Oracle Enterprise Manager 10g Grid Control Implementation Guide

Michael New

New York   Chicago   San Francisco
Lisbon   London   Madrid   Mexico City   Milan
New Delhi   San Juan   Seoul   Singapore   Sydney   Toronto

The McGraw·Hill Companies

Library of Congress Cataloging-in-Publication Data

New, Michael.
  Oracle Enterprise manager 10g grid control implementation guide / Michael New.
    p. cm.
  ISBN 978-0-07-149275-1 (alk. paper)
  1. Oracle (Computer file) 2. Computational grids (Computer systems)
  3. Database management. I. Title.
  QA76.9.C58N49 2009
  005.74—dc22
                                                                  2008047021

McGraw-Hill books are available at special quantity discounts to use as premiums and sales promotions, or for use in corporate training programs. To contact a special sales representative, please visit the Contact Us page at www.mhprofessional.com.

Oracle Enterprise Manager 10*g* Grid Control Implementation Guide

Copyright © 2009 by The McGraw-Hill Companies, Inc. (Publisher). All rights reserved. Printed in the United States of America. Except as permitted under the Copyright Act of 1976, no part of this publication may be reproduced or distributed in any form or by any means, or stored in a database or retrieval system, without the prior written permission of Publisher.

Oracle is a registered trademark of Oracle Corporation and/or its affiliates. All other trademarks are the property of their respective owners.

Screen displays of copyrighted Oracle software programs have been reproduced herein with the permission of Oracle Corporation and/or its affiliates.

1234567890   FGR FGR   0198

ISBN   978-0-07-149275-1
MHID   0-07-149275-5

Sponsoring Editor	**Technical Editors**	**Composition**
Lisa McClain	Phil Choi	International Typesetting
	Edward Whalen	and Composition
Editorial Supervisors	**Copy Editors**	**Illustration**
Janet Walden	Carolyn Welch	International Typesetting
Jody McKenzie	Bill McManus	and Composition
Project Manager	**Proofreader**	**Art Director, Cover**
Vastavikta Sharma, International Typesetting and Composition	Sanjukta Chandra	Jeff Weeks
Acquisitions Coordinators	**Indexer**	**Cover Designer**
Mandy Canales	Claire Splan	Pattie Lee
Jennifer Housh	**Production Supervisor**	
	George Anderson	

Information has been obtained by Publisher from sources believed to be reliable. However, because of the possibility of human or mechanical error by our sources, Publisher, or others, Publisher does not guarantee to the accuracy, adequacy, or completeness of any information included in this work and is not responsible for any errors or omissions or the results obtained from the use of such information.

Oracle Corporation does not make any representations or warranties as to the accuracy, adequacy, or completeness of any information contained in this Work, and is not responsible for any errors or omissions.

I dedicate this book to my daughter, Leah, and to my late brother, Robbie.

About the Author

Michael New is a senior information systems architect with 17 years of experience in the software industry. After receiving his B.S. from M.I.T. in Aeronautics and Astronautics, he worked for six years as a software engineer, principally on the GPS satellite program at Rockwell. For the past 11 years, Michael has worked as an Oracle core and Oracle Applications DBA. He began as a full-time DBA, moved into consulting with an Oracle partner, then served as a Technical Manager for Oracle Consulting before launching his own firm, New DB Solutions. Michael has implemented the full array of Oracle products for clients in many industries, including the Oracle E-Business Suite, RAC, Data Guard, RMAN, Application Server, and Grid Control (in which he is a recognized authority). Michael has authored several white papers on database administration, has been a presenter at Oracle OpenWorld, and has been a technical editor for a McGraw-Hill Professional book on Oracle 10g. Michael lives in New York City with his five-year-old daughter, Leah. You can reach him at Michael@NewDBsolutions.com or through his web site, www.newdbsolutions.com.

About the Technical Editors

Phil Choi worked at Oracle for several years documenting the Enterprise Manager suite of products. Before Oracle, he worked in a variety of industries, including consulting, software, and e-commerce. Phil has yet to put his Mechanical Engineering degree to good use, but the degree in Creative Writing has come in handy during his attempts to write short stories and essays for publication. Phil (and his dog Ajax) lives and works in San Francisco.

Edward Whalen is the founder and Chief Technology Officer of Performance Tuning Corporation. During his career, he completed many complex consulting projects and continues to write books and technical papers on various Oracle and Microsoft technologies. Ed has published more than eight books on SQL Server and Oracle, most recently *Microsoft SQL Server 2005 Administrator's Companion and Oracle Database 10g Linux Administration*. Ed is recognized as a leader in database design and performance optimization and provides critical services to companies worldwide.

Contents

Acknowledgments xiii
Introduction xv

PART I
Install Grid Control

1 Overview of the Grid Control Architecture 3
    What Is Grid Computing? 4
    What Is Grid Control? 7
    Grid Control Components 10
        Grid Control Console 13
        Oracle Management Agent 14
        Oracle Management Service 15
        Oracle Management Repository 17
        Data Flow Between Grid Control Components 19
    Grid Control, Database Control, and AS Control 22
        Grid Control vs. Database Control 23
        Grid Control vs. AS Control 24
        Agents for Grid Control, Database Control, and AS Control 26
        Grid Control, Database Control, and AS Control: All Together Now 26

2 Grid Control Preinstallation 29
    Key Architectural Design Decisions 33
        Decide How Many Grid Control Environments to Build 35
        How Many Regional Sites Are Required? 36
        Make Required Installation Choices 41
    Network Configuration Steps 47
        Set Up Host Name Resolution 49
        Fully Qualify Host Name References 49
        Host Name Constraints 52
        Use Static IP Addresses 53
        Connectivity Checks 53

Hardware Requirements 55
    Hardware Installation Requirements 56
    Hardware Operating Requirements 57
Software Requirements 60
    Verify Certification Information 60
    Create Required OS Groups and Users 61
    Create Required Directories 65
    Stop Database Listeners Using Port 1521 70
    Synchronize OS Timestamps/Time Zones 70
    Confirm Platform-Specific Software Requirements 72

3 Grid Control Installation 75
Gather Needed Installation Information 77
Address Installation Bugs 77
Initialize the Oracle User Environment 83
    Set Up X Server (UNIX Only) 83
    Set/Unset OS Environment Variables 85
Running the Installer with Desired Options 87
    Silent Installation Method 87
    Static Ports Feature 88
    Starting and Monitoring the Installation 89
Install Grid Control Using a New Database 91
    Specify Installation Type 91
    Specify Installation Location 91
    Language Selection 93
    Specify Inventory Directory and Credentials 93
    Product-Specific Prerequisite Checks 94
    Specify Configuration 95
    Specify Optional Configuration 95
    Specify Security Options 97
    Oracle Configuration Manager Registration 99
    Summary 100
    Install 101
    Configuration Assistants 101
    End of Installation 103
Install Grid Control Using an Existing Database 103
    Steps to Install the OMR Database 105
    Run the Installer—GC Existing Database Option 111
    Install Standalone Agents on Dedicated OMR Nodes 114
Log in to the Web Console 114
    Install the Grid Control Security Certificate in Your Browser 115
Install an Additional OMS 116

Contents **vii**

4 Grid Control Post Installation 119
   Patch Grid Control 120
      Apply the Latest Grid Control Patch Set 121
      Apply Latest Database Patch Set Certified for OMR 123
      Apply the Latest EM Critical Patch Update 124
      Apply Required Grid Control One-off Patches 125
   Oracle Management Service Configuration 129
      Reduce Grid Control Logging 130
      Set Up Oracle User Environment on OMS Hosts 135
      Modify the Default Console Timeout 138
      Configure OMS for Failover and Load Balancing 138
      Tune OMS Thread Pool Size 139
      Add OMR Alias to OMS tnsnames.ora 140
      Back Up Critical OMS Files 140
   Oracle Management Repository Configuration 141
      Set Up Oracle User Environment on OMR Nodes 141
      Confirm That Listeners Load-Balance Across RAC Nodes 141
      Configure *i*SQL*Plus Access in Grid Control 142
      Modify the Data Retention Policy 148
      Modify Job Purge Policy 150
      Secure the emkey 150
      Set Up Grid Control Auditing 151
      Configure and Tune the OMR Database 152
      Install Oracle Configuration Manager 152

5 Preinstall Standalone Management Agents 155
   An Overview of Common Agent Preinstallation Steps 157
   Agent Key Installation Decisions 158
      Select Agent Installation Methods 158
      Use an Existing User or Create a Separate Agent User 162
      Install a Local or Cluster Agent 163
      Configure a Server Load Balancer First 165
      Use an LDAP or Local Agent User 165
      Choose to Secure or Not Secure the Agent 166
   Agent Hardware Requirements 166
      Agent Disk Space Requirements 166
      Agent CPU Requirements 167
      Agent RAM Requirements 168
      Summary of Agent Hardware Requirements 168
   Agent Software Requirements 168
      Initialize the oracle User Environment 168
      Confirm No Existing Agent Is Installed 169
      Change Response File if Using an SLB 170
      Prepare for Cross-Platform Agent Installation 171
      Prepare Targets for Discovery 173
      Gather Needed Agent Installation Information 173

6 Install Management Agents via Agent Deploy 175
Agent Deploy Installation Overview 176
Install Required Packages 178
Configure SSH User Equivalence 178
    Set Up SSH Server (SSHD) on Windows 179
    Back Up Current SSH Configuration 179
    Set ORACLE HOME on OMS Host 179
    Run sshConnectivity.sh Script 180
    Verify SSH User Equivalence Is Configured 184
Set Up Time Zone for SSH Server 184
Validate All Command Locations 185
Modify Agent Deploy Properties File for SLB 187
Choose Inventory Location 187
Verify Agent User Permissions 188
Prepare for Cross-Platform Agent Push 189
Include Additional Files (Optional) 189
Run Agent Deploy Application 190
    Fresh Install 192
    NFS Agent 195
    Upgrade Agent 196
Agent Deploy Post-Installation 197
    Non-NFS Agents 198
    NFS Agents 198
Troubleshoot Agent Deploy 198

7 Install Management Agents Locally 201
nfsagentinstall Installation 202
    Configure Shared Storage 205
    Install Master Agent on Shared Storage 209
    Install NFS Agents 211
agentDownload Installation 213
    Prepare for agentDownload Installation 214
    Copy agentDownload Script from OMS to Target Host 217
    Execute agentDownload on Target Host 219
Agent Cloning Installation 223
Silent Agent Installation 225
Interactive Agent Installation 228
    Install Required Packages 228
    Run the Interactive Installer 229
Agent Post-Installation Steps 236
    Set Up Agent User Environment 237
    Confirm Agent Is Working 238
    Refresh Host Configuration (If Needed) 240
    Run agentca for Cluster Agent (Windows Only) 240
    Confirm Agent Restart on Reboot Is Configured 240
    Back Up the Agent 241

8 Install Grid Control Clients 243
  Install and Configure the EM Java Console 244
    Determine Whether You Need the EM Java Console 245
    Installation Steps for the EM Java Console 246
    Start the EM Java Console 249
    Configure Change Manager 250
  Install Adobe SVG Viewer 262
  Install EM Command Line Interface 263
  Install Oracle Configuration Manager Client 264

9 EM Login and Component Control 267
  Log in to EM 269
    Web Console Login to Grid Control 269
    EM Java Console Login 272
    AS Console Login 273
    iSQL*Plus Login 274
    Metrics Browser Login 276
  Control EM Components 278
    Control iSQL*Plus Server 278
    Control the Management Repository 279
    Control the AS Console 279
    Control the Management Service 280
    Control the Management Agent 286
    Windows EM Services 290
    EM Startup/Shutdown Order 292

PART II
Configure and Maintain Grid Control

10 General Console Configuration 297
  Introduction 299
    Tour of the Setup and Preferences Menus 299
    Follow an Event from Trigger to Notification 303
    Reasoning Behind this Chapter's Menu Navigation Order 305
  Set Up Administrators 307
    Create Roles 307
    Create Administrators 313
    Set Administrator Preferences 317
    Set Preferred Credentials 320
  Configure Notifications 330
    Define Notification Methods 330
    Set Notification Schedules 334
    Create and Subscribe to Notification Rules 339
  Enable Patching Features 356
    Complete the Patching Setup 356
    Schedule the RefreshFromMetalink Job 362

11	Configure Target Monitoring	365
	Remove Targets from Monitoring	367
	Ensure Agents Are Started	367
	Back Up targets.xml File	367
	Confirm No Targets Are in Blackout	367
	Remove Targets from Grid Control	368
	Discover and Configure Targets for Monitoring	370
	Three Paths to Discover/Configure a Target	371
	Agent Configuration	376
	ASM Discovery and Configuration	383
	Database Discovery and Configuration	386
	Listener Discovery and Configuration	407
	Host Additional Monitoring	409
	OracleAS Discovery and Configuration	411
	Enter Target Properties	413
	Grant Management Pack Access	414
	Management Options: Packs, Plug-ins, and Connectors	414
	Schedule Blackouts for Planned Downtimes	421
	Set Blackouts in the Console	421
	Set Blackouts at the Command Line	427
12	Configure Group and Service Monitoring	429
	Group Configuration	430
	Create a Group	431
	Group Features	435
	Service Configuration	438
	SLA Objectives for Services	439
	Key Elements of Services	441
	Configure a Service	445
13	Tune Metrics and Policies	489
	Change Default Metrics and Policies	492
	Change Default Metrics	492
	Change Default Policies	501
	Enable Metric Baselines	504
	Metric Baseline Concepts and Terms	506
	Create Metric Baselines	509
	Disabling Metric Baselines	514
	Add Corrective Actions	515
	Create Corrective Actions	516
	Add Corrective Actions to Metrics or Policies	518
	Implement User-Defined Metrics	521
	Create an OS User-Defined Metric	522
	Create a SQL User-Defined Metric	527

		Use Metric Snapshots	529
		Create Metric Snapshots	530
		Apply Metric Snapshots	532
		Leverage Monitoring Templates	534
		Create Monitoring Templates	536
		Maintain Custom Metrics when Applying Monitoring Templates	540
		Apply Monitoring Templates	541
		Compare Settings Between Targets and the Template	544
14		**Backup and Recovery of Grid Control**	**549**
		Introduction to Grid Control B&R	551
		Grid Control B&R Concepts	551
		Resetting Agents	554
		Recommended Grid Control Backups	557
		OMR Database Backup and Recovery	560
		Back Up the OMR Database in Grid Control	561
		Recover the OMR Database	597
		Leverage Flashback Technology for OMR Database Recovery	606
		Manage OMR Database Backups in Grid Control	608
		Grid Control Software Backup and Restoration	612
		OMR Database Software Backup and Restoration	613
		OMS Software Backup and Restoration	615
		Agent Software Backup and Restoration	619
15		**Configure Grid Control for High Availability and Disaster Recovery**	**623**
		Grid Control High Availability Recommendations	625
		OMS HA Recommendations	627
		OMR Database HA Recommendations	633
		Agent HA Recommendations	636
		Grid Control Disaster Recovery Recommendations	637
		OMR Database DR Recommendations	639
		OMS DR Recommendations	644
		Agent DR Recommendations	648
16		**Securing Grid Control Data Transfer**	**651**
		Enable EM Framework Security	656
		Enable Framework Security Between Agents and OMS	656
		Secure Repository Data Transmissions	663
		Secure Console Connections	675
		Enforce HTTPS Between the Browser and AS Console	676
		Enforce HTTPS Between Browser and OHS	677
		Security Considerations for Console Web Cache Access	679
		Limit GC Console Access to Certain Clients	680

17 Maintain and Tune Grid Control 685
Perform Routine Grid Control Maintenance Tasks 686
    Weekly Online Maintenance Tasks 687
    Monthly Offline Maintenance Tasks 693
Gather and Document Grid Control Metrics 699
    Types of Grid Control Metrics to Gather 699
    Specific Grid Control Metrics to Gather 703
    Procedure to Evaluate Grid Control Metrics 708
Tune Grid Control to Reduce Bottlenecks 709
    Reduce High CPU Utilization 710
    Resolve Loader Backlog 715
    Minimize Rollup Delays 719
    Evaluate Job, Notification, and Alert Metrics 722
    Tune I/O Bottlenecks 723
    Improve Poor Console Performance 727
    Platform-Specific Tuning Recommendations 734

Appendixes Online at www.OraclePressBooks.com
A Agent Subsystems
B Certification and Platform-Specific Installation Requirements
C Configure and Tune a Database for the Management Repository
D Set Up SSH User Equivalence
E Configure Grid Control Automatic Startup
F RMAN Scripts from Oracle-Suggested Backup Strategy in Grid Control
G Configuring OAS Using Oracle Net Manager
H Tools for Grid Control Performance Metrics
I How to Install the Grid Control Security Certificate in Your Browser

Index 737

Acknowledgments

This book sprung from material for a Grid Control "Deep Dive" class that I developed and taught to Oracle Consulting personnel at the behest of Dennis Horton, a Senior Director at Oracle. Without his impetus, I would never have tackled this book.

Thanks to the Oracle Press team at McGraw-Hill Professional: Lisa McClain, Mandy Canales, Jennifer Housh; the copyedit team, notably Vastavikta Sharma, Carolyn Welch, and Bill McManus; and the Illustration and Production departments for their work in producing this book. I am very grateful to Ed Whalen and Phil Choi for their roles as technical reviewers. Ed introduced me to the McGraw-Hill team by bringing me in as a technical editor for his book, *Oracle Database 10g Linux Administration* (Oracle Press, McGraw-Hill Professional, 2005). Later, he told me about McGraw-Hill's desire to publish a Grid Control book. After I signed up to write this book, Ed provided his company's computer resources as a platform for most of the Grid Control installation and configuration steps shown in the book. I am also largely indebted to Phil Choi, formerly the lead editor at Oracle for Enterprise Manager products, for extensively editing the book's first draft. Phil taught me a great deal about technical writing.

I greatly appreciate Oracle's help in supplying and validating technical content as needed. Thanks to Oracle Support in answering some tough questions along the way, particularly Mathieu Hornsperger for explaining the inner workings of Grid Control security. Narain Jagathesan and Anirban Chatterjee in the Enterprise Manager development group at Oracle provided much of the detailed information about the Management Agent given in Appendix A (available online at www.OraclePressBooks.com).

I'd also like to acknowledge my Oracle mentors: Howard Ostrow, a great all-around technical resource and RAC expert; Matthew Burke, a DBA master who tutored me for four years in DBA paths less traveled and encouraged me to present my first white paper; and Ashok Nagabothu, who took me, when a fledgling DBA, under his wing.

I am grateful to my family and friends for seeing me through a trying time personally while writing this book. My brother Jon, his wife Gina, their son Daniel, my mother, and good friends Adam Bierman and Brad Stonberg were very supportive. I am indebted to Amy Seawright, erstwhile companion, who brought me back to life before my mistress, Oracle, reclaimed it. Finally, I thank my daughter, Leah, and my muse for resurrecting me yet again, once and for all.

Introduction

This is the first book dedicated solely to the subject of implementing Oracle Enterprise Manager 10*g* Grid Control. It was only a matter of time before such a book was published. With the first release (10.1) of Grid Control roughly seven years ago, Oracle ushered in a true system management product. The latest release (10.2) of the product is enjoying large-scale adoption by companies looking to centrally manage their complete IT infrastructure—not just their Oracle systems, but non-Oracle ones as well. Firms around the world are significantly cutting IT expenditures by administering their data center components with Grid Control.

As you would expect from such a capable software product, the Enterprise Manager 10*g* Grid Control documentation library from Oracle Corporation is bulky, yet still does not contain solutions to certain common installation and configuration hurdles. Answers are found in supplemental material available on *OracleMetaLink*, the Oracle Technology Network (OTN), and elsewhere on the Web. (How much third-party material alone is out there? Google "Oracle Enterprise Manager 10g Grid Control" and you'll get 122,000 hits.) All in all, it's a lot of material to digest, especially if you're new to Grid Control.

Contrast a Grid Control novice's first installation using product and supplemental documentation with that of an expert who has performed many installations, having already waded through and assimilated the reference material and learned the bugs, workarounds, optimal installation order, shortcuts, tips, and tricks. The expert will likely implement Grid Control successfully, on time and under budget, whereas the novice may not be so fortunate. The aim of this book is to lend an expert hand to a DBA who is not intimately acquainted with Grid Control so that he or she can implement it according to best practices, as quickly and painlessly as possible.

This book distills what I've learned from completing many Grid Control engagements over the past seven years. Each implementation was unique: many were large installations at Fortune 500 companies, while others were smaller, proof-of-concept assignments; the platforms were different, as were the high availability requirements. Yet, these engagements also shared many commonalities: I ran across same issues over and over again—some generic, some platform-specific—and found myself returning to the same reference material for tried-and-true solutions. I documented the process for customers and gathered Grid Control collateral from colleagues. When I was with Oracle Consulting, I funneled this knowledge and experience into a week-long Grid Control "Deep Dive," a hands-on laboratory course that I developed and taught to Oracle consultants. More than anything else, it was this trial by fire in teaching fellow consultants—the toughest students—that motivated me to want to share my methods with a wider audience.

This book is a "best-of" compilation of all these Grid Control experiences, a how-to manual on which novice and aspiring expert can both rely. I provide step-by-step instructions on how to install, configure, maintain, and tune Grid Control releases 10.2.0.3 and 10.2.0.4 for all operating systems on which it is certified. The book does not cover how to upgrade an existing OEM 9.2 or 10.1 installation to the latest 10.2 version. I concentrate on how to *implement* Grid Control 10.2, not how to use it ad hoc (such as to tune managed targets). While configuring Grid Control is the book's central focus, I use Grid Control to configure itself whenever possible to demonstrate its features. By turning Grid Control on itself, so to speak, you will learn how to use it. For example, the Chapter 14 material on backup and recovery of the Oracle Management Repository Database applies equally to target databases.

The appendices for the book are available at www.OraclePressBooks.com on either the Downloads page or the book product page. The material there lends itself better to soft copy because you can copy the provided commands, scripts, etc. for setting up Grid Control to your clipboard.

This book is intended primarily for Oracle DBAs with at least a few years of experience who want to install and configure Grid Control to administer their IT infrastructure, as well as for system and network administrators assisting in the process. In addition, the book is suitable for IT managers responsible for a Grid Control project or just interested in gaining insight into the product's capabilities to administer both Oracle and non-Oracle products in their data centers. To effectively capitalize on the book, those in management roles should have at least a basic knowledge of Oracle technologies.

Conventions

Throughout the book, I refer to two main operating systems: Windows and UNIX. The term "UNIX" refers to all variants of that OS, including Linux. UNIX syntax is used unless the equivalent Windows syntax widely differs. The most prevalent difference is between the UNIX forward slash (/) and the Windows backslash (\) in directory specifications.

Following are the conventions used throughout the book:

Term	Meaning
<variable>	Text in italicized angle brackets represents a variable. Substitute a value for the variable text and do not type the angle brackets.
[parameters]	Text in square brackets represents one or more optional parameters for commands. Enter any optional parameters and do not type the square brackets.
\	In UNIX, the backslash continuation character at the end of a line permits entering a single command on multiple lines. Alternatively, on any platform, you can omit the backslash and type the entire command on a single line, provided it adheres to OS-specific restrictions on the maximum number of characters allowed per line. In this book, a backslash at the end of a line signifies to choose one of these methods to enter the command.

PART
I

Install Grid Control

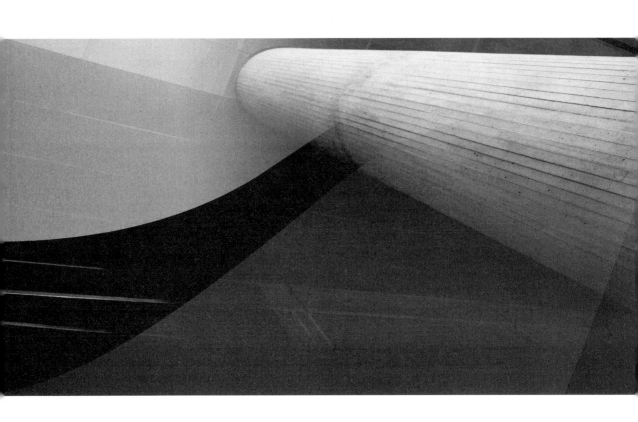

CHAPTER 1

Overview of the Grid Control Architecture

If this book were a novel, the first sentence would aspire to emulate the masters. Something along the lines of "All Children, except one, grow up."[1]; "It was the best of times, it was the worst of times…"[2]; or "As I walked through the wilderness of this world…"[3]. But this isn't a novel. So let me begin more humbly.

This chapter takes a high-level look at the architecture of Oracle Enterprise Manager 10g Grid Control (referred to throughout the book as Grid Control or GC), and includes descriptions of its core components. We'll "kick the tires," so to speak, then open the hood of the Grid Control engine, stick our heads in, and take a closer look at its parts and inner workings. I'm not asking you to remove these parts, spread them across the floor of your garage, and disassemble them one by one. However, near the end of the chapter, I will trace how data flows between all parts of the GC engine.

TIP
Most of this chapter deals with GC architecture, delving deeper into the topic than you might expect for a first chapter. My rationale is to establish a solid foundation you can refer back to as you progress through the book. If you're in a rush, skim the chapter and move on to Chapter 2, which covers the pre installation process.

This chapter answers the following questions:

- What is grid computing?
- What is Grid Control?
- What are the main GC components?
- What are the subcomponents that make up each component, and what are their functions?
- What protocols do components use to communicate with one another?
- What wiring enables communication between components for Console requests, Agent uploads, and alert notifications?
- What is the difference between Grid Control, Database Control, and Application Server Control?

What Is Grid Computing?

Before we begin an in-depth discussion of Grid Control architecture, we need to step back and ask this question: What is grid computing? Any description of Grid Control would be remiss without mentioning its namesake, "the grid." Introduced in the late 1990s, the grid refers to grid

---

[1] *Peter Pan* by J.M. Barrie, 1911.
[2] *A Tale of Two Cities* by Charles Dickens, 1859.
[3] *The Pilgrim's Progress* by John Bunyan, 1678.

computing, the next generation of computing that allows you to virtualize IT resources, including data, processing, servers, storage capacity, databases, and network bandwidth, as though they were utilities. The term was first coined to equate an IT department's delivery of services with an energy company's grid that supplies electrical power to its customers. In such a "utility computing" environment, applications rely on a grid infrastructure just as appliances rely on electricity from outlets. You don't know where the generators are or how the electric company does its wiring. You just want to plug in an appliance and get electricity. The same applies to an application—you want it to work just like a utility.

One of the most salient examples of grid computing is the Search for Extra-Terrestrial Intelligence At Home (SETI@home), a computing project that harnesses the processing power of over five million idle, Internet-connected computers to analyze signal data from radio telescopes in an attempt to detect intelligent life beyond Earth. Grid computing is based on an open set of standards and protocols, such as the Open Grid Services Architecture (OGSA), which describes a Service Oriented Architecture (SOA), enabling communication across heterogeneous, geographically dispersed sites. With grid computing, organizations optimize data and computing resources by pooling and sharing them across networks. Grid computing is another step in the evolution of IT developments, such as the Web, peer-to-peer computing, distributed computing, and virtualization technologies,[4] but it is distinguished from these technologies in the following ways:

- Like the Web, grid computing conceals complexity, but unlike the Web, which mainly enables communication, grid computing enables collaboration to achieve common business goals.

- Like peer-to-peer computing, grid computing allows individual users to share files, but grid computing also allows many-to-many sharing—not only file sharing but all computational resources as well.

- Like distributed computing, grids bring computing resources together, but grids can be geographically distributed and heterogeneous.

- Like virtualization technologies, grid computing virtualizes IT resources but on a much larger scale in terms of computing power and storage.

A grid strategy provides the following features:

- Scalable, highly available database clustering
- Load balancing
- Automatic self-tuning and self-management
- Storage virtualization
- Resource virtualization
- Grid management software

---

[4] MCADCafe, April 5, 2004, Jeff Rowe.

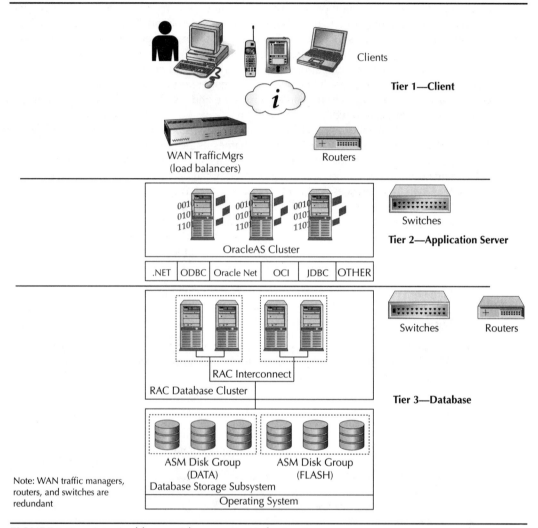

FIGURE 1-1. *A typical large grid computing infrastructure*

Figure 1-1 illustrates a typical real-world large grid computing architecture using Oracle technology.

As shown in the figure, enterprise grid computing provides highly available database services to applications by means of a flexible component provisioning (i.e., deployment) topology. This topology integrates all IT resources—servers, storage, databases, applications, and networks—to provide provisioning on demand. These resources can be divided into three tiers. Let's cover each tier from the user end to the back end.

- **Tier 1—Client** The client tier is the entry point for all application client connections. These are the client PCs, PDAs, ATMs, and other acronyms all requesting something that only IT can deliver—data.

- **Tier 2—Application Server** The Application Server tier consists of application servers and web servers, which Oracle provides in Oracle Application Server (OracleAS) 10g. Figure 1-1 depicts a three-node OracleAS Cluster, with each node containing an application server and a web server. This cluster communicates with the database cluster via various protocols and specifications, including Open Database Connectivity (ODBC), Java Database Connectivity (JDBC), and Oracle Call Interface (OCI).
- **Tier 3—Database** The database tier is comprised of two main parts:
    - **Database cluster(s)** The emergence of database clustering is made possible economically by the recent introduction of small multiprocessor hosts, low-cost blade servers, and the open source Linux operating system. Each Real Application Clusters (RAC) cluster presents a single database spanning all nodes in that cluster. In Figure 1-1, the cluster is composed of a four-node RAC database. (RAC nodes are shown in groups of two because you typically add two nodes at a time when more database server resources are required.) RAC offers the flexibility to add or remove nodes as capacity demands fluctuate. RAC also provides hardware failure protection; it can fence off failed nodes, redistribute processing evenly across the surviving nodes, and reconnect clients transparently to the remaining nodes.[5]
    - **Database storage subsystem** The storage subsystem is an independent collection of low-cost disk devices. The example in Figure 1-1 uses the Oracle solution of Automatic Storage Management (ASM) disk groups. Each disk group (two in this case, named DATA and FLASH) is composed of multiple disk devices (three in this example). ASM partitions the disk space and evenly distributes the data storage throughout the storage array. If you add or remove storage arrays, ASM "automagically"[6] redistributes the data storage.

On all tiers, the switches, routers, and WAN traffic managers (which also provide load balancing) are redundant. Figure 1-1 does not include a disaster recovery (DR) site, but such a site is essential to any highly available architecture. The DR site would be similarly configured, although perhaps scaled down.

Many books have been written about grid computing, and this cursory discussion on the subject certainly cannot do it justice. It is simply intended to introduce you to the concept of the grid architecture so that you can place Grid Control in its proper context. I should note that, while a grid strategy is undoubtedly the right technological direction for companies to gravitate towards, your firm doesn't need to be fully or even partially exploiting grid technology to take advantage of Grid Control.

What Is Grid Control?

With its usual uncanny foresight into the direction of IT technology, Oracle changed the name of Oracle Enterprise Manager 9*i* to Oracle Enterprise Manager 10*g* Grid Control when this new version of the product was first introduced in 2003. At that time, most corporate IT professionals were still relatively unversed in grid computing concepts, and their IT sites were anything but Grid compliant—most sites still aren't, though the trend of computing as a commodity has

---

[5] Transparent Application Failover (TAF), an Oracle database feature, only functions when applications are enabled for it.

[6] My mentor and friend, Howard Ostrow, a RAC expert for Oracle Consulting Services, is fond of this phrase in describing ASM's amazing capabilities.

certainly caught on. To address the challenges of companies whose business needs change faster than their IT departments can adapt, Oracle introduced the three products we've spoken of to enable an Oracle grid solution: Oracle Application Server 10*g* (Tier 2), Oracle Database 10*g* and the accompanying ASM solution (Tier 3), and Grid Control to manage the entire lot. Given that these three products together are integral to implementing a grid strategy, it should come as no surprise that the "*g*" in "10*g*" in each of the product names stands for "grid." Grid Control manages the grid and, as you will see in this chapter, is architected itself in a "grid-compliant" manner; that is, it is built on the same three tiers and, yes, it even monitors itself.

Grid Control is an Oracle Web application that centralizes management of your enterprise IT grid (or even non grid) environment. Grid Control allows administrators to monitor the entire technology stack in the grid, including both Oracle and non-Oracle components. This flagship systems management product can manage over 100 types of Oracle and non-Oracle targets (system components) configured in any architecture. These components include routers, load balancers, switches, business applications, application servers, databases, networks, storage, and hosts, most of which inhabit the example in Figure 1-1. If a component within the grid goes offline or begins to experience performance issues, Grid Control automatically triggers an alert to notify an administrator so that he or she can take appropriate action. Grid Control is also capable of automatically replying to alerts with pre defined "fixit" jobs.

Here are just some of Grid Control's many features and how you can use them:

- Administer the database server tier (Tier 3 in Figure 1-1) using RAC management

- Monitor Automatic Storage Management (ASM) instances, Oracle's solution to a shared array of disks in a database storage subsystem (Tier 3 in Figure 1-1)

- Manage application and web servers (Tier 2 in Figure 1-1) as part of providing service level management, which includes setting business service goals for transaction performance and business processes, monitoring and tuning service performance and usage, diagnosing root causes of problems, and reporting

- Monitor non-Oracle components such as WAN traffic managers (in Tier 1)—for example, the F5 BigIP Local Traffic Manager

- View overall system and service status at a glance using the home page and dashboards akin to monitors at a hosting site that allow drill-down to analyze the root cause of a problem

- Add alerts, security policies, and thresholds for all component types, and set rules for alert notifications

- Manage many databases, hosts, and other components as one using templates

- Bare metal provision new operating systems using pre tested software image libraries, and deploy new databases and application servers

- Manage the life cycle of the Oracle and OS patch process with direct connections to Oracle*MetaLink*, proactive patch notifications, and automated distribution

- Automate tasks at the operating system and database levels using the job subsystem

- Perform configuration management to capture your current configuration, analyze it against referenced, saved, or live configurations, and drive to a certified configuration

- Generate easy out-of-box reports for users, managers, and executives using graphical report creation tools with secure publishing capabilities

Grid Control gives you the ability to take control of a heterogeneous IT site that employs hardware platforms, network solutions, and databases from many vendors. Until recently, IT professionals—from the CIO down to the DBA—have only dreamed about centralizing control of all IT resources. One of Grid Control's main objectives is to proactively monitor all systems so that DBAs can spend more time participating in high-level activities, such as planning the future technology direction of their IT departments. Back in the day before GUI applications, DBAs monitored systems at the command line with custom scripts. Grid Control standardizes such monitoring with out-of-box metrics to alert IT staff of problems with system resources and critical applications.

Making Sense of Grid Control Versioning

Oracle Enterprise Manager (OEM) 10g Grid Control versioning is confusing because it is not consistent. The following list shows the versioning scheme for all releases of the product since 10.1.0.2:

Full Name of Release	Release Family	Release Number
OEM 10g Grid Control Release 4 (10.2.0.4.0)	EM 10gR4	10.2.0.4
OEM 10g Grid Control Release 3 (10.2.0.3.0)	EM 10gR3	10.2.0.3
OEM 10g Grid Control Release 2 (10.2.0.2.0)	EM 10gR2	10.2.0.2
OEM 10g Grid Control Release 2 (10.2.0.1.0)	EM 10gR2	10.2.0.1
OEM 10g Grid Control Release 1 (10.1)	EM 10gR1	10.1.0.6
OEM 10g Grid Control Release 1 (10.1)	EM 10gR1	10.1.0.5
OEM 10g Grid Control Release 1 (10.1)	EM 10gR1	10.1.0.4
OEM 10g Grid Control Release 1 (10.1)	EM 10gR1	10.1.0.3
OEM 10g Grid Control Release 1 (10.1)	EM 10gR1	10.1.0.2

As you can see, all versions of the first major release, Grid Control 10.1, are known as Release 1. However, the next major release, 10.2, actually refers to both 10.2.0.1 and 10.2.0.2 as EM 10gR2. Then, as of EM 10gR3, Release x describes the latest release family or patch set of 10.2. In other words, Release 3 (EM 10gR3) is 10.2.0.3, and Release 4 (EM 10gR4) is 10.2.0.4. This book covers these latest two releases which, combined, have been ported to every major platform. EM 10gR4 has already been ported to most major platforms, and EM 10gR3 is offered for the remaining ones.

Grid Control Components

Oracle Enterprise Manager 10g Grid Control is Oracle's solution for managing your complete IT environment—Oracle and non-Oracle products alike. Grid Control gathers information about your enterprise computing environment, consolidating it into a central Repository. Grid Control displays this information to database administrators from its Web Console and sends them alerts on threshold conditions of interest. Administrators can then use this information to ask GC to perform tasks for the computing systems it monitors.

Grid Control is built upon the Oracle technology stack: an Oracle database acts as the back end, while an Oracle Application Server is the glue of the middle tier. Let's start by examining the main components of Grid Control.

The basic GC topology consists of four core components, as depicted in Figure 1-2: the Grid Control Console, the Oracle Management Agent, the Oracle Management Service, and the Oracle Management Repository. Each component can be separated by a firewall.

NOTE
Throughout the book, I use the term "Grid Control infrastructure" or "Grid Control framework," to refer to the three principal Grid Control architectural components, namely, the Oracle Management Repository, Oracle Management Service, and the Oracle Management Agent, which is installed on each target host. Although the Grid Control Console is indispensable to this framework, it is a "client."

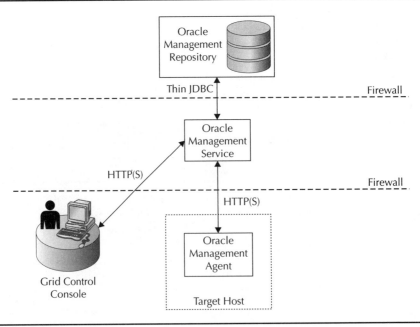

FIGURE 1-2. *Topology of Grid Control core components*

Chapter 1: Overview of the Grid Control Architecture 11

Following is a brief description of each core component. See Chapter 2 for a complete list of supported browser versions for the Console, and certified platforms for the GC framework, and check the Oracle*Metalink* Certify tab for the most up-to-date information.

- **Grid Control Console** The Grid Control Console (Web Console or Console) is a browser-based Console in which administrators can centrally manage their entire computing environment. If you had to choose one image to define the big picture for Grid Control, it would have to be the Console home page.

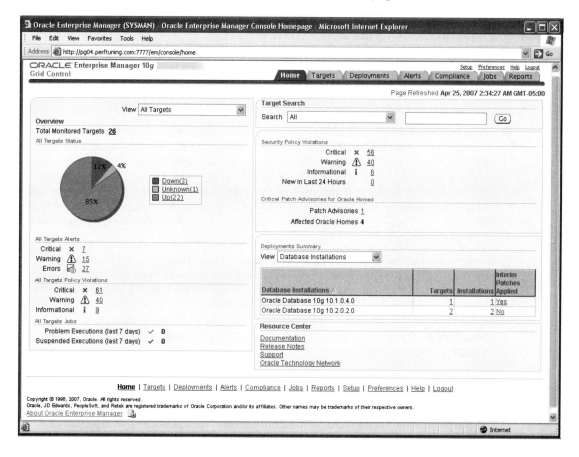

For now it's enough to know that you're looking through the front windshield of the Grid Control vehicle to get an overall view of your IT infrastructure. From the Console home page, you can turn down (i.e., drill down) any street of a specific managed target (system) within your IT infrastructure that Grid Control administers. The GC Console is certified to run on all the popular browsers, including Internet Explorer, Netscape, Mozilla, Firefox, and Safari. You don't install the Console; the Management Service renders it. You just open a browser and connect to the Console via the GC login URL. Administrators can open as many Console connections as the GC site can accommodate from a performance standpoint, which is usually a very large number.

- **Oracle Management Agent** The Oracle Management Agent (OMA or Agent), installed on each managed host, monitors that host and all targets on it, communicates information about these targets to the Oracle Management Service (OMS), and maintains the host and its targets as directed by administrators via their Console input. Targets are Oracle and non-Oracle products installed on a host. Grid Control monitors roughly 100 different target types out-of-box. For GC sizing purposes, each instance of a particular target type counts as a monitored target. Example target types are Database Instance, Listener, Oracle Application Server, and Host.

  You can also choose from among dozens of licensable management options or "add-ons" to extend the list of non-Oracle products that GC can monitor. Grid Control monitors itself, so an Agent also runs on each node hosting the Oracle Management Service and the Oracle Management Repository. Each managed host runs one and only one Agent. You can have as many Agents as you can scale Grid Control to support. Oracle certifie the Agent on a large number of 32-bit and 64-bit host platforms, including AIX, Linux, HP-UX, Solaris, and Windows.

- **Oracle Management Service** The Oracle Management Service (OMS) is a Java 2 Platform Enterprise Edition (J2EE) middle-tier application that renders the user interface for the Console. (J2EE is an environment for developing and deploying enterprise applications). The Agent uploads target data to the OMS which then processes this data before uploading it to a data store, the Oracle Management Repository. The Grid Control middle-tier contains three elements: Oracle Application Server Containers for J2EE (OC4J), Oracle HTTP Server (OHS), and Web Cache. The OHS deploys the Management Service J2EE Web application.

  You must install the OMS on at least one host, but you can also install it on more hosts as needed to support your environment for scalability or high availability. Each Management Service should reside on its own host. The OMS and Oracle Management Repository can reside on the same host, but for performance reasons, Oracle does not recommend this configuration for a production Grid Control environment, unless it is small (see Chapter 2 for details). All physical OMS hosts logically provide the generic Oracle Management Service.

  At the time of this writing, the OMS was certified to run on AIX, HP-UX, Linux, Solaris, and Windows.

- **Oracle Management Repository** The Oracle Management Repository (OMR or Repository) is the data store for Grid Control created in an Oracle database, and is located in the SYSMAN schema. This schema contains information about all Grid Control targets, administrators, and managed applications. The Repository organizes this data so that the OMS can retrieve and display it in the Console for administrators. A Grid Control site uses just one central Management Repository database. It can be a single-instance or RAC Oracle 10*g* or Oracle 9.2 database, although Oracle 10*g* is highly recommended.

In a nutshell, both the Console and the Agent communicate with the OMS. The following summarizes how these Grid Control components interact with each other (you can think of them as ad hoc and batch connections, respectively):

- The Grid Control administrator requests content in the Console over HTTP(S) in a browser session, which the OMS renders. The OMS then retrieves the data for the request from the Management Repository and displays it in the Console.

- Agents upload information to the OMS over HTTP(S), and the OMS uploads this data via JDBC to the OMR. The OMR sends data back to the OMS over JDBC, which is relayed to the Agent via a built-in HTTP listener.

Now that you are familiar with the main components of the Grid Control architecture, let's stop the engine, hoist Grid Control up on a lift, and take the mechanic's flashlight tour. We would do well to examine these components more closely and identify their subcomponents. Such scrutiny will help illuminate how the components interact, and will allow you to trace the data flow between them for each type of Grid Control communication.

Grid Control Console

The Web Console provides the user interface to the Grid Control product. From any location with Web browser access, you can log in to the Grid Control Console and centrally manage your entire enterprise Grid environment. Because the Console interface is rendered in HTML, it uses HTTP (or HTTPS if secured), making it lightweight, easy to access, and firewall friendly. You can offer Console login through a portal. Wireless devices can also access a critical subset of Enterprise Manager functionality through its wireless component, EM2Go. EM2Go requires no additional setup, and is available using a wireless personal digital assistant (PDA), a combination of Bluetooth PDA and mobile phone, a PDA using a Wireless Local Area Network (WLAN), or Pocket PC Emulator.

The usual way to log in to the Grid Control Console is at http[s]:<*OMShost*>:<*port*>/em. The URL protocol (HTTP or HTTPS) depends on whether you configure Secure Sockets Layer (SSL) encryption for browser access to Console communications. (SSL is an Internet protocol that secures data transmission using public-key cryptography). You can log in via Web Cache or directly to the OHS. The URL port differs depending upon both the method of login (Web Cache or OHS) and the OMS platform (UNIX or Windows). Web Cache page performance is a little better, but it is useful to know how to bypass it, such as when trying to isolate Web Cache as the source of a Console login problem.

In addition to the Web Console, there are several optional client components that work with Grid Control, and whose installation is covered in Chapter 8:

- **Enterprise Manager 10*g* Java Console**   The EM Java Console is a thick-client supplied with the Oracle 10.2 Client software, containing select functionality still not converted to the Grid Control Web Console (thin-client).

- **Adobe SVG Viewer**   This plug-in is required to display certain graphs in the Grid Control Web Console, which prompts you to install the plug-in if needed.

- **EM Command Line Interface**   The EM CLI is for administrators desiring to access Web Console functions from an OS shell either interactively or directly from scripts, or to set up workflow for business processes.

- **Oracle Configuration Manager Client**   OCM collects host configuration information to upload to Global Customer Services for analysis to better serve their customers. OCM is bundled with Grid Control 10.2.0.2 and can also be downloaded from OracleMetaLink as a standalone install kit.

See Chapter 8 for details on these optional Grid Control clients, including how to install them.

Let me speak briefly about the EM 10g Java Console, as many older DBAs are familiar with the OEM 9i Java Console. Almost all of the capability of the OEM 9i Java Console has been converted to HTML within the 10g Grid Control Web Console. To execute the few leftover 9i Java Console fat client features not yet available in the Grid Control 10g Console, such as management of Oracle Advanced Replication, you must download the Oracle10g Client software and install the Enterprise Manager 10g Java Console nondefault component. The EM Java Console only runs in standalone mode, optionally with its own standalone Repository; it does not connect to the Management Service or to the Management Repository.

Oracle Management Agent

An Management Agent must be installed on a host for Grid Control to monitor targets on that host. The Agent is the distributed portion of the Enterprise Manager framework, and is implemented in the C programming language for performance and resource reasons. It is a multithreaded process that uses Oracle core libraries, the Oracle Call Interface (OCI) and Oracle Secure Sockets Layer (SSL) to secure it by default.

You run one and only one Management Agent on each host that has targets to be managed. A properly installed Agent automatically (i.e., out-of-box) starts monitoring itself, its host, any Grid Control components (OMR, OMS) installed along with the Agent on the host, and certain Oracle target types (databases, listeners, and Oracle Application Servers) already installed on the host. (The Agent and host are targets in their own right; Grid Control treats them as any other target.) This automatic target discovery begins as soon as the Agent is installed and starts up. Those targets not automatically discovered or installed subsequent to the Agent can be manually discovered in the Console.

Targets can also be remote from the Agent to allow monitoring of targets without operating systems, such as firewalls and routers. In addition to the many built-in target types that Grid Control manages Oracle and its partners have built dozens of add-ons for non-Oracle products. Oracle-built add-ons are available for third-party products such as Microsoft SQL Server, Microsoft Active Directory, F5 BigIP, NetApp Filer, Dell PowerEdge Server, BEA WebLogic, and IBM WebSphere Application Server. Partnerbuilt add-ons include Citrix Presentation Server and Egenera pServer.[7] For a complete list of all add-ons, see http://www.oracle.com/technology/products/oem/extensions/index.html on the Oracle Technology Network.

NOTE
Grid Control can also monitor Oracle Collaboration Suite (OCS). Although Oracle uses OCS internally, this product does not have a sufficient market presence to warrant coverage in this book.

The Management Agent uses default monitoring and data collection levels to provide out-of-box monitoring and management information about all discovered targets. The Agent immediately uploads metric alerts and periodically uploads management information to the Management Service. The Agent also performs tasks on behalf of the Management Service, from running a job (a unit of work defined to automate administrative tasks such as backups or patching) to setting blackouts (suspending data collection on one or more targets to perform scheduled maintenance).

The Management Agent stores its configuration and management information about targets in flat OS files located under the Agent home directory. These files hold both unchanged, administrative

---

[7] pServer is Egenera's term for a virtual server.

data and dynamic alert and metric data that the Agent generates from monitoring its targets. See Appendix A (online at www.oraclepressbooks.com) for a complete treatment of Agent files and directories and for a description of all Agent subsystems and how they interact. This appendix is meant primarily as a referral source when needing to delve into the internals of the Agent subsystems for troubleshooting purposes. However, it will also give you a general feel for how the Agent organizes itself at the file level to monitor targets. It will probably not make much sense to you until after you install all Grid Control components, including standalone Agents on target hosts, which isn't covered until Chapters 5, 6, and 7.

Oracle Management Service

The Oracle Management Service is the Grid Control middle-tier that renders the user interface for the Console. The OMS is a critical component of any Grid Control implementation. The Management Service not only receives upload information from Management Agents, but also sends data to and retrieves data from the Management Repository. The OMS also renders this data in the form of HTML pages, which are requested by and displayed in the client Console. In addition, the Management Service performs background processing tasks, such as delivering notifications and dispatching Grid Control jobs. The OMS is deployed on the Oracle Application Server. The EM10*g* installation begins by installing an instance of the installation type, "Oracle Application Server J2EE and Web Cache." The OMS is deployed on its own OC4J container in this Application Server instance. Understanding the architecture of the OMS requires a grasp of each of its elements. The Management Service for EM10*g* Release 2 and higher comes bundled with Oracle Application Server Release 10.1.2,[8] which consists of the following subcomponents:

- Oracle Application Server Containers for J2EE (OC4J)
- Oracle HTTP Server (OHS)
- OracleAS Web Cache

NOTE
The OMS is technically part of the OC4J, but throughout this book the entire GC middle-tier is collectively referred to as the OMS.

You can deploy multiple Management Services, each on a separate host. However, the OMS on each host includes all subcomponents, and points to the same Repository database for a given Grid Control site. Each OMS communicates via Java Database Connectivity (JDBC) to the Management Repository. (JDBC is a standard Java interface for connecting from Java to relational databases.) For GC 10.2.0.3 or lower, Oracle does not support a Management Service installation on nodes running any kind of clusterware, including Oracle Real Application Clusters (RAC). This means that, unless running GC 10.2.0.4 or higher, it is not supported to install the OMS on a host containing a RAC database that houses the Management Repository.

The following sections describe each GC middle-tier subcomponent.

Oracle Application Server Containers for J2EE (OC4J)

The OMS sits within an Oracle Application Server 10*g* Release 2 (10.1.2) environment, thereby benefiting from Oracle Process Manager and Notification Server (OPMN) control. (OPMN allows you to manage all OMS processes.) The other two OMS subcomponents, Oracle Application

---

[8]EM10*g*R1 is bundled with Application Server 9.0.4.

Server and Web Cache, also integrate with OPMN for process management, death detection, and failover for OC4J and OHS processes. As such, you should control all Management Service subcomponents using either the Oracle Application Server command-line tool, *opmnctl*, or EM Application Server Control, as discussed in Chapter 9.

Oracle HTTP Server

Oracle HTTP Server (OHS) is the web server for Oracle Application Server, and is based on the proven code base of Apache 1.3 Web server. Oracle HTTP Server, as part of Oracle Application Server, offers the following functionality to Grid Control[9]:

- **Security**
  - Permits Grid Control to take advantage of the SSL encryption capabilities of Oracle HTTP Server to enable HTTPS security between browsers and the Grid Control Console.
  - Allows GC administrators to define directives to limit access control (for example, by domain name and IP address).
  - Provides the opportunity to integrate Grid Control with Oracle Application Server Single Sign-On (SSO) to authenticate GC administrators.

- **High availability and scalability**   Provides these features through the use of multiple Oracle Management Servers, each with its own Oracle HTTP Server.

- **Virtual directories**   Make URLs available for certain Grid Control functionality. For example, the Oracle HTTP Server includes the Management Service virtual directory *http://<OMShost>:4889/agent_deploy/* from which hosts to be managed can pull the *agentDownload* script and response file needed to install the Agent via this method.

- **PL/SQL stored procedures**   Provide access to PL/SQL code stored in the Management Repository database through the Oracle module *mod_plsql*.

- **PL/SQL Server Pages**   Allow Grid Control HTML pages to use PL/SQL code as the scripting language. HTML is translated into a stored procedure that uses the *mod_plsql* module to send the output to the Grid Control Console.

- **Server Side Include**   Affords an easy way to add static and dynamic content across the Grid Control site (for example, the uniform static header and footer information that appears on most Console pages).

- **Perl**   Supplies some dynamic content to Grid Control. The Oracle Application Server 10.1.2 bundled with Grid Control uses Perl 5.6.1.

- **Dynamic Monitoring Service (DMS)**   Collects runtime performance metric statistics for both Oracle HTTP Server and OC4J processes. With this data, you can locate bottlenecks and tune application servers accordingly, though you should not tune the HTTP Server bundled with GC. However, DMS does supply Grid Control's many HTTP Server metrics, including statistics on active connections, memory, process and CPU usage, request failures, and error rates.

---

[9] Source: *Oracle HTTP Server Administrator's Guide 10g Release 3 (10.1.2)*, Chapter 1. I've adapted this material to explain how Grid Control takes advantage of generic Oracle HTTP Server functionality.

OracleAS Web Cache

OracleAS Web Cache (Web Cache) is a reverse proxy server accelerator, which is a server that appears as a content server to clients outside the firewall, but actually relays requests to back-end servers behind the firewall, then delivers retrieved content back to the client. This provides the following benefits for Grid Control[10]:

- **Performance**   Reduces response times to Console requests by storing frequently accessed URLs in memory to avoid processing duplicate URL requests on the Oracle HTTP Server and Management Repository database. Not only does Web Cache cut down on the number of URL requests, its external cache also increases static throughput (logos, menus, tabs, and header/footer information) of the Grid Control Web site, processing these requests 10 to 100 times faster than application-specific object caches. Note that over 90 percent the data on GC Console pages is dynamically generated, and does not benefit from increased Web Cache throughput.

- **Scalability**   Allows for many more concurrent Console connections, thereby reducing Oracle HTTP Server errors. The increased throughput itself also provides scalability.

- **Cost savings**   Provides cost savings for Grid Control implementations because of the increased performance and scalability. Administrators need fewer Oracle Management Servers (i.e., fewer HTTP Servers) for the same Console load.

OracleAS Web Cache behavior varies depending on whether a request from a GC administrator results in a cache hit or a cache miss. Here's how it works. An administrator logged into the GC Console through Web Cache sends an HTTP or HTTPS request for content. Web Cache acts as a virtual server on behalf of the Oracle HTTP Server for its Management Service. If any part of the content is in its cache (such as from a previous navigation to this page), Web Cache sends this part of the content directly to the Console browser and stores a copy of the page in cache. This is a cache hit. If Web Cache does not have the requested content or if the content is stale or invalid, it hands off the request to the Oracle HTTP Server, which requests this data over HTTP or HTTPS from the Management Service. This is a cache miss.

Oracle Management Repository

The Oracle Management Repository is the comprehensive source for all the management information for Grid Control. It consists of schema definitions, stored procedures, and RDBMS jobs within an Oracle database, all owned by the SYSMAN database user. A particular Grid Control implementation employs only one Management Repository. You can install multiple Management Services, but each must use this central Repository as their data store.

All Grid Control Console administrators share the same Repository information based on the priviledges granted to them. Information in the Repository includes:

- Configuration details about the managed targets
- Managed target availability information

---

[10]Source: *Oracle Application Server Web Cache Administrator's Guide 10g Release 2 (10.1.2)*, Chapter 1. I've adapted this material to explain how Grid Control takes advantage of generic Web Cache functionality.

- Historical metric data and alert information
- Target response time data
- Inventory information on patches and products installed

This information allows administrators to manage their complete application stack databases, application.

The Management Repository can reside on either the same host as a Management Service or on a dedicated host. You make this decision on the first screen of the Grid Control installer, where you select one of the first two Installation Types: Enterprise Manager 10*g* Grid Control Using a New Database or Enterprise Manager 10*g* Grid Control Using an Existing Database. When you choose the first installation option, the installer creates a new 10.1.0.4 database, and lays down a new Repository, OMS, and Agent, all on the local host. As for the second installation option, the OMS can be placed on either the OMR host or a separate dedicated host—you run the installer on the host where you want the OMS to reside. Oracle supports running the Repository in either a single-instance or Real Application Clusters (RAC) database, Release 9.2 or 10*g*. You must locate the Repository in an Enterprise Edition database and must select the Partitioning Option because GC uses partitioned tables to store management data.

Now let's examine the Repository schema, including the schema owner, tablespaces, and objects within that schema. SYSMAN is both the schema owner and the default Super Administrator account. You cannot remove or rename the SYSMAN account. It can be used to:

- Create GC privileges and roles
- Perform the initial GC setup, such as creating administrator accounts and notification rules
- Discover new targets
- Create generic jobs to run on all databases or hosts

The Grid Control installer creates two default tablespaces to hold its objects:

- **MGMT_TABLESPACE**   This tablespace holds all management information except configuration management data.
- **MGMT_ECM_DEPOT_TS**   This tablespace stores configuration management data, which includes support for Binary Large Objects (BLOBs). Aside from the benefit of logically isolating configuration data in a separate tablespace, it is a best practice for performance and tuning reasons to locate BLOBs in a separate tablespace.

As for the schema objects, an open Repository schema is the key to one of Grid Control's most important architectural features: extensibility. An open schema means that you can customize the use of Repository data if the standard configuration does not meet your requirements. The Repository tablespaces contain base tables and indexes that begin primarily with MGMT_ and other data types, including over 3000 database *views*. (Views are virtual object tables that offer specialized or restricted access to the data and objects in a database. Views provide the flexibility to look at the same relational or object data in more than one way.[11]) The views, whose names start

---

[11] *Oracle Database Concepts 10g Release 2 (10.2)*, Chapter 27.

mostly with "MGMT$", are particularly handy for mining Repository information that you want to process further. The views are comprehensive so that you can avoid having to directly access the base tables. The inherent advantage to these views is they insulate custom applications from underlying changes to the base SYSMAN schema due to new releases or patching.

CAUTION
Views supplied with Enterprise Manager 10g Release 10.2 are "provisional." This means that while Oracle fully supports these views for Release 2, they may change in subsequent releases and are not guaranteed to be backward compatible.

Following are the categories of GC views that provide access to metric, target, and monitoring data stored in the Repository:

- Monitoring
- Inventory
- Policy definition
- Policy association
- Policy violation
- Management template
- Job
- Application Service Level Management
- Configuration
- Oracle home patching
- Linux patching
- Security
- Configuration management
- Database cluster
- Storage reporting

Views provide the basis for passing alerts on to other System Management products, for customizing new add-ons, reporting, performing historical analysis, and data computation.

Data Flow Between Grid Control Components

Now that you know all GC components and subcomponents, it's time to connect the dots between components to understand how they interact, even down to the subcomponent level as is required to do justice to the Grid Control architectural model. Figure 1-3, 6 diagrams the internal workings of this engine to show how data flows between components for both Console communications and Agent uploads of metric data and alerts.

20 Oracle Enterprise Manager 10g Grid Control Implementation Guide

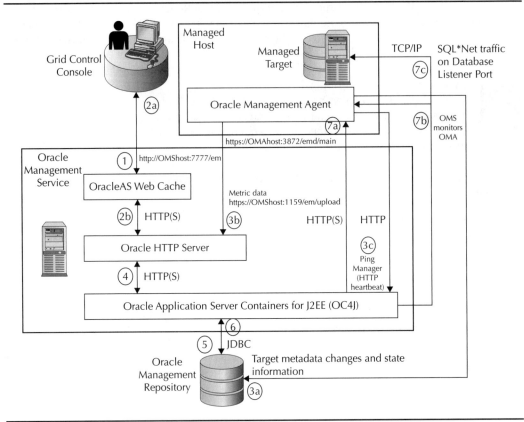

FIGURE 1-3. *Data flow for interactions between Grid Control components*

Let's follow the numbers and arrowed lines in this figure to visualize the flow of Console and management data requested from one component or subcomponent to another. Default ports are listed in the figure. However, an administrator can customize all ports within the GC infrastructure as a specific site's environment requires.

1. An administrator logs in to the Console at http[s]://<OMShost>:<port>/em. Port 7777 shown in the figure is the Web Cache nonsecure login port for UNIX.

2. The administrator requests content in the Console, prompting two possible responses, depending on whether the requested content is found in Web Cache (this is irrespective of whether you log in via Web Cache):

    a. If some of the requested content is in its cache (such as from a previous navigation to this page), Web Cache sends this content directly to the Console browser.

Chapter 1: Overview of the Grid Control Architecture 21

    b. If Web Cache does not have any of the requested content or if the content is stale or invalid, it hands off the request to the Oracle HTTP Server (2b), which requests this data over HTTP or HTTPS from the Management Service (4). The OMS then requests this data via JDBC from the Repository (5), and, upon receipt, renders the content in the Console from (6) to (4) to Console.

3. An Agent communicates with both its OMS and the Repository. The Agent uploads its information directly to the Repository or through the Oracle HTTP Server of the OMS to the Repository, depending on the criticality of the Agent communication as defined by the Agent channels (see "How Agent Files Are Uploaded" in the online Appendix A). The Agent also sends a periodic heartbeat called an *Agent ping* to its OMS indicating that it is available. Here are the specifics:

    a. The Agent connects directly to the Management Repository to report target metadata changes (contained in A channel files) and state information (stored in B channel files).

    b. The Agent uploads metric data (C and D channel files) to the Oracle HTTP Server URL (unlike administrator requests, which first go through Web Cache if they're logged in that way). The default HTTP Server URL is https://<OMShost>:1159/em/upload for secure HTTPS uploads. (The Agent uses http://<OMShost>:4889/em/upload for unsecured HTTP uploads and wallet requests to secure the Agent; even when the Agent is already secured, the unsecured port must still be accessible to allow an Agent to re-secure if necessary.)

    c. The Agent heartbeat sent to the OMS indicates that the Agent is running.

4. The Oracle HTTP Server uploads Agent data over HTTP or HTTPS to the OMS.

5. The OMS uploads Agent data via JDBC to the Repository.

6. If the Repository needs to send data to the Agent, it uses JDBC to send information to the OMS to relay to the Agent via its built-in HTTP listener.

7. The OMS communicates to the Agent in several ways:

    a. The Management Service forwards data directly to the Agent over HTTP. As mentioned earlier, no Oracle HTTP Server is required on the Agent host because the Agent software includes a built-in HTTP listener listens on *http://<OMAhost>:3872/emd/main*. The OMS also submits jobs and other management tasks through this URL.

    b. The OMS uses the Agent URL to monitor the status of the Agent. (If the OMS-to-Agent communication is not successful, the OMS pings the host on which the Agent resides with ICMP Echo requests.)

    c. The OMS sends SQL*Net traffic over TCP ports to target database listeners on Agent hosts, such as when patching database targets through Grid Control.

Now that you've looked more closely at the internal workings of the Grid Control engine, you'll have a better appreciation for it when you fire up the "stock" engine in Chapter 3. With enough detailed instructions, even a novice mechanic can install the base GC product. You'll

> **Grid Control Weaves into a Grid Infrastructure and Monitors Itself**
> Now that you have studied the GC components, you can plug them into a grid infrastructure and use Grid Control to monitor that infrastructure, including its own components. Yes, Grid Control can actually monitor itself. Think about it. The Repository is housed in a database, which can be one of the databases supported by the RAC Cluster in Tier 3 (refer back to Figure 1-1). The application tier (Tier 2) can host the OMS. The Console is a browser on a client workstation, like any other client in Tier 1. Agents on all grid hosts monitor all grid components as targets, including targets without operating systems (switches, routers, load balancers, etc.) monitored by remote Agents, and the Grid Control components and subcomponents themselves.[12] While it is the best practice to weave Grid Control into a grid computing environment, Grid Control monitors itself irrespective of your adherence to a grid computing architecture.

have plenty of time before the tougher high availability custom job in Chapter 15 rolls into your garage.

Grid Control, Database Control, and AS Control

For EM 10g, the term "Enterprise Manager" encompasses Grid Control, Database Control, and Application Server (AS) Control. The word on the street is that the release of EM 11g is taking an ambitious "EM Everywhere" approach to eliminate the distinctions between the "Controls". For EM 10g, however, we must maintain this distinction.

Think of Database Control as Grid Control for just one database. Therefore, the database-related material in this book, with few exceptions, applies equally to Database Control as it does to Grid Control. Application Server Control administrative functionality, on the other hand, is not encompassed in Grid Control, but is part of the Oracle Application Server. Grid Control does monitor AS Control, and simply provides the administrative link to AS Control for the AS targets it monitors. While this book touches on Application Server Control as it relates to Grid Control, for more in-depth coverage of AS Control, consult the Oracle documentation for Oracle Application Server or refer to the excellent book *Oracle Application Server 10g Administration Handbook*, McGraw-Hill/Osborne, 2004.

NOTE
This book narrows the scope of Enterprise Manager coverage to just Grid Control. It does not address Database Control or Application Server Control per se, although most of the target database-related material applies just as much to Database Control as it does to Grid Control.

What differentiates Grid Control from Database Control and from Application Server Control?

---

[12] How can Grid Control send you notification that the Repository or OMS are down when they must be up to send such notification? For the answer, see the section "Enable Notification if Grid Control Goes Down" in Chapter 11.

Grid Control vs. Database Control

What is the difference between Oracle Enterprise Manager 10*g* Database Control and Oracle Enterprise Manager 10*g* Grid Control? Database Control is a web-based application (like Grid Control) installed with Oracle Database 10*g*. From the Database Control Console, you can only administer the accompanying single instance or RAC Oracle Database and ASM instances the database uses for storage.[13] The Database Control architecture is very similar to that of Grid Control but on a smaller scale. The equivalent of the Grid Control Central Agent and OMS are rolled into the same OC4J application, and the Repository is located in the target database itself. This means that Database Control does not work if the target database is down, except that it can start the database. The Database Control home page has the same look and feel as the Grid Control home page. To take an example, here is a home page for a cluster database named *ptc*.

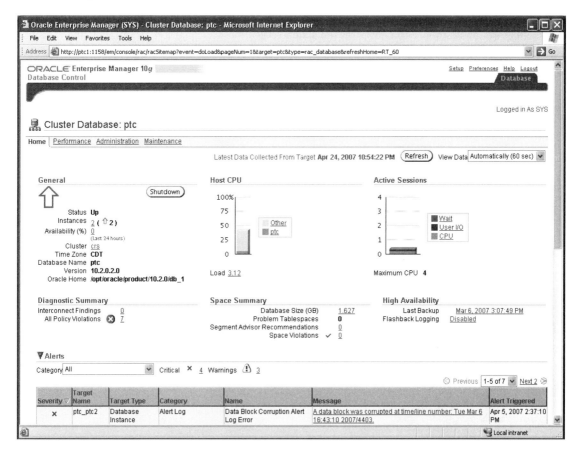

It is clear from this home page that this Database Control Console only manages this particular cluster. By contrast, from the Grid Control Console, you can centrally manage multiple databases

---

[13] Database Control also gathers some host statistics for the server on which the database resides. Therefore, you can also monitor and manage the host with Database Control.

and many other types of targets as well. It is best practice to run either Grid Control or Database Control, but not both, although you can run both concurrently.

Grid Control vs. AS Control

How is Grid Control different from Application Server (AS) Control? In short, as just one of many target types it monitors, Grid Control *monitors* the Oracle Application Server, whereas AS Control *administers* the Oracle Application Server.

Grid Control Monitors Oracle Application Server

While Grid Control administers all Oracle products in your enterprise, it only monitors the Oracle Application Server—both the AS bundled with Grid Control and any separate AS targets that Grid Control monitors. Monitoring AS instances in the Grid Control Console encompasses real-time monitoring, alerting, and historical data collection, just as with any other Grid Control target. For example, the Application Servers tab within the GC Console reveals the status, alerts, policy violations (noncompliance with a desired state for security, configuration, or storage), and resource usage of all Application Servers and their subcomponents, as shown here.

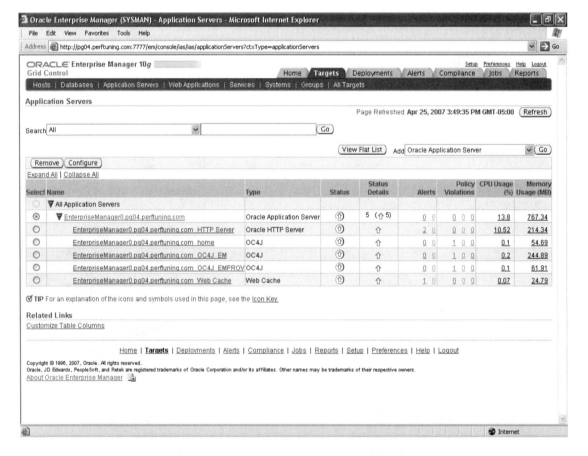

AS Control Administers Oracle Application Server

By contrast, use the standalone AS Control Console to administer AS subcomponents. To get a better understanding of what it means to administer the Application Server, take a look at the AS Control Console home page, accessible directly at http://<AShost>:1156 on UNIX or http://<AShost>:18100 on Windows, or through the Grid Control Console on the Application Server home page using the Administer link.

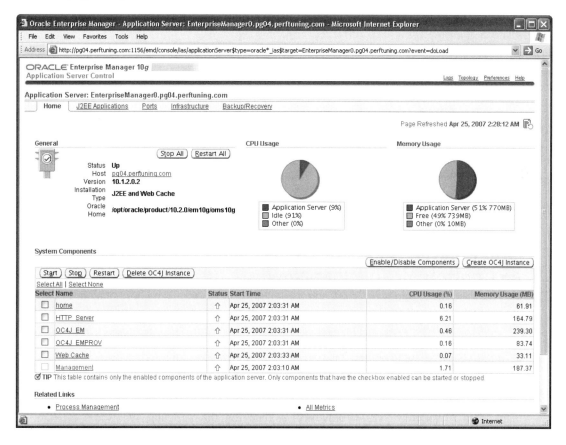

From the AS Console home page, you can control OMS and non-OMS Application Server subcomponents—the Oracle HTTP Server, OC4J instances, and Web Cache. This control is from A to Z; you can stop, start, restart, enable, or disable subcomponents. You can also gather information about the management software itself by clicking the Management link at the bottom of the System Components section on the Application Server home page. The AS Console has four other tabs: J2EE Applications (which allow you to view response times for OC4J instances), Ports (which you can change through the UI), Infrastructure (including Identity Management and OracleAS Farm Repository Management), and Backup/Recovery (of AS data and configuration files). Without going into more detail, as Application Server Control is beyond the scope of this book, I hope I've given you just enough information to make the distinction clear between Grid Control and Application Server Control.

Agents for Grid Control, Database Control, and AS Control

Now that we are clear that Grid Control, Database Control, and Application Server Control are all distinct, you're probably not surprised to hear that separate Agents exist for each. The Grid Control central Agent on each managed host (including OMS and OMR hosts) monitors all targets on that host and uploads target information to the SYSMAN schema in the Management Repository.

NOTE
In this book, Management Agent *or* Agent *refers to the* central *Grid Control Agent.*

The Database Control Agent, on the database host, similarly stores data in a SYSMAN schema, but this schema is local to its associated Oracle database. In contrast to both Grid Control and Database Control Agents, the Application Server Control Agent, bundled with the Oracle Application Server, only communicates real-time data internally to the Application Server Control, as there is no Repository for historical data. All three Management Agents are distinct. They run from their respective home directories—the Grid Control Agent from its own Agent home, the Database Control Agent from the database home, and the Application Server Control Agent from the Application Server home including that for the OMS.

Grid Control, Database Control, and AS Control: All Together Now

Let's build on the topology of Grid Control core components shown earlier in Figure 1-2 by adding Database Control and Application Server Control to the mix, as represented in Figure 1-4 (I've omitted protocols for clarity).

I have also included four typical managed targets: a 10g Database Server, a 10g Application Server, an 8i/9i Database Server, and a third-party application. For good measure, the figure illustrates two OMS hosts, with two of the central Agents uploading management data to one of the OMS hosts, and the remaining two Agents uploading to the other OMS host.[14] In addition, a Grid Control Console communicates with an OMS, which also receives Agent uploads of target metric data. Although each OMS only receives data from the particular Agents that report to it, each OMS uploads this data to a central Management Repository, which in turn makes information from all Agents available to each OMS. This allows any GC Console to administer or monitor all Grid Control targets. On the other hand, a particular Database Control Console is in contact with just one database, and likewise, a particular AS Control Console connects with just one Oracle Application Server. The Database and AS Consoles are privy only to local database and Application Server data, respectively. All three Consoles can operate concurrently, but (as already stated) it is best to shut down Database Control when using Grid Control, as Grid Control can administer each of the databases to which a particular Database Console has access.

---

[14] In a high availability configuration, as discussed in Chapter 15, rather than assigning Management Agents to particular Management Services, you would use a hardware server load balancer to virtualize the Management Services and shared storage to allow for failover between them.

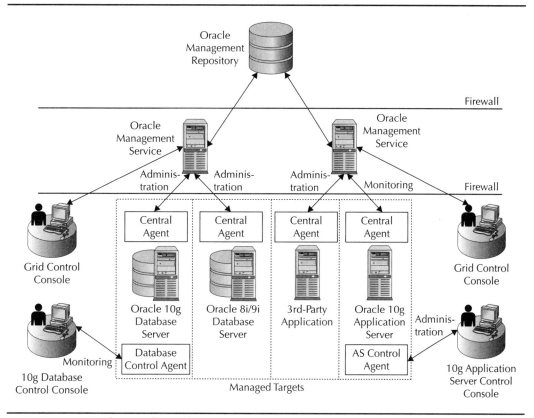

FIGURE 1-4. *Grid Control, Database Control, and AS Control*

Each Application Server Control Console should remain available, however, to administer a particular Application Server, because Grid Control can only monitor an Application Server. I hope all this makes the distinction clear between Grid, Database, and AS Control.

Summary

Here is a synopsis of what was covered in this introductory chapter:

- We would be remiss if we made no mention of grid computing technology, which Grid Control was designed to administer. I began this chapter by presenting a brief overview of grid computing.

- Next, I provided a brief history of Grid Control and highlighted several product features.

- From there, I introduced you to the four main Grid Control components: the Console, Agent, OMS, and Repository.

- I then broke down the main GC middle-tier components into subcomponents, and examined the interaction between all GC components. You saw how components work together to satisfy Console requests and upload metric data from the Agent on a target host through the OMS to the Repository, and how the OMS in turn delivers data back to the Agent.

- I ended this chapter with an ancillary but critical explanation of how Grid Control differs from Application Server Control and Database Control. This delineation helps further refine the scope of the book to just Grid Control and Application Server Control as it relates to Grid Control.

Now that you understand the basic architecture and concepts of Grid Control, we can proceed to Chapter 2, where we'll meet in the garage for preinstallation. Then, in Chapter 3, we'll install Grid Control and get this thing on the road.

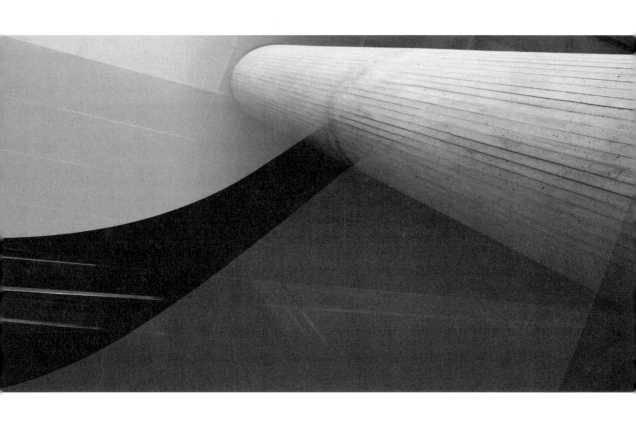

CHAPTER 2

Grid Control Preinstallation

In Chapter 1, you got a feel of the workings of the Grid Control "engine" by becoming familiarized with its components (and subcomponent) and examining how they work together to process Agent management data and Console requests. In this chapter, we prep the engine, and in Chapter 3, we begin building it. The preinstallation steps discussed here are necessary to install a basic GC environment, which consists of an Oracle Management Repository (OMR) database (single-instance or RAC) servicing one or more Oracle Management Service (OMS) hosts that in turn talk to Oracle Management Agents (OMAs) on each target host. You will prepare to install the OMR database and one OMS either on the same host or on separate hosts, in close proximity to one another (i.e., in the same data center and preferably on the same subnet). I also provide sizing guidance and instructions for deploying additional Active OMSs on separate hosts.

NOTE
This chapter is for setting up Active/Active OMS and Repository (i.e., RAC) installations. See Chapter 15 now if interested in implementing an OMS or OMR Active/Passive configuration using Cold Failover Cluster (CFC) as each requires a special Grid Control installation procedure.

Let's take a look at how the preinstallation tasks are organized in this chapter. They fall into the following four major categories:

- **Architectural design**   The first preinstallation step is to make the minimum number of design decisions required to install a basic GC environment, which can serve as a common denominator for any high availability and disaster recovery topology to be implemented when we get to Chapter 15 (except CFC configurations, which you must implement when installing GC).

- **Network configuration**   After fleshing out the basic architecture, confirm the network configuration, including host naming rules and constraints and connectivity checks between GC component nodes.

- **Hardware requirements**   Provide the hardware resources (disk space, RAM, swap, and CPU speed) for OMR and OMS hosts to meet Grid Control installation and operating requirements.

- **Software requirements**   Confirm that GC meets all certification standards, create the necessary OS groups, users, and directories, stop listeners on port 1521, synchronize OS timestamps/time zones, and satisfy all platform-specific software requirements.

The preinstallation tasks in this chapter apply to the hosts where the OMR, OMS, and chain-installed Agents will be installed, as well as to hosts where standalone Agents will be installed.

Stage the Grid Control Software
OMR, OMS, OMA

In the inimitable IT multitasking fashion, I advise that you begin downloading and staging the GC software while carrying on with the remaining preinstallation steps. You'll need to stage the distribution soon enough anyway to run the prerequisite checker, as described shortly. To stage the GC software, you must obtain it, then set up access to run the installer on the chosen host. The method of access varies depending upon whether the target system is local or remote. There are two ways to obtain the Oracle Enterprise Manager 10*g* Grid Control distribution software for a particular platform: download it from OTN or order the Grid Control Media Pack from the Oracle Store. (Even if you order the Media Pack, you still need to download the latest Patch Installer, which is not included.)

- **Download from OTN** Download the latest full GC software release for your platform from the Oracle Technology Network (OTN) Web site at http://www.oracle.com/technology/software/products/oem/index.html. Make sure to select the link under the Full Installers section for the latest full installation available for your platform.[1] The Full Installers contain the OMR, OMS, OMA, and Management Packs, whereas the Patch Installers only contain the OMS, OMA, and Management Packs. No platforms currently offer Full Installers for the latest GC release. Therefore, you need to download the Full Installer archive for an earlier release, and the Patch Installer archive for the latest release (applied in Chapter 4). The Full Installer archives consist of one or more zip files, depending upon the platform, ranging from 1.3G on Windows to 4.3G on HP-UX Itanium. Download all zip files and unzip them into the same staging directory (it creates its own directory structure).

- **Order the Media Pack** Alternatively, order the latest Enterprise Manager 10*g* Grid Control Media Pack for a particular platform from the Oracle Store for a nominal fee. To order the Media Pack, go to http://oraclestore.oracle.com, choose your country, click the Media Packs link on the left navigation pane, select the platform, select the Enterprise Manager Media Pack, click Add to Cart, Checkout, and follow the checkout procedures. The Media Pack for each OS contains the base EM distribution and the Monitoring plug-ins (discussed in Chapter 11). When you receive the Media Pack, you can either run the DVD-ROMs directly or stage them on disk.

Whether you download the distribution software or order the Media Pack, you also need to download the Agent software for managed host platforms other than your OMS platform. Download this Agent software under the Mass Agent Deployment section of the OTN site indicated above for Enterprise Manager.

Once you obtain the GC software from either one of these sources, you can access the software via several methods, depending on whether the target system is local or remote.

(Continued)

---

[1] I assume here that you are doing a fresh GC installation. If you need to upgrade a GC 10.2.0.X or 10.1.0.X installation, see the latest Patch Installer README or release notes for your platform.

If the target system is local, access the GC software directly on that system. On Windows platforms, no further action is required to run the installer locally. On UNIX platforms, run the installer either from an Xwindows session or from a UNIX workstation.

If the target system is remote, access the GC software from a remote DVD drive or use remote access software, as follows:

- Access the GC software from a remote DVD drive (UNIX only). If the system where EM is to be installed does not have a DVD drive, you can share a remote DVD drive.

- Access the GC software using remote access software. If you don't have physical access to a remote system where installing GC, but this system has a local hard drive, you can install GC using remote access software such as Microsoft built Remote Desktop Connection, VNC from RealVNC Ltd., or Hummingbird Exceed. Start the remote access software on both your local computer and the remote system and install GC in one of the following ways:

  - Copy the GC software to a hard drive and install from that hard drive. To do this, share the hard drive, then using the remote access software, map a drive letter on the remote system to the shared hard drive.

  - Access GC software from a DVD drive in your local computer. Insert the DVD into a drive on your local computer and share the DVD drive. Use the remote access software to map a drive letter on the remote system to the shared DVD drive. You are now set up to run the Oracle Universal Installer (OUI), as described in Chapter 3, from the shared hard drive on the remote host using the remote access software.

On most UNIX systems, a DVD disk mounts automatically when inserted into the disk drive. In case the DVD is not automatically mounted, the following lists the command to set the mount point for each UNIX platform (execute as *root*):

AIX	HP-UX	Red Hat Linux	SUSE Linux	Solaris
/usr/sbin/mount -rv cdrfs <DVD device name> <disk mount point directory>	/usr/sbin/mount -F cdfs -o rr <DVD device name> <disk mount point directory>	mount -t nfs <hostname>:/mnt/<DVD_path>	mount -t nfs <hostname>:/media/<DVD_path>	/usr/sbin/mount -r -F hsfs <DVD device name> <disk mount point directory>

See the platform-specific GC documentation for more detailed instructions on mounting the product disks.

How do chain-installed Agents (as referred to in the Oracle EM documentation) differ from standalone Agents?

- **Chain-installed Agents**   You do not explicitly install these Agents; their installation is bundled with that of an OMS . The Agent accompanies the OMS, "tags along for the ride," so to speak, to monitor the OMS as well as the Repository (if installed on the same host) and any other host targets. Chain-installed Agents don't have any distinct prerequisites beyond those for the OMS they accompany.

- **Standalone Agents**   You explicitly install these Agents on each target host (except on OMS hosts where Agents are chain-installed) using one of six methods covered in Chapters 5, 6, and 7. Some of the preinstallation steps for the OMS also apply to standalone Agents, as indicated in this chapter.

Standalone Agent and chain-installed Agent installations, once completed, are identical. Each plays the role of a central Agent that GC needs to manage a particular target host.

In this chapter, I annotate all preinstallation tasks with OMR, OMS, and/or OMA (for any Agent installation method) to indicate on which host(s) to execute the task. Complete all OMR and OMS requirements now on the OMR node(s) and OMS hosts, respectively. (You can repeat the OMS tasks covered in Chapter 15 for any additional OMS hosts you may install then for high availability reasons.) You can either complete the standalone OMA requirements on all target hosts concurrently, or wait and circle back to them when you get to Chapter 5, which lists additional prerequisites for standalone Agents. Most administrators are anxious to install the OMR and OMS and don't want to concurrently perform certain preinstallation steps for standalone Agents on all target hosts. If you feel this way, then, as the wizard says in *The Wizard of Oz*, "Pay no attention to that man behind the curtain," i.e., "Pay no attention to those standalone Agent requirements in Chapter 2." (My quote doesn't have the same ring to it, but you get the idea.) You can certainly wait to take care of these Agent requirements all at once when you get to Chapter 5. Don't worry; I will remind you.

NOTE
Again, the OMA annotations below indicate standalone (i.e., central) Agent preinstallation steps for any Agent method. You can elect to postpone these steps until Chapter 5, as they are not required for the Agent chain-installed with the OMS in this chapter.

While many of the preinstallation steps apply generically to all operating systems, the commands to verify or fulfill some of these steps differ by platform. Where the syntax varies, I provide the variations for Windows and all flavors of UNIX.

Key Architectural Design Decisions

Grid Control preinstallation starts with design. This book splits that design into two stages: build a stock GC engine (described in this chapter), then "sup' up" this engine to meet high availability (HA) and disaster recovery (DR) requirements (covered in Chapter 15).

Here are a handful of few reasons to justify this two-tiered approach for those IT professionals who are skeptical by nature (i.e., for all you IT professionals):

- It's a daunting task for a Grid Control newbie to tackle a full-blown implementation using all the HA and DR bells and whistles. On the other hand, it's very attainable for an experienced DBA not completely familiar with Grid Control to make a few easy preliminary design decisions to implement the base product according to step-by-step instructions. Supplying such instructions is the major intent of this chapter.

- Oracle shops with little or no production monitoring in place understandably want to get GC up and running as quickly as possible to start monitoring their systems. It's better to install the base GC product quickly to reap the benefit of immediate out-of-box monitoring than to continue without monitoring while you design and implement a high availability GC solution.

- You can benefit from GC data about itself to better compute its availability and performance as measured against SLA requirements. From this data, you may determine that the additional OMS host or RAC Repository node you planned to add isn't necessary after all.

Even a stock GC engine has built-in options. Choose those options now that you absolutely need, and leave the high-end stuff for later. You don't need to know the inner workings of a sports car engine to buy one and drive it off the lot. You only need to answer three key questions now to install a basic GC system in Chapter 3:

- How many Grid Control environments should you build?
- How many regional sites are required?
- What installation choices should you make?

Before addressing these questions, let me define what I mean by a GC site as opposed to a GC environment. A particular GC environment is comprised of one or more sites.[2] A Grid Control environment is a system consisting of one or more coupled OMS and OMR hosts (single-instance or RAC) that manage Agents on target hosts. Targets on these hosts can be production, nonproduction, or both, depending on whether you decide to monitor production and nonproduction targets with separate GC environments. Regardless of this decision, the targets for a particular environment are logically related in that you want to manage all of them with just one GC system, if possible. Grid Control sites are mini-environments (each with an independent set of OMS, OMR, and OMA hosts) that are logically coupled with other sites to provide the geographical coverage necessary to service a particular environment. Network limitations or other considerations require you to install multiple sites. The sites are logically, not physically, related so they are really environments in their own right.

---

[2] In this chapter, I distinguish between a GC "environment" and a GC "site." However, in other chapters, I use the terms interchangeably.

Decide How Many Grid Control Environments to Build

OMR, OMS

Before you can prepare OMS and OMR hosts for installation, you need to select the host or hosts where these components will reside. The first question that springs to mind is whether this first (and perhaps only) GC environment will monitor production targets, nonproduction targets, or both. Most companies want to monitor nonproduction targets, too, so they must choose one of the following GC instance management strategies:

- Build at least two GC environments, one for production targets and one or more environments for nonproduction targets
- Build one GC environment for both production and nonproduction targets

NOTE
Your instance management strategy may call for multiple nonproduction environments, such as testing, pre-production, and development. However, most companies do not run more than one nonproduction environment.

A GC implementation is itself a production target if it monitors production targets. As such, best practice is the first option: configure a production GC environment to monitor only production targets. This prevents nonproduction targets from adversely impacting a production environment. "Shorts" can occur anywhere in the "wiring" that connects components: from the Agent to the OMS, and from the OMS to the Repository:

- **Agent to OMS**   On a host with both production and nonproduction targets, the Agent-to-OMS upload process can stall on a file containing nonproduction data that must be synchronously loaded, i.e., data that blocks all other Agent uploads. These queued Agent uploads could contain critical production alerts whose notification would be delayed by a "stuck" nonproduction data upload.

- **OMS to OMR**   The collection of nonproduction target data can cause a Repository database deadlock or other unexpected conditions, thereby causing an OMS-to-Repository loader backlog of Agent data.[3]

A prerequisite for isolating production and nonproduction GC management is that target hosts exclusively contain either production or nonproduction targets. Only one Agent can run on a given host (perhaps a virtual host), and this Agent can only belong to one GC environment. If, as a rule, production and nonproduction targets are allowed to "cohabitate" on the same servers, you may as well let one GC environment manage both types of targets.

---

[3] A salient case in point, since fixed in EM 10.1.0.4, was Bug 3785480 where a metric query that took a long time to execute due to a row exclusive blocking lock on a Repository table resulted in a backlog on the OMS tier of Agent files to be uploaded to the Repository. Some Oracle customers opted to completely shut down all production Agents until the bug was fixed.

CAUTION
Oracle does not support running more than one Agent on a single host, except in Active-Passive environments that run third-party Cold Failover Cluster (CFC) software.[4] While it may be feasible to do so, such as by changing the default ports of the second Agent installation, it makes no sense as a way to separate production from nonproduction environments. If a host contains both production and nonproduction targets, you've already intertwined these two environments in that they both must contend for common hardware resources. Installing separate Agents would only compound the problem since multiple Agent installations would have to vie for the same hardware resources.

This advice that, with regards to production and nonproduction targets, "neither the twain shall meet" also applies to production OMS and OMR hosts and nonproduction targets. In fact, try to avoid placing even production targets on hosts running production OMS or OMR components, as it's not possible to predict possible resource contention. Any target that consumes undue resources on a GC infrastructure host can cripple or even bring down that host, thereby jeopardizing the monitoring of all targets.

If you've decided on separate production and nonproduction GC environments, it's standard practice to build the nonproduction environment first, just as it would be for any software application. If you plan to monitor all targets with one environment, then consider this chapter the beginning of a production build. Either way, you need to earmark one to three servers for your installation. Decide now whether these servers are for a production or a nonproduction GC environment. In the following two sections, you will determine the number of servers required and the GC components to be installed on each server.

How Many Regional Sites Are Required?
OMR, OMS

A question that goes hand in hand with whether to build separate production and nonproduction environments is "How many sites are geographically required for a particular GC environment?" More aptly put, what is the fewest number of GC sites you can get away with for a given environment? For instance, if all production targets are on the east coast of the U.S. and all nonproduction targets are on the west coast, network bandwidth constraints due to geography may require building two regional sites, one for each environment.

Here I define a GC "site" as a logical entity. One or more GC sites comprise a particular GC environment. Companies deploying Grid Control install one or more regional sites to provide complete geographical coverage for an environment, which is defined in terms of managing production targets, nonproduction targets, or both. Sites are logically related, but are physically no different from environments. GC sites/environments are physically unrelated to other sites/environments; that is, all components—the constituent Agents and OMSs reporting to a Management Repository—for a particular site/environment are separate from any other site/environment. The difference is purely on a logical level.

---

[4] In CFC environments, you must install one Agent to each physical node of the cluster and another Agent using the logical host name of the cluster.

Barring the instance management strategy dictating the number of environments, Oracle's direction, and that of the IT industry in general, definitely favors a consolidation strategy of using just one site to manage each environment. The technology direction for Grid Control consolidation is no exception. Grid Control has been implemented globally, but only at firms with fast networks. To calculate whether you can get away with one global site for a particular environment, you need to know the network requirements between GC components.

It turns out that network performance between the OMS and OMR dictates how many GC sites each environment requires. Fortunately, Grid Control network error handling is robust, allowing it to tolerate network glitches and outages between components. However, Agent/OMS and Console/OMS network issues impact GC performance much less than network performance between OMS and OMR tiers. A particular Agent to OMS link may even be severed for a time without being noticed at the system level. Of course, certain administrators may be left hung out to dry if the network connection slows between the Console and a particular OMS. However, a network problem between OMS and OMR tiers has more global implications, because it could significantly reduce overall system performance. Console connection performance would deteriorate, along with monitoring (alerts and notifications), job execution, and every other GC function.

NOTE
It is the minimum network performance requirements between OMS and OMR hosts that usually dictate how far you can "spread" an installation, not any inherent limitations of Grid Control, which can scale for hundreds of administrators and tens of thousands of targets.

Figure 2-1 shows the minimum bandwidth and maximum latency requirements between OMS, OMR, and OMA tiers. And Table 2-1 summarizes these requirements for easy reference. OMR to OMR storage requirements are also given, but typically the OMR host is physically coupled with its storage devices.

Both Figure 2-1 and Table 2-1 show that the minimum bandwidth and maximum latency requirements between OMS and OMR hosts are much more stringent (10× or so) than those between OMS and OMA hosts. (Ideally, OMS / OMA latency should be less than 1ms.) Due to the necessarily tight network coupling between OMS and OMR tiers, most sites run at least one OMS host from within the same data center as the OMR host, usually placing one server next to the other (or these days, one blade on top of the other). Corporate Local Area Networks (LANs) typically run Gigabit Ethernet (GigE) or higher, so they are almost always fast enough to meet the 1Gbps minimum bandwidth requirement between OMS and OMR. Table 2-2 lists LAN bandwidths for the most prevalent LANs.

OMS / OMR hosts should perform well if they're on a GigE LAN. But can you meet GC network requirements (particularly between OMS and OMR) when using multiple OMSs geographically separated and connected over a Wide Area Network (WAN)? Companies with computing resources located around the country (or even around the world) often rely on a WAN to connect LANs at remote office locations, and secure that WAN connection over a VPN tunnel. To establish whether the WAN is fast enough to support the OMS locations you are contemplating, let's look briefly at typical WAN speeds. How do they fare in relation to GC network requirements?

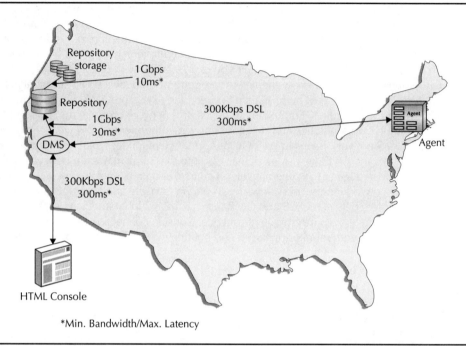

FIGURE 2-1. *Bandwidth and latency specifications between Grid Control tiers*

WAN bandwidths vary widely by provider, depending upon the underlying device technology. Most U.S. national and international firms run over at least a DS-1 (Tier 1) or DS-3 (Tier 3) WAN. Based on the broadband transmission rates shown in Table 2-3, DS-1 at 1.544Mbps provides substantially more than the required 300Kbps minimum bandwidth that Agent (and Console) connections with the OMS must meet.

However, not even DS-3 transmission speeds of 44.736Mbps would be sufficient to support OMS / OMR network bandwidth requirements, which are nearly 23 times higher (at 1Gpbs). These requirements would necessitate an OC-48 (2.488Gbps) or higher WAN connection. Many firms cannot justify the expense of such a high-speed broadband network. Those companies that

Communication between GC Components	Minimum Bandwidth Required	Maximum Latency Allowed
Console <-> OMS	300Kbps	300ms
OMA <-> OMS	300Kbps	300ms
OMS <-> OMR	1Gbps	30ms
OMR <-> OMR Storage	1Gbps	10ms

TABLE 2-1. *Minimum Network Requirements between Grid Control Components*

LAN Device	Bandwidth
Fast Ethernet (100base-X)	100Mbps
FDDI	100Mbps
Gigabit Ethernet (1000base-X)	1Gbps
Myrinet 2000	2Gbps
Infiniband 1X	2.5Gbps
10 Gigabit Ethernet (10Gbase-X)	10Gbps

TABLE 2-2. *LAN Bandwidths for Different LAN Devices*

can justify the expense can only do so for high-bandwidth networks between primary and disaster recovery (DR) sites, and they lease lines from telecommunications giants like Sprint, Cingular, and AT&T for such purposes. IT departments would be hard-pressed to justify vastly increasing their network budgets just to satisfy a single global Grid Control implementation. System Management Products like Grid Control are usually the orphan step-child of IT departments, and, from a budgetary perspective, must often ride on the coattails of other IT expenditures.

As for GC latency requirements, WANs almost always meet the 300ms latency requirements for Agent / OMS or Console / OMS communications. Even dial-up modem users over the Internet (WAN) typically experience only 150-200ms latency. However, while corporate LANs invariably meet the 30 millisecond (ms) OMS / OMR maximum latency, most WANs do not. Again, the exception is a company with a leased line between production and disaster recovery (DR) data centers. They may very well meet this 30ms latency requirement, provided their DR facility is close enough to the primary facility. (The speed of light is the limiting factor in meeting a 30ms latency over thousands of miles).

WAN Connection	Transmission Speed
DS-1 (Tier1)	1.544Mbps
E-1	2.048Mbps
DS-3 (Tier 3)	44.736Mbps
OC-3	155.52Mbps
OC-12	622.08Mbps
OC-48	2.488Gbps
OC-192	9.953Gbps

TABLE 2-3. *Broadband Transmission Rates*

> **Can You Run an Active OMS from a DR Location?**
> As explained in Chapter 15, you can configure a standby OMS host at a DR location and leave it down but ready for service to connect to a standby Repository in case of failover. However, you can also run an active OMS (or an Active/Passive OMS) at a DR location, although other DR hosts remain in standby mode. However, on failover you'd need to point this active OMS located at the DR site to the standby Repository. Wait until Chapter 15 to worry about DR and high availability considerations. For now, you just need to evaluate whether your DR and primary data centers are close enough to place an active OMS at each location.

So what does this all mean in terms of the fewest number of sites you can get away with for a particular Grid Control environment? Companies running a VPN over a WAN between offices can use a single GC site for nationwide or even worldwide coverage of each environment (which may just be one environment if it monitors both production and nonproduction targets). Agents and Console connections can span large distances to OMS hosts. However, these OMS hosts must be placed geographically close enough to the OMR node(s) to meet the networking requirements of 1 gigabit per second (Gbps) bandwidth and 30ms latency. This usually means placing one or more OMS hosts alongside the OMR node(s), and if an additional OMS host is required for scalability reasons, placing it at a remote location (perhaps at the DR data center) within 2,790 miles of the production location. This distance can be as the crow flies, or as the satellite flies (i.e., it can be from a Low Earth Orbit (LEO) satellite network typically 500 to 1,500 miles from earth). It so happens that 2,790 miles is the exact distance between New York City and Los Angeles (though why you'd want to place a DR data center at either location is beyond me). How did I come up with 2,790 miles?

The speed of light is 186,000 miles per second. If the second OMS host were separated from the Repository host by 2,790 miles, it would take at least 15 milliseconds for each packet to be sent between hosts. Taking packet acknowledgment into account, the latency doubles to 30 milliseconds.[5]

TIP
To "string out" your environment, you can geographically separate the first OMS host from the OMR host. You can then place one or more OMS hosts on the other side of the OMR host from the first OMS host.

With this background, you are now ready to start sketching a picture of your proposed Grid Control topology. Take out a piece of paper and a pencil, even crayons if you want—they're usually closer at hand, thanks to my four-year-old daughter. Let's paint by numbers now. You might not need to do this if your site is small and all servers are located in the same data center.

---

[5] Thanks to Ed Whalen, one of the technical editors of this book, for the general idea behind this calculation. See his excellent book, *Oracle Database 10g Linux Administration*, The McGraw-Hill Companies, Inc., 2005, Chapter 13.

However, if you have a disaster recovery site located elsewhere, you'll probably want to go through this exercise.

1. The first fork in the road is to decide how many GC environments are required. This depends upon whether you will use separate GC environments to monitor production and nonproduction targets, and upon your corporate IT management structure.

    - If your company routinely co locates production and nonproduction targets on the same host, then you should probably build one GC environment (consisting of one or more sites as geographically required) to monitor both production and nonproduction targets.

    - On the flip side, if your company runs independent regional data centers, perhaps across the country or in different countries, and corporate policy dictates that each data center be run independently, you may need one GC environment per data center (if you're monitoring production and nonproduction targets together) or even two environments per data center (one for production and the other for nonproduction targets).

    - If your corporate network (intranet) is slow, geographical considerations alone favor building two regional environments, as dictated by network bandwidth constraints.

2. For each GC environment, draw a crude map to indicate where your GC infrastructure and managed targets will geographically reside. Draw a box on this map to represent each data center, and indicate whether it is a primary or DR location.

3. Figure out how many Grid Control sites you need for each environment, given Grid Control network bandwidth and latency requirements.

4. For each environment and site, position OMR and OMS in these data center boxes where you propose to locate the hosts for these GC components. You need at least one OMR and can have multiple OMSs. Draw a line between each OMS and its associated OMR. Draw OMAs to represent where your target hosts are located and connect them to the OMS(s), either each OMA to a particular OMS (manual load balancing), or each OMA to multiple OMSs (which requires a server load balancer).

Now go show that pretty picture to your network administrator to confirm your bandwidth and latency estimates between all components. If the NA makes you sign your drawing in big block letters and hangs it on the lunchroom refrigerator, take out another piece of paper and draw a more realistic picture.

Make Required Installation Choices

OMR, OMS

The installation choices you need to make include answering the following basic questions regarding Grid Control installation options and topologies:

- Should you install Grid Control using a new database for the Repository, or should you install Grid Control using an existing database for the Repository?

    - If you're using an existing database, should you use a single-instance or RAC Repository database?

    - If you're using a RAC database, how many nodes will you need?

- How many OMS hosts should you deploy?

Tackling these questions requires an understanding of the GC installation options and possible topologies (i.e., defined here narrowly as the host or hosts where GC components will reside). When you kick off the GC installer, the first screen shot asks you to choose from one of four installation types:

Installation Type	Nickname
Enterprise Manager 10g Grid Control Using a New Database Installation Type	GC New Database Option
Enterprise Manager 10g Grid Control Using an Existing Database Installation Type	GC Existing Database Option
Additional Management Service Installation Type	GC Additional OMS Option
Additional Management Agent Installation Type	GC Additional OMA Option

These installation types have long-winded names and are referenced many times throughout this book, so I use the above nicknames for them. When you launch the Grid Control installer for the first time, you must select either the GC New Database Option or the GC Existing Database Option (after preparing an existing database). These options are mutually exclusive, and you can only select one of these options *once* for a particular GC environment. The reason is that both options install an OMR either in a new or in an existing database, and only one OMR can exist in a particular GC environment or site. After you select either the first or second option and complete the installation, you can deploy more OMSs by running the GC Additional OMS Option (the third installation type) multiple times, once for each additional OMS host you want to install[6]. The following table summarizes the GC components that each installation type deploys.

Installation Type	Database Created for OMR?	OMR Installed?	OMS Installed?	OMA Installed?
GC New Database Option	Yes	Yes	Yes	Yes
GC Existing Database Option	No	Yes	Yes	Yes
GC Additional OMS Option	No	No	Yes	Yes
GC Additional OMA Option	No	No	No	Yes

As the table shows, the first three installation types install an OMS and chain-installed Agent, and only the GC New Database Option creates a database.

---

[6] See Chapter 7 for instructions on the fourth installation type, which is the interactive installation of standalone Agents.

The remainder of this section offers a concrete rationale to help you decide what GC installation type to employ, whether to use a RAC Repository, and if so, how many RAC nodes to use, and how many additional OMS hosts to install, if any.

Choosing a New or Existing Database for the Repository

The GC New and Existing Database Options are mutually exclusive, and you can only select one of these two options *one time* to build a particular GC site. As I just mentioned, the reason is that both options install a Repository in either a New or Existing Database, and only one Repository can exist in each EM ite. The option you choose determines the GC components installed (at least one component is deployed on the local host), whether you can install them on more than one host, and whether you can use a RAC database for the OMR. Following is a description of these first two options and how they impact component locations on host(s) earmarked for GC installation.

GC New Database Option This option lays down all three components (OMR, OMS, and chain-installed OMA) on the same host, creating a new single-instance Oracle 10g Release 1 (10.1.0.4) database to house the OMR.

CAUTION
Except for small deployments of fewer than 1,000 targets, Oracle recommends locating the OMR and OMS components on separate, dedicated hosts for performance reasons. The OMR Database should be dedicated to the Repository and not shared with other applications. From a licensing perspective, you can use the OMR Database to host an Oracle Recovery Manager (RMAN) repository[7], but this violates both Grid Control and RMAN best practices.

GC Existing Database Option This option installs an OMS and chain-installed Agent on the local host where the installer is run, and installs an OMR on a preexisting database. You can install the OMR in a single-instance or Real Application Clusters (RAC) database. This presents several variations. You can

- Install an OMS and chain-installed Agent on one node, and install the OMR in a preexisting database on another node or nodes. This variation has two prongs:
    - The preexisting database can be a single-instance database.
    - The preexisting database can be a RAC database. Typically, it is a two-node RAC, but you can start with a single-node RAC and add additional nodes later to scale out.
- Install all GC components on the same host as in the first option above[8] (i.e., install an OMS and chain-installed OMA on the same host as that running an existing database, and install the OMR in this database).

---

[7] See Oracle*MetaLink* Note 394626.1.
[8] The only reason to select this variation is to install a 10.2 or 9*i* database (which is not recommended) rather than a 10.1.0.4 database installed with the GC New Database Option.

TIP
In Chapter 3, you will learn how to install an OMS on a Repository RAC node. This configuration, supported as of GC 10.2.0.2, is not ideal because the OMR and OMS must compete for resources. However, it may be the only alternative, particularly in a nonproduction environment where server real estate is tight, but where you want to mirror a RAC OMR production environment.

The major Grid Control architectural "fork" in the road is whether to choose the GC New Database Option or the GC Existing Database Option. If I were at that fork, and you stopped to ask me which way to go, my answer would have to be conditional. Here are the most succinct directions I could think to give you (I call it fork #1 because there's one more fork to come):

"**Fork #1:** Take the New Database road if all of the following are true: both the OMR and OMS are certified on your platform, your host meets the combined OMR and OMS hardware requirements, you don't need high availability for the OMR, and you monitor fewer than 1,000 targets. Otherwise, take the Existing Database road."

This answer is densely packed, so let's break out a "map" to give you more details on each one of these four conditions.

Selecting the GC New Database Option

Below is a description of each of the four conditions you should satisfy before choosing the GC New Database Option.

- **If both the OMR and OMS are certified on your platform**   The GC New Database Option installs the Repository and OMS on one server, so both components must be certified on whatever operating system that server is running. The OMS can run on most Windows 32-bit platforms and on many UNIX-based 32-bit and 64-bit platforms. However, the OMR database can also run on 64-bit Windows platforms and on additional UNIX 32-bit and 64-bit platforms. See the section "Verify Certification Information" later in the chapter for all OMS-certified platforms.

- **If your host meets the combined OMR and OMS hardware operating requirements**   If OMR and OMR components reside on the same host, which is applicable only for small deployments of fewer than 1,000 targets, the combined hardware requirements are two 3GHz CPUs, 8GB RAM, 50GB total disk space, 1.85GB temporary disk space, and 4.5GB swap space. If your host meets these combined hardware requirements, read on about using the GC New Database Option. (Otherwise, you must use the GC Existing Database Option.)

- **If you don't need high availability for the OMR**   If you select the GC New Database Option, it creates a new single-instance Repository. If you need high availability (HA) for the OMR first install a RAC Database or configure the OMR for use with Cold Failover Cluster, which both require selecting the GC Existing Database Option. (You could use the GC New Database Option to install a single-instance database, then RAC it, but it is easier to pre install a RAC database off-the-bat and use the GC Existing Database

Option.) If you only need data availability, called cold failover—not high availability—for the Repository, install a single-instance database with the GC New Database Option, and then build a physical standby database for the Repository later using Oracle Data Guard, which also supplies data protection, disaster recovery, and other benefits (see "OMR Database DR Recommendations" in Chapter 15).

- **If you will monitor fewer than 1,000 targets** Whereas the previous point addresses the need for OMR high availability, this point speaks to OMR capacity. Generally speaking, a single-instance Repository, installed using the GC New Database Option, should be able to manage fewer than 1,000 targets, defined as a "small" GC site. A multinode RAC Repository, which requires selecting the GC Existing Database Option, will provide the immediate scalability to monitor more targets. (See the next section for how many RAC nodes you'll need.)

Remember that targets are not just monitored hosts, but all Oracle and non-Oracle products installed on these hosts, such as database instances, listeners, Application Servers, E-Mail servers, and Net App Filers. There are over 100 different target types, and each instance of a particular target type counts as a monitored target.

If your site does not meet all four of these conditions for using the GC New Database Option, plan to use the GC Existing Database Option, described next.

CAUTION
iSQL*Plus access in Grid Control requires running iSQL*Plus server from any database home. The OMR database home is the logical place for such a home, but the GC New Database Option doesn't install iSQL*Plus, and you can't install it afterward, either by rerunning the GC installer or the 10.1 database installer. However, you can install iSQL*Plus in the OMR database home via the GC Existing Database Option. This should not be a major factor in deciding which GC installation type to choose, as you can configure any target database server containing iSQL*Plus to provide iSQL*Plus access in the Console to all target databases.

Selecting the GC Existing Database Option

While the GC New Database Option installs the Repository in a new single-instance database on the *same* host as the OMS, you use the GC Existing Database Option to preinstall either a single-instance or RAC Repository on a *different* host than the OMS host.[9] The only major decision you need to make when using the GC Existing Database Option is whether to use a single-instance or RAC Repository.[10] Remember: you've already taken the Existing Database road, and now you come to another fork. If, a la Twilight Zone, I magically beat you to this new fork in the road before you

---

[9] You can use the GC Existing Database Option to install the Repository on the same host as the OMS, but this would only be used to preinstall Database version 10.2 or 9.2 (which is not recommended), rather than install Database version 10.1 with the GC New Database Option.

[10] Again, please consult Chapter 15, if you're interested in using an Active/Passive configuration for multiple Repository or OMS nodes.

had a chance to drive there, and you were so amazed that you stopped again for directions, I would tell you this:

> "**Fork #2:** Howdy. Hope you know you're on to the GC Existing Database road right now. At this fork, you can either take the single-instance database road or the RAC database road. If you *do* need high availability for the OMR, or if you will monitor *more* than 1,000 targets, take the RAC database road." Otherwise, you can take the single-instance database road."

Finally, how many RAC nodes are required? That's the easy part. No matter what the GC deployment size is, you shouldn't need more than a two-node RAC cluster to satisfy requirements for OMR high availability and GC site size. The following table shows this and also quantifies the number of targets that constitute a small, medium, or large GC site.

Size of GC Site	Number of OMR Nodes (SI or RAC)
Small <1,000 targets	1-node, single-instance (or single-node RAC to scale out later)
Medium 1,000 to 10,000 targets	2-node, RAC
Large >10,000 targets	2-node, RAC

Remember that Oracle does not support customers running the OMS on any RAC Repository node until 10.2.0.2, and it's not recommended in any case, except perhaps for nonproduction sites.

Selecting the GC Additional OMS Option

After your initial Grid Control installation using either the GC New Database or GC Existing Database Options, you can install additional OMSs on separate hosts by running the GC Additional OMS Option. You must decide how many additional OMS hosts to deploy, if any, and where to locate these hosts geographically. Both decisions depend on the size of your GC site and on high availability (HA) requirements. I'll use the same definition for a small, medium, and large GC site in establishing sizing requirements. The following table indicates that the total number of OMS hosts needed is as easy as one, two, three (unless you need HA for the OMS component):

Size of GC Site	Number of OMS Hosts
Small <1,000 targets	1 (2 if HA required)
Medium 1,000 to 10,000 targets	2
Large >10,000 targets	3

I will defer comprehensive coverage of Grid Control HA considerations until Chapter 15. It should suffice for now to know that providing HA for an OMS in an Active/Active configuration (where both OMS hosts operate independently and concurrently) means installing an additional OMS, even for a small GC site by selecting the GC Additional OMS Option. If you're considering configuring the OMS in an Active/Passive Cold Failover Cluster (CFC), you must follow a special installation procedure where you actually install both the initial and second OMS concurrently. (Again, see Chapter 15 for details.) If you're not sure what distinguishes an OMS Active/Active from an Active/Passive configuration, the short answer is that an Active/Passive configuration requires third-party cluster software and shared storage, most commonly NFS. Unless you possess this hardware and storage, you will be using an Active/Active OMS configuration (a best practice anyway, particularly when configured with a Server Load Balancer).

There you have it. At this point, hopefully you've earmarked whether you're installing a production or nonproduction GC environment, how many servers you need for the GC installation, and which installation options and component topology you will select. Now, let's learn how to use the prerequisite checker to determine whether your chosen environment meets certain platform-specific requirements. Then, we'll cover all network configuration, hardware, and software requirements in detail.

Network Configuration Steps

The network configuration steps for Grid Control involve setting up host name resolution, adhering to certain host naming rules and constraints, using a static IP address for all GC hosts, and confirming connectivity, most importantly between GC hosts. If you abide by the rules described below, you will avoid possible problems at installation time, and even more painfully, down the road. While it is not very difficult to deinstall and reinstall GC or any of its components after a failed installation, the cost of doing so rises exponentially once you start using and relying upon the product for your monitoring solution.

You may be questioning why I place GC network configuration steps ahead of all other requirements. The main reason for getting these network steps out of the way now is that it may take some time for your network administrator to make the network changes, such as fully qualifying the host name or ensuring that it is registered properly in the Domain Name System (DNS). Your company may have host name naming standards that conflict with those presented here, and these conflicts may take time to resolve. At larger corporations, even the simplest network changes may take considerable time to approve and implement.

One of those network changes to request from your network administrator is to open any necessary firewall ports between hosts to contain GC components. The easiest way to ensure installation success is to first disable Internet access on all GC hosts (if necessary for security), then open all ports on firewalls between these hosts. You can re secure firewalls after securing traffic between GC components (see Chapter 16). If you disable outgoing Internet access on the OMS before installation, afterward you'll need to re-enable such access so the OMS can search for and download patches. For information on specific firewall and proxy server configuration requirements, see Chapter 6 of Oracle Enterprise Manager Advanced Configuration 10g Release 4 (10.2.0.4.0).

Take Advantage of the Standalone Prereqchecker

The *Prereqchecker* (prerequisite checker) is a routine that automatically validates whether a GC site meets certain (but not all) platform-specific hardware and software prerequisites for OMS and standalone OMA hosts based on the installation type chosen. The GC installer automatically runs the Prereqchecker, or you can run the Prereqchecker in standalone mode. When you run the GC installer and choose an installation type on the first screen, the appropriate Prereqchecker routines automatically run and report their results on the Product-Specific Prerequisite Checks screen. You invoke the Prereqchecker in standalone mode by running the GC installer with the appropriate options, including one to denote the desired installation type.

> **TIP**
> *Although the Grid Control installer runs the Prereqchecker for you, I recommend running it in standalone mode. It's easier to let the Prereqchecker verify requirements and report all results, rather than your having to resolve unmet requirements on the fly during an installation.*

Use the standalone Prereqchecker in one of the following ways as part of your strategy to confirm all preinstallation requirements:

- Start with the standalone Prereqchecker to get an exception report, so to speak, of platform-specific shortcomings, including hardware resource issues, missing packages, insufficient kernel parameters, and the like. Manually redress these shortcomings, and rerun the standalone Prereqchecker to verify that you took care of them. Then using this book as a guide, manually verify your site's compliance with the remaining prerequisites that the Prereqchecker doesn't evaluate.

- Manually certify that your site meets all prerequisites; then use the standalone Prereqchecker to double-check that you've satisfied at least the subset of requirements the Prereqchecker verifies. I personally prefer this method, given that the Prereqchecker does not comprehensively check all GC installation prerequisites and that it's buggy.[11]

If you choose the first option, run the standalone Prereqchecker now according to the instructions in Appendix B (online at www.oraclepressbooks.com) under the section "How to Run the Prereqchecker in Standalone Mode." If you select the second option, wait to run the standalone Prereqchecker until you've come to the end of this chapter. (I'll remind you then.) Also, see "The Prereqchecker" section in Appendix B for a list of the checks performed for each GC installation type (see Table B-1; for now ignore the Additional OMA column), its bugs and quirks, and for sample Prereqchecker output.

---

[11] See online Appendix B, "Prereqchecker Bugs and Quirks," for a complete list.

Set Up Host Name Resolution
OMR, OMS, OMA
The Grid Control installer will fail if you do not first configure name resolution[12] on hosts where OMR, OMS, or standalone OMA components are installed. You can configure host names to resolve through the local *hosts* file on each server, through the Domain Name System (DNS), or both.

> **NOTE**
> *In this section and throughout the book, the term "local* hosts *file" or "*hosts *file" is used to signify* /etc/hosts *on UNIX and* C:\WINDOWS\system32\drivers\etc\hosts *on Windows.*

On UNIX, to determine the order of priority for name resolution, check the "hosts" entry in the */etc/nsswitch.conf* file as follows:

```
grep -i hosts /etc/nsswitch.conf
```

The *hosts* file takes precedence over DNS if the output of this command contains "files" before "dns", similar to the following:

hosts: files   [NOTFOUND=continue UNAVAIL=continue] dns

Many sites configure their servers with such *hosts* file priority to minimize network overhead. Conversely, DNS takes precedence over the *hosts* file if the output from the command above contains "dns" before "files."

- If the *hosts* file is configured to take precedence (or is used as DNS backup), verify that it contains at least two entries: one for the localhost loopback address and the other for the IP address to hostname mapping. For example, the *hosts* file must contain at least the following two lines:

  127.0.0.1 localhost localhost.<localdomain>
  <IP address>   <hostname>.<domain>   <hostname>   <alias>

- The *localhost* entry is a loopback address, and the first alias for it, "localhost," should not be fully qualified.
- If DNS is set to take precedence (or is backup to the *hosts* file), ensure that you configure both forward and reverse *nslookup* for each host name running a GC component (see the section "Fully Qualify the DNS Lookup" later in the chapter).

Fully Qualify Host Name References
OMR, OMS
OMR and OMS host names (and target hosts running standalone Agents, if possible) should be consistently fully qualified wherever they are referenced. A fully qualified host name contains the short host name and domain name (such as *omshost.mycorp.com*) as opposed to just the short host name (*omshost*). Host names are populated in GC configuration files, and using a

---

[12] Name resolution refers to finding an IP address that corresponds to a given host name.

mixture of fully qualified and not fully qualified host names can cause unexpected results, such as GC installation errors, and GC Console or Application Server Console browser pages not found. Not using the fully qualified host name can also cause confusion in that each host across your enterprise may not be uniquely identified. An example where using the short name may be confusing (at best) is when GC manages two target hosts with the same short name (say, *host1*) in two different countries (such as the United States and the United Kingdom), each with a distinct domain name, (for example, *us.mycorp.com* and *uk.mycorp.com*). Grid Control assigns unique names to each target in the Agent's *targets.xml* file, but you are only inviting trouble that these hosts will somehow be mixed up.

NOTE
Throughout this book, fully qualified host names are used in all explanations and examples.

The main places where host name references ought to be fully qualified are as follows (perform all checks on the local host for itself):

- In the *hostname* command output
- In the local *hosts* file
- In the DNS

Let's look at each of these host name settings in particular.

TIP
If your company's network policy is not to use the fully qualified host name everywhere, at least try to be consistent in not using it anywhere (i.e., use the short host name). This may not be possible, though, as a DNS usually returns the fully qualified host name, whereas network policy or application requirements may dictate using a short name in the hostname *output and/or as the first alias in the local* hosts *file. The best advice I can give you is to lobby to change this network policy, which is inherently inconsistent.*

Fully Qualify the *hostname* Command

Many UNIX networking programs use the *hostname* command to identify the machine. Verify that the *hostname* command returns the fully qualified host name, as follows:

 `hostname`

<hostname>.<domainname>
    On UNIX only, verify that the domain name has not been set dynamically by executing the following command:[13]

 `domainname`

---

[13] This command displays the name of the Network Information Service (NIS) domain. NIS is Sun Microsystems' "Yellow Pages" (YP) client-server directory service protocol for distributing system configuration data such as host names between computers on a computer network. Source: Wikipedia.

Verify that no results are returned—the *hostname* command should statically supply both host name and domain name.

If the *hostname* output is not fully qualified for the OMS or OMR hosts, make this change permanent so that it persists across reboot, provided your company network policy allows it. You must have root privileges on the server to change a host name. Temporarily changing the hostname is also helpful when installing GC on a server that requires reboot for the host name change to take effect, but that can't be rebooted at the time. You can temporarily change a UNIX hostname by issuing the following command as *root*:

    hostname <hostname>.<domainname>

Do not temporarily change the host name unless you plan to permanently change it by rebooting at the earliest opportunity. A temporary host name change would revert back to the short name after reboot, which would be out of sync with the host name stored in the GC configuration files.

Fully Qualify the First Alias in a *hosts* File

As already mentioned, if a server uses a local *hosts* file for name resolution (even as a backup method behind DNS), this file should include at least the following two lines:

    127.0.0.1 localhost localhost.<localdomain>
    <DNS IP address>   <hostname>.<domainname>   <hostname>

As for the host name, the first alias (after the IP address) should be fully qualified; not doing so can cause your GC installation to fail when the Web Cache Configuration Assistant runs.[14] For example, on a UNIX host named *omshost.mycorp.com*, the following two lines should be listed in */etc/hosts*:[15]

    127.0.0.1 localhost localhost.mycorp.com
    138.2.4.84 omshost.mycorp.com omshost

CAUTION
When installing Grid Control on SUSE Linux (SLES9), Bug 3843195 causes the bundled Oracle Application Server 10g installation to throw a JAVA error. The workaround is to remove or comment out all IPv6 entries of the following form in /etc/hosts:

    #::1   localhost ipv6-localhost ipv6-loopback
    #fe00::0   ipv6-localnet

Fully Qualify the DNS Lookup

If a GC host relies upon DNS for name resolution, even as a backup method to a local *hosts* file, a forward or reverse name server lookup (*nslookup*) of the short name or fully qualified host name

---

[14] Oracle*MetaLink* Note 351841.1.

[15] For RAC database nodes, whether for a target or a Repository, in addition to the public IP address entry, you must also include entries for private and VIP addresses. See the Oracle Clusterware documentation for details.

should return the fully qualified host name. Try the following commands on each GC host, substituting the host name on which you are executing the commands:

```
nslookup
<hostname>.<domainname>
<hostname>
<IP address>
```

Following is the command output from a forward and reverse *nslookup* of the host name *omshost.mycorp.com* (input is in **bold**):

```
$ nslookup
> omshost.mycorp.com
Server:         138.2.203.16
Address:        138.2.203.16#53

Name:   omshost.mycorp.com
Address: 138.2.5.85
> omshost
Server:         138.2.203.16
Address:        138.2.203.16#53

Name:   omshost.mycorp.com
Address: 138.2.4.84
> 138.2.4.84
Server:         138.2.203.16
Address:        138.2.203.16#53

84.4.2.138.in-addr.arpa name = omshost.mycorp.com.
```

Host Name Constraints

OMR, OMS, OMA

The host names where GC components are installed must adhere to certain naming rules that vary depending upon the component:

- OMR, OMS, and OMA host names cannot be *localhost, localhost.localdomain,* or an IP address, but must be valid host names[16]. A valid host name is one listed in the local *hosts* file, or for an OMR or OMA host, a virtual host name as discussed next.

- You can use a virtual host name for an OMA or OMR host, but not for an OMS host. A virtual host name is anything other than the physical host name of the server, as indicated by the *hostname* command output. To use a virtual name, specify the ORACLE_HOSTNAME argument when executing the Grid Control Oracle Universal Installer (OUI). The OUI executable name is different on UNIX than on Windows.

  On UNIX the OUI syntax to use a virtual host name is:

  `./runInstaller ORACLE_HOSTNAME=<virtual host name>`

  On Windows the syntax is:

  `setup.exe ORACLE_HOSTNAME=<virtual host name>`

---

[16] The Prereqchecker confirms this requirement. In fact, it's the only network-related prerequisite check performed.

- An OMA or OMR host can have multiple aliases registered with the DNS under a single IP address. If this is the case, set the ORACLE_HOSTNAME argument to the alias you want to use as the host name. You can still set ORACLE_HOSTNAME for the OMS installation, as long as the host name defined is a physical host name, not a virtual host name.

- An OMR, OMS, or OMA host can be multihomed[17], meaning that it can contain multiple network (NIC) cards, each with its own IP address and host name. If you don't specify ORACLE_HOSTNAME as an argument to an OMS or standalone OMA installation using the OUI on a host with multiple NIC cards, then the OUI assigns the first multihomed entry in the *hosts* file as the host name.

Use Static IP Addresses

OMR, OMS

The hosts on which you are installing the OMS and OMR must have unique static IP addresses, not Dynamic Host Configuration Protocol (DHCP) assigned addresses. For hosts running standalone Agents, unique static IP addresses are recommended but not required.[18]

Connectivity Checks

OMR, OMS, OMA

Perform the following tests to test connectivity to the SMTP Server, the Internet, and to a Proxy Server (if used). Also check connectivity to Oracle*MetaLink* and check the account credentials Grid Control will use to download patches and patch information.

SMTP Server Connectivity

OMS, OMA

The GC installer prompts you to enter an SMTP Server name and e-mail address where the OMS is to send alert notifications for the default GC administrator called SYSMAN. It is optional to enter this information during the installation but highly recommended because it immediately enables notifications for out-of-box alert thresholds.

Test SMTP Server   If you enter an SMTP server, the installer must confirm this server is reachable from the OMS host before you can advance to the next screen. If the SMTP server is not reachable, you must uncheck the box to enable e-mail notifications. For this reason, it is wise to confirm before starting the installer that all OMS hosts can reach this SMTP server.

---

[17] An OMR host can also be multihomed, but the ORACLE_HOSTNAME argument is not used to specify the OMR host name (the first multihomed entry in the *hosts* file is used).

[18] You can run Grid Control on your personal computer (PC) if it has a static IP address. Given that most personal computers run on DHCP-assigned networks, an alternative is to run the Grid Control Vmware bubble. Unfortunately, it is only available for download from an internal Oracle site, the Oracle Fusion Factory at http://ff.us.oracle.com/vmware. Your local Oracle sales rep or on-site Oracle Consultant would be more than happy to demo this bubble (and any others you may be interested in), and perhaps even leave it for you to take it out on an unaccompanied test drive. Even the bubble needs a static IP address, which you provide by adding the Vmware image IP address to the *hosts* file on your PC. This bubble runs on Red Hat Linux 3.0 and requires the following minimum hardware: a Pentium IV 2.6Ghz or Centrino 1.4Ghz processor, 2GB of RAM, and 24GB of free disk space.

In addition, it is important that the Monitoring Agent (the Agent chain-installed with the initial OMS) and any additional Agents desired be able to send an out-of-band (OOB) notification through this SMTP server. An *OOB notification* is an SOS call sent directly through an SMTP server (not through an OMS) to a designated e-mail address to relay that no OMS hosts are online to receive Agent alerts or data uploads. You will configure the OOB notifications themselves in Chapter 11 in the section "Enable Notification if Grid Control Goes Down." For now, just make sure that the Monitoring Agent, and, preferably, at least one target host per subnet, can ping the SMTP Server:

```
ping -c 4 <SMTP_Server>
```

Test Notification E-mail Address Next, from each OMS host, make sure the SMTP server can relay notifications to the e-mail address you want to associate with the SYSMAN account (and to any other e-mail addresses where you want to receive notifications). This is also important to check in advance because the GC installer does not confirm receipt to the e-mail address you supply. The e-mail address can be a generic administrator e-mail account, to which you can assign one or more administrators, or a distribution list, as you see fit. Send a test e-mail to this address via a telnet connection to the SMTP server on port 25 from the command line. Following is the output from such a telnet session on the host *omshost*. In this session, we send a test e-mail from sender *EnterpriseManager@oracle.com*, which must be an e-mail account that your SMTP server recognizes, to the same e-mail address. Input commands are shown in **bold**:

```
$ telnet mail.mycorp.com 25
Trying 138.1.161.138...
Connected to bigip-mail.mycorp.com (138.1.161.138).
Escape character is '^]'.
220 rgmsgw02.mycorp.com ESMTP Mycorp Corporation Secure SMTP Gateway
Switch-3.1.7/Switch-3.1.7 - Ready at Wed, 5 Apr 2006 11:43:43 -0600 -
Unauthorized Usage Prohibited.
helo
250 rgmsgw02.mycorp.com Hello omshost.mycorp.com [138.2.4.61], pleased to meet you
mail from: EnterpriseManager@mycorp.com
250 2.1.0 EnterpriseManager@mycorp.com... Sender ok
rcpt to: EnterpriseManager@mycorp.com
250 2.1.5 EnterpriseManager@mycorp.com... Recipient ok
data
354 Enter mail, end with "." on a line by itself
Subject: SMTP Test
Hello this is an SMTP test for Grid Control
.
250 2.0.0 k35HhhOm026568 Message accepted for delivery
quit
221 2.0.0 rgmsgw02.mycorp.com closing connection
Connection closed by foreign host.
```

Confirm that you receive this test e-mail at *EnterpriseManager@mycorp.com* in a timely manner.

Using this telnet method, also confirm that Monitoring Agent OOB notifications will succeed by sending a test e-mail from each Monitoring Agent to the e-mail address you plan to designate for OOB notifications.

OMS MetaLink Connectivity
OMS
The OMS must be able to connect to Oracle*MetaLink* at http://oracle.com/support/metalink/index.html in order to locate and download patches. Open a browser session from the OMS and confirm Internet access to this URL. If the OMS must go through a proxy server to reach the Internet, this confirms that the browser's proxy server information is correct. Ensure that you provide this same proxy server information during the installation.

Also verify the Oracle*MetaLink* credentials that the OMS will use for patch-related functionality. I recommend creating a new Oracle*MetaLink* account name for Grid Control to use exclusively, such as "*mycorp*_OEM."

If you cannot install a browser on the OMS to check Oracle*MetaLink* connectivity and credentials, at least check that the OMS can access Oracle*MetaLink* by pinging it:

```
ping -c 4 updates.oracle.com
```

Be aware that this ping will fail if your network administrator disabled ping access from inside the firewall to external sites.

Connectivity Between All GC Hosts
OMR, OMS, OMA
The OMS needs to be able to communicate in both directions with the OMR host via JDBC, and with OMA hosts via HTTP and HTTPS. If a firewall separates the OMS from the OMR or, more likely, if a firewall separates the OMS from the OMA, check that your company's proxy server is configured as required to allow communications between these GC components. First, from the OMS host, ping the proxy server between OMS and OMR, if one exists:

ping -c 4 <proxy_server>

Second, from the OMS, ping the OMR and the OMA to confirm that they are reachable. Also ping in the reverse direction from the OMR to the OMS and from the OMA to the OMS. Specify fully qualified host names when executing all *ping* commands.

> **TIP**
> *A check ("✔") symbol next to a subheading either under hardware and software requirements below or in Appendix B (online) indicates that the Prereqchecker confirms this prerequisite for you. Table B-1 in Appendix B summarizes all Prereqchecker actions.*

Hardware Requirements
Requirements for *installing* Grid Control versus *operating* Grid Control in production are vastly different. This section covers both, in that order. The installer Prereqchecker confirms that your system possesses the minimum resources just to install Grid Control (at least in my opinion). These resources include hardware requirements such as CPU, RAM, and disk space. Whether the hardware has enough horsepower to go live with a GC production site supporting all your managed targets for at least the short term ... well, that's another story.

First, let's spell out the minimum hardware requirements to get through a GC installation. This is useful to know if you don't yet have the hardware resources to put Grid Control into production but want to get started and test-drive its functionality.

Hardware Installation Requirements ✔

OMR, OMS, OMA

If your production Grid Control environment requires high availability HA (addressed in detail in Chapter 15), then you need to design and build in fault-tolerant storage up front for all GC framework components—OMR database, OMR software, OMS software, and OMA software. The topic of fault-tolerant storage falls outside the scope of this book. However, you'll find an excellent treatment on best practices for configuring HA storage in Section 2.1 of Oracle Database High Availability Best Practices 10g Release 2 (10.2). This section identifies the following six elements of a fault-tolerant database storage subsystem, but the last three of these elements apply equally to storage for all GC framework components, as indicated in the table below:

Element of Fault-Tolerant Storage	Applies To
Evaluate database performance requirements and storage performance capabilities	OMR Database
Use Automatic Storage Management (ASM) to manage database files	OMR Database
Use a simple disk and disk group configuration	OMR Database
Use disk multipathing software to protect from path failure	All GC Framework Components
Use redundancy to protect from disk failure	All GC Framework Components
Consider HARD-compliant storage	All GC Framework Components

Table 2-4 lists the hardware installation requirements on a per-host basis when installing Grid Control using the first three of four GC GUI installation types (all except Additional Management Agent).[19] Remember, these are only installation minimums, not operating requirements (listed next). All requirements are per host. The GC Existing Database Option is represented in two rows of the table. The first row is for the host where the OMS and chain-installed OMA are installed, and the second row is for the host where the OMR is installed.

---

[19] The requirements are specified as parameters in the file *oraparam.ini* located in the *install* directory at the top level of the EM software distribution. The total disk space installation footprint is broken out by OS, and is for all component Oracle homes installed for the indicated installation type. The installation prerequisite checks verify all requirements except those for the OMR host as specified in the row GC Existing Database Option (OMR Host), as this host is different from the installation host.

Installation Type (Component Host)	Total Disk Space				Physical RAM	Temporary Disk Space			Swap Space	
	AIX	HP-UX	Linux Solaris	Windows	All Platforms	Linux, Solaris, Windows	AIX	HP-UX	AIX, HP-UX, Linux, Windows	Solaris
GC New Database Option[20] (OMR/OMS/OMA Host)	9.0GB	9.0GB	4.5GB	4.2GB	512MB	250MB	1.3GB		250MB	500MB
GC Existing Database Option (OMS/OMA Host)	5.0GB	5.0GB	2.5GB	2.5GB	512MB	250MB	1.3GB		250MB	500MB
GC Existing Database Option (OMR Host)	4.0GB	4.0GB	2.0GB	1.7GB	960MB	250MB	0GB		250MB	500MB
Additional OMS (OMS/OMA Host)[21]	2.0GB	2.75GB	2.0GB	2.5GB	512MB	250MB	1.3GB		250MB	500MB

TABLE 2-4. *Hardware Installation Requirements per OMR/OMS Host(s) Broken Down by Installation Type*

Hardware Operating Requirements

OMR, OMS

Following is a summary of the hardware operating requirements for each OMR host (Table 2-5) and OMS host (Table 2-6), including number of CPUs,[22] RAM, total disk space, temporary disk space, and swap space, needed to sustain the indicated GC site sizes. Because the units of these specifications are on a per-host basis, the number of OMR and OMS hosts listed in earlier tables is listed again for the sake of completeness in Tables 2-5 and 2-6.

> **NOTE**
> *Again, these requirements assume the OMR and OMS components reside on different hosts, which Oracle recommends for performance reasons. However, if these components reside on the same host, which is acceptable for small deployments of fewer than 1,000 targets, the combined OMR/OMS requirements are two 3GHz CPUs, 8GB RAM, 50GB total disk space, 1.85GB temporary disk space, and 4.5GB swap space.*

---

[20] These requirements also apply to the GC Existing Database Option if the OMR and OMS are installed on the same host.

[21] The Prereqchecker does not verify some of these installation requirements, including physical RAM, temporary space, and swap space.

[22] Expect diminishing performance returns from OMR or OMS hosts running multicore microprocessors. In other words, don't expect 2× the performance from a dual-core device over a single-core processor.

Deployment Size	Number of OMR Hosts	Number of CPUs (speed)	Physical RAM	Total Disk Space	Temporary Disk Space	Swap Space (UNIX Only)
Small <1,000 targets	1-node, Single-Instance	2 (3GHz)	4GB	40.0GB	1.6GB	4.0GB
Medium 1,000 to 10,000 targets	2-node, RAC	4 (3GHz)	8GB	80.0GB	1.6GB	8.0GB
Large >10,000 targets	2-node, RAC	8 (3GHz)	12GB	300.0GB	1.6GB	12.0GB

TABLE 2-5. *Hardware Operating Requirements per Host Dedicated to the OMR*

Deployment Size	Number of OMS Hosts	Number of CPUs (speed)	Physical RAM	Total Disk Space	Temporary Disk Space	Swap Space (UNIX Only)
Small <1,000 targets	1	2 (3GHz)	4.0GB	10.0GB	250MB	500MB
Medium 1,000 to 10,000 targets	2	4 (3GHz)	4.0GB	10.0GB	250MB	500MB
Large >10,000 targets	3	4 (3GHz)	4.0GB	20.0GB	250MB	500MB

TABLE 2-6. *Hardware Operating Requirements per Host Dedicated to the OMS*

The range of values correlates with the deployment size (i.e., number of targets) and significantly exceeds the minimum installation hardware requirements above.[23] Notice that as Grid Control scales upward, each OMR node needs proportionally more CPU and RAM than each OMS. This is partially explained by the fact that more OMS than OMR nodes are needed as the deployment size increases.

The commands to check CPU, RAM, swap space, disk space, and temporary disk space requirements for all supported operating systems are listed in Table 2-7.

---

[23] The disk space allotments for both OMR and OMS are generous compared with those specified in the Oracle EM documentation, and allow for growth over a site's lifetime. The OMR disk space is sized to handle expected log and trace files, growth in the OMR database tables as well as in an external Change Manager schema, and a Flash Recovery Area large enough to hold two days of on-disk RMAN backups (and archive logs), a block change tracking file, and 24 hours of flashback data. (Chapters 3 and 4 detail OMR database configuration.) The OMS disk space is double that recommended in the EM documentation based on real-world experience. It allows for centralized storage of patches, preferably on a shared drive accessible by all OMS hosts. This allows administrators to deploy patches when logged into the Console via any OMS.

	AIX	HP-UX	Linux	Solaris	Windows
Number (speed) of CPUs[24]	lscfg[25] | grep 'proc[0-9]' | awk 'END {print NR}' top	ioscan -C processor	cat /proc/cpuinfo	prtconf -v | head	Right-click Computer Name, select Properties, and then select the General tab. You can also get the number of CPUs from environment variable NUMBER_OF_PROCESSORS.
Physical RAM	bootinfo -r lsattr -EHl <resource_name> such as mem0	grep "Physical:" /var/adm/syslog/syslog.log	grep MemTotal /proc/meminfo	/usr/sbin/prtconf | grep "Memory size"	Right-click Computer Name, select Properties, and then select the General tab.
Swap Space	swap -s lsps -a (for paging info)	/usr/sbin/swapinfo -a	grep SwapTotal /proc/meminfo	/usr/sbin/swap -s	Right-click Computer Name, select Properties, and then select the Advanced tab. Under Performance, click Settings, which brings up Performance Options. On the Advanced tab under Virtual memory, Total paging file size for all drives is listed.
Total Disk Space	df	bdf	df	df	Right-click the local disk where GC software will be installed.
Temporary Disk Space[26]	df /tmp	bdf /tmp	df /tmp	df /tmp	Defaults to C:\Documents and Settings\<user id>\Local Settings\Temp. Right-click drive and select Properties.

TABLE 2-7. *Commands to Check Hardware Requirements on All Certified Platforms*

Because of the minimum requirement of 4GB of RAM for the OMR, I highly recommend running the OMR on a 64-bit platform. On a host dedicated to an Oracle database, such as the GC Management Repository, most of the RAM is usually assigned to the Oracle System Global Area (SGA). So even the smallest of GC sites running an OMR on a host with 4GB of RAM requires more than 2GB SGA. On all 32-bit platforms, attaining an SGA above 1.7GB involves implementing the Very Large Memory (VLM) option for that platform, as discussed in Chapter 17 (see Platform-Specific Performance Recommendations). The VLM option introduces complications of its own on all 32-bit platforms. By contrast, if you run the OMR on a 64-bit platform that natively supports large SGAs, you avoid these VLM complications. (Linux 64-bit platforms require implementing HugePages, which is technically a VLM option, but measurably improves performance.) Also, memory performance on 64-bit platforms is far superior to that of 32-bit platforms running VLM, not to mention the superior performance that 64-bit platforms afford in every other regard.

---

[24] For existing databases, you can also look at the initialization parameter CPU_COUNT that Oracle automatically determines. However, this parameter is not always accurate for several reasons. Dual-CPUs may not be accounted for. Also, beware of Bug 2735470 fixed in Oracle 10.1.0.1 that causes Oracle to ignore any CPUs you add after installing the Oracle database.

[25] *lscfg* is included in the AIX package called Monitor.

[26] You can supersede the default temporary directory using environment variables TEMP and TMP (set both to be sure).

Software Requirements

This section details the following software requirements[27] for OMR, OMS, and standalone OMA hosts:

- Verify certification information
- Create required OS groups and users
- Create required directories
- Stop database listeners using port 1521
- Synchronize OS timestamps/time zones
- Confirm platform-specific software requirements

All supported GC platforms must meet these generic prerequisites (all requirements except platform-specific ones). Such requirements include creating the requisite OS groups, users, and directories that the GC installer needs.

NOTE
If you are using a preexisting database to house the OMR, you have probably already satisfied the OMR host generic software requirements, entitled Create Required OS Users and Groups and Create Required Directories, as they are also Oracle Database preinstallation steps.

In addition to generic requirements, the section "Confirm Platform-Specific Software Requirements" at the end of this chapter lists unique requirements for each platform and OS version, such as required Solaris 9 patches that apply only to this operating system.

Verify Certification Information

OMR, OMS, OMA

One of the GC software preinstallation steps is to check that all components (OMS, Repository, Agents, Console) and Monitoring Plug-ins are certified on the intended platforms and for applicable O/S, database and browser, and third-party hardware and software versions. To verify that your intended GC environment is supported, consult the "Certification Information" section in Appendix B (online), which also explains the reasons behind, implications of, and recommendations related to the certifications. Appendix B contains certification information for Oracle Enterprise Manager 10g Grid Control Release 3 (10.2.0.3) and Release 4 (10.2.0.4.0), the latest two 10.2 releases available. I've limited the data to these two releases because, between them, Grid Control runs on all major operating systems found in today's data centers. Some certification information

---

[27] I define software requirements much more broadly than does Oracle Enterprise Manager Grid Control Installation and Basic Configuration. In Chapter 1 of that Oracle manual, the section entitled Enterprise Manager Software Requirements limits software to Oracle software that GC depends upon, such as the Oracle database software required for the Repository when choosing the GC Existing Database Option. In this section, software requirements also include software configuration steps and platform-specific software requirements.

may have changed since the book's publication. Therefore, double-check everything against the Oracle Enterprise Manager 10g Grid Control Certification Checker, Oracle*MetaLink* Note 412431.1. This note is accessible from the Certify tab on the Oracle*MetaLink* home page and provides an up-to-date matrix of certification data.

Create Required OS Groups and Users
OMR, OMS, OMA
Create the identically named OS users and groups below on all hosts housing GC components, if these users don't already exist. Follow a consistent naming convention to simplify administration.

- Oracle inventory group (*oinstall*)
- OSDBA group (*dba*)
- OSOPER group (*oper*). This group is optional.
- Oracle software owner (*oracle*). This owner is also used on Windows platforms.
- Unprivileged user (*nobody*). This user exists on UNIX platforms only.

For UNIX platforms in particular, also use the same user ID (uid), group ID (gid), and supplemental group name for the *oracle* user across all servers where GC components are installed. Hosts containing GC components may require consistent user and group names to one degree or another, as shown in these examples:

- Choose the same Agent user and groups that own (or predominantly own) existing Oracle targets to be monitored. Additional configuration is sometimes required for Agents to monitor targets installed by a different user (see Chapter 5).
- All RAC Repository nodes require the same Oracle users, user IDs, groups, and group IDs named above.[28]
- A Repository Data Guard configuration requires or recommends the same Oracle users, user IDs, groups, and group IDs named above on all primary and standby nodes.

Create Oracle Inventory Group (oinstall)
If you install Grid Control on a UNIX system with no existing Oracle products,[29] the OUI prompts you to create a new Central Inventory to store information about the GC software (see Figure 3-4 in Chapter 3) and any other Oracle software later installed on the host. You must enter the inventory directory path (the standard name to use for a Central Inventory is $ORACLE_BASE/*oraInventory*) and OS group name (typically named *oinstall*) with write permission to the inventory directory. The OUI then creates the Central Inventory in the specified location. The OUI also creates a Central Inventory pointer file (called *oraInst.loc*) in a default location to point to the inventory path. The following table

---

[28] Oracle Database Oracle Clusterware and Oracle Real Application Clusters Installation Guide 10g Release 2 (10.2), Chapter 2 (for all UNIX platforms).

[29] Technically, the OUI prompts you to create a new Central Inventory if no Central Inventory pointer file *oraInst.loc* exists on the host.

shows the default location of the *oraInst.loc* file (and the *oratab* file discussed in the section "Create Oracle Base Directory" later in the chapter) for all UNIX platforms:

Platform	Location of Configuration Files *oraInst.loc* and *oratab*
AIX	/etc
HP-UX	/var/opt/oracle
Linux	/etc
Solaris	/var/opt/oracle

The inventory pointer file consists of two lines following the example format here:

inventory_loc=/u01/app/oracle/oraInventory
inst_group=oinstall

The parameter *inventory_loc* is set to the specified inventory path, and the parameter *inst_group* is set to the specified inventory group owner.

First, determine if there is already an Oracle inventory group by checking whether the *oraInst.loc* file exists in the default location for your platform. If the *oraInst.loc* file exists, verify that the group name defined by the parameter *inst_group* is an existing OS group in the */etc/group* file, as follows (assuming *oinstall* is the group name):

```
grep oinstall /etc/group
```

If the Oracle inventory group doesn't exist, login as *root* and create it as follows, which varies by platform:

- On HP-UX, Linux and Solaris, enter:

  ```
  /usr/sbin/groupadd -g 501 oinstall
  ```

  This example assumes you are creating the *oinstall* group with group id (gid) 501. Specify the same group name and gid on all hosts containing GC software.

- On AIX, enter:

  ```
  smit security
  ```

  Choose the appropriate menu commands to create the *oinstall* group, and press F10 to exit.

On Windows platforms, the Grid Control OUI does not prompt you to create a new Central Inventory, but creates one in the default location *%SystemDrive%:Program Files\Oracle\Inventory*. Windows does not use an inventory pointer file, but instead specifies the inventory location in the registry key HKEY_LOCAL_MACHINE\SOFTWARE\ORACLE\INST_LOC. Windows creates an OS group called *OSDBA* to own the Central Inventory.

Create OSDBA Group (dba)

This step is only necessary if you plan to select the GC New Database Option, which creates a new Repository database. In this case, you must create an OSDBA group if one doesn't already exist. This group identifies operating system user accounts that have database administrative privileges (the SYSDBA privilege) on the Repository Database. The standard name for this group is *dba*. Check for the existence of the OSDBA group by doing the following:

- If an OSDBA group does not exist, create an OSDBA group using the same command syntax listed above for the Oracle inventory group.

- If an OSDBA group already exists, you can use this group after installation to log in to the Repository Database with operating system authentication. On the other hand, if you don't want the users in the OSDBA group to have the SYSDBA privilege on the Repository Database, create a new OSDBA group now to give a different group of OS users the SYSDBA privilege.

Create OSOPER Group (oper)

This step is only required if you plan to select the GC New Database Option, which creates a new Repository database. If so, create an OSOPER group only if you need a group of OS users with a limited set of DBA privileges (SYSOPER operator privileges). For most GC environments, an OSOPER group is not required. The default name for this group is *oper*. If you need an OSDBA group, check for its existence and create it using the same command syntax listed above for the Oracle inventory group. As with the OSDBA group, you can create a new OSOPER group if one already exists to give a different group of OS users (other than the existing OSOPER group) operator privileges for your GC installation.

Create Oracle Software Owner (oracle)

The GC software owner is typically the user *oracle*, and will be so throughout this book. This user should have the Oracle inventory (*oinstall*) group as its primary group and the OSDBA (*dba*) group as a supplemental (or secondary) group.[30]

CAUTION
If other Oracle software is already installed on a server, and the Oracle inventory group (oinstall) owns the inventory, then this oinstall group must be the primary group of the OS user installing the GC software.

You can create a different user name other than *oracle*, if one already exists, to assign a different user ownership for your GC installation. Check for the existence of this user as follows:

```
id oracle
```

If the *oracle* user exists, then the command output will look something like this:

```
uid=200(oracle) gid=500(oinstall) groups=501(dba),502(oper)
```

If the user exists, then decide whether to use the existing user or create another user. To use the existing *oracle* user, confirm as shown above that the primary group is the Oracle inventory group (*oinstall*), and that it is a member of the OSDBA group and, optionally, the OSOPER group.

---

[30] Some sites do not use the *oinstall* group at all, preferring only to use the *dba* group. This will also work, though it is not a best practice. The primary group is the first group listed for "groups=" when you use the *id* command, and is specified by the *useradd* or *usermod* "-g" option. Supplemental groups are defined by the "-G" option.

To create a new Oracle software user, either because it doesn't exist or because you require a new one, use the following command syntax as *root*, depending upon the platform:

- On HP-UX, Linux, or Solaris, to create an *oracle* user with the properties shown above (assuming the *oinstall* and *dba* groups already exist with the gid's listed):

  ```
  /usr/sbin/useradd -u 200 -g oinstall -G dba,oper -s /bin/bash -m oracle
  ```

  where

  - -u is the numerical value of the user's ID
  - -g is the group name of the user's initial login group
  - -G is a list of supplemental groups of which the user is a member

  Set the password for the *oracle* user as follows:

  ```
  passwd oracle
  ```

  <Enter password>

- On AIX, enter the following command to create the *oracle* user:

  ```
  smit security
  ```

  Choose the menu commands to create the *oracle* user as follows:

  In the Primary GROUP field, specify the Oracle inventory group, *oinstall*. In the Group SET field, specify the OSDBA group *dba*, and, if required, the OSOPER group *oper*. Press F10 to exit.

  Then set the *oracle* user's password:

  ```
  passwd oracle
  ```

  <Enter password>

  If the *oracle* user exists, but its primary group is not *oinstall* or it is not a member of the OSDBA or OSOPER groups, then enter the following command to modify it (not all *usermod* options below may be necessary):

  ```
  /usr/sbin/usermod -g oinstall -G dba,oper -u 200 oracle
  ```

Create Unprivileged User (nobody)

The unprivileged user called *nobody* must exist only if you plan to select the GC New Database Option, which creates a new Repository database. If so, verify that the *nobody* user exists. The *nobody* user must own any external jobs (extjob) executable after the installation. To see whether the *nobody* user exists, enter the following command:

```
id nobody
```

If the *nobody* user does not exist, then enter the following command to create this user on HP-UX, Linux, or Solaris:

```
useradd nobody
```

On AIX, enter the following command to create this user:

```
smit security
```

Enter the *nobody* user name, and leave the Primary GROUP and Group SET fields blank.

Create Required Directories
OMR, OMS, OMA

It is best practice that all GC directory structures adhere to the Optimal Flexible Architecture (OFA) guidelines, and that you consistently specify these three directories across all GC hosts. Create or choose in advance paths for three such directories that the GC software must use:

- Oracle base directory
- Oracle inventory directory
- Grid Control parent directory

Using the same directory structure on all hosts allows you to easily find the home or inventory, hard-code Agent home directory paths into generic Agent scripts (if absolutely necessary), and document your Agent installations.

Here is a brief summary of the tasks to perform in this section. Before running the GC installer, ensure the Oracle base directory exists and has permissions for the *oracle* user to create the GC Oracle homes below it. Also, decide upon Oracle inventory and GC parent directory specifications, and back up any preexisting Oracle inventory directory.

CAUTION
The Grid Control installer and Console cannot process symbolic links or spaces, so don't use them in directory paths, in configuration files, or in the User Interface.

Create the Oracle Base Directory

The Oracle base directory, defined by the OS environment variable ORACLE_BASE, is the top-level directory for all Oracle software installations. OFA guidelines recommend the following path:

ORACLE_BASE=<mount_point>/app/oracle

where <mount point> is the file system to contain the GC software.

This directory specification assumes that the *oracle* user is the owner of the Oracle software. You can either use the same Oracle base directory for all installations or separate Oracle base directories for each Oracle product. If a different OS user owns each Oracle product on the same server, then you can create a separate Oracle base directory for each user. Hopefully, you are following best practices, dedicating OMS and OMR hosts to your GC infrastructure, with no other products installed on these hosts. In this case, there will be no preexisting Oracle base directory. However, if this is not the case, or you're not sure whether an Oracle base directory already exists, you can usually determine what it is by any existing Oracle inventory or Oracle home directories. Directory paths that end with the Oracle software owner are likely candidates for existing Oracle base directories. Check both *oraInst.loc* and *oratab* files for such directory paths. For example, the *inventory_loc* value in the *oraInst.loc* file is usually an Oracle base directory. For the setting used in the earlier section "Create Oracle Inventory Group (oinstall)",

inventory_loc=/u01/app/oracle/oraInventory

the parent directory, /u01/app/oracle, is probably an Oracle base directory. Similarly, if you're using the oracle user to install the GC software, and an oratab file exists and contains the following two lines,

prod:/u02/app/oracle/product/10.2.0/db_1:Y
*:/u03/app/iasora/product/10.2.0/as10g/:N

then /u02/app/oracle would be a viable Oracle base directory to use. Ideally, you should place this directory on a different file system than the operating system, and make sure it has enough free disk space to satisfy Grid Control as specified in the section "Hardware Requirements" earlier in this chapter. Later in this chapter, when you set up the environment for the oracle user to run the installer, you will be reminded to set the ORACLE_BASE environment variable accordingly.

To create the Oracle base directory with the correct owner, group, and permissions, enter the following commands (you probably need to be logged in as root):

```
mkdir -p /u01/app/oracle
chown -R oracle:oinstall /u01/app/oracle
chmod -R 775 /u01/app/oracle
```

Choose the Oracle Inventory Directory

If you install Grid Control on a UNIX system with no existing Oracle products, the OUI prompts you to enter the Oracle inventory directory path and creates the inventory in the specified location with the proper owner, group, and permissions. The installer also creates an oraInst.loc file in the default location for your platform to point to the inventory directory (refer back to the earlier section "Create Oracle Inventory Group (oinstall)" for more background). Do the following now to make sure the OUI can perform these steps when you run the GC installer in Chapter 3:

1. If the oraInst.loc file does not exist in the default location for your platform, the GC installer will create this file, provided the GC OS user has write permissions to do so (confirm this now).

2. If the oraInst.loc file exists in the default location:

   - Ensure root owns this file with 644 permissions, and that inventory_loc points to the desired location for the Grid Control inventory.

   - Ensure the GC user (oracle) and/or group (oinstall) own the inventory directory with 664 permissions so that the GC installer can update the inventory. If needed, recursively grant such access to the inventory directory as follows:

     chown -R oracle:oinstall <inventory_loc>

     chmod -R 664 <inventory_loc>

On Windows hosts, you must use a Central Inventory. However, on UNIX systems, you can manage inventories on hosts where GC components, including Agents, are to be installed using one of two approaches:

- A Central Inventory
- A separate inventory for Grid Control

Use a Central Inventory Installing Grid Control (or any Oracle product) into an existing registered Central Inventory is the Oracle best practice. It is also a Grid Control best practice to install the OMS and Repository Database on a dedicated host or hosts. In such a case, no Oracle inventory will exist when installing the OMS or OMR. However, on target hosts where you will deploy Agents, a Central Inventory will probably exist for Oracle products that Grid Control is to manage.

Using a Central Inventory to contain both GC software and any other Oracle software on a host (rather than using a separate inventory for GC) greatly reduces inventory administration. For example, after installation, inventory information for GC and target software alike will automatically appear in the Console under the Deployment tab without additional GC configuration required (described in Chapter 11 in the section "Enable Multi-Inventory Support"). One possible downside of using a Central Inventory is that, although it's a best practice, it can cause deinstallation, upgrade, and patching problems for Oracle products (including GC components) sharing that inventory, but only if you don't take proper precautions to back up the inventory before running the OUI.

Unless you have a good reason to use a separate inventory for GC installations, I would recommend using a Central Inventory. If a Central Inventory exists on a host where you plan to install a GC component (likely just the Agent), do the following to validate and back up the inventory on each of these hosts:

1. Confirm the OUI version, accessibility, and validity of the Central Inventory. Run the following command as the owner of the GC component to be installed:

   ```
   opatch lsinventory
   ```

   - Confirm from the value of the OUI version that the preexisting Central Inventory was created with a version 10.2 Oracle Universal Installer (the same OUI version the GC installer uses). While a preexisting inventory created by a 10.1 installer may work, you're asking for trouble, based on my experience. Unless you want to upgrade your OUI to 10.2, I'd suggest using a separate inventory for Grid Control in this case.

   - Verify that the existing Central Inventory is not corrupted and is accessible (i.e., can be run) by the user who owns the GC component to be installed. Confirm that no errors appear in the output from the command above, and that the last output line says "OPatch succeeded."

2. Prepare to install any GC component on the host by backing up the preexisting Central Inventory and the corresponding *oraInst.loc* file pointing to it. If the GC installation fails, you can simply restore the backup of the preexisting Central Inventory rather than having to detach GC from it using the OUI.

   - On Windows, backup Central Inventory at *%SystemDrive%:Program Files\Oracle\Inventory* to another location.

   - On UNIX, backup the inventory as follows:

     cd <inventory_loc>/..

     cp -Rp <inventory_loc> <backup_loc>

Use a Separate Inventory for Grid Control Your site may maintain separate inventories for each Oracle product installed on a particular host. In this case, you may prefer to register the GC installation in a new separate inventory in one of the following ways, as illustrated in Figure 2-2.

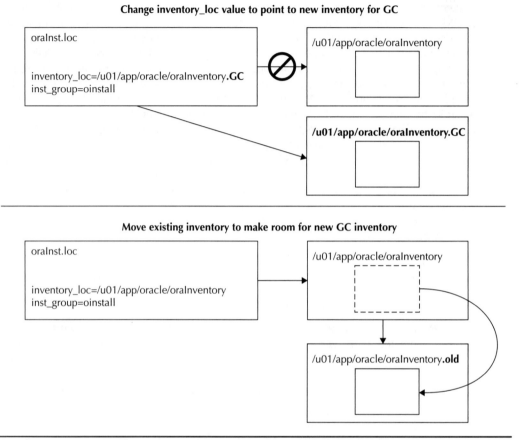

FIGURE 2-2. *The two different methods to use separate Oracle Inventories*

Here is a description of the two methods for moving the inventory:

- **Change inventory_loc value to point to new inventory for GC**   Change the *inventory_loc* value in the inventory pointer file *oraInst.loc* to point to a new location. In Figure 2-2, we are changing *inventory_loc* from */u01/app/oracle/oraInventory* to */u01/app/oracle/oraInventory .GC*. First back up the existing default *oraInst.loc* file to a meaningful name indicating the corresponding Oracle product already installed (or "pound out" the current entry in the file by entering a "#" as the first character of the line). Then change the *inventory_loc* value in the default *oraInst.loc* file to point to a new inventory path for the GC installation.

- **Move existing inventory to make room for new GC inventory**   Move the existing inventory directory containing information on existing Oracle products and specified by *inventory_loc* to an alternate location, and use the existing *inventory_loc* location for the new GC Central Inventory. In Figure 2-2, we move the inventory directory */u01/app/ oracle/oraInventory* to */u01/app/oracle/oraInventory.old*.

If you decide to use a separate inventory for each product, you will later need to configure Grid Control to discover each inventory located in a nondefault location (other than that pointed to by the default *oraInst.loc* file). This configuration process is covered in Chapter 11 in the section "Enable Multi-Inventory Support."

Choose the Grid Control Parent Directory

For all GC components on all hosts, use the same GC *parent directory* specification if possible (the directory that the GC installer prompts for that contains all the GC Oracle homes as subdirectories). The parent directory should be a subdirectory of the Oracle base directory. You do not need to pre create this parent directory; the GC installation will create it as long as the *oracle* user has read-write permissions and the *oracle* group has read permissions (750 on UNIX) on the lowest existing directory (typically ORACLE_BASE) comprising the parent directory.

If you do pre create a parent directory, the Prereqchecker warns that it exists (the installer still continues); it checks that the path doesn't contain a space; and it performs an Oracle Home Compatibility check that the component oracle.rsf.oracore_rsf is not present.

An example parent directory used throughout this book is */u01/app/oracle/product/10.2.0/em10g*. Under this parent directory, the GC installer creates a separate Oracle home for each core component installed under the following subdirectories (created only if that component is installed):[31]

Grid Control Component	Subdirectory Created below Parent Directory
OMR	db10*g*
OMS	oms10*g*
OMA	agent10*g*

The Grid Control parent directory, if pre created, should be empty, and once installed, should only contain the GC Oracle home subdirectories listed above (post-installation; do not create additional subdirectories under this directory).

If you select the GC New Database Option (which installs a new database for the Management Repository), the installer lays down all three Oracle homes under the parent directory. If you use the GC Existing Database Option (where a Management Repository is installed in an existing database), the installer creates two subdirectories under the parent directory: *agent10g* and *oms10g*. Both the GC New and Existing Database Options hard-code these subdirectories. When you have a choice of home directories, as when installing standalone Agents on target hosts, for consistency use the same OMA Oracle home as the hard-coded one for the chain-installed Agent (all the way down to the *agent10g* subdirectory)[32] across all target nodes. As with the Grid Control installation of the OMS and chain-installed Agent, most standalone Agent installation methods create the subdirectory *agent10g* under the Parent Directory specified.

The same applies if you choose the same Oracle home path for the OMR database software (even if it's on a different host than the OMS) when using the Existing Database option as is hard-coded when using the New Database option. This ensures consistency for all GC Oracle homes across all hosts containing GC components.

Remember not to use symbolic links in any directory specifications—Grid Control cannot parse them.

---

[31] The *agent10g* subdirectory is also used when installing standalone Agents on target hosts using the GC Additional OMA Option.

[32] See Chapters 5, 6, and 7 for more information on installing standalone Agents.

Stop Database Listeners Using Port 1521
OMR
If you choose the GC New Database Option, this creates a database and starts a default Oracle Net listener on port 1521 and the IPC key value EXTPROC. However, if you're sharing the OMR database host with another Oracle product running a listener, and if it uses the same port or key value, the installer can only configure the new listener, not start it. Before kicking off the installer, shut down any existing database listeners on port 1521 to guarantee that the new database listener for the Repository database starts during the installation. You can determine whether an existing listener process is running, the listener name, and its Oracle home directory by executing the following command as the *oracle* user:

```
ps -ef | grep tnslsnr
… /u01/app/oracle/product/10.2.0/db_1/bin/tnslsnr LISTENER -inherit
```

If output similar to the following is returned, the listener is running. Here, */u01/app/oracle/product/10.2.0/db_1* is the Oracle home directory where the listener is installed[33] and the listener name is the default, LISTENER. Then, get the TCP/IP port and IPC key value from the *listener.ora* file or from a listener status command. If the listener uses port 1521, change it to another port number, or if needed, change the OMR port number after installation.[34]

Synchronize OS Timestamps/Time Zones
OMR, OMS, OMA
Ensure that the timestamp on the OMS host is synchronized with those on all other GC hosts, including the Repository and Agent hosts. Synchronized timestamps are particularly important for the following:

- When viewing GC log files in the Console or at the OS level
- Between RAC Repository nodes to prevent interconnect problems
- On Cluster Agent nodes to avoid Agent installation errors such as "/bin/tar: ./Apache/Apache/conf/dms.conf: time stamp 2006-03-21 23:27:52 is 21 s in the future"

You can make minor changes in time on RAC nodes, in the seconds range. To make larger time changes, shut down the instances on all RAC nodes. This avoids false evictions, especially if you've applied any 10g low-brownout patches for Bug 4896338, which allow for very low miscount settings. DBMS_SCHEDULER jobs will be affected by time changes, as they use actual clock time rather than System Change Numbers (SCNs). Apart from these issues, the Oracle server is immune to time changes (e.g., transaction and read consistency operations will not be affected).[35]

---

[33] This may not be the Oracle home if TNS_ADMIN is set; in this case, get the Oracle home from the *oratab* file.

[34] This requires changing the port in the OMR listener.ora file and in the $OMS_HOME/sysman/config/emoms .properties file on the OMS for the properties *oracle.sysman.eml.mntr.emdRepPort* and *oracle.sysman.eml.mntr .emdRepConnectDescriptor*.

[35] Note 200346.1 (Oracle internal only).

The recommended approach to synchronize timestamps is to run the Network Time Protocol (NTP) on all GC hosts.[36] All Grid Control and target hosts should use the same NTP server. (Even if you're running the NTP daemon, spot-check that timestamps are synchronized across GC hosts by manually running the *date* command as simultaneously as possible on each host.) To check that the NTP daemon is running, execute the following command:

```
/sbin/service ntpd status
```

You can also check for the presence of the ntpd process, which will return output similar to that shown here:

```
ps -ef | grep ntpd | grep -v grep
ntp       1420     1  0 Sep29 ?        00:00:03 ntpd -U ntp -p /var/run/ntpd.pid
```

If the NTP daemon is not running, start it as *root*:

```
/sbin/service ntpd start
```

For Linux hosts, you can add the "-x" flag to this ntp daemon startup command to prevent the clock from going backwards.

CAUTION
If you plan to use a RAC Repository database, you must run the NTP or an equivalent service to keep the clocks synchronized across RAC nodes. Otherwise, you may experience unexpected OMS shutdowns, and data collection timestamps may be incorrect (Bug 4500241).

In addition to timestamps, set the OS time zone (and update the hardware clock) as desired for Grid Control on the OMS, OMR, OMA, and target servers. It is particularly important to set the OS time zone correctly on the node(s) to contain the OMR database, as the database time zone defaults to the OS time zone. If the OS time zone is not a valid Oracle Database time zone, the latter defaults to UTC.[37] For instructions on changing the OS time zone, consult your platform-specific documentation.

TIP
If your GC implementation spans multiple time zones and is centrally managed, consider using the UTC time zone for all Grid Control hosts, including managed target hosts. This will help you avoid having to convert time zones when examining log files, and the like.

---

[36] Each machine has a different clock frequency and, consequently, a slightly different time drift. NTP computes this time drift within approximately 15 minutes and stores this information in a drift file. NTP adjusts the system clock based both on this known drift and on a chosen time-server for all RAC nodes.

[37] Oracle*MetaLink* Note 330737.1 provides a list of the time zones that the 10*g* Oracle Database supports; this list is the same one the GC Agent supports.

As an example, on Linux you can change the system time zone in a few short steps. Logged in as *root,* first check which time zone your machine is currently using by executing the following command, whose output will be in the format shown here:

```
/sbin/clock
Mon 11 Dec 2006 05:58:34 AM EST   -0.097270 seconds
```

In this case, EST is the current time zone. Say you wanted to change the time zone to UTC. Back up the previous time zone configuration, which for Linux is stored in */etc/localtime.* (Don't try to read this file, it contains binary data.)

```
cp /etc/localtime /etc/localtime.bk
```

Next, change to the directory */usr/share/zoneinfo.* Here you will find a listing of time zone regions, both files and directories. If appropriate, change to the directory named after the appropriate region. If you live in the United States and want to use local time, you would change to the directory *US* or *America* (the latter is also valid for Canada and the other Americas) and choose your local time zone. In this example, you would not need to change directories because *UTC* is a file under */usr/share/zoneinfo.* Copy the appropriate time zone to */etc/localtime,* as shown in this example:

```
cd /usr/share/zoneinfo
cp UTC /etc/localtime
```

If you have the utility *rdate,* update the current system time with this new time zone by executing the following:

```
/usr/bin/rdate -s time.nist.gov
```

Finally, set the hardware clock by executing:

```
/sbin/hwclock --systohc
```

Confirm Platform-Specific Software Requirements ✔
OMR, OMS, OMA

OMR, OMS, and standalone OMA hosts have platform-specific requirements over and above the generic software requirements already mentioned. These requirements and the commands to manually check them are listed in the section "Platform-Specific Software Requirements" in Appendix B (online).[38] Use this section of the appendix now to confirm and redress any outstanding prerequisites for your platform. This section begins by pointing you to the Release Notes to learn

---

[38] Package and kernel requirements in the online Appendix B of this book are interpreted from the Oracle manual, Enterprise Manager Grid Control Installation and Basic Configuration, Appendix B: Platform-Specific Package and Kernel Requirements, cross-referenced with the platform-specific Oracle Enterprise Manager Grid Control Quick Installation Guides. The generic installation guide does not make it clear whether the requirements apply to the OMS, OMR, standalone OMA, or all of them; the platform-specific guides are clearer but still hazy on this point. Appendix B of this book clarifies the requirements for each GC component.

> **Don't Forget to Run the Standalone Prereqchecker**
> Reminder: If you haven't yet run the Prereqchecker in standalone mode, run it now [see "The Prereqchecker" section in Appendix B (online) for instructions] to double-check that you've satisfied at least the subset of requirements that the Prereqchecker verifies.

about platform-specific installation issues. The body of the section offers a breakdown by platform of required OS packages, patches, kernel parameters, and group and user rights.

Summary

This chapter is long because preparation is the key to installing Grid Control successfully and efficiently. You begin by downloading and staging the GC software. At the same time, you make some basic architectural design decisions, then address prerequisites related to network configuration, hardware, and software (both generic and platform-specific). Now you're ready to move on to Chapter 3, where you actually install Grid Control. Prepare to reap the benefits of this thorough preinstallation groundwork.

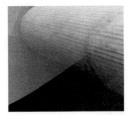

CHAPTER 3

Grid Control Installation

This chapter gives you directions from start to finish on how to install the Grid Control Repository, OMS, and chain-installed[1] Agent, using the first three installation types, which you may recall from Chapter 2 are as follows:

Installation Type	Nickname
Enterprise Manager 10*g* Grid Control Using a New Database Installation Type	GC New Database Option
Enterprise Manager 10*g* Grid Control Using an Existing Database Installation Type	GC Existing Database Option
Additional Management Service Installation Type	GC Additional OMS Option

Chapters 5, 6, and 7 cover the fourth installation type, the GC Additional OMA Option, as well as the other methods for installing standalone Agents.

To cover these three installation types, I've organized the chapter as follows:

- Detail the information you must gather as input to the installer
- Provide direction on how to consult platform-specific release notes to address installation bugs
- Describe how to initialize the OS environment for running the GC installer
- Explain three ways of running the GC installer—interactively, silently, or via the static ports feature—and how to monitor the installation
- Supply detailed instructions and screen shots for the three installation types when you run the GC installer interactively:
  - For the GC New Database Option, I show all screen shots the installer presents. Roughly 75 percent of these screen shots also appear when you select the GC Existing Database and GC Additional OMS Options.
  - For the GC Existing Database Option, I explain how to create, configure, and tune a new database to become the "existing" database, and show two screen shots unique to this installation type (rather than presenting duplicate screen shots to those for the GC New Database Option).
  - For the GC Additional OMS Option, no screen shots are provided since they are a subset of those for the GC Existing Database Option. I explain what information the installer does not request compared with the GC Existing Database Option.
- Direct how to log in to the GC Console as a preliminary check to ensure the GC installation is successful (before installing any additional OMS hosts).

Let's get down to the first order of business. I strongly advocate that you determine your input to all installation questions *before* launching the GC installer. Such a fact-finding session accelerates the

---

[1] See the introduction of Chapter 2 for an explanation of this term.

installation process and prevents you from having to reinstall Grid Control because of insufficient preparation. Consultants and/or in-house IT staff who hold a single meeting to gather all required installer input exhibit more professionalism than those who start the installation blindly, only to have to repeatedly return to others for input.

Gather Needed Installation Information
OMR, OMS
The installation information requested when you're choosing any of the first three installation types differs by installation type as detailed in Table 3-1. Following are descriptions of the columns to help you navigate through the table. Starting with the top row, consult the columns in the order in which they are presented in the table:

- **Columns 1 to 6** Find the column for your chosen installation type (columns 1, 2, 3) to determine whether it will display the Installation Screen Title (column 4) requesting the Input Field (column 5) described here (column 6).

- **Column 7: Recommended Input** Where applicable, I recommend an input value or choice based upon Grid Control best practices and Optimal Flexible Architecture (OFA)[2] standards.

- **Column 8: Is Input Mandatory or Optional?** This column shows whether the requested input is mandatory or optional. I recommend entering all optional information now, but you can also configure it through the GC Console after the installation completes.

- **Column 9: Figure Number** These figures are the installation screen shots themselves found in the later sections for the GC New and Existing Database Options. These sections provide in-depth coverage of each input field. You don't need to reference these figures now unless you need more detail than that provided in the table.

After documenting the information you'll enter when running the Grid Control installer for your specific installation type, you can move on to evaluating installation snags you may encounter, which depend upon the GC platform used, as described next.

Address Installation Bugs
OMR, OMS, OMA
This book is intended to serve as a major source of information to navigate the Grid Control installation and configuration process. To make this experience as painless as possible, I try to minimize the need to bounce between this book and the Oracle GC documentation. You cannot ignore the Oracle GC documentation altogether. Release notes, which contain information on installation bugs, documentation errata, and the like, accompany each new software release for a specific OS, leaving us with a wide array of current release notes that would be impossible to incorporate into this book. To complicate matters, given that all platforms require the Full Installer for releases 10.2.0.1 through 10.2.0.3, then Patch Installers for 10.2.0.3 (the latest release available on some platforms) or 10.2.0.4, you must consult the respective release notes for both the Full Installer and Patch Installer needed at your site.

---

[2] For more information on OFA standards, see Appendix D of the Oracle Database Installation Guide 10*g* Release 2 (10.2) for your Repository platform.

Applies to GC New Database Option	Applies to GC Existing Database Option	Applies to GC Additional OMS Option	Installation Screen Title	Input Field	Field Description	Recommended Input	Is Input Mandatory or Optional?	Figure Number
Yes	Yes	Yes	Specify Installation Type	Choose GC New Database, Existing Database, or Additional OMS Option	Choose desired installation type	Site specific	Mandatory	3-1
Yes	Yes	Yes	Specify Installation Location	Parent Directory	Creates Oracle homes db10g (only for GC New Database Option), oms10g and agent10g subdirectories under this base directory	<mount point>/app/oracle/product/10.2.0/em10g	Mandatory	3-2
Yes	Yes	Yes		Product Languages	Language Selection defaults to English	Site specific	Mandatory	3-3
Yes	Yes	Yes	Specify Inventory directory and credentials	Full path of Inventory directory	(UNIX only) Screen appears only if this is the first Oracle installation on this host[3]	<ORACLE_BASE>/oraInventory where <ORACLE_BASE>=<mount point>/app/oracle	Mandatory	3-4
Yes	Yes	Yes		OS Group name	(UNIX only) OS group must have write permission to Inventory directory	oinstall or dba	Mandatory	
Yes	Yes	Yes	Product-Specific Prerequisite Checks					
Yes	No	No	Specify Configuration	Repository Database Name	New database name. Recommend using fully qualified database name.	emrep.<domainname>	Mandatory	3-5
Yes	No	No		Repository Database File Location	Installer appends database name to the File location specified	<mount point>/oradata/ or if using ASM, +DATA/<ORACLE_SID>/datafile/	Mandatory	3-6

Installation Screen	Field	Description					Value	Mandatory/Optional	Page
	OSDBA group OSOPER group	(UNIX only) Installer user must be a member of both groups to grant SYSDBA and SYSOPER permissions for creating the new database.	Yes	No	No	Yes	dba dba	Optional	
Specify Repository Database Configuration	Database Hostname Port Service/SID SYS Password	Use fully qualified hostname Listener port For single-instance use DB_NAME; for RAC use one INSTANCE_NAME (both nonqualified) Enter SYS password.	No	Yes	Yes	No	<hostname>.<domain> site-specific (default 1521) <DB_NAME> for single-instance, <INSTANCE_NAME> for RAC Site specific	Mandatory	3-14
	Management Tablespace Location Configuration Data Tablespace Location	Specify full path for datafiles. Cannot use Oracle Managed Files (OMF).	No	No	Yes	No	<full path>/mgmt.dbf <full path>/mgmt_ecm_depot1.dbf[4]	Mandatory	
Specify Optional Configuration	Configure e-mail Notification? e-mail Address SMTP Server	Check the box to specify e-mail address[5] and SMTP Server name for SYSMAN alert notifications.	Yes	No	Yes	No	Check the box	Optional	3-7
	Configure MetaLink? Username, Password	Check the box to enter MetaLink credentials to enable GC to search for and download patches.[6]	Yes	No	Yes	No	Check the box	Optional	

TABLE 3-1. *Grid Control Installation Information Needed Based on Installation Type (Continued)*

[3] On Windows, the Inventory location defaults to <*SystemDrive*>:*Program Files**Oracle**Inventory*\.

[4] If you're using ASM, <*full_path*> is +DATA/<*ORACLE_SID*>/datafile.

[5] You will likely receive multiple notifications following installation, triggered from out-of-box alerts. So initially, enter the e-mail address of the GC administrator responsible for tuning alerts.

[6] I recommend creating a dedicated MetaLink account not tied to a person so that it remains open indefinitely and so that the account password is not inadvertently changed.

Applies to GC New Database Option	Applies to GC Existing Database Option	Applies to GC Additional OMS Option	Installation Screen Title	Input Field	Field Description	Recommended Input	Is Input Mandatory or Optional?	Figure Number
Yes	Yes	Yes		Configure Proxy? Proxy Server, Port Do Not Proxy For Proxy Username, Password, Realm	Check the box if OMS must go through a firewall to connect to Agents or to MetaLink. Specify fully qualified hostname and port for proxy server. Enter internal address(es) not to proxy for, comma-separated such as *.us.mycorp.com,*.mycorp.com. Enter username, password, and realm for proxy server if it requires authentication.	Site specific	Optional	
Yes	Yes (under Specify Passwords screen)	Yes (under Specify Repository Database Configuration screen)	Specify Security Options	Agent Registration Password Require secure communication for all Agents?	Specify registration password for Agents to securely communicate with OMS. Check the box to require that the OMS communicate only with secure Agents.	Enter a password Check the box	Mandatory	3-8
Yes	No	No		Use same or different passwords for accounts? SYS, SYSTEM, SYSMAN, DBSNMP	Use same or different passwords as dictated by security policies. Passwords(s) must be five-characters minimum plus one number and start with a letter.	Site specific	Mandatory	

Yes	Yes	Oracle Configuration Manager Registration	Enable OCM, CSI, MetaLink Account Username, Country Code, Connection Settings, Test Registration	Register for OCM by supplying all fields to provide Global Customer Service with system configuration information.	Check Enable Oracle Configuration Manager and complete remaining fields	Optional	3-9
No	Yes (under Specify Repository Database Configuration screen)	Specify Passwords	Agent Registration Password Require secure communication for all Agents?	Specify a registration password for Agents to securely communicate with OMS. Check the box to require that the OMS communicate only with secure Agents.	Enter password Check the box	Mandatory	
No	Yes (under Specify Repository Database Configuration screen)		SYSMAN Password	Enter new password for SYSMAN (user should not exist).	Site specific	Mandatory	
Yes	Yes	Summary	Click Install	Summarizes installation based on all input	Click Install if all options are correct	Mandatory	
Yes	Yes	Install (Execute Configuration scripts)	root password	Need root password to run scripts orainstRoot.sh and allroot.sh.	Site specific	Mandatory	

TABLE 3-1. *Grid Control Installation Information Needed Based on Installation Type (Continued)*

Release notes for the base 10.2.0.1 release on OTN are available in the Oracle Enterprise Manager Documentation Library at http://www.oracle.com/technology/documentation/oem.html. Release notes for the latest Patch Installers are bundled with the platform-specific GC software that you already staged (see Chapter 2, "Stage the Grid Control Software"). These release notes are also put out as a separate Oracle*MetaLink* consolidated document called Oracle Enterprise Manager Grid Control Release Notes for *<platform>* 10*g* Release *x* (10.2.0.*x*.0), where *x* is the latest release family. (This consolidated document also contains the Grid Control List of Bugs Fixed for that platform.) Before installing Grid Control, inspect the Installation and Upgrade Issues under the Known Issues section of both the Full and Patch Installer release notes for bugs and workarounds.

To give you an idea of the types of installation bugs you may run across, following are some Linux bugs I've encountered (and workarounds) in various GC 10.2.0.x releases.[7] (Most of my Grid Control installation testing has been on the Linux platform.) Some of these bugs are documented in the GC 10.2.0.x release notes for Linux, and others are referenced in Oracle*MetaLink* notes cited in the footnotes for this chapter.

- The Grid Control OUI prompts you to install Oracle Configuration Manager (OCM), if desired. On UNIX, if the *oracle* user isn't authorized to use CRON, OCM software is installed but OCM setup fails, as revealed only in the OUI installation log files. The reason is that the OCM setup process needs to add an entry to the *oracle* crontab to run data collections every 15 minutes. Do one of the following, depending on whether you can authorize *oracle* to use CRON:

    - Enable *oracle* to use CRON by adding the *oracle* user to */var/adm/cron/cron.allow* file (which requires *root* permission). Be sure to unset CCR_DISABLE_CRON_ENTRY before running the OUI as it is also documented that setting this variable to any value enables it.

    - As a workaround to not using CRON, set the environment variable CCR_DISABLE_CRON_ENTRY to 1 before launching the Grid Control OUI. Schedule OCM to run using a scheduling method other than CRON or run OCM manually whenever desired, as follows (ORACLE_HOME is the home in which OCM is running):[8]

        ```
        $ORACLE_HOME/ccr/bin/emCCR start
        $ORACLE_HOME/ccr/bin/emCCR collect
        $ORACLE_HOME/ccr/bin/emCCR upload
        $ORACLE_HOME/ccr/bin/emCCR status
        $ORACLE_HOME/ccr/bin/emCCR stop
        ```

- Before installing Grid Control on Linux x86, make sure to unset all NLS-related OS environment variables (NLS_LANG, ORA_NLS, ORA_NLS33, ORA_NLS10, etc). Setting NLS environment variables can cause the Grid Control installation to fail.[9]

---

[7] Some of these bugs may have been fixed since the publication of this book. To confirm this, consult the latest GC release notes for Linux.

[8] See Note 369619.1.

[9] See Oracle*MetaLink* Note 272493.1. This was confirmed for GC 10.1.0.2 on Linux x86, but later releases on other platforms may be affected as well.

- The following bugs are specific to SUSE Linux Server Enterprise 9 (SLES 9) as the installation platform for the OMS:
    - /usr/lib/libdb.so.2 is not available on SLES 9, which causes problems for the Grid Control Oracle Universal Installer (OUI). The solution is to create the following symbolic link before launching the OUI:

        ```
        ln -sn /usr/lib/libdb.so.3 /usr/lib/libdb.so.2
        ```

    - If you plan to select the GC New Database Option on SLES 9, first manually create an empty /etc/oratab file. The installer reports an error if this file is missing and does not create a new oratab file.

    - During the execution of *root.sh*, the installer tries to create a hard link from */etc/rc3.d/S99gcstartup* to the file */etc/init.d/gcstartup*. However, on SLES 9 the */etc/rc3.d* directory does not exist. The equivalent directory is */etc/init.d/rc3.d*. Therefore on SLES, create the following hard link manually after the Grid Control installation completes:

        ```
        ln /etc/init.d/rc3.d/S99gcstartup /etc/init.d/rc3.d/gcstartup
        ```

    - No Grid Control components restart automatically on reboot. You must manually start all EM components, including the OMR database and listener if created using the GC New Database Option, the OMS, and chain-installed OMA. A workaround is to implement the *orarun* script available from SuSE.[10]

Initialize the Oracle User Environment
OMR, OMS, OMA
Before running the Grid Control installer, you need to initialize the environment from which you will launch the OUI as the *oracle* user (or other user chosen to own the GC software). This initialization involves setting up X Server for UNIX platforms, then setting or unsetting the appropriate environment variables.

Set Up X Server (UNIX Only)
To run the Grid Control Oracle Universal Installer (OUI) interactively on UNIX, you need to set up X Server on the installer host. If you are on a UNIX platform, you can either use a PC with X Server software installed or a UNIX workstation.

NOTE
You can also run the installer silently, in which case you don't need to install X Server libraries or run X Server. See the section "Silent Installation Method" later in the chapter for more information.

If you're using a PC with X Server software, start the X Server on the PC. If using a UNIX workstation, start a terminal session such as an X terminal (xterm) to the remote UNIX system where you want to install the software. Either way, log in directly as the *oracle* user—if you use *su* or *sudo* to log in as *oracle*, it's likely the X Server software won't work.

---
[10] See Oracle*MetaLink* Note 266049.1.

> ### UNIX Shell Variations
>
> Initializing the environment on UNIX platforms varies according to the UNIX shell you use. There are four main UNIX shells in use today. The following list identifies the shell startup script name, shows how to run the startup script, and provides the command syntax for setting environment variables for each shell. (Note that I use the Bash shell in examples throughout this book.)
>
> | | Bash Shell (bash) | Bourne Shell (sh) | Korn Shell (ksh) | C shell (csh or tcsh) |
> | --- | --- | --- | --- | --- |
> | User's Startup Script for Shell | .bash_profile | .profile | .profile | .login |
> | How to Execute Startup Script from User's Home Directory | . ./.bash_profile | . ./.profile | . ./.profile | source ./.login |
> | Syntax to Set x=1 (Set Environment Variable x to "1") | x=1; export x | x=1; export x | x=1; export x | setenv x 1 |

NOTE
Leave the session open where you're logged in directly as the Oracle user, as you will be initializing the environment and launching the Grid Control installer in this session.

Allow the remote UNIX host to display X applications on the local X Server by executing the following command on the remote host (use the fully qualified host name of the remote host):

    xhost <remote_hostname>

To set the environment as indicated below, first determine the default shell for the *oracle* user:

    echo $SHELL

Open the shell startup file for the *oracle* user in a text editor, such as *vi*. Refer back to the list in the "UNIX Shell Variations" sidebar above for the shell startup file name. The example here assumes a Bash shell startup file is used.

    vi .bash_profile

Enter or edit the following line to set the default file mode creation mask to 022 in the shell startup script. This ensures that the Oracle installation lays down files with 755 permissions:

    umask 022

Save the shell startup file and exit the editor. Then run the shell startup script (again, we use the Bash shell in examples throughout this book):

    . ./.bash_profile

Enter the following command to direct X applications like the OUI to display on your local PC:[11]

```
DISPLAY=<local_system>:0.0; export DISPLAY
```

where *<local_system>* is the host name or IP address of your local PC.

Test Xwindows functionality by verifying that when you run the *xclock* command on the UNIX host, a clock appears on your X Server display. The location of *xclock* varies by UNIX platform as follows:

Platform	xclock Command
AIX	/usr/bin/X11/xclock
HP-UX	/usr/bin/X11/xclock
Linux	/usr/X11R6/bin/xclock
Solaris	/usr/openwin/bin/xclock

Set/Unset OS Environment Variables

The following environment variables should be set or unset for GC installation as noted here.

Variable Name	Value	Description
CCR_DISABLE_CRON_ENTRY	1	Set only if you want the GC OUI to install Oracle Configuration Manager and the *oracle* user isn't authorized to use CRON.
LANG, NLS_LANG, and other NLS variables	Unset	These variables might be incompatible with the OS default locale and OMR database character set.[12]
OPATCH_NO_FUSER	Unset (for GC 10.2.0.2 or higher)	Opatch (called by OUI) uses the *fuser* executable to detect active processes. Unset if *fuser* is present in /sbin or /usr/sbin or is in the PATH.
	FALSE (for GC 10.2.0.1 or lower)	This is a workaround for Bug 4898745 (fixed in GC 10.2.0.2) to bypass the OPatch active process checks.
ORACLE_BASE	Set	Set to */u01/app/oracle*, for example. This is not required, but the OUI picks up this location to pre populate the EM Parent and oraInventory directories.

---

[11] You do not need to set the DISPLAY environment variable for GC silent installations.

[12] This is a recommendation. However, the installation will fail if you don't unset NLS variables on Linux for GC 10.1.0.2, and perhaps for other releases and platforms. See Oracle*MetaLink* Note 272493.1.

Variable Name	Value	Description
ORACLE_HOME	Unset	The OUI prompts for a parent directory, which contains multiple Oracle homes.[13]
ORACLE_SID	Unset	OUI will prompt you to enter the OMR database name.
PATH	Do not include Oracle paths. Include *fuser* executable in PATH if it's not in /sbin or /usr/sbin.	This variable should not contain any paths for installed Oracle products, but only standard OS paths for all users.
TMP and TMPDIR (UNIX), TEMP, TMP (UNIX and Windows)	Unset (set if required)	OUI uses this directory for executables and link files. Unset if free space in temporary directory is at least 250MB and won't be purged. Otherwise set to an alternate directory that meets these specifications.
TZ	Unset	Set desired time zone at OS level instead.

As this table shows, you should set ORACLE_BASE (not required but recommended), and set OPATCH_NO_FUSER and temporary directory variables only as needed. Unset all other Oracle-related variables not specifically mentioned.

On UNIX, consult the list in the "UNIX Shell Variations" sidebar above for the syntax to set or unset variables in the various UNIX shells. To confirm that the UNIX environment has been set correctly, enter the following commands as the *oracle* user:

```
umask
env | sort
```

Verify that the *umask* command displays a value of 22, 022, or 0022 and that the environment variables you set in this section have the correct values.

On Windows, open a command window and use the following syntax to set or unset variables, respectively:

```
set <variable_name>=<value>
set <variable_name>=
```

Leave the command shell open to execute the GC installer (*setup.exe*) later, thereby allowing the OUI to pick up all environment variable settings. To verify whether the Windows environment is set correctly, execute the following command from the command shell:

```
set
```

---

[13] Multiple Oracle homes are installed under this parent directory for the OMR, OMS, and OMA, depending on the installation type chosen.

Running the Installer with Desired Options

Before kicking off the Grid Control installation, it's always a good idea to consult the Readme instructions for your platform that accompany the Full Installer version you staged in Chapter 2. The *README.txt* is located in the top level directory of the distribution. Scan it and you'll see that for a fresh installation of Grid Control, it says to "Run runInstaller (or setup.exe for Windows) to start installation." Let me improve the Readme by adding, "with the appropriate parameters." Two options that require parameters are the silent installation method and the static ports feature.

This section is laid out as follows:

- Silent Installation method
- Static ports feature
- Starting and monitoring the installation

Both options have limited usefulness, so I explain them only briefly. I also explain how to start and monitor your installation, which is the same regardless of the installation type you select.

NOTE
If you are planning to install the OMS or OMR in an Active/Passive Cold Failover Cluster (CFC) environment, see the following sections in Chapter 15 for alternate Grid Control installation instructions: "Install OMSs in an Active/Passive Configuration" and "Install the Repository in a Cold Failover Cluster." If a server load balancer is available now for use with Grid Control, also see "Use a Server Load Balancer" in Chapter 15, as it requires a different approach to GC installation.

When reading about the silent installer and the static ports feature, for sake of completeness, I provide the syntax for their respective parameters. However, don't kick off the installer until you get to the appropriate section for your installation type.

Silent Installation Method

If you have good reason not to run the Grid Control installer interactively, run it silently. Silent installations work for all four installation types. What reason might you have for using the silent installer?

- If you can't run X Server on the host where running the installer. The X Server libraries may not be installed, or your corporate security policy may not allow you to run X Server on the host in question.

- If you are installing many GC environments. The silent installer eliminates the need to interact with the installation. This may reduce installation time appreciably when building multiple Grid Control environments, such as at each corporate satellite office or in a classroom setting. (In my opinion, the silent installer adds unnecessary complexity when laying down just one or two GC environments.)

- If you are installing many standalone Agents on target hosts. (The fourth installation type, the GC Additional OMA Option, is not covered in this chapter. I detail Agent deployment using the silent installer in Chapter 7, along with the other methods of installing a standalone Agent.)

- If you want to better guarantee the reproducibility of the installation, such as to ensure identical production and test environments, use the silent installer in conjunction with the static ports feature. This is particularly useful if multiple administrators are running the installer.

To perform a silent installation, start by editing a response file template for your chosen installation type, supplying the same information that the OUI screens request. The response file takes the place of entering this information interactively. The four response file templates, one for each installation type, are located in the *Disk1/response* directory below the top level directory of the GC distribution. To run the Grid Control installer silently, you would simply launch the GC installer executable used for interactive installation with the following parameters, one of which specifies the edited response file (remember, don't actually run the installer yet):

On UNIX

`./runInstaller-responseFile <responsefile path> <optional_parameters> \`

`-silent -waitForCompletion`

On Windows

`.\setup.exe -responseFile <responsefile path> <optional_parameters> -silent`

For complete instructions on running Grid Control silent installations of the first three installation types (that deploy an OMS), see Chapter 4 of Oracle Enterprise Manager Grid Control Installation and Basic Configuration 10*g* Release 2 (10.2).

Static Ports Feature

I can think of two principal reasons for considering the static ports feature.

- If you're installing an additional OMS. Using static ports is the only way to make the port configuration identical across OMS servers. For some reason, if you don't use static ports, Grid Control only assigns some—not all—default ports for additional OMSs, even if these ports are available.
- If one or more default OMS ports are not free for any OMS. However, this should not be the case if you dedicate the OMS host to Grid Control.

You can use the static ports feature with either the interactive or silent installer. The silent ports feature allows you to specify custom port numbers for the Console, OMS, and chain-installed Agent components to override default port numbers that the installer otherwise assigns.[14] Post installation, you can reconfigure any GC component to use a different port, but the static ports feature makes it easier to use custom ports up-front. The static ports feature relies upon your editing a template file called *staticports.ini* located in the *Disk1/response* directory below the top level directory of the GC distribution. (This directory is also where the silent installation response file templates are found.) Example custom port entries in the file are as follows:

Enterprise Manager Central Console Port=5345
Web Cache HTTP Listen port=7799

---

[14] See Oracle*MetaLink* Note 353736.1 for a listing of the default EM 10.1 and 10.2 ports.

When specifying the static ports option to the installer, it uses the values from the file instead of the default port numbers. The EM documentation states that default port numbers are used for any components not explicitly listed in the *staticports.ini* file. However, my own experience with this feature suggests otherwise. I highly recommend that you specify all components explicitly in this file, assigning each component a default or custom port number, whichever the case may be. The *staticports.ini* file uses the same general format as the file *$OMS_HOME/install/portlist.ini*, which the GC installation creates. Therefore, if you install Grid Control, with either default or custom ports, and you wish to use the same port numbers to install an additional OMS, use the ports listed in the *portlist.ini* file from the first installation in the *staticports.ini* file for additional OMS installations. If copying and pasting entries from *portlist.ini*, comment out the header information with a hash mark ("#") or remove it altogether.

When you finish editing the *staticports.ini* file, you are ready to install Grid Control with the static ports feature. When the time comes to run the installer, either interactively or silently, enter the following at the command line:

```
./<runInstaller or setup.exe> -staticPortsIniFile <path>/staticports.ini
```

For more details on the static ports feature, see Chapter 4 of Oracle Enterprise Manager Grid Control Installation and Basic Configuration 10*g* Release 2 (10.2).

Starting and Monitoring the Installation

You are ready now to *learn* how to start and monitor your installation. Then you can go to one of the following sections, depending on the installation type you want to choose, and actually kick off the installer:

- Install Grid Control Using a New Database
- Install Grid Control Using an Existing Database

After choosing one of these installation types, continue with the remaining sections:

- Log in to GC Web Console
- Install an Additional OMS (if applicable)

When launching the Grid Control installer, you will run it interactively or silently using either default or static ports. Most of you will be installing Grid Control interactively rather than with the silent installer. You will probably not need the static ports feature unless you're installing an additional OMS. Whatever installation type you choose, remember to start the installer in the same command shell you opened and initialized in the earlier section "Set/Unset OS Environment Variables." You will execute one of the following commands, depending on the platform (with the appropriate parameters for a silent installation and/or the static ports feature):

On UNIX

```
cd
./<path>/runInstaller <parameters>
```

On Windows

```
cd
.\<path>\setup.exe <parameters>
```

The *<path>* varies based on the location of the installation software:

- If running the OUI from DVD, *<path>*=*<DVD>*/Disk1. When starting the installer from a command line, you should not be in or below the DVD mount point directory, as the installer process will tie up that directory, prohibiting you from ejecting the DVD when prompted to insert Disk 2.

- If installing from a staged location on disk, *<path>*=*<stage_directory>*/Disk1. You can also dot execute the installer from within its own directory, i.e., "*./runInstaller*" or ".\setup.exe".

Regardless of the installation method you opt for—interactive or silent—at some point the installer will take over and the process behind the scenes will become identical. That point for interactive installations is when you click "Install" after completing the interview process in the OUI screens (with or without using the static ports feature), or if running the installer silently, after launching it. The installation time for the base GC release will vary according to your architectural choices, the installation type you select, the existing database in which you're installing the OMR, and the available hardware resources. Following is a rough estimate of the total running time for each installation type:

Installation Type	Installation Running Time (Approximate)
GC New Database Option	2 hours
GC Existing Database Option	1.5 hours
GC Additional OMS	1 hour

Don't be too hasty to terminate an installation that may appear to be hung because of the limited output from the Java installer. Don't suffer in silence; you can find out what's happening behind the scenes by tailing the installation log files using the command *tail -f <logfile>* on UNIX systems (for Windows a UNIX style *tail* command is also available in the Windows 2003 Resource Kit). Here are some notable log files:

- **<oraInventory_dir>/logs/installActionsYYYY-MM-DD_HH-MI-SS-[A-P]M.log** This main installer log file is also echoed to *stdout*. It is useful to consult this log file to review previous installer activity.[15]

- **$OMS_HOME/sysman/install/log/emca_repos_createHH_MI_DD.log** This is a useful log file for a particularly long operation summarized (with spelling error to boot) in the Java output as "Operation Creating OMS Respository [sic] is in progress."

---

[15] On UNIX platforms, the directory *<oraInventory_dir>* is specified by the Inventory pointer location file, *oraInst.loc*, if it exists. If it doesn't exist, you specify *<oraInventory_dir>* when installing Grid Control on the screen "Specify Inventory directory and credentials." On Windows platforms where no Oracle products have been installed, the inventory directory is located under *<ORACLE_BASE>\oraInventory* if ORACLE_BASE is set in the environment where the installer was launched. If at least one Oracle product was previously installed (depending upon the product), *<oraInventory_dir>* can be *<SystemDrive>:\Program Files\Oracle\Inventory*.

- **$ORACLE_HOME/cfgtoollogs/cfgfw**   These are log files for each Configuration Assistant, located under the database Oracle home.
- **cfmLogger_YYYY-MM-DD_HH-MI-SS-[A-P]M.log**   All configuration logs are appended to this log file.

NOTE
The installer is touted to have a resumability feature if it unexpectedly terminates. The installation should pick up where it left off, but don't count on this feature working. In testing, I was not able to demonstrate this functionality. The installer appeared to start from scratch and when I specified the same parent directory used in the failed installation, it complained about the existence of the previous EM installation and asked me to remove it.

Install Grid Control Using a New Database

If you want to install Grid Control using the GC New Database Option, it's finally time to kick off the installer. Go ahead, don't be shy. I'll wait. (See the earlier section "Starting and Monitoring the Installation.")

The GC New Database Option is the default installation type selected when you run the installer. This option lays down all Grid Control components on a single host. A new Oracle Database 10g (10.1.0.4) Enterprise Edition is created with a Repository, and the OMS and tag-along OMA are deployed. Following are the screen shots presented when you choose this installation type.

NOTE
Many of these screens below from the GC New Database Option also appear in the other two Grid Control installation types. Therefore, for reference purposes each screen is labeled "New," "Existing," and/or "Add'l" to indicate which installation types display the screen.

Specify Installation Type
As shown in Figure 3-1, select Enterprise Manager 10g Grid Control Using a New Database, and click Next.

Specify Installation Location
Use the screen shown in Figure 3-2 to specify a parent directory under which the installer will create three subdirectories, one for each Oracle home: *agent10g*, *db10g*, and *oms10g*. This directory cannot be a symbolic link. The base directory of the parent directory will default to ORACLE_BASE/GridHomes if ORACLE_BASE is defined. Change the directory specification as desired. In this book, */u01/app/oracle/product/10.2.0/em10g* is used as the parent directory. It is better to let the installer create the parent directory for you, as it will use the desired permissions. The products are installed in English by default. Click Product Languages to install in different language(s). Otherwise, click Next. The next screen shows the Product Languages screen.

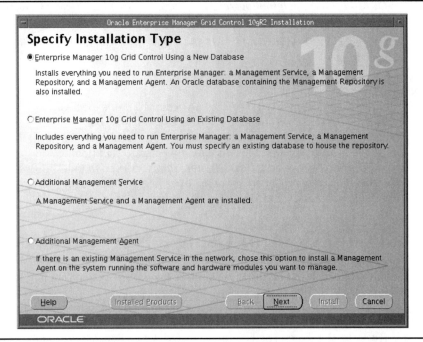

FIGURE 3-1. *Specify Installation Type: New, Existing, Add'l*

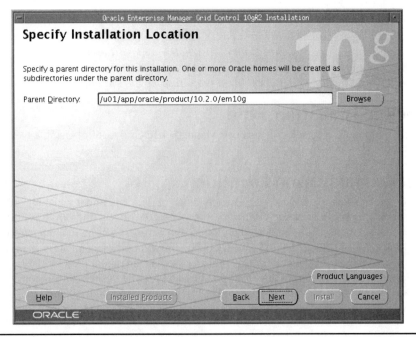

FIGURE 3-2. *Specify Installation Location: New, Existing, Add'l*

NOTE
Screens such as the one above and throughout the book are taken from Grid Control installations on UNIX, so they contain UNIX style file paths (forward slashes). If installing Grid Control on Windows, use Windows-style drive specifications (such as C:) and file paths (backslashes).

Language Selection

This screen (see Figure 3-3) provides 14 languages from which to choose. You can choose multiple languages, not for the installation itself, but for subsequent use of the product. The installer installs text in the selected languages and the fonts required to display these languages. Select the language(s) desired and click OK. Then click Next on the Specify Installation Location screen.

Administrators must add the desired languages in their browsers used for Console access to display the GC web pages in these languages. Taking Internet Explorer as an example, select Tools, Internet Options, click Languages…, click Add…, select the additional languages, and then click OK three times.

Specify Inventory Directory and Credentials

Enter the path to the inventory directory, as shown in Figure 3-4. This path does not need to exist, but the *oracle* user must have permissions to create it. In this illustration, the *oinstall* group is used as the Operating System group name. You can also use the *dba* group, depending on your site's Oracle standards. This screen will not appear on Microsoft Windows platforms, where the inventory location defaults to *<SystemDrive>:\Program Files\Oracle\Inventory*. Click Next.

FIGURE 3-3. *Language Selection: New, Existing, Add'l*

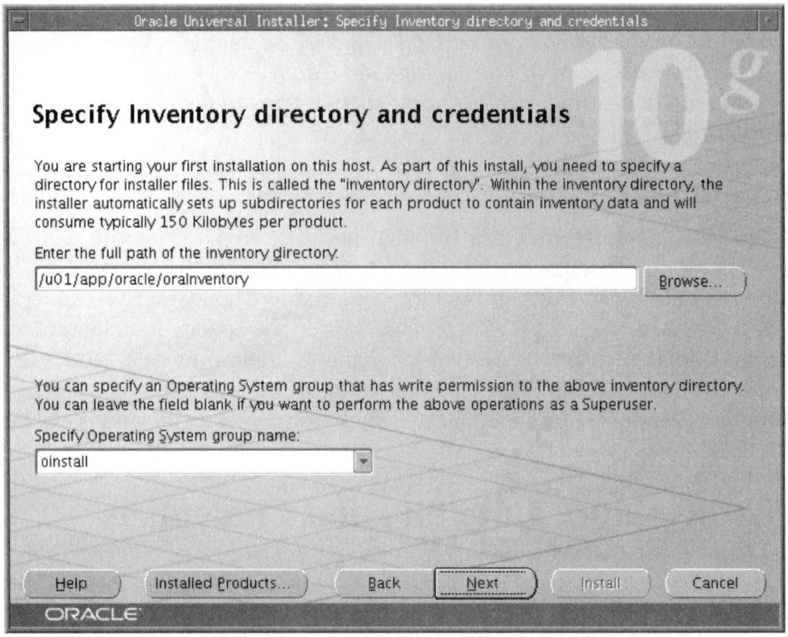

FIGURE 3-4. *Specify Inventory Directory and Credentials: New, Existing, Add'l*

If on the above screen you receive the error, "OUI-10182: The effective user ID does not match the owner of the file, or the process is not the super-user, the system indicates that superuser privilege is required," it is probably due to insufficient privileges to create the oraInst.loc file in the platform-specific default location. To resolve, either pre create this file or grant the OS user running the installer (or the OS group you specify on this screen) write permissions to the directory where oraInst.loc is created. This should enable you to continue without having to restart the installation.

Product-Specific Prerequisite Checks

Figure 3-5 shows the prerequisite check, type (Automatic, Optional, or Manual), and status (Succeeded, Warning, Skipped, or Failed).

- Click Next if all prerequisite checks are successful, which should be the case if you ran the Prereqchecker in standalone mode without error as instructed in Chapter 2.

- Let the prerequisite checks finish, then click each failed prerequisite check to view its corresponding details at the bottom of the screen. These details show expected vs. actual results, error messages, and instructions to resolve them. Fix the failed checks manually if possible, and if fixed, select the check and click Retry. After resolving all checks as best you can, click Next. It is best to resolve all errors before continuing. However, you can click Yes to ignore the errors and attempt to proceed with the installation, although the installer may not allow you to continue, depending upon the severity of the error.

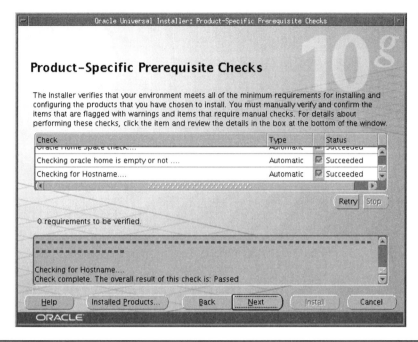

FIGURE 3-5. *Product-Specific Prerequisite Checks: New, Existing, Add'l*

Specify Configuration

Enter the desired Repository database name, datafile location, and OS groups in the screen shown in Figure 3-6. Fully qualify the database name you choose. The Groups Specification section is grayed out for Windows platforms. Click Next.

Specify Optional Configuration

Figure 3-7 is *optional* in that you don't need to complete it for the installation to succeed and all fields are disabled by default. However, you must specify these input fields if you want Grid Control to send you e-mail alerts or help you patch Oracle software. "Optional" refers to whether you want the installer to configure this functionality now or subsequently perform this configuration yourself through the Console. The intent of making this configuration optional is so you don't have to delay a GC installation just because your SMTP server, Oracle*MetaLink* credentials, or proxy server are not ready, or because an OMP or OMS server are not on the network yet. But if these external elements are ready (and they will be if you followed the preinstallation steps in Chapter 2), why not let the GC installer configure them rather than having to do this yourself after the installation?

- **Configure E-mail Notification** Checking the box and entering the required information here configures a global e-mail notification method, not only for the built-in SYSMAN account, but for any GC users you later create. SYSMAN is the owner of the Repository database schema and a super administrator (e.g., given full privileges) who is automatically configured to receive certain out-of-box alerts.

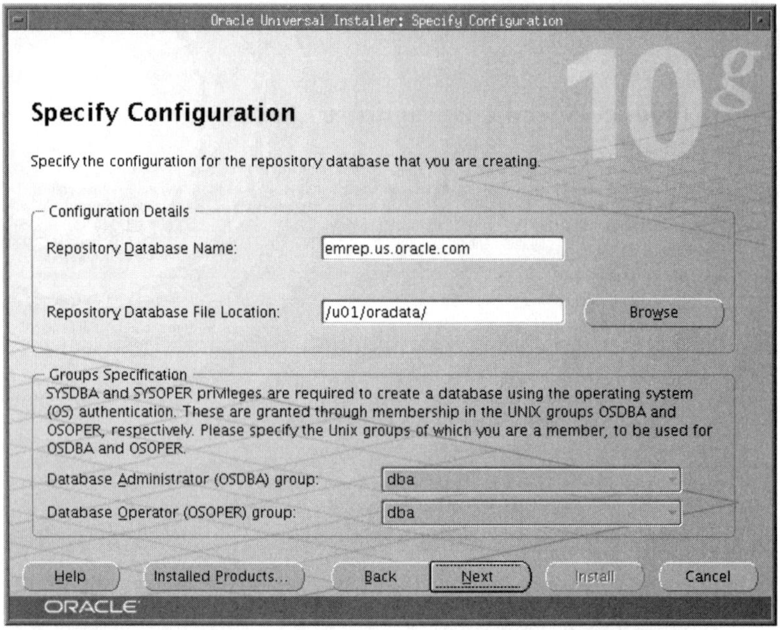

FIGURE 3-6. *Specify Configuration: New*

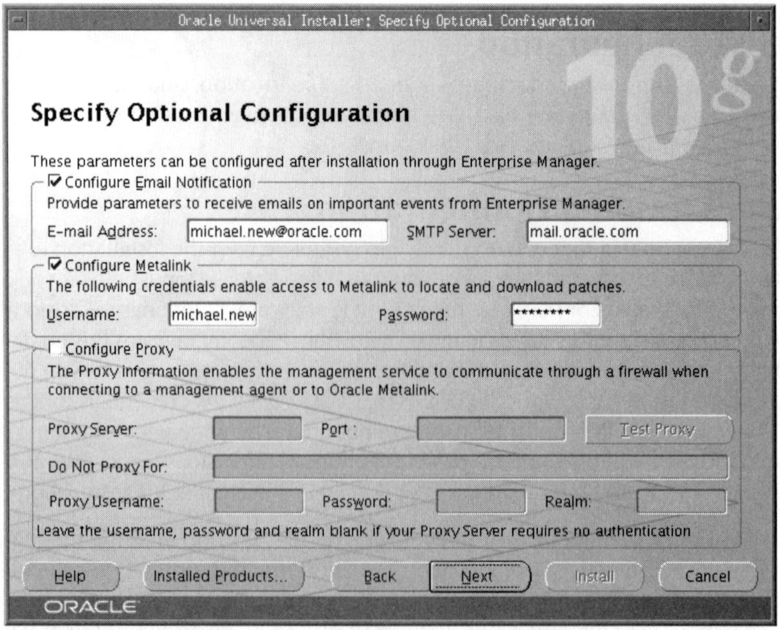

FIGURE 3-7. *Specify Optional Configuration: New, Existing*

- **E-mail Address**   Enter the e-mail address to associate with the SYSMAN account. The installer does not verify that this e-mail address is valid, so you should confirm this yourself.

- **SMTP Server**   The installer checks that the SMTP server is reachable, and doesn't let you continue unless you enter a valid SMTP server name or uncheck the box, Configure E-mail Notification.

■ **Configure MetaLink**   Check this box to enter a valid Oracle*MetaLink* username (an e-mail address) and password for Grid Control to use when searching for and downloading patches. The installer does not verify the validity of these credentials. Ideally, you should set up a dedicated account for Grid Control that is not tied to a particular person.[16] This will prevent connectivity problems due to a user password change or to an account that is closed when someone leaves the company. After the installation, you can also enter these credentials through the Console by navigating to Patching Setup under Setup.

■ **Configure Proxy**   Check the box if the Management Service must go through a firewall to connect to Oracle*MetaLink* or to an Agent on a target host. If you click Test Proxy, the installer confirms whether the proxy server is reachable.

- **Proxy Server**   Specify the fully qualified hostname for the proxy server.

- **Port**   Enter the port for the proxy server, typically port 80 or 8080.

- **Do Not Proxy For**   Enter internal address(es) not to proxy for, separated by commas but no spaces, such as *.us.mycorp.com, .mycorp.com*.

- **Proxy Username, Password, Realm**   If the proxy server requires authentication, enter a username, password, and realm for the proxy server.

Click Next to advance to the next screen.

Specify Security Options

In Figure 3-8, you specify an agent Registration password to enable Agents to communicate securely with the OMS via HTTPS as well as HTTP. Also, check the box "Require Secure Communication for all agents" to lock the OMS, thereby insisting upon such secure communication. Not checking the box configures the OMS in an unlocked state, i.e., both secure and unsecured Agents can talk to the OMS. On the bottom half of the screen, you can choose to use different passwords for all database accounts or to use the same password, as your company's security policy directs. The password for *ias_admin* (the administrative user for the Oracle Application Server Console) is set to that for SYSMAN. Both Grid Control administrator and database passwords must abide by the following restrictions (unless indicated otherwise for one or the other):[17]

- Passwords must be between 5 and 30 characters long.

- Passwords must start with an alphabetic character—they cannot start with a number.

- GC administrator passwords must include at least one integer.

---

[16] To create a MetaLink account, go to http://metalink.oracle.com, click Register for MetaLink under First-Time Users, and complete the interview process. You need a valid Customer Support Identifier (CSI) number to create an account.

[17] This password scheme for GC administrators is admittedly the weakest in the database network, due to Grid Control Bug 4113146.

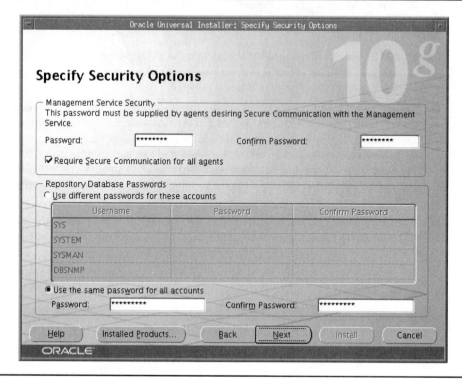

FIGURE 3-8. *Specify Security Options: New, Existing, Add'l*

- GC administrator passwords can only contain alphanumeric characters and "_" (underscore).
- Database passwords cannot use Oracle reserved words,[18] and I discourage their use for GC administrator passwords as well.

Click Next.

TIP
You may be tempted to leave the OMS unlocked if all Grid Control components are behind the corporate firewall. Do not let this lull you into a false sense of security. Roughly 70 percent of security threats come from within (from hackers, internal theft, and the like). A fundamental security concept is that of multiple layers of protection. Avail yourself of this simple security layer and lock the OMS. The only cost is Agent Registration password management.

---

[18] For a list of Oracle reserved words, see Table G-1 in Appendix G of *Oracle Enterprise Manager Grid Control Installation and Basic Configuration 10g Release 4 (10.2.0.4.0)*.

Oracle Configuration Manager Registration

As of GC 10.2.0.2, the OUI is bundled with Oracle Configuration Manager (OCM), which is installed in the OMS and chain-installed Agent homes, collects configuration information for these homes and the host, and uploads it regularly to Global Customer Support (GCS) ("Oracle Support") to help improve its technical support service. Completing the screen shown in Figure 3-9 is more convenient than manually installing and configuring OCM later. (This is covered in Chapter 8 under the heading, "Install Oracle Configuration Manager Client." However, as this section points out, you must use the Command Line Interface to install and configure the OCM in the OMR Database home.)

- **Enable Oracle Configuration Manager** If you check this box to enable OCM now, you must complete all fields on this screen. When you click Next, the OUI prompts you to click Accept License Agreement.

- **Customer Identification Number (CSI)** Enter your Customer Support Identifier (CSI) here. If registration is successful, the Oracle*MetaLink* administrator of this CSI will receive an e-mail confirmation that OCM has been enabled on this server.

- **Metalink Account Username** Enter the username of the administrator associated with this CSI, which must match Oracle's records.

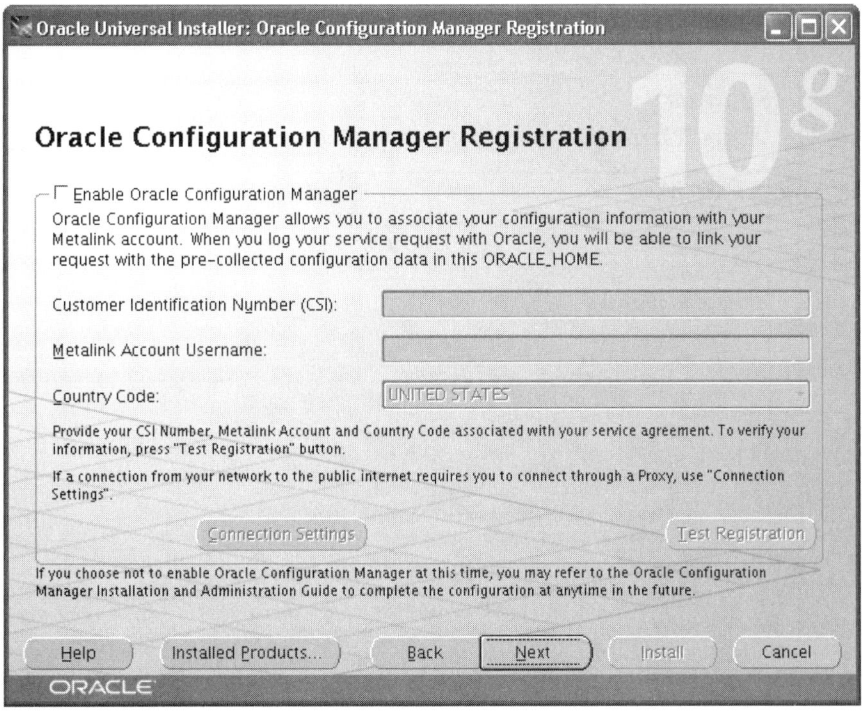

FIGURE 3-9. *Oracle Configuration Manager Registration: New, Existing, Add'l*

- **Country Code** Enter the country code for the MetaLink Username. To confirm, log in to Oracle*MetaLink* as this user and see the Profile section under the Licenses link for the Country Code associated with the CSI.

- **Connection Settings** Click this button if you must connect from this host through a proxy server to reach the Internet. You are prompted for the name and port of the proxy server, and if the proxy server requires authentication, the proxy username and password.

- **Test Registration** Click this button to test the connection between the host and the OCM service. You cannot proceed past this screen until the information has been verified, so you may as well test the registration using this button. If the test fails, and you cannot obtain the correct OCM information at this time, uncheck Enable Oracle Configuration Manager and configure OCM manually later on as instructed in Chapter 8.

A log of all OCM installation steps and errors is located under the directory *$OMS_HOME/ccr/log*. Click Next to continue.

Summary

Verify the options you chose in Figure 3-10. If everything is correct, click Install.

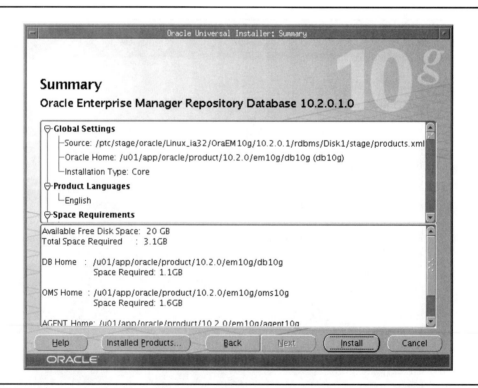

FIGURE 3-10. *Summary: New, Existing, Add'l*

Chapter 3: Grid Control Installation **101**

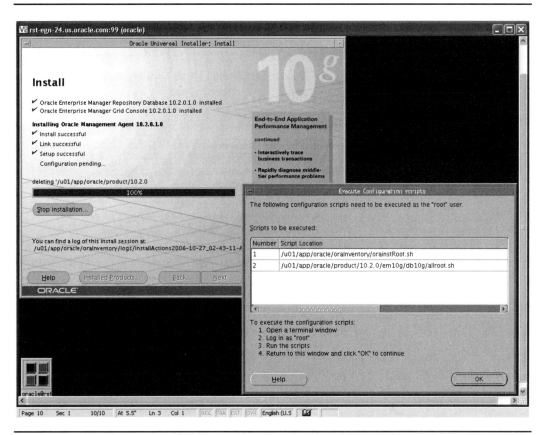

FIGURE 3-11. *Install and Execute Configuration Scripts: New, Existing, Add'l*

Install

An installation screen (see Figure 3-11) shows a progress bar. Approximately 30 minutes into the installation, the Execute Configuration Scripts dialog box pops up for UNIX platforms (it does not appear on Windows platforms). If this is the first Oracle installation on the server, you are prompted to log in to the OMS host in a separate session as *root* and run *orainstRoot.sh* under the Inventory directory. The installer always prompts you to execute *allroot.sh* under the database home as *root*. This script in turn executes three *root.sh* scripts, one for each of the Oracle homes installed. Provide default answers to all questions these shell scripts ask by hitting RETURN at each prompt. Return to the dialog box and click OK to continue.

Configuration Assistants

The screen in Figure 3-12 appears when the assistants begin running. The screen shows the name, status, and type (Recommended or Optional) of each configuration assistant. If an assistant fails, you can select that assistant, click Stop, fix the problem, then click Retry. Depending on the severity of the error, and the dependency of one assistant on another, the installer may not give

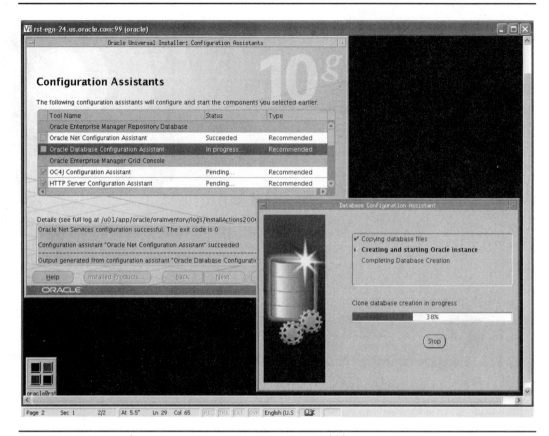

FIGURE 3-12. *Configuration assistants: New, Existing, Add'l*

you the chance to continue to the next assistant. In this case, you must either click Cancel, or manually kill the installer. Regardless of how many configuration assistants fail, if the installer gets through all of them, it will not report the installation as having failed at the end. This is misleading, as failure of any of the assistants leaves the installation in a very questionable state. You should seriously think about de installing Grid Control even if only one assistant fails. It is better to address the problem and re install cleanly, or GC may not function properly, either immediately or down the line after much post configuration effort.

Once you input the information on the installer screens and click Install, it takes about half an hour for the installer to lay down the OMR, OMS, and OMA software trees. The configuration assistants then run for another 1.5 hours or so when you select the GC New Database Option. Most Configuration assistants run for less than 5 minutes, except that for the Oracle Database (which runs for about 15 minutes), and the OMS Configuration (which runs for about 30 minutes).

Click Next if all configuration assistants succeed, or one or more fail and you are willing to take a big chance (in which case you should consider another line of work).

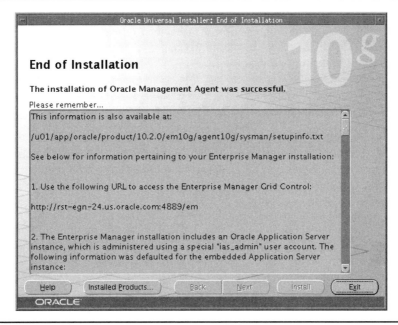

FIGURE 3-13. *End of Installation: New, Existing, Add'l*

NOTE
You can also run a configuration assistant in standalone mode for some OUI errors, as instructed in the error message. However, it is typically easier and definitely more reliable to de install Grid Control entirely, resolve the problem that caused the failure, and rerun the installer from the beginning.

End of Installation
This screen (Figure 3-13) appears with the URL for how to access the Console, the Repository connection details, and the path to a file containing the Release Notes. Click Exit. A dialog box asks you to confirm if you really want to exit. Click Yes.

Install Grid Control Using an Existing Database
Before you can install Grid Control using an existing database, you must first install and configure a new database (or configure an existing database)[19] to contain the Management Repository.

---

[19] Instead of building a new database for the OMR, you could use a preexisting database, but this would likely occur only in an upgrade scenario when using an OMR database from an earlier GC release. GC upgrades are outside the scope of this book, so configuration of a preexisting database is not covered here.

In this section we build a new database,[20] which then becomes the "existing" database where the Repository is later installed via the GC Existing Database Option. You need to decide whether this database will be single-instance or RAC; Chapter 2 provides a clear basis for this decision. If you haven't decided, do so now by revisiting Chapter 2, as the installation procedures for single-instance vs. RAC databases vary considerably.

To pre create a database to house the OMR, consult the official Oracle Database installation documentation, supplemented in this section with instructions to ensure the database supports the OMR, to streamline the installation process, and to guarantee compliance with 10g database best practices. It is outside the scope of this book to provide instructions on how to build a single-instance or RAC database for each operating system and database version certified with the Repository. The procedure differs considerably by platform, predominantly in the preinstallation steps. Other installation steps are more generic; but none of them are GC-specific, so repeating them here would add little value to understanding Grid Control. However, this section does cover all GC-specific requirements for the OMR database. The section stands on its own, e.g., you don't have to rely upon the Oracle Enterprise Manager documentation set.[21] It's enough that you have to consult the Oracle Database documentation, informed by the content in this section.

TIP

Though I cannot cover Oracle Database requirements for each database version and platform, OracleMetaLink Note 401705.1 contains a comprehensive reference list of MetaLink articles detailing OS requirements for all Linux x86 and x86-64 platforms.

I would be remiss if I didn't at least identify the specific Oracle database manuals for you to reference. The Oracle Database platform-specific installation documentation, available on OTN, for single-instance and RAC databases, is as follows:

- If you're installing a single-instance database, complete the preinstallation, installation and post installation tasks listed in the *Oracle Database Installation Guide* or *Quick Installation Guide* for your platform and database release.

- If you're installing a RAC database, follow the tasks on each RAC node as outlined in the *Oracle Database Oracle Clusterware and Oracle Real Application Clusters Installation Guide*[22] for your platform and release. You generally need to complete Parts I, II, and III, which contain the following material:

  - **Part I**   Installation Planning and Requirements
  - **Part II**   Preinstallation Procedures
  - **Part III** Installation and Post installation

---

[20] Here, I am speaking of creating a new database to use with the GC Existing Database Option, as opposed to the GC installer creating a new database using the GC New Database Option.

[21] Some of the database installation and configuration points in this chapter differ from or do not appear in the Oracle EM documentation, but are found in Oracle*MetaLink* notes or based upon experience in the field.

[22] This is a single Oracle guide, though the length of its title may suggest otherwise. In the Books tab of the Oracle Database Documentation Library, these books are listed with an even longer title of *Oracle Clusterware and Oracle Real Application Clusters Installation and Configuration Guide* for each platform.

Steps to Install the OMR Database

To put a Grid Control spin on this Oracle documentation as it relates to the Repository database, let's divide the OMR database installation into four main steps to clarify what choices to make (steps apply to both single-instance and RAC unless otherwise noted):

- Install Oracle Clusterware software (RAC only)
- Install Oracle ASM software (10g only)
- Install Oracle Database software
- Create the Oracle Database

Notice that you will install three separate Oracle homes if using RAC, or two homes if not using RAC: the Clusterware home (RAC only), the ASM home, and the database software home.

TIP
I strongly suggest installing the latest certified 10gR2 database Patch Set for all Oracle homes to take advantage of the new 10.2 features and latest Patch Set bug fixes. While you can run the Repository in an Oracle 9.2 database, there should be no reason to do so. You should be able to use a 10g database as long as you can dedicate a database to the OMR (i.e., the database doesn't also need to support a third-party application not certified on Oracle 10g). Nevertheless, if you use an Oracle 9.2 database for the OMR, the syntax for most of the commands and initialization parameters in this section still apply.

Following are pointers on how to navigate the four installation steps above, all of which are covered generically in the Oracle database documentation. For illustrative purposes in this discussion, it is assumed you are installing Oracle 10g Release 2 (10.2.0.3) for all Oracle homes, which was the latest 10.2 Patch Set certified for the Repository at the time this book was published. Be sure to check the Grid Control Certification Matrix on Oracle*MetaLink* for the up-to-date certification information.

Install Clusterware Software (RAC Only)

To install a RAC database for the Repository, begin by installing the base Clusterware 10.2.0.1 release according to the Oracle Clusterware and Oracle Real Application Clusters and Configuration Guide for your particular platform. Then apply the latest Clusterware Patch Set for your platform that is certified for the OMR (currently 10.2.0.3) by consulting the release notes that accompany that Patch Set. There is not much else to say about the Clusterware with respect to supporting the GC Repository database. It is the only truly generic installation step with respect to Grid Control of the four steps covered here.

Install ASM Software (10g Only)

Rather than using cooked file systems for your Repository database files, consider configuring disks for use with Oracle Automatic Storage Management (ASM), which is an Oracle 10g best practice for both single-instance and RAC database storage. ASM is an integrated file system and volume manager for Oracle datafiles, spfiles, control files, online redo logs, archive logs, change tracking files, Data Guard configuration files, flashback logs, Data Pump dump files, and RMAN backups. ASM provides optimized I/O performance comparable to that of raw devices coupled with the easy management of a file system. The overall steps to install ASM 10.2.0.1 are to run the 10.2 database installer, choose Configure ASM, then patch ASM to 10.2.0.3. It is best practice to install a separate ASM home than that for the Oracle Database software. Among other reasons, this allows you to patch the ASM and the database software homes independently.

CAUTION
Make sure the Oracle Clusterware is running on all nodes before launching the OUI to install ASM. This enables you to perform a Cluster Installation of ASM on all cluster nodes.

The ASM home is identical to the database home. As such, you install it by running the Oracle Server 10.2 OUI. Here are a few key installation points to install the ASM home and configure the ASM instance to support the Repository database (screen titles are the section subheadings):

Installation Type   Select Custom.

Available Product Components   Select the following components:

- Oracle Database 10*g*
- Enterprise Edition Options
    - Oracle Advanced Security (optional)[23]
    - Oracle Net Services (Oracle Net Listener)
    - Oracle Call Interface

Create Database   Select Configure Automatic Storage Management (ASM).

Oracle Net Configuration Assistant   Check "Perform typical configuration" to create an ASM listener on each node using default port 1521 and prefix LISTENER. Before continuing, shut down any processes, such as other listeners using port 1521.

Create ASM Instance   An ASM best practice is to choose "Create server parameter file (SPFILE)" rather than creating an IFILE, and for Server Parameter Filename, specify a shared raw device (15M is sufficient).

ASM Disk Groups   Create separate DATA and FLASH disk groups (these are arbitrary names which I use in the remainder of the chapter) for the database datafiles and Flash Recovery Area (FRA), respectively. These disk groups must be sufficiently large—taking into account the desired redundancy or mirroring level—to support the hardware operating requirements listed in Chapter 2.

When the 10.2.0.1 ASM installation finishes, apply the latest database Patch Set (currently 10.2.0.3) to the ASM home.

See Oracle*MetaLink* Note 370915.1 for an excellent FAQ on ASM, including useful ASM commands and debugging tips.

---

[23] Oracle Advanced Security (OAS) is an optional feature that must be licensed separately. To evaluate whether you need OAS for GC, see "EM Framework Security" in Chapter 16.

Install Database Software

Whether using ASM or not, it is recommended that you install the latest Oracle 10g Database software certified for the Repository (currently the 10.2.0.3 Patch Set) in a separate Oracle home from the ASM home. You can streamline the installation process by patching the database software *before* creating the database itself, which is the subject of the next heading. This installation order lets you circumvent the database post-installation steps for individual components—standard advice for building any Oracle Database. For those less experienced DBAs, the way this plays out is to use the OUI to install the base database software distribution (10.2.0.1) but decline to create the database, exit the installer, patch the software to the latest 10.2 Database Patch Set, then run the Database Configuration Assistant (DBCA) in standalone mode to create the database.

Make the specific choices below when installing the Oracle Database 10.2.0.1 software (screen titles are the section subheadings):

Installation Method (Windows Only)   Select Advanced.

Installation Type   Select Custom.

Specify Home Details   Enter the following for Destination Name and Path to parallel those the Grid Control installer uses for the OMS and Agent.[24]

- **Name**   Specify *db10g*.
- **Path**   Enter a path of the form *<mount_point>/app/oracle/product/10.2.0/em10g/db10g*, such as */u01/app/oracle/product/10.2.0/em10g/db10g*.

Specify Hardware Cluster Installation Mode   (RAC database only). Select "Cluster Installation" (rather than Local Installation) and check all Node Names.

Available Product Components   Choose the following product components:

- **Oracle Advanced Security (optional)**[25]   Select. This allows you to secure the OMR database at a later time. For details on configuring OAS, see the section "Secure the Repository Database" in Chapter 16.
- **Oracle Partitioning**   Select. This is a requirement for the OMR, which creates nearly 1,000 partitioned tables and roughly 700 indexes in the SYSMAN schema for storing management metrics.
- **Oracle Enterprise Manager Console DB**   De select. This is Database Control, which you do not want to select for the OMR database; Grid Control will manage this and all other target databases. If you do not de select this component, the later Grid Control installation will fail. DB Control installs configuration files and Repository objects using Enterprise Manager Configuration Assistant (EMCA). Grid Control checks that neither exists because it needs to install its own version of these configuration files and Repository objects in the same locations.

---

[24] The home names and lowest path that the Grid Control installer uses are "oms10g" for the OMS and "agent10g" for the Agent.

[25] Oracle Advanced Security (OAS) is an optional feature that must be licensed separately.

- ***i*SQL*Plus**   Select. This provides *i*SQL*Plus access in Grid Control to the OMR database and all target databases (configured in Chapter 4). Be sure to choose the *i*SQL*Plus component now because you can't install it after patching the Database software.

- **Other Components**   Select Oracle Spatial, Oracle OLAP, Data Mining Scoring Engine, and Oracle XML Development Kit. The GC New Database Option installs these components with its database software. Although the OMR does not actually rely on these components, I recommend selecting them to standardize all your GC installations irrespective of the installation type you select.

Create Database   Choose "Install database Software only." First upgrade to the latest database Patch Set (done next) before creating the database.

When the 10.2.0.1 OUI installation completes, apply the latest 10.2 database Patch Set to the database home. This is the same Patch Set you applied to the ASM home as described above.

Create the Oracle Database

At this point, you have installed the Clusterware (if using RAC), the ASM home and ASM instance, and the Oracle Database home, and all homes are at the latest available Patch Set level. The last step is to actually create a RAC or single-instance database for the Repository by running the Database Configuration Assistant (DBCA), a GUI tool that guides you through the database creation process. At the end of the Oracle Database software installation, I instructed you not to create a database, which would have launched DBCA. Instead, I advised you to install the database software only so that you could apply the latest database Patch Set before creating the database. Now that the database software is patched, you are ready to launch DBCA as a standalone tool. For specifics on running DBCA in standalone mode to create a RAC database, see Chapter 6 of the Oracle Database Oracle Clusterware and Oracle Real Application Clusters Installation Guide for your platform. To create a single-instance database for the OMR using DBCA, consult Chapter 2 of the guide called Oracle Database 2 Day DBA 10*g* Release 2 (10.2). Once you kick off DBCA to create either a RAC or single-instance database, you can also get online help by clicking Help on any DBCA installation screen.

TIP
Don't forget to start the Clusterware, node apps, and the ASM instance on all cluster nodes before launching DBCA to create a RAC database for the OMR. Similarly, before launching DBCA to create a single-instance database, start ASM, if used. For a single-instance database, ASM, in turn, requires Cluster Synchronization Services (CSS), which are automatically installed with ASM and started as root using "crsctl start css."

Follow the documentation cited above to run DBCA in standalone mode on the OMR node, or on one of the cluster nodes, informed by the following installation choices (screen titles are the section subheadings and installation options are capitalized):

Welcome   Choose Oracle Real Application Clusters Database or Single-Instance Database, as already decided.

Operations   Choose Create a Database.

Node Selection   Choose all cluster nodes.

Database Templates   Select the Custom Database template.

Fine-Grained Access Control (FGAC)   Select FGAC, which is required to create the OMR.

Management Options   De select Configure the Database with Enterprise Manager. This option is intended for installing a target database to point to an *existing* Grid Control installation. Here, by contrast, you are creating a database to host the OMR component for a *new* GC installation, which, in its course, automatically configures the OMR Database as a Grid Control target.

Storage Options (10*g* Only)   Select Automatic Storage Management (ASM) if it's installed.[26] As already mentioned, using ASM as the storage mechanism is a database best practice for both single-instance and RAC databases. If you select ASM, the next installation screen allows you to choose the DATA disk group created earlier as the database storage location.

Database File Locations   Select Use Oracle-Managed Files.[27] OMF is a database best practice for the generic database datafiles. Unfortunately, the GC OUI does not allow you to specify an OMF file format for the GC tablespaces it creates. For the Database Area, enter the DATA disk group as "+DATA".

Recovery Configuration   Note the following about the Recovery Configuration options:

- **Specify Flash Recovery Area (10*g* only)**   Check this option as a database best practice. If using ASM, specify the FLASH disk group created earlier as the FRA and allocate a FRA size of two, preferably three times the database size. (After installing Grid Control, which creates two tablespaces, the database size will be approximately 2.5GB.)

- **Enable Archiving**   Do not check this box. It will speed up the Grid Control installation process. Instead, for recovery purposes you can take a cold backup of the database once it's built, as recommended later in this chapter. Chapter 4 reminds you to enable archiving at the end of the OMR database configuration process.

Database Content   Under the Database Components tab, I recommend selecting the same components that the GC New Database Option installs, which are Oracle Data Mining, Oracle Text, Oracle OLAP, and Oracle Spatial. Also, click the Standard Database Components button and select all three components: Oracle JVM, Oracle XML DB, and Oracle Intermedia. Components not required and which should be grayed out (i.e., nonselectable) anyway are Oracle Ultra Search, Oracle Label Security,[28] Sample Schemas, and Enterprise Manager Repository. The only component that you must be careful *not* to select is the Enterprise Manager Repository, as the GC Existing Database Option must install the Repository.

Initialization Parameters   It is easier to wait until after DBCA creates the database to change the bulk of the initialization parameters the OMR requires or recommends. This is because the DBCA user interface (UI) is cumbersome, whereas, once the database is created, you can change

---

[26] The other two storage choices for RAC databases are Raw Devices or Cluster File System, and choices for single-instance databases are Raw Devices or File System.

[27] The other option is to Use Common Location for all Database Files, which is a non-OMF selection.

[28] The Oracle EM 10.1 documentation incorrectly listed Oracle Label Security as a requirement, but it is not required either for EM 10.1 or 10.2.

all initialization parameters more quickly by scripting the process, for example. A few parameters, however, should be set within DBCA, as they are difficult to change after DBCA creates the database. These parameters are as follows (**bolded** text indicates tab names on the Initialization Parameters screen):

- **Sizing**   Ensure Block Size is set to 8192 bytes, the Oracle-recommended value for an OMR database. Set DB_BLOCK_SIZE now at database creation time, because changing it requires recreating the database.

- **Character Sets**   Make sure the chosen database and national character sets store the main language used by GC administrators and additional language groups desired. A good choice is the Database Character Set Use Unicode (AL32UTF8) and the accompanying National Character Set "AL16UTF16 - Unicode UTF-16 Universal character set."[29] Also, select your desired Default Language and Default Date Format.

- **Database Storage**   It's easier to specify the correct redo log sizes for DBCA to create than it is to re create these redo logs afterward. Redo log sizing depends upon the anticipated size of your GC environment. For redo log recommendations, see Table B-2 in Appendix B (online at www.oraclepressbooks.com). At this time, don't worry about the number of redo log groups, the SGA size, or tablespace, datafile, or tempfile specifications. In the next section, "Configure/Tune a Database for the OMR—Part 1 of 2," we use SQL*Plus to change them as required for Grid Control.

Creation Options   The following provides information about the Creation Options:

- **Create Database**   Check this box, particularly for a RAC database. If you don't check this box to let DBCA create a RAC database in the GUI, and rely instead on database creation scripts (see next bullet point), then you will need to perform additional manual steps not documented here or in the Oracle documentation to complete the RAC installation process.

- **Generate Database Creation Scripts**   I recommend selecting this option. If a single-instance database installation fails, you can run the scripts (rather than re running DBCA and re entering all required information), or create an identical GC environment on another host, if needed. The scripts also serve as a record of the installation process.[30]

This concludes the database installation process for a single-instance or RAC database to house an OMR installed using the GC Existing Database Option.

Configure/Tune a Database for the OMR—Part 1 of 2

Perform the steps in the section "Configure/Tune a Database for the OMR—Part 1 of 2" in Appendix C (online), which lists database configuration and tuning tasks to complete before launching the installer and selecting the GC Existing Database Option. As I've already mentioned, you

---

[29] These are both Unicode character sets, which store Western and non-Western characters. See Oracle Database Administrator's Guide 10g Release 2 (10.2), Manually Creating an Oracle Database, Considerations Before Creating the Database, Table 2-1.

[30] There is a bug in the init.ora script created by selecting Generate Database Creation Scripts. Trying to create a 10.2 database with these scripts produces the error "ORA-01678: parameter db_file_name_convert must be pairs of pattern and replacement strings." Resolve this error by unsetting its value.

could have performed most of the tuning steps (such as the initialization parameter changes) when running the Database Configuration Assistant (DBCA). However, it's quicker to do all configuration and tuning through SQL*Plus once you've created the database to house the OMR.

After you run the Grid Control installer as described in the next section, and patch Grid Control according to the instructions in Chapter 4, I will refer you back to the online Appendix C to complete the steps in the section "Configure/Tune a Database for the OMR—Part 2 of 2." The reason for the two parts is that, by and large, the steps in Part 2 (such as enabling ARCHIVELOG mode), if implemented before installing and patching Grid Control, would reduce the performance of the Grid Control installation and patching process.

Run the Installer—GC Existing Database Option

Like the GC New Database Option, the GC Existing Database Option installs an OMS, OMR, and chain-installed OMA, but with two differences. The principal difference is that the GC Existing Database Option doesn't create a database for the OMR but installs the OMR into an existing qualified database. The second distinction is that the GC Existing Database Option can install an OMR, OMS, and chain-installed OMA on the same host, or it can install the OMR on a separate host from the OMS and chain-installed OMA. (The GC New Database Option can only install GC components on the same host.) When you choose the GC Existing Database Option, the installer creates the OMR in an existing single-instance or RAC database you point to, and installs the OMS and chain-installed OMA on the local host. The GC installer is not what you would call "RAC-savvy" (as I will explain shortly), but you can circumvent the few RAC-related gotchas by following the instructions in this section.

TIP
Now would be a good time to take a cold backup of the existing database, particularly if you did not enable archiving yet, which I recommended to keep the GC installation time to a minimum. A backup will preserve the database creation, configuration, and tuning you just completed, providing a way to recover and restore to this point should the Grid Control installer fail or other unrecoverable error occur.

Let's kick off the Grid Control installer now (for instructions see the section above, "Starting and Monitoring the Installation"). The GC Existing Database Option is very similar to the GC New Database Option. The obvious difference is that on the Specify Installation Type screen, you select Enterprise Manager 10*g* Grid Control Using an Existing Database.

NOTE
To anticipate all screen shots you will encounter, see earlier Table 3-1, which lists the screens that appear when you choose the GC Existing Database Option (column 2), and which references the figure number of the screen (column 9). Each screen is also labeled as to which Grid Control installation type(s) render it (the label is "Existing" for the GC Existing Database Option).

The other differences are fewer screens because some only apply to the GC New Database Option and two additional installation screens for the Existing Database option. One screen shot requests that you enter a password for the new SYSMAN user, which is simple enough and not shown here. The other screen shot is entitled Specify Repository Database Configuration (see Figure 3-14).

This screen can trip you up. Let's review the information it requests, field by field, observing the following guidelines:

- **Database Hostname** Use the fully qualified hostname of the existing database. If you're using a RAC Repository, enter one of the node names.

- **Port** Enter the database listener port (the default port is 1521).

- **Service/SID**
  - For single-instance OMR, enter the DB_NAME.
  - For a RAC Repository, rather than the DB_NAME, specify the INSTANCE_NAME running on the node entered above for Database Hostname. (After the Grid Control installation is complete, as Chapter 4 instructs, you will need to modify the Management Service connection string from INSTANCE_NAME to DB_NAME to take advantage of client failover to the other RAC instance in the event of a RAC host outage.)

  The example INSTANCE_NAME in the screen shot is *emrepa1* to exemplify a GC environment with multiple sites. The DB_NAME for this site is *emrepa* and for additional sites would be *emrepb*, *emrepc*, etc.

- **SYS Password** Enter the current SYS password.

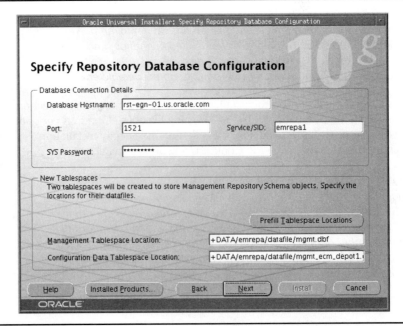

FIGURE 3-14. *Specify Repository Database Configuration: Existing, Add'l*

- **Tablespace Locations**   Specify the full path for the datafiles. You cannot specify Oracle-Managed Files (OMF) for these GC tablespaces (though you could convert them to OMF post-installation, if desired). So if you're using ASM, when prompted for the GC tablespace locations, specify a non-OMF string with the full path for the datafiles. It is recommended that you use the default datafile names, in which case the full path when using ASM would be of the form +*DATA/<database_name>/datafile/<datafile_name>*. For example, a good naming convention for these datafiles is as follows:

    Management Tablespace Location: +*DATA/emrep/datafile/mgmt.dbf*
    Configuration Data Tablespace Location: +*DATA/emrep/datafile/mgmt_ecm_depot1.dbf*

    When you install the Management Repository into an existing database (including a RAC database), the installer does not allow you to set the size of these required EM tablespaces. The default initial datafile extents assigned are not usually sufficient for a production environment, so the installer sets AUTOEXTEND ON to allow these datafiles to grow. However, these default extent sizes become a problem when storage for the database (usually a RAC database) is on a raw device, because the AUTOEXTEND ON feature does not work for raw devices. If the database used for the Management Repository is configured with raw devices, there are two options for increasing the size of the Repository for each of the two EM tablespaces:

    - Pre create multiple raw partitions, setting the first partition size equal to the default size of the tablespace as defined by the installation process. The second partition allows for growth.

    - Alternatively, pre create a large partition sized to allow for EM growth, create a tablespace using the default size, create a dummy object that will increase the size of the tablespace to the end of the raw partition, and then drop the dummy object.

    Regardless of which option you choose, when using raw devices disable the default AUTOEXTEND ON space management for all objects. (Do not disable AUTOEXTEND ON when using ASM-managed storage for the Repository database.)

Trick to Installing the OMS on a RAC Node

It is unsupported to install an OMS on a RAC node prior to EM10g Release 10.2.0.2. However, you must still know the trick to installing a 10.2.0.3 or 10.2.0.4 OMS on a RAC node.

On the Specify Repository Database Configuration screen; if you specify the *local* INSTANCE_NAME or local node where running the GC installer, you get the following error due to Bug 4745547: "You do not have sufficient privileges to write to the specified path for tablespaces." The workaround is to specify a remote RAC node name and INSTANCE_NAME. The error is due to ASM, not RAC. If the OMS and the Repository Database are on the same box, then Oracle tries to check the path writability of the tablespace locations entered. For ASM disks, the path writability returns false (patch writability cannot be used for ASM disks), the installation fails at this validation, and it cannot proceed. The reason for the success of this workaround is that this check is not performed if the database is perceived to be on a remote node. The only anomaly with this configuration is that some of the OMS ports the GC installer selects are nondefault.

Install Standalone Agents on Dedicated OMR Nodes

You don't need to explicitly install an Agent on the OMS host. The Grid Control installation does this for you with the chain-installed Agent, regardless of the option selected—either using the GC New or Existing Database Options, or when installing an Additional Management Service.

By contrast, you *do* need to install a standalone Agent on an OMR host when it is remote to an OMS installed using the GC Existing Database Option. This option lays down an OMS and chain-installed OMA on the local server, and creates the Repository through a SQL*Net connection in an existing single-instance or RAC database on a local or remote node(s). But the GC Existing Database Option does not install a Management Agent on each remote OMR node; you must install an Agent on each OMR node in a separate process, using any standalone installation method. See Chapters 5, 6, and 7 for instructions on installing a standalone Agent via one of six available installation methods.

NOTE
You don't have to install standalone Agents right now on dedicated OMR node(s). You can wait until you begin installing standalone Agents in Chapter 6. Whenever it is, I suggest that you deploy the first standalone Agents on OMR nodes. Monitoring the OMR database and node(s) is paramount, as all other target monitoring relies on the OMR database remaining operational.

Log in to the Web Console

Now that you've installed Grid Control using either the GC New Database Option or GC Existing Database Option, you probably want to log in to the Console before proceeding further, just to make sure the GC installation was successful, or at least successful enough for you to get in the front door, so to speak. If the GC installation was successful, all GC processes required for login will automatically start, allowing you immediate access to the Console as the built-in SYSMAN administrator, and with the password specified during the installation. The following table lists the URL variations to log in to the Grid Control Console.[31]

Login Method	Protocol Used	Port for UNIX OMS Host	Port for Windows OMS Host	Grid Control Console Login URL
Web Cache	HTTP	7777	80	http://<*OMShost*>:<*port*>/em
Web Cache	HTTPS[32]	(must configure)	(must configure)	https://<*OMShost*>:<*port*>/em
HTTP Server	HTTP	4889	4889	http://<*OMShost*>:<*port*>/em
HTTP Server	HTTPS	1159	1159	https://<*OMShost*>:<*port*>/em

---

[31] If you used the static ports feature, your port numbers may be different.

[32] This Web Cache HTTPS URL does not work out-of-box after you install Grid Control. See "Configure/Enforce HTTPS Between Browser and Web Cache" in Chapter 16 for details on how to configure this access. By default, upon installation, traffic between browser and Console via Oracle HTTP Server is already secured over HTTPS on port 1159.

As you see, Console login is via OracleAS Web Cache or the Oracle HTTP Server (OHS), over a secure SSL (HTTPS) or nonsecure (HTTP) connection. You can log in to the GC Console via Web Cache, which in turn accesses OHS in the background, or you can bypass Web Cache and log in directly via OHS. It is recommended that you log in via Web Cache (either over HTTP or HTTPS as your security requirements dictate) because page performance is usually slightly better. But it's useful to know how to bypass Web Cache, such as when you're trying to isolate a Console login problem. The URL protocol varies (HTTP or HTTPS) depending on whether you use SSL encryption for browser <-> Console communications. The URL port also differs by platform (UNIX or Windows) for Web Cache login.

Console access via Web Cache doesn't help performance much when compared with that via OHS because the information provided in Grid Control is largely dynamic. Web Cache does cache some information that doesn't change (such as icons, headers, and footers), but this is a small percentage (typically 1 percent to 5 percent) of the data returned to the Console. (You can sometimes distinguish the cached data by the way it sometimes loads before the rest of a page.) The Console must display target information obtained in real-time for accuracy, so this information cannot be cached. Therefore, you receive only a slight benefit to caching Grid Control data in Web Cache.[33] I still prefer to log in via Web Cache because any performance improvement, however little, is better than nothing. Also, Web Cache is used to monitor the built-in EM Website service (see Chapter 12).

For now, ensure that you can log in as SYSMAN via all configured methods shown in the table above. Once logged in, you're in a whole new world of IT management, but the goal here is just to make sure things appear in order. If you can see all the OMS subcomponents on the Targets tab under the All Targets subtab, and if most of these targets are showing the correct UP status (a green arrow pointing up), then this is a good enough spot check for now.[34] However, knock yourself out and tool around in the Console until you're satisfied the Console is running smoothly. Don't get too carried away, though. From Chapter 10 forward, we spend most of our time in the Console configuring Grid Control.

Install the Grid Control Security Certificate in Your Browser

Each time you log in to the Console using HTTPS, a security alert dialog box appears in your browser informing you that the certificate of the Grid Control site is not trusted. Your browser is simply letting you know that it does not accept the Grid Control certificate for the purposes of identifying the Console web site itself. In other words, your browser does not assume on your behalf that the Grid Control application certificate authority (CA) root certificate is trustworthy. This root certificate is a trusted certificate included in the OHS wallet,[35] which also contains the OMS server's certificate and private key. Your browser prompts you to set up a trusted relationship with this CA root certificate as required by Secure Sockets Layer (SSL) to encrypt browser <-> OHS communication.

---

[33] Several analysts in the Oracle Support EM group told me that some customers disable Web Cache to conserve OMS resources. If you ask these analysts why Oracle supplies Web Cache with Grid Control, their answer is because it's bundled with OracleAS, which is the framework for the OMS middle tier.

[34] Don't be surprised if the database and ASM targets (except a new Management Repository database) show a DOWN status. You must configure them by entering the password for the DBSNMP and SYS users, respectively, as described in Chapter 11.

[35] The OHS wallet, generated during Grid Control installation, is located in the directory *$OMS_HOME/Apache/Apache/conf/ssl.wlt/default* as defined by the value for *SSLWallet file:* in *$OMS_HOME/Apache/Apache/conf/ssl.conf*. For more information on this wallet, see "Configure/Enforce HTTPS Between Browser and Web Cache" in Chapter 16.

Appendix I contains the procedures to install the Grid Control certificate in Internet Explorer as well as to suppress the Security Information dialog box. The procedure to install the certificate for other certified browsers (including Netscape, Mozilla, Firefox, and Safari) is different, but the basic idea is the same: to install and accept the certificate forever. (See your browser documentation for the certificate installation procedure.) After installing the certificate for *most* browsers, you won't receive any further security alerts when logging in to the Console on your client workstation using this browser. However, I've found that in Netscape Navigator a dialog box called Security Error: Domain Name Mismatch appears every time, even after accepting the certificate forever, which is particularly bothersome.

TIP

I recommend using Microsoft Internet Explorer for the Web Console, as it allows you to use web application features without additional configuration required by other browsers. These features include service tests and the Transaction Recorder. (See "How to Configure a Service" in Chapter 12 for details on these features.)

Install an Additional OMS

As the name implies, this option installs a second (third, fourth…) OMS and chain-installed Agent on the local host where running the GC installer. The prerequisite for choosing this option is to first install an OMR, OMS, and chain-installed OMA, i.e., first run the GC installer and choose either the GC New Database Option or the GC Existing Database Option. The option you choose dictates where you locate an additional OMS and chain-installed Agent, as follows:

- For either option, install an OMS and chain-installed Agent on a system not running any other Grid Control components, or

- Install an OMS and chain-installed OMA on a node running an existing single-instance or RAC[36] Repository database, where the primary OMS was installed on a separate host. This is not a likely scenario.[37]

The procedures for the GC Additional OMS Option and GC Existing Database Option are very similar. The only difference is that the former only updates the SYSMAN schema to reflect that another OMS exists, whereas the latter creates the Repository in an existing database in the SYSMAN schema. OMSs in an Active/Active configuration are independent of each other, except as relates to the Shared Filesystem Loader (if configured).[38] However, do not mistake independence for ignorance. As fellow members of the GC infrastructure, OMSs are aware of each other in that they all appear as targets in the GC Console. This "independent awareness" has its benefits. For instance, you can shut down OMS #1 for maintenance and administrators can access the Console via OMS #2.

---

[36] Again, Oracle does not support installing an OMS on a host running Clusterware until GC 10.2.0.2.

[37] In this scenario, you would install the first OMS on a different host than the OMR, then install a second OMS on the same host as the OMR. It is more likely that you would do the reverse, i.e., install the first OMS on the same host as the OMR (using the GC New Database Option), then install a second OMS on a different host than the OMR (via the GC Additional OMS Option).

[38] See Chapter 15 for more information on both the Shared Filesystem Loader and OMS Active-Passive configurations.

NOTE
The screen shots above with the label "Add'l" appear when choosing the GC Additional OMS Option. For a complete list of screen shots for this option, see column 3 of Table 3-1.

As contrasted with the GC New or Existing Database Options which install the initial OMS, the following screens do not appear when installing an Additional Management Service:

- **Specify Repository Database Configuration**   The installer does not prompt you for GC tablespace information because the other options create the required tablespaces.

- **Specify Optional Configuration**   The installer does not prompt for e-mail notification or Oracle*MetaLink* credentials because you enter them when installing Grid Control using the other options.

As a general guideline, the number of required Repository processes increases by 100 for each additional OMS. My recommended setting of 2000 for PROCESSES (see Appendix C, online) should be sufficient for multiple (i.e., three or four) OMSs.

TIP
When installing an additional OMS, use the static ports feature to ensure that the port configuration on all OMS servers is identical (see the section "Static Ports Feature" earlier in the chapter). Perplexingly, Grid Control does not use all the same default ports for additional OMSs, even if these ports are available.

Summary

This chapter focused on installing the Grid Control infrastructure using the GC New or Existing Database Options, and on installing additional Management Services (beyond the first) using the GC Additional OMS Option. The chapter began with gathering installation information, addressing installation bugs, initializing the OS environment for the GC installer, and running the GC installer with any desired options, silently or using the static ports feature.

In the section on the GC New Database Option, I presented screen shots and instructions to complete them. Following this section is a major section on the GC Existing Database Option devoted to creating, configuring, and tuning a new database to house the Repository. In this section, you learned which screen shots are unique to the GC Existing Database Option (not presented by the GC New Database Option), and the corresponding input required. I concluded the chapter by providing guidance when choosing the GC Additional OMS Option. Because this procedure is almost the same (only with several fewer screen shots) as the one for the GC Existing Database Option, it suffices to point out which screen shots do not appear.

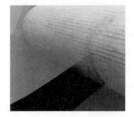

CHAPTER 4

Grid Control Post Installation

In Chapter 3 you installed the latest Grid Control Full Installer release for your platform, including the Oracle Management Repository (OMR), one or more Oracle Management Services (OMSs), and tag along Oracle Management Agents (OMAs). In this chapter, you will perform the following post installation tasks:

- Patch the Grid Control environment
- Configure the OMS host(s)
- Configure and tune the OMR node(s) and OMR database

See the beginning of each section for a summary of the specific tasks to accomplish in that section. Let's begin with the first task: patching Grid Control to bring it up to the latest patch level.

Patch Grid Control

Following are the four categories of Grid Control patches presented in this section, some of which you may already have applied:

- Apply the latest Grid Control Patch Set
- Apply the latest Database Patch Set certified for the OMR
- Apply the latest EM Critical Patch Update (CPU)
- Apply required Grid Control one-off patches

NOTE
Most Grid Control patching functionality requires one of the OEM Provisioning Packs. For more information, see Enterprise Manager Licensing Information, available on OTN in the Reference section of the EM Documentation Library.

Grid Control can stage patches for itself, and in some cases can actually patch itself. The limitation is that Grid Control can only apply patches that rely upon the *opatch* executable, which updates the OUI inventory for the product being patched. As for the patches covered here, it's efficient to stage almost all of them in the Console, and I illustrate the navigation path to do so. This way, you learn how to *use* Grid Control while *configuring* Grid Control. I also point out how you can apply one of these patches (the CPU) if running multiple OMS hosts.

It gets a little reflexive here, but think of it this way: Grid Control can patch itself like a doctor can patch (or "operate" on) himself or herself. A doctor can stitch his or her own leg and even perform minor surgery under local anesthesia, but certainly can't operate on himself or herself under general anesthesia. For Grid Control, local anesthesia to self-patch is provided by multiple OMSs. You keep one OMS running while patching another (check first that the patch Readme allows for this). Patching an OMR database with Grid Control is almost out of the question, since even a *one-off* database patch through *opatch* requires shutting down the database, which shuts down Grid Control. The upshot is that the OMR database and at least one OMS must remain running when applying a patch through the Console.

CAUTION
It is not supported to change the Application Server configuration forming the Grid Control middle-tier outside of applying Grid Control patches to it.

Let's begin the Grid Control patching process now. It's always best to test these patches in a nonproduction environment before putting them into production. However, that ship sailed at the beginning of Chapter 2 when you decided how many GC environments to build:

- If you built a production GC environment to manage both production and nonproduction targets, then take a good backup (as noted below) and hope for the best.

- If you opted to build both a production and nonproduction environment, then you've already installed the nonproduction environment. If the production hardware is ready, consider starting the production build now so as to begin production monitoring as soon as possible. To do so, double back and complete to the instructions in Chapters 2 and 3 on the production hardware. Then you can return to this chapter and patch nonproduction, with the assurance that these patches will not affect production.

Apply the Latest Grid Control Patch Set

Regardless of whether you used the GC New or Existing Database Option to install Grid Control, you should download and apply the latest release of the GC Patch Installer[1] for your platform (currently 10.2.0.3 or 10.2.0.4, but referred to here simply as the "GC Patch Set"). You must be on GC release 10.2.0.1 or higher to apply the latest GC 10.2.0.x Patch Set. If you were lucky enough when installing Grid Control to be on an OMS platform for which the latest GC release was available in a Full Installer version, then you can skip this section. GC Patch Sets are cumulative. For instance, you can directly apply the GC 10.2.0.4 Patch Set without first applying the 10.2.0.3 Patch Set. The latest GC Patch Set fixes many bugs in the base 10.2.0.1 release that provide critical functionality. GC Patch Sets also introduce new functionality and add-ons. One new feature as of the GC 10.2.0.2 Patch Set is Oracle Configuration Manager (OCM), a separate product that extracts GC configuration information and uploads it to Oracle Support. (OCM makes opening Service Requests easier because you don't need to manually supply GC configuration details.)

TIP
You can apply the GC Patch Set to many Agents at one time through the Grid Control Console. For details, see the GC Release Notes for your platform. By contrast, you cannot patch OMS hosts through the Console, as the Patch Set requires shutting down all *OMS hosts to patch any* one *of them.*

---

[1] A GC Patch Installer release contains an OMS, OMA, and Management Packs, but no OMR or OMR Database.

Let me summarize the GC OMS patching process detailed in the Release Notes to clarify and supplement them in a few places. (I do not go into detail on applying the GC Patch Set to Agents, as the Release Notes are very clear on this subject.)[2]

1. **Download the Patch Set software and Readme for your OMS platform** to a shared location amongst OMS hosts, if possible. You may have already done so, as recommended in Chapter 1. (While downloading, you can proceed to the next step.) The *Readme.txt* file is located under the *doc* directory in the software distribution. You can also download the Readme separately for certain platforms from the OTN Enterprise Manager Downloads page at http://www.oracle.com/technology/software/products/oem/index.html under the Patch Installers section via the README link. See the Readme for general GC Patch Set directions and consult the Release Notes included with the Patch Set for specific instructions.

2. **Take a cold backup of your entire GC installation**, preferably via server-wide backup. This includes all OMS homes, the OMR Database, all OMR homes, and any standalone Agents already installed, (While backing up, you can go to the next step, which requires some reading.) The most reliable mechanism for de installing GC Patch Sets is to restore from backup. See "Backups Before and After a Configuration Change" in Chapter 14 for details.

3. **Make a Go/No-Go decision now on whether to apply the Patch Set**. Read the Known Issues[3] section of the GC Patch Set Release Notes to determine whether any bugs are showstoppers. You may decide that the bugs are such that you'd rather wait for the next GC Patch Set, though this is unlikely. (You won't have the option of waiting if applying GC 10.2.0.4, as this is the terminal 10.2 release.) These bugs were discovered even before the Patch Set was released. Generally speaking, the bugs should be fewer and less loathsome than those in the previous Patch Set. Workarounds are available for some of these bugs. Flag any "must-have" workarounds. You'll need to implement them after applying the Patch Set (done in step 10 below of this procedure).

4. **Perform tasks specified in the Preinstallation section of the Release Notes**. If you already began configuring Grid Control in the Console or have patched it since initially installing it, see the Release Notes for possible preinstallation steps. However, if you have not used or patched Grid Control yet (i.e., you're following instructions in this book), there is only one possible preinstallation task, depending on how you installed Grid Control.

    ■ If you selected the GC New Database Option, or if you chose the GC Existing Database Option using Oracle Database 10.1.0.4 or 10.1.0.5, apply patch 4329444 to the OMR database.

    ■ If you selected the GC Existing Database Option and used Oracle Database 10.2 for the Repository, there are no preinstallation tasks for the GC Patch Set.

5. **Set a side-wide blackout to suppress alert notifications for all GC components**. To set this blackout, log in to the Console with OPERATOR privilege on all targets, click the Setup link at the top right of the home page, click the Blackouts tab on the left navigation page, and click Create. On the Properties page, select "all target types" in the Type

---

[2] Perhaps i goes without saying that you only need to patch the OMR database once, regardless of the number of OMS hosts you installed.

[3] I have a less euphemistic name for the Known Issues section: New Bugs with this New Patch Set.

field under Available Targets and click Move All to move all targets to Selected Targets. Advance to the Schedule page, enter immediately for the Start Date, and Indefinite for the Duration of the blackout.

6. **Shut down all OMS services associated with the Repository** by executing *$OMS_HOME/opmn/bin/opmnctl stopall* on each OMS host. However, leave the Repository database and listener running.

7. **Set ORACLE_HOME to the OMS home in the environment on each OMS host**. On each OMS you intend to patch (starting with the first OMS installed), log in to an XWindows session as *oracle* and set the ORACLE_HOME to the OMS home in the environment. This ensures that the OUI automatically picks up the ORACLE_HOME for the OMS. (You can also pass the ORACLE_HOME as an argument to the installer executable—see the next step.)

8. **Apply the GC Patch Set to each OMS host**. When you apply the Patch Set to the first OMS host, it also patches the associated Repository, whether located on the local OMS host or remote to it. To launch the installer, change directories to *3731593/Disk1/* in the Patch Set distribution, and execute *runInstaller* on UNIX or *setup.exe* on Windows.[4] Specify ORACLE_HOME=<OMS_HOME> as an argument to the installer executable if you did not set it in the environment in the previous step.

9. **Complete any applicable GC Patch Set post installation tasks** for each OMS, if applicable; some of these tasks depend upon whether metadata from certain platforms (such as Solaris) is present in the OMR, and others redefine the host metrics called Generic Log File Monitoring and Log File Pattern Matched Line Count.

10. **Implement any "must-have" workarounds** for each OMS that you identified in Step 3 above from the Known Issues section of the GC Patch Set Release Notes.

11. **Apply the Patch Set to all tag-along and standalone**[5] **Agents** (ideally in one operation) according to the instructions in the GC Patch Set Release Notes.

Apply Latest Database Patch Set Certified for OMR

The Oracle Management Repository is currently certified on Oracle Databases 9.2.0.6+, 10.1.0.4+, and 10.2.0.2+ (i.e., "+" means "and subsequent Patch Sets in the series"), both standalone and RAC. Actually, Database 11.1.0.6 is also certified, but I can only recommend using it with EM 11*g*, which is being developed on Database 11*g*. As I recommended in Chapter 2, you should seriously consider running Database 10.2 to take advantage of its latest features, and running the latest 10.2 Patch Set available for your platform to incorporate the most recent bug fixes. If you installed Grid Control using the GC Existing Database Option according to the instructions in Chapter 3, then you already patched the OMR database to the latest 10.2 release. However, if you selected the GC New Database Option, the GC installer laid down Oracle Database 10.1.0.4. In this case, I suggest

---

[4] You can also run the Patch Set silently using a response file.

[5] Chapter 7 reminds you to do this in the "Agent Post-Installation Steps" section following standalone Agent installations.

upgrading the OMR database and software to 10.2.0.1, then applying the latest 10.2 Patch Set for your platform.

- For instructions on upgrading Oracle database and software from 10.1.0.4 to 10.2.0.1, see the Oracle Database Upgrade Guide 10*g* Release 2 (10.2) available on the Oracle Technology Network (OTN) under the Database link below the Documentation tab.

- To patch the Oracle Database and software from 10.2.0.1 to the latest release for your platform, see the Oracle Database Patch Set Notes 10*g* Release 2 for that release and operating system. You can download the Readme and Patch Set from Oracle*MetaLink*.

While Grid Control can stage the latest Oracle Database 10.2 Patch Set, it cannot apply the Patch Set because it requires shutting down the OMR database.

Apply the Latest EM Critical Patch Update

Oracle highly recommends that customers apply the latest Critical Patch Update (CPU) for EM Grid Control. A CPU is a cumulative set of patches that address both critical security vulnerabilities and nonsecurity bugs on which these security fixes depend. CPUs are put out every quarter (in January, April, July, and October) and, like Patch Sets, are rigorously tested as a collection, so they are more reliable than one-off patches.

To determine the CPU requirements and patch number to apply, see the Critical Patch Update *<Month Year>* Advisory on Oracle*MetaLink*. This document contains a link to a document called the Critical Patch Update *<Month Year>* Availability for Oracle Server and Middleware Products. This document, in turn, contains two sections:

- Minimum Requirements specify minimum Patch Set requirements for the Oracle Application Server and the Oracle Database components of EM. These components must be on the minimum Patch Set levels stated here to apply the CPU.

- Oracle Enterprise Manager Patch Availability shows the CPU patch numbers to apply for the OMS/OMR and for the OMA (separate patch numbers) based on the EM release and platform (UNIX or Windows).

You can apply the CPU manually or use Grid Control to stage and apply some of the CPU to certain GC components, provided you comply with all CPU Readme instructions. For example, applying the Enterprise Manager CPU to an OMS requires shutting down that OMS. Thus, to apply the CPU to a nonrunning OMS via Grid Control requires logging in to the Console through a separate, running OMS. If you have at least two OMS hosts, say OMS #1 and OMS #2, patch them as follows, beginning with OMS #1:

1. Shut down OMS #1.
2. Log in to the Console via OMS #2, and click Patch Advisory under Deployments.
3. On the Patch Advisories tab, under Interim Patches to Apply, click the CPU patch number, select the OMS #1 home under Affected Oracle Homes, and click Patch. Follow the Patch Wizard to stage and apply the patch. The only step in the Wizard that is not self-explanatory

is on the Stage or Apply page. In the Pre-Patch and Post-Patch sections, you need to enter the following commands to shut down and restart OMS #1, which you are patching:

```
%oraclehome%/opmn/bin/opmnctl stopall
%oraclehome%/opmn/bin/opmnctl startall
```

Enter the *%oraclehome%* variable above literally, as it represents the OMS #1 home you are patching. Complete the remaining Patch Wizard pages and submit your request. Grid Control submits a job to apply the CPU to the selected OMS #1 Oracle home. Click the View Job button for details on the job's progress.

4. When finished, shut down OMS #2, and start and log in to the Console via the now patched OMS #1.

5. Repeat this patching procedure for OMS #2.

Apply Required Grid Control One-off Patches

One patching step many GC administrators overlook is applying any one-off patches or *one-offs* that are functionally required or needed to implement workarounds.

One-offs are patches that address individual bugs, some of which resolve significant functional bugs or deficiencies. Oracle creates one-offs as interim patches to apply on top of a specific (hopefully the latest) GC Patch Set, and rolls these one-offs into the next GC Patch Set. Many of the one-offs are intended for the OMR database. The proactive way to determine which one-offs you functionally require is to scan their descriptions (and Readme files if needed), then choose those one-offs that provide fixes you cannot live without. The reactive way to choose needed one-offs is to wait until Grid Control fails in some unacceptable way, then look for a one-off that fixes that failure.

TIP
I recommend the reactive approach of not applying GC one-off patches unless absolutely necessary. Oracle apparently agrees with this conservative approach, as evidenced by the following statement placed in nearly every database one-off Readme: "You must have NO OTHER PATCHES installed on your Oracle Server since the latest Patch Set (or base release x.y.z if you have no Patch Sets installed)."

You can identify an up-to-date list of one-off patches for a specific GC Patch Set, then download these patches directly from Oracle*MetaLink* and stage them manually or download and stage them in Grid Control.

NOTE
See "Perform OMR Partition Maintenance" in Chapter 17 for specific GC and OMR Database one-off patches that may be required, depending on the versions you are running.

Let's take a closer look at how to download and stage one-off patches, both in Oracle*MetaLink* and in the Console. I use the Linux x86 platform in the examples.

Downloading One-off Patches in Oracle*MetaLink*

To download a patch in Oracle*MetaLink* and stage it manually, follow these steps:

1. Download a one-off patch by clicking the Patches & Updates tab, then click the Advanced Search link. Complete the requested fields as shown below by clicking the Flashlight icon, selecting the appropriate values for your site, and then clicking Go. Make sure the Release field reflects your currently *installed* GC release.

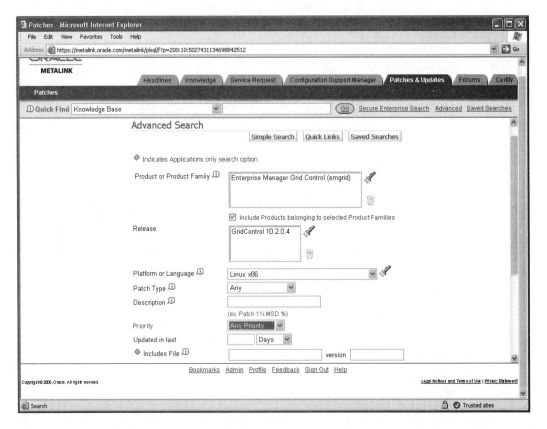

2. Click the View Readme icon for each patch you are interested in evaluating and then click the Download Now icon to download the patch.
3. Manually unzip the patch to stage it.

Downloading and Staging One-off Patches in Grid Control

To download and stage any patch (including a one-off) in the GC Console, do the following:

1. Go to the Deployments tab, and click Patch Oracle Software under the Patching section.

2. The Patch Wizard leads you through the staging/patching process. On the Select Patch page, click Search Criteria.

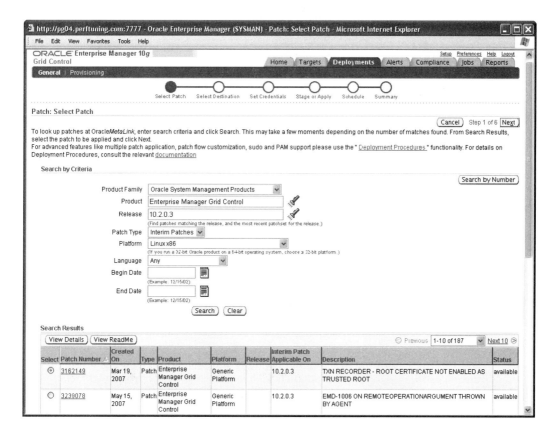

3. Enter the required fields, and click Search. The patches matching the criteria are returned as shown above. Under Search Results, select the desired patch. If needed, click View Details (which lists a patch description, whether you've already downloaded it to Grid Control, etc.) or click View ReadMe. To stage the patch, click Next

4. On the Select Destination page, select the Destination Type called Host and Directory.

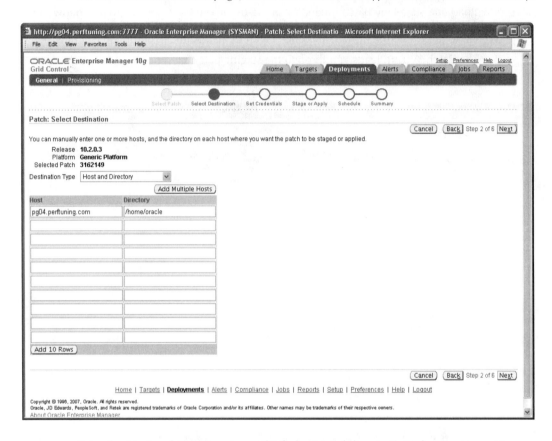

Enter the Host and Directory (or multiple hosts and directories) where you want to download and stage the patch, and click Next. When you specify */home/oracle* as the Directory, the patch directory structure GC creates looks like this:

```
pg04.perftuning.com > pwd
/home/oracle/EMStagedPatches/3162149
pg04.perftuning.com > ls -la
total 2080
drwxr-----   3 oracle    dba              96 Jun 27 13:24 .
drwxr-----   3 oracle    dba              96 Jun 27 13:23 ..
drwxr-xr-x   4 oracle    dba              96 Mar 20 02:05 3162149
-rw-r-----   1 oracle    dba         1056947 Jun 27 13:23 p3162149.zip
```

As shown above, Grid Control creates and appends the subdirectory EMStagedPatches/<*patch number*> to the Directory you specify, and then places the patch zip file in this subdirectory alongside another <*patch number*> subdirectory where it unpacks the patch file. Click Next.

5. Complete the remaining Patch Wizard pages, which are self-explanatory, and submit your request for hand-off to the GC job system to execute. Click the View Job button to ensure that the patch was staged successfully.

The advantages of using the Console rather than Oracle*MetaLink* for patching is that GC stages patches and offers the benefits of its Patch Cache (available on the Deployments tab by clicking View Patch Cache).

The Patch Cache keeps a record in the Repository of all downloaded patches, and allows you to apply a patch to multiple destinations in the same operation while only having to stage it once, which is particularly useful when patching multiple Agents.

I am limiting this discussion to staging a one-off patch, not applying it, because most one-offs require shutting down all OMS hosts (prohibiting Console access) and/or stopping the Agent on the target host (preventing patch job submittal). See the patch Readme for directions on how to apply a staged one-off patch outside of Grid Control.

Oracle Management Service Configuration

Following are suggested OMS configuration changes to make. While none of these changes is strictly mandatory, I highly recommend all of them for tuning and customizing your OMS configuration.

- Check whether certain Grid Control log files are growing very large.
- Set up the *oracle* user environment on OMS hosts
- Modify the default Console timeout
- Configure the OMS for failover and load balancing (applies only to a RAC OMR database)
- Tune the OMS thread pool size

> **UNIX Scripting Technique to Modify Configuration Files**
>
> Some of the following post installation steps call for you to change OS configuration files. You can manually edit them with a text editor such as *vi* in UNIX or Notepad in Windows. Alternatively, you can employ the UNIX shell techniques below to append lines to a file or to make substitutions in a file. You can also use these techniques on Windows if running a UNIX shell on Windows, some of which are available as freeware.
>
> To append lines to a file called *filename*, use the following inline scripting technique:
>
> ```
> cat >> filename <<EOF
> <insert text to append>
> EOF
> ```
>
> To make substitutions in a file, use the UNIX command, *sed*, in conjunction with the *vi* substitute command, *s*. For instance, to change all occurrences of SCOTT to TIGER in *filename* and to save these changes in *newfilename*, issue the following command (file entries changed are case sensitive):
>
> ```
> sed 's/SCOTT/TIGER/g' filename > newfilename
> ```
>
> The more files you need to change, the more time you will save using these techniques. As described in later chapters, you may need to make changes to certain configuration files, such as the Agent *emd.properties* file. If you are monitoring scores of target hosts, rather than manually changing each file, you can save a lot of time by writing an OS script to make the required changes and executing this script in one operation on all target hosts using one of two Grid Control eatures: "The Execute Host Command Feature" in Chapter 9, or the GC job system. If you need to bounce a component (such as an Agent) to make the change effective, you can write a script that changes the configuration file, then restarts the component. For an excellent example, see Oracle*Metalink* Note 560905.1.

- Add an OMR alias to the OMS *tnsnames.ora* file
- Back up critical OMS files

Reduce Grid Control Logging

Now that you have patched Grid Control, it's time to check that you are not hitting a particular Grid Control bug described below that causes an Oracle Notification Server (ONS) log file to grow very large within hours. In addition, because some OracleAS default logging levels on OMS hosts cause certain logs to grow large over time, we learn below how to restrict or disable logging and/or rotate and purge logs to reduce their footprint on the OMS filesystem. Without such measures, the OMS filesystem could eventually fill up, which would halt most GC operations, notably uploads of Agent metric alerts and data.

Check for Abnormal ons.log Growth

The Oracle Notification Server (ONS) running in the OMS home utilizes the ports defined in the *$OMS_HOME/opmn/conf/ons.config* file. The OMR Database listener, which is an ONS client, also uses the ONS ports identified in its own *$ORACLE_HOME/opmn/conf/ons.config* file. When Grid Control installs OracleAS and Oracle Database 10*g* on the same server, particularly when upgrading a Grid Control 10.1 installation to 10.2, the installer sometimes mistakenly configures identical ONS ports in both homes, which creates an operational conflict when both the OracleAS ONS and 10*g* RDBMS listener services are running. In these instances, ONS log files called *$OMS_HOME/opmn/logs/ons.log.** or *$ORACLE_HOME/opmn/logs/ons.log.** (where a timestamp is appended) begin growing exponentially large to gigabytes in size due to repeating messages of the following form[6]:

Local connection 0,127.0.0.1,6100 missing form factor

If you encounter these error messages, disable the OMR Database listener from subscribing to the Oracle Notification Server (ONS). To implement this solution, rename the OMR Database *$ORACLE_HOME/opmn/conf/ons.config* file so that the OMR Database listener cannot reference this file, then restart the OMR Database listener. Grid Control will not experience any problems when using this workaround.

Reduce OMS Host Logging

The Oracle Application Server (OracleAS) bundled with Grid Control contains the following subcomponents that generate their own log files:

- Oracle HTTP Server
- OC4J (applications that run in an OC4J instance within OracleAS):
    - *EM* application
    - *home* J2EE application
- Oracle Process Manager and Notification Server (OPMN)
- Web Cache

Some of these log files, such as access logs and trace logs, are verbose and devoid of diagnostic information, but can grow to hundreds of megabytes or gigabytes in size over time and consume inordinate OMS filesystem space. You can reduce the amount of disk space these log files consume on the OMS filesystem by rotating and purging rotated log files, reducing logging levels, or disabling logging altogether. Some OracleAS subcomponents include a facility to rotate and purge log files. Other subcomponents can only rotate logs, but you can purge them easily enough at the operating system level. Table 4-1 below contains details of OracleAS subcomponent logging in Grid Control, including log file names, descriptions, configuration files controlling logging behavior, and logging strategy needed, if any.

Follow the instructions below to reduce OMS host logging for the Oracle HTTP Server and OC4J subcomponents.[7] Default OPMN and Web Cache logging should be sufficient, but I provide the logging directories and configuration files for these subcomponents in case you're interested in controlling their logging as well.

---

[6] Bug 4621067. Also see Note 284602.1 for details.

[7] The procedure here for reducing HTTP and OC4J logging is based on Note 339819.1.

Subcomponent of OracleAS	Log File Location	Description of Log File	Configuration File That Controls Logging Behavior	Logging Strategy
Oracle HTTP Server (OHS)	$OMS_HOME/Apache/Apache/logs/error_log.<time>	Errors accessing OHS	$OMS_HOME/sysman/config/httpd_em.conf (and $OMS_HOME/sysman/config/httpd_em.conf.template)	Reduce, Rotate, Purge
	$OMS_HOME/Apache/Apache/logs/access_log.<time>	Access to the OHS	$OMS_HOME/sysman/config/httpd_em.conf (and $OMS_HOME/sysman/config/httpd_em.conf.template)	Reduce, Rotate, Purge
OC4J	$OMS_HOME/j2ee/OC4J_EM/log/OC4J_EM_default_island_1/default-web-access.log	Access to *OC4J_EM* web site	$OMS_HOME/j2ee/OC4J_EM/config/default-web-site.xml	Rotate/Purge or Disable
	$OMS_HOME/j2ee/home/log/home_default_island_1/default-web-access.log	Access to *home* web site	$OMS_HOME/j2ee/home/config/default-web-site.xml	Rotate/Purge or Disable
OPMN	$OMS_HOME/opmn/logs	Logs of OPMN-managed OracleAS processes.	$OMS_HOME/opmn/conf/opmn.xml	None Needed
Web Cache	$OMS_HOME/webcache/logs	Web Cache access and error logs	$OMS_HOME/webcache/webcache.xml	None Needed

TABLE 4-1. *OMS Host Logging Details*

Reduce Oracle HTTP Server Logging   The OHS log file names that grow large over time are:

- $OMS_HOME>/Apache/Apache/logs/error_log.<time>
- $OMS_HOME>/Apache/Apache/logs/access_log.<time>

To best approach here is to reduce the amount of non essential OHS logging in these log and access files and to rotate logs. To implement this approach, make equivalent changes as described below to the pair of Apache configuration files, *httpd_em.conf* and *httpd_em.conf.template* in the *$OMS_HOME/sysman/config/* directory. You must change both files the same way because *emctl secure lock/ unlock* operations regenerate *httpd_em.conf* from scratch using *httpd_em.conf.template*. You then register these changes with the Distributed Configuration Management (DCM) repository and bounce OHS to effect the changes. If you don't use register changes with DCM, manual changes to the configuration files would be overwritten the next time the configuration were resynchronized.

Here is the procedure to reduce OHS logging and rotate logs:

1. Edit both *httpd_em.conf* and *httpd_em.conf.template* as follows:

    a. Add the following entries for each VirtualHost to signify not to log Agent uploads or frequently used, non critical URLs:

    ```
    SetEnvIF Request_URI .*/em/upload.* no-log
    SetEnvIF Request_URI .*/em/dms.* no-log
    SetEnvIF Request_URI .*/em/cabo.* no-log
    SetEnvIF Request_URI .*/em/images.* no-log
    ```

b. Instruct OHS to rotate its log files by modifying the following lines:

Change the lines:

ErrorLog <OMS_HOME>/Apache/Apache/logs/error_log

TransferLog <OMS_HOME>/Apache/Apache/logs/access_log

  to:

ErrorLog "|/<OMS_HOME>/Apache/Apache/bin/rotatelogs <OMS_HOME>/Apache/Apache/logs/error_log 86400"

TransferLog "|/<OMS_HOME>/Apache/Apache/bin/rotatelogs <OMS_HOME>/Apache/Apache/logs/access_log 86400"

Above, the units for 86400 are in seconds, meaning to rotate the logs every 12 hours. Adjust this rotation period as desired.

2. Implement the above changes by registering them with the DCM repository and bouncing OHS:

```
$OMS_HOME/dcm/bin/dcmctl updateconfig -ct ohs -v -d
$OMS_HOME/opmn/bin/opmnctl restartproc process-type=HTTP_Server
```

3. Schedule an operating system job to regularly purge older rotated logs. For example, on UNIX, add the following entries to the *crontab* file for the *oracle* user on the OMS host:

```
0 4 * * * find <OMS_HOME>/Apache/Apache/logs/error_log.* -ctime +5
-exec rm {} \;

0 4 * * * find <OMS_HOME>/Apache/Apache/logs/access_log.* -ctime +5
-exec rm {} \;
```

The commands in the entries above run every day at 4 AM to remove all rotated Apache logs older than 5 days. Change the time these commands run and rotation period as desired.

Reduce or Disable OC4J Logging When you install Grid Control, default access logging levels are used for the *OC4J_EM* instance where the OMS is deployed (identified by process-type "OC4J_EM") and for the default *home* OC4J instance (shown as the process-type "home" in *opmnctl status* output). The respective OC4J log file names for each OC4J instance are:

- $OMS_HOME/j2ee/OC4J_EM/log/OC4J_EM_default_island_1/default-web-access.log
- $OMS_HOME/j2ee/home/log/home_default_island_1/default-web-access.log

In a Grid Control environment, you can either rotate and purge OC4J access logs or disable such logging altogether. It is preferable to rotate/purge OC4J logs, provided you have enough disk space, rather than disable it altogether, as logs can be useful in debugging OC4J access problems in Grid Control. A log file of the format *default-web-access.log.** (with a timestamp appended) for the OC4J_EM and *home* instances exists in respective directories and *$OMS_HOME/j2ee/OC4J_EM/log/* and *$OMS_HOME/j2ee/home/log/*. Logging behavior for these log files is controlled by their respective configuration files, called *default-web-site.xml*, located in directories *$OMS_HOME/j2ee/OC4J_EM/config/* and *$OMS_HOME/j2ee/home/config/*. To rotate/purge or disable OC4J access logging, make the changes in the respective subheadings below to both configuration files.

Rotate and Purge OC4J Logs

To rotate OC4J logs and purge them periodically, do the following:

1. Instruct both *OC4J_EM* and *home* instances to rotate both logs daily by editing their respective configuration files *$OMS_HOME/j2ee/OC4J_EM/config/default-web-site.xml* and *$OMS_HOME/j2ee/home/config/default-web-site.xml*. Find the following line:

   <access-log path="../log/default-web-access.log" />
   Append the **boldfaced** text shown below to this line:

   <access-log path="../log/default-web-access.log" **split="day"** />

2. Purge rotated OC4J logs at the OS level, as required when rotating Apache logs (see above). OracleAS does not include a facility to purge rotated logs. In the UNIX shell, schedule the following OS commands to run in the *crontab* file for the *oracle* user on the OMS host:

   0 4 * * * find *<OMS_HOME>*/j2ee/home/log/default-web-access.log.* -ctime +5 -exec rm {} \;

   0 4 * * * find *<OMS_HOME>*/j2ee/OC4J_EM/log/default-web-access.log.* -ctime +5 -exec rm {}

   These commands, scheduled to run every day at 4 AM, remove all rotated OC4J logs older than 5 days. Adjust run times for these commands and rotation periods as preferred.

Disable OC4J Logging

As an alternative to rotating and purging OC4J logs, you can disable OC4J logging altogether as follows:

1. Remove all existing OC4J logs for both the *OC4J_EM* and *home* instances:

   ```
   rm $OMS_HOME/j2ee/OC4J_EM/log/default-web-access.log.*
   rm $OMS_HOME/j2ee/home/log/default-web-access.log.*
   ```

2. Edit the configuration files *default-web-site.xml* located in both the *$OMS_HOME/j2ee/OC4J_EM/config/* and *$OMS_HOME/j2ee/home/config/* directories. In this configuration file, change the line of the following form:

   <web-app application="em" name="em" load-on-startup="true" root="/em" />

   so that it reads as follows (append instructions in **boldface** to turn off logging):

   <web-app application="em" name="em" load-on-startup="true" root="/em" **access-log="false"** />

3. Register the above changes with the Distributed Configuration Management (DCM) repository and bounce all OC4J instances:

   ```
   $OMS_HOME/dcm/bin/dcmctl updateconfig -ct oc4j -v -d
   $OMS_HOME/opmn/bin/opmnctl restartproc ias-component=OC4J
   ```

4. Log in to the Console, then check that your logon did not generate any OC4J access logs, as follows:

   ```
   ls -la $OMS_HOME/j2ee/home/log/default-web-access.log.*
   ls -la $OMS_HOME/j2ee/OC4J_EM/log/default-web-access.log.*
   ```

Set Up Oracle User Environment on OMS Hosts

Grid Control administration is easier when you can specify a GC component home directory path with an environment variable, such as *$OMS_HOME*. Administrators often want to use such environment variables to change to a particular GC home directory or in generic scripts. Admins must also frequently run certain executables in these GC homes to control GC components. (Chapter 9 covers GC component control in depth.) One such executable is *$OMS_HOME/opmn/bin/opmnctl*; it is used to start and stop all OMS subcomponents. It is convenient to be able to call GC executables without having to specify their full path. Adding the paths to these executables to the PATH environment variable provides this convenience. For example, adding *$OMS_HOME/opmn/bin* to the PATH allows you to simply enter *opmnctl stopall* to stop the OMS rather than needing to type *$OMS_HOME/opmn/bin/opmnctl stopall*. (Notice how an environment variable, *OMS_HOME*, is used to add the path to the *opmnctl* executable to the PATH—these things build on one another.)

Below is a script listing in UNIX format of suggested environment variables to initialize and directory paths to add to the PATH for the *oracle* user on an OMS host, which may also contain an OMR database and Oracle Clusterware to run the database. Take a quick look at the script listing first to see what I'm talking about.

```
# Oracle homes defined and executables added to PATH:
# Oracle Home    Variable         Executable
# -----------    -----------      ---------------------------------
# OMS home       OMS_HOME         dcmctl, opmnctl
# CRS home       ORA_CRS_HOME     cluvfy, srvctl
# OMA home       AGENT_HOME       agentca, emctl
# OMR db home    ORACLE_HOME      isqlplus, lsnrctl, opatch, sqlplus
# ASM home       ORA_ASM_HOME     asmcmd
#
#Other variables defined:
#ORACLE_BASE, ORACLE_SID, LD_LIBRARY_PATH, SHLIB_PATH, TZ[8], CV_NODE_ALL
#
ORACLE_BASE=/u01/app/oracle #example value
export ORACLE_BASE
#
#OMS - Define OMS_HOME. Add opmnctl,dcmctl paths to PATH
OMS_HOME=$ORACLE_BASE/product/10.2.0/em10g/oms10g #example value
PATH=$PATH:$OMS_HOME/opmn/bin:$OMS_HOME/dcm/bin
export OMS_HOME PATH
#
#CRS - Define CV_NODE_ALL, ORA_CRS_HOME. Add cluvy,srvctl paths to PATH
CV_NODE_ALL=node1,node2 #example value
ORA_CRS_HOME=/u01/crs #example value
PATH=$PATH:$ORA_CRS_HOME/bin
export CV_NODE_ALL ORA_CRS_HOME PATH
#
#OMA - Define OMA_HOME. Add agentca,emctl paths to PATH
AGENT_HOME=$ORACLE_BASE/product/10.2.0/em10g/agent10g #example value
```

---

[8] The *TZ* environment variable is for the tag-along Agent. For more detail, see the section "Set Up Agent User Environment" in Chapter 7.

```
PATH=$PATH:$AGENT_HOME/bin
export AGENT_HOME PATH
#
#Database - Define ORACLE_SID,ORACLE_HOME,LD_LIBRARY_PATH
#Add isqlplus, lsnrctl, opatch, sqlplus paths to PATH
ORACLE_SID=emrep #example value. For RAC OMR, set to SID of instance, i.e., emrep1
ORACLE_HOME=$ORACLE_BASE/product/10.2.0/em10g/db10g #example value
LD_LIBRARY_PATH=$ORACLE_HOME/lib #to locate 64-bit shared libraries
SHLIB_PATH=$ORACLE_HOME/lib32 #to locate 32-bit shared libraries
PATH=$PATH:$ORACLE_HOME/bin:$ORACLE_HOME/OPatch
export ORACLE_SID ORACLE_HOME LD_LIBRARY_PATH PATH
#
#ASM - Define ORA_ASM_HOME. Add asmcmd path to PATH
ORA_ASM_HOME=$ORACLE_BASE/product/10.2.0/asm #example value
PATH=$PATH:$ORA_ASM_HOME/bin
export ORA_ASM_HOME PATH
#
#Set TZ - time zone for Agent
TZ=UTC #example value
export TZ
umask 022
```

Do the following to implement this script listing on your OMS hosts:

- On UNIX systems, add these lines to the login script for your selected UNIX shell (see Chapter 3, Table 3-2, for login script names for the most popular UNIX shells).

- On Windows platforms, set these environment variables (except don't enter the *umask* command, which is UNIX-specific) for the *oracle* user:

    - Log in as the *oracle* user, click the Start button, select Control Panel, and then select System.

    - In the System Properties window, click the Advanced tab, and then click the Environment Variables button.

    - Under User variables for the *oracle* user, click New, then enter the variable name and variable value and click OK. Repeat this procedure for all variables to be defined. Add the PATH variable, defining all values separated by a semicolon.

    - Click OK to close the Environment Variables screen, then OK again to close the System Properties screen.

You can also create a logon script on Windows,[9] but setting user environment variables is more reliable, in my experience.

Table 4-2 summarizes the syntax differences between UNIX and Windows platforms for elements used in shell scripts for each platform.

---

[9] To create a Logon script in Windows, navigate to System Tools, Local Users and Groups, and Users. Right-click the *oracle* user and select Properties. Click the Profile tab. In the User Profile section, fill in values for Profile path (such as *C:\app\oracle*) and Logon script (such as *profile_oracle.bat*). Create this Logon script under the Profile path specified, in which you define all environment variables and the PATH as listed above. Log out and log in again for variable settings to take effect, or execute the script manually.

Script Element	UNIX	Windows	Example for Windows
Comment line	#	REM (must be the first characters)	REM This is a comment.
Define environment variable	<variable name>=<value>	set <variable name>=<value>	set ORACLE_BASE=C:\app\oracle
Use a defined variable in a value	$<variable_name>	%<variable name>%	OMS_HOME=%ORACLE_BASE%\product\10.2.0\em10g\oms10g
PATH separator	: (colon)	; (semicolon)	PATH=%PATH%;%AGENT_HOME%\bin
Export a variable	export <variable_name>	Not applicable	Not applicable
Unset a variable	unset <variable name>	set <variable name>=	set ORACLE_HOME=

TABLE 4-2. *Syntax Differences Between UNIX and Windows Shell Scripting*

Note the following points regarding the initialization listing:

- For OMS hosts that do not contain an OMR database, remove OMR variable definitions and PATH references, such as for CRS, OMR, and ASM variables. (If the OMR is on a different host than the OMS, wait to configure the OMR host environment until you're instructed to do so in the "Oracle Management Repository Configuration" section later in the chapter.)
- ORACLE_HOME is the OMR database home.
- Replace example values above with your site's actual values.
- Some GC executables are located in two Oracle homes, such as *emctl*, which occurs in both the OMS and Agent homes. If you run *emctl* without qualification, the PATH ordering executes it from the Agent home, not the OMS home.[10] *$OMS_HOME/bin* is omitted from the PATH to eliminate confusion. To control the OMS, use *opmnctl*, which is in the PATH, or change the directory to *$OMS_HOME/bin/*, and run "./emctl".
- Define *ORACLE_SID* as follows:
    - For a single-instance Repository, set ORACLE_SID to the database name (equal to the DB_NAME initialization parameter).
    - For a RAC Repository, set ORACLE_SID to the specific INSTANCE_NAME running on that node. As an example, for a two-node (*node1*, *node2*) RAC OMR database named *emrep*, set ORACLE_SID to *emrep1* on *node1* and to *emrep2* on *node2*.

---

[10] Accidentally executing *emctl* from the Agent home to control the OMS will return a usage error at best. On the flip side, you cannot control the Central Agent by running *emctl* from the OMS home (it acts upon the AS Control Agent instead).

Later steps in this chapter reference the variables listed above. Therefore, on each OMS host, after adding the appropriate lines above to the *oracle* user login script on UNIX, initialize the environment again with these new variables (on UNIX, log out and log in again as *oracle*, and on Windows, open a new command window).

NOTE
The OMS properties file, emoms.properties, referenced several times in this OMS configuration section, is located under the directory $OMS_HOME/sysman/config/.

Modify the Default Console Timeout

By default, GC logs out sessions that have been inactive for 45 minutes. To change this timeout according to administrator preference and your company's security policy, add the following parameter to *emoms.properties*:

oracle.sysman.eml.maxInactiveTime=<time_in_minutes>

Changes to the OMS properties file do not take place until you restart the OMS. Instructions to bounce the OMS appear below at the end of the section "Tune OMS Thread Pool Size," as all steps in between also require changing the *emoms.properties* file.

TIP
When you bounce the OMS, the entries you add to the emoms *.properties file may move to different line numbers, so don't be surprised if new lines you append to this file are reshuffled. Look elsewhere in the file and you'll find them.*

Configure OMS for Failover and Load Balancing

When you use a RAC cluster to provide high availability for the Management Repository, you should configure the OMS to take advantage of redundancy in the repository provided by the RAC nodes. You accomplish this by changing the Oracle Net connect string the OMS uses to communicate with the Repository. This connect string is the variable *oracle.sysman.eml.mntr .emdRepConnectDescriptor* in the *emoms.properties file* on all OMS hosts. It already contains the required failover and load balancing parameters.[11] You should only need to change the parameter SERVICE_NAME to point to the service name for the database rather than to the SID of the specific instance you specified when installing the OMS. This allows the OMS to connect to any available instance in the RAC configuration. Make the change as shown below in **bold** below to alter the connect string.

from:

oracle.sysman.eml.mntr.emdRepConnectDescriptor=(DESCRIPTION\=(LOAD_BALANCE\=on) (FAILOVER\=on)(ADDRESS_LIST\=(ADDRESS\=(PROTOCOL\=TCP)(HOST\=<*OMR_node1*>)

---

[11] A bug that plagued earlier versions of EM 10*g* did not provide failover or load balancing parameters in the connect string. In this case, you need to add these parameters as well as a second connect string for the second RAC node not specified in the EM installation. For more information on both FAILOVER and LOAD_BALANCE parameters, see Chapter 13 of Oracle Database Net Services Administrator's Guide 10*g* Release 2.

(PORT\=1521))(ADDRESS\=(PROTOCOL\=TCP)(HOST\=<*OMR_node2*>)(PORT\=1521))
(CONNECT_DATA\=(**SID**\=<*INSTANCE_NAME*>)(FAILOVER_MODE\=(TYPE\=select)
(METHOD\=basic)))))

to:

oracle.sysman.eml.mntr.emdRepConnectDescriptor=(DESCRIPTION\=(LOAD_BALANCE\=on)
(FAILOVER\=on)(ADDRESS_LIST\=(ADDRESS\=(PROTOCOL\=TCP)(HOST\=<*OMR_node1*>)
(PORT\=1521))(ADDRESS\=(PROTOCOL\=TCP)(HOST\=<*OMR_node2*>)(PORT\=1521))
(CONNECT_DATA\=(**SERVICE_NAME**\=< *SERVICE_NAMES*>)
(FAILOVER_MODE\=(TYPE\=select)(METHOD\=basic)))))

The backslash ("\") characters are required to escape the equal signs ("="). Irrespective of whether the Repository is in a RAC database, the value for SERVICE_NAME in the connect string must match that for SERVICE_NAMES (a database initialization parameter), which should include the domain name (i.e., emrep.mycorp.com). Therefore, the value for SERVICE_NAME *must* include the domain name. If these parameters do not match, you get the following error message when attempting to log in to the Console: "503 Service Unavailable - Servlet error: Service is not initialized correctly. Verify that the repository connection information provided is correct."

In addition, when running a RAC Repository, add the following property to the emoms.properties file on all OMS hosts to switch on JDBC Fast Connection Failover (FCF) for the OMS so that loader connections are monitored at run time and rebalanced across RAC nodes:

em.FastConnectionFailover=true

Do not change the values for the following two variables:

oracle.sysman.eml.mntr.emdRepSID=<*INSTANCE_NAME*>

oracle.sysman.eml.mntr.emdRepServer=<*RAC_nodename*>

These variables are set to the instance name and node name for the RAC node you specified during the GC installation of the OMS.

Putting this *emoms.properties* change into effect requires restarting the OMS. You can wait until after completing the step "Tune OMS Thread Pool Size" below (if required), as it also calls for an OMS bounce.

Tune OMS Thread Pool Size

The OMS loader processes the data that continuously arrives from all Agents, and uploads this data to the OMR. Configure the number of Loader Threads to keep pace with the incoming Agent data volume. Per OMS, the default is one loader thread and the maximum is ten loader threads. Each loader thread you add to an OMS host typically increases the overall CPU utilization on that host by 2% to 5% in a large GC site with many Agents. While a small GC deployment may not need more than one loader thread, you may as well increase loader threads to more fully utilize the CPUs on OMS hosts. Here is a simple guideline for determining how many loader threads to configure on each OMS:

Loader threads per OMS host = 2 x (total #CPUs[12] on all OMR nodes) / #OMSs

---

[12] If you're using dual-core CPUs on OMR nodes, count each dual-core processor as 1.5 CPUs. The OMS application code is not specifically optimized for multicore devices.

For a RAC Repository, add up all CPUs across cluster nodes. Take the case of a medium-size GC deployment with two OMS hosts and a two-node RAC Repository with four CPUs per node. This works out to eight loader threads per OMS host:

Loader threads per OMS host = [2 x (4 + 4)] / 2 = 8

Expect diminishing returns as you add loader threads. However, OMS performance should incrementally increase as you add loader threads until you start approaching CPU saturation. To add more loader threads, set the following configuration parameter in the *emoms.properties* file:

em.loader.threadPoolSize=n

where 'n' is a positive integer from 1 to 10. The thread pool size defaults to 1 if you specify a value other than 1 to 10.

Bounce the OMS as follows to put this change and any previous *emoms.properties* changes into effect:

```
$OMS_HOME/bin/emctl stop oms
$OMS_HOME/bin/emctl start oms
```

Add OMR Alias to OMS tnsnames.ora

In order to apply many of the Grid Control one-off patches from the Console (as referenced in their Readme files), you need to create a new *$OMS_HOME/network/admin/tnsnames.ora* file on all OMS hosts, and add an alias for the Repository database. Use the same alias (or aliases in the case of a RAC Repository) found in the *tnsnames.ora* file on the OMR database node(s). This alias is also useful when you want to perform ad hoc queries via SQL*Net from the OMS host to the OMR database (if on separate tiers or if on the same tier but the environment is set for the OMS). If you plan to configure RMAN in Grid Control, add a Recovery Catalog alias at the same time as adding an OMR database alias to the *tnsnames.ora* file on the OMS host.

Back Up Critical OMS Files

At the end of this chapter, after performing all post installation tasks, you will be at a good stopping point to begin regular Grid Control backups, starting with a complete cold backup of your Grid Control environment installed to that point. The sidebar at the end of this chapter called "Begin Regular Grid Control Backups" will remind you to take that cold backup.

Right now, however, I suggest backing up the following indispensable OMS files:

- **$OMS_HOME/webcache/internal.xml** It is possible that the Web Cache may fail to start because of OMS server failure or forced reboot. The problem can result from a corruption in the *internal.xml* file, and the solution is to restore this file. Therefore, it is a good idea to back up this file immediately after installing the OMS.

- **$OMS_HOME/sysman/emd/targets.xml, $AGENT_HOME/sysman/emd/targets.xml** Back up the *targets.xml* files under both OMS and Agent homes, which contain Agent and host target information for the Application Server Agent and Central Agent, respectively. The Agent and host targets are added to both files during EM installation. While you can rediscover these targets using the Agent Configuration Assistant (*agentca*), sometimes *agentca* produces errors or does not work as expected. For this reason, back up both *targets.xml* files so that you can restore them if either file is corrupted or accidentally removed. (This recommendation also applies to the Agent *targets.xml* file installed on target hosts not containing an OMS.) It is also useful to have a backup of these files for reference in case you need to remove and rediscover any OMS or Agent targets. You may have to do this to resolve a status mismatch where the actual status of a target is UP but the Console shows its status as DOWN.

- **$OMS_HOME/sysman/config/emoms.properties**   Back up the OMS properties file, as it contains the encrypted SYSMAN password. I have heard of cases where this file is zeroed out if the disk on which it resides fills up. The line containing the encrypted SYSMAN password is of the following form:

  oracle.sysman.eml.mntr.emdRepPwd=4f6373387e40367843dfe6271b381ed6

Oracle Management Repository Configuration

Following are the OMR configuration changes to perform. Technically, none of these changes are mandatory, but will improve the functionality, ease of use, security, and performance of your GC environment.

- Set up the *oracle* user environment on OMR nodes
- Confirm that listeners load-balance across RAC nodes
- Configure *i*SQL*Plus access in Grid Control
- Modify the data retention policy
- Modify job purge policy
- Secure the emkey
- Set up Grid Control auditing
- Configure and tune the OMR database
- Install Oracle Configuration Manager

Set Up Oracle User Environment on OMR Nodes

Below is a listing in UNIX format of suggested environment variables to initialize (both GC home directories and other variables like *ORACLE_SID*) and directory paths to add to the PATH for the *oracle* user on an OMS host, which may also contain an OMR database.

The section earlier in the chapter, "Set Up Oracle User Environment on OMS Hosts," contains a UNIX style script listing of OMS, OMR, and OMA variables to set in the *oracle* user environment for a host that has all three of these components. On OMR nodes that don't contain an OMS, you don't need to use the OMS variable specifications and PATH references in this listing. When you've finished making changes, re-initialize your OS environment by logging out and logging in again as *oracle*. This makes it easier to accomplish some of the remaining steps in this chapter, and my instructions reference these variables, including the OMR ones.

Confirm That Listeners Load-Balance Across RAC Nodes

The management services rely on server side connect time load balancing on a RAC OMR database to evenly distribute connections between RAC nodes. To distribute the load optimally for a RAC Management Repository, ensure that the PREFER_LEAST_LOADED_NODE_<*listener_name*> property is set to ON (the default value) in the *listener.ora* file on each RAC node. (I'd recommend setting this property explicitly as a reminder not to set it differently.) With this property setting, a listener selects an instance in the order of least loaded node, then least loaded instance. In the case of dedicated OMR nodes, this setting will likely not change anything, given that the least loaded node will probably contain the least loaded instance. However, if you share OMR node(s) with other applications or databases, this property setting will maximize overall OMR database performance by taking into account non-GC loads and directing GC traffic to the least loaded RAC node.[13]

Configure *i*SQL*Plus Access in Grid Control

Each database target home page in the GC Console has an *i*SQL*Plus link, which provides access to that database via a web version of SQL*Plus.[14] However, in an out-of-box GC installation, clicking on this link returns a big red error, with the tip that for Grid Control, *i*SQL*Plus requires manual configuration.[15] The reason is that *i*SQL*Plus Application Server is not shipped with the OMS but is a component of the Oracle Database software.[16] This requires configuring an OMS host to point to a (10*g*) database software home[17] containing an *i*SQL*Plus Server.

NOTE
*You don't have to configure iSQL*Plus access if you don't plan on using it. As the length of this section implies, iSQL*Plus configuration, while not difficult, takes some effort.*

Which Oracle Database home should run *i*SQL*Plus Server for Grid Control? It seems logical to choose an OMR database node. The OMR database is a central GC component upon which all targets rely, so why not rely upon an OMR node for *i*SQL*Plus Server access to all targets? You can do just that if you installed Grid Control using the GC Existing Database Option. (In Chapter 3, I advised you to select *i*SQL*Plus as a component of the OMR database software installation.) However, as I cautioned in Chapter 2, the GC New Database Option does not install *i*SQL*Plus in the OMR database home, and you can't even install it after the fact using the Oracle Database 10.1 installer. In this case, I'd run *i*SQL*Plus Server in a database home for a target you're unlikely to remove in Grid Control, as dropping this database target would disable Console *i*SQL*Plus access to *all* target databases.

Regardless of which Grid Control option you chose, you can configure *i*SQL*Plus for high availability and load balancing. If the OMR is in a RAC database with multiple OMS hosts, you can configure each OMS host to use an *i*SQL*Plus Server on a different OMR node. Similarly, if the OMR is in a single-instance database with multiple OMS hosts, you can configure each OMS host to use an *i*SQL*Plus Server on a different target database node.

To demonstrate this high availability *i*SQL*Plus architecture in Grid Control, consider a two-node RAC OMR database called *emrep*, with the *emrep1* instance on *OMRnode1* and the *emrep2* instance on *OMRnode2*. In addition, assume there are two OMS hosts, *OMShost1* and *OMShost2*. Figure 4-1 graphically depicts half of the high availability architecture, which involves *OMRnode1* and *OMShost1*. (The other half of the architecture is the same, except it consists of *OMRnode2* and *OMShost2*.)

---

[13] See Oracle*MetaLink* Note 226880.1 for more background on server side connect-time load balancing and on testing it with client side connect-time load balancing (for which the OMS is already configured).

[14] See the section "*i*SQL*Plus Login" in Chapter 9 for login instructions.

[15] For Database Control, the *i*SQL*Plus Server is started by default. However, you may recall from Chapter 1 that it is redundant to run both Database Control and Grid Control.

[16] You must manually select the *i*SQL*Plus component, as no database installation type installs it by default. This was one of the components you were instructed to select in Chapter 3 when installing database software for the OMR to use with the GC Existing Database Option.

[17] I have not tested running *i*SQL*Plus from an Oracle 9*i* Database home. I'd suggest running *i*SQL*Plus from a 10*g* database home, provided you have one.

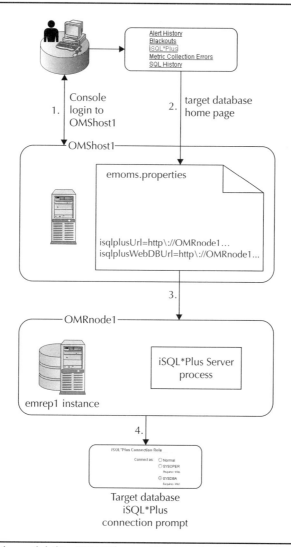

FIGURE 4-1. *A high availability iSQL*Plus architecture in Grid Control (Continued)*

In this figure, *OMShost1* is configured to use the *i*SQL*Plus Server on *OMRnode1*. To illustrate how this architecture works (and how *i*SQL*Plus works in general), let's look at a Console connection through *OMShost1* (the numbering in Figure 4-1 corresponds to that used here):

1. Log in to the Console through *OMShost1*.
2. Navigate to the home page for a target database and click the *i*SQL*Plus link.
3. *OMShost1* is configured to use the *i*SQL*Plus Server on *OMRnode1* from the Oracle home for the OMR database instance *emrep1*. Specifically, in the *emoms.properties*

144 Oracle Enterprise Manager 10g Grid Control Implementation Guide

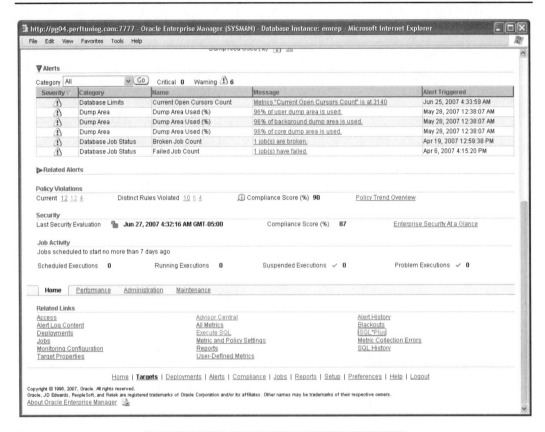

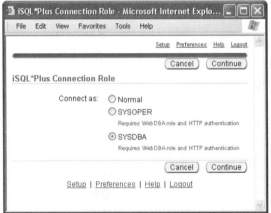

FIGURE 4-1. *A high availability iSQL*Plus architecture in Grid Control (Continued)*

file on *OMShost1*, the isqlplusUrl[18] parameter (for Normal connections) and the isqlplusWebDBUrl[19] parameter (for SYSDBA and SYSOPER connections) both point to the *iSQL*Plus* Server on *OMRnode1*.

4. Care of the *iSQL*Plus* Server on *OMRnode1*, a new window opens prompting you to select the *iSQL*Plus* Normal, SYSDBA, or SYSOPER connection role. Enter login credentials for a database user on the target database, and the *iSQL*Plus* Workspace (for Normal access) or *iSQL*Plus* DBA Workspace (for SYSDBA or SYSOPER) is displayed for your pleasure.

Following are the steps to configure Grid Control *iSQL*Plus* access in nonsecure mode[20] for the Normal connection role, followed by additional steps to configure *iSQL*Plus* connectivity via SYSDBA or SYSOPER.

Configure *iSQL*Plus* Normal Access

Perform the following tasks on at least one OMS host to configure *iSQL*Plus* in Grid Control for the Normal connection role. If configuring *iSQL*Plus* for high availability, repeat these steps for each OMS host that will provide *iSQL*Plus* access.

1. Decide which Oracle Database home will run the *iSQL*Plus* Server for this OMS. As just discussed, this choice depends on the GC installation type you used to install Grid Control:

   - If you used the GC Existing Database Option, choose a node containing an OMR database home. For *iSQL*Plus* high availability when running a RAC OMR database and multiple OMS hosts, earmark a particular RAC node to run *iSQL*Plus* Server for the OMS you are presently configuring.

   - If you used the GC New Database Option, select a database home for a database unlikely to be removed as a GC target. For *iSQL*Plus* high availability when running multiple OMS hosts, earmark a specific database node to run *iSQL*Plus* Server for the OMS you are currently configuring.

2. Configure the OMS host to provide *iSQL*Plus* Normal access. Edit the *emoms .properties* file on the OMS host and replace placeholders %ISQLPLUS_PROTOCOL%, %ISQLPLUS_HOST%, and %ISQLPLUS_PORT% with the protocol, host, and port, respectively, for the *iSQL*Plus* Server servicing this OMS host. These placeholders occur in the following two lines in the *emoms.properties* file:

   ```
   oracle.sysman.db.isqlplusUrl=%ISQLPLUS_PROTOCOL%\://%ISQLPLUS_
   HOST%\:%ISQLPLUS_PORT%/isql
   plus/dynamic

   oracle.sysman.db.isqlplusWebDBAUrl=%ISQLPLUS_PROTOCOL%\://%ISQLPLUS_
   HOST%\:%ISQLPLUS_PORT
   %/isqlplus/dba/dynamic
   ```

---

[18] This URL is http://*OMRnode1*:5560/isqlplus/dynamic. HTTP port 5560 is the default port.
[19] This URL is http://*OMRnode1*:5560/isqlplus/dba/dynamic.
[20] If you restrict Console access to HTTPS as described in "Secure Console Connections" in Chapter 16, you must configure iSQL*Plus for secure mode. For instructions, see Enabling SSL with *iSQL*Plus* in the Configuring SQL*Plus chapter of the SQL*Plus User's Guide and Reference.

As previously stated, the backslash ("\") characters are escape characters. You only need to edit the first line for *i*SQL*Plus Normal access, but you may as well edit both lines in case you want to configure SYSDBA or SYSOPER access next or later on. For both lines:

- Replace "%ISQLPLUS_PROTOCOL%" with "http"
- Replace "%ISQLPLUS_PORT%" with "5560"
- Replace "%ISQLPLUS_HOST%" with "<isqlplus_node>" where <isqlplus_node> is the fully qualified database node where the *i*SQL*Plus Server for this OMS host will run.

3. Bounce the OMS for the above *emoms.properties* changes to take effect:

   ```
   $OMS_HOME/bin/emctl stop oms
   $OMS_HOME/bin/emctl start oms
   ```

4. Start the *i*SQL*Plus Server on the OMR host servicing this OMS host as follows:

   ```
   $ORACLE_HOME/bin/isqlplusctl start
   ```

5. You can test the configuration thus far by clicking the *i*SQL*Plus link to one of these databases, but only if you've already discovered a database in the Console.[21] See the section "*i*SQL*Plus Login" in Chapter 9 for login instructions. Test only the Normal connection role at this point, as connecting via SYSDBA or SYSOPER requires additional configuration, described next.

Configure *i*SQL*Plus SYSDBA/SYSOPER Access

To configure *i*SQL*Plus for SYSOPER or SYSDBA connectivity, you must set up the WebDBA user to authenticate using Java AuthoriZationN (JAZN).[22] This involves adding credentials to an OMS Application Server authentication file by setting up an OC4J user manager. OC4J can use either the XML-based provider type (*jazn-data.xml*) or the LDAP-based provider type (Oracle Internet Directory). Instructions here are for the XML-based provider.[23] You can perform this setup either in the JAZN shell or at the OS level. The commands are shorter if you invoke the JAZN shell, but OS commands are useful in scripting the *i*SQL*Plus setup process, so I provide the syntax for both in this section.

Before we get started configuring *i*SQL*Plus SYSDBA/SYSOPER access in Grid Control, I'd like to provide some background on the very first step: to navigate to the correct directory before running the JAZN commands. This directory differs depending on whether the *i*SQL*Plus database home is a local or a Cluster installation.

- For a local database installation, you must change to the directory *$ORACLE_HOME/oc4j/j2ee/isqlplus/application-deployments/isqlplus*.
- For a Cluster database installation, you must change to the directory *$ORACLE_HOME/oc4j/j2ee/isqlplus_<isqlplus_node>/application-deployments/isqlplus*.

---

[21] If you're using a RAC OMR and have not installed any standalone Agents yet (which is perfectly acceptable), this *i*SQL*Plus test must wait until after you install Agents on the RAC nodes and discover the OMR database in Grid Control.

[22] JAZN is Oracle's implementation of the Java Authentication and Authorization Service (JAAS), a Java package that enables applications to authenticate and enforce user access controls. It implements a Java version of the standard Pluggable Authentication Module (PAM) framework.

[23] For information on how to set up the LDAP-based provider, see the Oracle9*i*AS Containers for J2EE documentation.

Changing to the correct directory is required to change the *jazn-data.xml* configuration file located in the *config* subdirectory below it. A Cluster Oracle home with the *i*SQL*Plus component installed initially contains only the local directory. However, starting the *i*SQL*Plus Server on a Cluster installation creates the Cluster directory.

CAUTION
*In case you didn't catch that, in a Cluster installation, it is the process of actually starting the iSQL*Plus Server that creates the Cluster directory $ORACLE_HOME/oc4j/j2ee/isqlplus_<isqlplus_node>. You will be looking for this directory in vain until you start the iSQL*Plus Server.*

When configuring *i*SQL*Plus for Grid Control in a Cluster installation, be careful not to mistakenly change to the directory for a local installation. You must be in the directory containing the RAC node name (isqlplus_<*isqlplus_node*>). If you are not in the correct directory, you'll be configuring the XML file in the wrong directory and *i*SQL*Plus will completely ignore your changes.

Now let's get down to configuring *i*SQL*Plus access for an OMS host to provide the SYSDBA or SYSOPER connection roles. In addition to performing the tasks for Normal access as described in the previous section, carry out the following steps on the database node servicing the OMS host that you're configuring:

1. Navigate to the correct directory for setting up an OC4J user manager. As already explained, you must change to one of two directories, depending upon whether the *i*SQL*Plus database home for this OMS host is a local or a Cluster installation.

    - For a local installation:
    ```
    cd $ORACLE_HOME/oc4j/j2ee/isqlplus/application-deployments/isqlplus
    ```
    - For a Cluster installation:
    ```
    cd $ORACLE_HOME/oc4j/j2ee/isqlplus_<isqlplus_node>/application-deployments/isqlplus
    ```

2. Set JAVA_HOME in the OS environment to the location of the Java Development Kit (JDK), which must be version 1.4 or later. The JAVA_HOME location is typically the database *$ORACLE_HOME/jdk* directory.
    ```
    JAVA_HOME=$ORACLE_HOME/jdk; export JAVA_HOME
    ```

3. If you prefer to use the JAZN shell rather than OS commands to complete this configuration, start the JAZN shell on the database node as follows:
    ```
    $JAVA_HOME/bin/java
    -Djava.security.properties=$ORACLE_HOME/sqlplus/admin/iplus/provider -jar
    $ORACLE_HOME/oc4j/j2ee/home/jazn.jar -user "iSQL*Plus DBA/admin" -password
    <admin_password> -shell
    ```

    where <*admin_password*> is the password for *admin*, the *i*SQL*Plus DBA realm administrator user, which is set to *welcome* by default. This command should return the following prompt:
    ```
    JAZN:>
    ```

4. (Optional) Change the default JAZN password, either in the JAZN shell or at the OS command line (both accomplish the same thing):

   - In the JAZN shell:

   `JAZN:> setpasswd "iSQL*Plus DBA" admin welcome <new_password>`

   or

   - At the OS command line:

   ```
   $JAVA_HOME/bin/java -
   Djava.security.properties=$ORACLE_HOME/sqlplus/admin/iplus/provider -jar
   $ORACLE_HOME/oc4j/j2ee/home/jazn.jar -user "iSQL*Plus DBA/admin" -
   password <password> -setpasswd "iSQL*Plus DBA" admin welcome
   <new_password>
   ```

5. Grant the webDba role to the *admin* user in the JAZN shell or at the OS command line:

   - In the JAZN shell:

   `JAZN:> addrole "iSQL*Plus DBA" webDba`

   or

   - At the OS command line:

   ```
   Djava.security.properties=$ORACLE_HOME/sqlplus/admin/iplus/provider -jar
   $ORACLE_HOME/oc4j/j2ee/home/jazn.jar -user "iSQL*Plus DBA/admin" -
   password <password> -grantrole webDba "iSQL*Plus DBA" admin
   ```

6. Exit the JAZN shell (if logged in) and bounce the *i*SQL*Plus process on the database node to put the above changes into effect:

   ```
   JAZN:> exit
   $ORACLE_HOME/bin/isqlplusctl stop
   $ORACLE_HOME/bin/isqlplusctl start
   ```

7. Test that you can now connect via *i*SQL*Plus to a target database as SYSDBA or SYSOPER. See the section "*i*SQL*Plus Login" in Chapter 9 for connection instructions.

Table 4-3 lists additional JAZN and OS commands that may be of use in maintaining your *i*SQL*Plus configuration:

Modify the Data Retention Policy

The Repository database stores Grid Control data in the SYSMAN schema according to a default data retention policy. This data includes metrics, alerts, and other indicators of past target health. The data retention policy differs for database vs. Application Server data and for each metric *period*. Metric periods range from raw data typically collected in 5-minute increments, to data rolled up over one hour and over one day. The retention policies for all metric periods are shown in Table 4-4. *RT* indicates Application Server policies, and the value for *<datatype>* is either DOMAIN, IP, or URL.

Do not change the default data retention policy unless required, as it can adversely impact performance and scalability. Some companies must nevertheless increase the retention policy to enable management tools to extract OMR data, or to meet stringent auditing, historical drilldown, or reporting requirements. Take, for instance, a company that has to track metrics for all database targets over a rolling 3-year period. The default retention value for 1-day metrics is 365 days.

Step	JAZN Command	OS Command
To list users in the *i*SQL*Plus *DBA* realm	listusers "*i*SQL*Plus DBA"	Djava.security.properties=$ORACLE_HOME/sqlplus/admin/iplus/provider -jar $ORACLE_HOME/oc4j/j2ee/home/jazn.jar -user "*i*SQL*Plus DBA/admin" -password <password> -listusers "*i*SQL*Plus DBA"
To add a new user to the *i*SQL*Plus *DBA* realm	adduser "*i*SQL*Plus DBA" <username> <password>	$JAVA_HOME/bin/java -Djava.security.properties= $ORACLE_HOME/sqlplus/admin/iplus/provider -jar $ORACLE_HOME/oc4j/j2ee/home/jazn.jar -user "*i*SQL*Plus DBA/admin" -password <password> -adduser "iSQL*Plus DBA" <username> <password>
To remove a user from the *i*SQL*Plus *DBA* realm	remuser "*i*SQL*Plus DBA" <username>	$JAVA_HOME/bin/java -Djava.security.properties= $ORACLE_HOME/sqlplus/admin/iplus/provider -jar $ORACLE_HOME/oc4j/j2ee/home/jazn.jar -user "*i*SQL*Plus DBA/admin" -password <password> -remuser "iSQL*Plus DBA" <username>
To revoke a user's webDba role	revokerole webDba "*i*SQL*Plus DBA" <username>	$JAVA_HOME/bin/java -Djava.security.properties= $ORACLE_HOME/sqlplus/admin/iplus/provider -jar $ORACLE_HOME/oc4j/j2ee/home/jazn.jar -user "*i*SQL*Plus DBA/admin" -password <password> -revokerole "iSQL*Plus DBA" <username>

TABLE 4-3. *Useful JAZN and OS Commands to Maintain an iSQL*Plus Configuration*

Table Name Where Data Is Stored	Parameter in MGMT_PARAMETERS That Sets Retention Value	Default Retention Value
MGMT_METRICS_RAW	mgmt_raw_keep_window	7 days
MGMT_METRICS_1HOUR	mgmt_hour_keep_window	31 days
MGMT_METRICS_1DAY	mgmt_day_keep_window	365 days
MGMT_RT_METRICS_RAW	mgmt_rt_keep_window	24 hours
MGMT_RT_<datatype>_1HOUR	mgmt_rt_hour_keep_window	7 days
MGMT_RT_<datatype>_1DAY	mgmt_rt_day_keep_window	31 days
MGMT_RT_<datatype>_DIST_1HOUR	mgmt_rt_dist_hour_keep_window	24 hours
MGMT_RT_<datatype>_DIST_1DAY	mgmt_rt_dist_day_keep_window	31 days

TABLE 4-4. *The Default Grid Control Data Retention Policy*

(*One-day* metrics are 24 1-hour metrics that are rolled up, i.e., averaged.) Therefore, at a minimum, the company would need to increase the one-day metrics retention period to 1095 days. This company might similarly decide to triple the default retention values for other metrics periods. The only limitation is disk space the Repository database consumes as the management tables grow.

To change default retention period(s), insert the relevant parameter(s) shown in Table 4-4 into the MGMT_PARAMETERS table. None of these parameters exists in MGMT_PARAMETERS by default, and any parameter not inserted into the table "defaults" to its default value, for lack of a better word. To increase the raw data retention period from 7 to 14 days, as an example, execute the following statement in SQL*Plus as the SYSMAN user:

```
INSERT INTO MGMT_PARAMETERS (PARAMETER_NAME, PARAMETER_VALUE)
VALUES ('mgmt_raw_keep_window', '14');
```

Modify Job Purge Policy

The default Grid Control policy is to purge completed jobs older than 30 days. Your company policy may dictate that job information be kept for longer than this. You cannot change the job purge policy from within Grid Control, but you can change it from within SQL*Plus. To change the job purge policy to 60 days, for example, log in to SQL*Plus as the SYSMAN user and execute the following two procedures:

```
EXEC MGMT_JOBS.DROP_PURGE_POLICY('SYSPURGE_POLICY');

EXEC MGMT_JOBS.REGISTER_PURGE_POLICY('SYSPURGE_POLICY, 60, null');
```

Secure the emkey

An *emkey* is an encryption key used to encrypt and decrypt sensitive data in Grid Control, such as host and database passwords, and consists of a random number generated during the installation of the Repository. When you install the first OMS or an additional OMS, an emkey is copied from a table in the Repository database to an *emkey.ora* file stored in the *$ORACLE_HOME/sysman/config* directory of the OMS. During startup, the OMS checks the status of the *emkey.ora* file. If it has been properly configured, the OMS uses it to encrypt and decrypt data.

After the emkey has been copied to the *emkey.ora* file, you should remove the emkey from the Repository database, as it is not considered secure. If the emkey is not removed, data such as database passwords, server passwords, and other sensitive information can be easily decrypted. Remove the emkey from the Repository database by executing the following command on any OMS host:

```
$OMS_HOME/bin/emctl config emkey -remove_from_repos
```

> **NOTE**
> *You should remove the emkey from the OMR database only if you are not planning to install an additional OMS now or in the near future. If you remove the emkey and later decide to install an additional OMS, you can copy the emkey from an existing OMS back to the Repository database using $OMS_HOME/bin/emctl config emkey -copy_to_repos.*

To confirm the status of the emkey and whether it was removed from the Repository, execute:

```
$OMS_HOME/bin/emctl status emkey
```

You will see the following output if the emkey is removed from the Repository:

Oracle Enterprise Manager 10g Release 10.2.0.0.0
Copyright (c) 1996, 2005 Oracle Corporation. All rights reserved.
The Em Key is configured properly.

If the emkey is still present in the Repository, the output of the above status command is:

Oracle Enterprise Manager 10g Release 10.2.0.0.0
Copyright (c) 1996, 2005 Oracle Corporation. All rights reserved.
The Em Key is configured properly, but is not secure. Secure the Em Key by running
"emctl config emkey -remove_from_repos".

It is essential that you back up the *emkey.ora* file, given that all encrypted data would become unusable if this file was lost or corrupted on all OMS host(s). You can back up the *emkey.ora* file from any OMS host, as it is identical across all OMS hosts. Take several kinds of backups to allow for multiple recovery options:

- Backing up the OMS home as suggested in the "Begin Regular Grid Control Backups" sidebar at the end of this chapter will back up the *emkey.ora* file itself.
- Back up the *emkey.ora* file to another machine not running a Grid Control component so that you can quickly restore this file in an emergency.
- Backup the *emkey.ora* to disk and store the disk off-site.

Set Up Grid Control Auditing

Now, shortly after installing Grid Control, is the best time to set up Grid Control auditing, not after you've experienced a security breach. You can't prevent or track down a suspicious operation unless auditing procedures are in place to capture information about that operation. This is why I remind you now to set up the auditing you require rather than waiting until the Grid Control security chapter (Chapter 16) near the end of this book. IT shops with stringent security policies may need to jump directly to Chapter 16 now to institute all best-practice security measures covered there, in addition to auditing.

Even after implementing all security measures in Chapter 16, you want to know when these measures are either under attack or have been breached. Auditing allows you to capture all changes in Grid Control or a subset of changes for certain operations you specify. You can mine audit data to send real-time custom notifications of suspicious activity through Grid Control. This data can also help you track down an intruder after the fact. Grid Control security features, including those for auditing, are compliant with standards set in the Sarbanes-Oxley Act of 2002 (SAS 70).

You execute particular audit APIs using the MGMT_AUDIT_ADMIN package in the Repository Database to choose the audit level—all operations, no operations, or selected operations. These APIs also let you specify which operations to audit and how often to purge audit data. By default, no operations are audited. The specific Grid Control operations you can audit include logon, logoff, and changes in users, passwords, privileges, roles, or jobs. Grid Control audits basic information for each operation chosen, including details on the user name and client, the operation performed, and the OMS host used. You can examine all audit operation data using the view MGMT$AUDIT_LOG.

For all details on setting up Grid Control auditing, see Oracle Enterprise Manager Advanced Configuration 10*g* Release 4 (10.2.0.4.0), Chapter 5, Section 5.5. This section is well written, and contains tables of audit fields and operation codes. Forgive my sending you to the EM documentation, but in this case, there is no sense in just reproducing the material in that section.

> **Begin Regular Grid Control Backups**
> Now that you've completed the OMR and OMS installation and post installation process, it would be a shame to suffer a hardware or software failure, or a human error that sent you reeling back to the beginning of this great adventure.[24] Now is the ideal time to begin a regular schedule of Grid Control online hot backups, as described in the section "Daily or Weekly Server-Level Backups" in Chapter 14.

Configure and Tune the OMR Database

A few tasks remain to configure and tune the OMR database to comply with GC best practices. If you installed Grid Control using the GC New Database Option, you have more steps to complete:

- If you installed Grid Control with the GC New Database Option, complete all steps in Appendix C (online at www.oraclepressbooks.com) under both the following sections:
  - Configure/Tune the OMR Database (Part 1 of 2)
  - Configure/Tune the OMR Database (Part 2 of 2)
- If you installed Grid Control with the GC Existing Database Option, and you followed the instructions in Chapter 3 (which told you to perform the steps in section "Configure/Tune a Database for the OMR - Part 1 of 2," of Appendix C), then the only steps remaining for you to complete are those in Appendix C under the section "Configure/Tune a Database for the OMR - Part 2 of 2."

You may justifiably ask why you'd need to make any OMR database changes if the GC installer creates this database for you when selecting the GC New Database Option. Why wouldn't the installer take care of everything? The answer is that the OMR database installed with the GC New Database Option will serve Grid Control well but will serve it better if you carry out all configuration and tuning recommendations. The GC New Database Option creates an OMR database that complies with all OMR requirements but not with all OMR recommendations to optimize its performance.

Install Oracle Configuration Manager

(Optional) The Oracle Configuration Manager (OCM) is an optional product separate from Grid Control that helps Oracle Support analyze a reported problem with an Oracle product. You can install OCM in the Oracle home for a database, application server, and OCM will upload configuration information for the product on a regular basis to Oracle Support.

If you choose the GC New Database Option, the Grid Control installer prompts you to install OCM, if desired. In this case, OCM is deployed in the Oracle home for the OMR database, OMS, and chain-installed Agent.

However, if you install Grid Control using the GC Existing Database Option, Grid Control does not install the database used for the Repository; rather, you must pre-install a database for this purpose. In this case, if you want to install Oracle Configuration Manager (OCM) in the

---

[24] Lou Reed, loosely quoted.

Repository Database home, you must do so outside of the Grid Control OUI. (For the various OCM installation options, see Oracle Configuration Manager Installation and Administration Guide 10g). Whatever method you employ to install OCM outside of Grid Control, afterward you must complete the following two steps as outlined in the above-referenced OCM guide:

- **Instrument the OMR database for general OCM collection**   This step is required for OCM to monitor any database, not just the OMR database. See Section 2.3.2 for details.

- **Configure OCM to collect Repository specific information**   This additional configuration is necessary to instruct OCM to collect information for a Grid Control Repository. See Section 2.3.2.2 for instructions.

Summary

This chapter described OMS and OMR post installation tasks to perform after installing Grid Control in Chapter 3. I divided these tasks into three broad categories:

- Patch the Grid Control environment
- Configure the OMS host(s)
- Configure and tune the OMR node(s) and OMR database

Each category contains numerous steps, summarized at the beginning of each section, then lovingly spelled out for you.

If I could express my aim in this chapter as a negative, it would be to not let you get too ahead of yourself. Many administrators experience the IT version of a maiden drive in a new sports car when they first log in to the Console, which opens up a new world of IT management possibilities. Administrators understandably want to remain in the Console to start managing targets. After all, that's what Grid Control is for. The Oracle Enterprise Manager documentation doesn't help in this respect. In Oracle Enterprise Manager Grid Control Installation and Basic Configuration 10g, chapters on Agent installation and Console target configuration immediately follow those on Grid Control installation. This chapter has no equivalent in the Oracle EM documentation. My rationale for the chapter is that the three most important things in IT are preparation, preparation, and preparation. Those who ignore patching or configuring Grid Control pay the price later with more complicated patch set preinstallation and post installation steps to perform, more bugs to fix, and more false alert notifications to address. In this chapter I have tried to channel any would-be "Console euphoria" into marching orders for patching and configuring the product.[25]

I do have good news, however, if you're revving the gas of the Grid Control engine. You don't have to wait any further to get started on installing Agents. Chapter 5 leads you through the standalone Agent preinstallation process. Again, there's no equivalent chapter in the Oracle EM documentation. If you're disappointed that it's about Agent *preinstallation* and not Agent *installation*, fret not: two more Agent installation chapters immediately follow, covering all six Agent installation methods, thereby finishing off Agent installation.

---

[25] I don't actually ask you to begin Console target configuration until Chapter 10.

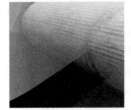

CHAPTER 5

Preinstall Standalone Management Agents

After presenting an overview of the Grid Control architecture in Chapter 1, Chapters 2 through 4 discuss installation of the Management Repository and one or more Management Services (each with a chain-installed Agent). What remains is to deploy standalone Management Agents on hosts you want to monitor. Standalone Agents are the last required piece of the infrastructure puzzle, and an important piece. Without an Agent on a host, you cannot use Grid Control to manage that host or its services and applications.

NOTE
Throughout this chapter, the host on which you are installing a standalone Agent is referred to interchangeably as the remote host, the Agent host, the target host, or the destination host.

Installation of standalone Agents is broken down into three chapters:

- This chapter covers preinstallation tasks for all installation methods.
- Chapter 6 is dedicated to the Agent Deploy installation method, as distinguished from *Agent deployment,* which refers to all Agent installation methods.
- Chapter 7 provides instructions on the remaining local installation methods.

Why does it take three chapters to cover standalone Agent installation? The main reason is that there are so many different ways to install Agents—six to be exact. Agent Deploy in particular allows you to automate the mass deployment of Agents to multiple hosts in a scalable fashion and as such requires an entire chapter because of the many prerequisites. Other methods are available to give administrators a choice that best suits their environment. Mirroring the length of coverage on Agent installation is the length of time it takes to deploy Agents on hosts across your IT environment. Agent deployment is usually the most time-consuming part of a Grid Control installation. The principal reason is that target hosts usually far outnumber OMS and Repository nodes. In addition, you may need to employ multiple installation methods on different target platforms and versions, and each target host can present unique configuration problems. The order for deploying standalone Agents is as follows:

1. First, install an Agent on each node running an OMR database that does not also contain an OMS. (Hosts running OMS components already contain a chain-installed Agent.) The scenario here is that you create either a single-instance or RAC database to house the OMR, then install Grid Control using the Existing Database option, locating the OMS on a *different* host from the OMR host. It makes sense to install Agents first on OMR host(s) before installing Agents on other target hosts. Grid Control needs to know about problems with its infrastructure before it can effectively manage target hosts.

2. Next, install Agents on all (other) target hosts you want Grid Control to manage. An Agent is the only Grid Control component required on target hosts.

An Overview of Common Agent Preinstallation Steps

Let's start with a high-level look at common preinstallation tasks for all Agent deployment methods. Coverage of these tasks is split between Chapter 2 and this chapter.

You may remember that Grid Control prerequisites discussed in Chapter 2 apply to all GC components, including the standalone Agent. The Agent requirements presented in that chapter are a subset of those for the OMR and OMS. Under each subheading in Chapter 2, *OMS*, *OMR*, and *OMA* denote the components to which the preinstallation steps apply. Following are subheadings from Chapter 2 that apply to standalone Agents. If you did not complete these standalone Agent preinstallation requirements back in Chapter 2, it's time to complete them now:

- Stage the Grid Control Software
- Network Configuration Steps
    - Set Up Host Name Resolution
    - Fully Qualify Host Name References
    - Host Name Constraints
    - Use Static IP Addresses
    - Connectivity Checks
- Software Requirements
    - Verify Certification Information
    - Create Required OS Groups and Users
    - Create Required Directories
    - Synchronize OS Timestamps/Time Zones
    - Confirm Platform-Specific Software Requirements

Again, complete these steps now before proceeding further.

Following are additional standalone Agent preinstallation steps covered in the remainder of this present chapter. These steps are not required for OMS or OMR hosts.

- Agent key installation decisions
- Agent hardware requirements
- Agent software requirements

Let's delve into these additional Agent preinstallation requirements now.

Agent Key Installation Decisions

You need to make a handful of decisions before installing Management Agents:

- Select Agent installation methods
- Use an existing user or create a separate Agent user
- Install a local or cluster Agent
- Configure a server load balancer first
- Use an LDAP or local Agent user
- Choose to secure or not secure the Agent

Some administrators are not aware that they are making all of these decisions because they don't know of the alternative choice for each bullet point above. This section brings these decisions to the surface and offers direction on how to make them.

Select Agent Installation Methods

You may recall from running the Grid Control GUI installer that a standalone Agent—called an *Additional Management Agent*—is the fourth installation type offered. Including this interactive method, standalone Agents can be installed six ways. Why so many? Well, each method takes a different tack, one of which may be more appropriate than another for a particular host or set of hosts in your environment. Below are the Agent installation methods available in GC Release 2, with my overall ranking from most to least preferred. My rating gives points for automation, efficiency, speed, number of bugs, ease in scripting, and ability to upgrade:

- **Agent Deploy** Also known as the *Agent Installation* feature, this is a web-based Agent push from the Console capable of mass deploying or upgrading standalone local, cluster, or NFS Agents.

- **nfsagentinstall** NFS Agents share the Agent home on a Network File System (NFS) such as a Net App Filer disk, among target hosts, on which state and log information is locally stored. I use the moniker *nfsagentinstall* (the script name that installs NFS Agents) to distinguish it from an Agent Deploy installation of NFS Agents.

- **agentDownload** This is a text-based script that runs on target hosts, pulling the Agent installation from the OMS by calling the silent installer script under the covers.

- **Agent Cloning** After installing a "master Agent" on a target host via any other method, you can clone this Agent to multiple hosts running on the same platform. To clone, you simply zip the Agent home, transfer the zip file to another host running the same O/S, unzip it, and manually run the Agent Configuration Assistant (*agentca*) to discover targets on that host, specifying an Agent Cloning option to update the Oracle inventory with the Agent installation information.

- **Silent Installation** Installs an Agent locally using a response file. The default response file is pre-populated to install Agents on the same platform as the OMS. You must edit this response file to deploy Agents on other platforms.

- **Interactive Installation** This method uses the Grid Control OUI to install Agents by selecting the Additional Management Agent Installation Type.

Following are a few general thoughts regarding Agent installation methods:

- The Oracle documentation refers to all but the interactive method as "mass deployment" methods, but I would reserve this description only for Agent Deploy.
- I do not consider *nfsagentinstall*, *agentDownload*, Agent Cloning, and silent installation methods as mass deployment methods, given that their installation scripts must run on each local target host. They are "automated" methods because you can automate local Agent installations by employing wrapper scripts, cron jobs, or O/S system features. However, such automation requires additional programming that is not necessary with Agent Deploy.
- All methods except Agent Deploy are "local" methods, because they are executed on the local Agent host. These methods are covered in Chapter 7.
- Resultant Agent installations are identical for all methods except NFS Agent, which uses a different Agent architecture as briefly described above.
- The Select Agent Installation Methods heading above uses the plural, "methods," because you may decide to choose more than one method to install Agents across all hosts.
- Keep the number of Agent installation methods to a minimum to avoid the learning curve required for each method, which sometimes also varies by platform.
- Ideally, you would select one method of Agent installation, but this isn't always possible. Differing standards between IT departments may restrict choice in installation methods (e.g., prohibiting NFS technology would rule out NFS Agents, and disallowing Xwindows on target hosts would rule out interactive installations).
- Start by eliminating unsuitable Agent installation methods to narrow the field, such as crossing NFS Agents off the list for RAC database environments.
- Analyze your environment against Agent features and limitations to select one (or more if required) installation methods.
- Consider not only which methods will install Agents in the least amount of time, but also the amount of upkeep required. In other words, give serious consideration to Agent Deploy, as it is the only method that can upgrade multiple Agents in one operation.
- As the number of target hosts decreases, the complexity of the Agent Deploy installation process (see Chapter 6) begins to overshadow its benefits and other methods become more appealing.

To help you select the most appropriate Agent installation method, let's compare and contrast their relative merits.

Agent Deploy

- Agent Deploy pushes Agents in one operation to multiple hosts *of the same operating system*. This means that you are limited to installing batches of Agents, at least one batch per O/S.
- Sites with tens of hosts on the same O/S should seriously evaluate Agent Deploy for time savings from an installation and maintenance perspective.
- Sites with less than ten or so hosts on the same O/S may want to consider using methods other than Agent Deploy.
- Only Agent Deploy allows you to deploy or upgrade standalone local or NFS Agents en masse (i.e., in one installation process).
- Agent Deploy on Windows is complicated and buggy. You should avoid this combination, in my judgment.

NFS Agents

- Whether or not you use NFS Agents depends a lot on your company's current use of NFS technology. If you're not already using Net App Filer or Network Attached Storage (NAS) drives, then NFS Agents may not be your best choice. Some companies simply mount a network share, but the share should be redundant to avoid a single point of failure if the host or storage device malfunctions.

- NFS Agents are very attractive to many sites that want to cut down on disk space requirements or reduce maintenance costs (only the master NFS Agent needs to be upgraded).

- NFS Agent installation is not supported on hosts running Clusterware or Oracle Cluster File System (OCFS).

agentDownload

- You can't use *agentDownload* if there is a proxy server between Agent and OMS hosts, or if the *wget* utility is not available, as is often the case on AIX.

- *agentDownload* ignores prerequisite checks and redirects output and errors to a log file, which administrators tend to overlook, compared with *stdout* (standard output)[1] and *stderr* (standard error)[2] to the terminal, which the other methods provide. Check the log file or consider running the Prereqchecker available with the Full Installer for each unique OS to confirm that all prerequisites are met.

- As with Agent Deploy, *agentDownload* lets you centralize the installation of Agents from the OMS across all platforms; but you must download a platform-specific version of the *agentDownload* script from the OMS.

- *AgentDownload* can deploy Agents in secure web environments that allow HTTP access to Agent hosts but that restrict other types of access. Restrictions can be due to firewalls, or lack of an ftp server or NFS mount to copy or share Agent installation files, respectively.

- It is a silent, scripted method with text-only output, so requires no graphical (i.e., Xwindows) setup, which is also true of the silent installation method.

Agent Cloning

Agent Cloning is a time saving installation method when you have lots of Agent one-off patches to apply; after patching just one Agent, you can clone it to other Agent hosts.

Silent Installation

Silent installation, like *agentDownload*, is a useful method to automate deployment on destination hosts that do not run Xwindows software.

Interactive Installation

Sites with fewer than ten or so hosts running on the same platform will probably find interactive installation quite sufficient.

---

[1] Stdout is the stream where a program writes its output data, typically the text terminal that initiated the program.

[2] Stderr is another output stream that programs use to output error messages, typically the text terminal that started the program.

Agent Installation Method	Can Automate	Installs Cluster Agent	Installs Through an SLB	Automatically Runs Required *root* Scripts	Secures the Agent Automatically	Agent Restarts on Boot	Installation Problems
Agent Deploy	Yes	Yes	Yes[3]	Yes (if option is chosen)	Yes (if option is chosen)	Yes	Doesn't update Inventory or *oratab*
nfsagentinstall	Yes	No	Yes	Yes	No	Yes[4]	Doesn't update Inventory. Typo in *oratab* entry.
agentDownload	Yes	Yes	Yes	No	Yes[5]	Yes (as of 10.2.0.3)	None
Agent Cloning	Yes	No	Yes	No	No	No	A little buggy
Silent Installation	Yes	Yes[6]	Yes	No	No	Yes	None
Interactive Installation	No	Yes	Yes	No	Yes	Yes	None

TABLE 5-1. *Features of All Agent Installation Methods*

Table 5-1 above summarizes the features and limitations of all Agent installation methods, characterizing their relative strengths more quantitatively.

Below are some finer points and conclusions to draw from each column of this table of features:

- **Can Automate** Agent Deploy is the only true mass deployment installation method. All remaining methods except interactive installation can be automated using a wrapper script, remote commands, etc.

- **Installs Cluster Agent** Choose a cluster Agent over local Agent installation for RAC targets. It is much easier to install and upgrade a cluster Agent (once you complete the required preinstallation steps) because you can do so in one operation.

- **Installs Through an SLB** For GC releases prior to 10.2.0.4, you cannot install Agents using Agent Deploy when accessing the Console via a server load balancer (SLB)[7] located in front of your OMSs. All other installation methods can install through an SLB. For GC 10.2.0.3 or earlier, you can install Agents using Agent Deploy through an SLB by logging in to the Console using the URL for a specific OMS host rather than for the SLB. This bypasses an SLB configured to load balance Console logins.

---

[3] Using Agent Deploy behind an SLB is only possible as of Grid Control 10.2.0.4.

[4] This is true only for *nfsagentinstall* from 10.2.0.3 or later master Agents.

[5] *agentDownload* automatically secures the Agent as of 10.2.0.3 if you enter the Agent registration password when prompted or if you preset AGENT_INSTALL_PASSWORD.

[6] You can only install a cluster Agent using silent installation if the target host is on the same platform as the OMS.

[7] A server load balancer is either a hardware device such as an F5 BIG-IP or less desirably, a dedicated software program that you employ in GC HA environments to spread Agent uploads and Console logins amongst two or more OMS hosts.

- **Automatically Runs Required *root* Scripts**   Agent Deploy runs *root.sh* and *orainstRoot.sh* scripts as *root*, thereby avoiding the need to locally log in to each target host during installation.

- **Secures the Agent Automatically**   Agent Deploy can automatically secure Agents, critical functionality for the same reason as being able to run *root* scripts—it avoids the need for local login. For *agentDownload*, the only local installation method offering this feature, it's more of a luxury than a necessity, as you're already logged in locally and can easily run *emctl secure agent*, manually or in a script.

- **Agent Restarts on Boot**   Agent Cloning is the only installation method that does not attempt to address Agent automatic restart on boot. Although the remaining installation methods attempt to put in place such an Agent restart mechanism, none actually succeed in configuring the Agent to restart on boot on UNIX platforms. Appendix E (online at www.oraclepressbooks.com) describes this Agent restart mechanism and how to finish enabling it.

- **Installation Problems**   All installation problems are minor. For bugs, see the Agent Deployment section of the Release Notes for your platform.

Use an Existing User or Create a Separate Agent User

The Oracle Enterprise Manager documentation set does not provide much guidance, if any, on whether to create a separate Agent user to install the Agent software on managed hosts or use an existing owner of Oracle software (typically *oracle*) that's already installed.[8] Yet this is one of the first dilemmas confronting administrators who are sorting through how to install Agents, so let me weigh in on this important decision.

Using a different user for the Agent than that for other managed targets on a given host can complicate the OS environment. A new Agent user adds system administrative overhead and may require additional Oracle Inventory management (see the section "Choose the Oracle Inventory Directory" in Chapter 2). For these reasons, I recommend creating a separate Agent user *only under the following circumstances*:

- If the security policy at the site dictates that you install a separate Agent user.

- If you want to completely decouple the GC infrastructure from the targets it manages. This may be security-related or to avoid any possible corruption to managed targets, such as to their Oracle inventory, assuming you are using a central Inventory. (Note that you can use a central Inventory or separate Inventories for each Oracle product, whether you install a new Agent user or co-opt an existing user.)

- If the *oracle* user must prompt for user input on login. Many sites running multiple databases on a single host add logic to the *oracle* user login script that prompts you to select from a menu of database environments to initialize. Such logic would prohibit *oracle* from being the OS preferred credential user because GC cannot supply input to *stdin* (standard input).[9]

---

[8] The only guidance I found was in Meta*Link* Note 273579.1, which states that "you can use the same [Agent] owner as existing installations."

[9] Stdin is data (often text) going into a program. In this case, the program is a user login script, which prompts you to enter input from the text terminal that started the program. Grid Control's Execute Host Command cannot provide input to a login script for the OS user specified in Preferred Credentials.

- If you want to use a unique login environment for the Agent user. For example, you may want to use the Execute Host Command functionality in Grid Control to run OS scripts (or commands) across multiple target hosts in one operation. In such a case, all scripts can rely on variables set in the login environment rather than in each script.

- If for security or other reasons, the *oracle* user cannot be granted *read* access to the *$OAS_HOME/sysman/emd/targets.xml* file belonging to an Oracle Application Server (OAS) target. The Agent user requires *read* access to this file to discover the OAS target. You can grant such *read* access either through owner or group permissions. The default OAS owner and group are *iasora:dba*, so granting the *dba* group (of which *oracle* is a member) *read* access to the OAS *targets.xml* file avoids the need to create a separate Agent user.

- If you want to use the Agent Deploy or *nfsagentinstall* methods for mass-installing Agents and either the *oracle* password, username, user ID (*uid*), group name, or group ID (*gid*) are different across hosts. These installation methods require that this user information be the same across all target hosts where deploying the Agent. A separate Agent allows you to assign this user information uniformally on all target hosts.

This last point deserves a little more coverage. Most Oracle products are installed using the *oracle* user. However, some Oracle shops use specific users for each Oracle product installed, or use the default *iasora* user for Oracle Application Server installations. If the *oracle* user exists on all hosts to be monitored, it is not difficult to change the user ID and/or group ID so that they are identical. Typically, the showstopper to using an existing *oracle* user for Agent Deploy or *nfsagentinstall* installations is that you cannot set the *oracle* password identically across hosts, usually for security reasons. This is particularly true at larger Oracle sites where different teams of DBAs are assigned different databases. (Ironically, these larger sites are where these methods are most suitable to mass deploy Agents across many hosts.) Whatever constrains you from using an existing *oracle* user as the Agent user, you can still install Agents via Agent Deploy or *nfsagentinstall* by creating a new Agent user with the same group ID and password.

NOTE
If you install a separate Agent user and want to use the same user on all hosts, including OMS hosts, then you must install Grid Control as this user.

If you decide to create an Agent user on all target hosts that is different from the existing owner of Oracle products (typically *oracle*), then create such a user now. You can call this user *gridcontrol, oemagent, agentem*, or any other name. (I would not use "10g" in the name, as you will one day be upgrading this Agent to 11g.) Use the same *uid* and *gid* for this user on each host. (You can also create an LDAP user if desired. See "Use an LDAP or Local Agent User" below.) For instructions on creating an Agent user, see the section "Create Required OS Groups and Users" in Chapter 2.

Install a Local or Cluster Agent

You can install an Agent on a cluster node (either an Oracle RAC cluster node or a node running third-party clusterware) using the Agent Deploy, *agentDownload*, silent, or interactive methods.

What Is an EMSTATE Directory?

EMSTATE (also known as the "state" directory) refers to a local subdirectory under the Agent home that certain Agent installations create to store configuration files (upload, state, etc.) and log files for the local host. An *EMSTATE* directory pertains to local Agents on RAC nodes and cluster Agents, but it also relates to NFS Agents and Database Control Agents. The *EMSTATE* directory paths differ for these Agents in the following manner:

- **Local Agent on a RAC node**   *EMSTATE* is the Agent home subdirectory *$AGENT_HOME/<nodename>.<domain>*, created only on the local node.

- **Cluster Agent**   *EMSTATE* is also the Agent home subdirectory *$AGENT_HOME/<nodename>.<domain>*. Subdirectories named after each node are created under the AGENT_HOME on *all* nodes, but cluster Agents only use the *EMSTATE* subdirectory named after the local node—the other subdirectories remain empty.

- **NFS Agent**   Each NFS Agent host shares common NFS-mounted Agent binaries. When deploying NFS Agents on each host, you specify a local *EMSTATE* directory of your choice, and locate it on local storage for security purposes.

- **Database Control Agent**   The *EMSTATE* directory is the Agent home subdirectory *$AGENT_HOME/<nodename><ORACLE_SID>*. (This differs from the *EMSTATE* directory for a Grid Control Agent, which is not tied to a particular database so it does not contain *<ORACLE_SID>* in its directory path.)

To determine the location of the *EMSTATE* directory once an Agent of any type is installed, execute the following command (example output is from one of the example GC cluster Agent nodes referenced below, called *ptc1.perftuning.com*):

```
emctl getemhome
Oracle Enterprise Manager 10g Release 10.2.0.4.0.
Copyright (c) 1996, 2006 Oracle Corporation.  All rights reserved.
EMHOME=/opt/oracle/product/10.2.0/em10g/agent10g/ptc1.perftuning.com
```

For each of these methods, the choice is to deploy either a local Agent on each node individually or a cluster Agent on multiple nodes in a single installation by choosing the Cluster Installation option. The terms *cluster Agent* and *RAC Agent* are synonymous, and equally misleading, in my view. These terms refer to the RAC target on which the Agents exist, not to the Agents themselves. The Agents are not part of an Oracle Clusterware installation, which binds hosts so that they operate as a single system. The Agents on each RAC node are installed on local cooked file systems, and are entirely separate (i.e., each node has its own Agent, which acts independently). One thing cluster Agents do have in common with RAC databases is that, like Clusterware, they are installed via the Secure Copy (SCP) Program across all specified RAC nodes in a single installation process. This feature makes cluster Agents more attractive than local ones because you can install and upgrade them in one operation. The resulting Agent installation is the same for both local and cluster Agent deployments.

Configure a Server Load Balancer First

If you plan to use a server load balancer (SLB) to front-end multiple Active/Active OMSs, but did not configure an SLB when first installing Grid Control, consider doing so now, particularly if you will be installing many standalone Agents. The section "Use a Server Load Balancer" in Chapter 15 refers you to Oracle*MetaLink* Note 353074.1 for instructions to configure an SLB. If you configure an SLB before installing standalone Agents, you don't need to complete Part II of the above-referenced Note, called Configuration Steps on Management Agent Side. Instead, when installing Agents with an SLB already in place, you just need to specify the SLB hostname, not an OMS hostname, so that the Agent uploads to the SLB rather than to a particular OMS.

Use an LDAP or Local Agent User

If you cannot install an Agent as an existing local user (typically named oracle) that owns current Oracle software on a target host, then you must create a new Agent user for this purpose. Instead of creating a local Agent user on each target host, consider installing Agents using a Lightweight Directory Access Protocol (LDAP) operating system user. This can be a great boon when you have many target hosts to manage because you only need to create one LDAP user rather than multiple local users. While the official Oracle documentation does not address whether you can use an LDAP user for the Agent, I have found empirically that you can do so *except* in the following cases:

- When using the Agent Deploy method
- When using a cluster Agent (whether installed interactively or through Agent Deploy)

You cannot use an LDAP user in these cases because they require setting up SSH user equivalence[10] (from the OMS to the target host in the first case and between RAC nodes in the second case for interactive installations), which in turn necessitates installing a local Agent user.

An LDAP user has the following additional limitations, irrespective of the Agent installation method employed:

- To control the Agent (start, stop, etc.), you must run *sudo* commands as a local user on the target host. Typically the local user is either your personal user account or the *oracle* account that owns target Oracle software. Sudo commands require that you fully qualify all executables. Example *sudo* command syntax to start an Agent that an LDAP user installs is:
sudo -u *<ldapagent_user>* /*<mount_point>*/app/product/10.2.0/em10g/agent10g/bin/emctl start agent
- You cannot manage Oracle Application Server or Collaboration Suite targets, because the owner of this software must be the Agent user and cannot be an LDAP user.

You need to weigh the benefits against the drawbacks of using an LDAP Agent user. If you decide on an LDAP Agent user, there are two requirements that the system administrator must satisfy. First, grant the OS *dba* group (to which DBAs belong) sudo privileges to this user. Second, ensure that the LDAP Agent user id (uid) and *dba* group id (gid) are each consistent across hosts using the LDAP Agent. See your LDAP vendor's documentation for instructions on how to create an LDAP user.

---

[10] SSH user equivalence allows you to connect via the Secure Shell (SSH) protocol from one host to another without specifying a password.

Choose to Secure or Not Secure the Agent

Best practices recommend securing all Agent communications with the OMS to ensure the privacy of transmitted data. Secure Socket Layer (SSL) is used to encrypt Agent data sent over HTTPS. If you specified during EM installation that the OMS must communicate only with secure Agents (see Chapter 3, Figure 3-8), then you must secure the Agent during installation. Some Agent installation methods secure the Agent if you preset the AGENT_INSTALL_PASSWORD environment variable before running the particular installation executable. (For instructions on how to set environment variables for the most common UNIX shells, see the "UNIX Shell Variations" sidebar in Chapter 3.) On Windows, open a command window and use the following syntax to set or unset variables, respectively:

set <variable_name>=<value>
set <variable_name>=

Post-installation, you can manually secure an Agent by running the following command on the Agent host:

$AGENT_HOME/bin/emctl secure agent [-reg_passwd <password>]

If you don't use the *reg_passwd* argument, you are prompted for the Agent registration password.

Agents always deploy on a nonsecure port (default 4889), which must remain accessible throughout an Agent's lifetime to allow it to secure or re-secure if necessary. You may need to re-secure an Agent for several reasons such as to change an Agent registration password or to reconfigure an Agent for communications with a server load balancer. By default, unsecured HTTP uploads and wallet requests to secure an Agent take place on http://<*OMShost*>:4889/em/upload, and secure HTTPS uploads take place on https://<*OMShost*>:1159/em/upload.

Agent Hardware Requirements

The Oracle Enterprise Manager documentation does not clearly state any specific hardware operating requirements for the standalone (i.e., non OMS chain-installed) Management Agent[11]. However, the Agent, like any software application, consumes hardware resources, including disk space, CPU, and RAM.

Agent Disk Space Requirements

Disk space requirements for the Agent software home are much easier to define than those for CPU and RAM. Both the interactive and Agent Deploy installation processes (except for NFS Agents) check that the minimum required disk space is available. Table 5-2 lists these minimums for each certified Agent platform as well as the total disk space operationally needed. Here's how I derived the totals.

In addition to the minimum space for Agent home software, you need to allocate disk space for collection, state, upload, log, and trace files in the *$AGENT_HOME/sysman/emd* directory. The upload manager supports a default maximum size of 50MB in the *upload* directory before

---

[11] The disk space requirements for just the initial Agent software home are actually buried in Appendix C (Agent Deploy Application—Installation Prerequisites) of the Enterprise Manager Installation and Basic Configuration Guide.

Platform	Minimum Disk Space Required for Agent Software (GB)	Total Disk Space to Allocate for Agent (GB)
AIX	1.50	2.00
HP-UX	1.50	2.00
HP Itanium	2.00	2.50
Linux x86	0.45	0.75
Linux x64	0.75	1.25
Solaris	0.85	1.25
Windows	0.45	0.75

TABLE 5-2. *Minimum and Maximum Disk Space Requirements for Agent Installation*

temporarily disabling collections, logging, and tracing. (This size is modifiable in the Agent properties file *emd.properties*, but most sites shouldn't need to change the default.) Factor in another 50MB for remaining files (collection, state, log, and trace files). Finally, add 20% free buffer space, which leaves room to upgrade. Totaling this up, you get:

- Total Agent Disk Space = (*<Agent home software>* + *<collection, state, upload, log files>*) * 1.20

- Total Agent Disk Space = (*<Agent home software >* + 100MB) * 1.20

Table 5-2 does the math for you and lists the total disk space to allocate for each platform, rounding up to the nearest 250MB.

Agent CPU Requirements

As for CPU requirements for the Agent, a good guideline from real world experience is that the Agent typically consumes approximately 1% to 2% CPU on average[12], and can peak as high as 10% CPU. This peak is borne out by the default Agent metric thresholds used to self-monitor the Agent out-of-box. The default CPU Usage (%) metric thresholds are 10% Warning and 20% Critical CPU consumption. This CPU consumption is measured as a percentage of CPU time at any given moment in time of the Agent process and any of its child processes. (The Agent occasionally creates child processes when evaluating a metric or running a job.) The Agent must reach these percentage levels at least four times in 1038 seconds (about 17 minutes) to trigger an alert. System administrators do not find it heartwarming that monitoring software itself can chew up 10% of the CPU cycles, but the Agent typically will not consume more than 2% CPU. Still, this discussion should convince you not to run an Agent on a severely CPU-bound machine.

---

[12] Obviously, this percentage is inversely proportional to CPU speed. An Agent workload that consumes 10% CPU on a 1GHz system would consume much less CPU on a 4GHz system. Empirical data is the best indicator of an Agent's CPU drain.

Agent RAM Requirements

As for RAM requirements, again the Agent metrics shed some light on anticipated usage. The Resident Memory Utilization (%) metric, which represents the amount of physical memory used by the Agent and all of its child processes, has a default 20% Warning threshold and 30% Critical threshold. The Resident Memory Utilization (KB) metric has thresholds of 128MB Warning and 256MB Critical. It is not uncommon for these metrics to fire, so you should plan on parting with at least 128MB RAM and as much as 256MB RAM for the pleasure of the Agent's company. Again, this is not real popular with system administrators.

> **TIP**
> *I would tactfully suggest that you not mention these worst-case CPU or RAM figures to SAs. First monitor the Agent consumption yourself. You may be pleasantly surprised. Then tighten Agent metric thresholds enough to ensure you learn about any Agent hogs before your SA does.*

Summary of Agent Hardware Requirements

To summarize, set aside the following minimum and maximum disk space, CPU, and RAM resources for the Agent on target systems running at 1GHz CPU or higher, as shown next:

Resource	Minimum	Maximum
Disk Space (see Table 5-2 for each platform's requirements)	0.75GB (Linux x86)	2.50GB (HP Itanium)
CPU	2%	10%
RAM	128MB	256MB

To conclude this discussion on Agent hardware resources, I'd like to point out that Agent Deploy operations take up both CPU and RAM resources on the involved OMS hosts. For example, an Agent Deploy installation on six to nine target hosts consumed an average of 15% CPU and two Gigabytes of memory on a Linux OMS host with two Pentium 4 3.00GHz processors.[13]

Agent Software Requirements

Following are additional standalone Agent software requirements over and above those listed in Chapter 2.

Initialize the oracle User Environment

Execute the same steps now as described in the subheading of the same name, which is located in Chapter 3, and note the following additional points about setting OS environment variables (shown in capital letters below):

- If you have installed a separate Agent user, there probably won't be any Oracle variables to unset.

---

[13] From 10g R2 Management Agent Deployment Best Practices, an Oracle White Paper, June 2006, Section 5.3.2.

- If you predetermined to secure all Agent/OMS traffic (refer back to the earlier section in this chapter "Choose to Secure or Not Secure the Agent"), then for the *agentDownload* and interactive installation methods, optionally set AGENT_INSTALL_PASSWORD to the value of the Agent Registration password specified during the GC installation. (This avoids your being prompted for the password when installing the Agent.) You can set this variable in a specific session (or in the Agent user's shell startup script, though this is not a good security practice).

- If the *listener.ora* directory is not in the default location, you can specify the nondefault directory path location with TNS_ADMIN. This did not work in GC 10.1, but was fixed in GC 10.2 (see Bug 3488126).

- For cluster Agent installations, set the CLUSTER_NAME environment variable to the cluster you want GC to manage. Set ORA_CRS_HOME (for 10*g* cluster targets) or CRS_HOME (for 9*i* cluster targets). Also, make sure the Clusterware is started.

For cluster Agent installations on GC 10.2.0.2 and earlier using any installation method, log in to each target cluster node and make sure any environment variables that contain either a semicolon (";"), a space character (" "), or a double colon ("::") are temporarily unset, whether Oracle-related or not (see Bug 5208716 and Oracle*MetaLink* Note 352140.1). To determine whether any such variables exist, execute the *set* command to view all environment variable settings, back up the *oracle* user login script, then edit this script and temporarily append commands to unset any such variables. Typical variables that contain semicolons, spaces, and double colons are ADJREOPTS, BASH_VERSION, CONTEXT_FILE, CONTEXT_NAME, customfile, compareVersion, IFS, LESSOPEN, LS_COLORS, PROMPT_COMMAND, SSH_CLIENT, SSH_CONNECTION, PS1, PS2, PS3, PS4, ver, ver1, ver2, and VNCDESKTOP. Note that unsetting the variable PROMPT_COMMAND will cause no prompt to be displayed, but this is only temporary. You can restore the login script from backup as soon as the cluster Agent installation completes. If you neglect to unset any such variables, the cluster Agent installation will fail with the following error:

Failed to copy file *<AGENT_HOME>*/sysman/install/hosttgt.out to destination.

If you get this error, rather than reinstalling these Agents, a workaround is to execute the following command on each cluster node to reconfigure the Agent:

```
$AGENT_HOME/oui/bin/runConfig.sh ORACLE_HOME=$AGENT_HOME MODE=perform
ACTION=configure
```

Confirm No Existing Agent Is Installed

You may have inherited an existing GC environment in some stage of installation, and you may not be sure whether an Agent is installed on a particular host. You can only install one Agent per host, except on Active/Passive clusters where you install one Agent per hardware node and one for each virtual IP address the cluster software generates (see the section "Configure the Agent in Active/Passive Environment" in Chapter 15). The Agent Deploy method confirms as a prerequisite check that only one Agent is deployed per host, but no other installation method confirms this. Here are a few ways to check for the presence of an existing Agent:

- If the Grid Control installation is running, log in to the Console and check whether the host has been discovered (navigate to the Host home page).

- Check under the *ORACLE_BASE* directory for an Agent home.

- For UNIX only, look at all existing Oracle Inventories. First check the *inventory_loc* value in the *oraInst.loc* file for the directory path to the default Inventory. Then you can either run the OUI and select Installed Products to see if the Agent is installed, or look at the *inventory.xml* file under the *<inventory_loc>/ContentsXML/* directory. The Agent home will be contained in a line of the following form:

    <HOME NAME="agent10g" LOC="/u01/app/oracle/product/10.2.0/em10g/agent10g" TYPE="O" IDX="3"/>

- For UNIX only, the Agent home might be defined in the *oratab* file (not all Agent installation methods add such an entry). Look for a home named after the Agent, such as the following:

    *:/u01/app/oracle/product/10.2.0/em10g/agent10g:N

- Look for any Agent processes currently running on the system. On UNIX, run the following command to see if the Agent process or Agent watchdog process exists (example output is shown below for a running Agent):

    ```
    ps -ef | grep emagent

    oracle    30272     1  1 16:12 pts/3    00:00:00
    /u01/app/oracle/product/10.2.0/em10g/agent10g/perl/bin/perl
    /u01/app/oracle/product/10.2.0/em10g/agent10g/bin/emwd.pl agent
    /u01/app/oracle/product/10.2.0/em10g/agent10g/sysman/log/emagent.nohup

    oracle    30289 30272 25 16:12 pts/3    00:00:02
    /u01/app/oracle/product/10.2.0/em10g/agent10g/bin/emagent
    ```

- Look in the PATH for any Agent home references.

Change Response File if Using an SLB

An Agent installation response file, named *agent_download.rsp*, is provided in the OMS home under *$OMS_HOME/sysman/agent_download/<version>/* for several installation methods (including Agent Deploy, *agentDownload*, and silent installations) to consult when noninteractively installing the Agent. The values in the original response file, *agent_download.rsp*, are populated at the time of OMS installation to allow you to deploy Agents against that particular OMS host and platform. However, in environments with multiple OMS hosts virtualized by a server load balancer (SLB), you want target hosts to be able to point to any available OMS host, rather than to a particular one. This is especially useful for load balancing purposes when OMS resources are limited, either because of deploying many Agents simultaneously, or because of high user load. All Agent installation methods except Agent Deploy on Grid Control 10.2.0.3 or earlier allow you to go through an SLB rather than a particular OMS host to deploy Agents. On each OMS, you simply need to edit *agent_download.rsp*, and modify the following two lines to point to the SLB host name and SLB port, respectively:

s_OMSHost="<SLBhost>"
s_OMSPort="<SLBport>"

Prepare for Cross-Platform Agent Installation

Regardless of the Agent installation method, by default you can only install an Agent on host platforms matching that of the OMS host. For instance, if your OMS is installed on a Solaris host, you can only install Solaris Agents by default. To install cross-platform Management Agents (i.e., Agents on a different operating system than that of the OMS host), you must go to the Mass Agent Deployment Downloads page on the Oracle Technology Network (OTN) web site and download the Agent kits, available in zip format, for these other platforms. These Agent kits have the naming convention *<platform>_Grid_Control_agent_download_<version>.zip*. Once downloaded, you must unpack an Agent kit, depending upon the Agent installation method employed (detailed below), either into the Agent download directory[14] of the OMS home, defined as *$OMS_HOME/sysman/agent_download/<version>/<platform>*, or to any location accessible by the Agent host, to make it available for use by your particular installation method(s). Table 5-3 shows all *<platform>*

Platform Name	Description
linux	Linux
x86_64	Linux X86_64
ppc64	Linux on power PC
linux390	z/Linux
ia64linux	IA64 Linux
solaris	Solaris 64 bit
solarisx86	Solaris-x86
hpux	HP-UX
decunix	HP Tru64 UNIX
hpunix	HP 64 bit
hpi	HPUX-Itanium
aix	AIX5L
macosx	MACOSX
vms	VMS
windows_ia64	Microsoft Windows (64-bit IA)
windows_x64	Microsoft Windows (64-bit AMD64)
win32	Microsoft Windows 32-bit

TABLE 5-3. *Platform Names in Agent Download Directory*

---

[14] Do not confuse the term "Agent download directory" with the *Agentdownload* installation method. The Agent download directory contains installation software used by all Agent installation methods to install cross-platform Agents (i.e., Agents on a platform other than the OMS). The *Agentdownload* installation method is just one of these installation methods.

values that are part of the Agent download directory path, and the corresponding platforms these values represent.

The *<version>* subdirectory below which the Agent kit is unzipped allows you to stage installers for different Agent versions alongside each other.

To stage cross-platform Agent software, download the Agent Zip file for the alternate platform from the Oracle Technology Network (OTN) at http://www.oracle.com/technology/software/products/oem/htdocs/agentsoft.html, and then stage the Agent kit in one of the following two locations, depending upon the installation method you plan to use for this cross-platform Agent installation:

- For Agent Deploy or *agentDownload* installations, stage the Agent kit on the OMS host under the Agent download directory, *$OMS_HOME/sysman/agent_download/<version>/<platform>*, as follows:

    - Copy the downloaded zip file to *$OMS_HOME/sysman/agent_download/<version>/* (first create this directory if it doesn't exist).
    - cd *$OMS_HOME/sysman/agent_download/<version>*
    - Unzip the Agent software with the command:

        $OMS_HOME/bin/unzip <platform>_Grid_Control_agent_download_<version>.zip

- For *nfsagentinstall*, interactive, silent, or Agent Cloning installations, stage the Agent kit to local disk on the Agent host, or to any location readable by the Agent host by simply unzipping the zip file at that location. The UNIX installer is located under *<unzipped_location>/<platform>*/agent/*runInstaller* (on Windows the executable is called *setup.exe*).

Back up and replace the existing response file, *agent_download.rsp*, located under the directory *$OMS_HOME/sysman/agent_download/<version>/* with the response file found at the top-level directory of the Agent kit, named *agent_download.rsp.bak*. Then change the response file to reflect either the OMS host and port, or the server load balancer (SLB) host and port, if Agent deployments will go through an SLB. Note that the response file is version-specific, but not platform-specific. That is, the response file is located in the *<version>* directory and above the *<platform>* directory, so it is independent of the Agent platform. This is why you need to copy the platform-specific version of the response file to this location under *<version>*. First, back up the *agent_download.rsp* file installed on each OMS, then replace it with the platform-specific version of the response file:

cd $OMS_HOME/sysman/agent_download/*<version>*

mv agent_download.rsp agent_download.rsp.orig

mv agent_download.rsp.bak agent_download.rsp

Next, modify *s_OMSHost="%s_hostname%"* and *s_OMSPort="%s_OMSPort%"*, replacing the dummy values with either of the following:

- If the OMS does not use an SLB, use the OMS host name and nonsecure port 4889, respectively. Although the OMS is secure by default, the *agentDownload* script itself is downloaded unsecured.

- If the OMS uses an SLB and if you are not using Agent Deploy with Grid Control 10.2.0.3 or earlier, use the SLB host name and SLB port, respectively.

CAUTION
You may have already changed this response file according to the previous section to reflect the use of an SLB (see the earlier section "Change Response File if Using an SLB"). However, you need to redo the steps from this previous section if you're doing a cross-platform Agent installation.

Any additional cross-platform Agent installation steps required by a particular installation method are delineated in Chapters 6 and 7 for that method.

Prepare Targets for Discovery

Before installing a standalone Agent, perform the following on each destination host to maximize the success of the target auto-discovery process:

1. Check that the *oratab* file on the destination host contains all databases and Oracle homes that you want Grid Control to monitor. The reason for this check is that the *oratab* file is parsed during the Agent target discovery process.
2. If the host is a RAC node, start the Oracle Clusterware.
3. Prepare the inventory on the target host. Refer back to the "Choose the Oracle Inventory Directory" section in Chapter 2.

Gather Needed Agent Installation Information

Table 5-4 lists common installation information required to install the Management Agent regardless of the method used. For some fields, suggested values are provided, which comply with values used throughout the book.

Information Required	Suggested Value or Recommendation
Parent Directory	<mount_point>/app/oracle/product/10.2.0/em10g
Inventory Directory	<mount_point>/app/oracle/oraInventory
Operating System Group Name	oinstall (or dba if this is your standard)
Management Service Host Name	Use fully qualified hostname
Management Service Port (Agent always installs on nonsecure port)	4889
Password for Agent user (typically the *oracle* user)	N/A
Password for *root* (to run *orainstRoot.sh* and *root.sh*)	N/A
Agent Registration password	Set AGENT_INSTALL_PASSWORD environment variable if running installation locally
Inventory Pointer File Location file (*oraInst.loc*)	Leave in default location for your platform (see "Create Oracle Inventory Group (oinstall)" in Chapter 2)

TABLE 5-4. *Required Agent Installation Information*

Summary

There is much to consider when preparing for installation of standalone Management Agents on destination hosts. I logically divided the steps between Chapter 2 and this chapter. Chapter 2 contains common prerequisites for all GC components, including standalone Agents, while this chapter provided additional preinstallation requirements exclusive to standalone Agents. Accordingly, I began this chapter by referring you back to specific tasks in Chapter 2 pertaining to standalone Agents. If you hadn't already completed these tasks (which is a likely possibility for the reasons stated in that chapter), I instructed you to double back. This chapter then guided you through some key Agent installation decisions and detailed the additional hardware and software requirements specific to standalone Agents.

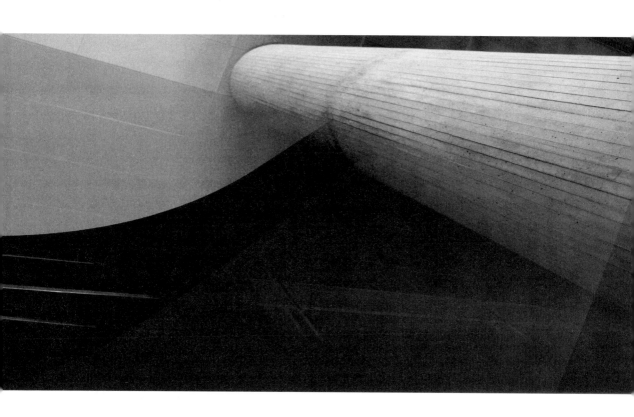

CHAPTER 6

Install Management Agents via Agent Deploy

he last chapter supplied preinstallation steps for installing standalone Management Agents on managed hosts using any method. In the next two chapters you will learn how to install Agents using each method. To review, the six methods are as follows:

- Agent Deploy
- *nfsagentinstall*
- *agentDownload*
- Agent Cloning
- Silent Installation
- Interactive Installation

This is the overall ranking recommended in the last chapter for mass or automated deployment of standalone Agents. Predictably, this order is roughly from most to least complex. As with most software technology, the more automated and capable the program, the more complicated it is to implement. This chapter begins with Agent Deploy, the most difficult installation method, and in the next chapter we proceed south in difficulty to cover the remaining five installation methods. This chapter becomes a little treacherous at times, but it's well worth the trouble to mass-deploy Agents if your site contains many managed host targets. If Agent Deploy is your only installation method, there's no need to tap the other Agent installation methods covered in Chapter 7, and you could even skip that chapter if you like.

NOTE
Going forward, the Agent owner is assumed to be the oracle *user. If you decided to create a separate Agent user in Chapter 5 (see the section "Use an Existing User or Create a Separate Agent User"), perform all steps as that user instead. Steps requiring* root *access are indicated as such.*

Agent Deploy Installation Overview

Agent Deploy is by far the most labor intensive of all the Agent installation methods. You should therefore aim to minimize the overall number of *iterations* for the set number of Agents you are deploying. By "iterations" I mean the number of times you must run through all Agent Deploy installation steps. Two main factors influence the overall number of Agent Deploy iterations required:

- The number of OMS hosts where performing the steps
- The number of Agent batches deployed

First, let's look at the number of OMS hosts. Most of the installation steps are performed on the OMS host side. If you install multiple OMS hosts, you need to decide which OMS host(s) to configure for Agent Deploy. The answer is to choose at least one OMS host, or if you desire high availability for the Agent Deploy application, to configure at least two OMS hosts. As Chapter 5 points out, the Agent Deploy application does not work through a server load balancer (SLB) if running Grid Control 10.2.0.3 or earlier. Therefore, if you're on GC 10.2.0.3 or earlier and are using an SLB to virtualize the connection to multiple OMSs, you must access the Agent Deploy application by logging in to the Console using the URL that directly accesses one of the OMS hosts configured for Agent Deploy, rather than using the URL for the virtual Management Service. This simply means logging in to the Console using http://<OMShost1>:7777/em rather than http://<SLBhost>:<SLBport>/em, for example. As of GC 10.2.0.4, you can run Agent Deploy via SLB login after specifying the SLB host and port in an Agent Deploy configuration file (called *$OMS_HOME/sysman/prov/agentpush/agentpush.properties*) on each OMS and after performing OMS configuration steps for all OMSs.

The number of Agent batches deployed is the second factor that affects the overall number of Agent Deploy iterations. As mentioned in Chapter 5, Agent Deploy pushes Agents in one operation to multiple hosts of the same operating system. Thus, you are limited to installing batches of Agents, *at least one batch* per OS. Strive to make it *just one batch* per OS. In other words, deploy as many Agents as possible per batch. Why? Because each batch has common tasks required before running the Agent Deploy API in the Console (called *pre-API* in this chapter) that do not increase with the number of Agents installed. One such task is during SSH user equivalence setup when you run the *sshConnectivity.sh* script and specify a list of target hosts as one of its arguments:

./sshConnectivity.sh -hosts "<space separated host list>" -user oracle

If you had 20 Linux x86 Agents and deployed them all at once, you'd only have to execute the above command (and other pre-API steps) once. However, if you deployed them in two batches of 10 Agents per batch, you'd have to run the command (and other steps) twice.

Now that you understand the macroeconomics of Agent Deploy—which OMS host or hosts to configure it for, and how to reduce the number of installation "iterations" required—let's get into the microeconomics, e.g., the installation process itself. This process is so lengthy that it requires its own "table of contents" to keep track of it. Table 6-1 provides this TOC of required steps, and shows on which hosts you need perform them.

Agent Deploy Installation Step	Tasks Involve OMS Host, Agent Host, or Both?
Install Required Packages	AGENT
Configure SSH User Equivalence	AGENT, OMS (see subheadings)
Set Up Time Zone for SSH Server	AGENT, OMS
Validate All Command Locations	OMS
Modify Agent Deploy Properties File for SLB	OMS
Choose Inventory Location	AGENT
Verify Agent User Permissions	AGENT
Prepare for Cross-Platform Agent Push	OMS
Include Additional Files (Optional)	OMS
Run Agent Deploy Application	OMS (through the Console)
Agent Deploy Post-Installation	AGENT

TABLE 6-1. *Agent Deploy Installation Steps*

In this table, the shorthand I use signifies the following:

- **AGENT**   Perform this step on all remote Agent hosts.
- **OMS**   Perform this step on the local OMS host (to simplify discussions here, I assume you are configuring only one OMS host), or
- **AGENT, OMS**   Perform some of the substeps on remote Agent hosts and other substeps on the local OMS host.

These shorthand tags for the hosts on which each step must be executed are also listed under each subheading below. As you may notice, the actual running of the Agent Deploy application is the second to last step, and is preceded by a long list of preparation tasks. You must be very methodical when using Agent Deploy. Building it is like assembling a racecar engine—you must bring out your A-game mechanical skills to build it properly. If you do, the result is a high performance mass-deployment of Agents, and the ability to upgrade them later by simply running the Agent Deploy application (i.e., because you've already completed the pre-API steps, which are a one-time setup). However, if you don't assemble this Agent Deploy engine carefully, it will likely crash spectacularly on the track.

NOTE
Before commencing with these Agent Deploy tasks, remember to complete all preinstallation steps covered in Chapter 5.

Install Required Packages
AGENT
Appendix B (online at www.oraclepressbooks.com) lists the packages required by the Agent Deploy application (and other Agent installation methods) for certain target Agent platforms and their specific OS versions, as well as the commands to check for the existence of these packages.[1] Install any missing packages now. See the platform-specific documentation for instructions, or ask your system administrator, if you have that luxury.

Configure SSH User Equivalence
AGENT, OMS
Before installing the Agent using Agent Deploy, you need to set up SSH (Secure Shell) user equivalence from the OMS to the target nodes. SSH user equivalence allows you to connect via SSH from the OMS host to Agent hosts without specifying a password. The Agent Deploy application uses SSH and SCP to connect to the target hosts and copy the installation files, respectively. SSH user equivalence need only be set up in one direction: *from* the local OMS *to* the remote target nodes (not between target hosts or from target hosts to the OMS). Without this equivalency, the Agent Deploy application would be halted by password authentication requests.

---

[1] Just in case new platforms are certified with the Agent after this book is published, let me specify which platforms (and OS versions) do *not* require any additional packages (and therefore are *not* listed in Table 6-2): HP Tru64, Linux Itanium (SLES 10), Linux x64 (all OSs), z/Linux (all OSs), Linux on Power (all OSs), HP-UX PA-RISC (11.31), HP-UX Itanium (all OSs), Solaris SPARC (Solaris 10 Local container), Microsoft Windows x86 (all OSs *except* Win NT 4.0), Microsoft Windows x64 (all OSs), and Windows Itanium (all OSs).

Following are the steps to configure SSH user equivalence from the OMS to target hosts as a prerequisite for installing standalone Agents with Agent Deploy:

- Set up SSH Server (SSHD) on Windows
- Back up current SSH configuration
- Set ORACLE_HOME on OMS host
- Run *sshConnectivity.sh* script
- Verify SSH user equivalence is configured

Set Up SSH Server (SSHD) on Windows
AGENT, OMS
While SSH is usually installed by default on UNIX and Linux hosts, Windows systems do not usually provide SSH access.[2] However, the Cygwin Suite does. Cygwin Suite consists of two parts: a collection of tools that provide a Linux look and feel for Windows and a DLL (*cygwin1.dll*). This DLL acts as a Linux API emulation layer, providing substantial Linux API functionality, including SSH Server. Whether you want to use Agent Deploy with a Windows OMS host or deploy Agents to Windows target hosts, you must install the Cygwin Suite on all Windows hosts involved in the Agent Deploy operation.

Appendix D (online) contains instructions for installing the Cygwin Suite, whether on a Windows OMS host or Windows target host where you're deploying the Agent. (See the section "How to Set Up SSH Server (SSHD) on Windows".) Perform all steps at this time, either as the Grid Control software owner (if on the OMS host) or as the Agent user (if on a target host).

Back Up Current SSH Configuration
AGENT
First, back up any currently configured SSH user equivalence on each target host, because the *sshConnectivity.sh* script run from the OMS host to each target host (see below) might break any current user equivalence. Oracle RAC nodes will already be configured with SSH user equivalence between nodes—this is actually a RAC installation requirement. Backing up the current SSH configuration ensures that, if you do break the existing SSH equivalence, you can restore it by appending the contents from each original SSH configuration file to the newly created replacement SSH configuration file. Backing up the SSH user configuration on each node is as simple as backing up the *.ssh* directory, typically installed under the *oracle* user *$HOME* directory. Execute the following on target Agent hosts. On Windows, first start a Bash shell and substitute the hard coded path in double quotes for *$HOME*:

```
cd $HOME
cp -rp .ssh .ssh.bk
```

Set ORACLE HOME on OMS Host
OMS
The next step in setting up SSH user equivalence is to log in to the OMS and point the ORACLE_HOME environment variable to the OMS home. This is a requirement of the *sshConnectivity.sh*

---

[2] For those who may be wondering, Active Directory equivalence does not handle pushing out Agents via Agent Deploy from a Windows OMS server to Windows target hosts. You must set up SSHD on Windows as described here.

script to tell it where the OMS is located. The procedure to set ORACLE_HOME to the OMS home is slightly different on UNIX than it is on Windows:

- On UNIX, execute the following on the local OMS host:

  ```
  export ORACLE_HOME=$OMS_HOME
  ```

- On Windows, the syntax for the ORACLE_HOME value depends upon the Cygwin version you are using. Start a Bash shell and determine the version using the *uname* command as follows:

  ```
  bash-3.2$ uname
  CYGWIN_NT-5.2
  ```

  - If you are running Cygwin version 5.2, due to a bug (Bug 5233554), you must specify a Cygwin style (UNIX) path enclosed in double quotes:

    ```
    bash-3.2$ export ORACLE_HOME="C:/app/oracle/product/10.2.0/em10g/oms10g"
    ```

  - If you are running any version of Cygwin other than 5.2, use a Windows style path enclosed in double quotes, such as:

    ```
    bash-3.2$ export ORACLE_HOME="C:\app\oracle\product\10.2.0\em10g\oms10g"
    ```

Now you can execute the *sshConnectivity.sh* script as done in the next section. Change to the directory where the script is located:

```
cd $ORACLE_HOME/sysman/prov/resources/scripts
```

If you are on a Windows 10.2.0.3 OMS host, you must first convert this script from UNIX to Windows format (see Bug 5613495).[3] Specifically, the script converts from the UNIX new line character /n to the Windows new line character /r/t. Back up the script, then convert it to Windows format as follows:

```
bash-3.2$ C:\cygwin\bin\dos2unix.exe sshConnectivity.sh
```

Run sshConnectivity.sh Script

AGENT, OMS

Grid Control provides a script to set up user SSH equivalence from the OMS to target hosts. Before GC 10.2.0.2 and later versions were ported to all GC-certified platforms, the script to establish SSH user equivalence depended on the GC release and OMS platform used, as shown here:

Grid Control Release	OMS Platform	Script Name
10.2.0.2 or later	UNIX, Windows	sshConnectivity.sh
10.2.0.1	Windows	sshUserSetupNT.sh
10.2.0.1	UNIX	sshUserSetup.sh

---

[3] If you don't convert the *sshConnectivity.sh* script to Windows format, you will get the following errors:
: command not foundh: line 24:
: command not foundh: line 25:
./sshConnectivity.sh: line 63: syntax error near unexpected token `elif'
'/sshConnectivity.sh: line 63:` elif [$j = "-obPasswordfile"]

All scripts are on the OMS host under *$OMS_HOME/sysman/prov/resources/scripts/*. However, now that GC is available on all certified platforms for version 10.2.0.2 or later, I only discuss the *sshConnectivity.sh* script here.

This script confirms network connectivity between the OMS host and each of the remote hosts, and then asks multiple times for the user's password on the remote hosts. To test that SSH equivalency is successfully created, the script finishes by running the *date* command remotely on the target hosts via SSH. If the output is the date, then the configuration succeeded. However, if you are prompted for a password once more, then SSH connectivity was not successfully established.

Here is the syntax showing all options for the *sshConnectivity.sh* script, including those for both UNIX and Windows:

./sshConnectivity.sh
-hosts "<space separate host list>" | -hostfile <path of cluster configuration file>
-user <remote user> [-shared] [-verify] [-confirm] [-remotePlatform <platform id>]
[-localPlatformGrp <unix | win>] [-advanced] [-exverify] [-asUser <local user>] [-asUserGrp <group>]
[-sshLocalDir <local dir>] [-logfile <logfilename>] [-homeDir <user home dir>] [help]

Each option is explained in Table 6-2.

Option	Description	Example Value for Option
-hosts	A space separated remote hosts list. Use the exact same host names here (preferably FQDN) as in the Agent Deploy application, or installation will fail (Bug 4679997).	-hosts host1.domainname host2.domainname host3.domainname
-hostfile	You can specify the host names either through the *-hosts* option or by specifying the absolute path of a cluster configuration file (a text file optionally created for a Clusterware installation). The first column in each row of this file is used as the host name.	-hostfile /home/oracle/Agenthosts where this file contains: host1 10.1.0.0 - host2 10.1.0.1 - host3 10.1.0.2 -
-user	The user on remote hosts.	-user oracle
-shared	Takes no argument. Used when installing an NFS Agent across remote hosts that share a master NFS Agent home directory.	-shared
-verify	Takes no argument. Checks if SSH user equivalence is set up by running the *date* command on each remote host. (No SSH configuration is performed; it is only verified.) If you are prompted for a password, this means that SSH user equivalence is not set up correctly.	-verify

TABLE 6-2. *Options for the sshConnectivity.sh Script (Continued)*

Option	Description	Example Value for Option
-confirm	Takes no argument. Allows you to establish SSH user equivalence without being prompted to remove *write* permissions for *group* and *world* for the user home directory and the ~/.ssh directory on the remote hosts. This option is misleadingly named, as it does *not* confirm the permissions changes.	-confirm
-remotePlatform	You must specify this option if the remote platform is not the same as the local platform. Specify the platform ID of the remote platform; see *platforminfo.properties*.[4]	-remotePlatform 46
-localPlatformGrp	*unix* or *win*. The default for this option is *unix*. Specify this option if the local OMS platform is Windows and use the argument *win*.	-localPlatformGrp win
-advanced	Takes no argument. Sets up SSH user equivalence between all remote hosts. Use to preserve equivalence between RAC nodes.	-advanced
-exverify	Takes no argument. Does exhaustive verification for all hosts, including SSH equivalence from local to all remote hosts (standard), from each remote host to itself, and to all other remote hosts. Use in conjunction with *-advanced* option.	-exverify
-asUser	User, such as SYSTEM, for which SSH equivalence needs to be set on the local machine.	-asUser SYSTEM
-asUserGrp	Group to which the specified asUser belongs.	-asUserGrp root
-sshLocalDir	Windows directory path where keys should be generated on the local machine for asUser.	-sshLocalDir "C:\cygwin\.ssh"
-homeDir	Windows-style full path of the home directory of the current user	"-homeDir "C:\Documents and Settings\oracle"
-logfile	The absolute path of a logfile to create. Will contain additional details not displayed to stdout. Will append current date and time to logfile name in the form logfilename2007-03-09_10-33-35-PM.	-logfile "./log_ssh"
help[5]	No value required Takes no argument. Lists Usage and examples.	help

TABLE 6-2. *Options for the sshConnectivity.sh Script*

---

[4] The *platforminfo.properties* file is located under *$OMS_HOME/sysman/prov/resources/*. The platform IDs for each Agent platform are **AIX=610**, AIX 5L (64-BIT)=212, **HP-UIX=2**, HP-UX Itanium=197, HP-UX PA-RISC=289, HP Tru64=87, **IBM Power Based Linux=227**, Linux x64 (AMD64/EM64T)=226, Linux x86=46, Linux Itanium=214, Linux on POWER=249, **Solaris SPARC=453**, Solaris SPARC (64-BIT)=23, Solaris x86=173, Solaris x86_64=267, Windows Itanium (64-BIT)=208, Windows x64=233, Windows x86=912, z/Linux=211.

[5] Do not use a dash for this option.

Note the following about running the script *sshConnectivity.sh*:

- Answer *no* when you are asked if you want to specify a passphrase for the private key the script will create for the local host. Otherwise, you may be asked to supply the passphrase indefinitely; this is contrary to the message you would receive were you to enter a passphrase: "The estimated number of times the user would be prompted for a passphrase is 8. In addition, if the private-public files are also newly created, the user would have to specify the passphrase on one additional occasion."

- As required for SSH to work, the script turns off X11 Forwarding at the Agent user level by appending the following lines to the end of *~/.ssh/config* on each remote host:

    ```
    Host *
    ForwardX11 no
    ```

    If this *config* file already exists, the *sshConnectivity.sh* script first backs it up to *$HOME/.ssh/config.backup* before changing it. Even if X11 Forwarding is turned off at the system level, it is overridden by the Agent user-level setting, so regardless of the system-level setting, you don't need to make any changes to the *config* file once *sshConnectivity.sh* runs.[6]

- On Linux RAC 10.2 clusters, the Agent Deploy installation may fail due to a lost SSH connection. To prevent this, confirm that LoginGraceTime in /etc/ssh/sshd_config is not set to zero (0). The zero value gives an indefinite time for SSH authentication. Use the default value of 120 seconds or more to force the SSH server to disconnect after this time if you have not successfully logged in. You must bounce the SSH daemon to effect this change.

- For some platforms and EM versions, I've observed that the *sshConnectivity.sh* script does a *cat* of the *known_hosts* file to itself, duplicating all entries, and thus doubling its size every time the script runs. While this does not break any preexisting SSH equivalence, optionally you can restore the original *known_hosts* file after running the script.

Example Uses of sshConnectivity.sh
OMS
Let's take a bunch of examples to illustrate how to use these options to establish SSH user equivalence for various combinations of OMS and Agent platforms.

- To set up SSH equivalence if both the local OMS platform and the remote nodes are UNIX hosts of the same operating system and all users are local (nonshared):

    ./sshConnectivity.sh -hosts "*<space separated host list>*" -user oracle

- To establish SSH equivalence between remote NFS-shared Agent homes if both the local OMS platform and the remote nodes are UNIX hosts of the same operating system:

    ./sshConnectivity.sh -hosts "*<space separated host list>*" -user oracle -shared

---

[6] As an example, on Linux x86 platforms, X11 Forwarding is enabled by default at the system level via the line "X11Forwarding yes" in the */etc/ssh/sshd_config* file. You don't need to disable X11 Forwarding at the system level, but you can do so by commenting out this line, which requires bouncing the SSH daemon.

- To set up SSH equivalence if the local OMS platform is Linux and the remote nodes are Solaris hosts:

    ./sshConnectivity.sh -hosts "<space separated host list>" -user oracle -remotePlatform 453

- To configure SSH equivalence if the local OMS platform is Windows, and the remote platform is Solaris SPARC (you must execute the following within the Cygwin Bash shell or the command will fail):

    bash-3.2$ ./sshConnectivity.sh -hosts "<space separate host list>"
    -user oracle -remotePlatform 453 -asUser SYSTEM -asUserGrp root
    -sshLocalDir "C:\cygwin\.ssh" -localPlatformGrp win

To get an idea of the actual output you should expect from executing the *sshConnectivity.sh* script, see the section in the online Appendix D entitled "Output From Setting Up SSH User Equivalence."

Verify SSH User Equivalence Is Configured

Finally, verify that you successfully configured SSH user equivalence by running the *sshConnectivity.sh* script with the same options used to configure SSH, but add the *-verify* option. As an example, for local and remote UNIX hosts of the same platform where all users are nonshared, run the following command on the OMS host against all remote hosts:

./sshConnectivity.sh -hosts "<space separated host list>" -user oracle -verify

For sample output of running the *sshConnectivity.sh* script with the *-verify* option, see the section in the online Appendix D entitled "Output From Verifying Setup of SSH User Equivalence."

Set Up Time Zone for SSH Server
AGENT, OMS

Agent Deploy requires that the time zone be set correctly for SSH connections from the OMS to remote hosts. The time zone may already be set for non-SSH connections (such as telnet and ftp), but it is probably not set for SSH access by the Agent Deploy application. You can set the time zone for SSH connection to remote hosts in one of two ways:

- Use the *-z* option in the Additional Parameters field on the Agent Deploy application to pass the correct time zone (see the section "Run Agent Deploy Application" later in the chapter for directions).

- Set the TZ environment variable for the SSH daemon on each Agent host.[7] First test whether TZ is already set by running the following command from the OMS to each remote host:

    ssh -l oracle -n <Agent_host> 'echo $TZ'

---

[7] If the TZ environment variable is set for the SSH daemon on the remote host, this TZ value takes precedence over the value specified by the *-z* option

If TZ is not displayed, set TZ as follows in the *rc* file of the shell that the Agent user employs on the remote hosts. For the Bash shell, add the following line to the *.bashrc* file in the home directory of the Agent user on each remote host:

export TZ="*<timezone>*"

For the CSH shell, add the following line to the *.cshrc* file in the home directory of the Agent user on each remote host:

setenv TZ *<timezone>*

TIP
On Red Hat using the Bash shell, I have observed that setting TZ in .bashrc does not set TZ for the SSH server. Instead, you must set TZ in /etc/profile.d/bash.sh, which requires root access.

Now bounce the SSH daemon on all remote hosts to effect this setting of the TZ environment variable. You can do this in an SSH session, or in an Xwindows session to protect against possible lockout if the SSH daemon doesn't restart:

```
/etc/init.d/sshd restart
```

Note that you cannot use the *service sshd restart* command because it doesn't set the TZ. Retest to see if the SSH server on each Agent host can now access the TZ variable by executing the following from the OMS:

```
ssh -l oracle -n <Agent_host> 'echo $TZ'
```

Validate All Command Locations
OMS

Agent Deploy uses certain commands when executing its application programming interfaces (APIs), such as the *zip* executable on the remote hosts and *ping* on the local OMS host. Properties files, located under *$OMS_HOME/sysman/prov/resources/* and described in Table 6-3, provide the default locations for these commands, which vary by platform. Additional properties files related to tuning and ignoring messages are also listed.

These properties files are listed in order of ascending precedence. In other words, a value specified for a property listed in *userPaths.properties* overrides a value for the same property in previous files.

You must validate all Agent Deploy command locations specified in these properties files by executing the script *validatePaths* as shown below. You must first set JAVAHOME, which is not mentioned in EM documentation but is required nevertheless:

```
cd $OMS_HOME/sysman/prov/resources/scripts
export JAVAHOME=$OMS_HOME/jdk
./validatePaths
```

186   Oracle Enterprise Manager 10g Grid Control Implementation Guide

Properties File Name	Description	Example Entry
platformInfo.properties	Contains a list/mapping of platform IDs and files to be loaded for each platform. The files in turn specify the paths for the commands.	#unix -1=Paths.properties, sPaths.properties, userPaths.properties
Paths.properties	Contains command arguments Agent Deploy must pass every time the commands listed here are executed.	ZIP_ARGS=-rv
sPaths.properties	Contains the paths for all commands to be executed, regardless of platform.	SSH_PATH=/usr/bin/ssh
ssPaths_<platform>.properties	A platform-specific file that contains commands to be executed for that platform. For Windows, ssPaths_msplats.properties contains paths to the Cygwin binaries.	SSH_PATH=/usr/local/bin/ssh

userPaths.properties	Lists all variables used to specify command paths. You must uncomment variables you want to use, and specify appropriate values.	#PING_PATH=
		ZIP_PATH=/home/oracle/bin/zip
		#TAR_PATH=

system.properties	Contains properties to tune the Agent Deploy application. Notable properties are:	
	Number of parallel threads created to execute commands.	oracle.sysman.prov.threadpoolsize=8
	The maximum number of threads, which can increase dynamically depending on workload.	oracle.sysman.prov.threadpoolmaxsize=32
ignoreMessages.txt	If any command returns a banner, security alert, or non-error, inform SSH to ignore the output by appending it to this file.	Warning: No xauth data; using fake authentication data for X11 forwarding.

TABLE 6-3.   *Descriptions of Properties Files for Agent Deploy*

You may get an error such as:

```
The following executables were not found at their default locations. Please
provide the locations in userPaths.properties.

SUDO_PATH=/usr/local/bin/sudo
```

Change the appropriate properties file to resolve the error and rerun *validatePaths* to confirm it completes without error.[8]

Modify Agent Deploy Properties File for SLB
OMS
As mentioned in Chapter 5, all Agent installation methods except Agent Deploy 10.2.0.3 or earlier allow you to go through a Server Load Balancer (SLB), provided you modify the Agent response file *$OMS_HOME/sysman/agent_download/<version>/agent_download.rsp* to account for this SLB virtualization (see "Change Response File if Using an SLB").

Prior to Grid Control 10.2.0.4, if you log in to the Console via an SLB, the Agent Deploy application in the UI fails. In order to use Agent Deploy for GC 10.2.0.3 or earlier, you need to log in to the Console directly via one of the OMSs (i.e., using a URL containing the OMS host name) rather than accessing the Console via the SLB URL. Beginning with Grid Control 10.2.0.4, Oracle addresses this limitation. You can access the Console via the SLB URL, then run the Agent Deploy application if you specify SLB information in the Agent Deploy properties file, called *$OMS_HOME/sysman/prov/agentpush/agentpush.properties*, on every OMS host (and if you configure all OMS hosts for Agent Deploy). Uncomment the following two properties in this file and specify the SLB host name and SLB port, respectively:

oracle.sysman.prov.agentpush.slb.host=<*SLBhost*>

oracle.sysman.prov.agentpush.slb.port=<*SLBport*>

It is not necessary to bounce the OMS service after making these changes, as Agent Deploy parses this properties file in real time.

Choose Inventory Location
AGENT
You should have already decided back in Chapter 2 (in the section "Choose the Oracle Inventory Directory") where to create the inventory for Agent installations: either in an existing Central Inventory or in an alternative (i.e., separate) location. But how do you accomplish this with Agent Deploy? There are three scenarios for Agent Deploy: use an existing Central Inventory, specify a different (separate) inventory location, or don't specify an inventory at all:

- To use an existing Central Inventory on remote hosts for the Agent Deploy installation, ensure that the user specified in the OS Credentials field of the Agent Deploy application has read, write, and execute permissions to the Central Inventory directory. Agent Deploy will use this Central Inventory by default. (Remember that the *inventory_loc* variable in the *oraInst.loc* file defines the Central inventory path.) Back up the existing Central Inventory before running Agent Deploy so that you can restore it if the Agent installation fails.

---
[8] I have observed that sometimes you cannot resolve an error according to these instructions. If the error relates to a non essential command, such as *sudo* in this example, you can carry on and use Agent Deploy by not using the functionality associated with this command. In the case of *sudo*, you can avoid an error by not selecting *Run root.sh* under the "OS Credential section. Alternatively, you can try resolving the error outside the framework of the *validatePaths* script. In this example, you can create a symbolic link from where the *sudo* command is actually located on Linux, */usr/bin/sudo*, to where *validatePaths* expects to find *sudo*, under */usr/local/bin/sudo*. The command would be *ln -s /usr/bin/sudo /usr/local/bin/sudo*. Don't expect any miracles, however, when working outside the box. This workaround still doesn't allow *root.sh* to run.

- To use a different inventory location, either because you cannot grant the Agent user permissions on the existing inventory, or because you want to use a separate inventory for the Agent installation (for reasons discussed in Chapter 2), you can specify the path to an alternative inventory location by using the *-i <inventory_location>* option in the Additional Parameters section of the Agent Deploy application (see "Run Agent Deploy Application" later in the chapter).

- If no Oracle installation exists on remote hosts, no inventory pointer file (or inventory) yet exists. In this case, if you don't use the *-i* option in the Additional Parameters "section" (the subheading on the Agent Deploy application page), Agent Deploy will create a new inventory and inventory pointer file in the Agent user's home directory. This is not desirable because it's non-standard; it's best to specify the standard pointer file location, such as */etc/oraInst.loc*.

Verify Agent User Permissions

AGENT

In order for the Agent Deploy installation to work and, once installed, to manage certain targets, the Agent user must have certain permissions, depending upon targets you want to monitor and Agent Deploy options you want to use:

- Confirm that the Agent user on all remote hosts has write permissions to the following directories, depending upon the type of Agent installation:

    - If installing a new standalone Agent, the Agent user (through Agent Deploy) must be able to create a new Agent installation base directory (or write to this directory if it already exists). This means that the Agent user must have write permissions on the lowest created subdirectory of the Agent installation base directory. For instance, if you are installing an Agent under the installation base directory */u01/app/oracle/product/10.2.0/em10g/*, and only */u01/app/oracle/* has been created, then the Agent must have write permissions on */u01/app/oracle/*.

NOTE
One Agent Deploy prerequisite check is that the Agent home directory itself must be empty. The Agent home directory is the Agent installation base directory appended by agent10g. *In the above example, the Agent home directory would be* /u01/app/oracle/product/10.2.0/em10g/agent10g/.

- If upgrading an existing Agent, the Agent user must be able to write to an existing Agent base directory.

- If installing a new NFS Agent, the Agent user must be able to create a new installation base directory, a new EMSTATE directory, and must be able to read from the NFS location.

- If you wish to manage an installation of Oracle Application Server (OAS) or Oracle Collaboration Suite, verify that the group used to install these products is the same as the Agent user's group.

- Confirm that the Agent user has *sudo* privileges to execute the *root.sh* script on remote UNIX hosts. Automatically running *root.sh* at the end of the Agent installation is one of the Agent Deploy options; it must be run one way or another after the installation completes in order to discover all the targets. (You can choose not to select this option, and instead manually execute *root.sh* locally on each target host at the end of the installation, but this would be cumbersome if you are installing many Agents.) You can test remote *root.sh* execution by running the following command from the OMS node:

    ```
    ssh -l oracle -n <Agent host> 'sudo -u root whoami;hostname'
    ```

    This should return:

    ```
    root
    <Agent_host>
    ```

Prepare for Cross-Platform Agent Push
OMS
As explained in Chapter 5 under "Prepare for Cross-Platform Agent Installation," a *cross-platform* Agent installation is where the OMS host and target hosts run on different operating systems. A *cross-platform Agent push* is the same thing, but it is specifically for the Agent Deploy installation method. In such a case, you cannot select other OSs in the drop-down menu for the Platform field in the Agent Deploy application (this field is covered below in "Run Agent Deploy Application"). To accomplish a cross-platform Agent push using the Agent Deploy application for Enterprise Manager 10.2.0.2 or later, apply patch 5455350 to the OMS. After you apply the patch, all platforms you stage in the Agent download directory will appear in the drop-down menu of the Agent Deploy UI. Remember that you can only install Agents on one platform at a time (i.e., in one Agent Deploy "batch" or operation at a time).

In addition to applying this patch, for GC 10.2.0.1 on UNIX platforms, set the AGENT_INSTALL_PASSWORD environment variable in the Agent user's login script (Bug 5124872).

Include Additional Files (Optional)
OMS
If you want Agent Deploy to copy additional files (such as scripts) to the Agent hosts, perform the following steps on the OMS host(s) that will render the Console from which you will run Agent Deploy:

1. Copy all files into the *$OMS_HOME/sysman/agent_download/<version>/<platform>/addons* directory that you want to deploy along with the Agent.
2. Create a file called *set_perms.sh* in this directory that sets permissions on all files to be copied to the Agent hosts.
3. Execute the following command on the OMS host(s) to extract all files into the newly installed AGENT_HOMEs after the deployment process and to run *set_perms.sh* to set the desired file permissions:

    $OMS_HOME/jdk/bin/jar cvfM \

    $OMS_HOMEsysman/agent_download/<version>/<platform>/agent_scripts.jar *

Run Agent Deploy Application

OMS (through the Console)

Finally, after much painstaking preparation, you are ready to strap yourself in and fire up the Agent Deploy application "engine." Agent Deploy is a J2EE web-based application accessible through the GC Console. The application checks SSH connectivity, performs prerequisite checks on the remote hosts, and displays failures/warnings and how to correct them. The user can ignore nonfatal warnings and proceed with the installation.

To run the Agent Deploy application, navigate to the Deployments tab and select the Install Agent link under Agent Installation. This will load the application and present the following three types of Agent installations:

- Fresh Install
- Add Host to Shared Agent
- Upgrade Agent

Here is a screenshot of this choice of installation types, with a brief description that the application provides for each type.

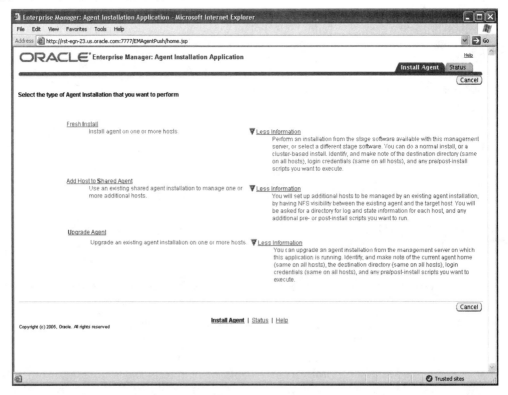

Once you choose an installation type, an application screen loads requesting your input. Each type has a different application screen, but these screens share common fields that I've distilled below. I list both the section names and field(s) located under each section. This does away with the need to present this same material for each installation type. Following this coverage of the common fields, I cover the unique sections of each installation type. Many of the common fields are self-explanatory. But they contain enough intricacies and quirks to warrant the descriptions provided. Each field name is followed by the word "Mandatory" or "Optional" in parentheses to indicate whether or not the field is required. (The application screen itself designates mandatory fields with an asterisk next to the field name.) Without knowledge of this content, you could expect some trial and error to finesse your way successfully through an Agent Deploy installation. Like most things, it's not difficult, *once* you know how to do it.

- **Hosts**
    - **Platform**   (Mandatory) The platform of the Agents to be installed. Each operation or "batch" can only deploy to one platform. By default the drop-down menu will consist solely of the OMS platform installed.[9]
    - **Provide Host List**   (Mandatory) List the host names or IP addresses where Agents are to be installed, comma-delimited, and no spaces allowed. Specify the same host name format (fully qualified or not) used to execute *sshConnectivity.sh*.[10]
- **OS Credential**
    - **Username, Password**   (Mandatory) These credentials—both username and password—must be the same for all Agent hosts specified and must match those used when running *sshConnectivity.sh*.
    - **Run root.sh**   (Optional) If the user has *sudo* privileges on the remote hosts, then check this box to run *root.sh* at the end of the installation. Otherwise, you need to run *root.sh* manually on each target host (which obviously reduces the "mass" in mass-deployment).
- **Management Server Security**
    - **Management Server Registration Password**   (Mandatory) Provide the Agent Registration password to configure Agents to run in secure (HTTPS) mode.
- **Additional Scripts**
    - **Pre-Installation Script File**   (Optional) You can run a custom script, optionally as the *root* user, before installing Agents. Designate a local directory path on Agent hosts where the script is located (i.e., the script must exist locally on all Agent hosts). An example use is to script any required Agent preinstallation steps (see Chapter 5).
    - **Post-Installation Script File**   (Optional) You can also run a custom script, optionally as the *root* user, after installing Agents. Designate a local directory path on Agent hosts where the script is located. An example use is to script any required Agent post-installation steps (see Chapter 10, "Agent Configuration").

This concludes coverage of the common fields found in the application screens for all installation types. Now we review the unique sections and/or fields found in each installation type, beginning with a Fresh Install. Again, all fields are notated as either mandatory or optional.

---

[9] To add additional platforms to the pull-down menu from which to select, see the section above "Prepare for Cross-Platform Agent Push."

[10] The button Get Host Names From File brings up the local hosts file on your Console workstation, not the OMS hosts file, as you might expect.

Fresh Install

Select the Fresh Install type to deploy either a local or Cluster Agent, or to install a new Cluster Agent on a recently added cluster node. If you're extending a cluster, you must specify all nodes of the existing cluster as well as the new cluster node name(s).

When you choose the Fresh Install type, the New Agent Installation: Installation Details screen loads, presenting the following additional sections:

Source Software

Source Shiphome Directory
- ● Default, from Management Server location.
  The staged software should be available at <OMS_HOME>/sysman/agent_download/<version>/<platform>
- ○ Another Location
  Shiphome Location [_____]
  This is the full path of the products.xml of the source software. Example: /net/host1/shiphomes/linux/agent/stage/products.xml

Version
Select the agent source software version.

Version [10.2.0.3.0 ▼]
This list is determined by the staged software available on Management Server.

Hosts
The list of hosts, specified here, should be separated by comma, spaces or newline. You can use either the Name, or IP Address to specify Hosts.

Platform [Windows NT, Windows 2000 ▼]
This list is determined by the staged software available on Management Server.

* Provide Host List [_____]

Example: host1,host2,168.78.5.118, ...

[Get Host Names From File]

☐ Cluster Install
All specified hosts must be in the same cluster. Regardless of what the cluster node list looks like, the application will install agent only on hosts specified.

Following is an explanation of each section and recommended values for each field:

- **Source Software, Source Shiphome Directory** (Mandatory) You have two choices here: "Default, from Management Server location" or "Another Location." I highly recommend using the default location, which is the Agent download directory of the OMS, *$OMS_HOME/sysman/agent_download/<version>/<platform>*. There are bugs associated with using an alternate location,[11] and your OMS home should have plenty of disk space to stage Agent software for any additional platforms needed, provided you complied with the OMS disk space requirements listed in Chapter 2. Remember that you can only deploy Agents for operating systems whose installation files are placed in the Agent download directory. For details, see Chapter 5, "Prepare for Cross-Platform Agent Installation."

- **Version** (Mandatory) Multiple versions only appear if you first perform the steps listed in Chapter 5 under the heading "Prepare for Cross-Platform Agent Installation." Otherwise, only the current GC version is selectable. If you are deploying cross-platform, you will be able to select from all combinations of versions and platforms contained in the Mass Agent

---

[11] For Agent 10.2.0.1, the Alternate Location simply doesn't work (Bug 4576575). For Agent 10.2.0.2, several bugs exist (Bug 5178469 and Bug 5183333) with workarounds that you can avoid simply by using the default location. Agent Deploy is complicated enough already—why make it more difficult?

Deployment software you downloaded and staged on the OMS. Select a version, and the Platform field pull-down menu (under Hosts) will offer all platforms ported for that version.

- **Hosts** (Mandatory) The following three additional item appear in the Fresh Install application screen:

  - **Cluster Install** (Mandatory for cluster Agent installation) Check this box to install a RAC Agent, or leave it unchecked to install a non-RAC Agent on a RAC or non-RAC node. A non-RAC Agent can still discover cluster targets. If you're installing Agents on cluster nodes, it is best practice to check the Cluster Install box to install a cluster Agent.

  - **Cluster Node List** (Mandatory for cluster Agent installation) This field is not applicable for non-cluster Agent deployments, but it is mandatory if you're installing a cluster Agent. Use the Populate Defaults button to automatically populate the cluster node list, or less desirably, type in a comma-separated list yourself. The cluster nodes should not be fully qualified, and no spaces are allowed in the list. You should list all cluster nodes. (A new cluster installation should deploy Agents on all cluster nodes even if you don't specify them all, but don't rely on this; it's better to specify all cluster nodes.) For Oracle Clusterware, the cluster node list is the same as the host list indicated in the Provide Host List field described above (except that the cluster node list is not fully qualified). By contrast, for vendor Clusterware, the node name may be different from the machine name you entered in the Provide Host List field. You can establish the Oracle Clusterware node names by executing *olsnodes* (Oracle 10*g*) or *lsnodes* (Oracle 9*i*) on one of the cluster nodes. These executables are located in the *<ORA_CRS_HOME>*/bin or *<CRS_HOME>*/bin directory, respectively.

NOTE
The recommendation not to use the canonical (fully qualified) host name in this field is an exception to my general recommendation to use the canonical host name.

  - **Cluster Name** (Optional) Entering a cluster name is optional. For 10*g* Clusters, the current cluster name will be used, unless you override it with the name specified here. This field is useful for standardizing the cluster naming convention within GC when CRS cluster names are lacking in this regard, especially if two or more clusters on different hosts have the same name and you'd like to distinguish them in the Console by names you can choose. If you don't know the cluster name, use one of the following commands, depending upon the Oracle Clusterware version.
    For Oracle 10*g*:

    ```
    $ORA_CRS_HOME/bin/cemutlo -n
    ```

    For Oracle 9*i*:

    ```
    $CRS_HOME/bin/cemutls -n
    ```

For 9.2 clusters using third-party vendor cluster software, if you do not specify a cluster name, the name will default to *<first_node_name>_cluster*.

Destination

* Installation Base Directory [   ]
The Oracle home directory will be created as a sub-directory under this directory.

Port
Specify appropriate port for Agents

Port [   ]
If not specified, a free port will be selected within the default range.

Additional Parameters
Additional parameter to agent installation can be provided here.

Additional Parameters [   ]
Example: -i /home/oraInst.loc

- **Destination, Installation Base Directory** (Mandatory) Specify the absolute path for the Installation Base Directory that Agent Deploy will create on all target hosts. Confirm that the Agent user has write permissions to this base directory. The *agent10g* subdirectory is created under this directory, resulting in an Agent Oracle home of *<Installation Base Directory>/agent10g*. You do not need to pre-create the *agent10g* subdirectory, but if it does exist, it must be empty.

- **Port** (Optional) Enter the target host port for all installed Agents to use to communicate with the OMS. Each Agent Deploy operation must install all its Agents on the same destination port. The default port is 3872. If a port is not available on one or more destination hosts, the application follows certain rules to choose a free port:

    - If you specify a port and it is not available on the first target host specified in the Provide Host List field, the Agent Deploy application prompts you for an alternative port.

    - However, if the specified port is available on the first target host listed, but is in use on other target host(s), the installation will fail on these other host(s).

    - If you don't specify a port, and the default port is not available on the first target host listed, the application chooses a free port in the range of 1830-1849.

    While these rules may be interesting reading from an academic perspective, you don't have to worry about them as long as you predetermine whether the default port is available on all target hosts. If it is, do not specify a port so that this default port will be used. If the default port is not free on one or more hosts, choose an alternative port that is free on *all* hosts. To determine which ports are currently in use, execute one of the commands below, depending on the host platform.

    On UNIX:

    `netstat -a | grep <port>`

    On Windows:

    `netstat -a | findstr <port>`

    You should also consult the */etc/services* file on UNIX platforms and the *%SystemRoot%\system32\drivers\etc\services* file on Windows platforms for all port associations with services. Some of these services may be in use (such as *ftp* on port 21 and *telnet* on port 23), but perhaps not at the time you run the *netstat* command. Checking the *services file* ensures that these ports are not reserved for any services that may later be started.

- **Additional Parameters**   (Optional) Pass any desired additional parameters. It may sound strange, but the list of parameters differs entirely depending upon the Source Software location you specify:

    - When using the default location, use the parameters supported by the *agentDownload* script listed in Table 7-4 in Chapter 7. The Agent Deploy application employs the *agentDownload.<platform>* script behind-the-scenes to perform the installation.

    - When using Another Location, which is highly discouraged, use the parameters that the Oracle Universal Installer (OUI) supports; in this case *agentDownload* deploys the Agent by running the OUI in silent mode. See Table 7-7 in Chapter 7 for a list of these parameters.

NFS Agent

Choosing the Add Host to Shared Agent installation type installs NFS Agents to reference an existing master Agent installation. You can install multiple NFS Agents through Agent Deploy or one at a time by running the *nfsagentinstall* script locally on a target host. However, you must first install the master Agent on an NFS drive via any method except *nfsagentinstall*. For details, see the section "nfsagentinstall Installation" in Chapter 7.

CAUTION
Don't try to install the NFS master Agent using Agent Deploy. You must always begin an NFS Agent installation by installing the master Agent via any method except the nfsagentinstall *script. (If you use Agent Deploy, you must choose the Fresh Install type, not the NFS Agent type.) Then you can use Agent Deploy to install additional NFS Agents that use this master Agent.*

When you choose this installation type under Agent Deploy, the Agent NFS Installation: Installation Details screen loads. Most options are the same as for a Fresh Install. Two additional sections appear over and above those common sections listed above:

- **Destination**   (Mandatory) These are the locations of the NFS master Agent and the target host EMSTATE directories.

    - **NFS Agent Location**   (Mandatory) The location of the existing master Agent Oracle home, such as */nfs1/agent10g*. This home must exist on a shared NFS drive and must be mounted on each target host you specify for installation.

- **State Directory Location**   (Mandatory) The location of the EMSTATE directory you desire to create on all target hosts—for example, */u01/app/oracle/product/10.2.0/em10g/agent10g*. Unlike other installation methods, an NFS Agent installation does not use a parent directory, so it does not append *agent10g* to the path entered. You must specify *agent10g* explicitly in the path, if this is your desired EMSTATE directory.[12]
- **Port**   (Mandatory) The port specification is mandatory for this installation type, whereas it is optional for a Fresh Install. Default port 3872 is pre-filled. Specify this port, assuming it is available on all hosts where you are deploying the NFS Agent. If the default port is not free on one or more hosts, choose an alternative port that is free on all hosts. (See the write-up of the Port field in the "Fresh Install" section earlier for how to establish whether a port is free on both UNIX and Windows platforms.)

Upgrade Agent

The ability to upgrade an Agent through Agent Deploy is one of the prime benefits of investing in configuring the Agent Deploy application. I grant that you need to jump through more than a few hurdles to use Agent Deploy. But once you set it up, expect the extended mileage of a vintage Bimmer (BMW). One caveat is that you need to upgrade each cluster node separately, despite the fact that you were able to install a Cluster Agent on all cluster nodes in one operation.[13]

Selecting the Upgrade Agent type brings up the Agent Upgrade: Installation Details screen, with the following sections or fields in addition to those common sections listed above.

---

[12] Using *agent10g* as the final subdirectory in the Agent home path is recommended to match the location of non-NFS Agents, such as the Agent automatically installed with the OMS, at the very least. In this latter case, *agent10g* is hard-coded.

[13] Source: Appendix I of the 10*g* R2 Management Agent Deployment Best Practices, An Oracle White Paper, June 2006.

- **Source Agent Information, Existing Agent Home** (Mandatory) Supply the Agent home of the Agent you want to upgrade, including the *agent10g* subdirectory appended to the end. Example: */u01/app/oracle/product/10.2.0/em10g/agent10g*. This field appears to be redundant, as the Installation Base Directory field under the Destination section (discussed below) already specifies the Agent home to upgrade.

- **Version** (Mandatory) As described for the Version field in the Fresh Install installation type, this section only appears if you first perform the steps listed in Chapter 5 under the heading "Prepare for Cross-Platform Agent Installation." Otherwise, only the current Grid Control version is selectable. If you are deploying cross-platform, you will be able to select from all combinations of versions and platforms contained in the Mass Agent Deployment software. Select a version, and the Platform field pull-down menu (under Hosts) will offer all platforms ported for that version.

- **Hosts, Cluster Upgrade** (Optional) Check this box if you are upgrading a cluster Agent, specify Platform and all nodes in the cluster under Provide Host List. If upgrading non-RAC Agents, do not check this box. (In this case, see descriptions of Platform and Provide Host List above in the introduction to this section, "Run Agent Deploy Application.")

- **Destination, Installation Base Directory** (Mandatory) Specify the absolute path for the Installation Base Directory that, when the *agent10g* subdirectory is appended to it, corresponds to your current Agent Oracle home of *<Installation Base Directory>/agent10g*.

- **Additional Parameters** (Optional) Pass any desired additional parameters, as already described for the same field in the "Fresh Install" installation type above. To summarize, if you specify the default Source Software location, use the parameters supported by the *agentDownload* script listed in Table 7-4 of Chapter 7. If you're using Another Location, which you should avoid if at all possible, use the parameters in Table 7-7 that the Oracle Universal Installer (OUI) supports.

Agent Deploy Post-Installation
AGENT
Perform the steps in the section "Agent Post-Installation Steps" in Chapter 7 for generic standalone Agent post-installation procedures, which include upgrading the Agent to the latest patch set, setting up the Agent OS environment, and determining whether the Agent is functioning properly. Agent Deploy also has a couple of additional post-installation steps, as described next.

Non-NFS Agents

If deploying a non-NFS Agent, and it is the first Oracle product installed on a destination host, and you didn't specify the inventory option (-*i* for a Default Source Shiphome Directory) in the Additional Parameters section, then Agent Deploy creates the inventory and *oraInst.loc* file, the inventory pointer file, in the Agent user's home directory. In this case, after installing the Agents, you may want to manually move the inventory on the Agent hosts to the default location *$ORACLE_BASE/oraInventory/*, particularly if you plan to share the inventory with other Oracle products installed later. If you move the inventory, also move *oraInst.loc* to the default location, which is either */etc/* (on AIX and Linux) or */var/opt/oracle/* (on HP-UX and Solaris). Don't forget to change the *inventory_loc* parameter in *oraInst.loc* to point to the new inventory location. There should be no repercussions to moving the inventory and inventory pointer file. You will actually benefit by moving the inventory pointer file to the default location because you won't need to manually configure Grid Control to discover the inventory located in a nondefault location (other than that pointed to by the default *oraInst.loc* file). This manual configuration process is covered in Chapter 11 under "Enable Multi-Inventory Support (UNIX Only)."

NFS Agents

If deploying an NFS Agent, and it is the first Oracle product installed on a target host, you need to manually reevaluate the Inventory collection in Grid Control and upload it into the Management Repository. To do so, execute the following command from the EMSTATE directory after the installation completes:

<EMSTATE_DIR>/bin/emctl control agent runCollection *<NFShost>*:host Inventory
where *<NFShost>* is the NFS host name.

Troubleshoot Agent Deploy

The Agent Deploy application runs in a different OC4J (Oracle Containers For Java) container than Enterprise Manager. Therefore, you deal with Agent Deploy problems separately from EM. If an application error occurs, stop and restart the OC4J instance in which Agent Deploy is running as follows:

```
OMS_HOME/opmn/bin/opmnctl stopproc process-type=OC4J_EMPROV
OMS_HOME/opmn/bin/opmnctl startproc process-type=OC4J_EMPROV
```

   Do not attempt to click the Status tab to monitor the status of the Agent Deploy prerequisite checks while they are running, as this will cause an application error. Rather, tail the application logs. Each log and its description is listed in Oracle Enterprise Manager Grid Control Installation and Basic Configuration 10*g* Release 4 (10.2.0.4.0), Appendix F.2 (online).

Summary

Agent Deploy installations are not plug and play, so to speak. This is why it is crucial to consider what Agent installation method(s) are most appropriate for your Grid Control environment. If you do commit to Agent Deploy, you are committing to a rather lengthy set of installation steps. For this reason, at the start of the chapter I pointed out two main factors that influence how many times you must run through the Agent Deploy installation process. The first factor is the number of OMS hosts on which you intend to configure Agent Deploy: at least one OMS host in any case, and two OMS hosts for Agent Deploy high availability. The second factor is the number of Agent batches deployed. Striving to deploy all Agents of the same OS in one batch or API operation minimizes the total number of batches required.

Once you scope the magnitude of the overall installation effort, you must carry out the formidable list of tasks, performed on OMS and/or Agent hosts, as follows:

- Install Required Packages   (AGENT)
- Configure SSH User Equivalence   (AGENT, OMS)
- Set Up TimeZone for SSH Server   (AGENT, OMS)
- Validate All Command Locations   (OMS)
- Modify Agent Deploy Properties File for SLB   (OMS)
- Choose Inventory Location   (AGENT)
- Verify Agent User Permissions   (AGENT)
- Prepare for Cross-Platform Agent Push   (OMS)
- Include Additional Files (Optional)   (OMS)
- Run Agent Deploy Application   (OMS)
- Agent Deploy Post-Installation   (AGENT)

All tasks except the final two are pre-API tasks. Grid Control somewhat automates the lengthiest of these pre-API tasks, "Configure SSH User Equivalence," with its *sshConnectivity.sh* script, located in the OMS home. On Windows, however, SSH user equivalence requires installing and configuring the Cygwin Suite, which takes a little effort. The *validatePaths* script, also located in the OMS home, performs the task, "Validate All Command Locations." These are platform-dependent commands the Agent Deploy API employs behind the scenes, which are defined in properties files located under the OMS home. With the exception of these two steps, you must "drive a stick" to perform the remaining pre-API steps for Agent Deploy. Luckily, these steps are not very labor-intensive.

After completing these pre-API tasks for the OMS and target hosts, you don't need to worry about them ever again when you're installing and upgrading Agents on these target hosts. You can kick off the Agent Deploy application and usually coast downhill from there. It is relatively painless to complete the fields on the "one-pager" Agent Deploy application screen in the Console (a long page, I grant you). The probability that Agent Deploy will successfully mass-deploy Agents is directly proportional to how thoroughly you perform the pre-API steps.

Post-installation for Agent Deploy in particular (as opposed to the general post-installation tasks discussed in Chapter 7 that are common to all Agent installation methods) involves just a few steps at most. For non-NFS Agents, you may want to move the inventory if it was installed under the Agent home, and for NFS Agents you need to force a reevaluation of the Inventory collection in Grid Control.

We now round the corner into our final chapter—Chapter 7—of three chapters on Agent installation. This chapter covers the five remaining Agent installation methods. You could say from this chapter division alone that Agent Deploy is roughly as powerful as the five other installation methods combined. But if you're just going to the grocery store, don't take your Ferrari—take the mini van. One or more of these five installation methods may amply suffice for your environment, particularly if your Agent count is not terribly high.

CHAPTER 7

Install Management Agents Locally

The last chapter supplied the steps to use the Agent Deploy method to mass-deploy standalone Management Agents on target hosts. In this chapter you learn how to install Agents using the remaining five methods. To review, the six methods are as follows:

- Agent Deploy
- *nfsagentinstall*
- *agentDownload*
- Agent Cloning
- Silent Installation
- Interactive Installation

The remaining five installation methods covered here are less complicated to use than Agent Deploy. However, we start with the second most complicated method, *nfsagentinstall*, and then work our way downward in complexity, roughly speaking.

As in Chapter 6, the Agent owner in this chapter is assumed to be the *oracle* user. For help deciding which user to choose for the Agent, see the section "Use an Existing User or Create a Separate Agent User" in Chapter 5. If you decide to create a separate Agent user, perform all steps as that user instead. Steps requiring *root* access are indicated as such.

NOTE
Before installing the Agent using any method covered in this chapter, first complete the preinstallation tasks detailed in Chapter 5, which apply to all methods. I do not remind you to do this again in the section for each method.

nfsagentinstall Installation

Grid Control Release 2 introduced a way to install a *master Agent* on a shared Network File System (NFS), then to share the NFS bits across multiple target hosts of the same platform that run *NFS Agents*. Each host shares the same NFS-mounted binaries and also runs its own local Agent processes. Host-specific configuration, log, state, collection, and upload files reside locally in an EMSTATE directory on each target host.[1]

NFS Agent installation is not supported on a host running clusterware of any kind—either Oracle or third-party clusterware. It is not just that the NFS Agent home cannot reside on clustered storage. You cannot install an NFS Agent anywhere on a host running clusterware, even if you locate the Agent home on local storage. This restriction does not relate to the location of the Agent home, but rather to the fact that an NFS Agent is not certified to manage Oracle or third-party cluster technology, including Oracle Clusterware, Oracle Automatic Storage Management (ASM),

---

[1] This is not a hard and fast requirement. You could place the EMSTATE directory on shared storage too, but it is not recommended for several reasons. Principal among them are security concerns that target information for the entire enterprise would reside in one location on shared storage. It makes more sense to place each host's EMSTATE directory on local storage, as this directory contains information about local targets for that particular host.

or Oracle RAC databases. The Agent Deploy, *agentDownload*, silent, and interactive methods provide an option to install a cluster Agent.

NFS Agent installation is supported on both UNIX and Windows platforms. All hosts running NFS Agents that use the same master Agent must run on the same operating system and version. However, you can run separate groups of NFS Agents, one for each OS and version, where each group relies upon a different master Agent for that particular OS and version.

DBAs sometimes find the idea of NFS Agents conceptually difficult. This is because the Agent is not a single entity, like an Agent installed using the other installation methods. Instead, the Agent is distributed for each host, consisting of a master Agent on shared storage and NFS Agents on each target host, including on the master Agent host. The term "master Agent" is somewhat misleading, implying that the Agent on a master Agent host is vastly different from NFS Agents on other hosts. In reality, the choice of the master Agent host is arbitrary. It is simply any target host from which you install a standalone Agent on shared storage. You could run this Agent installation from any target host that can mount the NFS drive, although it does make more sense to choose a host geographically closest to the NFS Server and on the fastest network connection. Once this master Agent is installed, the method for creating NFS Agents is identical on both the master Agent host and all other NFS Agent hosts. The only stipulation is that you must run the *nfsagentinstall* script (or Agent Deploy, selecting the NFS Agent type) first on the master Agent host (thereby creating an NFS Agent on that host), before running this script on the other NFS Agent hosts.

It is key to your understanding of NFS Agents to distinguish the "master Agent host" from the "master Agent." The master Agent host is the host from which you mount an NFS share to install the master Agent. The master Agent is like any other Agent, except that it is installed on shared storage and never started explicitly, which I explain below. Once you install this master Agent, the master Agent host loses its master status, falling back to become the first NFS Agent host. It's true that this host gets preferential treatment for being the first host on which you install an NFS Agent, but that's its only distinction, beside the fact that it should also be the only host with write permissions on the NFS share. (Again, stay tuned, as I explain more about this additional distinction below.)

Hopefully, you now have a basic understanding of the concept of an NFS Agent. To flesh out this understanding, we need to get down to the specifics of deploying NFS Agents using *nfsagentinstall*. Let's break down the NFS Agent installation process into bite-size pieces to debunk the perceived complexity of this very efficient installation method. There are three such pieces:

- **Configure shared storage** Configure and mount an NFS share to the master Agent host and all target hosts where installing NFS Agents. If you're not using a dedicated highly available NFS server, the master Agent host itself can also serve as the NFS server, but for production systems this is only recommended if you protect the NFS share with hardware mirroring (and replication too, preferably).

- **Install master Agent on shared storage** Install the master Agent on an NFS share from the master Agent host using any method **except** *nfsagentinstall*. After the installation completes, shut down the Agent processes on the NFS share, and never start them again from the NFS share home. Instead, you will start local NFS Agent processes, which call the binaries on the NFS share behind the scenes.

- **Install NFS Agents** Install the first NFS Agent on the master Agent host, then install NFS Agents on all remaining destination hosts. All installations can be via the *nfsagentinstall* script or Agent Deploy (see Chapter 6, "Run Agent Deploy Application"). Start each NFS Agent from the EMSTATE ("state") directory.

We now delve into each of these main steps in enough detail to allow you to deploy NFS Agents. An example goes a long way. Let's take the case of installing three NFS Agents on a Linux x86 platform using *nfsagentinstall*. The first step, Configure Shared Storage, differs most by UNIX platform, and is completely different on Windows platforms, with respect to OS commands, file entries, and configuration file names. However, the remaining two steps are much the same for all flavors of UNIX and Windows. As Figure 7-1 depicts, you log in to *masterhost* and install the master Agent as you would any other Agent, but on shared storage located on an NFS server called *nfsserver*. (Remember that the choice of *masterhost* is arbitrary—it can be any target host on which you want to deploy an NFS Agent.) Next, use *nfsagentinstall* to deploy an NFS Agent on *masterhost* and two additional NFS Agent hosts, *nfshost1* and *nfshost2*. The NFS share installation directory is */nfs1/agent10g*. NFS Agent configuration, state, collection, and upload files are installed locally

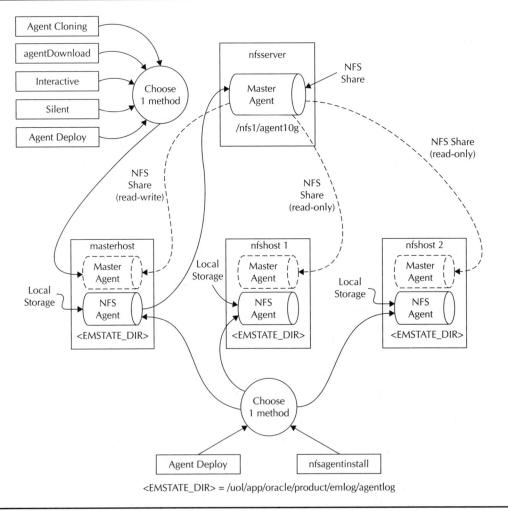

FIGURE 7-1. *Depiction of a typical NFS Agent installation using nfsagentinstall*

Description of Variable	Variable Name	Sample Value for Variable
NFS Server	<NFS_SERVER>	nfsserver
NFS master Agent base installation directory	<NFS_MASTER_BASEDIR>	/nfs1
NFS master Agent installation directory	<NFS_MASTER_HOME>	/nfs1/agent10g
NFS master Agent host	<NFS_MASTER_HOST>	masterhost
NFS Agent installation directory	<EMSTATE_DIR>	/u01/app/oracle/product/10.2.0/em10g/agent10g[2]
NFS Agent host name(s)	<NFS_AGENT_HOST>	nfshost1, nfshost2

TABLE 7-1. *NFS Agent Variables and Values Used in the Examples Herein*

on target hosts in the EMSTATE directory */u01/app/oracle/product/10.2.0/em10g/agent10g/*. However, these NFS Agents share the same NFS-mounted Agent binaries.

Table 7-1 summarizes all variables and values used in the example related to host names and directories.

Let's start with the first step, to ready the shared storage for master Agent installation. Each command or file entry is listed below in two formats: with variables, then side-by-side with substituted values listed in the table. Enter each command or file entry on one line, even though they may be shown on multiple lines.

Configure Shared Storage

Configuring shared storage involves two basic steps:

- Allocate a highly available NFS share
- Mount this share on each host where NFS Agents are to be installed

Allocate the Shared Storage

The first step is to have your system administrator (SA) set up a shared drive to house the master Agent Oracle home. The master Agent is supported on a Net App Filer disk or other Network Attached Storage (NAS) drive, but is not supported on Storage Area Network (SAN) devices,[3] or on any kind of clusterware, including Oracle Clusterware, the Oracle Cluster File System (OCFS) Version 1 or Version 2, or third-party clusterware. Confirm that the desired master Agent installation directory is on a supported, shared, highly available device, and that it has enough free disk space to accommodate a normal Agent installation. See Table 5-2 in Chapter 5 for the Agent's total disk space requirements on each platform. Only the master Agent host is granted read-write access to this share. NFS Agent hosts only need read-only access.

---

[2] Notice that I used best practices by specifying an <EMSTATE_DIR> directory that matches the Agent home specification format recommended throughout this book for any Agent installation method.

[3] See Oracle*MetaLink* Note 353836.1.

I hope you noticed the above words, "highly available." At larger Oracle shops, NFS servers are typically clustered for high availability. It is critical to recognize that the NFS master Agent storage is a single point of failure, as all NFS Agents rely upon the master Agent. Best practices dictate not installing NFS Agents unless you can install the master Agent on a highly available NFS server. DBAs at smaller companies without the luxury of an NFS server sometimes make the mistake of staging an NFS share from one of the Agent hosts (usually the master Agent host), which is almost certainly not highly available. Some DBAs also mistakenly believe that installing the master Agent as a cluster Agent could solve this HA problem, but this is not supported, as already mentioned.[4]

CAUTION
It is best left to an experienced system administrator to set up NFS servers and mounts, as they carry an inherent security risk and warrant many other administrative considerations. NFS administration tasks require Sys Admin privileges (i.e., root *access on UNIX) in any case.*

If your company employs NFS technology, then it's probable that only system administrators (SAs) and network administrators have access to these NFS servers, and equally probable that DBAs have no such access. Even so, you may benefit from learning the specific steps to stage an NFS share, in this case, on Linux x86. You may wear both DBA and SA hats, may be an SA interested in a real example, or, most likely, are a DBA who wants to provide specific directions to your SA or set up NFS Agents in a test environment. Whatever your position, the steps can be summarized as follows: On the staging server where the master Agent is to be installed, create the mount point directory for the NFS share, set up the share to be exported, export it, then make the export persistent across reboot. You must be the *root* user to perform all steps, unless otherwise noted:

1. Create a local mount point directory on the NFS server for the NFS share with the correct permissions:

mkdir -p <NFS_MASTER_BASEDIR>	mkdir -p /nfs1
chown oracle:dba <NFS_MASTER_BASEDIR>	chown oracle:dba /nfs1
chmod 750 <NFS_MASTER_BASEDIR>	chmod 750 /nfs1

2. Set up the share to be exported on the NFS server. To do so, as *root* first create or edit the file */etc/exports*, which contains the access control list for filesystems allowed to be exported to NFS clients. Add the following line to this file to allow the NFS master Agent home to be exported to the master Agent host with read-write access and to all NFS Agent hosts with read-only access:

<NFS_MASTER_BASEDIR> <NFS_MASTER_HOST>(rw,async) <space-separated list of NFS Agent hosts (ro,async)>	/nfs1 masterhost(rw,async) nfshost1(ro,async) nfshost2(ro,async)

---

[4] If you were cheeky enough to try running a master Agent on a RAC cluster, you'd find that the *nfsagentinstall* script throws an error preventing the NFS Agent from starting. You can't install NFS Agents on an Active/Passive cluster because NFS Agents are not supported on clusterware of any kind, including Active/Passive clusters.

Entries may be continued across newlines using a backslash. Each space-separated NFS client that is allowed to mount the filesystem is immediately followed by a parenthesized, comma-separated list of export options for that client. Do not use a space anywhere within these options or the share will not be created.

Notice a few options in the example entry above. First, see how only the master Agent host, *masterhost,* is provided read-write (rw) access, and how each NFS Agent host is only granted read-only (ro) access, as stated above. Also, observe how all hosts list the *async* option (the default is *sync*), which allows the NFS server to violate NFS protocol and reply to requests before any changes made by that request have been committed to NFS disk storage. Using the *async* option improves performance without incurring the usual associated cost of possible data corruption if an NFS server crashes. (*masterhost* does not need to write to the NFS share, except to upgrade or patch the master Agent, and a backup before such changes should amply protect the NFS share.)

3. Export the above share on the NFS server:

   `/usr/sbin/exportfs -a`

4. Make the share persistent across reboot on the NFS server. Add the line below to the filesystem table, called */etc/fstab* on Linux and on some UNIX platforms, and */etc/vfstab* on other UNIX platforms. The filesystem table is a configuration file containing information about where and how to mount a machine's partitions and storage devices. The *async* option is used again here to allow for improved NFS server performance.

   <NFS_MASTER_BASEDIR> <mount point> <filesystem type>   /nfs1 /nfs1 nfs
   <mount options> <dump options> <filesystem check options>   auto,nouser,async
                                                                rw 2 1

See your UNIX operating system specific documentation for details on all configuration files and commands, specifically the *man* pages for *mount, fstab, nfs, exports, mountd, nfsd,* and *nfs*.

Mount the NFS Share on Target Hosts

As *root*, the next step is to mount the NFS share on each NFS Agent host, starting with the master Agent host, and to make this share persistent across reboot. While you as a DBA probably don't have access to any NFS servers, you may very well have *root* access on the NFS Agent hosts, including the master Agent host.[5] If you have *root* access, you can perform the following steps that require such access; otherwise, ask your SA to do it.

1. Create a local directory for the NFS share on the master NFS Agent host and on each host where you will deploy an NFS Agent.[6] For administrative clarity, find a mount point name that is available on all NFS Agent hosts. It is logical to use the same directory specification that you used for the NFS server mount (*/nfs1*), assuming this mount point is not already defined on an NFS Agent host. (Check the */etc/fstab* file on each NFS Agent

---

[5] You could also mount the share in the unlikely event that your SA was feeling generous and specified the *user* option in */etc/fstab* for the NFS device—the default is *nouser*, which means only *root* can mount the device.

[6] Create the directory either as *root* or *oracle*, depending on the permissions required for the mount point chosen. If you must be *root* to create the directory, which is the case here because we're creating the *nfs1* directory below the root ("/") mount point, grant *oracle* ownership and permissions to this directory afterward as shown below.

host to see if this mount has been assigned yet.) If the mount point is already taken, use a different mount point, but use it consistently across each NFS Agent host if possible to standardize the installations. Here we assume /nfs1 is available on all NFS Agent hosts. As *root*, create the local directory for the NFS share and grant *oracle* permissions to this directory:

mkdir -p <NFS_MASTER_BASEDIR>	mkdir -p /nfs1
chown oracle:oinstall <NFS_MASTER_BASEDIR>	chown oracle:oinstall /nfs1
chmod 755 <NFS_MASTER_BASEDIR>	chmod 755 /nfs1

2. Mount the NFS share to this local directory using the following command executed as *root*:

mount -t nfs <NFS_SERVER>:/nfs1 /nfs1	mount -t nfs nfsserver:/nfs1 /nfs1

3. Make the NFS share persistent across reboot by adding the following line to /etc/fstab as *root* (options vary as shown below depending upon whether you are on the master Agent host or on NFS Agent hosts):[7]

- On the master Agent host *only*, mount the NFS filesystem read-write on reboot:

<NFS_SERVER>:<NFS_MASTER_BASEDIR> <mount point> <filesystem type> <mount options> <dump options> <filesystem check options>	nfsserver:/nfs1 /nfs1 nfs auto,nouser,async,bg,hard,nointr,rsize=32768,tcp,vers=3,timeo=600,actime=0 rw 2 1

- On each NFS Agent host, mount the NFS filesystem read-only on reboot:

<NFS_SERVER>:<NFS_MASTER_BASEDIR> <mount point> <filesystem type> <mount options> <dump options> <filesystem check options>	nfsserver:/nfs1 /nfs1 nfs auto,nouser,async,bg,hard,nointr,rsize=32768,tcp,vers=3,timeo=600,actime=0 ro 2 1

4. Check that you can write a test file to the NFS share while logged in as the *oracle* user on the master Agent host:

touch <NFS_MASTER_BASEDIR>/testfile	touch /nfs1/testfile

---

[7] See the table in Oracle*MetaLink* Note 359515.1 for mount options to use for other operating systems. In this Note, use the mount options listed in the table's second column, labeled "Mount options for Binaries."

5. Log in as *oracle* to each NFS Agent server, including the master Agent host, and confirm that you can read the test file created above on the NFS share (the first command), but that you cannot write to the NFS share (the second command):

   cat <NFS_MASTER_BASEDIR>/testfile   cat /nfs1/testfile
   touch <NFS_MASTER_BASEDIR>/testfile2   touch /nfs1/testfile2

   The first command should succeed. The second command should return the following error:
   `touch: creating 'testfile2': Read-only file system`

Install Master Agent on Shared Storage

Now that you have configured shared storage for all NFS Agent hosts, it's time to install the master Agent from the master Agent host in the NFS directory *<NFS_MASTER_HOME>*, which in this example is */nfs1/agent10g*. This is required before deploying NFS Agents on all remaining target hosts, including the master Agent host, as shown in the next and last section. To install the master Agent on shared storage, follow these steps:

1. Specify a local Inventory directory on the master Agent host—it cannot be on shared storage.

   - If this is not the first Oracle product installation, check that the Inventory directory specified by the *inventory_loc* in the */etc/oraInst.loc* file is set to a nonshared directory:

     cat /etc/oraInst.loc
     inventory_loc = <non_NFS_directory>
     inst_group = oinstall

   - If this is the first Oracle product installation, pre-create an *oraInst.loc* file, specifying a non-NFS directory.

2. Confirm that the userid (uid), group id (gid) and credentials (both username and password) of the user installing the NFS master Agent match those of the user installing each NFS Agent. This is required because the NFS Agents need permission to execute the NFS master Agent executables, such as *emctl* and *nmo*.

3. Decide upon an installation method to install the master Agent on the NFS share created above. You can choose from any of the six standalone Agent deployment methods **except** *nfsagentinstall*. For an overview of each method, see Chapter 5. The rationale for deciding upon an installation method is provided in Chapter 5 under the heading "Select Agent Installation Methods." If you use Agent Deploy here (choosing the Fresh Install type), you will probably also want to use Agent Deploy to install the NFS Agents as described in the next section.

4. Log in as *oracle* to the master Agent node and install the master Agent on the NFS share created above using the standalone Agent deployment method you've just chosen. Follow the procedure discussed in this book for your chosen standalone installation method. I refer you to Chapter 6 for Agent Deploy procedures, and to the corresponding sections in this chapter for *agentDownload*, Agent Cloning, silent, or interactive

installation methods. Whatever method you choose, install the Agent home in the NFS share directory, which in this example is /nfs1/agent10g. Remember that most Agent installation methods append *agent10g* to the path specified for the Agent. For the particular installation method you choose, use the installation options, if available, that prevent the master Agent from starting and that disable target auto-discovery when the installation completes. Table 7-2 indicates these options for each installation method for which they are available (the *nfsagentinstall* method is not applicable because you cannot use it to install a master Agent).

If you choose a method that does not allow you to suppress Agent startup, then shut down the Agent after the installation completes. Here is the command to shut down the master Agent, if it was started by the installation:

<NFS_MASTER_HOME>/bin/emctl stop agent   /nfs1/agent10g/bin/emctl stop agent

You should never again start the Agent from the NFS shared home. As described in the following section, you will only start NFS Agents, including that on the master Agent node, by using the *emctl* executable located in the local EMSTATE/bin directory on each target host.

If the master Agent is started, even momentarily, and auto-discovery occurs, this Agent will temporarily show up in the Console UI along with the targets on the master Agent host. When you install the NFS Agent on the master node (as described below), an additional Agent and associated targets will appear. So you will temporarily see two Agents and duplicate targets on the master Agent host. The master Agent will show "Agent Unreachable" because you will have just shut it down, and the NFS Agent will show status UP. Ignore this, as within minutes after the NFS Agent is installed, the master Agent and targets it discovered "disappear" from the UI, leaving you with just the NFS Agent and its targets on the master Agent host.

Installation Method	Option To Prevent Agent Startup	Option To Disable Target Auto-Discovery
Agent Cloning	None	Do not execute $AGENT_HOME/bin/agentca -f
Agent Deploy	-t	-d
agentDownload	-t	-d
Interactive Installation	None	None
nfsagentinstall	N/A	N/A
Silent Installation	startAgent=false (command-line) b_startAgent=false (response file)	doDiscovery=false (command-line) b_doDiscovery=false (response file)

TABLE 7-2. *Installation Options to Prevent Agent Startup and Auto-Discovery After Deployment*

Install NFS Agents

Now that you have installed the master Agent on a shared drive, you need to install an NFS Agent on the local drive for the master NFS Agent host and on each NFS Agent host. You can either install one NFS Agent at a time using the *nfsagentinstall* script, available in the master Agent home, or deploy multiple NFS Agents in one operation using Agent Deploy as described in Chapter 6 under the heading "NFS Agent." (Remember that NFS Agent hosts must be running the same OS (including version) as the master NFS Agent host.) The NFS Agent on the "master" Agent host is identical to any other NFS Agents you deploy (regardless of installation method used).

Complete the following steps to install an NFS Agent using the *nfsagentinstall* script on the master NFS Agent host or any target host. (This is the second time you are running an installation on the master NFS Agent host—the first time was in the section above, "Install Master Agent on Shared Storage.") These instructions will simply refer to installing the NFS Agent, with the understanding that the host on which it is being installed is either the master Agent host or a regular NFS Agent host, the steps being the same. In this example, the master NFS Agent host is <NFS_MASTER_HOST> and the NFS Agent hosts are *nfshost1* and *nfshost2*.

1. Ensure you have sufficient local disk space for the EMSTATE directory of an NFS Agent. While installation requires only 1MB of disk space, upload files can consume up to 50MB by default, based on the maximum amount of XML upload file data that Agents are preconfigured to hold. Factoring in space for log and collection files, 100MB of space is more than sufficient to keep an NFS Agent operational.

2. Confirm that the Oracle Inventory is not in a shared location, as this Inventory contains installation information unique to each host. This will definitely be a problem if the NFS Agent user's home directory is shared, because the default Inventory location for an NFS Agent is the *oraInventory* subdirectory directly below the Agent user's home. The Inventory may be shared for other reasons as well. For whatever reason, if the Inventory is shared, change the value of *inventory_loc* in the *oraInst.loc* file to a nonshared, local directory:

   - If the NFS Agent installation is not the first Oracle product installation, the *oraInst.loc* file will already exist and may point to a shared location. If necessary, change the existing *inventory_loc* value in this file to a local drive specification.

   - If the NFS Agent installation is indeed the first Oracle installation, the *oraInst.loc* file will not yet exist. In this case, you need to manually create this file in the platform-specific location—under */etc/* for AIX and Linux, and under */var/opt/oracle/* for HP-UX and Solaris. The *oraInst.loc* file is composed of two lines in the following format (assuming ORACLE_BASE is */u01/app/oracle* and the Inventory owner is *oinstall*):

   ```
   inventory_loc=/u01/app/oracle/oraInventory
   inst_group=oinstall
   ```

3. Execute the *nfsagentinstall* script to install the NFS Agent. Following is the syntax for this script and the corresponding values from the example we've been studying:

   cd <NFS_MASTER_HOME>/sysman/install   cd /nfs1/agent10g/sysman/install
   ./nfsagentinstall -s <EMSTATE_DIR> -p <PORT>   ./nfsagentinstall -s /u01/app/oracle/product/10.2.0/em10g/agent10g -p 3872

Observe the following regarding the *nfsagentinstall* script:

a. Use the actual Agent home for <EMSTATE_DIR>, i.e., include *agent10g* in the directory specification. Unlike most other installation methods, the *nfsagentinstall* script does not use a parent directory, but uses <EMSTATE_DIR> as the argument to the *-s* option to define the Agent home.

b. Use the *-p* option, which denotes the port on which you want the Agent to run. This option is optional for Enterprise Manager 10.2.0.2 onwards, but if you don't specify a port and default port 3872 is being used, the script will select the next available port number in the range 1830 to 1849. Thus, for sake of consistency, it is best to enter a specific port number available on all NFS Agent hosts. Use default port 3872 if possible.

c. Run <EMSTATE_dir>/root.sh as *root* when prompted near the end of the installation, specifying the <EMSTATE_DIR> as an argument to this command (this is the only *root.sh* script in any Enterprise Manager product that employs an argument):

```
cd <EMSTATE_DIR>                    cd /u01/app/oracle/product/10.2.0/em10g/agent10g
su                                  su
./root.sh <EMSTATE_DIR>             ./root.sh /u01/app/oracle/product/10.2.0/em10g/agent10g
```

There are a few bugs related to *root.sh*. The first bug is that in EM 10.1 only, *root.sh* is not be installed on any host except in the NFS Agent home on the master Agent host. In this case, copy *root.sh* from the master Agent host to the other NFS Agent hosts. The second bug is that *root.sh* encloses the Agent home in brackets "< >" in the *oratab* file (located under /etc/ or /var/opt/oracle/, depending on the platform). Remove these brackets from /etc/oratab.

d. Address the following error, which should occur on all hosts but the master host at the end of the installation:

Error encountered executing command <EMSTATE_DIR>/bin/emctl reload. reload failed with ret=-2

To resolve this error, secure the Agent from the NFS Agent home as follows:

```
<EMSTATE_DIR>/bin/emctl              /u01/app/oracle/product/10.2.0/em10g/
secure agent <agent_registration_    agent10g/bin/emctl secure agent
password>[8]                         <agent_registration_password>
```

3. Run the following commands as the *oracle* user to collect information about the host's inventory, but only if this is the first Oracle installation on the host. *nfshost1* is used as

---

[8] The *emctl* executable in the EMSTATE directory points to the Agent binary in the NFS-mounted master Agent home.

the <NFS_AGENT_HOST> in the example syntax. (This step is not required on the master Agent node as the master Agent installation already collects Inventory information.):

cd <EMSTATE_DIR>/bin   cd /u01/app/oracle/product/10.2.0/em10g/
                                           agent10g/bin

./emctl control agent runCollection   ./emctl control agent runCollection
<NFS_AGENT_HOST>:host   nfshost1:host Inventory
Inventory

4. Confirm that the NFS Agent is started using the *EMSTATE/bin/emctl status agent*. Remember never to start the NFS Agent using the *emctl* executable from the NFS share directory (*/nfs1/agent10g/bin* in this example).[9]

agentDownload Installation

The *agentDownload* method of installation uses a command-line script called *agentDownload.<platform>* on UNIX and *agentDownload.vbs* on Windows. The OMS platform-specific version of the script is provided as part of the OMS installation, and is located under the directory *$OMS_HOME/sysman/agent_download/<version>/<platform>/*. The script is also available for all other certified Agent platforms as part of the Mass Agent Deployment kit downloadable from the Oracle Technology Network (OTN) at the following URL:

http://www.oracle.com/technology/software/products/oem/htdocs/agentsoft.html

The Readme file for the Agent kit is available from this URL as well.

NOTE
For discussion purposes in this section, the placeholder agentDownload is generically used to signify agentDownload.<platform> and agentDownload.vbs scripts.

Whereas Agent Deploy employs push technology to deploy Agents from the OMS host to multiple Agents in one operation, *agentDownload* uses pull technology, requiring local execution of the script on each target Agent host (except when installing a cluster Agent). To summarize the installation procedure, first you copy the *agentDownload* script from the OMS host to the target host. Then you execute *agentDownload* on the target host. The script uses the *wget* utility to download the Oracle Universal Installer (OUI) and the installation response file, *agent_download.rsp* from the OMS home via HTTP to the Agent host. Then the script performs a local, silent OUI installation using values supplied by the response file. The script uses *wget* again to download the installation code from the OMS via HTTP. This is why *agentDownload* does not work when a proxy server separates the OMS and Agent hosts.

---

[9] Don't try and get fancy and rename or remove the *emctl* executable from the master Agent home on the NFS share in an attempt to prevent users from running it. While you should never directly call this executable, the local *<EMSTATE_DIR>/bin/emctl* executable does call it.

Now that we have a conceptual feel for the *agentDownload* installation method, let's discuss specifics on how to use it to deploy Agents. The installation procedure consists of three basic steps:

- Prepare for *agentDownload* installation
- Copy *agentDownload* script from OMS to target host
- Execute *agentDownload* on target host

Throughout I use the variables in the following table:

Variable Name	Example Value(s)
<platform>	linux, solaris, win32
<version>	10.2.0.4.0
agentDownload (in text references) and <agentDownload> (in command syntax)	agentDownload.linux (for UNIX), agentDownload.vbs (for Windows)

Prepare for agentDownload Installation

Preparing for *agentDownload* installation involves several steps, which can be broken down into preparing the *agentDownload* script, and, for Windows, confirming that the proper version of Windows Script Host is installed.

Prepare agentDownload Script and Response File

Edit each original platform-specific script on each OMS host that will play a role in an *agentDownload* installation:

1. **Make changes required for SLB (only if OMS and Agent are on same platform)** If OMS and target platforms are the same *and* if you are using a server load balancer (SLB) to virtualize the Management Service, make the following two changes in the *agentDownload* script (if OMS and target platforms are different, skip to step 2):

   Replace %s_OMSHost% with the load balancer host name.
   Replace %s_OMSPort% with the load balancer port.

2. **Download and stage Agent software and prepare scripts (cross-platform only)** If OMS and target platforms are the same, skip to step 3. If OMS and target platforms are different, you must download the Agent software from a Mass Agent Deployment distribution. In this case, you need to stage the software and prepare both the *agentDownload* script and the Agent response file as follows:

   - **Download and stage the Agent software** Download the software listed under Mass Agent Deployment on OTN under http://www.oracle.com/technology/software/products/oem/index.html. To stage the Agent software, copy the downloaded zip file to $OMS_HOME/sysman/agent_download/<version>/, change to that directory, and unzip the file (see *instructions.txt*, which is included in the distribution).

- **Prepare the *agentDownload* script** The OMS host and port will not be instantiated in the downloaded script, only in the *agentDownload* script from the initial GC installation. Therefore, you need to make the following two changes in the *agentDownload* script:

    Replace %s_OMSHost% either with the OMS host name or with the load balancer host name if you're using an SLB.
    Replace %s_OMSPort% either with the OMS port, which defaults to 4889, or with the load balancer port if you're using an SLB.

- **Prepare the *agent_download.rsp* file** If no *agent_download.rsp* file exists under *$OMS_HOME/sysman/agent_download/<version>/* on the OMS host, then use the Agent response file, *agent_download.rsp.bak*, from the EM software distribution. In other words, do the following (again, only if there's no response file on the OMS):

    cd $OMS_HOME/sysman/agent_download/<version>
    mv agent_download.rsp.bak agent_download.rsp

3. **Grant execute privileges to *agentDownload* script** Grant execute privileges to the *agentDownload.<platform>* script:

    cd $OMS_HOME/sysman/agent_download/<version>/<platform>
    chmod 755 <agentDownload>

4. **Fix *agentDownload* bug** Optionally, add the code shown in **bolded italics** below between and including **Customization begins** and **Customization ends** lines to the CleanUp function of the *agentDownload* script on the OMS (surrounding code is shown in normal font). This removes an extraneous empty directory, named *agentDownload10.2.0.x.0Oui*, which is left over by the script after it successfully installs the Agent. This directory is used temporarily to house the downloaded installer files from the OMS, and is created in the same directory where the *agentDownload* script is downloaded:

```
...
    echo "Removed ${InstallerLocalStage}/Translations" |tee -a $LogFile
  fi

# Customization begins
  if [ -d "${InstallerLocalStage}" ]
  then
    rmdir ${InstallerLocalStage}
    echo "Removed ${InstallerLocalStage}" |tee -a $LogFile
  fi
# Customization ends

  if [ -d $ORACLE_HOME ]
  then
...
```

5. **Configure wget** As mentioned, the *agentDownload* script uses the *wget* utility on the Agent host to download the OUI and response file, then the installation binaries, from the OMS host. If *wget* is not installed,[10] and you get clearance from your SA, download it from the Web and install it. GNU *wget* is available for most platforms at http://www.gnu.org/software/wget/wget.html. The method of configuring *wget* varies for UNIX vs. Windows:

   - Windows does not support *wget* by default, but a Windows version of this utility is available at http://gnuwin32.sourceforge.net/packages/wget.htm. Install *wget*, preferably in a directory already specified in the PATH on all Agent hosts where *agentDownload* will run. *<SystemDrive>:\Windows\system32* will almost certainly be in the PATH already. (The default *wget* installation directory, *C:\Program Files\GnuWin32*, likely will not be in the PATH).

   - On UNIX, ensure the *wget* executable is located either in */usr/local/bin/*, */usr/bin/*, or is in the PATH on all Agent hosts where *agentDownload* will run. You can confirm whether *wget* is in the PATH of the Agent user by typing:

     ```
     which wget
     ```

     This command should return the path of this executable. If *wget* is not in either directory, or in the PATH, then do one of the following, depending upon the situation:

   - Modify the *agentDownload* script on each OMS to look for the location of *wget* in that directory.[11] For example, assuming *wget* is located in */usr/sbin*, then add the text in **bolded italics** below. (This solution has the advantage of requiring only one change on the OMS host, rather than having to make a change on every target host, which both remaining solutions require.)

     ```
     # check if /usr/local/bin/wget exists use that
     if [ -f /usr/local/bin/wget ]
     then
          WGET="/usr/local/bin/wget --dot-style=mega --verbose --tries=5"
     else
          if [ -f /usr/bin/wget ]
          then
              WGET="/usr/bin/wget --dot-style=mega --verbose --tries=5"
     else
          if [ -f /usr/sbin/wget ]
          then
              WGET="/usr/sbin/wget --dot-style=mega --verbose --tries=5"
          fi
     fi
     ```

---

[10] Many system administrators (SAs) regard *wget* as a security risk, so be sure to clear the use of this utility with them. If an SA objects to installing *wget* on a permanent basis in a central location, such as in */usr/local/bin* or */usr/bin*, download *wget* to a directory where you have permissions, alter the *agentDownload* script as shown in the previous example to search for *wget* in that directory, then add the directory to the PATH of the Agent user.

[11] You can also create a soft link on Agent hosts from the existing *wget* location to */usr/local/bin* or */usr/bin*, but this is a less desirable solution because you'd have to do so on each Agent host.

- Change the PATH of the Agent user on every target host to include the directory where *wget* is located.
- Change the location of *wget* on every target host to */usr/local/bin*, */usr/bin*, or to some other directory already in the PATH of the Agent user.

6. **Confirm that the jar executable is in the PATH**   The *agentDownload* script uses the *jar* executable to unjar the OUI installer that will be downloaded from the OMS. Confirm that the *jar* executable file is in the PATH for the *oracle* user on the target host by executing the following command:

   ```
   which jar
   ```

   This should return */usr/bin/jar* or other (nonstandard) path. Any location is fine, as long as this executable is in the PATH. The *agentDownload* script itself uses the above "*which*" command to determine the location of the *jar* executable.

Confirm WSH Version on Target Host (Windows Only)

If the target host where you are deploying the Agent is running on a Windows platform, the command-line version (*Cscript.exe*) of Windows Script Host (WSH) version 5.6 is required to run *agentDownload*. (There is a Windows-based version called *Wscript.exe*, but it is not required.) A script host is a program that provides an environment to execute scripts in a variety of languages that use different object models to perform tasks. WSH provides command-line switches for setting script properties. You run scripts with Windows Script Host at the command prompt by typing **Cscript.exe** followed by the name of a script file, which in this case is *agentDownload.vbs*.

Windows Script Host is bundled with each Windows OS version certified to run the Agent. To display your installed version of WSH, type **Cscript**[12] at the command prompt and then press ENTER. If you have WSH 5.6 installed, you should see output similar to this, followed by usage information:

```
C:\Windows> Cscript
Microsoft (R) Windows Script Host Version 5.6
Copyright (C) Microsoft Corporation 1996-2001. All rights reserved.
```

Windows Vista and 2008 come with the latest WSH version 5.7. See Table 7-3 for the Cscript version included with other Windows OSs and URLs to download the latest versions on these OSs.[13]

Copy agentDownload Script from OMS to Target Host

Now that you have corrected *agentDownload* at the source location on the OMS host, download this script to each target host, perhaps to the Agent user's home directory, because you know this user has read, write, and execute privileges in this directory. For a cluster Agent installation, you only need to download *agentDownload* to one cluster node. Use whichever method you'd like to place the *agentDownload* script on the destination host.

---

[12] The *Cscript.exe* executable is located under %SystemDrive%:\Windows\system32\, which is in the PATH. So, you should not need to enter the full path to this executable.

[13] See http://msdn.microsoft.com/en-us/library/9bbdkx3k.aspx for more information on WSH.

Operating System	Windows Script Included	Latest Available	Download Upgrade From
Windows Server 2003	5.6.0.8515	5.7.0.16535	http://www.microsoft.com/downloads/details.aspx?FamilyID=F00CB8C0-32E9-411D-A896-F2CD5EF21EB4&DisplayLang=en
Windows XP	5.6.0.6626	5.7.0.16535	http://www.microsoft.com/downloads/details.aspx?familyid=47809025-D896-482E-A0D6-524E7E844D81&displaylang=en
Windows 2000	2.0 (also known as WSH 5.1)	5.7.0.16535	http://www.microsoft.com/downloads/details.aspx?familyid=c03d3e49-b40e-4ca1-a0c7-cc135ec4d2be&displaylang=en

TABLE 7-3. *Where to Download Latest Available Windows Script Host Version for some OSs*

The most straightforward way is to employ an OS command (*ftp, rcp, scp*) to copy the appropriate *agentDownload* script, named "*agentDownload.<platform>*" from the OMS directory $OMS_HOME/sysman/agent_download/<version>/<platform>/ to the target host.

The alternative way to download the *agentDownload* script from the OMS to each target host relies on OMS functionality. The OMS installation creates a web server alias called *agent_download* at the URL http://<OMShost>:4889/agent_download/ that maps to the physical directory *$OMS_HOME/sysman/agent_download/* in the OMS Application Server home. Use this web server alias to download the platform-specific script from the OMS to the target host either from the command-line using *wget*, or from within a browser window:

- Log in to the target host as *oracle* and pull the script from the OMS host to the current directory using the *wget* utility at the command-line as follows:
  - For UNIX:

    wget http://<OMShost>:4889/agent_download/<version>/<platform>/agentDownload.<platform>
  - For Windows:

    wget http://<OMShost>:4889/agent_download/<version>/win32/agentDownload.vbs
- Download the script by accessing the URL http://<OMShost>:4889/agent_download/.

TIP
You need to append the slash to the above web server URL for this alias to work.

The script is actually located below this URL under the subdirectories *<version>/<platform>*. To download the script using a Mozilla browser for example, log in as the *oracle* user on the target host and open a browser window in an Xwindows session by typing **mozilla**, entering the above URL, clicking the *<version>*, then the *<platform>* directory to drill down to and select the *agentDownload* script, selecting File, right-clicking and selecting Save Page As…, then saving under the *oracle* user's home directory (change Files of type from Text File to All Files).

Execute agentDownload on Target Host

Now it's time to actually execute the *agentDownload* script. Log in to the target host as the *oracle* user in a command-line session using *telnet* or *ssh*, for example. Running the *agentDownload* script does not require an xterm session.

TIP
As of GC 10.2.0.2, the Agent uploads by default to a secure OMS. However, agentDownload *does not secure the Agent by default until GC 10.2.0.3, and only if Agent and OMS are on the same platform. The* agentDownload *script accomplishes this by storing the Agent Registration password in an encrypted format. As a catch-all to guarantee the Agent is automatically secured (for cross-platform deployments, for example), set the AGENT_INSTALL_ PASSWORD environment variable on the Agent host before running* agentDownload, *as instructed in Chapter 5.*

As mentioned, the *agentDownload* syntax is different for UNIX than it is for Windows because you must call the Windows version using *Cscript.exe*, although the options are the same:

On UNIX platforms

./agentDownload.*<platform>* [-bcdhilmnoprtuxN]

On Windows platforms

Cscript.exe agentDownload.vbs [-bcdhilmnoprtuxN]

You can use *agentDownload* to install an Agent on a single node, on each RAC cluster node individually, or on all nodes of a RAC cluster in one operation. The *agentDownload* script has command options to install all three variations of Agents, and offers more options to control other installation aspects. All options are shown alphabetically in Table 7-4. Example syntax for these parameters is shown in the right-most column, including sample values for parameters that require them.

Let's take a look at the *agentDownload* command syntax for the following three variations of Agent deployment, using both generic and example values to better illustrate this syntax:

- Install an Agent on a single node
- Install an Agent on each RAC cluster node individually
- Install a cluster Agent on a RAC cluster

Parameter	Parameter Value Required	Description	Example Value for Parameter
-b	Yes	Installation base directory. The script appends *agent10g* to this directory specification.	-b /u01/app/oracle/product/10.2.0/em10g
-c	Yes	Cluster node list. Used during installation only. Specify public node names in double-quotation marks, separated by commas—no spaces are allowed.	-c "node1,node2,node3"
-d	No	Do not initiate automatic target discovery	-d
-h	No	Display and describe all options you can use with this script	-h
-i	Yes	Inventory pointer location file	-i /etc/oraInst.loc
-l	No	Specify to install only on the local node of the cluster	-l
-m	Yes	Specify OMS host name for downloading the Agent installation code	-m omshost
-N	No	Do not prompt for Agent Registration password	-N
-n	No except for 9*i* clusters	Tells the Agent to discover an existing cluster. Required for 9*i* clusters. Optional for 10*g* clusters, and can provide a cluster name[14] as an argument to override the default cluster name (*crs*) assigned during the Clusterware installation.[15]	-n crs
-o	Yes	Specify the old ORACLE_HOME during an upgrade	-o /u01/agent10g

TABLE 7-4. *Parameters Available for agentDownload Script*

---

[14] To determine the cluster name of an existing cluster in Oracle10*g*, execute *$ORA_CRS_HOME/bin/cemutlo -n*, and for Oracle9*i*, execute *$CRS_HOME/bin/cemutls -n*.

[15] For Oracle9*i* clusters, you must set the CLUSTER_NAME environment variable to the cluster name for Grid Control to discover that cluster. It is good practice to set the CLUSTER_NAME variable for Oracle10*g* to override the default cluster name. This avoids the problems associated with multiple clusters of the same default name (see Oracle*MetaLink* Note 429068.1 for one of these problems).

Parameter	ParameterValue Required	Description	Example Value for Parameter
-p	Yes	File location for static port file for Agent. Template file is located under <DVD>/response/staticports.ini. Copy file to hard disk and edit local copy.	-p /home/oracle/staticports.ini
-r	Yes	Specify port for connecting to the OMS host. The default port is 4889.	-r 4889
-t	No	Do not start the Agent after installation or upgrade	-t
-u	No	Use to upgrade the Agent	-u
-x	No	Enable shell debugging	-x

TABLE 7-4. *Parameters Available for* agentDownload *Script (Continued)*

Install an Agent on a Single Node

To install a local Agent on a single node, execute the following command:

./<agentDownload> -b <base_dir> -i   ./agentDownload -b /u01/app/oracle/product/10.2.0/
<Inventory_pointer_file>   em10g -i /etc/oraInst.loc

This will install the agent in the directory */u01/app/oracle/product/10.2.0/em10g/agent10g* and will lay down the Inventory in the directory defined by *inventory_loc* in */etc/oraInst.loc*. The *-i* option is not mandatory, but allows you, if desired, to keep the Agent Inventory separate from the other Inventory of installed Oracle products on a destination host. The Inventory option also reminds you at the very least to make sure you are using the correct Inventory pointer location file.

Install an Agent on Each RAC Cluster Node Individually

You can install a local Agent on each node of a RAC cluster (as in this section), or on all RAC nodes in a single installation operation (as in the next section). The Agent will use the cluster name specified during the Oracle Clusterware installation, unless you override it by specifying a new cluster name after the *-n* option (for Oracle10*g* only), or by setting the CLUSTER_NAME environment variable before running *agentDownload* (for Oracle10*g* and required for Oracle9*i* if *-n* is not used).

To install an Agent on each node of a RAC cluster, so that the Agent is capable of managing that cluster, run the following command on each cluster node:

./<agentDownload> -b <base_dir> -i   ./agentDownload -b /u01/app/oracle/
<Inventory_pointer_file> -n [<clustername>]   product/10.2.0/em10g -i /etc/oraInst.loc -n

Install a Cluster Agent on a RAC Cluster

To install a cluster Agent on all nodes of a RAC cluster in one operation, for example on a two-node UNIX cluster consisting of *node1* and *node2*, enter the command below from just one of the cluster nodes:

./<agentDownload> -b <base_dir> ./agentDownload -b /u01/app/oracle/
-i <Inventory_pointer_file> -n product/10.2.0/em10g -i /etc/oraInst.loc -n -c
[<clustername>] -c "node1,node2, ..." "node1,node2"

This is the same syntax as the command above, but in addition the *-c* option is specified to list all cluster nodes.[16]

Final Installation Steps for All Three Variations

Regardless of which *AgentDownload* variation you choose, you must run *root.sh* on each RAC node, even for cluster Agent installations. Near the end of a UNIX installation, the following screen output is displayed:

To execute the configuration scripts:

1. Open a terminal window
2. Log in as "root"
3. Run the scripts
4. Return to this window and click "OK" to continue[17]

However, further screen output immediately follows, then the installation completes and a command prompt is returned, which is confusing to say the least. At this point, change directories to the Agent home directory, log in as *root*, and run *root.sh*:

cd <base_dir>/agent10g cd /u01/app/oracle/product/10.2.0/em10g/agent10g
su su
./root.sh ./root.sh

The final step for all variations of the *agentDownload* process is to check the installation log files. The *agentDownload* script performs all prerequisite checks, but bypasses displaying the output of these checks during the installation. The script still records prerequisite check output in the installer logs written to the installation base directory, which in the example above is

---

[16] You can install a cluster Agent on chosen nodes of a RAC cluster, but this would not be a good practice. Either install an Agent on each cluster node, or install a cluster Agent on all cluster nodes.

[17] You can ignore step 4 from the above output, which was intended as a GUI installation instruction and is incorrectly displayed here by the agentDownload script.

/u01/app/oracle/product/10.2.0/em10g. After the *agentDownload* script completes, check these log files for installation errors.

Agent Cloning Installation

Agent Cloning is an easy way to deploy non-NFS, non-RAC Agents to multiple destination hosts. This method relies upon a working Agent configuration to serve as the source Agent. You can install this source Agent using any non-NFS or non-RAC method, excluding Agent Cloning itself; this means using *agentDownload*, interactive, silent, or Agent Deploy[18] methods. Once you install the source Agent, patch it as required. You only need to patch one Agent for each target platform (both OS and version) in your environment; then you can clone that Agent to all other target hosts on that platform. The principal advantage of Agent Cloning over other installation methods is that you can clone the Agent in a patched state to avoid having to patch all target Agents in the same manner. The more the source Agent has been patched, and the more Agents you need to clone, the more time you save by not having to patch these cloned Agents.

The Agent Cloning process is rather simple, both conceptually and in practice. Here is one admittedly long sentence describing the broad strokes. Install and patch the source Agent; compress and copy it to another host; clone it using the *runInstaller* command with the appropriate options; secure the cloned Agent; run the Agent Configuration Assistant (*agentca*) to discover targets on the destination host; then finally, run *root.sh* against the cloned Agent.

Let's go through the particulars of Agent Cloning on a UNIX host running the Bash shell. The example is to install a source Agent on *<OMAhost1>* into Agent home */u01/app/oracle/product/10.2.0/em10g/agent10g*, then clone this Agent to destination host *<OMAhost2>*, which must run on the same OS and version. For clarity in specifying all Agent commands, assume that *$AGENT_HOME* is defined on the source host. Here are the specifics:

1. Install a source non-NFS, non-RAC Agent on target host *<OMAhost1>* using the interactive, silent, *agentDownload*, or Agent Deploy methods. For instructions on using the first three methods, see the appropriate sections in this chapter. To use Agent Deploy, see Chapter 6 for installation instructions.

2. Patch and certify that the source Agent works on *<OMAhost1>* as required. As for patching the Agent, see the first section in Chapter 4 entitled "Patch Grid Control" for recommendations on the types of Agent patches to consider. These include the latest EM patch set and Critical Patch Update (CPU), and any one-off patches functionally required, over and above the latest EM patch set. Once the Agent is patched, certify at the very least that it can upload management data to the OMS by doing a manual upload:

   ```
   $AGENT_HOME/bin/emctl upload agent
   ```

   You may want to do more extensive testing by configuring all targets on the source Agent host to confirm that the Agent is working as expected and that you are receiving alerts.

---

[18] It's unlikely that you'll choose Agent Deploy for a single Agent installation, as it's the lengthiest installation method of all and lends itself best to mass Agent deployment.

3. Use a Zip utility or the UNIX *tar* and *compress* commands (as shown below) to compress the source Agent home and copy it to the destination host. On source Agent host *<OMAhost1>*, change directory to the Agent home, then create a compressed tar ball named *agent.tar.Z* of the Agent home (the name is arbitrary):

   ```
   cd $AGENT_HOME
   ```

   ```
   tar cvf - * | compress > agent.tar.Z
   ```

4. Create the Agent home directory on the destination host *<OMAhost2>*. For consistency, use the same Agent home directory for all cloned Agents as for the source Agent:

   ```
   export AGENT_HOME=/u01/app/oracle/product/10.2.0/em10g/agent10g
   mkdir -p $AGENT_HOME
   ```

5. Use any method (*rcp*, *scp*, or *ftp*, for instance) to copy the Agent archive from source to destination host (here I assume *scp* is configured between them). Run the following commands on destination host *<OMAhost2>*:

   ```
   cd $AGENT_HOME
   scp <OMAhost1>:/u01/app/oracle/product/10.2.0/em10g/agent10g/
   agent.tar.Z ./.
   ```

6. Uncompress and untar the file created in step 3 on destination host *<OMAhost2>*:

   ```
   zcat agent.tar.Z | tar xvf -
   ```

   or if the *zcat* command is not installed, use *uncompress*:

   ```
   uncompress agent.tar.Z | tar xvf -
   ```

7. Run the Agent installer at the command line on *<OMAhost2>* to clone the Agent on this host. In the newly unpacked Agent home, change directories to the *$AGENT_HOME/oui/bin* directory, and execute *runInstaller* with the following options (enter the command on one line or on multiple lines using the "\" continuation character as shown below):

   ```
   cd $AGENT_HOME/oui/bin
   ./runInstaller -clone -forceClone   \
   ORACLE_HOME=$AGENT_HOME   \
    ORACLE_HOME_NAME=agent10g -noconfig -silent
   ```

8. Execute the following command on destination host *<OMAhost2>* to secure communications between the cloned Agent and the OMS or SLB (the destination Agent is configured to upload to the same OMS or SLB as the source Agent):

   $AGENT_HOME/bin/emctl secure agent [*<agent_registration_password>*]

   If you don't specify the Agent Registration password at the command line, you are prompted to enter it.

9. Run the Agent Configuration Assistant (*agentca*) on destination host *<OMAhost2>* to configure the Agent and discover its targets:

   ```
   $AGENT_HOME/bin/agentca -f
   ```

10. Finally, to complete the cloning process, run the *root.sh* script as *root* on destination host *<OMAhost2>*:

    ```
    cd $AGENT_HOME
    su
    ./root.sh
    ```

    When prompted, press ENTER to accept all defaults.

This concludes installation of an Agent using the Agent Cloning method. As you can see, the installation process is very straightforward, and easily reproducible. As all commands are executed at the command line, you can script the entire installation process using shell or Perl commands, for example. The same applies to Agent Cloning on a Windows platform.

Silent Agent Installation

Silent Agent deployments offer two principal advantages over interactive installations. First, you can automate the installation process, as it does not require user intervention. Second, silent installations on UNIX platforms do not require Java, so you can deploy Agents on UNIX hosts where X Server is not installed or running. Silent installations use a response file to install Agents. The EM Grid Control software distribution, available on DVD from the Oracle Store or downloadable from OTN, provides a template response file called *additional_agent.rsp* that you must edit for your particular environment before using. There are similar response files for the other three main options to install Grid Control using a New or Existing Database, and to install an additional Management Service. All response files, including that for the Agent, are located in the EM software distribution under the directory *<top-level directory>/response/*.

This template file contains variable-value combinations. Some variables have actual values assigned, such as *b_silentInstall=true*. Other variables are assigned the placeholder value *<Value Required>* for a mandatory value or *<Value Unspecified>* for an optional value.

This brief background should be sufficient to begin the Agent silent installation process, which is quite straightforward. Perform all steps as the Agent user on the target host, unless otherwise specified.

1. Stage either the EM Grid Control DVD or the installation software downloaded from OTN to a local or shared directory on the target host. For detailed instructions, see the text box at the beginning of Chapter 2 entitled "Stage the Grid Control Software."

2. Copy the Agent response file from the EM software distribution location, *<top-level_dir>/response/additional_agent.rsp*, to a central staging location where you can edit the file and substitute common values applicable to all target hosts. The resulting file will be a customized template, if you will, for your GC environment. A convenient central location for this customized template is under *$OMS_HOME/sysman/agent_download/<version>/* on an OMS host.

3. Edit *additional_agent.rsp* in the staging location, substituting for the *<Value Required>* or *<Value Unspecified>* placeholders as shown in Table 7-5. As we've done earlier in this chapter, we use the example of a two-node cluster Agent installation on *node1* and *node2*, where the Agent is to communicate with an OMS host, *<OMShost>*. The columns in this table (from left to right) provide the exact syntax for specifying variable name

Variable_name=<value>	Description of Variable	Mandatory or Optional	Common or Target-Specific Value
UNIX_GROUP_NAME="oinstall"	UNIX group to set for the Inventory directory. Defaults to *inst_group* value in *oraInst.loc*.	Optional	Common
BASEDIR="/u01/app/oracle/product/10.2.0/em10g"	Installation base directory (installer appends *agent10g*)	Mandatory	Common
CLUSTER_NODES="node1,node2"	Cluster node names on which to install a cluster Agent.	Mandatory (for cluster Agent installation)	Target-specific
REMOTE_NODES="node1,node2"	Remote node names on which to install the Agent. Use for cluster Agent installations.	Mandatory (for cluster Agent installation)[19]	Target-specific
LOCAL_NODE="node1"	Local host where installing the Agent	Mandatory	Target-specific
b_doAgentConfig=true	Controls whether Agent Configuration Assistant runs. Set to false if installing a master NFS Agent.	Mandatory	Common
ORACLE_HOSTNAME="<hostname_alias>"	Allows use of an alternate Agent host name than is returned by the *hostname* command, such as for virtual host.[20]	Optional	Target-specific
sl_OMSConnectInfo={"<OMShost>","4889"}	OMS host name and port (or SLB and port) with which Agent communicates	Optional	Common (if using SLB or single OMS)

TABLE 7-5. *Description of Variables and Values in additional_agent.rsp File*

---

[19] The response file lists the value for REMOTE_NODES as <*Value Required*>, but it is <*Value Unspecified*> unless installing a cluster Agent.

[20] For more on virtual hosts, see the section in Chapter 15 entitled "Configure the Agent in Active/Passive Environment."

and value,[21] a description of the variable's function, whether it is mandatory (*<Value Required>*) or optional (*<Value Unspecified>*), and whether the value is *target-specific* or whether you can likely use a *common* value for all installations. In this step, fill in only the common values—wait until after copying this template response file to each target host in the next step to enter target-specific values.

If you're installing an Agent from a Mass Agent Deployment distribution, the Agent response file is called *agent.rsp* and is located in the directory *<platform>/agent/install/ response/*. In this case, you cannot install a cluster Agent or use an alternate host name. For this reason, there are far fewer variables. Following are the variables, all of which are mandatory (i.e., can't be commented out) and require target-specific values (or, in the case of the last variable, must be set to null):

Variable_name=<value>	Description of Variable
ORACLE_HOME="/u01/app/oracle/product/10.2.0/em10g"	Installation base directory (installer appends *agent10g*)
ORACLE_HOME_NAME="OraEM10g"	The Oracle home name used in creating folders and services
b_upgrade="False"	Whether this is an upgrade of an existing installation
s_oldAgentHome=""	Oracle home of Agent installation being upgraded (if any)

4. Copy the response file you just edited from the OMS staging location to the host where you are performing the silent Agent installation. If you are installing a cluster Agent, copy the response file to just one of the nodes. The Agent user's home directory is a good place to copy the file, as the Agent has full privileges in this directory. The following command illustrates using *scp* (if configured) to copy the response file from the OMS host, in the directory suggested above, to the Agent user's home directory on the target host. (If *scp* is not configured on both OMS and target hosts, use *ftp* or *rcp* to copy the response file.) Execute this command from the OMS host to push the response file to the Agent host:

    scp $OMS_HOME/sysman/agent_download/*<version>*/additional_agent.rsp *<OMAhost>*:/*<OMA_home_directory>*/.

    Once you copy the template response file to a local host, edit this file and input all target-specific values identified in Table 7-5.

5. Log in to a command-line session on the Agent host as the desired Agent owner. Run the same GC *runInstaller* or *setup.exe* executable (for UNIX or Windows, respectively) used to install the OMS from the *install* directory where the EM DVD or downloaded installation software was staged in step 1 above.[22] However, specify the *-silent* and *-responseFile* options as follows (shown for a UNIX host):

    ./runInstaller -silent -responseFile $HOME/additional_agent.rsp

---

[21] Comma-separated strings cannot contain spaces. Strings must be enclosed in double quotes, but boolean values do not require quotes.

[22] Prior to EM 10.2, running the silent installer on UNIX required that Xwindows libraries be installed, although you did not need to actually start an xterm session to run the silent installer. As of EM 10.2, Xwindows libraries are no longer required.

6. If this is the first Oracle product installation on the target host, you are prompted to run *orainstRoot.sh* as *root* (assumed here to be in the OFA-compliant location, ORACLE_BASE. However, the *oraInventory* directory will be in the location defined by *inventory_loc* in the *oraInst.loc* file.):

   ```
   cd $ORACLE_BASE/oraInventory
   su
   ./orainstRoot.sh
   ```

7. Run *root.sh* as the *root* user from the Agent home directory when prompted (the value shown for AGENT_HOME below is that listed in Table 7-5):

   ```
   cd /u01/app/oracle/product/10.2.0/em10g/agent10g
   su
   ./root.sh
   ```

8. Stop and secure the Agent (either supply the Agent Registration password on the command line or wait to be prompted for it):

   ```
   emctl stop agent
   ```

   emctl secure agent [<agent_registration_password>]

For a log file of the installation, see *<oraInventory_dir>*/logs/silentInstall*<RRRR-MM-DD_HH_MI_SS[AM|PM]*.log. Check this log file for errors.

Interactive Agent Installation

The interactive Agent installation is perhaps the easiest to accomplish, provided you don't have Xwindows display problems firing up the installer. This section presents the steps to interactively install one Management Agent at a time, or multiple Agents in one operation on all RAC cluster nodes by selecting the Cluster Installation mode.

Install Required Packages

Table 7-6 lists the packages required for certain platforms when using the interactive installation method to install the Agent, as well as the commands to check for the existence of these packages. Platforms not listed do not require any additional packages. These packages relate to your Xwindows display and are a very small subset of those packages required for the Agent Deploy installation method. You can use the GC installer itself to check that all packages are present. The Oracle Universal Installer (OUI) executable is called *runInstaller* on UNIX and *setup.exe* on Windows. On the chosen node, and from the location specified in the next section, run the installer with the *-executeSysPrereqs* parameter (referenced in Table 7-7):

- On UNIX, enter: ./runInstaller -executeSysPrereqs
- On Windows, enter: ./setup.exe -executeSysPrereqs

Operating System	Platform	Package(s) Required	Command to Check for Packages
HP-UX	HP-UX 11.11 PA-RISC2.0	X11MotifDevKit version 0.0, X11MotifDevKit.MOTIF21-PRG version 0.0	/usr/sbin/swlist -l product I grep X11
Linux x86 (32-bit)	Red Hat Enterprise Linux AS/ES 3.0	openmotif21-2.1.30-9[23]	rpm -qa I grep <package name>
	Red Hat Enterprise Linux AS/ES 4.0	openmotif21-2.1.30-11,[24] openmotif-2.1.30MLI4 version 0.0[25]	rpm -qa I grep <package name>
	SUSE Linux Enterprise Server 8	openmotif version 2.1.30-11, openmotif-2.1.30MLI4 version 0.0	rpm -qa I grep <package name>

TABLE 7-6. *Package Requirements for Interactive Installation of the Agent*

Run the Interactive Installer

The source software location varies from which you run the installer to deploy the Agent interactively:

- If you're installing an Agent on the same platform as the OMS host, use the GC software distribution that you used to install the OMS. This distribution is available on DVD or you can download it from the Oracle Technology Network (OTN). For detailed staging instructions, see the text box called "Stage the Grid Control Software" at the beginning of Chapter 2. The installer executable is located under the top directory of this distribution.

- If you're installing a cross-platform Agent, download the Agent kit according to the instructions in Chapter 5 under the section "Prepare for Cross-Platform Agent Installation". The installer executable is located under the directory *<unzipped_directory>/<platform>/agent/*.

Following are instructions, including screen shots, for installing a cluster Agent interactively. The screen shots are from an installation on a two-node Linux cluster running on Red Hat 4 (x64). The cluster node names are *ptc1* and *ptc2*, and the Agent home directory specified is */opt/oracle/product/10.2.0/em10g/agent10g*.[26] A Cluster installation is the same as that of a single Agent, except the latter will not display the screen in step 5 below entitled Specify Hardware Cluster

---

[23] This package is listed incorrectly without the "-9" as "openmotif21 version 2.1.30" in 10*g* R2 Management Agent Deployment Best Practices, An Oracle White Paper, June 2006.

[24] This package is listed incorrectly as "openmotif version 2.1.30-11" in 10*g* R2 Management Agent Deployment Best Practices, An Oracle White Paper, June 2006.

[25] This package appears to be unavailable on the Red Hat Network, and Agent installations work without it.

[26] The path */opt/oracle* is not the default Optimal Flexible Architecture (OFA)-compliant ORACLE_BASE directory—*/<mount_point>/app/oracle* is. Whatever ORACLE_BASE value you choose for the Agent, it's good practice to standardize on it across all hosts in your environment, regardless of UNIX flavor. Except for mount point differences, you can even use the same ORACLE_BASE value on Windows hosts.

Installation Mode. **Bolded** titles for these steps match the OUI screen titles and **bolded** substeps match the OUI screen field names.

1. If installing a RAC Agent, make sure all RAC nodes meet both of the following requirements (if you don't, the screen shown in step 5 that permits a cluster installation will not appear):

   a. Ensure the Oracle Clusterware is started.

   b. Make sure the *oraInst.loc* file points to the Inventory for the Oracle Clusterware installation (and any other Oracle installations sharing that Inventory) or specify the *-crslocation* option to the installer (see step 2 below).

2. On one of the cluster nodes, run the OUI from the GC installation software distribution.[27] To start the GC installer interactively, log in as the *oracle* user to own the GC software, then execute one of the following commands, depending upon the platform:

   ■ On UNIX, enter: ./runInstaller [options]

   ■ On Windows, enter: .\setup.exe [options]

   The above "dot execute" syntax (".\ or ./") assumes you are running the installation from the hard drive where the software is staged. If installing from DVD, do not start the installer below the mount point directory (as shown above), as you may not be able to eject the DVD disk. In other words, type the path to the installer, such as *<DVD>/Disk1/runInstaller*. You can use any of the installation options listed in Table 7-7,[28] although no options need to be specified in this example.

   If you need to specify proxy information, do so as follows when invoking the OUI:

   [./runInstaller | ./setup.exe] oracle.sysman.top.agent:s_proxyHost="*<proxy_host>*"
   oracle.sysman.top.agent:s_proxyPort="*<proxy_port>*"

3. **Specify Installation Type** Choose the last of the four options shown in Figure 3-1 in Chapter 3, which is to install an Additional Management Agent:

4. **Specify Installation Location** This screen is the same as appears in Figure 3-2 when installing Grid Control.

   ■ **Parent Directory** Enter the parent directory (or installation base directory) for the Agent. The installer creates the *agent10g* subdirectory below this base directory. Enter */opt/oracle/product/10.2.0/em10g* for this example.

   ■ **Product Languages** Click this button if you want to install the Agent in a language other than English. The Language Selection screen appears, allowing you to choose one or more languages for the Agent. The language of the Agent installation itself does not change, but remains English. The installer lays down text in the selected languages and the fonts required to display these languages in the Console.

---

[27] Don't forget to follow all preinstallation steps in Chapter 5, including "Initialize the oracle User Environment." I remind you again because these steps are particularly important for an interactive install.

[28] This information comes from Appendix D, Table D-2, Oracle Enterprise Manager Grid Control Installation and Basic Configuration 10g Release 4 (10.2.0.4.0).

Parameter	Description
-cfs	Oracle home specified is on the cluster file system. This is mandatory when -local is specified so that OUI can register the home in the Inventory
-clusterware oracle.crs, <crs_version>	Specifies version of the installed Oracle Clusterware
-crslocation <path>	For cluster installs, specifies the path to the CRS home location. Specifying this overrides CRS information obtained from the Central Inventory.
-debug	Provides debug information from the OUI
-executeSysPrereqs	Executes system prerequisite checks only
-force	Allows silent mode installation into a nonempty directory
-help	Displays usage description of all parameters
-ignoreSysPrereqs	Ignores the results of the system prerequisite checks[29]
-invPtrLoc <full path of oraInst.loc>	Points to a nonstandard Inventory location (UNIX only)
-jreLoc <location>	Path where the Java Runtime Environment (JRE) is installed, without which the OUI cannot run. JRE is included in the EM distribution and its location specified in *oraparam.ini*, so this parameter should not be required.
-local	Installs on the local node only. Used to perform a local installation on a cluster node.
-logLevel <level>	Filters log messages with a lesser priority level than <level>. Valid options are: severe, warning, info, config, fine, finer, and finest.
-paramFile <location of file>	Location of *oraparam.ini* file. The default location is <platform>/agent/install/.
-printmemory	Logs debug information for memory usage
-printtime	Logs debug information for time usage
-responseFile <full path to response file>	Response file and path to use
-sourceLoc <full path to products.xml>	Software source location
-updateNodeList	Updates the node list for this home in the OUI Inventory

TABLE 7-7. *Parameters that the OUI Supports*

---

[29] This parameter is useful when you have already run *-executeSysPrereqs* and know that the target host has met all prerequisites.

5. **Specify Hardware Cluster Installation Mode** Choose a cluster Agent if you're installing on a RAC node. If the installer is run on a RAC cluster, the following screen should appear, offering the Cluster Installation or Local Installation radio buttons to install a cluster Agent on all nodes or a local Agent on one of the nodes, respectively:

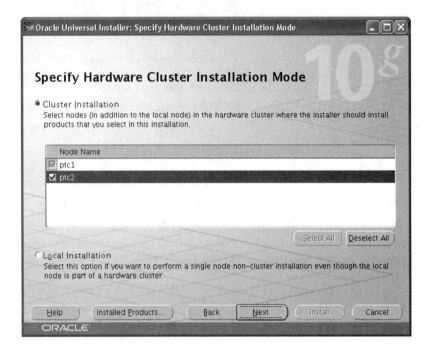

In this example, we are installing a cluster Agent (the recommended approach for RAC targets), so choose the Cluster Installation radio button. Under Node Name, select all nodes in the hardware cluster (in addition to the local node, which is grayed out because you must install the Agent locally).

6. **Specify Inventory directory and credentials** Choose the Inventory location. If this is a UNIX host and is the first Oracle product being installed, which is not the case in this example, then a screen by this name will appear, requesting the desired path to the new Inventory the installer will create (typically *$ORACLE_BASE/oraInventory*, although you need to substitute the actual ORACLE_BASE value here) and the Inventory group (typically *oinstall* or *dba*):

   - **Enter the full path of the inventory directory** Enter the full path to the desired central *oraInventory* directory.

   - **Specify Operating System group name** Specify the group owner you want for the Inventory, typically *oinstall* or *dba*.

7. **Product-Specific Prerequisite Checks** Resolve any failed prerequisite checks. This screen, similar to the one shown in Figure 3-5 of Chapter 3, displays the name, type

(Automatic, Optional, or Manual), and status (Succeeded, Warning, Skipped, or Failed) of all prerequisite checks:

- If all prerequisite checks are successful, click Next. These prerequisite checks should all succeed if you ran the PrereqChecker in standalone mode as instructed above in the section "Install Required Packages."

- If a check does not succeed, click Stop to halt the remaining checks. Click the failed prerequisite check to view its corresponding details at the bottom of the screen. These details show expected vs. actual results, error messages, and instructions to resolve them. Fix the failed check manually if possible, select the check, and click Retry. After resolving all checks as best you can, click Next. An error message appears if any recommended prerequisite check fails. Click No to rerun the check or Yes to ignore the error and proceed with the installation.

TIP
If you're installing the Agent on RHEL3, you can ignore the following error in Product-Specific Prerequisite Checks as the later package indicated is not available on this OS:
"Checking for openmotif-21-2.1.30-11; found openmotif-2.2.3-9. RHEL4.1. Failed <<<<"

8. **Specify Oracle Management Service Location**   This screen provides information for the OMS host with which the Agent is to communicate.

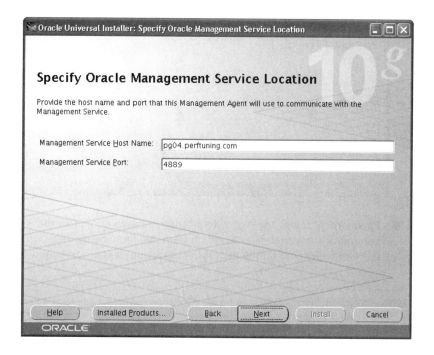

- **Management Service Host Name**   Use the fully qualified host name of one of the OMS hosts, preferably the closest host geographically to the Agent host.

- **Management Service Port**   Enter the port number for the Management Service. The default nonsecure port number is 4889. Even though the OMS is secure and locked by default, you must enter the nonsecure port number to connect via HTTP to receive the certificate. Then the Agent can begin communicating with the OMS securely over HTTPS.

    If you use multiple OMS hosts behind a Server Load Balancer (SLB), rather than entering a specific OMS and port, enter the SLB host name and port in these fields.[30]

9. **Specify Agent Registration Password**   If you didn't set the environment variable AGENT_INSTALL_PASSWORD before launching the interactive installer, the following screen will appear to allow you to enter this password:

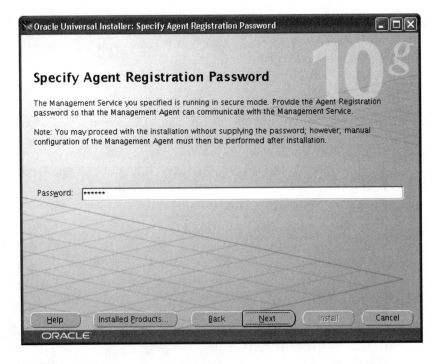

- **Password**   Provide the Agent Registration password to allow the Agent to communicate with the OMS, which is enabled for SSL by default. This is the password specified during the OMS installation. If you don't know the password,

---

[30] Alternatively, specify one OMS host and port for the installation. Then, after the Agent is installed, edit the Agent properties file, *$AGENT_HOME/sysman/config/emd.properties*, change the host name and port specified for REPOSITORY_URL to those of the SLB, and restart the Agent.

ask the administrator who installed GC.[31] If you can't obtain a valid password at this time, leave this field blank, but in order for the Agent to communicate with the OMS, you must later secure the Agent by issuing the command *emctl secure agent.*

10. **Oracle Configuration Manager Registration**   This screen is the same one shown in Figure 3-9 of Chapter 3 when installing Grid Control. Enable OCM if desired. As of GC 10.2.0.2, the OUI is bundled with Oracle Configuration Manager (OCM). If you're only installing a handful of Agents, this is much more convenient than manually installing and configuring OCM later (covered in Chapter 8 under the heading, "Install Oracle Configuration Manager Client").

    - **Enable Oracle Configuration Manager**   Check this box if you elect to enable OCM now. In this case, you must complete all fields on this screen. When you click Next, the OUI prompts you to click Accept License Agreement.

    - **Customer Identification Number (CSI)**   Enter your Customer Support Identifier (CSI) here, which is dubiously labeled "Customer Identification Number." The Oracle*MetaLink* administrator of this CSI will receive a confirmation e-mail that OCM has been enabled on this server.

    - **Metalink Account Username**   Enter the username of the administrator associated with this CSI.

    - **Country Code**   Enter the country code for the MetaLink username. To confirm, log in to Oracle*MetaLink* as this user and see the Profile section under the Licenses link.

    - **Connection Settings**   Click this button if you must connect from this host through a proxy server to reach the Internet. You are prompted for the name and port of the proxy server, and proxy username and password if the proxy server requires authentication.

    - **Test Registration**   Click this button to test the connection between the host and the OCM service. You cannot proceed past this screen until the information has been verified, so you may as well test the registration using this button. If the test fails, and you cannot obtain the correct OCM information at this time, uncheck Enable Oracle Configuration Manager and configure OCM manually later on (again, see Chapter 8). A log of all OCM installation steps and errors is located under the directory *$AGENT_HOME/ccr/log.*

11. **Summary**   Review the Summary screen, similar to that for any of the first three GC installation types (see Figure 3-10 in Chapter 3), to confirm all installation options selected. These options include Global Settings (including the Oracle home and cluster nodes selected), Product Languages, Space Requirements, and New Installations.

    Verify the choices you made and click Install to begin the Agent installation.

---

[31] If the administrator doesn't recall the password, a super administrator can add a new password by navigating to Setup: Registration Passwords, then clicking Add Registration Password.

12. **Execute Configuration scripts** After the installation runs for a while, this popup will appear. Run the listed configuration scripts as *root* on each node where a RAC Agent was installed:

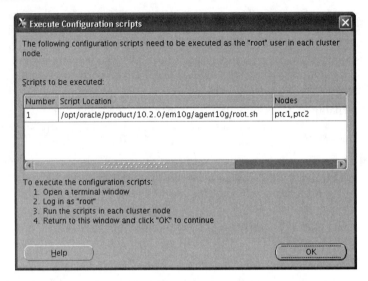

The script *orainstRoot.sh* will only appear if this is the first Oracle product installation on these nodes. The Nodes column will not appear if the installation is not on a RAC cluster, or if it is on a RAC cluster but you chose the Local Installation mode in step 5 above. Execute *oraInstRoot.sh* (if indicated) and *root.sh* on each cluster node as the *root* user, one node at a time.

When running the scripts, hitg the ENTER key to accept the default answers. Then return to the above window and click OK to continue.

13. **End of Installation** The screen displays whether the installation of the Agent was successful, and provides the location of the Release Notes, which are under $AGENT_HOME/relnotes/README_EM.htm. Click Exit, as this is the end of the installation and click Yes to confirm that you want to exit from the installer.

14. Secure the Agent on remote nodes if you performed a cluster Agent installation. You must secure the Agent on all nodes except the local node from which you ran the interactive installer, which should already be secure:

```
cd $AGENT_HOME/bin
./emctl stop agent
./emctl secure agent
./emctl start agent
```

This completes the interactive installation of an Agent.

Agent Post-Installation Steps

Return to Chapter 4, which covers Grid Control post-installation tasks, and perform all Agent-related patching tasks mentioned in the first section, called "Patch Grid Control".

After completing Agent patching, perform the Agent post-installation steps below regardless of the Agent installation method used, unless otherwise noted. These steps are as relevant for Agent Deploy (covered in Chapter 6) as they are for the other installation methods.

- Set up Agent user environment
- Confirm Agent is working
- Refresh host configuration (if needed)
- Run *agentca* for cluster Agent (Windows only)
- Confirm Agent restart on reboot is configured
- Back up the Agent

Set Up Agent User Environment

It is convenient to be able to initialize the Agent user environment as follows:

- Define the environment variables ORACLE_BASE (to reference the Inventory), AGENT_HOME, and TZ.
- Add *$AGENT_HOME/bin* to the PATH before a target database's *$ORACLE_HOME/bin* so you can run the Agent *emctl* rather than that for DB Control and the Agent Configuration Assistant executable (*agentca*) without path qualification.

Define these variables in the login script for the Agent user on each target host (assumed to be the *oracle* user in this chapter). The login script name varies by shell: it is named *.bash_profile* for the bash shell, *.profile* for the Bourne or Korn shells, and *.login* for the C shell. Below is an example of these variable definitions for an Agent initialization script on UNIX.

```
umask 022
ORACLE_BASE=<ORACLE_BASE> #placeholder
#ORACLE_BASE=/u01/app/oracle #example value
export ORACLE_BASE
#
#OMA
AGENT_HOME=<AGENT_HOME> #placeholder
#AGENT_HOME=$ORACLE_BASE/product/10.2.0/em10g/agent10g #example value
#For NFS Agents set AGENT_HOME to the EMSTATE directory.
#For cluster Agents, set AGENT_HOME to that of local cluster node
#i.e., should end with "<OMAhost>.<domain>" such as:
#AGENT_HOME=$ORACLE_BASE/product/10.2.0/em10g/agent10g/node1.yourcorp.com #example value
PATH=$AGENT_HOME/bin:$PATH
export AGENT_HOME PATH
#
TZ=<TZ>
TZ=UTC #example value
export TZ
#
# Large AIX sites only: Set data segment size and disable loading of
# runtime libraries in kernel space; also change Threadscope context from
# default Processwide 'P' to Systemwide 'S' to reduce mutex contention.
LDR_CNTRL="MAXDATA=0x80000000"@NOKRTL
AIX_THREADSCOPE=S
```

To set the Agent environment variables on Windows platforms, do the following:

1. Log in as the Agent user, click the Start button, select Control Panel, then System.
2. In the System Properties window, click the Advanced tab, and click the Environment Variables button.
3. Under User variables for the *oracle* user, click New, enter the variable name and value, and click OK. Repeat this procedure for all variables to be defined. Add the PATH variable, defining all values separated by a semicolon.
4. Click OK to close the Environment Variables screen, then OK again to close the System Properties screen.

Execute the new or revised *oracle* user login script to initialize the environment on all Agent hosts. To initialize the environment for the Bash shell, for example, "dot" execute the login script as the *oracle* user (the dot refers to the PERIOD, which is the first character):

```
. $HOME/.bash_profile
```

An alternative way to execute the login script is to log out and log in again as *oracle*.

The remainder of this chapter assumes the Agent environment is set up as described above. For example, commands referenced later assume AGENT_HOME is defined and that *$AGENT_HOME/bin* is in the Agent user's PATH so that you do not need to specify the path to the *emctl* executable.

Confirm Agent Is Working

On the target host, the *emctl* executable located under the Agent home starts, stops, and displays the Agent's status. This utility also performs many other Agent functions. However, for now, you likely need to know only how to query an Agent's status, and if necessary to stop or start it. The Agent should already be started if the installation completes successfully, regardless of the Agent deployment method used. In the unlikely event of an Agent startup problem, see the basic Agent troubleshooting steps below.

The *emctl* executable is located in the *bin* directory under the Agent home. This location varies depending upon the Agent type as shown in Table 7-8.

Execute *emctl* from the appropriate Agent home to confirm that the Agent is running as expected. The *emctl* syntax for basic Agent control is simple:

```
emctl [ stop | start | status ] agent
```

Agent Type	Location of *emctl* Executable
non-cluster Agent	$AGENT_HOME/bin
cluster Agent	$AGENT_HOME/<*OMAhost*>/bin
NFS Agent	<EMSTATE_DIR>/bin

TABLE 7-8. *Location of Agent emctl Executable Shown by Agent Type*

The following sample output is from the example cluster Agent installation performed above in the section "Interactive Agent Installation" (the important lines are **bolded**):

```
Oracle Enterprise Manager 10g Release 4 Grid Control 10.2.0.4.0.
Copyright (c) 1996, 2007 Oracle Corporation.  All rights reserved.
Agent is already running
[oracle@ptc1 bin]$ emctl status agent
Oracle Enterprise Manager 10g Release 4 Grid Control 10.2.0.4.0.
Copyright (c) 1996, 2007 Oracle Corporation.  All rights reserved.
---------------------------------------------------------------
Agent Version          : 10.2.0.4.0
OMS Version            : 10.2.0.4.0
Protocol Version       : 10.2.0.4.0
Agent Home             : /opt/oracle/product/10.2.0/em10g/agent10g/ptc1.perftuning.com
Agent binaries         : /opt/oracle/product/10.2.0/em10g/agent10g
Agent Process ID       : 29824
Parent Process ID      : 29772
Agent URL              : https://ptc1.perftuning.com:3872/emd/main
Repository URL         : https://pg04.perftuning.com:1159/em/upload
Started at             : 2008-04-06 13:30:47
Started by user        : oracle
Last Reload            : 2008-04-10 20:37:20
Last successful upload                       : 2008-04-10 22:12:29
Total Megabytes of XML files uploaded so far :    49.37
Number of XML files pending upload           :        0
Size of XML files pending upload(MB)         :     0.00
Available disk space on upload filesystem    :    11.59%
Last successful heartbeat to OMS             : 2008-04-10 22:13:08
---------------------------------------------------------------
Agent is Running and Ready
```

Here is the meaning of these **bolded** items:

- **OMS Version** confirms that the correct OMS version is shown.
- **Agent URL and Repository URL** protocols should be the same, i.e., HTTP or HTTPS. If the Agent URL protocol is not secure (HTTP) and the Repository URL protocol is secure (HTTPS), then secure the Agent with the command *emctl secure agent*.
- **Last successful upload** should have a current timestamp. If not, manually attempt to upload Agent data with the command *emctl upload*. You should see the following output:

  ```
  [oracle@ptc1 bin]$ emctl upload
  Oracle Enterprise Manager 10g Release 4 Grid Control 10.2.0.4.0.
  Copyright (c) 1996, 2007 Oracle Corporation.  All rights reserved.
  ---------------------------------------------------------------
  EMD upload completed successfully
  ```

- **Agent is Running and Ready** should be shown. If it is not shown, and only copyright information is listed, it is likely due to one of three causes:
    - The Agent is not secure. Secure it with the command *emctl secure agent*.
    - The Agent is not running. Start the Agent with the command *emctl start agent*.

- The OMS port is not correct. Verify whether the Agent is connecting to the correct OMS port, and that the wallet URL and time zone (TZ) are correct in the *$AGENT_HOME/sysman/config/emd.properties* file, as shown here:

REPOSITORY_URL=https://<*OMShost*>:<*port*>/em/upload	Where the default secure port is 1159
emdWalletSrcUrl=http://<*OMShost*>:<*port*>/em/wallets/emd	Where the default unsecured port is 4889
agentTZRegion=<*timezone*>	Where time zone should match that set for the OMS

To troubleshoot the above problems, you may need to consult the Agent log files located in the *$AGENT_HOME/sysman/log/* directory.

After confirming the Agent is running as expected according to the above command, *emctl status agent*, log in to the Console and see whether at least the Agent and other targets appear under the Targets tab, All Targets subtab. Hopefully, the Agent also automatically discovers any single-instance databases, Oracle Application Servers, or other auto-discoverable targets on that host (discussed in Chapter 11), but this is not required for the Agent installation to be considered successful. You can always manually discover these additional targets if at least the Agent target exhibits an UP status.

Refresh Host Configuration (If Needed)

Shortly after Agent installation, the host configuration should appear in the Console on the host home page, and should reflect the OS and platform. However, particularly for NFS Agent installations, the host configuration may not be immediately available. GC automatically refreshes the configuration every 12 hours, but you can manually do so by clicking on the Host, selecting the Configuration subtab, then clicking Refresh. Confirm that the host configuration is populated.

Run agentca for Cluster Agent (Windows Only)

After installing a cluster Agent on a Microsoft Windows platform, only the host and cluster targets are discovered, regardless of the installation method employed. To discover all remaining targets, you must run the Agent Configuration Assistant (*agentca*) on each node of the cluster as follows:

```
$AGENT_HOME/bin/agentca -d -n <cluster name> -c <node name>
```

After running this command, you may need to resecure the Agent on all nodes using the command *emctl secure agent*.

Confirm Agent Restart on Reboot Is Configured

It is important that an Agent automatically start when a managed host is rebooted to ensure continued monitoring of all targets on that host. To that end, the GC installation process creates mechanisms on all platforms to automatically start the Agent (and other installed GC components) on server reboot. However, two problems exist with this Agent automatic startup process.

First, the Agent Cloning installation method does not automatically configure the Agent restart mechanism, as indicated in Table 5-1 in Chapter 5. All other installation methods configure this mechanism, with the caveat that *agentDownload* and *nfsagentinstall* do so, but only in GC 10.2.0.3 or later.

The second problem is that, even for Agent installation methods that do configure automatic Agent restart on reboot, the UNIX. Agent restart scripts are flawed and must be modified to enable Agents start on boot, although Windows Agents do start on reboot.

See Appendix E (online at www.oraclepressbooks.com) for instructions to configure and verify automatic startup on both Windows and UNIX platforms to ensure this important GC high availability feature works as intended.

Back Up the Agent

It is a good idea to begin regular backups of all Agent installations now that you have installed the m. If possible, take a baseline server-level hot tape backup of each Agent host. Less ideally, back up the Agent software home installation. See "Agent Backup and Restoration" in Chapter 14 for specific backup instructions. In a pinch, if you just can't back up the Agent homes at this time, at least back up the *$AGENT_HOME/sysman/config/targets.xml* file, which contains Agent and host target information. Target definitions are added to this file during Agent installation. While you can rediscover these targets using the Agent Configuration Assistant (*agentca*), sometimes this script encounters an error or does not work as expected. For this reason, back up the *targets.xml* file so you can restore it if this file is corrupted or accidentally removed. It's also useful to have a backup of the *targets.xml* file for reference in case you need to remove and rediscover any Agent targets. This may be required to resolve a status mismatch where the actual status of the target is UP, but the Console shows its status as DOWN.

Summary

This chapter spells out the steps to deploy Agents using local installation methods, which comprise all methods except Agent Deploy. We began with the most conceptually tricky method for DBAs to wrap their heads around, *nfsagentinstall* installation. You will likely need to work closely with your system administrator in configuring the shared storage, but you will find that *nfsagentinstall* is a reliable, space-efficient way to deploy Agents, particularly in large EM environments where you can plug in to existing NFS technology upon which your company already relies. The remaining installation methods, *agentDownload*, Agent Cloning, silent installation, and interactive installation, provide complementary installation features. Each installation method suits a particular context. Your mastery of all methods and their relative strengths provides the flexibility needed to suit any GC environment. Whatever method you choose, all roads lead to common Agent post-installation steps, which include setting up the OS environment, confirming the Agent is working, refreshing the host configuration if needed, running agentca for Windows cluster Agents ensuring Agent restart on system boot for high availability, and backing up the Agent home.

Each one of these installation methods has its own personality, strengths, and quirks. However, all methods lead to the same goal of deploying a working Agent able to monitor a host and all targets on that host, and to upload this target management information to the Management Service.

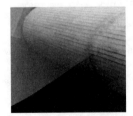

CHAPTER 8

Install Grid Control Clients

he only required client component for Console display is a certified browser. However, as mentioned in the section "Grid Control Console" in Chapter 1, you should evaluate four additional optional client installations for their relevance to your site:

- **Enterprise Manager 10g Java Console**   This thick-client is supplied with the Oracle 10g Client software; it contains select functionality such as Change Manager that is still not converted to the Grid Control Web Console (thin-client).

- **Adobe SVG Viewer**   This plug-in is required to provide certain graphics in the Grid Control Web Console, such as a topology graph for system targets. You are prompted to install the Adobe SVG Viewer if your Web Console request relies on it.

- **EM Command Line Interface**   The Enterprise Manager Command Line Interface (EM CLI) is for administrators who want to access a subset of Web Console functionality directly from scripts or interactively from an OS shell.

- **Oracle Configuration Manager Client**   OCM collects host configuration information and uploads it to Oracle Support for analysis to better service their customers. OCM is bundled with Grid Control as of GC 10.2.0.2 and is also available as a standalone install kit.

First, let's address the EM Java Console (as distinguished from the EM Web Console thin-client browser). Then, we'll investigate the remaining optional client installations.

Install and Configure the EM Java Console

As you learned in Chapter 1, Grid Control employs a web-enabled, thin-client Console to manage all targets. This Web Console is the point of entry to access most GC features. I also mentioned in Chapter 1 that the Enterprise Manager 10g Java Console, which is a thick-client available as part of the Oracle 10g Client software, provides some functionality that the Web Console still lacks. Oracle Enterprise Manager (OEM) 9i relied completely upon a thick-client EM Java Console. The Oracle EM development team converted most of the OEM 9i Java Console functionality to the Web Console in EM 10.1, and in EM 10.2 they inched ever closer to a full conversion of all Java Console features to the Web Console. However, the EM team is still working to convert some remaining EM Java Console database management features.

Hard-core OEM 9i Java Console buffs unfamiliar with Grid Control 10g no longer need to wonder where the equivalent 10g Java Console features went. These features didn't disappear—they were converted to the 10g Web Console. For example, the 10g Java Console does not display a database instance's session lock information, which was available in the 9i Java Console. This is because the OEM 9i Diagnostics Pack that provided this information no longer exists in the EM 10g Java Console. You can now manage locks using standard GC 10g Web Console functionality via the Instance Locks link on the Database Performance page.[1]

---

[1] See Oracle*MetaLink* Note 356223.1.

Feature	Offered in EM 10g Java Console?	Offered in GC 10g Web Console?
Capture	Yes	Yes
Compare	Yes	Yes
Reverse Engineer Schema	Yes	Yes
Versioning	Yes	Yes
Alter	Yes	No
Propagate	Yes	No
Synchronize	Yes	No

TABLE 8-1. *Change Management Feature-by-Feature Comparison of Java Console vs. Web Console*

Determine Whether You Need the EM Java Console

The EM 10.1 documentation appropriately covers the Java Console, but perplexingly, all mention of the Java Console disappears in the Oracle EM 10.2 documentation set. The hole in the EM documentation is that it does not identify which database features remain the sole domain of the EM 10g Java Console compared with the most recent GC Web Console version, 10.2.0.4. Filling this hole is our first order of business, as only then can you decide whether you even need to install the EM 10g Java Console. If you don't need to install it, you can skip to the section "Install Adobe SVG Viewer" later in this chapter.

Let's look at the short list of remaining functionality the 10g Java Console alone possesses to help you decide whether you even need to install this thick-client. The Java Console we need to compare to the GC 10g Web Console is now version 10g, as it is part of the 10g database client software distribution. (If you didn't catch that little switcheroo, when OEM 9i was replaced by OEM 10g, the Java Console jumped ship from OEM 9i to the Oracle 10g Database client.) The remaining management areas where Java Console is the only game in town (i.e., no Web Console functionality exists) are Workspace Manager, Advanced Replication, Advanced Queues, Spatial, Failsafe (Windows only), LogMiner Viewer, Text Manager, Enterprise Security Manager, Wallet Manager, Policy Manager, and certain Change Manager features. Of all Oracle products listed above, Change Manager is perhaps the most popular, so let me at least tell you which Change Manager features are found in the Java Console, Web Console, or both. Table 8-1 compares features afforded by the EM 10g Java Console and Grid Control 10g Web Console as they relate to Change Manager.[2]

As Table 8-1 shows, the Web Console possesses all Change Management utility features of the Java Console except the capability to alter and propagate schemas, and synchronize schema changes. The Change Manager modules that provide this functionality will not be integrated into the EM Web Console until Grid Control 11g.[3] Do you need these three Change Manager features

---

[2] Based on Oracle*MetaLink* INTERNAL Note 371834.1.
[3] See Oracle*MetaLink* Note 414411.1.

or any of the above Oracle products right now? If not, congratulations. You can skip to the next major section, "Install Adobe SVG Viewer." If you do need one or more of these products now to manage any of your EM targets, read on—it's not difficult to set up the 10*g* Java Console. If you're not sure whether you need one of these products, don't install the Java Console until you *are* sure. You can always install the Java Console later. Even better, by that time, you may have upgraded to GC 11*g* and the Change Manager module you need will have already been integrated into the GC 11*g* Web Console.

You may find certain Java Console features have overlapping functionality in the Web Console, depending on the Grid Control release you're running. If you find that a particular feature is available in both the Java Console and Web Console, use the Web Console. Except when a function is missing in the Web Console, it is strongly recommended over the Java Console for several reasons. One reason is that all changes in a target's management data due to Web Console actions are stored in the GC Management Repository, whereas the Java Console's standalone repository, which is not integrated with the GC Repository, stores management data only from certain applications, such as from Change Manager. Another reason to use the Web Console when offered the choice is to avoid missing out on new feature updates and bug fixes. (When Enterprise Manager 11*g* is released, the Java Console will no longer be required nor new Oracle Database client releases offered because its remaining unique features will be folded in to the 11*g* Web Console.)

For more background information about the Java Console, and its products, including Change Manager, see the Online Help.

Installation Steps for the EM Java Console

If you need the functionality cited above, you'll need to install the 10*g* Java Console, which is part of the Oracle 10*g* Client software. You must install it on all computers where you want to have thick-client access to the EM Java Console.

TIP

A more efficient way than installing the Oracle 10g Client software on every administrator's workstation is to install it once on a server (such as an OMhost) where every admin has access. An admin can then connect to the server and, if it's running on UNIX, start the Java Console in an X Server session. The limitation is that only one admin at a time will be able to connect to the Java Console.

Following are the steps to install the Database Client 10gR2. Notable screen shots are shown below. **Boldface** text matches the OUI screen titles and field names and user input for these fields is *italicized*.

1. Before starting the installer, back up the Central Inventory (if one exists) on the host where the Java Console is to be installed. (For background on the Central Inventory for the various platforms, see the section "Create Oracle Inventory Group (oinstall)" in Chapter 2.) Backing up the Inventory allows you to restore it in case the Java Console installation is not successful, without having to run the OUI installer to remove the Inventory in the GUI.

2. Unset any Oracle-related environment variables except ORACLE_BASE, as these variables may point to other installed Oracle products, thereby confusing the 10.2 Client installer.

3. Run the Oracle 10*g* Client 10.2.0.x OUI installer from the software distribution media or from disk, if copied there:
   - On UNIX, enter: ./runInstaller
   - On Windows, enter: .\setup.exe

4. **Select Installation Type—Oracle Client 10.2.0.x.0**   Choose either Administrator or *Custom* (as shown here).

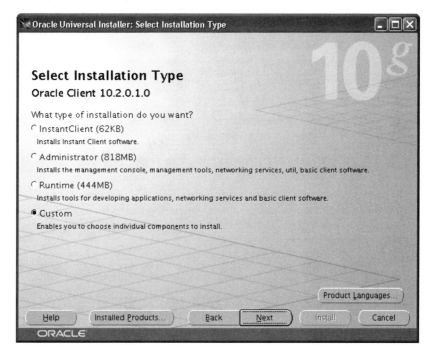

5. **Specify Home Details—Destination**   Enter the name and path of the Oracle home. The following values comply roughly[4] with Optimal Flexible Architecture (OFA) best practices.
   - **Name**   *OraClient10g_home1*
   - **Path**   */opt/oracle/product/10.2.0/client_1*

---

[4] I say "roughly" because OFA guidelines actually specify */<mount_point>/app/oracle/product/10.2.0/client_1*, but the only mount point available on the host used here was */opt* (as is sometimes the case on Solaris).

248 Oracle Enterprise Manager 10g Grid Control Implementation Guide

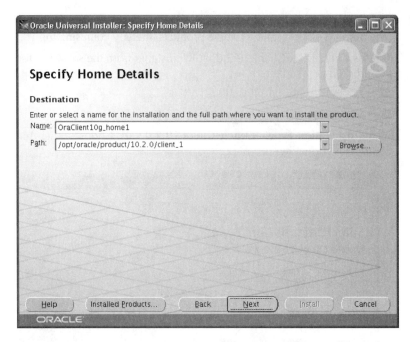

6. **Available Product Components—Oracle Client** This screen appears only if you select the Custom installation type in step 1. Select Enterprise Manager 10g Java Console 10.2.0.x.0 as one of the installation components of the 10.2 Client software. It is not selected by default.

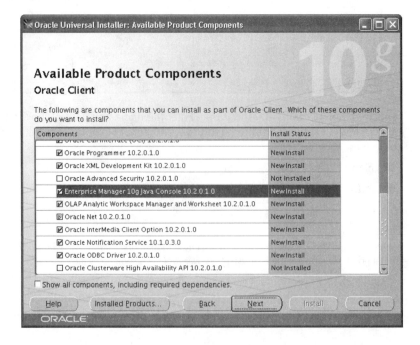

Platform	Environment Variable	Required Value
AIX	LIBPATH	$ORACLE_HOME/lib32:$ORACLE_HOME/lib:$LIBPATH
HP-UX	SHLIB_PATH	$ORACLE_HOME/lib32:$ORACLE_HOME/lib:$SHLIB_PATH
Linux	LD_LIBRARY_PATH	$ORACLE_HOME/lib:$LD_LIBRARY_PATH
Solaris	LD_LIBRARY_PATH	$ORACLE_HOME/lib32:$ORACLE_HOME/lib:$LD_LIBRARY_PATH
Tru64	LD_LIBRARY_PATH	$ORACLE_HOME/lib:$LD_LIBRARY_PATH

TABLE 8-2. *Required Shared Library Path Environment Variable Setting to Start the Java Console*

7. Select all other desired components[5] but exclude the following components if not needed, as they require separate licensing and additional configuration:

   ■ Oracle Advanced Security 10.2.0.x.0[6]

   ■ Oracle Clusterware High Availability API 10.2.0.x.0

8. Respond to the remaining screens, **Product-Specific Prerequisite Checks**, **Summary**, and **Execute Configuration scripts**, to finish installing the Java Console on the chosen host.

Start the EM Java Console

Start the Java Console and ensure that it works properly by connecting to a database to be managed. If you plan to use Change Manager, you may as well connect to the database planned for the standalone repository (assumed here to be the OMR Database) by following these steps:

1. Set the environment variables required to open the Java Console.

   ■ Set the ORACLE_HOME environment variable to the Oracle Client home.

   ■ If the Oracle Client is on a UNIX platform, set the shared library path environment variable to include the directories listed in Table 8-2 above, which lists the variable name and value (the directories) for each UNIX platform.

2. Start the Java Console.

   ■ If you're running a Windows client, click Start, Programs, select the Oracle 10*g* Client home, whose default name is *Oracle – <Oracle Home Name>*, and select Enterprise Manager Console.

---

[5] A discussion of all available Oracle Client components is outside the scope of this book.

[6] Install Oracle Advanced Security if you plan to enforce data encryption and/or data integrity on the Database server side as suggested in Chapter 16 (see the section "Secure Repository Data Transmissions").

- On either a UNIX or a Windows client (at the command line), start the Java Console as follows (where ORACLE_HOME is the Oracle 10*g* Client home specified with Unix syntax):

  `$ORACLE_HOME/bin/oemapp console`

3. Add the desired database. The first time you log in, a dialog box appears to add a database to the tree. Enter the hostname, port, SID, and desired Net Service Name for the OMR Database. The Java Console will then open.

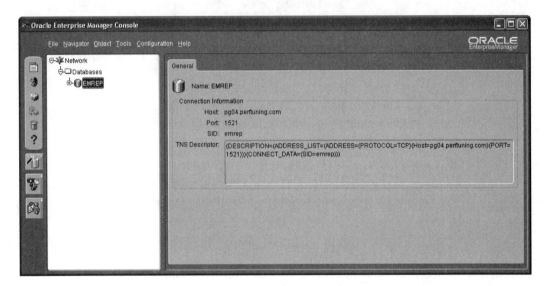

If you plan to use Change Manager, leave the Java Console open so that you can use it to create the standalone repository, as described in the next section "Configure Change Manager".

Configure Change Manager

Now that you've installed the Java Console, you may also need Change Manager functionality in particular. If so, read on. Otherwise, skip to the next section ,"Install Adobe SVG Viewer." It is important to realize that the Enterprise Manager 10*g* Java Console, even though it contains the words, "Enterprise Manager," does not integrate into the Grid Control framework (i.e., OMS, OMR, and OMA) at all. The Java Console does not use any OMS application processing or storage in the Management Repository database. Instead, the Java Console runs as a standalone thick-client or, when used with Change Manager, requires a standalone repository schema that's separate from the GC Management Repository's SYSMAN schema. Change Manager must store management information for the databases it accesses in this standalone repository schema. Although the schema is both physically and logically separate from the OMR SYSMAN schema, I suggest placing the standalone repository in the existing 10*g* OMR Database. Think of this solution as the next best thing to the Web Console offering all this functionality itself, which will eventually be the case. The Management Repository is already located in a centralized database for all GC web users, so why not let this database also play host to the standalone repository user (i.e., schema) as well? You may, of course, create or use a separate database for the standalone repository if you want.

Following is an overview of the Change Manager configuration process. The instructions assume you are creating the standalone repository in the OMR Database:

- **Create a tablespace for Change Manager**   The standalone repository needs to create schema objects in a tablespace that should possess certain attributes, so it is best to create a new dedicated tablespace distinct from the two OMR tablespaces.

- **Create a database user for Change Manager**   The standalone repository also requires certain roles and privileges, which you should assign to a new dedicated user to own the standalone repository schema, distinct from the SYSMAN user that owns the OMR schema.

- **Create a standalone repository for Change Manager**   Once you create this tablespace and user, use the Java Console to create the standalone repository for Change Manager.

Following are the details on these three steps. I list the SQL*Plus commands to accomplish the first two steps, and to demonstrate Grid Control functionality, I show the equivalent GC Web Console navigation route for these two steps (at the risk of confusing you, since this is tantamount to configuring the Java Console using the Web Console).

Regardless of your configuration method, your chosen database must be certified to host the standalone repository. Currently Oracle Databases 10.2, 10.1.x, 9.2.x, and 9.0.1.x are certified. If you choose to locate the standalone repository in the OMR Database, database versions eligible to host the standalone repository are further limited by those supported for the OMR, which are currently Oracle Database 10.2.0.2+, 10.1.0.4+, and 9.2.0.6+ (where "+" means "or later").

Create a Tablespace for Change Manager

Following are the steps to create a tablespace for Change Manager. I show the SQL*Plus commands first, then the equivalent GC Console navigation process.

Create a Tablespace in SQL*Plus   Here is the procedure to create a tablespace for Change Manager in SQL*Plus:

1. Verify that the database hosting the standalone repository has object support enabled. This is the default for all supported database versions, but you can double-check that object support is enabled by ensuring that VALUE='TRUE' in the following query:

    ```
    SELECT * FROM V$OPTION WHERE PARAMETER = 'Objects';
    ```

2. Create a tablespace named OEMREPOS_STANDALONE (though you can choose any name) for the standalone repository as shown here. If you're not using Oracle Managed Files (OMF), then specify a datafile name in single quotes after the DATAFILE keyword:

    ```
    CREATE TABLESPACE OEMREPOS_STANDALONE
    DATAFILE SIZE 32M EXTENT MANAGEMENT LOCAL
    SEGMENT SPACE MANAGEMENT AUTO;
    ```

3. Set the datafile to AUTOEXTEND ON with an appropriate NEXT and MAXSIZE. You must specify a datafile name whether or not you're using OMF. If you're using OMF, specify the OMF-created datafile name, which you can get from querying DBA_DATA_FILES.

   ALTER DATABASE DATAFILE '<datafile_name>' AUTOEXTEND ON NEXT 5M MAXSIZE 2000M;

Create a Tablespace in Grid Control   As an alternative to using SQL*Plus, you can create a tablespace for Change Manager using Grid Control. Here's how:

1. Log in to the Console, click the Targets tab, and then click the Databases sub tab. Choose the Repository database, then the Administration page. The following illustration shows the Administration page for a Database Instance.

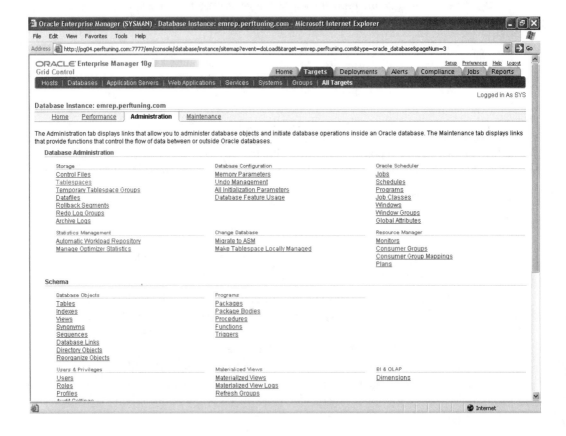

Chapter 8: Install Grid Control Clients 253

2. Click the Tablespaces link under the Database Administration heading.

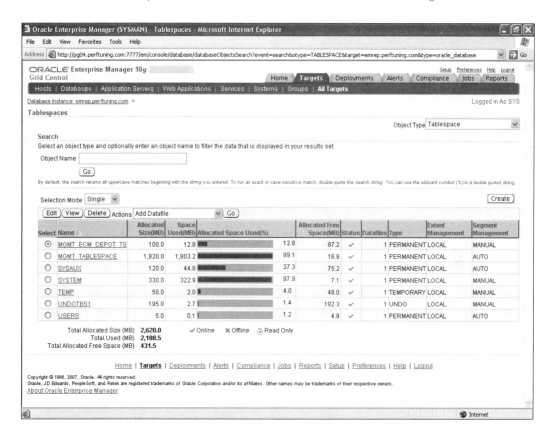

3. Click the Create button to create the tablespace.

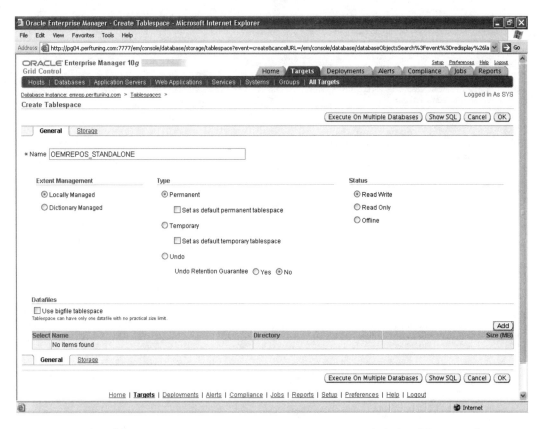

4. Enter the tablespace name OEMREPOS_STANDALONE, and click Add to specify a datafile. Enter the desired datafile attributes, which here include enabling autoextend, and click Continue.

5. Click the Storage tab. Enter the desired attributes, which here are the defaults of choosing automatic extent allocation and segment space management, and enabling logging. Click OK to create the tablespace.

Create a Database User for Change Manager

Create a database user to own the standalone repository schema and grant the necessary privileges to that user. The username OEMREPOS_STANDALONE specified below matches that chosen for the standalone repository tablespace name, but you can choose any username you like.

Create a Database User in SQL*Plus   Execute the following steps to create a database user for Change Manager in SQL*Plus:

1. Create the standalone repository database user with the following quotas:

   ```
   CREATE USER OEMREPOS_STANDALONE IDENTIFIED BY <password>
   DEFAULT TABLESPACE OEMREPOS_STANDALONE
   TEMPORARY TABLESPACE <temp_tablespace_name>
   QUOTA UNLIMITED ON OEMREPOS_STANDALONE;
   ```

   If the database is Release 10.1 or earlier, grant this user unlimited quota on TEMP:[7]

   ```
   ALTER USER OEMREPOS_STANDALONE QUOTA UNLIMITED ON TEMP;
   ```

2. Grant the following system privileges and roles, which are required for all database versions certified to run the standalone repository as specified above):

   ```
   GRANT CONNECT TO OEMREPOS_STANDALONE;
   GRANT CREATE PROCEDURE TO OEMREPOS_STANDALONE;
   GRANT CREATE TRIGGER TO OEMREPOS_STANDALONE;
   GRANT CREATE TYPE TO OEMREPOS_STANDALONE;
   GRANT EXECUTE ANY PROCEDURE TO OEMREPOS_STANDALONE;
   GRANT EXECUTE ANY TYPE TO OEMREPOS_STANDALONE;
   GRANT SELECT_CATALOG_ROLE TO OEMREPOS_STANDALONE;
   GRANT SELECT ANY DICTIONARY TO OEMREPOS_STANDALONE;
   GRANT SELECT ANY TABLE TO OEMREPOS_STANDALONE;
   ```

   The above privileges are documented in the EM Grid Control 10.1 documentation.

3. Grant the following additional privileges to the standalone repository user, but only if that user is in an Oracle 10.2 Database:

   ```
   GRANT RESOURCE TO OEMREPOS_STANDALONE;
   GRANT ALTER SESSION TO OEMREPOS_STANDALONE;
   GRANT CREATE CLUSTER TO OEMREPOS_STANDALONE;
   GRANT CREATE DATABASE LINK TO OEMREPOS_STANDALONE;
   GRANT CREATE SEQUENCE TO OEMREPOS_STANDALONE;
   GRANT CREATE SYNONYM TO OEMREPOS_STANDALONE;
   GRANT CREATE TABLE TO OEMREPOS_STANDALONE;
   GRANT CREATE VIEW TO OEMREPOS_STANDALONE;
   ```

   You must explicitly grant these additional privileges in Database 10.2 because they were removed from the CONNECT role in this release.[8] (We grant the CONNECT ROLE to the standalone repository user in step 2 above.) If you don't grant these additional privileges, standalone repository creation will fail in an Oracle 10.2 Database with the following error:

   An unexpected exception occurred during a product action execution:

   oracle.sysman.vdb.VdbSQLException

---

[7] See Oracle*MetaLink* Note 307004.1. This is not supported in Oracle 10.2 - see Bug 3623814.
[8] See Bug 4287046.

Create a Database User in Grid Control Following are the equivalent steps to create a database user for the standalone repository using Grid Control:

1. Navigate to the Repository database Administration page, shown earlier. In the Schema section, click on Users.

2. Click Create to create the database user that will own the standalone repository schema.

3. This brings up the Create User page, consisting of a series of seven pages. Click the following pages, which allow you to enter the main user attributes and grant the necessary roles, system privileges, and quotas to this new user.

   a. **General**   Begin with this page.

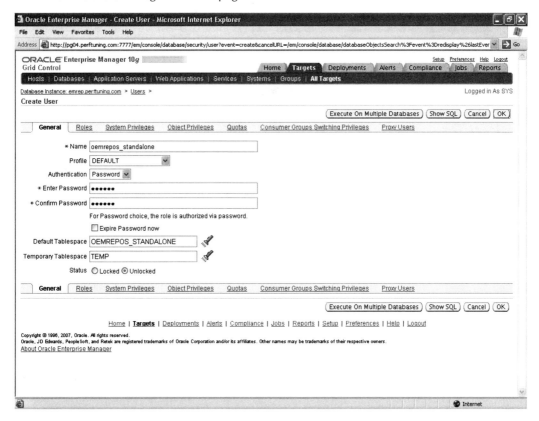

Fill in the fields as shown above for the main user attributes. For the user's default tablespace, select the OEMREPOS_STANDALONE tablespace created above.

b. Roles Click the Roles page, then click Edit List.

Move the SELECT_CATALOG_ROLE from Available Roles to Selected Roles, then click OK.

Chapter 8: Install Grid Control Clients   **259**

c. **System Privileges**   Click the System Privileges page, then click the Edit List button.

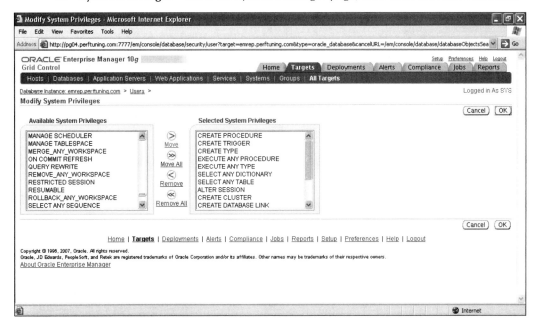

Move the privileges listed above in the section "Create a Database User in SQL *Plus" from Available System Privileges to Selected System Privileges. You can hold down the CTRL key to select multiple privileges to move. Click OK when you're finished with this page.

d. **Quotas**   Select the Quotas page to grant the required quotas to your new user.

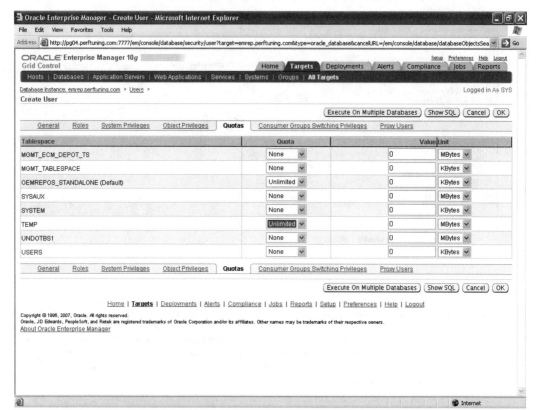

Select Unlimited quota for tablespace OEMREPOS_STANDALONE. and also for tablespace TEMP (as shown above) if the database release is 10.1 or earlier.[9] Click OK to create the user with the attributes you just entered.

Create a Standalone Repository for Change Manager

Now that you have created the tablespace and user for the standalone repository, it remains to let the EM Java Console create this repository for Change Manager to use. The procedure, provided below, is to start the Change Manager application, at which point the Java Console will prompt for the database user to assign as the owner of the repository schema. You will specify the OEMREPOS_STANDALONE user just created and the standalone repository will be created.

1. Return to the Java Console opened above (see the section "Start the EM Java Console") and start the Change Manager application in one of following three ways:

    - Right-click on the OMR Database instance name in the navigator panel, click Change Management, and then click any Change Manager application, such as Capture Database Objects.

---

[9] Ibid Footnote 7.

- On the Tools menu, point to the Change Management Pack and select Change Manager.
- Select the Change Management Pack drawer on the left-hand side (second icon from the bottom), and then click the Change Manager icon.

2. Close the dialog box shown here that appears on top of a Repository Login box.

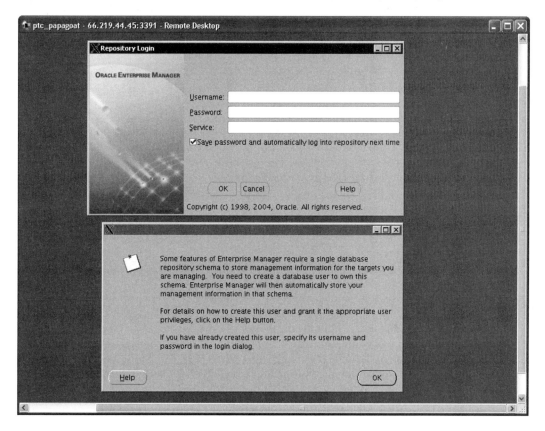

To close the dialog box, click on the close ("X") icon in the upper right-hand corner (the OK button doesn't seem to work).[10]

3. Enter credentials in the Repository Login box shown above for the database user OEMREPOS_STANDALONE created to own the standalone repository, and Service name for the OMR Database, check the box to save the password, and then click OK.

---

[10] If you cannot close the dialog box this way, kill the *oemapp* process, restart it, and then quickly hit ENTER to get past this dialog box.

4. Wait for the EM Java Console to create the standalone repository, which shouldn't take more than 30 minutes. When the dialog box reports that Create Repository processing is completed, click Close and confirm that the Change Manager application opens in a separate window as shown here.

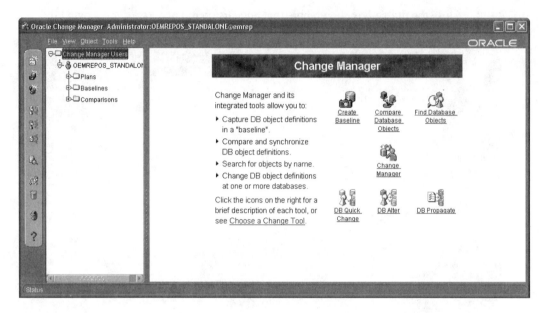

This completes the process of configuring a standalone repository for Change Manager. You can use this application with any accessible database, including the OMR Database itself and other GC target databases. To add new databases, click Navigator, Add Database to Tree, and enter the database connection information. All Change Manager management data gathered about discovered databases will be stored in the OEMREPOS_STANDALONE schema in the OMR Database. However, remember that this data will be entirely separate from the Grid Control management data stored in the SYSMAN schema.

Install Adobe SVG Viewer

You must install the latest Adobe SVG Viewer on Web Console client workstations that need to view certain Grid Control graphics. One such graphic is the topology for a particular system target.[11] (See Chapter 12 for more information on topologies.) You can install the SVG Viewer now on desired Console workstations, or wait until it's required, at which time Grid Control will prompt you to download this software. If you are a consultant, it's better to instruct administrators to install the SVG Viewer up-front so that the GC functionality it affords appears seamless.

---

[11] The navigation path for a system topology is to select the Targets tab, then the Systems sub tab. Choose a particular system target and then click on the Topology tab.

The SVG Viewer is available only for Windows, Mac, Red Hat Linux, and Solaris 8 platforms from http://www.adobe.com/svg/viewer/install/. This limited number of platforms for which Adobe SVG Viewer is available will not likely pose a problem, given that most admins run their Console browsers on Windows workstations.

Install EM Command Line Interface

The Enterprise Manager Command Line Interface (EM CLI) is for administrators who want to access a subset of GC Console functionality directly from scripts or interactively from an OS shell:

- **EM CLI offline use**   GC admins can use EM CLI for offline workflow scripts, meaning that the EM CLI client does work directly on behalf of an admin, backed by an OS user for security, after the admin performs a one-time interactive setup to give authority and credentials to EM CLI to execute its commands.[12]

- **EM CLI online use**   Admins can interactively perform GC Console-based operations like managing targets, jobs, groups, blackouts, notifications, and alerts. EM CLI can be a very efficient alternative to GUI entry when you have a large GC site with many targets and admins. A salient application of EM CLI is to streamline what might otherwise be an intensive data-entry process of setting preferred credentials for admins (see the section "Programatically Setting Preferred Credentials" in Chapter 10).

EM CLI is fully integrated with EM security and user administration functions, allowing administrators to carry out operations with the same security as provided by the GC Console. For example, EM CLI users will only be able to see and operate on targets for which they are authorized.

EM CLI allows you to access selected GC functionality from text-based consoles (shells and command windows) on a variety of operating systems. Currently, only OS shell and SQL scripts are certified for EM CLI, but certification is expected soon for SQL*Plus, Perl, and Tcl scripts. You also need EM CLI to develop a new Management Plug-in, specifically, to add a group of target definition files to a Management Plug-in Archive. For information on developing your own Management Plug-in to monitor a custom target type, see Oracle Enterprise Manager Extensibility 10*g* Release 2 (10.2) for Windows or UNIX.

Installing and configuring the EM CLI client is a short and easy process. Similarly to the EM Java Console, you can install the EM CLI client on an OMS host for administrators to share, or you can install it on a workstation with network access to the OMS. Go to the following OMS URL for a link to download the EM CLI kit:http://<*OMShost*>:7777/em/console/emcli/download. A screen shot of this URL page is shown on the following page.

---

[12] Source: Bug 4512405.

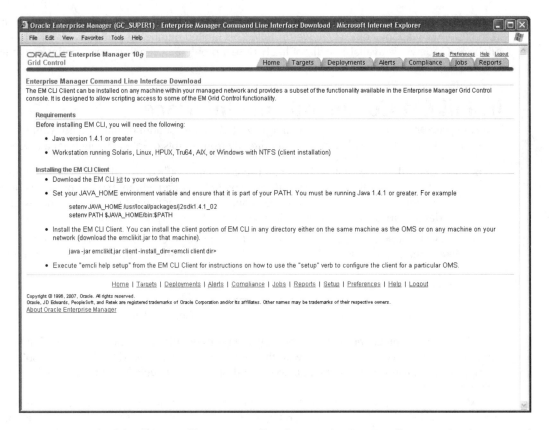

For a complete description of how to install and set up the EM CLI client, see Section 1.2.1 of *Oracle Enterprise Manager Command Line Interface 10g Release 2 (10.2)*.

Install Oracle Configuration Manager Client

Oracle Configuration Manager (OCM) is the data collecter that Software Configuration Manager (SCM) uses to gather host, Oracle Database, Oracle Application Server (including OMS), and e-Business Suite configuration information and upload it securely via HTTPS to Oracle Support. (SCM is the new support framework for collecting and centralizing configuration information. For details, see the Oracle Software Configuration Manager Getting Started Guide on Oracle*MetaLink*.) Once configured, OCM data is automatically uploaded to Oracle Support on a regular basis to help provide better service to customers. For instance, you can associate configuration data directly with a logged service request (SR), which reduces support resolution time and standardizes the SR process a bit more.

Although OCM shares some commonalities with the GC Management Agent in terms of how it functions, it is a completely different product intended primarily for Oracle Support to use. As such, OCM should be considered an optional (but recommended) installation for GC components and targets. OCM overlaps somewhat with GC functionality, such as in recommending patches and assessing system health. However, OCM also complements Grid Control by offering a direct upload capability to Oracle Support, and SCM framework features customize the Oracle*MetaLink* interface for your environment. For example, SCM provides relevant Dashboards, Headlines, and Knowledge Base Articles relevant to your specific Oracle product configurations.

The OCM client is installed into an ORACLE_HOME directory, and only collects configuration data related to that home and the host on which it is deployed. For help in quickly resolving Oracle Support service requests and in improving your Oracle*MetaLink* experience, I suggest installing OCM in the Oracle homes of all GC framework components (OMS, Agent, and OMR Database) as well as in the homes of all production GC target Databases, Oracle Application Servers, and e-Business Suite systems (although OCM for GC targets is not part of this chapter's scope). OCM is self-maintaining; that is, it checks if any software updates to the OCM client itself are available, and downloads and installs these updates automatically. OCM's footprint is also small enough to generally avoid raising concerns that it's consuming undue resources on GC or target servers that are not otherwise strapped for resources.

In general, there are two methods for installing OCM into an Oracle product's home: using the OCM Command Line Interface (which can be scripted) or using a product's OUI when OCM is bundled with that OUI. OCM is bundled with the GC OUI as of Release 10.2.0.2. The best practice is to install OCM in the Oracle homes of all GC framework components using the following installation methods, which vary by component:

- **Install OCM in OMS and chain-installed Agent homes**   You can install OCM in the homes of the OMS and chain-installed Agent using the OCM Command Line Interface or GC OUI (see Figure 3-9 in Chapter 3). I recommend the GC OUI method, as it does not require any manual steps and is therefore less susceptible to error.

- **Install OCM in standalone Agent homes**   You can install OCM in standalone Agent homes using the OCM Command Line Interface or the GC OUI (see the section "Run the Interactive Installer" in Chapter 7), depending on how you deploy Agents:

  - **Interactive Agent installation method**   When you're installing an Agent interactively, again, the best practice is to install the OCM bundled with the GC OUI, as it does not require any manual steps.

  - **Other Agent installation methods**   When you're not deploying Agents interactively, the most efficient way to install OCM in Agent homes is to mass-deploy it. Place the OCM distribution file in a central location under the OMS home, and schedule a Grid Control OS Script job to copy, unzip, and set up OCM on all Agent hosts. See Section 3.1.3 of Oracle Configuration Manager Installation and Administration Guide 10*g* Release 2 (10.2) for an example OS script to install OCM.

- **Install OCM in OMR Database home**   You can only use the OCM Command Line Interface (not the Grid Control OUI) to install OCM in the OMR Database home. Afterward, you need to run an additional shell script whose syntax is listed in Section 2.3.2.2 of Oracle Configuration Manager Installation and Administration Guide 10*g* Release 2 (10.2).

To download the Command Line Interface version of OCM, click on the Collector tab on the Oracle*Metalink* home page. Then, click "Download the configuration manager," select the platform, and click Download.

For detailed instructions on installing OCM using both the Command Line Interface and OUI, see the following sections of the Oracle Configuration Manager Installation and Administration

Guide 10*g* Release 2 (10.2), available on OTN as part of the Documentation Libraries for both Oracle Enterprise Manager 10.2 and Oracle Database 10.2:

- To install OCM using the Command Line Interface, see Section 2.2.3.
- To install OCM using the OUI (such as the GC OUI), see Section 2.2.4.

In addition, a good FAQ on OCM is available in Oracle*MetaLink* Note 369111.1.

Summary

This short chapter describes the installation procedures for all optional Grid Control clients. These clients, the EM Java Console (including Change Manager), the Adobe SVG Viewer, the EM Command Line Interface (EM CLI), and Oracle Configuration Manager (OCM), all play a part in realizing Grid Control's full capabilities. The EM Java Console, which includes the Change Manager application, supplies some vestiges of functionality still not available in the GC Web Console. The Adobe SVG Viewer provides the topology diagrams for certain targets. EM CLI affords a way to script Console tasks, which is especially useful in replacing repetitive, time-consuming Console navigation. The OCM is a strange bird in that it is not an exclusive Grid Control product. It is bundled not only with Grid Control 10.2.0.2 and subsequent patch sets, but also with the installers for other product releases and patch sets, and can be installed as a standalone application to boot. OCM is an important tool that Oracle Support uses to gather information about the Grid Control infrastructure and its targets in helping to resolve service requests.

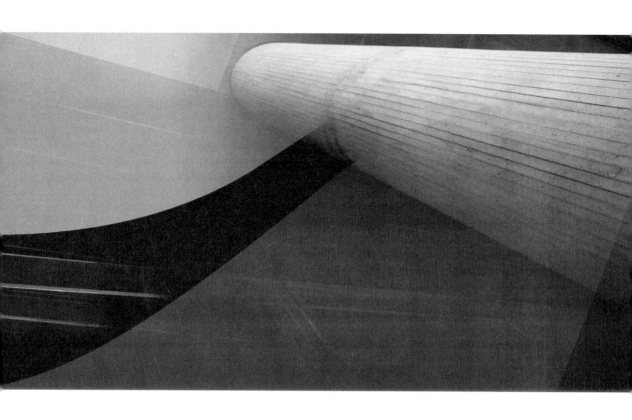

CHAPTER 9

EM Login and Component Control

his chapter serves as the central place to consult for how to log in to Enterprise Manager (EM) and control its component processes. If you want to know how to log in to EM via a particular entry point, or how to manage EM processes, it is convenient to be able to look for this information in one place. This chapter is that place.

I intentionally place this chapter after those in which you perform the configuration steps necessary to access the EM login entry points discussed here. You already learned at the end of Chapter 3 the different ways of logging in to the GC Web Console. However, you've configured additional EM login entry points that fall within the scope of this book[1], including login to the EM Java Console, to a target database through *i*SQL*Plus, and to the OMS AS Console. This chapter covers login steps for each of these EM access points.[2]

TIP
Instructions for login and control of the AS Console, though intended primarily for the OAS bundled with Grid Control, apply equally to OAS targets that Grid Control monitors.

Controlling EM processes is the second main subject of this chapter. Chances are that by now you've also needed to control at least one EM component, either to reboot a server, perform a cold backup, or resolve a failed EM process. The treatment here covers process control for the GC framework (OMS, OMR, and OMA), AS Control, and *i*SQL*Plus server. All command-line utilities for controlling component processes are equally applicable to UNIX and Windows platforms; however, on Windows these processes are also implemented as services, which allow for process management in the Services control panel. Therefore, I provide a table of all equivalent EM Windows services to the command-line utilities.

This chapter ends with a high-level overview of the order in which to start up and shut down EM. It takes into account how to flip the switch either way (i.e., starting or stopping) on all components you would find in a high availability EM environment, including multiple OMSs, a RAC Repository, and a standby database for the Repository. This is useful information for performing environment-wide cold EM backups and other maintenance operations. It is not a far leap from this high-level order to a usable sample script I supply for administrators to cleanly shut down and subsequently restart their EM infrastructures in an ad hoc fashion (e.g., outside the context of automatic startup on boot).

Some of the procedures for login and process control that are provided here are also touched upon cursorily in previous chapters as required to complete basic EM configuration. However, the more in-depth coverage here should shed some additional light on these procedures, even for those familiar with Enterprise Manager 10*g*. If you're new to EM10*g*, you will probably refer to this chapter multiple times when wanting to control or log in to Enterprise Manager.

---

[1] As I already mentioned in Chapter 1, Database Control is also part of Enterprise Manager, but is beyond the scope of this book.

[2] Actually, I present one login entry point not found in earlier chapters: the Metrics Browser, which is a troubleshooting tool that provides raw data collected by an Agent. However, fear not, as it is easy to enable and access.

Login Method	Protocol Used	Identifier in *portlist.ini*	Platform Installed (determines port)	Grid Control Console Login URL
Web Cache	HTTP	Web Cache HTTP Listen port	UNIX	http://<OMShost>:7777/em
Web Cache	HTTP	Web Cache HTTP Listen port	Windows	http://<OMShost>:80/em
Web Cache	HTTPS	None	UNIX, Windows	https://<OMShost>:<port>[3]/em
HTTP Server	HTTP	Enterprise Manager Central Console Port	UNIX, Windows	http://<OMShost>:4889/em
HTTP Server	HTTPS	Enterprise Manager Central Console Secure Port	UNIX, Windows	https://<OMShost>:1159/em

TABLE 9-1. *URL Variations to Log in to the Grid Control Console*

Log in to EM

Following are the multiple login entry points within Enterprise Manager, which include login to EM Consoles and to managed targets through GC Console links:

- Web Console login to Grid Control
- EM Java Console login to a target database
- AS Console login to the OMS Application Server or target OAS
- *i*SQL*Plus login to a target database
- Agent Metric Browser login

The following sections provide detailed login procedures for each of these EM entry points.

Web Console Login to Grid Control

As you learned in earlier chapters, the Web Console is the principal tool for managing your Grid Control environment. The URL for logging in to the Web Console is generically of the form:

http(s):<OMShost>:<port>/em

This URL varies depending on whether it is over a secure (HTTPS) or nonsecure (HTTP) connection and whether you communicate via Web Cache or Oracle HTTP Server (OHS), which determines the port. The port for Web Cache varies by platform. See Table 9-1 for these URL variations to log in to the GC Console. This table also lists the identifier in the *$OMS_HOME/install/portlist.ini* file associated with these port specifications.

The *portlist.ini* file is a static file. In other words, if someone changes the OMS port configuration after installation,[4] *portlist.ini* does not change accordingly. If you're not sure whether someone

---

[3] You choose a custom port if opting to configure Web Cache SSL access (see Chapter 16, "Security Considerations for Console Web Cache Access").

[4] Changing OMS ports post-install is not covered in this book. For such instructions, see Section 12.2.2 in Oracle Enterprise Manager Advanced Configuration 10g Release 4 (10.2.0.4.0).

changed any OMS ports, check the currently configured ports for OHS access (both secure and nonsecure) in the *$OMS_HOME/sysman/config/emoms.properties* file. Values for the following two properties are the current ports being used to access the Console via OHS:

Property in *emoms.properties*	Value for Property
oracle.sysman.emSDK.svlt.ConsoleServerPort	Port for nonsecure access via OHS
oracle.sysman.emSDK.svlt.ConsoleServerHTTPSPort	Port for secure access via OHS

Let's break down each piece of the above Web Console URL, from left to right:

- **HTTP vs. HTTPS** The URL protocol varies—HTTP (nonsecure) or HTTPS (secure)—depending upon whether you configure SSL encryption for browser to Console communications. When Grid Control is installed on any platform, SSL encryption is configured by default for OHS access, but not for Web Cache access.
- **OMS Host** You can log in directly to any configured active Management Service or through a server load balancer.[5]
- **Port** You can log in to the Console via OracleAS Web Cache or OHS by going through their respective access ports. Ports for OHS secure and nonsecure access (1159 and 4889 respectively) are the same for all platforms, but the Web Cache nonsecure port differs for UNIX and Windows (7777 and 80 respectively). Web Cache page performance can be slightly better, but it's useful to know how to bypass Web Cache to isolate a Web Cache Console login problem or to log in over HTTPS, which is only available via OHS out-of-box.

Regardless of the URL used to access the Web Console, the following screen appears for you to enter login credentials:

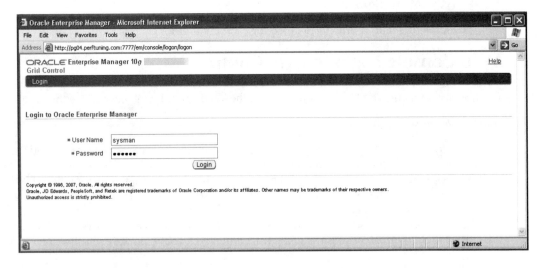

---

[5] In addition to a physical OMS host name, the name could be that of a virtual host when front ending multiple OMSs using a server load balancer. See the section "Use a Server Load Balancer" in Chapter 15.

The URL in this screen is for a nonsecure Web Cache login on a UNIX platform. Enter your GC administrator username and password. You can log in as the built-in SYSMAN user (as shown above) who owns the Repository schema, or by using any administrator name created after the installation. Once you log in, the home page appears, as shown in the section, "Grid Control Components" in Chapter 1. When you've finished with your work in the Console, you can log off either by clicking the Logout link available on the top right of most Console screens or by simply closing the browser window.

Web Console Login As a User Other Than DBSNMP

When you log in to Grid Control and navigate to a database target, you are automatically logged in as the DBSNMP monitoring user. Reconnecting to the database in Grid Control as a different user is easy, but obscure and not well documented. Here's how:

1. Navigate to the Performance page for a target database and click Logout. (When not on a database Performance page, Grid Control simply logs you out and displays a page with the Login button.)

2. The screen in Figure 9-1 appears.

3. You are given two selections:

    - Log out of database. Check "Display database login page after logout."

    - Log out of Enterprise Manager altogether.

    Make your selection and click OK.

4. A login page appears that allows you to log in as any database user with Normal, SYSOPER, or SYSDBA privilege.

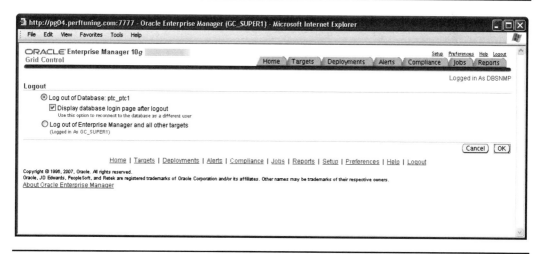

FIGURE 9-1. *The Logout page for a database*

EM Java Console Login

Introduced in Chapter 8, the EM Java Console is an optional thick-client that is part of the Oracle 10.2 Client distribution and offers a few database features that have still not been converted to the Grid Control Web Console. As Chapter 8 points out, use the Web Console rather than the Java Console, except when you need functionality that only the Java Console offers.

To log in to the Java Console, see the section in Chapter 8 entitled "Start the EM Java Console."

The first time you log in, a dialog box appears to add a database to the tree. You can manually enter the hostname, port, SID, and desired Net Service Name of a target database. Alternatively, you can populate the Navigator tree with multiple databases by reading database service names from the local *tnsnames.ora* file in the Oracle Client home. The Java Console will display a list of databases from this file. Whichever databases you select are added to the navigator tree.

You can also directly start a specific Java Console application from the command line by specifying the application as an argument to *oemapp*, as shown below. (A Windows style path is specified, as most admins run the Java Console on their workstations.)

```
%$ORACLE_HOME%\bin\oemapp <application_name>
where <application_name> = [ console | dbastudio | lmviewer | ocm |
                             ocmcli | sdoadvisor | txtmgr | worksheet ]
```

These arguments and their corresponding application names are listed in Table 9-2.

You can also log in to these applications once you are in the Java Console. As an example, start the Change Manager application from within the Java Console in any of the following ways:

- Select a database in the navigator panel, click the right mouse button, point to the Change Management menu option, then left-click Clone Table and Data, Compare Database Objects, or Capture Database Objects.

- On the Tools menu, point to Change Management Pack, then click Change Manager.

- Click the Change Management Pack drawer, then click the Change Manager icon.

oemapp Argument For Application	Application Name
console	Console login
dbastudio	DBA Studio (opens Console)
lmviewer	Oracle LogMiner Viewer
ocm	Oracle Change Manager
ocmcli	Oracle Change Manager Client Command-Line Interface (not a Java Console application)
sdoadvisor	Oracle Spatial Index Advisor
txtmgr	Oracle Text Manager
worksheet	SQL*Plus Worksheet

TABLE 9-2. *Arguments to Directly Open Java Console Applications*

AS Console Login

The Oracle Application Server Control Console (or "AS Console") is a web-based tool for managing any Oracle Application Server (OAS), and comes with the OAS product. There are two types of OASs in a Grid Control environment:

- OAS targets that Grid Control monitors
- The OAS bundled with the OMS middle tier

The procedure here to log in to AS Control applies to both of these OAS types. As discussed in Chapter 1 in the section "Grid Control vs. AS Control," you use the AS Console to *administer* both types of OASs and Grid Control to *monitor* them.

To log in to the AS Console for a given OAS, follow these steps. I provide generic instructions and also give details for an OAS bundled with an OMS. The AS Console must be started to log in (see "Control the AS Console" later for startup instructions).

1. Access the AS Console for an OAS instance using one of two methods:

    - Use the following URL:

        http://<OAShost>:<port>

        The AS Console port number is listed in the file <OAS_HOME>/sysman/setupinfo.txt. For the OAS bundled with Grid Control, <OAS_HOME> is $OMS_HOME. Use the following URLs to directly access the AS Console bundled with OEM 10g Grid Control Release 2 (10.2.0.x):[6]

        On UNIX: http://<OMShost>:1156

        On Windows: http://<OMShost>:18100

    - Use a link in the GC Console for the desired AS target. The OAS bundled with an OMS is named EnterpriseManager0.<OMShost>. Under the Targets tab and Application Servers subtab, select the name for the OAS, click the Administration page, then under Related Links, click the Administer link.

2. Whether you access the AS Console through the Console or by specifying the direct URL, a login box appears. Enter the built-in username *ias_admin* and the password. For OEM 10g Grid Control Release 2 (10.2.0.x), the password is that for SYSMAN, which the admin who installed Grid Control supplied.[7] Check the box "Remember my password" if desired, then click OK. This brings up the AS Control home page as shown in Chapter 1 under the section "AS Control Administers Oracle Application Server."

From the AS Control home page, you can administer the OMS middle tier subcomponents, which are Oracle HTTP Server, the OC4J EM application, and OracleAS Web Cache. You can also gather information about the management software itself by clicking on the Management link listed under System Components at the bottom of the page.

---

[6] In OEM10g Release 1 (10.1.0.x), the AS Console port is 1810 rather than 1156, so the URL is http://<OMShost>:1810.

[7] The default *ias_admin* password in EM10g R1 is *welcome1*. If you can't determine the password or it is corrupted, see Note 263023.1 for how to reset the password.

iSQL*Plus Login

For database targets in Grid Control, including the OMR Database itself, there is an *iSQL*Plus* link that allows you to log in over a web-based user interface with either DBA (e.g., SYSOPER, SYSDBA) or non-DBA credentials. However, to enable the *iSQL*Plus* link, you must configure and start an *iSQL*Plus* server according to the instructions in Chapter 4 (see "Configure *iSQL*Plus* Access in Grid Control"). Once *iSQL*Plus* is configured and started, you can log in to a target database using *iSQL*Plus* as follows:

1. Log in to Grid Control using HTTP rather than HTTPS, unless you configured *iSQL*Plus* to run in secure mode. (If you logged in to the Console via HTTPS but did not configure *iSQL*Plus* to run in secure mode, a screen would appear when logging in to *iSQL*Plus* stating that you were running in secure mode, but that *iSQL*Plus* was not configured to run in secure mode. If you get this screen, Click Yes to start *iSQL*Plus* in nonsecure mode.)

2. Click the Targets tab, All Targets subtab, and select the database instance name you want to access via *iSQL*Plus* (if a Cluster Database, you must select one of its instances):

3. At the bottom of the Home page for that database (or Performance, Administration, or Maintenance pages), click the *iSQL*Plus* link under Related Links.

TIP
*Rather than logging in to iSQL*Plus through Grid Control as just described, you can log in directly from a browser with or without SYSDBA/SYSOPER privileges using the following URLs, respectively:*
http://<iSQLPlus_node>:5560/isqlplus/dba/dynamic
http://<iSQLPlus_node>:5560/isqlplus/dynamic
*In Chapter 4, I suggest using one or more Repository Database nodes to host an iSQL*Plus server.*

4. The *iSQL*Plus* Connection Role popup appears, prompting you to connect as Normal, SYSOPER, or SYSDBA. Make your choice and click Continue.

   - If you choose to connect as NORMAL, continue to step 5, where you are prompted for a username and password.

   - If you choose to connect as SYSOPER or SYSDBA, you are prompted to enter the *iSQL*Plus* DBA realm administrator username and password. Enter the username *admin* and the password, which by default is *welcome*, but which you may have changed when configuring *iSQL*Plus*. Click OK.

5. You are prompted for *iSQL*Plus* login credentials and a connect identifier:

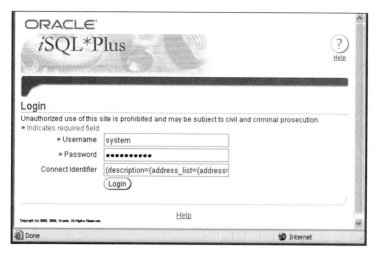

6. Enter a valid username and password. The connect identifier field contains a default value from information you enter when discovering the database. If connecting to a RAC database instance, correct the Connect Identifier if the *instance_name* keyword incorrectly specifies the *service_name*. Use the non-fully qualified *instance_name* on a particular RAC node. (Incidentally, the same solution is required when connecting to a RAC database instance through the Java Console.) Click Login.

7. The *iSQL*Plus* DBA Workspace window is displayed. In this workspace, you can execute ad hoc SQL, PL/SQL, and SQL*Plus statements, as well as load and save scripts. See the SQL*Plus User's Guide and Reference for the features offered by *iSQL*Plus*.

Metrics Browser Login

The Metrics Browser is an Agent troubleshooting tool accessed via a web browser that you can use to look at the raw data an Agent collects. The Metrics Browser is useful in troubleshooting metric errors because it allows you to evaluate an Agent's metric values real-time on a target host. In the case of the GC Metrics Browser, you can evaluate metric values even before they are uploaded to the OMS. You can actually use the Metrics Browser for Grid Control maintenance, as shown in Chapter 17 in the section "Fix Target Metric Collection Errors." There I offer an example of using an Agent's Metrics Browser to resolve a metric error.

Metrics Browsers exist for all three types of Enterprise Manager Agents—the central Grid Control Agent (GC Agent), Application Server (AS) Control Agent, and Database (DB) Control Agent. For a description of how these Agents differ, see the section in Chapter 1 "Agents for Grid Control, Database Control, and AS Control." As mentioned in that chapter, when running Grid Control, you do not need to—nor should you—run Database Control. Therefore, I don't discuss the Metrics Browser for the DB Control Agent. My comments here on Metrics Browsers apply equally to those for GC and AS Control Agents. Furthermore, these remarks encompass AS Metrics Browsers for both target OAS instances and for OASs bundled with Grid Control.

Following is the procedure for logging in to the Metrics Browser for a GC or AS Control Agent. All references to "Agent home" apply to that for the GC or AS Control Agent.

1. Enable the Metrics Browser (it is disabled by default). Simply uncomment the following line in the *emd.properties* file on the host under the *sysman/config/* directory of the Agent home:

   enableMetricBrowser=true

2. Reload the Agent. The above property is reloadable, meaning that you can reload the Agent rather than having to bounce it in order to put the property change into effect. Reload the Agent as follows using the *emctl* command under the *bin/* directory of the relevant Agent home:

   emctl reload agent

   Because reloading does not stop the Agent, it will not trigger any alert notifications of an Agent down or host unreachable.

3. Log in to the Agent Metrics Browser.[8] (Again, see Chapter 17 for an example.) Following are the default login URLs for both GC and AS Control Agents on UNIX and Windows:

Type of Agent	Platform	URL for Metrics Browser	Default Port
GC Agent	UNIX, Windows	https[9]://<Agent_host>:<Agent_port>/emd/browser/main	3872
AS Control Agent	UNIX	http://<AS_host>:<AS_port>/emd/browser/main	1157
	Windows	http://<AS_host>:<AS_port>/emd/browser/main	18120

---

[8] If the Metrics Browser is not enabled and you attempt to log in to its URL, your browser will report HTTP Error 403.

[9] The Central Agent is secure by default (i.e., it uses HTTPS). However, the Agent may be unsecure if you do not secure it on installation. This is allowed if you don't require secure communication for all Agents when installing an OMS or if you later unsecure an OMS using *emctl secure unlock*. If the Agent is not secure, you must connect to the Agent Metrics Browser via HTTP. For details, see the section "Enable Framework Security Between Agents and OMS" in Chapter 16.

4. After you've logged in to the Agent Metrics Browser, the following screen should appear:

TargetType(DisplayName)	TargetName(DisplayName)	BrokenReason
oracle_emd(Agent)	pg04.perftuning.com:3872	
host(Host)	pg04.perftuning.com	
oracle_database(Database Instance)	emrep.perftuning.com	
oracle_listener(Listener)	LISTENER_pg04.perftuning.com	
oc4j(OC4J)	EnterpriseManager0.pg04.perftuning.com_home(home)	
oracle_apache(Oracle HTTP Server)	EnterpriseManager0.pg04.perftuning.com_HTTP Server(HTTP_Server)	
oracle_webcache(Web Cache)	EnterpriseManager0.pg04.perftuning.com_Web Cache(Web Cache)	
oracle_ias(Oracle Application Server)	EnterpriseManager0.pg04.perftuning.com	
oc4j(OC4J)	EnterpriseManager0.pg04.perftuning.com_OC4J_EMPROV(OC4J_EMPROV)	
oc4j(OC4J)	EnterpriseManager0.pg04.perftuning.com_OC4J_EM(OC4J_EM)	
oracle_csa_collector(CSA Collector)	pg04.perftuning.com_oms_csa_collector(pg04.perftuning.com_oms_csa_collector)	
oracle_emrep(OMS and Repository)	Management Services and Repository	
oracle_beacon(Beacon)	pg04.perftuning.com_beacon	

EMAgent:10.2.0.3.0

Note from this table that the Metrics Browser for the GC Agent is secure by default, whereas the Metrics Browser for the AS Control Agent is not secure. Note also that the port number in the login URL on UNIX is different from the one on Windows, but only for the AS Control Agent Metrics Browser.

5. Disable the Metrics Browser when you've finished troubleshooting a metric problem. This saves an unneeded expenditure of resources on the Agent host, even though the overhead of a Metrics Browser is low. To disable it, comment out the line that you uncommented above in the appropriate *emd.properties* file as follows:

   #enableMetricBrowser=true

6. Finally, reload the Agent so the change to disable the Metrics Browser takes effect:

   emctl reload agent

Default values for the Metrics Browser URL listed in the table above are fine, provided you don't inherit a system that your benefactor changed from the default configuration. Such changes can run the gamut from protocol to hostname (perhaps a virtual hostname now) to port for an Agent, OMS, or OAS target. There are two ways to confirm a Metrics Browser URL. The following values are for the GC Agent Metrics Browser shown above, which uses default ports:

- Examine the value for "Agent URL" listed in the output of *emctl status agent*. Following is an example default value for the GC Agent URL:

   Agent URL   : https://pg04.perftuning.com:3872/emd/main

   To connect to the Agent Metric Browser in this case, use protocol HTTPS, hostname *pg04.perftuning.com*, and port 3872.

- Check the EMD_URL parameter in the *emd.properties* file under the *sysman/config/* directory of the Agent home. In the above example for a GC Agent, the value for EMD_URL matches the value for Agent URL above, as it must for it to be working:

    EMD_URL=https://pg04.perftuning.com:3872/emd/main/

    The only difference between the two values above is the slash ("/") appended to EMD_URL parameter in the *emd.properties* file, which the internal Agent HTTP Listener requires [see "Agent Subsystems" in Appendix A (online at www.oraclepressbooks.com) for more on an Agent's inner workings]. However, this slash is immaterial when logging in to the Metrics Browser. That is, the Metrics Browser URL works, whether or not you append a slash to it. The values for either Agent URL or EMD_URL in this case confirm that the login URL for the Metrics Browser is: https://pg04.perftuning.com:3872/emd/browser/main.

Control EM Components

Now that you know all login entry points to Enterprise Manager, let's go over how to control EM components. I present instructions for each EM component in the suggested startup order, as follows:

- *i*SQL*Plus server
- Management Repository (OMR)
- AS Console
- Management Service (OMS)
- Management Agent (OMA)

This chapter assumes that all EM Oracle homes have been installed using the *oracle* user, and that, on hosts containing an OMS, Repository Database instance, or Agent, you've initialized the environment according to the instructions in Chapter 4 (for OMS and OMR) and Chapter 7 (for OMA) under section names beginning with "Set Up oracle User Environment." Such initialization allows you to call all executables described below, except $*OMS_HOME/bin/emctl*, without fully specifying their path.

Control iSQL*Plus Server

Execute the following command to start and stop the *i*SQL*Plus process on a database node hosting the *i*SQL*Plus Server (ORACLE_HOME below is the *i*SQL*Plus Server database home):

$ORACLE_HOME/bin/isqlplusctl start | stop

There is no option for the *isqlplusctl* command to check the status of the *i*SQL*Plus server. You must look at the process level. On UNIX, enter the following command to determine if the *i*SQL*Plus Server process is running (output will look something like the one shown here and will contain both "java" somewhere below the path $*ORACLE_HOME/jdk/bin* and "Djava" as an argument):

```
$ ps -aef | grep Djava | grep java | grep -v grep
  oracle 26634    1  0 19:53:10 pts/0    0:08
/u03/app/oracle/product/10.2.0/db_1/jdk/bin/IA64N/java -Djava.awt.headless=true -
Doracle.oc4j.localhome=/u03/app/oracle/product
```

On Windows, you can see the status of the *i*SQL*Plus service in the Services panel (see the section "Windows EM Services" later in this chapter).

Control the Management Repository

Control of the Management Repository Database depends on whether it is RAC or single-instance. Commands shown below are shown in order of startup; issue commands to shut down the Repository Database in reverse order.

Control a RAC Management Repository

For a RAC Repository Database, you'll need to start up the cluster, nodeapps, ASM (if used), and database, in that order (choose the "start" argument for all commands):[10]

1. First, as *root*, spin up the Clusterware, if it's not already started:

   /etc/init.d/crsctl start |stop| check crs

2. Then, as *oracle*, start the nodeapps[11] and ASM instance on each node, and then start the database. All commands can be run on any RAC node. (It is assumed that the directory where *srvctl* is located, *$ORA_CRS_HOME/bin,* is in the PATH as suggested in Chapter 4 in the section "Set Up oracle User Environment for OMR Nodes."):

   srvctl start |stop nodeapps -n <node_name>

   srvctl start |stop asm -n <node_name>

   srvctl start |stop database -d <database>

Control a Single-Instance Management Repository

For a single-instance Repository Database, log in to SQL*Plus with SYSDBA privileges and start up the database as follows:

```
SQL> startup;
```

To shut down the database, execute:

```
SQL> shutdown immediate;
```

Control the AS Console

Both chain-installed and standalone Agents are considered GC central Agents (GC Agents). The GC Agent chain-installed with the OMS is responsible for *monitoring* OMS targets, including the EM Application Server. This chain-installed Agent runs from its own Oracle home and is identical to standalone Agents on non-OMS boxes—it just monitors OMS targets rather than non-EM targets. However, each OMS server also runs an AS Agent under the OMS home that *administers* the EM Application Server. (For details on the difference, see "Grid Control vs. AS Control" in Chapter 1.) This AS Agent belongs to the EM Application Server 10.1.2, which is bundled with

---

[10] Single-instance databases can also run Clusterware and ASM, but this is not typically done for the Grid Control Repository Database.

[11] The nodeapps must run on each node to support a RAC database. Nodeapps include the Virtual IP (VIP) address, TNS Listener, Oracle Notification Services (ONS), and Global Service Daemon (GSD).

EM10*g* as its middle tier. The Application Server comes with an Application Server Control Console that provides Enterprise Manager features needed to administer the Application Server.

The Application Server Control command is called *emctl* and is located under the $OMS_HOME/bin/ directory. This executable is different from the *emctl* command made available with the central GC Agent under the $AGENT_HOME/bin/ directory.

CAUTION
Because the emctl executables under the Agent and OMS homes control different components, it is easy to execute the wrong one. Therefore, exercise caution when running these commands. I have found it best to list only the Agent emctl command (which is more heavily used) in the PATH so you can execute it without qualification. Then, if you need to run the OMS emctl command, fully qualify it or change directories to the $OMS_HOME/bin/ directory and dot execute it.

The AS *emctl* command starts both the Application Server Control Agent and Application Server Control, which are required for login to the Application Server Control Console.[12] (Each Application Server instance runs its own processes.[13] This is different from earlier versions of 9*i*AS where only one process ran on the server for all 9*i*AS instances.) Given that the AS and GC Agent *emctl* executables are different, it is not surprising that their syntax differs as well. Following is the syntax for the AS Control *emctl* command:

$OMS_HOME/bin/emctl start |stop| status iasconsole

By contrast, the command to control the GC Agent is

$AGENT_HOME/bin/emctl start |stop| status agent

Control the Management Service

As defined in Chapter 1, I refer to the entire Grid Control middle tier as the Oracle Management Service (OMS), which is a customary practice. To manage startup and shutdown of the OMS, you should understand that the OMS consists of three main subcomponents, which are all part of the Oracle Application Server distribution for Grid Control:

- Oracle HTTP Server (OHS)
- OMS applications, which are part of OAS Containers for J2EE (OC4J):
  - OC4J_EM   The main OMS application
  - OC4J_EMPROV   A new Grid Control 10.2 feature for provisioning and Agent Deploy
- OracleAS Web Cache 9.0.4

---

[12] The *$OMS_HOME/bin/emctl* command also controls the OMS, but this is covered later in the section "Control the Management Service."

[13] The AS *emctl* command used for EM is distinguished from that used to control 9.0.2 and 9.0.3 9*i*AS Instances, which is "emctl start |stop em".

There are three ways to control some or all of these subcomponents on any platform at the command-line level:

- **opmnctl**   This is the recommended command-line utility to detect and control all OMS subcomponents described above and others identified below, which are part of the Oracle Application Server bundled with Grid Control.

- **emctl**   This utility, located under the OMS home, starts the HTTP Server and OC4J_EM application, but it does not start OC4J_EMPROV or Web Cache. It is useful for putting changes to *emoms.properties* into effect.

- **EM 10g AS Console**   This browser-based utility provides equivalent functionality to the command-line utility, *opmnctl*, in controlling all OMS subcomponents.

NOTE
The Distributed Configuration Management Control (DCM) utility, <AS_HOME>/dcm/bin/dcmctl, is another command-line OAS tool. While dcmctl can control OHS and the OC4J_EM processes, it is not intended to manage the OMS, except to register manual changes to OHS configuration files with the DCM repository (such as in Chapter 4, "Reduce Oracle HTTP Server Logging"). The purpose of dcmctl outside the context of Grid Control is to administer and monitor AS clusters and farms and to deploy OC4J applications.

Let's review each method of controlling the OMS subcomponents.

opmnctl

Oracle Process Management Notification Control (*opmnctl*) is the recommended method to control all subcomponents of the OMS and additional OAS subcomponents that Grid Control does not require. The *opmnctl* executable is located in the *$OMS_HOME/opmn/bin/* directory. You don't need to qualify this directory if it's in your PATH, as recommended in Chapter 4, "Set Up *oracle* User Environment on OMS Hosts." Following are a few of the more useful *opmnctl* parameters:

- To get a detailed description of all *opmnctl* parameters or a particular parameter, execute the following:

    opmnctl usage [<parameter>]

- To start, stop, or get a status of all OAS subcomponents, execute one of the following commands, respectively:

    opmnctl startall | stopall | status

Below is a typical output of an *opmnctl status* command after executing *opmnctl startall*:

```
opmnctl status
Processes in Instance: EnterpriseManager0.pg04.perftuning.com
-------------------+--------------------+---------+---------
ias-component      | process-type       |     pid | status
-------------------+--------------------+---------+---------
```

```
DSA                 | DSA             | N/A   | Down
HTTP_Server         | HTTP_Server     | 13940 | Alive
LogLoader           | logloaderd      | N/A   | Down
dcm-daemon          | dcm-daemon      | N/A   | Down
OC4J                | home            | 13941 | Alive
OC4J                | OC4J_EMPROV     | 13942 | Alive
OC4J                | OC4J_EM         | 13944 | Alive
WebCache            | WebCache        | 13952 | Alive
WebCache            | WebCacheAdmin   | 13951 | Alive
```

OAS processes in **boldface** are required for access to the GC Console (including through Web Cache) and availability of all features, and they correlate with the OMS subcomponents already identified.

NOTE

For clarity in showing the processes started by opmnctl, all output in this section assumes the AS Console is not running (e.g., that you haven't executed $OMS_HOME/bin/emctl start iasconsole), as this starts the dcm-daemon.

Below is a brief description of all OAS processes, whether required or not. Grid Control functions properly with only required processes running; however, it's sometimes handy to start a few of the nonessential processes identified below, and you should at least be able to identify what they are. The format "*<ias-component>* (*<process-type>*)" is used to distinguish ias-components with the same name but with different process types.

- **DSA**   The OracleAS Guard server, a distributed server that runs on all the systems in an OracleAS Disaster Recovery (DR) configuration. OPMN does not start the DSA process by default because it is necessary only in the context of DR sites. DSA is not applicable to Grid Control, which runs multiple OASs (i.e., OMSs) independently. See "OMS Recommendations" in Chapter 15 for the recommended DR approach to the OMS.

- **HTTP_Server**   *Required.* The web server component of Oracle Application Server. Do not use *apachectl* to control Oracle HTTP Server.

- **LogLoader**   A compiler of log messages from log files you select into a single repository (the AS Metadata Repository) for viewing in the AS Console rather than at the OS level. Sometimes it's beneficial in Grid Control to start LogLoader for log readability and centralization when debugging infrastructure issues.

- **dcm-daemon**   A Distributed Configuration Management (DCM) process that enables OAS cluster-wide deployments. Grid Control does not need DCM because it uses independent nonclustered Application Servers. However, dcm-daemon starts by default when an administrator starts the AS Console.[14]

- **OC4J (home)**   The default OC4J container that comes with Oracle Application Server.

---

[14] dcm-daemon also starts after executing *opmnctl startall*, then running *dcmctl getstate*.

- **OC4J (OC4J_EM)** *Required.* The AS instance where the main Management Service application is deployed.

- **OC4J (OC4J_EMPROV)** *Required.* The AS instance where the EMPROV Management Service application is deployed. This is a new feature in Grid Control Release 10.2 used for provisioning and the Agent Deploy feature. This process is not required for other Grid Control functionality.

- **WebCache (WebCache)** *Required.*[15] Another entrée to the Console other than OHS, which marginally improves performance. Web Cache is also needed to configure End-User Performance Monitoring for the built-in EM Website service, which is given additional features in Chapter 12, "Manage Web Server Data Collection."

- **WebCache (WebCacheAdmin)** The Admin port, which, while not required, is used for administrator access to the OracleAS Web Cache Manager at http://<OMShost>:<WebCache_Admin_port> (port 9400 by default). The Web Cache Manager can be valuable in a Grid Control context to administer Web Cache, such as to configure HTTPS communication between client browsers and Web Cache.

TIP
Grid Control administrators who start LogLoader can configure it to do the following:

- *Mine logs in the AS Console UI rather than at the OS level, which for UNIX is generally in a less user friendly command-line text editor*

- *Search the Log Repository across all log files selected by date and time range, message types (error, warning, notification, etc.), and message text*

- *Review log histories that may otherwise have already been purged in the corresponding operating system log files*

- *Learn which log files are available and the name and path of each component's log file*

- *Define Log Loader Properties to control the behavior of LogLoader and the size of the Repository it updates*

LogLoader incurs unnecessary overhead if not tapped. Therefore, I recommend holding off on starting LogLoader unless needed for a specific troubleshooting situation, at which time you can retroactively configure it to view the desired logs.

To get a full listing of all OPMN subcomponent details, use the "-l" option:

```
opmnctl status -l
```

---

[15] Some—even certain EM Support analysts—say that Web Cache is not required for Grid Control. While you can log in via OHS and bypass Web Cache, it is integral to the complete monitoring of the EM Website service, which in turn is critical for GC self-monitoring. For example, the Web Cache URL is defined as the default home page for this service.

The output includes that shown above from the *opmnctl status* command (although only the *ias-component* column is shown below for clarity). The "-l" option also supplies the columns shown below:

```
Processes in Instance: EnterpriseManager0.pg04.perftuning.com
-------------+-----------+--------+---------+------
ias-component|       uid |memused |  uptime | ports
-------------+-----------+--------+---------+------
DSA          |       N/A |    N/A |     N/A | N/A
HTTP_Server  | 2135228907|  115576| 370:01:42|
http1:7778,http2:7200,https1:4444,https2:4443,https3:1159,http3:4889
LogLoader    |       N/A |    N/A |     N/A | N/A
dcm-daemon   |       N/A |    N/A |     N/A | N/A
OC4J         | 2135228908|   29288| 370:01:58| ajp:12501,rmi:12401,jms:12601
OC4J         | 2135228909|   32680| 370:01:58| ajp:12502,rmi:12402,jms:12602
OC4J         | 2135228914|  293556|  04:28:06| ajp:12503,rmi:12403,jms:12603
WebCache     | 2135228911|   25068| 370:01:57|
invalidation:9401,statistics:9402,https:4445,http:7777
WebCache     | 2135228912|    6184| 370:01:57| administration:9400
```

This extended information about all OMS component processes includes memory consumed, length of time started and ports assigned.

To control individual OPMN processes or ias-components (e.g., classes of components), run the following commands:

opmnctl startproc |stopproc process-type=<*process_type*>

opmnctl startproc |stopproc ias-component=<*ias_component*>

For example, to start OracleAS Web Cache (not Web Cache Admin), issue the following command:

```
opmnctl startproc process-type=WebCache
```

emctl

It is sometimes useful to be able to control just two subcomponents, the Management Service J2EE application (OC4J_EM process type) and OHS, as they allow access to the Console via OHS (rather than via Web Cache). The *emctl* command also allows you to implement changes made to the OMS properties file, *$OMS_HOME/sysman/config/emoms.properties*. An example of putting *emctl* to good use in this way is when you're changing OMS properties during OMS post-installation as described under the "Oracle Management Service Configuration" heading in Chapter 4.

The *emctl* executable is located in the *$OMS_HOME/bin/* directory. I do *not* assume that this directory is in your PATH, preferring it instead to contain the Agent *emctl* (see the section "Set Up *oracle* User Environment on OMS Hosts" in Chapter 4). Therefore, I fully qualify the *emctl* path below. Following are the basic *emctl* command options:

$OMS_HOME/bin/emctl start |stop |status oms

The *emctl status* command does not provide much detail, but only indicates whether the Management Service is running:

```
$OMS_HOME/bin/emctl status oms
Oracle Enterprise Manager 10g Release 3 Grid Control
Copyright (c) 1996, 2007 Oracle Corporation.  All rights reserved.
Oracle Management Server is Up.
```

Therefore, to see what processes are started compared with *opmnctl startall*, execute *opmnctl status* immediately after running *emctl start oms*:

```
$OMS_HOME/opmn/bin/opmnctl status
Processes in Instance: EnterpriseManager0.pg04.perftuning.com
-------------------+--------------------+---------+---------
ias-component      | process-type       |   pid   | status
-------------------+--------------------+---------+---------
DSA                | DSA                |   N/A   | Down
HTTP_Server        | HTTP_Server        |  11080  | Alive
LogLoader          | logloaderd         |   N/A   | Down
dcm-daemon         | dcm-daemon         |   N/A   | Down
OC4J               | home               |   N/A   | Down
OC4J               | OC4J_EMPROV        |   N/A   | Down
OC4J               | OC4J_EM            |  11124  | Alive
WebCache           | WebCache           |   N/A   | Down
WebCache           | WebCacheAdmin      |   N/A   | Down
```

This output shows what I've already stated: the *$OMS_HOME/bin/emctl start oms* command only starts the HTTP_Server and OC4J_EM process types (displayed in **boldface**). The four processes in *italics* are those that *emctl start oms* doesn't start but that *opmnctl startall* does start. You need to start the OC4J_EMPROV process-type to provision or use Agent Deploy. You also need to start the WebCache process type to log in to the Console via Web Cache. Because of these two limitations, I recommend starting the OMS middle tier with *opmnctl startall* to provide maximum Console functionality and login flexibility to administrators. I suggest you limit use of the *$OMS_HOME/bin/emctl* command to bouncing the Management Service when making OMS configuration changes. If OMS CPU or memory resources are very strained, you also have the option of using *emctl* on a short-term basis to start the minimal number of OAS processes necessary to use the Console. Of course, users would have to log in via OHS, and provisioning and Agent Deploy features would not be available. However, after resolving the OMS host resource constraint, you could resume using *opmnctl startall* to start those OAS processes that offer the full gamut of Grid Control features.

EM 10g AS Console

If desired, you can open a web browser and use the Application Server Console to manage a particular OMS. This is the GUI equivalent to using *opmnctl*. For directions and screen shots on logging in to the EM 10g AS Console, see the section "AS Console Login" earlier in the chapter. Once logged in, you can select those system components you'd like to start, stop, or restart under the System Components section (refer back to Chapter 1 under the section "AS Control Administers Oracle Application Server."). Remember, the AS Console is not available until you start it with the command *$OMS_HOME/bin/emctl start iasconsole*, as already mentioned.

Control the Management Agent

To monitor a target server, the Grid Control Agent must be running on that target. Remember from Chapter 1 that each target server runs a central Agent installed in a separate Oracle home from all target Oracle homes. (The OMS server also runs an AS Agent under the OMS home belonging to the EM Application Server. For more specifics, see the section "Control the AS Console" earlier in this chapter.)

To control an Agent, all platforms provide both the Agent command-line utility (*emctl*) and functionality afforded in the EM Web Console.

Control the Management Agent Using *emctl*

To control the Management Agent on any target host, including those containing the OMS and Repository, log in locally as the owner of the Agent Oracle home, assumed here to be *oracle*. Use the command-line utility, *emctl*, to manage the Agent. As recommended in Chapter 7 under "Set Up Agent User Environment," add the directory to the Agent *emctl* executable, $AGENT_HOME/bin, to the *oracle* user's PATH so that you don't have to qualify this executable.

> **NOTE**
> *The $AGENT_HOME in $AGENT_HOME/bin as defined in the PATH is the EMSTATE directory for a local Agent on a RAC node, a Cluster Agent, or an NFS Agent. For details, see "What Is an EMSTATE Directory?" in Chapter 5.*

To give you a flavor of the capabilities of this Agent command, the following is output from the *emctl* command executed without arguments, which provides all syntax variations:

```
$ emctl
Oracle Enterprise Manager 10g Release 3 Grid Control 10.2.0.3.0.
Copyright (c) 1996, 2007 Oracle Corporation.  All rights reserved.
     Oracle Enterprise Manager 10g Agent Commands:
         emctl start | stop agent
         emctl status agent
         emctl status agent -secure [-omsurl <http://<oms-hostname>:<oms-unsecure-port>/em/*>]
         emctl getversion
         emctl reload | upload | clearstate | getversion agent
         emctl reload agent dynamicproperties [<Target_name>:<Target_Type>]....
         emctl config agent <options>
         emctl config agent updateTZ
         emctl config agent getTZ
         emctl resetTZ agent
         emctl config agent credentials [<Target_name>[:<Target_Type>]]
 Blackout Usage :
         emctl start blackout <Blackoutname> [-nodeLevel]
[<Target_name>[:<Target_Type>]].... [-d <Duration>]
         emctl stop blackout <Blackoutname>
         emctl status blackout [<Target_name>[:<Target_Type>]]....
The following are valid options for blackouts
<Target_name:Target_type> defaults to local node target if not specified.
If -nodeLevel is specified after <Blackoutname>,the blackout will be applied to all
targets and any target list that follows will be ignored.
```

```
Duration is specified in [days] hh:mm
        emctl getemhome
        emctl ilint
        emctl start | stop | status subagent
Secure Agent Usage :
emctl secure agent <registration password> [-passwd_file <abs file loc>]
emctl unsecure agent
```

As this output shows, the *emctl* utility accomplishes many Agent functions. For example, it can secure the Agent, change its time zone, or control Agent blackouts. Following are the most important and frequently used *emctl* commands:

emctl stop | start | status | upload | reload | secure agent

- **stop, start** Stops or starts the Agent. There are two platform specific issues related to controlling the Agent:
    - If you have trouble stopping or starting the Agent on Windows NT (does not apply to any other Windows platform), attempt to shut down the Agent using e*mctl istop agent*, then restart the Agent in the Services control panel.[16]
    - The Agent may not start in IBM AIX environments with a hefty memory configuration where the Agent manages a large number of targets. To skirt this problem, set the following OS variables before starting the Agent:

        LDR_CNTRL="MAXDATA=0x80000000"@NOKRTL
        AIX_THREADSCOPE=S

        The LDR_CNTRL variable sets the data segment size and disables loading of runtime libraries in kernel space. The AIX_THREADSCOPE parameter changes AIX Threadscope context from the default Processwide 'P' to Systemwide 'S', which results in less mutex contention.[17]

- **status** Displays the status of the Agent. For sample output, see the section "Confirm Agent Is Working" in Chapter 7.

- **upload** Forces a manual Agent upload to the OMS of normal metric data that otherwise occurs automatically every fifteen minutes by default (determined by the value for UploadInterval in *$AGENT_HOME/sysman/config/emd.properties*). It is useful to be able to force an upload to test whether an Agent can fully communicate with the OMS. Just ensuring an Agent starts does not guarantee it can upload data.

- **reload** Reloading an Agent is used to effect the change of a reloadable Agent property in the *emd.properties* file. Reloading an Agent does not require shutting it down, and is similar to reloading a database listener.

- **secure** To secure a Management Agent, enter the Agent Registration password when prompted, or enter this password as an argument to the *emctl secure agent* command.

---

[16] Oracle Enterprise Manager Advanced Configuration 10*g* Release 4 (10.2.0.4.0), Section 2.1.2.

[17] Oracle Enterprise Manager Advanced Configuration 10*g* Release 4 (10.2.0.4.0), Section 2.1.1.

Control Management Agents in the Console

While the *emctl* utility provides complete control of the Agent, this control is limited to the local host on which the Agent runs. You can use the Console to your advantage to get a bird's eye view of all Agents. To see the status of all Agents, navigate to the Targets tab, All Targets subtab, and filter them by selecting Agent in the Search field, as illustrated here:

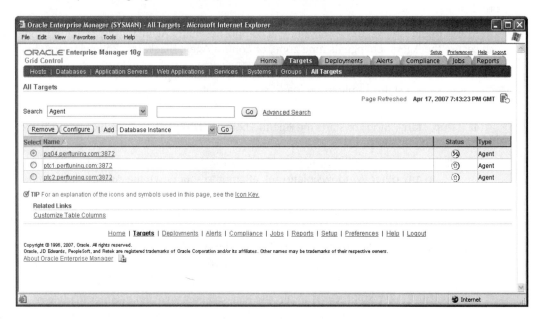

In this screen shot, you see that the Agent is unreachable on host *pg04*, but is running on *ptc1* and *ptc2*. Using a little ingenuity, you can use the Console to shut down any and all Agents in one operation, or control them (and the hosts on which they run) in other ways.

NOTE
Unlike the Agent, you cannot shut down the OMS or Repository Database in the Console, as both components must be running for the Console to function.

You can stop or restart many Agents in one operation within the Web Console by creating a Grid Control job, or ad hoc by using the Execute Host Command functionality under Related Links. (You cannot use either method to start Agents ad hoc in the Console, as an Agent must be running to run a job or execute commands for its host. However, you can create an OS script that stops an Agent, sleeps for a while, and then restarts the Agent.)

The Execute Host Command Feature Let's take a concrete example and shut down all Agents in one fell swoop using the Execute Host Command feature. To stop multiple Agents at once from within the Console, follow these steps (the hosts in this example are *ptc1* and *ptc2*):

1. Navigate to a Host home page and under Related Links, click Execute Host Command.

Chapter 9: EM Login and Component Control **289**

2. The Execute Host Command page appears.

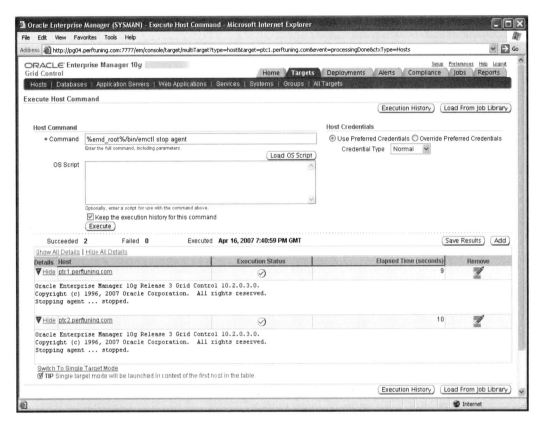

   a. Click Add to select all target hosts on which to run the command. A pop-up menu lists all discovered hosts. Select the desired hosts and click Select to close the pop-up menu. Adding hosts first allows you to select an OS script to load that is located on the selected hosts (as described next).

   b. Enter a single host command. Alternatively, enter the contents of an OS script, either directly or by clicking Load OS Script to choose its path on either your browser machine or the target machines. Click Help on this page to see which *target properties*, or system variables you can specify, such as *%emd_root%* (the Agent home path) shown above.

   c. Select the host credentials needed.

   d. Check "Keep the execution history for this command" if desired.

   e. Click Execute and a screen will appear stating that it is processing the Execute Host Command.

3. When finished, check that the Execution Status shows that the operation succeeded and optionally expand the details on each host, as shown above.

Essentially, anything you can script that affects the Agent and/or host can be accomplished through the Console. The caveat is that once the Agent is shut down through the Console (or even locally), you can no longer use the Console to contact that Agent ad hoc. So any Agent shutdown command that requires subsequent commands should be scripted, whether these commands are Agent-related or otherwise. An example of the usefulness of this concept is when you need to change an Agent property in the *emd.properties* file that requires a cold Agent start, such as *agentTZRegion*, which sets the Agent time zone. If you need to change Agent time zones across 100 UNIX hosts, rather than logging on to each host manually, you could save a lot of time by creating a script to run against all 100 hosts that shuts down the Agent, uses the *awk* or *sed* commands to edit the *agentTZRegion* line, and then restarts the Agent. Note 560905.1 contains an example script that changes an Agent configuration file using *awk* and restarts the Agent to put the change into effect, using the job system.

Windows EM Services

On UNIX and Windows, you can control all EM processes at the command-line, as just described. However, on Windows, you can also start and stop all EM processes in the Services panel. For these EM processes, Table 9-3 lists the executable syntax at the command line (for both Windows and UNIX), and the equivalent Windows service name to start or stop these processes. This table supplies additional service names (excluding those for OCFS) that you should start, and which may exist on OMR Database nodes if you preinstall a database to use with the GC Existing Database Option. In this table, ORACLE_HOME refers to the OMR Database home, and all commands should be run as *oracle* unless otherwise indicated.

Process Name	Command-Line Utility	Windows Service Name
TNS Listener (single-instance)	$ORACLE_HOME/bin/lsnrctl start|stop <listener_name>[18]	Oracle<HOME_NAME>[19]TNSListener
Oracle Database (single-instance)	$ORACLE_HOME/bin/ sqlplus '/ as sysdba'; startup|shutdown	OracleService<ORACLE_SID>[20]
AS Control	$OMS_HOME/bin/emctl start|stop iasconsole	Oracleoms10gASControl
OMS	$OMS_HOME/opmnctl/bin/ opmnctl startall|stopall	Oracleoms10gProcessManager
Agent	$AGENT_HOME/bin/emctl start|stop agent	Oracleagent10gAgent

TABLE 9-3. *EM Processes and the Commands and Windows Services That Control Them*

---

[18] If you're installing with the GC New Database Option, *<listener_name>* is *LISTENER*. This default value doesn't need to be specified.

[19] If you're installing with the GC New Database Option, *<HOME_NAME>* is *db10g*.

[20] If you're installing with the GC New Database Option, *<ORACLE_SID>* is *emrep*.

Process Name	Command-Line Utility	Windows Service Name
OMS Configuration Manager	(in crontab file on UNIX)	Oracleoms10gConfigurationManager
Agent Configuration Manager	(in crontab file on UNIX)	Oracleagent10gConfigurationManager
Additional Processes That May Exist When Preinstalling a Database for the GC Existing Database Option		
Clusterware	$ORA_CRS_HOME/bin/crsctl start|stop crs (as *root*)	OracleCRService, OracleCSService, OracleEVMService
TNS Listener and other nodeapps (RAC)	$ORA_CRS_HOME/bin/ srvctl start|stop nodeapps -n *<nodename>*	Oracle*<HOME_NAME>*TNSListener LISTENER_*<nodename>*
ASM Instance	$ORA_CRS_HOME/bin/ srvctl start|stop asm -n *<nodename>*	OracleASMService+ASM*<INSTANCE_NUMBER>*
*i*SQL*Plus Server[21]	$ORACLE_HOME/bin/ isqlplusctl start|stop	Oracle*<HOME_NAME>i*SQL*Plus
Oracle Database Instance (RAC)	$ORA_CRS_HOME/bin/srvctl start|stop database -d *<db_name>*[22]	OracleService*<ORACLE_SID>*

TABLE 9-3. *EM Processes and the Commands and Windows Services That Control Them (continued)*

To start or stop these services on Windows, click Start, Programs, Administrative Tools, and select Services. Right-click the desired service and click Start or Stop as desired.

When you deploy the Agent on a Windows system, three Agent-related services are created. In addition to the central Agent service listed in Table 9-3, two additional services are installed: *Oracle<HOME_NAME>SNMPPeerEncapsulator* and *Oracle<HOME_NAME>SNMPPeerMasterAgent*. These Simple Network Management Protocol (SNMP)[23] services play a role in integrating Grid Control with third-party management tools such as HP OpenView. Grid Control can use SNMP traps to notify third-party applications of metric alerts. You don't need to start either of these services if you're not integrating Grid Control with a third-party management tool. For more information on configuring SNMP on Windows and UNIX platforms to integrate with Grid Control, see Chapter 2 of the Oracle Enterprise Manager SNMP Support Reference Guide 10*g* Release 2.

---

[21] *i*SQL*Plus is present only when you install the Oracle Database Software and choose this component as recommended in Chapter 3 in the section "Install Database Software."

[22] More often than not, you want to stop the entire RAC database, which this command accomplishes. However, if you want to start or stop just one instance (ORACLE_SID) of a RAC database, issue the command *$ORA_CRS_HOME/bin/srvctl start|stop instance -d <db_name> -i <ORACLE_SID>*.

[23] Network management systems use the SNMP protocol to monitor products on the network for alert conditions. SNMP stores product management data in variables that system management products like Grid Control can query and push to other management systems.

EM Startup/Shutdown Order

Below is my recommended order to start all EM components and rationale for this order. (To stop EM components, use the reverse order.) When controlling your EM infrastructure, you need to take into account any high availability (HA) or disaster recovery (DR) architectures you've implemented (both are discussed in Chapter 15). At a high level, the order below addresses HA and DR best practices of running multiple OMSs, a RAC Repository Database, and a physical standby OMR Database in a Data Guard configuration as suggested in Chapter 15.

1. **Start *i*SQL*Plus Server**   Start *i*SQL*Plus Server to allow target databases to log in directly from a browser as well as through the *i*SQL*Plus link in Grid Control, once started.

2. **Start the OMR Database and related processes**   Following are a few points about why you start the OMR Database and related processes (listener, Clusterware, nodeapps, and ASM) at this point and the order in which to start them:

   - Start the standby OMR Database (and listener), if used, in managed recovery mode and related processes before starting the primary OMR Database and related processes.

   - Start the OMR Database listener before the OMR Database so that it can register more quickly with the listener if using automatic service management.

   - Start the Clusterware, nodeapps (which includes the listener), ASM, then the OMR Database, in that order, when using a RAC Repository, as with any RAC system.

   - Start the OMR Database before starting the OMS so that on startup the OMS can establish a connection to the Repository. If you start the OMS before the OMR, users attempting to log in to the GC Console will get a troubling error.[24]

   - Start the OMR Database before starting the Agents on OMR nodes to avoid alert notifications that the OMR Database is down.

3. **Start Application Server Control**   Start AS Control before starting the OMS to provide access to the AS Console so that you can start the OMS from the AS Console, if preferred, rather than using the *opmnctl* command-line utility.

4. **Start all OMSs**   Start each OMS before starting the Agent on the corresponding OMS host so that the Agent does not send an alert notification that the OMS is down. If running multiple OMSs, it doesn't matter which OMS you start up first.

5. **Start Agents on OMR/OMS hosts**   Start Agents on all OMR and OMS hosts to begin self-monitoring the Grid Control framework.

It is useful to script EM startup and shutdown when not rebooting for times when you need to perform GC maintenance, such as patching or cold backups. Appendix E (online) contains a sample script called *gcora* that starts or stops all EM components installed on a machine. This script calls executables at the command line, so it runs on UNIX or Windows.

---

[24] The error is "503 Service Unavailable. Service is not initialized correctly. Verify that the repository connection information provided is correct." If neither the OMR nor OMS is started, users get the generic browser error "The page cannot be displayed," which is a much less alarming message.

Startup/Shutdown Order with Standby OMR Database

If you're running the Repository in a Data Guard configuration, the primary and standby databases should be started and stopped in a particular order in relation to each other. Follow these three simple rules to maximize high availability using Data Guard:

- Always shut down the primary Repository Database before shutting down the standby Repository Database.

- Always start up the standby Repository Database in managed recovery mode (and start the primary and standby listeners) before starting the primary Repository Database.

- The primary and standby Repository Databases should be the last GC framework components shut down, i.e., you should shut down Agents on all OMR/OMS hosts, and stop OMS processes on *all* OMS servers before shutting down the primary, then the standby Repository Databases.

To remember the order in which to shut down and start the primary and standby databases, just remember that the standby should always be able to "see" the primary, so the standby must be shut down last and started first. If you violate any one of the above three rules, then the standby database will be down while the primary database is open and archiving. This can create an archive log gap, sometimes requiring manual recovery of the standby database. This recovery involves manually copying the archive logs from the primary to the standby server, then issuing the command RECOVER AUTOMATIC STANDBY DATABASE. This was definitely required with Oracle*9i* Data Guard, but 10*g* Data Guard usually takes care of standby database recovery automatically. While recovery is a simple process, the manual intervention required increases the chance of administrator error.

Summary

This chapter explained all the login paths into EM and how to control EM components. As the first part of this chapter demonstrated, all EM logins except through the Java Console are browser-based. Access to Grid Control is as simple as opening a browser and logging in to the Console, courtesy of the OMS layer. Once logged in, you can navigate to the links for *i*SQL*Plus and the EM Application Server Console, although you can also bring up both of these applications directly in a browser. The Metrics Browser is an Agent troubleshooting tool directly accessible through a web browser and which you enable on the spot for troubleshooting purposes, then disable when finished to conserve resources on the Agent host. The EM Java Console (and Change Manager in particular) is a different beast, requiring a separate client installation, as explained in Chapter 8. For those who installed the Java Console, explicit login directions were provided here for how to log in to the Java Console and to the Change Manager application within it.

Starting and stopping EM components was the second focus of this chapter. These EM components are *i*SQL*Plus, Application Server Control, and the three major Grid Control components: the Management Repository, OMS, and Agents. I gave you instructions for the command-line utilities available on both UNIX and Windows platforms to manage all components. In addition, you learned the alternate methods on Windows platforms to start and stop the associated services in the Services control panel. The chapter concluded with a high-level look at the order in which to start up and shut down the entire EM infrastructure. The order took into consideration high availability and disaster recovery configurations, using multiple OMSs, and a single-instance or RAC primary or standby Repository Database.

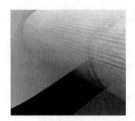

PART II

Configure and Maintain Grid Control

CHAPTER 10

General Console Configuration

This chapter marks the end of the Grid Control installation phase and the beginning of the configuration phase. Most Grid Control configuration takes place through the Web Console, known also as the User Interface (UI). Let me map out the overall strategy for this and subsequent chapters. First, I begin this chapter by laying the groundwork for what general Console configuration involves. Then, after we cover this introductory material, you perform general set up of administrators (or "admins"), configure *notifications* (events that admins elect to receive via e-mail and other defined methods),[1] and enable patching features. The following three chapters then concentrate on target management, including monitoring the most common target types (Chapter 11), creating and configuring groups and services (Chapter 12), and tuning the metrics and policies of all targets (Chapter 13).

NOTE
These general and target configuration changes are captured as metadata in the Repository and do not affect the targets themselves (except Agents on target hosts). Remember, the focus of the book is how to implement Grid Control, not how to use it to affect targets.

Let me trace out my approach for this chapter in more detail. It consists of the following major sections and subsections:

- Introduction
    - Tour the Setup and Preferences menus
    - Follow an event from trigger to notification
    - Understand the menu navigation order
- Set up administrators
    - Create roles
    - Create administrators
    - Set administrator preferences
    - Set preferred credentials
- Configure notifications
    - Define notification methods
    - Set notification schedules
    - Create and subscribe to notification rules
- Enable patching features
    - Complete the patching setup
    - Schedule the RefreshFromMetalink job

Let's dispense with the preliminaries first.

---

[1] Regardless of administrator notification preferences, all defined target events—metrics, policies, and job status changes—are reported in the Console.

Introduction

Most chapters in this book dive right into the meat of their titles without more than a few introductory paragraphs. However, this chapter demands more for two reasons: the chapter's length and the fact that this is your first full-on exposure to the UI. This section therefore contains preliminary material on general Console setup and is comprised of three parts:

- **Tour the Setup and Preferences menus** To experience the Setup and Preferences menus firsthand, click each menu item for a cursory view, and accompany this tour, as with a headset at a museum, by reading a short definition of each menu item. In this way, you learn the basics of Console terminology while experiencing the UI menus firsthand. I flag the few Setup menu items not receiving further coverage in this chapter, either because they're dependent on configuration performed in later chapters or are outside the scope of this book.

- **Follow an event from trigger to notification** To comprehend the interrelationships between menu items, trace an event from inception to notification using a flowchart. This is a framing mechanism to give grounding to subsequent discussions in this chapter on events, notification rules, and notification schedules.

- **Understand the menu navigation order** To follow how I derive a logical order for "circumnavigating" the Setup and Preferences menus, examine the interdependencies between menu items. It turns out that these interdependencies require toggling between the two menus. The resulting menu item order is the next best thing to a UI wizard to guide you through the general setup process.

Tour of the Setup and Preferences Menus

Let's begin with the two main menus for configuring Grid Control: the Enterprise Manager Configuration tab of the Setup menu (the Setup menu) shown in Figure 10-1 and the Preferences menu in Figure 10-2, whose links are both located at the top right of most Console pages. The numbering next to each menu item is their navigation order (more about that in a bit). The Setup menu allows you to configure and monitor Grid Control globally—that is, in a way that affects all admins. The Preferences menu is specific to each admin.

Following are short descriptions of all Setup and Preferences menu items. I encourage you to drill down into the menu items to get a feel for each item's UI. This will help cement in your mind what these menu items consist of.

The Setup Menu

The Setup menu items, which govern global GC behavior, are listed below. Remember that this chapter doesn't touch on every item in this comprehensive list; I cross-reference those chapters providing such coverage, and identify a few items that are beyond the scope of this book.

- **Roles** Roles allow you to group GC system and target privileges, and grant them to administrators or to other roles.

- **Administrators** Administrators are accounts created in the UI with which users log in to the Console to manage their GC environment. Each admin account is associated with certain system and target privileges, e-mail addresses, preferred credentials, notification rules, and a notification schedule. Using the OMR Database to store administrator properties, the OMS application provides the mechanism for interpreting and limiting administrator actions.

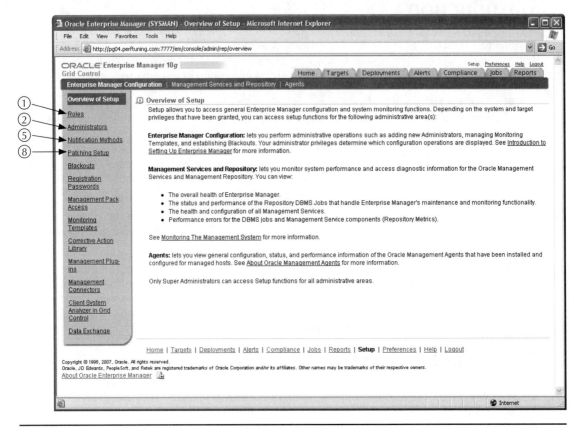

FIGURE 10-1. *The Setup menu*

- **Notification Methods**   Notification methods are global mechanisms defined to send notifications. Methods include SMTP Servers, custom OS scripts, PL/SQL procedures, and SNMP traps. Notification rules, defined per administrator, use selected notification methods to send administrators metric alerts, policy violations, or job status changes.

- **Patching Setup**   Patching Setup permits you to configure GC's patching features to stage and apply Oracle patches, patch sets, and critical patch advisories to targets on any Agent host. On this page you configure Oracle*MetaLink* credentials, a Patch Cache size, Internet proxy settings, offline patching, and patch validation (allowing only patches you approve to appear on the critical patch advisories page).

- **Blackouts**   Blackouts permit administrators to suppress data collection and DOWN notifications (indeed all notifications) for selected targets when purposefully brought down, such as for scheduled maintenance. Blackouts prevent off-duty DBAs from being paged unnecessarily, and allow GC to more accurately report on a target's performance, such as in meeting Service Level Agreements (SLAs). (Blackouts are covered in Chapter 11, "Schedule Blackouts for Planned Downtimes.")

Chapter 10: General Console Configuration **301**

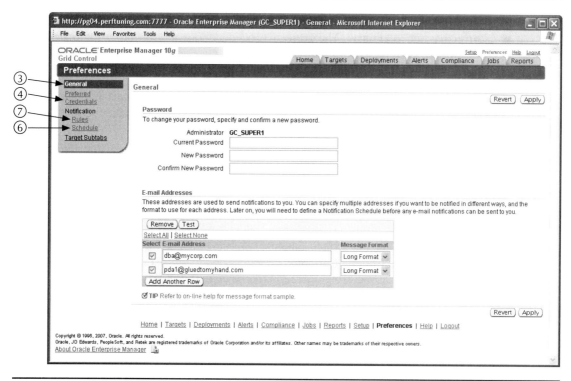

FIGURE 10-2. *The Preferences menu*

- **Registration passwords**   Registration passwords let administrators specify one or more passwords that Agents must use when first installed to communicate with any secure OMS in a GC environment. You can set multiple passwords with different expiration dates or set them to expire after one use. (This chapter does not further explain registration passwords; I refer to them throughout the book but don't explicitly cover their UI because it's so easy.)
- **Management Pack access**   Management Packs are licensed add-ons to Grid Control that considerably extend its functionality. On this page, you enable access to each pack on a target-by-target basis by acknowledging that you've licensed a pack for that target. Example key packs are the Database Diagnostics Pack, which supplies many advanced database metrics, and the Database Tuning Pack, which affords many of the performance features accessible as links on the Performance page for a database target. (For particulars, see the section "Grant Management Pack Access" in Chapter 11.)
- **Monitoring templates**   Monitoring templates let you standardize target monitoring by specifying monitoring settings once for a target, then applying them to multiple targets of that target type. Templates define metrics (including user-defined metrics), thresholds, collection schedules, corrective actions, and policies you can propagate to specified targets. You can later edit templates and reapply them. You can elect to prevent a target's settings for specific metrics from being changed when a monitoring template is applied

to the target. This allows you to use templates and still customize target monitoring. (For the minutiae, see the section " Leverage Monitoring Templates" in Chapter 13.)

- **Corrective Action Library**   The Corrective Action Library allows you to automate and reuse responses to alerts or policy violations. Corrective actions automatically execute to save admin time and improve the execution time of responses. A corrective action can run on the target where an alert or policy violation triggers, or can contain multiple tasks, each running on a different target. You can configure both critical and warning corrective actions and send notifications when they succeed or fail. (See the section "Add Corrective Actions" in Chapter 13.)

- **System Monitoring Plug-ins System**   Monitoring Plug-ins are target types that third-party vendors (or even in-house developers) provide to cover non-Oracle targets, in addition to Grid Control's predefined target types (some of which are also non-Oracle targets). You can define, search, import, export, or deploy monitoring plug-ins on this page. Microsoft SQL Server and F5 BIG-IP Local Traffic Manager plug-ins are a few examples. (See "System Monitoring Plug-ins" under "Grant Management Pack Access" in Chapter 11.)

- **Management Connectors**   Management Connectors integrate other system management products with Grid Control. Currently, Oracle has built four Management Connectors.[2] Two connectors, for Microsoft Operations Manager and BMC Remedy Help Desk, are bundled with EM 10gR3; the other two connectors, for Peoplesoft Enterprise HelpDesk and Siebel HelpDesk, are bundled with the products themselves. You must configure all connectors in Grid Control to integrate them. (See "Management Connectors" under "Grant Management Pack Access" in Chapter 11.)

- **Client System Analyzer in Grid Control**   The Client System Analyzer (CSA) allows web server administrators to collect and analyze information about client systems that connect to their web applications. End users can directly access CSA, or other applications can redirect users to it. GC comes preinstalled with an instance of the CSA application run by OMS web servers so you can collect client data without setting up a separate web application. Any client system that can access the GC Console URLs can use the CSA application to collect client data. (I do not cover CSA features in this book.)

- **Data Exchange**   Data Exchange provides a data transfer mechanism between GC and external monitoring systems, such as Oracle BAM Server. Grid Control can send target availability information, defined metrics, and alerts, and can receive externally monitored data such as business events and indicators. Currently, GC does not provide any data exchange hubs, but allows you to create them. (Data Exchange is outside the scope of this book.)

The Preferences Menu

Following are the Preferences menu items, configured for each administrator account:

- **General (Preferences)**   The General tab is where an administrator enters one or more e-mail addresses and message formats (long or short) for receiving notifications. An admin can choose from among the different e-mail addresses specified here when configuring his or her notification schedule.

---

[2] Oracle is also planning to release a connector for HP OpenView.

- **Preferred Credentials** Preferred Credentials simplify an administrator's Console access to targets by storing his or her login credentials in the Repository. With credentials set, an admin can access targets without being prompted to log in. Credentials are organized and summarized by target type. An admin can set default credentials for all targets of a particular type, can enter individual target credentials, or a combination of both, with individual credentials overriding default ones.

- **Notification Rules** Notification rules are the set of conditions that determine when a notification occurs. An admin can create rules and make them public so that other admins can subscribe to them (e.g., receive e-mail notification). There are thirteen built-in notification rules (such as Database Availability and Critical States). To see all aspects of a rule, click Edit and view each page. These aspects define metrics, policies, severity states, and job status changes that trigger notification, and for which targets. You can associate notification rules with multiple notification methods.

- **Notification schedule** A notification schedule determines when an admin receives notifications, and at which e-mail address(es). An admin can set up different notification e-mail addresses for every hour of every day. Schedules are based on a selected rotation frequency (or repeatable pattern) of one to eight weeks, and you can suspend notifications for a certain period as needed. This atomic level of configuration allows you to model almost any conceivable DBA schedule, however complex (such as when different DBAs are on call to handle a launch or maintenance period). Admin schedules can dovetail or overlap as desired, because each schedule is independent of the other.

- **Target Subtabs** Target Subtabs allow each administrator to customize the selection and ordering of Subtabs that appear below the Targets tab. An admin can add or remove subtabs as desired, such as adding a subtab for a group or removing a subtab for a target type he or she has no privileges on.

Follow an Event from Trigger to Notification

Now that you have traversed all menu items, particularly those related to notifications, you are in a better position to understand how an event occurrence trickles down to an e-mail notification (see Figure 10-3). This understanding will give you the big picture to paint the remaining material in this chapter by numbers.

Let's walk through the steps that determine whether an event occurrence (a metric alert, policy violation, or job status change) for a particular target becomes a notification for a specific administrator. To simplify the discussion, I exclude notifications of availability state transitions (such as from WARNING to CLEAR) and corrective actions and do not consider blackouts. (All target notifications and data collections are disabled while a target is blacked out. See "Schedule Blackouts for Planned Downtimes" in Chapter 11.)

The numbered steps below correspond to those in Figure 10-3. All steps are easy to grasp, except the first. So let me explain it.

Figure 10-3 is looking at events from the perspective of a GC admin who defines a metric, policy, or job for a target *and/or* for a rule that includes the target. So the first question—hence the first step—is whether an event actually fires for the target in question. I explain this anomaly in more detail later (in the section "Aligning Notification Rules with Target Metrics/Policies"), but a common configuration error is to define a particular metric threshold (for example) in a rule for a target but forget to define the metric for the target. In other words, a rule alone for a target does

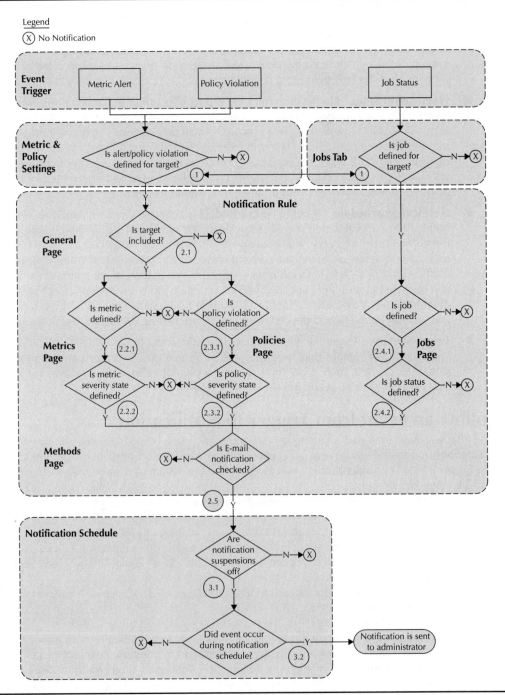

FIGURE 10-3. *The process from event occurrence to notification*

not trigger a notification; the metric also must be defined "upstream" for the target. This is what step 1 below is driving at.

1. The metric, policy, or job must be defined for this target for an event to occur. This initial check catches configuration mismatches where a rule for a target includes a metric, policy, or job not defined for the target.
2. This event is evaluated according to each of the administrator's notification rules, which have the following components:
    2.1. General—The rule must include the specific target.
    2.2. Metrics—If the event is a metric alert:
        2.2.1. this metric must be defined for the notification rule and
        2.2.2. the metric's severity state must be defined.
    2.3. Policies—If the event is a policy violation:
        2.3.1. this violation must be defined for the notification rule and
        2.3.2. the policy's severity state must be defined.
    2.4. Jobs—If the event is a job status change:
        2.4.1. this job must be defined for the notification rule and
        2.4.2. the job's status change must be defined.
    2.5. Methods—The E-mail Notification box must be checked to allow this admin to receive an e-mail notification of this event.
3. The event must occur at a time when the administrator's notification schedule permits the notification:
    3.1. The event must *not* occur during a period of suspended notifications.
    3.2. The event must occur during a time when the administrator is scheduled to receive notifications.

This diagram of the logic dictating whether an event becomes a notification shows just how configurable notifications are.

Reasoning Behind this Chapter's Menu Navigation Order

While both the Setup and the Preferences menus are organized into distinct categories, it's difficult for a first-time administrator to know where to begin configuring Grid Control after installing it. When you log in to the Console, no wizards offer to assist in the overall configuration.[3] It is safe to say that the necessary configuration order does not fall out by osmosis from the menu layout; that is, you cannot perform a successful configuration by following the menu order for Setup or Preferences (say, from top to bottom). Certain dependencies exist between items under the Setup and Preferences menus. You must take these dependencies into account when constructing a viable, efficient navigation order between menu items. The configuration order is somewhat flexible; there's more than one way to get through the menu

---

[3] By contrast, wizards or subtabs do exist within menu items themselves.

items and some ways are more circuitous than others. I derived the particular configuration order recommended in this chapter for addressing menu items by identifying all dependencies between any two items. Then, I lined up all items according to these dependency requirements so that you only have to visit each menu item once during configuration. Following are the dependencies between menu items. I only list the items and do not indicate which menu they belong to (Setup or Preferences):

- **Roles → Administrators**   Roles serve as templates for creating admins with similar privileges. If you first create roles that typify the types of privileges administrators share, then you can create administrators more quickly by granting them the appropriate role. This also reduces administrator maintenance, as changes to roles are propagated to all administrators already granted these roles.

- **Administrators → Preferred Credentials**   Preferred Credentials are unique to each administrator, so they cannot be set until you create an admin. As explained in the section "Create Administrators" later in this chapter, if certain admins share common preferred credentials, rather than using the UI to enter duplicate preferred credentials for each admin, use the EM Command Line Interface to programatically set preferred credentials for these admins.

- **Administrators → Notification Schedule**   A notification schedule is associated with a particular admin account, so you must create the account, then the schedule for it.

- **Notification Methods → Notification Rules**   At least one notification method must exist before you can define any notification rules. You can take care of this when installing GC by entering an SMTP server as the e-mail notification method. If you plan to use additional methods, create them first so you can apply the same method to multiple rules.

- **General Preferences → Notification Rules**   To receive e-mail notifications based on notification rules, you must enter at least one e-mail address under General Preferences. Otherwise, the Send Me E-mail box is grayed out under Notification Rules.

- **General Preferences → Notification Schedule**   You cannot define a notification schedule without adding at least one e-mail address under General Preferences for your admin account.

- **Notification Schedule → Notification Rules**   You can subscribe to receive e-mail notifications for every rule in Grid Control, but you won't receive one alert until you define a notification schedule to tell Grid Control when it's ok to send e-mails.

When installing Grid Control, if you don't specify SMTP Server information, there is an additional dependency:

- **Notification Methods → General Preferences**   Grid Control actually allows admins to enter e-mail addresses under General Preferences before a super administrator even sets up a mail gateway under Notification Methods. But no one will receive any e-mail notifications until you enter this gateway information. It's misleading to let users enter e-mail addresses and notification schedules thinking they'll receive alerts when you haven't globally configured an SMTP server to forward notifications.

In this chapter, I assume you enter an SMTP Server during GC installation, so this dependency does not factor into the order of steps presented here.

Based on the dependencies discussed above, following is one viable configuration order for menu items:

Roles → Administrators → General Preferences → Preferred Credentials → Notification Methods → Notification Schedule → Notification Rules → Patching Setup

The numbering next to specific menu items in Figures 10-1 and 10-2 reflects this ordering, and the chapter is organized accordingly. Such an order is not the only way to configure Grid Control, but it's one of the more efficient ways, as it avoids having to revisit menu items.

Set Up Administrators

With a whirlwind tour of the Setup and Preferences menus under your belt, a diagram to refer to on how notifications work, and an order in which to approach general Console configuration, we can now commence with this configuration. It begins with the four important elements of the Grid Control user model:

- Create roles
- Create administrators
- Set administrator preferences
- Set preferred credentials

This user model provides a mechanism for defining and assigning roles and privileges to an admin. The four-part course of action offers a logical progression for user configuration. Take note that page count coverage is heavily weighted in the fourth part, setting preferred credentials. However, this coverage is well deserved because you learn the terminology behind preferred credentials and some time-saving ways to set and verify them. Alas, the logical division of a chapter doesn't always coincide with its physical fleshing out in page count.

Create Roles

Roles are a collection of GC system and target privileges. Their primary benefit is to facilitate granting a particular set of shared privileges to multiple administrators. Without roles, a super administrator would have to individually grant each admin access to select targets from the existing pool of all targets. Roles ease user maintenance because changes to a role propagate retroactively, as it were, to admin accounts (and other roles) previously assigned that role.

Unlike administrators, who are actual Repository Database users, roles are not database roles; they offer specific privileges applicable only to GC management. Role definitions, like those for administrators, are stored only in SYSMAN tables in the Repository Database, not in target databases. Grid Control comes with a default role called PUBLIC, which is different from the database PUBLIC role,[4] although similar in function. The Grid Control PUBLIC role comes with no privileges, but it is assigned by default to all nonsuper administrators you create. PUBLIC is at your service. Its purpose is to store the least common denominator (LCD) of privileges you plan for all admins. I suggest

---

[4] For more on this enigmatic role, see Note 234551.1.

capitalizing on this utility. However, you can leave PUBLIC "underprivileged" (to borrow the societal euphemism for "no privileges at all"), at least for now, or even delete the PUBLIC role altogether if you see fit.

You can create roles based on any model that suits your IT environment. For example, roles can be based on geographic location (east coast, west coast), type of targets (production, nonproduction), privileges granted (read-only viewing of any target), line of business (executives, manufacturing), or any combination thereof.

That's enough background to get us started. Let me demonstrate how to create a role—in the spirit of learning by doing—called GC_VIEWALL; it will allow read-only access to all targets and the privilege to monitor GC performance. This is a good role to grant to IT managers. You may want such a role on your GC system. If not, you can still create the role to get a feel for the UI, then remove the role. Please take it in stride that three IT manager accounts—BGATES, MMOUSE, and SBUCKS—pre-date the role we're creating. These are "shell" accounts, created simply by specifying a username and password, and granting no roles (not even PUBLIC), system privileges, or target privileges. No malice implied—the lack of privileges is meant to show the utility of roles.

To create the GC_VIEWALL (or any) role, you must log in as a super administrator, SYSMAN in this case, and click the Roles link under the Setup menu.

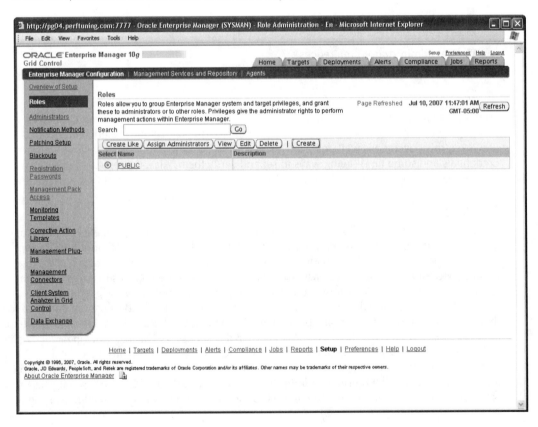

Here you catch your first glimpse of the default PUBLIC role just discussed—exciting stuff. Click the Create button, which launches a wizard consisting of the following seven steps.

Properties (Step 1 of 7)
The first step is to enter a name and optional description for the new role.

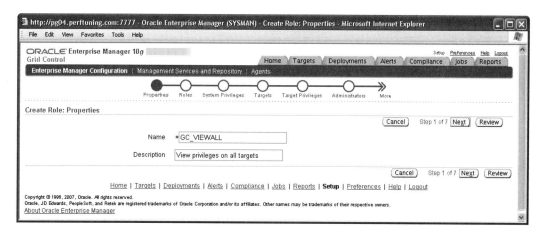

You cannot choose a role name that is already taken by an admin (or another role). You won't receive the ORA-20231 error message indicating as such until you click Finish. Also note that role names are displayed in uppercase when saving the role, and cannot be changed afterward. Click Next to advance to step 2.

Roles (Step 2 of 7)
The PUBLIC role and any previously created GC roles appear (see Figure 10-4). Add desired Available Roles to Selected Roles. For this example, don't select PUBLIC. Let's assume you do not plan to use PUBLIC to grant all admins certain privileges. In Figure 10-4, we don't need to grant any additional roles to the GC_VIEWALL role, which is fortunate, as there aren't any more roles to grant. Click Next to advance to step 3.

NOTE
The next three steps in creating a role—System Privileges (step 3), Targets (step 4), and Target Privileges (step 5)—are identical to steps to create a new administrator (as shown later). In fact, the only differences between the definitions for roles and admins are that admins can be assigned e-mail addresses and roles. Therefore, the "Create Administrators" section later in the chapter refers you back to the "Create Roles" section for these steps. Accordingly, descriptions for the following steps refer to roles "(or admins)."

310 Oracle Enterprise Manager 10g Grid Control Implementation Guide

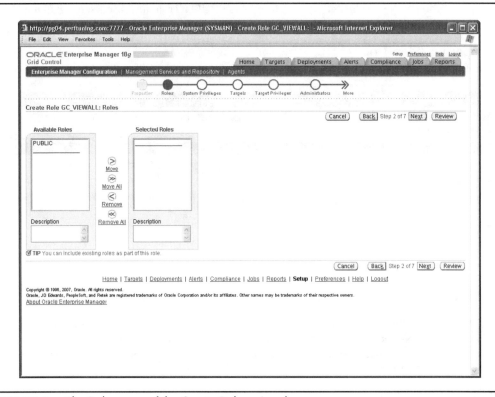

FIGURE 10-4. *The Roles page of the Create Roles wizard*

System Privileges (Step 3 of 7)

The System Privileges step offers five distinct system privileges to grant GC management features to a role (or admin) as described here.

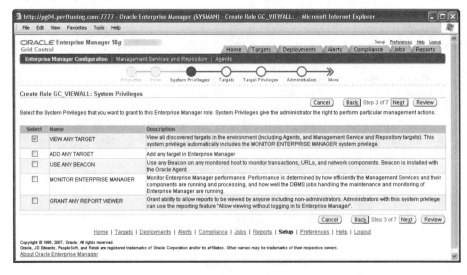

These GC privileges are similar in function to database system privileges such as CREATE USER. The analogy is that GC system privileges are application-wide (as database privileges are database-wide) and are not associated with specific targets (as database privileges are not related to specific objects). However, the limited number of GC system privileges limits their usefulness. The numbers alone speak for themselves—five GC privileges vs. over 150 DB10g system privileges. Check the system privileges to grant to the new role (or admin). In this example, we grant the VIEW ANY TARGET system privilege, which implicitly grants the MONITOR ENTERPRISE MANAGER system privilege as well, thereby fulfilling our system privilege objective for the role. Click Next to go to step 4.

Targets (Step 4 of 7)

The Targets step lets you select which of the current targets this role (or admin) has purview over.

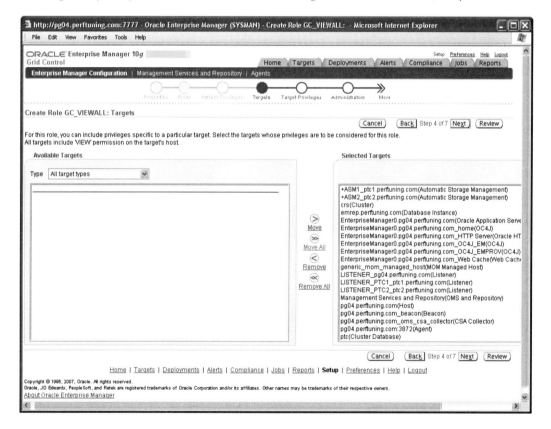

In Step 5, Target Privileges, you will assign specific target privileges to the targets you select here. Selecting a target implicitly provides the View target privilege on the host where the target resides. For instance, selecting a listener as a target grants the View privilege on the host where the listener resides. Use the Type field to select targets of a particular target type, such as Host, Database Instance, Agent, or OMS Repository. Selecting a group automatically includes all targets in that group, which is useful at large GC sites (see the section "Group Configuration" in Chapter 12). Select All Target Types here because in this example the GC_VIEWALL role provides read-only privileges to all targets. Click Next to advance to step 5.

Target Privileges (Step 5 of 7)

In the Target Privileges step, you grant target privileges to the role (or admin) for targets selected in the previous step.

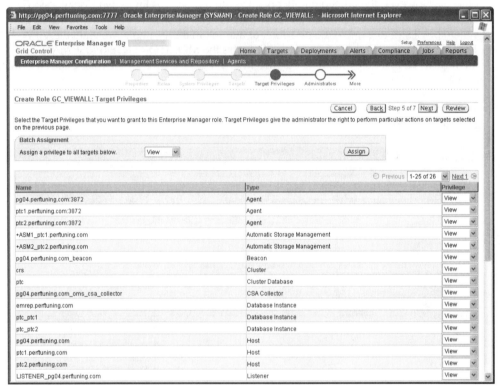

You can choose one of the following four target privileges for each target chosen on the previous page (in step 4) to give the role (or admin) certain rights:

- **None** Grants no target privileges.

- **View** This is the minimum privilege an admin needs to view a target in the Console. It allows viewing of target properties, inventory, and monitoring information, and is a handy privilege for IT managers who need full read-only access.

- **Operator** Permits startup, shutdown, and editing of target properties (all operations except to delete a target and set its credentials). The equivalent database entity for the Operator privilege is the SYSOPER role.

- **Full** Provides access to all operations for a target.

Target privileges are similar to database object privileges, such as the SELECT object privilege on a table or the EXECUTE object privilege on a procedure. However, as with GC system privileges, the low number of GC target privileges provides only limited functionality.

To grant a certain privilege to all targets, select it under the Batch Assignment section and click the Assign button. In this example, we choose the default View privilege for all targets. Click Next to go to step 6.

Administrators (Step 6 of 7)

This step allows you to select the admins to be granted the role. The UI is the same as shown in earlier Figure 10-4 for roles, except that here it allows you to select administrators. Here is where you can grant the role being created to existing admins. For the example, you imbue the three administrators, BGATES, MMOUSE, and SBUCKS, with all the role privileges just chosen. The SYSMAN administrator and other existing super administrators do not appear on this page, as they are automatically granted all GC privileges. Click Next to advance to the final step.

Review (Step 7 of 7)

The Review page allows you to review your choices from the previous steps. If you're satisfied with the choices you've made, click Finish. If you need to make some changes, click Back and change what you don't like. I only wish life were like that.

Create Administrators

Administrators are users created to manage the GC environment through the UI. Each admin account possesses its own preferences (found under the Preferences link), which include e-mail addresses, preferred credentials, notification rules, notification schedule, and target subtabs. (I cover all preferences in the remainder of this chapter). Each admin is also a Repository Database user, but is only granted CREATE SESSION through the MGMT_USER role. The reason for this is that the OMS uses the database user credentials only to authenticate GC Console login; thereafter, SYSMAN proxies for all admins.

Just as admins often need the same privileges, which roles can efficiently proliferate (see the earlier section "Create Roles"), certain admins may also share the same preferences, and particularly, the same preferred credentials.[5] Of all the Preferences menu items, the Preferred Credentials item has the most potential for manual input. This is because preferred credentials are set on a user and target basis, so they can be different for every normal and privileged username on every managed target. Unfortunately, EM10*g* doesn't offer a way in the UI to clone preferred credentials from one admin to another. This is a new feature planned for EM11*g* Release 2.[6] However, you can use the EM Command Line Interface (EM CLI) to set common credentials for administrators. (See "Programmatically Setting Preferred Credentials" later in the chapter.)

With an aim to capitalize on this new functionality in EM11*g* Release 2, I suggest creating a *template account* (my term for an account with a certain set of preferred credentials) in EM10*g* if multiple administrators share common preferred credentials.[7] Create these new administrators in the UI, but don't set their preferred credentials. Instead, write an EM CLI script to set the same preferred credentials for these administrators, and run the script against the template account as well. Then in EM11*g* R2, you can create a new admin with the same preferred credentials in the UI by copying the template account rather than using EM CLI.

---

[5] Data center security requirements are becoming more stringent every year, driven by world events—notably September 11—and by corporate dictates to comply with the Sarbanes-Oxley Act (SoX). Even so, it is still not uncommon for DBAs to use the same target login credentials, such as the *oracle* OS user and the SYS or SYSTEM database user.

[6] See Bug 6595613.

[7] Create a separate template account for each team of admins that share a unique set of preferred credentials. Then in EM11*g* R2, to create a new admin on a given team, you can clone that team's template account.

Following are some general recommendations for creating administrators to reduce the overhead of GC user management. Given that companies have varying security concerns, IT staff organizational structures, and the like, no definitive strategy exists for creating administrators. You need to decide which suggestions are apropos to your site.

- **Do not configure Preferences for the built-in SYSMAN account**   By default, the Administrators page lists three GC accounts: SYS, SYSTEM, and SYSMAN, all super administrators. You can't log in to the Console using SYS, and I don't advocate using SYSTEM, as it's a built-in database user.[8] SYSMAN, as the owner of the Repository, has unique system privileges, so cannot be duplicated (i.e., cannot serve as a template account). Therefore, unless you want all users to share the SYSMAN account for all time, and you're sure you'll never need to create more accounts,[9] don't configure additional preferences for SYSMAN.

- **Don't create accounts until adding all targets**   If possible, don't create new administrators (except perhaps a template account) until you first add and configure all common targets (see Chapter 11) and groups and services (discussed in Chapter 12). Only when all targets are discovered can you grant admins access to and privileges on these target.

- **Let users share GC accounts if they share target credentials**   If users share the same credentials for logging in to all targets, why create separate GC accounts for them? It's still not unheard of for all DBAs at a site to log in to targets with the same credentials. Users can share GC accounts even if their work schedules differ; the Notification Schedules UI allows them to independently enter individual e-mail addresses for their separate work hours. There are limitations to GC account sharing, however. If DBAs look after different targets, you'll need multiple accounts so you can assign each account different target privileges. Even then, however, you may not have to create a new GC account for every DBA. If teams of DBAs each manage certain targets using the same credentials, you can create one GC account per team.

- **Use EM CLI to set common preferred credentials**   If, for reasons just stated, you need to create separate GC accounts for users who log in to targets with the same credentials, it's tempting to hand over these accounts and let them "have at it." However, this places a burden on each admin to enter his or her credentials. Instead, consider writing an EM CLI script to set preferred credentials for these users. (Also, create a template account and set its credentials using the EM CLI script so that in EM11gR2, you can create new admins with the same credentials by cloning this account in the UI.)

- **Consider using generic names for accounts**   Many security policies call for individual user accounts and prohibit service accounts not associated with a particular person. The reason is that it is difficult or impossible to trace service account actions back to a particular user. If you're not bound by such security concerns, generic service account

---

[8] You can edit Grid Control privileges for SYSTEM without affecting its database privileges outside of Grid Control. However, even if you remove the GC SYSTEM user's super administrator access and all privileges, the user retains super administrator access.

[9] One thing you can be sure of in IT, as in life, is that you can't be sure of anything in IT.

Chapter 10: General Console Configuration 315

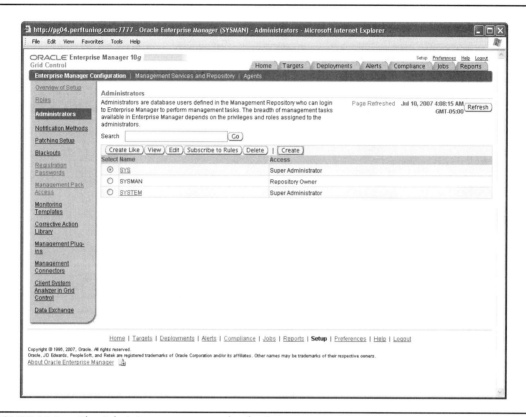

FIGURE 10-5. *The Administrators page under the Setup menu*

names (such as GC_SUPER1[10] used in this book) rather than those named after staff, may appeal to you. Administrator names, once created, cannot be changed, although you can delete them. If you follow this advice, you won't be stuck with defunct admin accounts named after former employees, or with having to create new accounts for employees taking over old positions.

Let's create a template account with super administrator privileges, called GC_SUPER1, an account I use throughout the rest of the book. A super administrator account is super easy to create—it's granted all privileges to all targets, so the Create Administrator wizard skips from the first step to the last. Don't be disappointed. You wouldn't learn anything new by creating a nonsuper administrator; the middle steps are identical to those for creating a role and the instructions for creating admins just refer you back to the "Create Roles" section above.

To create a super administrator (or any other admin) account, click Setup, then click Administrators (see Figure 10-5). Click Create. A wizard guides you through the interview process.

---

[10] Some sites may prefer a common prefix such as "GC_" to help easily distinguish all GC admins in the OMR Database from built-in database users. You can apply a consistent naming convention such as this to all types of names in Grid Control, such as for user-defined metrics, web applications, jobs, monitoring templates, etc.

Properties (Step 1 of 6)

The first step, the Properties page, appears.

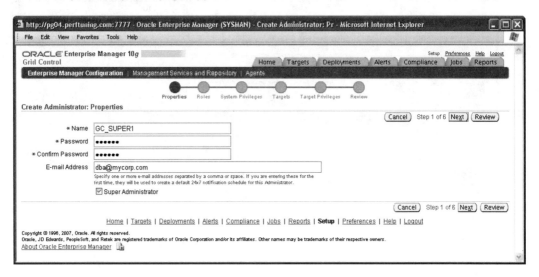

Enter the admin name, password, e-mail address, and check the Super Administrator box if applicable, as in this case. Before you click Next, note the following about this page:

- **Name**   Administrators are database users, so neither admin names nor passwords are case-sensitive. As with roles, you can't choose an admin name already taken by a role (or by another admin); names are displayed in upper case once they are created and cannot be changed afterward.

- **Password**   Passwords can only contain alphanumeric characters and the underscore character ("_").[11]

- **E-mail Address**   If you take my advice above and create templete accounts, let the user who maintains this template account also use it; therefore, enter the e-mail address for this user. If you want all notifications for the account sent to a distribution list, enter the distribution list e-mail address (such as *dba@mycorp.com* used here).

- **Super Administrator box**   Check the Super Administrator box if appropriate (as is here). If checked, when you click Next the wizard proceeds directly to Step 6 (Review). This is because super administrators are automatically granted complete access. Steps 2 through 5, which are Roles, System Privileges, Targets, and Target Privileges, respectively, are only for nonsuper administrators to define the level of their access.

Click Next.

Roles (Step 2 of 6)

If you did not check Super Administrator in the previous step, the wizard continues with this step to let you assign roles to the admin. The Roles page allows you to choose from Available Roles,

---

[11] See Bug 4113146.

as shown in Figure 10-4 earlier. As you learned earlier in the section "Create Roles," roles bundle GC system and target privileges you can grant to admins as here (or to other roles). If you need to assign system privileges, targets, or target privileges to the administrator you're creating, you have two choices (listed in the order I'd recommend):

- Pre-create one or more roles or add privileges to the PUBLIC role, then, when creating an admin, move desired roles from Available Roles to Selected Roles in this step. Skip steps 3, 4, and 5 for creating an administrator, as these are the privileges you grant through roles instead. Using roles is the most efficient way of granting privileges because you can reuse roles by assigning them to multiple admins and to other roles.

- Alternatively, do not use roles (i.e., do nothing in this step), but instead enter individual system and target privileges for each admin in steps 3, 4, and 5 next.

    When finished, click Next.

NOTE
The next three steps in creating an administrator—System Privileges (step 3), Targets (step 4), and Target Privileges (step 5)—are identical to the corresponding steps described earlier to create a new role. Therefore, I refer you back to the "Create Roles" section for a description of these steps.

System Privileges (Step 3 of 6)
See the System Privileges step in the "Create Roles" section. System Privileges are the same for roles as they are for administrators.

Targets (Step 4 of 6)
See the Targets step in the "Create Roles" section. Targets are the same for roles as they are for administrators.

Target Privileges (Step 5 of 6)
See the Target Privileges step in the "Create Roles" section. Target Privileges are the same for roles as they are for administrators.

Review (Step 6 of 6)
Look over your input and click Finish to create the template super administrator account.

Set Administrator Preferences

Once you create GC admins—whether template or individual accounts—use (or instruct them to use) the following pages on the Preferences menu to begin configuring preferences for their accounts:

- **General Preferences** On this page, an admin can change the account password, alter the assigned e-mail address (for example, add a PDA e-mail address or pager number), and choose the format (long or short) for notification messages sent to each address.

- **Target Subtabs** On this page, an admin can customize the subtabs appearing under the Targets tab, such as adding a particular group of interest, removing a subtab for a target type for which the account has no privileges, or reordering subtabs according to frequency of use.

- **Preferred Credentials** Each administrator account must have its own login credentials to access targets through the Console. I demonstrate the ins and outs of entering and verifying preferred credentials and explain a method to programmatically set credentials for admins en masse.

This initial access to the Console does not unduly tax a new administrator's GC skills. The user interface for these preferences, like that for many Console features, is intuitive. Nevertheless, you may prefer to perform these two steps for some—or even all—users, particularly any executives who rely on you to make their IT lives easier.

CAUTION
In the following sections relating to Preferences, remember to log in to the Console as the administrator (or super administrator) whose preferences you want to set.

Set General Preferences

The General page of the Preferences menu (shown in Figure 10-2 above) allows admins to specify one or more notification e-mail addresses (and a format for each address) for each hour of the day on their notification schedules. Administrators can set up e-mail addresses to use the Long Format or Short Format:[12]

- **Long Format** Long Format is an HTML e-mail meant for a mail reader such as Netscape Messenger or Microsoft Outlook. The long format contains full event details.

- **Short Format** The short format is a text e-mail meant for pagers or cell phones. Grid Control does not directly support sending pages (as EM 9*i* did) or SMS messages, but relies on third-party software to convert the e-mail to the appropriate format. The layout of the Short Format is controlled by two properties which, if used, should be added to the *$OMS_HOME/sysman/config/emoms.properties* files on all OMS hosts, as shown here:

Property Name	Default Value	Allowable Values	Description of Property
em.notification.short_format_length	155	1 to 155	The maximum character length of a short format e-mail
em.notification.short_format	both	subject body both	The format of the short e-mail in terms of how the message is split between subject and body[13]

---

[12] See Note 429292.1 and the Online Help from this page for details on short and long e-mail formats.

[13] If you're on GC 10.2.0.2 or 10.2.0.3, Bug 5409276 requires a patch to use this property.

These properties are not present out of box in the *emoms.properties* file, and default values are used for properties not added to this file.

For each admin account designated to receive notifications, including template accounts assigned to users, do the following to specify one or more e-mail addresses and their formats:

1. Log in to the Console as the admin whose general preferences you want to set.
2. Go to the General page under Preferences (refer back to Figure 10-2).
3. Click Add Another Row, enter the e-mail address, and select a Message Format (Long or Short).
4. Repeat step 3 for each e-mail address you want GC to be able to send concurrent notifications to.
5. Test that you will receive notifications at all e-mail addresses[14] by leaving all Select boxes checked and clicking the Test button.
6. Verify that the Test Results page shows "Test succeeded" for each e-mail address entered, and click OK.
7. Confirm that you receive a test e-mail at each specified e-mail address.
8. If you receive all test e-mails, click Apply. If not, do one of the following:
   - Resolve the problem (see "Test Notification E-mail Address" in Chapter 2 for the troubleshooting procedure), retest by repeating step 5 (select only the e-mail address that failed), step 6, and step 7, and then click Apply.
   - If you can't fix the problem for one or more e-mail addresses, select them, click Remove, then click Apply to add the e-mail address(es) that succeeded.

Customize Target Subtabs

By now you have seen and used target subtabs in the Console, which are the categories of targets that appear as minor tabs when selecting the main Targets tab. Target subtabs let you quickly access the target categories used most frequently. Administrators can use the Target Subtabs page under the Preferences menu (refer back to Figure 10-2) to select and order these subtabs below the Targets tab. Default subtabs are Hosts, Databases, Application Servers, Web Applications, Services, Systems, Groups, and All Targets. Many admins find the default target subtabs sufficient. However, each admin can add, remove, or reorder these subtabs as he or she prefers.

The ability to customize target subtabs is particularly applicable to large sites with many admins, each of whom is responsible for a small portion of all targets. Usually, such sites assign their admins targets of a particular type (database, host, etc). In these cases, admins with no privileges for targets of a particular type have no use for their corresponding subtabs. For instance, the account for a GC admin who manages five databases may be limited to full target privileges for just these databases. Without any other target privileges, this admin would see "No Targets found" under all default target subtabs except Hosts and Databases. (When granted privileges to a target, an admin is also given view privileges for the target's host.)

---

[14] You must have configured an outgoing Mail (SMTP) server during or after the Grid Control installation in order for the OMS to forward e-mail notifications. See the "Define Notification Methods" section below for how to configure a mail server post-installation.

To customize target subtabs for an administrator, do the following:

1. Log in to the admin account whose subtabs you want to customize.
2. Click Target Subtabs in the vertical navigation bar under the Preferences menu.

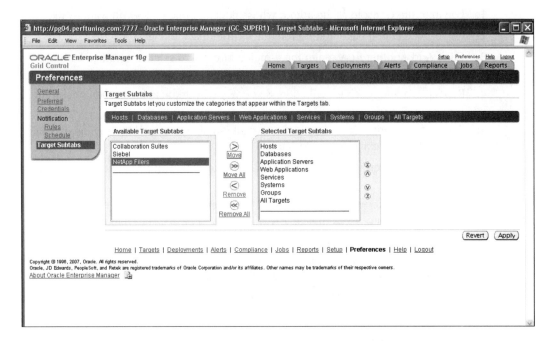

3. Select subtab names and add/remove them to/from the Selected Target Subtabs list as you prefer using the Move, Move All, Remove, and Remove All buttons.
4. Use the arrow buttons to organize the subtabs from top to bottom the way you wish them to appear from left to right in the Selected Target Subtabs list.
5. Click Apply when finished.

A popular change admins make is to remove the generic Groups subtab and add one or more groups they create as target subtabs. Adding a newly created group as a target subtab is one of the optional steps listed in the section "Create a Group" in Chapter 12.

Set Preferred Credentials

An administrator's preferred credentials simplify Console access to targets by storing target login credentials in the Repository. Grid Control uses these credentials so that admins can access targets in the UI without being prompted to log in every time. Many Console functions rely on preferred credentials, particularly the Related Links at the bottom of a target's home page. Like other preferences, admins set their own preferred credentials, allowing them to meet the most stringent corporate security standards, which require separate user accounts for each admin.

However, there is nothing to prevent admins who share target credentials from entering the same credentials for those targets.

I'd like to present the subject of preferred credentials using the following outline:

- Concepts for Preferred Credentials
- Enter Preferred Credentials
- Verify Default Credentials

Concepts for Preferred Credentials

To understand preferred credentials, you must understand the following terminology used in the Console screens. To flesh out these concepts, it helps while reading about them to look at a few Preferred Credentials pages. For your administrator account, click Preferences, Preferred Credentials, then click Set Credentials for both Host and Database Instance target types.

Normal vs. Privileged Credentials There are two types of preferred credentials: Normal and Privileged. Administrators need to enter Normal credentials at a minimum, and typically require both types of preferred credentials for accessing target databases and operating systems:

- For databases, Grid Control uses Normal credentials when a Console function does not require SYSDBA access or an admin chooses such credentials for a job; otherwise, GC uses Privileged credentials, such as to access a nonopen database, or to start up or shut down a database.

- For operating systems, Grid Control uses Normal credentials, unless a Console feature relies on *root* access (for UNIX) or on Administrators group access (for Windows) or unless a job specifies host Privileged credentials.

I recommend standardizing on usernames for database and OS preferred credentials for all target types that request them. The following table recommends normal and privileged usernames to specify for both database and OS credentials.

	Normal Credentials	**Privileged Credentials**
Database Username	DBSNMP	SYS
OS Username	*oracle*	*root* (on UNIX) *oracle*[15] (on Windows)

Target vs. Default Credentials There are two methods for setting preferred credentials: using target credentials and default credentials. Target credentials are specific to a target, whereas default credentials apply to all targets of a particular type. Certain rules govern the setting of preferred credentials:

- Credentials must be set using at least one of these methods.
- Individual target credentials, when set, override any default credentials set.
- Grid Control uses default credentials for a target if that target's credentials are not individually set.

---

[15] Use the *oracle* account (the owner of the Oracle target software) only if a member of the local Administrators group. Otherwise, use the Administrator user, the built-in account for administering the computer.

- Default credentials are useful for targets that share the same Normal or Privileged username and password, as this does away with having to enter individual target credentials.

- Target credentials can be tested before you save them; default credentials cannot be tested, but are only saved.

Let's take a look at how to enter preferred credentials, then discuss strategies for simplifying their management.

Enter Preferred Credentials

If you took my suggestion to drill down into some Preferred Credentials pages, then you know that the UI is straightforward. Remember to log in as the admin whose preferred credentials you want to set. If you're setting up a template account (such as the example GC_SUPER1 user I created above in the "Create Administrators" section), log in as that username. On the Preferences menu, click Preferred Credentials. The Preferred Credentials summary page appears, broken out by target type.

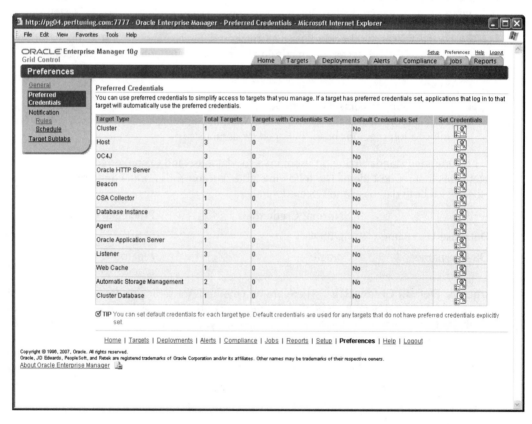

The row for each target type displays the following columns:

- **Total Targets**   The total number of discovered targets of that target type
- **Targets with Credentials Set**   The number of targets with Normal or Privileged credentials set under the Target Credentials section for that target type
- **Default Credentials Set**   Whether Normal or Privileged default credentials are set in the Default Credentials section for that target type
- **Set Credentials**   A link to set target and default credentials

Generally, you only need to set preferred credentials for the following target types, as you likely won't need to access the other target types through the Console:

- Agent (to configure plug-ins)
- Automatic Storage Management
- Cluster
- Cluster Database
- Database Instance
- Host
- Listener (if a listener password is configured)
- Oracle Application Server (to control it from AS Control)
- Web Cache

The target types Automatic Storage Management, Cluster Database, and Database Instance require both database and host credentials because GC must log on to the host before it can run the application, such as SQL*Plus, that the UI is implicitly calling. The other target types only ask for host credentials.

You can set preferred credentials in one or more of the following ways:

- Entering target credentials
- Entering default credentials
- Programmatically setting preferred credentials

You can use a combination of these methods, but keep in mind that individual target credentials override default credentials for a particular target type.

Entering Target Credentials   Follow these steps to enter preferred credentials for a particular target (the Repository Database *emrep* in this case):

1. Log in as the administrator whose target credentials you want to set, click Preferences, then click Preferred Credentials.

2. Click the Set Credentials icon for the target type whose credentials you want to set. (In the case of a Database Instance, clicking the Set Credentials icon brings up the Database and Cluster Database Preferred Credentials page.)

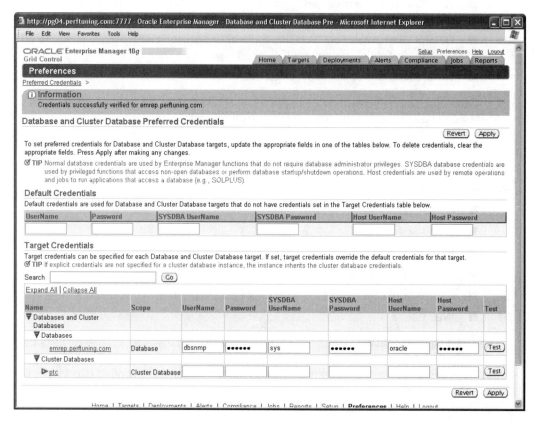

3. To set target credentials, under the Target Credentials section find the row for the target, which in this illustration is the Repository Database *emrep*. Enter normal and privileged (SYSDBA) database credentials, host credentials, and then click Test.

4. If successful, Grid Control returns the following informational message at the top of the screen: "Credentials successfully verified for <target_name>". If not successful, enter the credentials again and click Test. Click Apply to save verified credentials.

TIP
To return from the Preferred Credentials page for a target type to the summary page, click the Preferred Credentials link at the top left of the screen. This also brings back the Preferences vertical navigation bar.

Entering Default Credentials If credentials (usernames and passwords) are the same for at least two targets of a particular type, use Default Credentials to avoid having to enter credentials for each target, then enter credentials individually for remaining targets that don't share these default credentials. To enter default (then individual) credentials, follow the procedure below. (I use the Database Instance target type for this example as well):

1. Log in as the administrator whose default credentials you want to set, click Preferences, then click Preferred Credentials.
2. Click the Set Credentials icon for the target type you want to set. (Here, select the Database Instance target type.)
3. Enter credentials for UserName (DBSNMP in this case), SYSDBA UserName (SYS), and Host UserName in the Default Credentials section and click Apply.
4. If successful, GC displays the following message at the top of the screen: "Credential changes successfully *applied*."[16]
5. Enter credentials for individual targets with nondefault credentials under the Target Credentials section, click Test, and if verified, click Apply.

Note that the default credentials message differs from that for target credentials, which is "Credentials successfully *verified* for <target_name>". As cited earlier, the reason is that Grid Control does not verify default credentials for any targets. The section below called "Verifying Default Credentials" discusses two methods for verifying default credentials to overcome this "undocumented feature."

TIP
If you cannot successfully verify a credentials password, particularly for a database user, try resetting the password on the target host to the same password,[17] then test the target credentials again. This sometimes resolves the problem, although the reason remains a mystery to me.

Programatically Setting Preferred Credentials   It's great if you have the luxury of being able to use default credentials. However, most of you must comply with stricter security regulations, notably the Sarbanes-Oxley Act (SoX), prohibiting the use of default credentials; that is, of using the same OS and database passwords across multiple systems. Such regulations add administrative overhead by enforcing strong passwords and stipulating that users change production passwords at least every 90 days. Overworked DBAs (a redundant expression today)[18] bound by such regulations—even DBAs at "small" GC sites—are keenly interested in ideas to minimize preferred credentials maintenance. The idea that DBAs can set preferred credentials programmatically comes as music to their ears.

You can expedite the process of changing credentials on a recurring basis for each combination of administrator and target by scripting the process using the *set_credential* command available in the Enterprise Manager Command Line Interface (EM CLI). See Chapter 8 for an introduction to EM CLI, which refers you to the Oracle documentation for installing and setting it up.

---

[16] The italicized word is my emphasis.

[17] You can reset a UNIX OS password to the same password by logging in as *root* and using the "passwd <username>" command.

[18] For those aspiring wordsmiths out there, redundant expressions are called pleonasms.

Following is the procedure for using EM CLI to set or change preferred credentials:[19]

1. On any OMS host, log in as *oracle* and change directories to *$OMS_HOME/bin/*.
2. Execute the EM CLI command, specifying the *set_credential* verb as follows:

   ./emcli set_credential \
   -target_type=<*target_type*> \
   -target_name=<*target_name*> \
   -credential_set=<*credential_set*> \
   -user=<*administrator_user*> \
   -column="username:<*username*>;password:<*password*>"

   These options are described in Table 10-1.[20]

Option	Description (Table: Column)	Example Values
target_type	Type of target. (MGMT_TARGETS: TARGET_TYPE)	cluster[21] host oracle_database oracle_emd (Agent) oracle_listener osm_instance (ASM) rac_database
target_name	Name of target. (MGMT_TARGETS: TARGET_NAME)	<*db_name*>.<*db_domain*> <*fully_qualified_hostname*>
credential_set[22]	Credential set affected	HostCredsNormal HostCredsPriv DBCredsNormal DBCredsPriv
user	EM CLI administrator authorized to execute the EM CLI command	<*administrator_user*>
column	Username and password to set	"username:<*host_username*>;password:<*password*>" "username:<*db_username*>:password:<*password*>"

TABLE 10-1. *EM CLI Options for Setting Preferred Credentials*

---

[19] See Note 389754.1.

[20] Based on Oracle Enterprise Manager Command Line Interface 10*g* Release 2 (10.2) for Windows or UNIX, Chapter 2. See the *set_credential* description in the Verb Reference for more options.

[21] Bug 5382275 causes an error when specifying the *cluster* target_type.

[22] You must list the specified credential_set value exactly as shown under Example Values, in both uppercase and lowercase.

3. Confirm that the *emcli* command returns "Credentials for user <administrator_user> set successfully."

4. Verify that the first credentials you change/add appear in the Console. (Obviously, you won't manually check each credential you set as this defeats the purpose of programmatically setting them.) Log in to the Console as the admin <administrator_user>. Navigate to Preferences, Preferred Credentials, and select the target type specified by the <target_type> option. Confirm that the credentials you added/changed for the *target_name* are listed/correct.

Following are examples of the *set_credential* syntax to set/change host and database credentials:

- For administrator GC_ADMIN1, set/change Privileged host credentials on target host *host1.mycorp.com* to username root and password *newPass*.

    ```
    ./emcli set_credential \
    -target_type=host \
    -target_name=host1.mycorp.com \
    -credential_set=HostCredsPriv \
    -user=GC_ADMIN1 \
    -column="username:root;password:newPass"
    ```

- For super administrator GC_SUPER1, set/change Normal database credentials on target database *db1.mycorp.com* to username *dbsnmp* and password *newPass*.

    ```
    ./emcli set_credential \
    -target_type=oracle_database \
    -target_name=db1.mycorp.com \
    -credential_set=DBCredsNormal \
    -user=GC_SUPER1 \
    -column="username:dbsnmp;password:newPass"
    ```

What if You Don't Know the *root* Password?
Many DBAs at smaller shops perform both the database and system administration (SA) work for their database servers. However, at some Oracle shops, predominantly at large firms, there is a great divide between DBAs and SAs for reasons of security or corporate culture. Whatever the reason, DBAs are often not privy to the *root* password for their database hosts. If this is the case at your company, and you need *root* credentials to perform legitimate database functions in Grid Control, it's a reasonable request to ask an SA to enter the *root* credentials while you turn your back. Better yet, perhaps you can convince the SAs to use GC to manage their hosts, in which case they will gladly enter the *root* password for their own sake to take full advantage of all GC host features. The problem is that an SA needs to change the *root* preferred credentials every time the *root* password changes. To make the SA's job easier (and your job when changing database passwords), provide SAs with the automated solution of setting preferred credentials using the EM Command Line Interface (EM CLI). See the section above entitled "Programatically Setting Preferred Credentials."

Verify Default Credentials

You could always wait until GC needs to use default credentials and see whether they work. I'm not getting cheeky; this reactive approach is fine if you think it's fine. However, my style is to take a more proactive stance here, since so much GC functionality depends on correctly set credentials. There are two ways to verify the default credentials for a target type: manually and in a more automated fashion using Grid Control's own features.

Manually Verifying Default Credentials   The obvious way to manually verify default credentials is to enter those credentials for a specific target under the Target Credentials section (see "Entering Target Credentials" earlier). This method is constructive when only a few targets exist and you can quickly test all of them, or even when many targets exist but you're satisfied with spot checking a few targets because you're convinced they possess identical credentials. In our example above, testing credentials for the target database instance named emrep.perftuning.com would involve entering the same (default) credentials for this target and clicking Test. Grid Control would return "*Credentials successfully verified for* emrep.perftuning.com," which would tell you that the Default Credentials are correct for that target, assuming you entered the same password for both default and target credentials. Now you can remove the target credentials for emrep.perftuning.com and click Apply to revert to using only default credentials. Grid Control returns with an error that "*There are no credential changes to apply*", but nevertheless removes the target credentials and uses the default credentials.

Automatically Verifying Default Credentials   You can't test default credentials in the Preferred Credentials screen for a target without entering these credentials for the target. However, you can easily test that all target host or database credentials have been entered (and entered correctly) by using GC functionality to simultaneously execute host or database commands against all targets of each type where default credentials are set. (You can also use this method to verify target credentials rather than clicking the Test button in the Target Credentials section of the Preferred Credentials page for a target type.) These GC features are called Execute Host Command and Execute SQL, and they are available under Related Links on the respective Hosts and Databases summary pages and on individual host and database target home pages.

As for the Execute Host Command feature, do the following to execute a host command across all hosts where credentials are set (see "The Execute Host Command Feature" in Chapter 9 for a screen shot of its UI):

1. Log in as the administrator whose default credentials you already set.
2. Click Targets, Hosts, and under Related Links, click Execute Host Command.
3. The Execute Host Command page appears. In the Command field, enter a Host Command (such as *hostname* in this example*)*. If you're checking Privileged Credentials, preferably use a command that only the default Privileged host user has permissions to perform, such as "*/sbin/ifconfig -a*".
4. For Host Credentials, select Use Preferred Credentials.
5. For Credential Type, select Normal or Privileged to verify Normal or Privileged host credentials, respectively.

Chapter 10: General Console Configuration   329

6. Click Add to select those hosts whose credentials you want to verify. A pop-up menu lists all discovered hosts. Select the desired hosts and click Select to close the pop-up menu.

7. Click Execute and verify that the command completed successfully on all hosts by checking that Failed is 0 (zero)—displayed below the Execute button.

8. If Failed is not 0, see the Execution Status column for which hosts failed to execute the command. Then click Show under the Details column for the exact error message.

As for the Execute SQL feature, the procedure to test that you entered all database credentials correctly is very similar to that for hosts:

1. Log in as the administrator whose default credentials you already set.

2. Navigate to Targets, Databases, and under Related Links, click Execute SQL.

3. This brings up the Execute SQL page (shown with example selections and results).

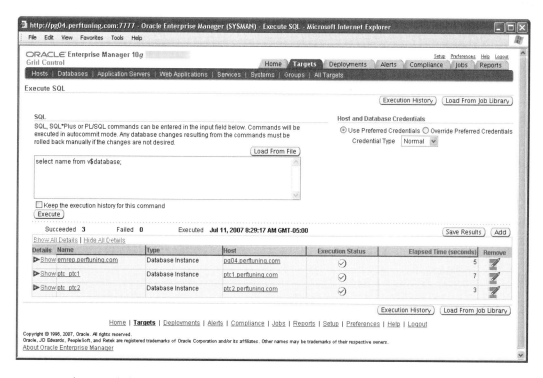

In the SQL field, enter a SQL command such as "*select name from v$database.*" (You don't need to terminate the command with a ";" or "/", although you can do so either on the same line or on a new line.) If you're checking Privileged Credentials, preferably use a command that only the default Privileged database user has permissions to perform, such as "*select * from dba_datafiles*".

4. For Host and Database Credentials, select Use Preferred Credentials.

5. For Credential Type, select Normal or SYSDBA to verify Normal or Privileged database credentials, respectively.

6. Click Add to select those databases whose normal credentials you want to verify. A pop-up menu lists all discovered databases. Select the desired databases and click Select to close the pop-up menu.

7. Click Execute and verify that the command completed successfully on all databases by checking that Failed is 0 (zero)—displayed below the Execute button.

8. If Failed is not 0, see the Execution Status column for which databases failed to execute the command. Then click Show under the Details column for the exact error message.

TIP
A great recurring job to schedule in Grid Control would be to test all preferred credentials defined for targets. Such a job would validate preferred credentials and alert you to unauthorized password changes.

Configure Notifications

Configuring notifications consists of three main steps:

- Define notification methods
- Set notification schedules
- Create and subscribe to notification rules

As I explained in the introduction, configuration dependencies require toggling between items on the global Setup menu and local Preferences menus for each administrator. After an admin sets his or her preferred credentials, a super admin needs to make sure any desired notification methods other than e-mail are configured under the global Setup menu. Then, each admin can complete the remaining two items under Preferences: Notification Schedule and Notification Rules.

Define Notification Methods

Notification methods define global means or mechanisms (such as e-mail) to contact administrators when events (metric alerts, policy violations, or job status changes) occur. At most GC sites, e-mail is a sufficient notification method. However, you can also define other notification methods, including SNMP traps for other system management products, OS scripts, PL/SQL procedures, and Remedy built-in Java callbacks. A super administrator only needs to set up a particular notification method one time. Admins then subscribe to *notification rules* (discussed later) that define, among other things, one or more of these notification method(s) to employ when sending event notices. At least one notification method must exist before you can define any notification rules. Notification methods are managed independently of notification rules, so you can apply the same method to multiple rules. Following is the Notification Methods page.

Chapter 10: General Console Configuration

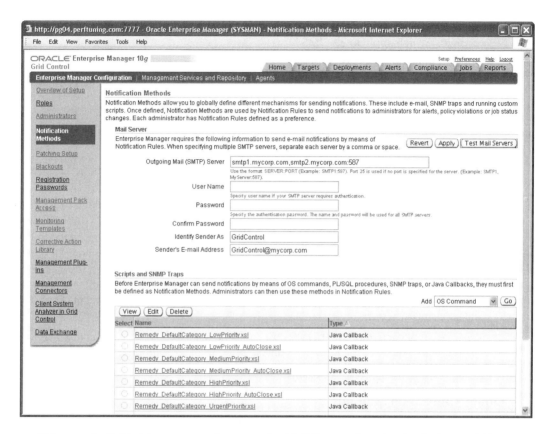

This page is divided into two sections: one for Mail Server, the other for Scripts and SNMP Traps.

Mail Server

E-mail is the predominant notification method used at GC sites to send alerts to administrators. A super administrator must specify one or more mail servers through which Grid Control can send such notifications. The Outgoing Mail (SMTP) Server field on this page will already contain an entry if you specified an SMTP server when installing Grid Control (see Figure 3-7 in Chapter 3). To provide high availability for notification delivery, you can enter additional mail gateways now (post-install), if your company uses them. Grid Control will attempt to deliver mail using the first listed mail server, and if unsuccessful, will use the next mail server listed. (Ask your network administrator whether your mail gateway is configured externally for high availability. If so, you may only need to enter a single virtual mail server name, which would forward the notification to a running mail server.)

Do the following to specify mail servers for all admins to use:

1. Log in to the Console as a super administrator and note which OMS host is used. If employing an SLB as a front end for multiple OMSs, log in directly through a particular OMS.

2. Click the Setup link at the top right and select the Notification Methods link.

3. In the Mail Server section of the Notification Methods page, complete the following fields, which apply to all OMS hosts:

- **Outgoing Mail (SMTP) Server**   Enter one or more mail servers in this field. Use the format "*<SMTPServer>:<port>*"and specify the port only if the mail server doesn't listen on default SMTP port 25. Separate mail server entries by commas or spaces. (The page above shows two mail servers, where the second one uses nondefault port 587.)

- **User Name / Password**   Enter a username and password if the SMTP servers require authentication. These credentials will be used for all SMTP servers.

- **Identify Sender As**   Enter the name of the e-mail account you want to appear as the sender of notification messages as shown in the From: field of the e-mails.

- **Sender's E-mail Address**   Specify the e-mail address through which to send e-mail notifications and to receive notifications of delivery problems. This must be a valid address from which the specified SMTP server(s) can send e-mail.

4. Click Test Mail Servers to verify that the OMS rendering the Console can use all specified mail servers to relay test messages from the sender's e-mail address. Test results are returned in the Console reporting on each SMTP server, and a test e-mail is sent through each SMTP server to the sender's e-mail address.[23] If the test succeeds, click Apply to save all changes.

5. Check that each additional OMS host can send test e-mails through the configured SMTP server(s). To do so, log in to the Console through each additional OMS (step 1) and repeat steps 2 and 4 above.

Other Notification Methods

In addition to the e-mail server notification method, you can define other methods to automate responses to target events, where both the targets and events are defined in a notification rule employing the method. Following are the additional notification methods available over and above e-mail notification:

- **OS scripts**   You can create OS scripts to automate event responses. OS scripts must exist in the same specified location on all OMS hosts. For example, you can automatically open an in-house trouble-ticket using an OS script.

- **PL/SQL procedures**   You can create PL/SQL procedures in the Repository Database to perform actions in response to events. For instance, you can notify a third-party application using a custom PL/SQL procedure.

- **SNMP traps**   SNMP traps can invoke and pass event data to SNMP-enabled third-party applications, such as HP OpenView.

- **Built-in Java callbacks**   You can also select in the UI from built-in Java callbacks to the BMC Remedy Help Desk to prioritize the urgency of notifications. Creating new Java callbacks is not supported as of the latest GC Release (10.2.0.4) available when this book was published, but Oracle plans to support them in EM11*g*.

---

[23] Check both Console output and e-mails received, as the Console output does not always seem to validate the specified e-mail address, only that the SMTP server can relay to the domain-part of that address.

Chapter 10: General Console Configuration **333**

These notification methods are similar to corrective actions in that they respond to an event occurrence, but differ from corrective actions in scope. Notification methods are performed for any event/target combination defined in a notification rule that employs the method, whereas corrective actions are applied to particular alerts or policies set for specific targets. As such, notification methods are more general in nature and are intended primarily for third-party integration with other applications.

To enter an OS command, PL/SQL procedure, or SNMP trap, select the appropriate choice in the Add field under the Scripts and SNMP Traps section. The following screens appear (with example input shown):

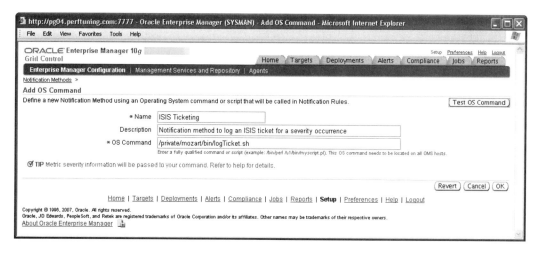

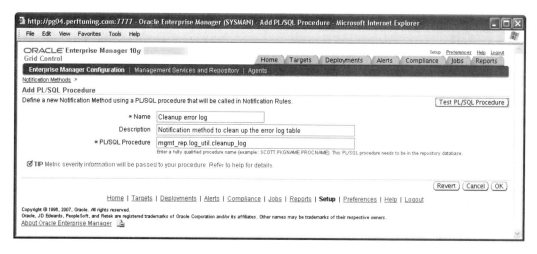

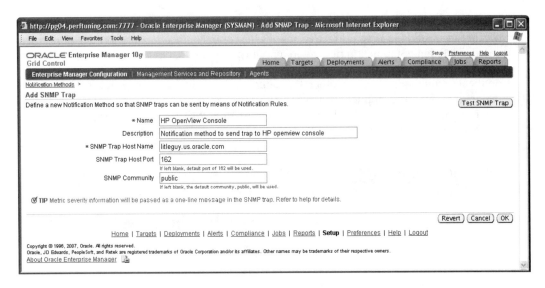

For all notification methods, a test button exists to test your setup. The input shown in the three screen shots above comes from the examples in the Oracle Enterprise Manager Online Help, which provides all details on how to set up these notification methods. For help, click the Help link (shown at the top right of these screen shots). Once you're in the Online Help, click the link, About Passing Metric Severity Information, for details on passing metric severity and policy violation information to a PL/SQL procedure or OS script. See Notification Examples for an example SNMP Trap and OS command sample script. For detailed information on SNMP traps, see Chapter 13 of Oracle Enterprise Manager Advanced Configuration 10g Release 4 (10.2.0.4.0).

Set Notification Schedules

A notification schedule defined for each administrator determines when and at what e-mail address(es) he or she is to receive GC notifications. Schedules are very configurable—you can specify a different notification e-mail address (or addresses) for every hour of every day, if warranted. You can also suspend notifications as needed, such as when an administrator is on vacation. A super administrator can set the notification schedule for any other admin, including another super administrator, but super administrators cannot change SYSMAN's schedule—only SYSMAN can. (SYSMAN's schedule is no different from a notification schedule for any other administrator; it dictates when the e-mail addresses for SYSMAN—if any are specified—would be sent alerts.)

The mechanics of specifying a notification schedule in the Console are straightforward enough. The schedule hinges on a *rotation frequency*, a concept not intuitively obvious to everyone. A rotation frequency is a repeatable pattern of the smallest time period you can define to model your desired notification schedule. Grid Control allows for a one- to eight-week rotation schedule. Most IT professionals—in the U.S., anyway—work a one-week (seven-day) rotation frequency, made up of a five-day workweek and a two-day weekend off.

To take a harder example, say you're a DBA firefighter who works roughly two 24-hour shifts per week, which fall on different days each week, but on the same days every four weeks. In this case, your work rotation frequency is four weeks, as this schedule repeats on a four-week basis.

Defining Notification Schedules

Let's illustrate how to set a notification schedule using the typical example of a one-week rotation frequency. A super administrator can define or edit a schedule for any other admin. In the following example, I use SYSMAN to edit GC_SUPER1's schedule. The GC_SUPER1 administrator who is on Eastern Standard Time wants to start receiving notifications on Monday July 23, 2007 by e-mail at *dba@mycrop.com* between 8AM and 5PM each Monday through Friday, and by e-mail and PDA at *pda1@gluedtomyhand.com* during off-hours.

1. First, confirm that all notification e-mail addresses are defined for the admin whose schedule you are defining. You can do this one of two ways:

   - Log in as the admin whose schedule you are defining/editing, go to the Preferences menu, click the General page as shown in earlier Figure 10-2, add a row for each desired e-mail address, and then click Apply.

   - Log in as any super administrator (SYSMAN here), navigate to the Administrators page under the Setup menu, select the administrator (GC_SUPER1), and click Edit.

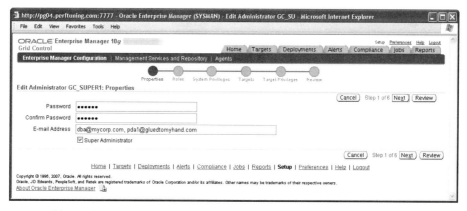

   Add any desired addresses, separated by commas, click Review, then click Finish.

2. As a super administrator, navigate to the Notification Schedule page by clicking on the Notification Schedule link on the Preferences menu (see Figure 10-6).

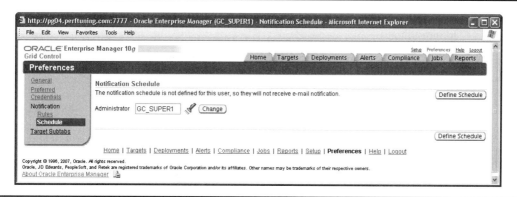

FIGURE 10-6. *The Notification Schedule page under the Preferences menu*

3. If necessary, click the flashlight icon next to the Administrator field and change the admin to that whose schedule you want to define/edit (GC_SUPER1 here).

4. Define (or edit) the time period of the schedule, including the rotation frequency, by clicking the Define Schedule button (if you're editing, the button is named Edit Schedule Definition). The following page appears.

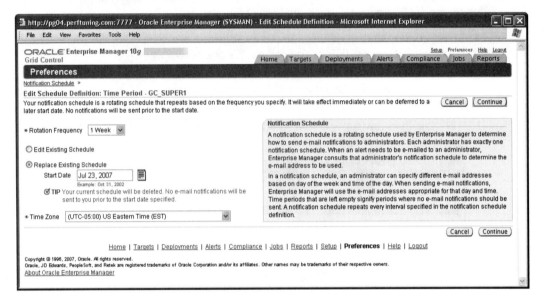

5. Enter the desired Rotation Frequency.

6. Click Edit Existing Schedule or Replace Existing Schedule, as appropriate.

7. Enter a Start Date of Monday of the current week, as this allows you to specify a schedule for the entire week.[24]

8. Choose the Time Zone in which this admin is located.

9. Click Continue.

10. Enter or edit the notification schedule on the page that appears (see below). The schedule defaults to 24×7 notification for the e-mail address(es) you specified when creating the administrator.

    You can take advantage of the Batch Fill-in functionality by filling in a block of time on given days to be notified at the same e-mail address(es). To do this, follow these steps (I call each iteration of steps 1 through 34 an *operation*):

11. Click the Flashlight next to the E-mail Addresses field and select one or more addresses you'd like to use for a particular block of time. (To remove all e-mail addresses during a block of time, clear the E-mail Addresses field select the same Start Time and End Time, all the Days of Week, and click Batch Fill-in.) Following is the result of all steps to set the example schedule. Here, the first e-mail address chosen is *dba@mycorp.com*.

---

[24] Grid Control schedules consider Monday the first day of the week.

Chapter 10: General Console Configuration **337**

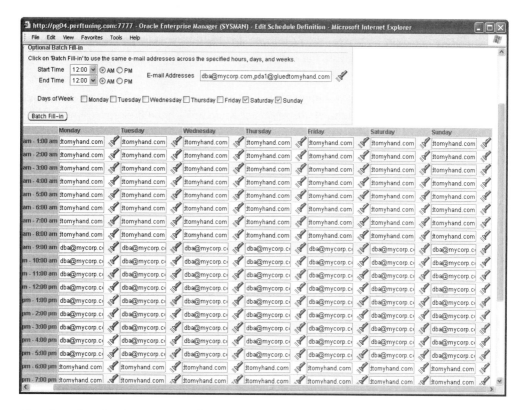

12. Complete the Start Time, End Time, and Days of Week fields for the first block of time where GC should send notifications to the specified e-mail address(es). In this case, Start Time is 8AM, End Time is 5PM, and Days of Week are Monday through Friday.

13. Click the Batch Fill-in button.

14. Repeat steps 1, 2, and 3 above for any other blocks of time desired. If, as in this example, one of these blocks of time crosses the midnight hour (from 5PM to 8AM), use two different Batch Fill-in operations for that block of time. In the first operation, start and end times are both before midnight (from 5PM to 12AM).[25] In the second operation, start and end times are both after midnight (from 12AM to 8AM). Here, then are the remaining Batch Fill-in operations for the example. For all operations, E-mail Addresses is *dba@mycorp.com;pda1@gluedtomyhand.com*).

    Start Time: 5PM, End Time: 12AM, Days of Week: Monday through Friday

    Start Time: 12AM, End Time: 8AM, Days of Week: Monday through Friday

    Start Time: 12AM, End Time: 12AM, Days of Week: Saturday and Sunday

15. Click Finish to save your changes.

---

[25] The default UI behavior only shows the beginning of each e-mail address field in the calendar, which in this example causes the entries for e-mail-only and e-mail/PDA addresses to appear identical. So for clarity, the screen shots here show the end rather than the beginning of the e-mail/PDA addresses to distinguish them from the e-mail-only addresses.

Suspending Notifications

Suspending notifications is easy. Say, for example, that the GC_SUPER1 administrator is on vacation from Monday, August 6 through Sunday, August 12, 2007. To suspend notifications for this time period, do the following:

1. As a super administrator (SYSMAN in this example), navigate to the Notification Schedule page by clicking on the Notification Schedule link on the Preferences menu (refer back to Figure 10-6).

2. If you are editing someone else's notification schedule, as in this case, click the flashlight icon next to the Administrator field and select the admin whose schedule you want to edit (GC_SUPER1 here).

3. Click the Suspend Notification Edit button as shown here.

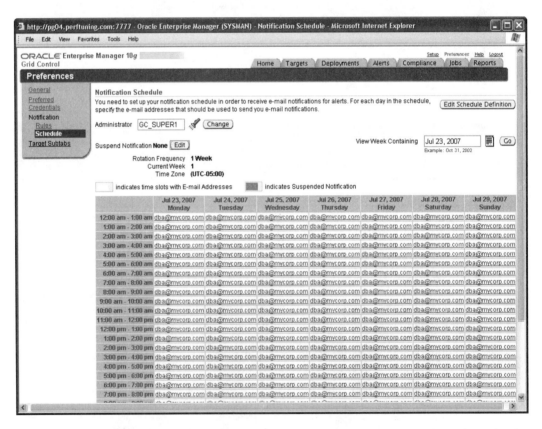

4. On the Edit Suspended Notification page that appears, enter or choose from the Calendar icon a Start Date (August 6, 2007) and an End Date (August 12, 2007) during which to suspend notifications for this admin, and click OK.

Create and Subscribe to Notification Rules

A notification rule defines a set of targets, target availability states, metrics, policies, job status changes, and notification methods that will produce a notification. Administrators can create their own notification rules or subscribe to public notification rules. Imagine a dashboard light for every possible change in a vehicle engine's condition. Notification rules allow you to select the dashboard lights (e.g., e-mails) that will appear in your vehicle to alert you of certain engine troubles. Grid Control collects data for events (metrics, policies) defined under Metric and Policy Settings for targets and for jobs defined on the Jobs tab. This collection is irrespective of whether the "downstream" notification rules include these metrics or policies. Like a vehicle's onboard computer, the Repository stores this data and the Console provides access to it—alerts on the Alerts tab, policy violations on the Compliance tab, and job status changes on the Jobs tab. Notification rules are a way of calling administrators' attention to those alerts and violations of interest so they don't have to constantly keep checking the Console for new events. If an admin subscribes to a notification rule that includes a particular event for a target, the admin receives notification of that event.

Figure 10-7 shows the Notification Rules page under the Preferences menu, and the 13 built-in rules (called *system-generated* rules in the Console and in the Online help).

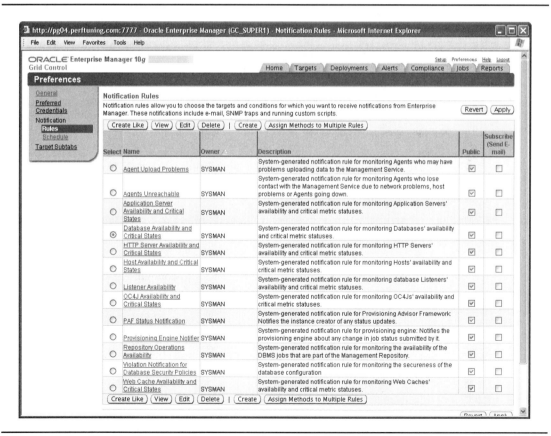

FIGURE 10-7. *The Notification Rules page*

On this page, an admin can create, edit, view, delete, or subscribe to notification rules. Each rule definition has six pages on which you accomplish the following functionality:

- **General** Define general attributes of the rule, such as name and target type, to which it applies. All notification rules, whether built-in or custom, relate to a particular target type, such as Agent, Application Server, Database, or Host.

- **Availability** Select whether the rule should notify you when a target goes down or restarts, and when a target goes down, whether to notify you if any defined corrective actions succeed or not. Also choose whether to be notified when metric errors are detected or resolved and when blackouts start or end.

- **Metrics** Choose specific metrics, their severity states, and corrective action states for which you want to receive notifications. Metrics are units of measurement such as host CPU Usage (%) used to establish a target's health. When a *metric threshold* (a boundary value against which a metric value is compared) is crossed, it triggers an alert in the Console. You can set notifications for three metric *severity states* (levels of alert severity):

    - *Warning* Requires attention in a specific area that still remains functional

    - *Critical* Demands immediate action in a particular area that is either completely or imminently nonfunctional

    - *Clear* Indicates when a Warning or Critical severity state is resolved

- **Policies** Select specific policies, their severity states, and corrective action states for which you want to receive notification. Policies define security, configuration, and storage best practices for certain types of targets. While policy violations are reported in the Console under the Compliance tab according to three levels of severity (Critical, Warning, and Informational), there are only two severity states for notifications: Violation (encompassing all three levels of severity) and Clear.

- **Jobs** Specify which jobs and status changes to notify on for the targets in the notification rule. Jobs are units of work defined to automate administrative tasks like backups or patching. You can be notified of the following job status changes: Scheduled, Started, Suspended, Succeeded, and Problem.

- **Methods** Choose the way(s) in which GC should notify you, including via e-mail (according to your notification schedule) and/or via advanced notification methods.

Let's look at each of these six pages in the context of creating a new notification rule. This will flesh out what notification rules consist of as well as how to configure them.

Creating Notification Rules

Following are the steps to create a new notification rule. I highlight best practices along the way as tips.

1. Log in as the administrator who will create the new notification rule.
    - To create a public notification rule, you can log in as any user; that is, as a super administrator or nonsuper administrator.

Chapter 10: General Console Configuration **341**

- To create a nonpublic notification rule, log in as the admin desiring sole access to the rule.[26]

TIP
When creating a public notification rule, log in as SYSMAN if possible because a public rule created using any other account is removed if that account is deleted.

2. Click Preferences and choose the appropriate link for your account.
   - If a nonsuper administrator, click My Rules under Notification on the vertical navigation bar.
   - If a super administrator, click Rules under Notification on the vertical navigation bar.
3. Create a new rule or a new rule based on an existing rule.
   - To create an entirely new rule, Click Create.
   - To create a new rule based on an existing rule, select the radio button in the Select column next to the model rule name and click Create Like.

NOTE
Given that admins share all public notification rules, leave the system-generated rules intact and use them only as templates to create new rules, so that other admins may do the same.

Let's say that SYSMAN wants to create a new notification rule based on the built-in rule, Database Availability and Critical States. Following are the specifications for the changes to be made to this built-in rule:

- General page
  - Name the Rule Database Availability and Critical/Warning States
  - Change the default description to indicate that it will notify on warning states as well as critical states
- Availability page
  - Notify when a target's availability changes (goes down or restarts)
  - Notify on both Problem and Succeeded corrective action states for target availability
- Metrics page
  - Add a new metric for notification called Archiver Hung Alert Log Error
  - For all metrics, add notification for Warning severity states in addition to the Critical severity states defined for the built-in rule

---

[26] SYSMAN has access to all rules, even nonpublic ones, and to all other Grid Control functionality.

- Policies page
  - Add a new policy called Default Passwords to notify on both Violation and Clear severity states and on both Succeeded and Problem corrective action states
- Jobs page
  - Notify on any jobs involving RMAN scripts with Suspended or Problem statuses
- Methods page
  - Do not send e-mail notification because, as is best practice, no admin in this example is associated with the SYSMAN account. Admins subscribing to the rule, once created, should subscribe to e-mail notification.

Following are explanations for each of the six pages comprising a notification rule. I examine all fields on these pages, even those you don't need to change to meet the example specifications.

General Page Use the General page to define the principal attributes for the notification rule.

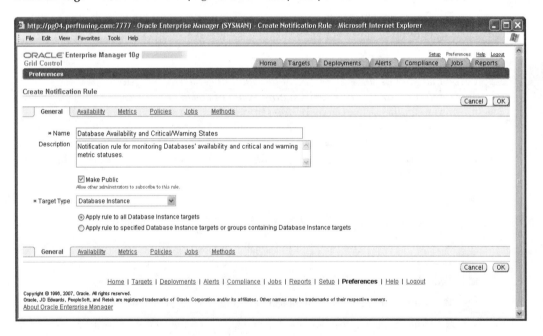

Enter or change the following fields and boxes (on this page, required fields are indicated by asterisks):

- **Name** (Required) Enter the name of the new rule.
- **Description** Enter a description for the rule. Default descriptions are provided for built-in rules.
- **Make Public** Leave this box checked.

Chapter 10: General Console Configuration **343**

TIP
Leave this box checked to allow other admins to subscribe to the rule, unless it should be restricted to the current admin for some reason. One reason would be that the admin defining the rule limits it to specific targets to which only he or she has privileges.

- **Target Type** (Required) Choose the target type for the rule's targets, defined next.
- **Apply rule to all/specified ... targets** Choose to apply the rule either to all targets of the selected type or to specified targets. This choice depends on your overall notification strategy. If you choose to apply the rule to specific targets, an Add button appears; click it to select the targets for which you want the rule to apply.

Availability Page Click the Availability page to select target availability states for which those subscribing to the rule should receive notification.

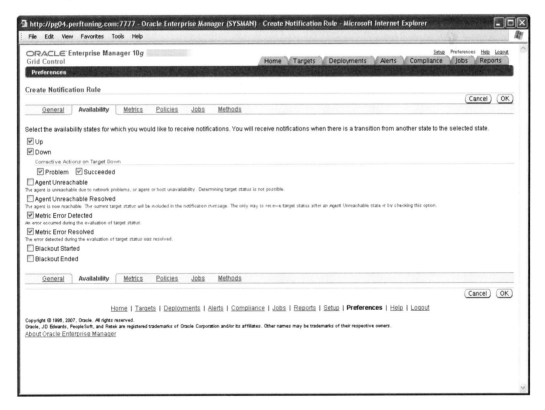

The boxes checked in this illustration are those generally recommended for all rules, except the Agents Unreachable rule as noted below. Nevertheless, let's describe each field, as you may want to use different settings for your own inimitable reasons.

- **Availability States** Check Up and/or Down, or neither, to receive notifications when a target's availability changes. All built-in rules are preconfigured with only the Down box checked. Leaving this box checked enables target down notifications even if you don't select any metrics on the Metrics page, covered next.[27]

TIP

Check the Up box (not checked by default) and leave the Down box checked to receive notifications when a production target's availability changes in either direction. By checking the Up box, the admin who manually restarts a target can indirectly convey to the other admins subscribing to the rule that the target has restarted.

- **Corrective Actions on Target Down** Select the Problem and/or Succeeded boxes, or both.

TIP

Select both boxes to enable notifications for corrective actions you've configured (or will configure) to restart downed targets included in the rule. You'll want to know if such corrective actions succeed or fail.

- **Agent Unreachable / Agent Unreachable Resolved** When an Agent or its host becomes unavailable, all monitored targets on that host are set to an Agent Unreachable state. Selecting either of these boxes for rules involving Agent or Host[28] target types would produce a flood of notifications, one for each target, were an Agent or host to go down or restart.

TIP

Do not check these boxes for any rule except the Agents Unreachable rule (for which these boxes are already checked by default), as the sole purpose of that rule is to monitor the status of all Agents and hosts.

- **Metric Error Detected / Metric Error Resolved** Check these boxes if you want the rule to notify that a metric error occurred or was resolved when checking a target's status.

TIP

Checking both boxes allows Grid Control to monitor itself; you are notified if a metric error blocks notification that a target has gone down. You can address other metric errors as part of weekly online housekeeping by reviewing the Errors subtab under the Alerts tab as described in Chapter 17, "Fix Target Metric Collection Errors".

---

[27] Checking the Up and Down "availability states" equates to requesting notification for a metric, which all target types have, called either Status or UpDown Status, and listed under Metrics and Policy Settings with Down hardcoded as the Critical threshold. On the Availability page for a rule, you should invariably choose Down at a minimum, as there is no more important notification than that a target has gone down.

[28] This contradicts the Oracle Enterprise Manager Online Help recommendation under "Collection Schedule: Metric Information" to create a notification rule for the Host target type that subscribes to the Agent Unreachable start and end conditions.

Chapter 10: General Console Configuration **345**

- **Blackout Started / Blackout Ended**   Generally, the targets included in a blackout don't correspond with those of a particular rule, unless the rule is designed specifically to report blackouts. While you cannot receive blackout notifications any other way, you can query all blackout information on the Blackouts page under the Setup menu. Administrators typically create blackouts to avoid notifications, so it defeats the purpose somewhat to configure notifications of the blackouts themselves.

TIP
Don't check these boxes unless, for each target defined in the rule, you want to receive a separate notification that a blackout has started or ended.

Metrics Page   Click the Metrics page (see Figure 10-8) to choose which metrics to associate with the rule.

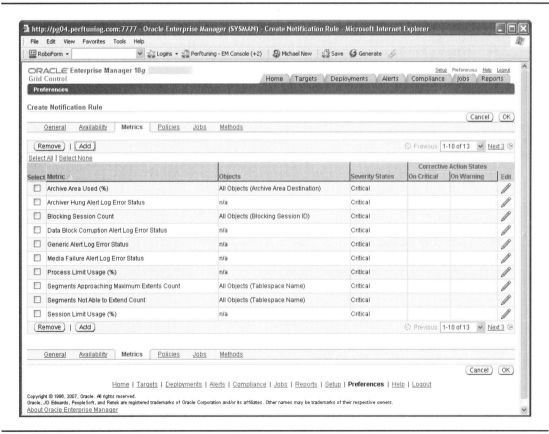

FIGURE 10-8.   *The Metrics page of a notification rule*

On this page you can add, edit, or remove metrics associated with this notification rule.[29] Clicking the Edit icon (pencil) for each metric permits you to change its severity and corrective action states. However, to assign the same states to multiple metrics in one operation, which you'll likely want to do, at least initially, it's much less painstaking to click the Add button.[30] This pulls up the Add Metrics page (see Figure 10-9).

Here, you can add new metrics, of course, but also change states for multiple metrics in one operation. The specification in the example calls for this; that is, to add the Warning severity state and all corrective action states for all metrics in the rule. The figure shows only the top and bottom of a very long page, because I've chosen "Show All ..." to select all metrics for this rule on one page. The page has three sections:

- **Metrics** Define metrics and specific objects (for metrics that allow multiple thresholds), if desired, to associate with the notification rule. (Objects apply to metrics that allow multiple thresholds, such as Tablespace Used (%), where you can set different thresholds for each object (i.e., tablespace) if you like. Because this is the Add Metrics page, all metrics appear, even those without thresholds. Metrics already associated with the rule are not checked, although this would be a nice feature to help keep track of the current metrics for the rule.[31] For this example, select those metrics you want to change by checking the 13 default metrics for this rule.

- **Severity States** Choose the severity states (Critical, Warning, and Clear) for which you'd like to receive notification on the metrics selected. For our example, click the Critical box to keep the original setting and click the Warning box because you want to add notifications for Warnings.

TIP
Choose at least Critical and Warning severity states for metrics defined in the rule to be notified of Critical and/or Warning thresholds you set. Setting critical and/or warning metric thresholds for a target has no apparent[32] purpose outside the context of triggering notifications when these thresholds are reached.

- **Corrective Action States** Select one or more corrective action states (Succeeded and Problem for Critical and Warning severity states) for which you want to receive notifications. For this example, select the Succeeded and Problem boxes for both Critical and Warning corrective actions.

Click Continue.

---

[29] The metrics listed are already selected for notification. In other words, you don't need to check a box under the Select column to add that metric. The boxes are only for removing metrics.

[30] In my opinion, there really should be an Edit button, as elsewhere in the UI.

[31] To keep track of the current Metrics tab settings, open the Metrics page for this rule in another session before editing the metrics. For IE, you can duplicate most UI pages by choosing File: New -> Window.

[32] I am reluctant to use this phrase "has no purpose" because DBAs are always adept at using a product's functionality in esoteric ways the product's designers never dreamed of.

Chapter 10: General Console Configuration 347

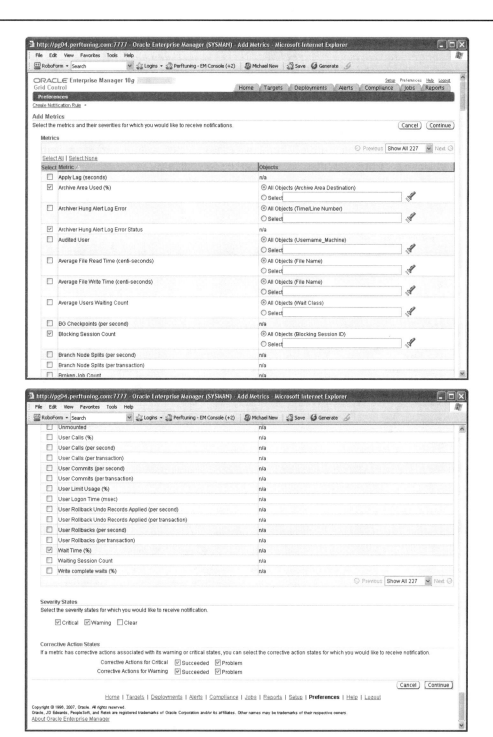

FIGURE 10-9. *The Add Metrics page for a notification rule*

TIP
Choose all corrective action states for the metrics defined in the rule, even if no corrective actions are defined. This enables notification for corrective actions defined later. If you want to know when a metric threshold is reached, chances are you also want the status of any configured corrective actions.

The technique of using the Add Metrics page to change states for multiple metrics only works for adding new Severity States and/or Corrective Action States, not for removing them.[33] After making all changes, Click Continue to return to the Metrics page

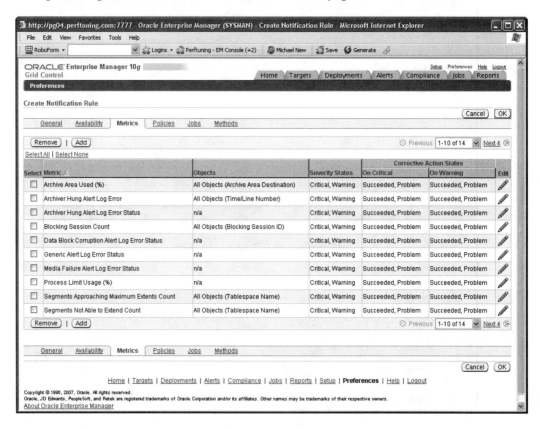

For all 13 metrics, you now see both Critical and Warning under the Severity States column, and all four Corrective Action States selected under the Corrective Action States columns.

Policies Page Click the Policies page to choose the policies to associate with the notification rule.

---

[33] This is because the boxes for these states are always unchecked on the Add Metrics page—even if they are already saved for metrics of a notification rule—so you can't uncheck them.

Chapter 10: General Console Configuration 349

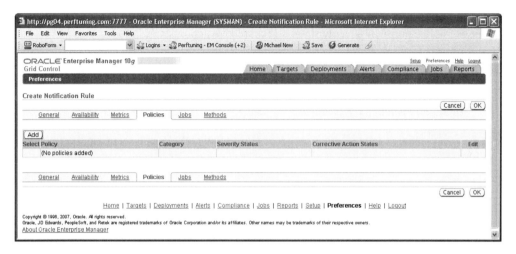

To add a new policy, click Add, which pulls up the Add Policies page.

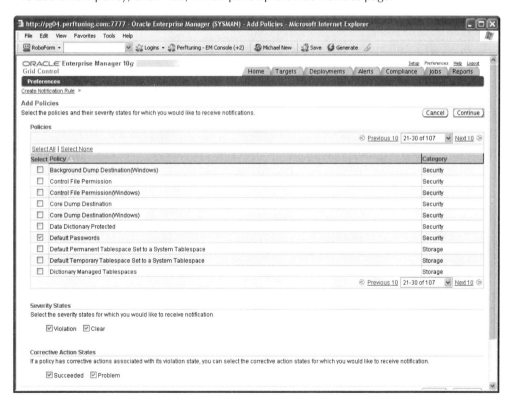

This is very similar to the Add Metrics page in earlier Figure 10-9. Both pages have the same general three sections. Except for the obvious difference in the first section, which lists policies rather than metrics, both pages have sections for severity states and corrective action states.

The Violation severity state includes all three levels of severity (Critical, Warning, and Informational) For example, choose all severity states and corrective action states.

TIP
Check all severity states and corrective action states for chosen policies. The reasons are the same as cited above (see Metrics page) for the equivalent metrics states.

Click Continue, which brings up a summary of all choices you made on the Policies page (not shown here).

Jobs Page   Click the Jobs page to define jobs for which you want the rule to notify when its status changes.

You can select Add Jobs by Criteria or Add Specific Jobs. Make your selection and click Go.

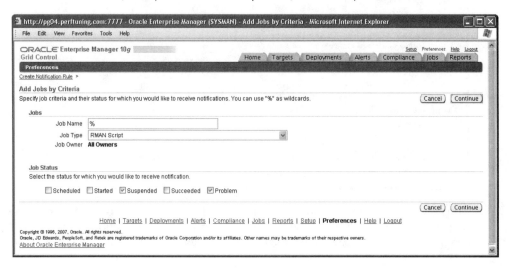

Chapter 10: General Console Configuration   **351**

Regardless of your selection, the page that appears (as shown above) contains two sections:

- **Jobs**   In the Job Name field leave "%" as a wildcard character if you want the rule to apply to all jobs of the Job Type selected.

- **Job Status**   For example, select Suspended and Problem. Click Continue. A new row appears on the Jobs page (not shown here) listing the new job.

TIP
Under Job Status, check at least the Suspended and Problem boxes to be notified of any job status exceptions.

Methods Page   Select the Methods page to specify the manner in which you want to be notified when a condition you defined on the other pages is satisfied.

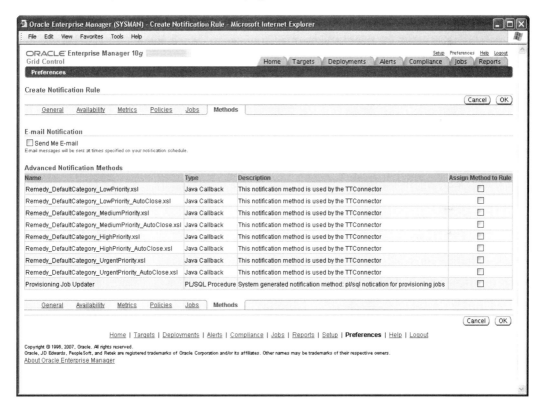

You can associate a rule with multiple notification methods—both e-mail and advanced methods; there is a section for each type of method:

- **E-mail Notification** Check the "Send Me E-mail" box if you want to enable e-mail notifications for the currently logged in administrator according to his or her notification schedule. Most admins want e-mail notification, whether this "e-mail" goes to an e-mail address, PDA, pager, or other device. For example, do not check this box because no admin uses the SYSMAN account; each has his or her own account.

TIP
If you're creating a nonpublic rule, check this box. If you're creating a public rule as SYSMAN, don't check the "Send Me E-mail" box if no administrator is associated with the SYSMAN account. Instead, advise administrators for whom the rule was intended to subscribe to it using one of the methods described in the next section.

- **Advanced Notification Methods** Choose any advanced notification methods you wish to assign to the rule. Unlike e-mail notification, which administrators subscribe to or not, advanced notification methods chosen for the rule form part of the rule's definition, so they carry over to any administrator subscribing to it.

Click OK to save the changes to this new rule made on all six pages.

Subscribing to Notification Rules

There are three methods to subscribe[34] to rules. One method permits administrators to subscribe to a particular rule when it's created or edited, as just shown on the Methods page above. The other two methods are for rules already created. These methods allow administrators to subscribe to multiple rules in one operation, and one of these methods allows super administrators to subscribe other admins to multiple rules in this fashion. Here are the three methods:

- **While creating/editing a rule** As the administrator creating/editing a particular rule, check the "Send Me E-mail" box on the Methods page as shown above to subscribe to it, and click OK.

- **On a notification rules page** An administrator can subscribe to multiple rules in one operation using the appropriate notification rules page on the vertical navigation bar. The exact page name differs, depending on whether you are a super administrator and, if not, whether the rule is public:

    - As a super administrator, use the Notification Rules page (refer back to Figure 10-7).

    - As a nonsuper administrator for public rules, use the Public Rules page; for nonpublic rules, use the My Rules page.

    For any of these notification rules pages, check boxes in the "Subscribe (Send E-mail)" column for those rules to which you want to subscribe, and click Apply.

- **On the Administrators page** A super administrator can subscribe other admins to multiple public notification rules in one operation. Click the Setup menu, then the Administrators page (refer back to earlier Figure 10-5). Select the desired admin and click Subscribe to Rules.

---

[34] You can also unsubscribe to rules using these same methods.

Chapter 10: General Console Configuration 353

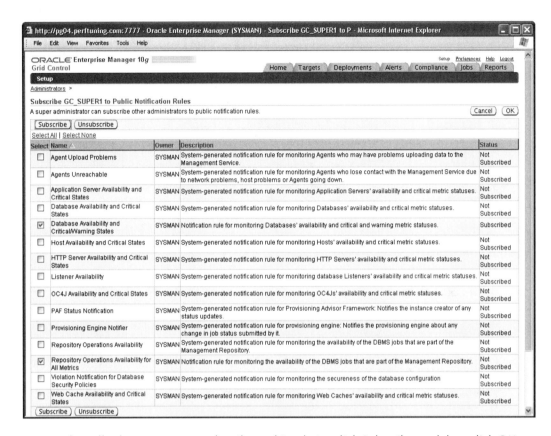

To select all rules you want to subscribe to this admin, click Subscribe, and then click OK to save your changes.

Aligning Notification Rules with Target Metrics/Policies
By design, out-of-box metrics with thresholds also appear as metrics in built-in notification rules, with few exceptions. Put another way, the metrics defined for each built-in rule are for the most part the same as the out-of-box metrics with thresholds defined for that rule's target type. For example, consider the out-of-box Metric and Policy Settings for a Database Instance (shown in Figure 10-10).

Those database instance default metrics with thresholds defined are, for the most part, the same metrics associated with the Database Availability and Critical States rule.[35] This correspondence, which is by design and true for all target types, stands to reason. A threshold for a target metric must be set to trigger an alert defined in a notification rule, and vice-versa (e.g., a notification rule must exist for a target metric to allow such a notification).

The problem is that this correspondence breaks down when it comes to the severity states of these metrics. Metrics for built-in notification rules only notify on Critical severity states (see Figure 10-8 above, for example),[36] yet Critical thresholds for many of these metrics are not

---
[35] Two exceptions are Process Limit Usage (%) and Wait Time (%), which are defined only for the rule.
[36] There's a reason the names for many built-in notification rules include the words "Critical States."

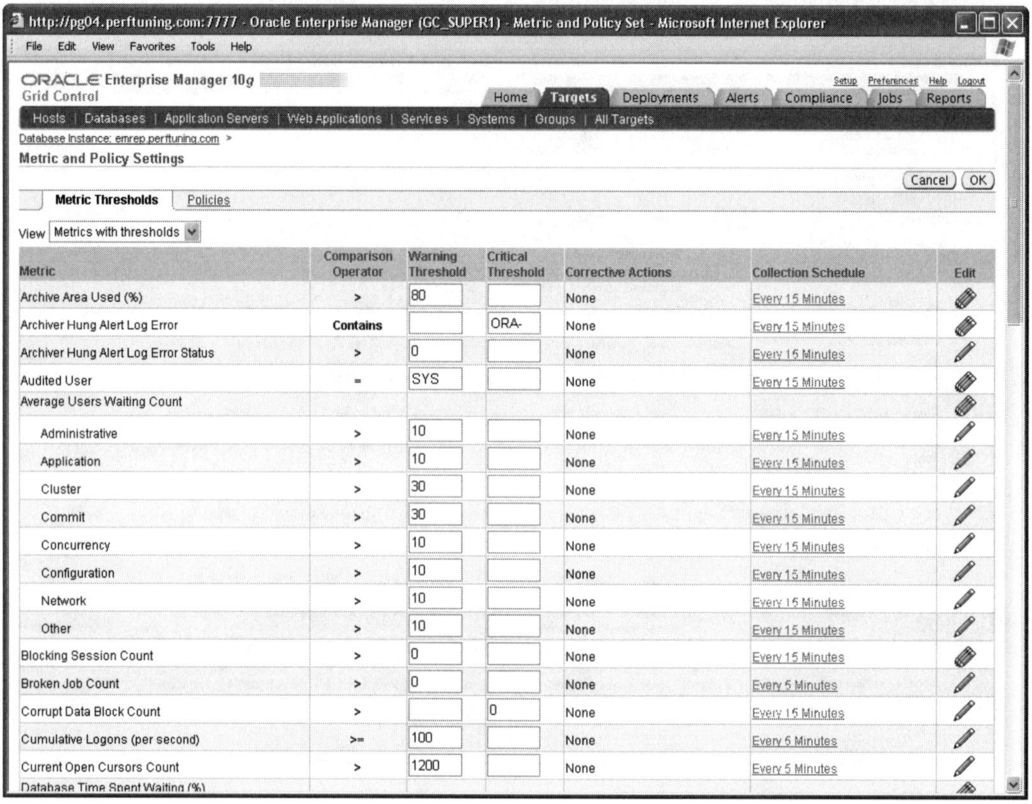

FIGURE 10-10. *The Metric and Policy Settings page for a Database Instance showing default thresholds*

set out-of-box (only Warning thresholds are). For example, only the Warning threshold is set for the Archive Area Used (%) metric shown in Figure 10-10, but the Database Availability and Critical States rule only defines the Critical severity state for this metric (and other metrics). This means that an admin subscribing to the Database Availability and Critical States rule would not be notified if the Archive Area Used (%) for any target database reached 100%, even though this would bring the database to a screeching halt. In general, admins have three options for addressing this mismatch between notification rules and defined target metrics:

- **Add critical metric thresholds to match critical notification rules**   Set critical thresholds for metrics on all targets, thereby "activating" corresponding critical notification rules. Choose this option when you want to be notified only when all metrics defined in rules reach critical levels. The warning metric thresholds defined will not trigger notifications because the built-in rules only notify on critical thresholds. (The other two options below address this mismatch.)

- **Add warning notification rules to match warning metric thresholds** Add the Warning severity state to metrics in notification rules. This option is preferable when you want to be notified of metrics reaching warning thresholds. (Critical thresholds defined for targets will also send notifications.)
- **Do both of the above** Set critical thresholds for metrics on all target databases and add the Warning severity state to metrics in notification rules. This option results in notifications for both critical and warning metric thresholds, using both of their corresponding notification rules.

When creating the example notification rule above, specifications call for choosing the second option, namely to add the Warning Severity State for all default metrics in the rule.

Notification Rules Recommendations

There are a variety of ways to configure notification rules to suit your company's notification requirements. Nevertheless, I have a few generic suggestions regarding notification rules:

- Because the GC infrastructure must remain available to ensure reliable alert notifications for all targets, use the Create Like feature to create a rule based on the Repository Operations Availability rule to notify on all available metrics for the target type OMS and Repository. Choose all severity states (Critical, Warning, and Clear) and all four corrective action states for each metric. Critical thresholds are set by default for all available metrics, so you don't need to add any metrics to Metric and Policy Settings for OMS and Repository targets.
- Enable database notifications for ORA- errors related to the flash recovery area (FRA) filling up (ORA-19815, ORA-16014, ORA-16038, ORA-19809, ORA-00312). If the FRA fills up, a database will hang. This is particularly critical for the Repository Database, as a full FRA would prevent all further alert notifications until the FRA was cleared. To add these notifications, ensure that the metric Generic Alert Log Error is one of the metrics for the rule you create based on the Database Availability and Critical States rule, and enable the metric for Critical, Warning, and Clear severity states. (Table 13-1 in Chapter 13 contains the syntax for the alert log errors to add to the Critical threshold value for the Generic Alert Log Error.)
- To prevent duplicate notifications for the same event, set all disk space notifications under rules relating to the Host target type rather than under rules pertaining to the Database target type. For example, set the Filesystem Space Available (%) metric in a rule based on the Host Availability and Critical States rule rather than setting the Dump Area Used (%) metric in a rule based on the Database Availability and Critical States rule. This would trigger only one notification rather than multiple simultaneous notifications for the same disk full condition (assuming background, core, and user dump areas were all located on the same disk, as is usually the case).
- Administrators who share GC accounts often want to receive notifications for targets according to different schedules, such as to be notified 24×7 for production targets and only during business hours for non-production targets. For these cases, you can create an administrator for each type of target (i.e., a prod admin and a non-prod admin), set the prod admin's schedule to 24×7 and the non-prod admin's schedule to business hours only, and create/assign separate notification rules for prod and non-prod targets.

An easier way is to use one admin, configure the account to receive notifications for all targets on a 24×7 basis, then set repeating blackouts for all non-prod targets during off-hours. This would require two repeating weekly blackouts: Monday through Friday starting when off-hours begin and for a duration covering until off-hours end; then Saturday and Sunday beginning at 12 AM for a 24-hour duration. However, there are two major drawbacks to using blackouts. You won't be able to configure these targets in GC while blacked out and no metrics will be collected during the blackouts.

Enable Patching Features

Enabling patching functionality in Grid Control is a lightweight, two-step process:

- Complete the Patching Setup
- Configure the RefreshFromMetalink job

Complete the Patching Setup

The Patching Setup page is for all admins, and therefore is located under the Setup menu.

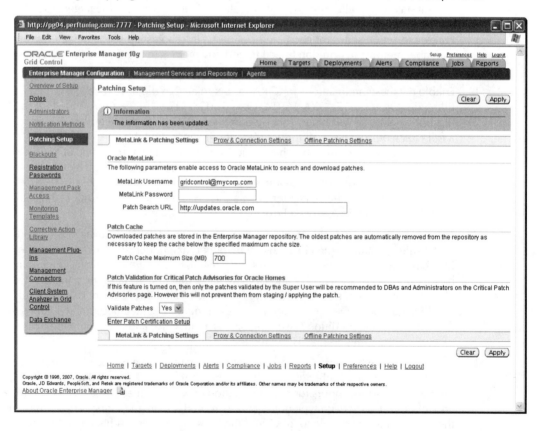

Chapter 10: General Console Configuration   **357**

On this page you specify parameters to enable GC's patching features. These features include staging and applying Oracle patches and patch sets to target hosts and collecting critical patch advisory information for managed targets. The Patching Setup page has three pages for configuring patching parameters:

- MetaLink & Patching Settings
- Proxy & Connection Settings
- Offline Patching Settings

Let's examine the parameters you can set on each of these pages.

MetaLink & Patching Settings

Use the MetaLink & Patching Settings page to configure Oracle*MetaLink* credentials, the *Patch Cache* (a location in the Repository for downloading patches), and the *patch validation feature*, which allows super administrators to validate patches appearing on the Critical Patch Advisories page. As shown above, the MetaLink & Patching Settings page has three sections corresponding to these features:

- **Oracle*MetaLink***   Grid Control uses the credentials listed here to search for Oracle patches an administrator requests and to collect Oracle critical patch advisory information. All credentials fields will be specified if you entered credentials during GC installation on the Specify Optional Configuration page (see Chapter 3, Figure 3-7). The Patch Search URL is pre-filled by the GC installation software with the correct Oracle*MetaLink* web site address at the time. You may need to change this URL in the future if Oracle changes their web site location.

TIP
Create a dedicated MetaLink account not tied to a particular administrator. This allows you to keep the account open indefinitely and reduces the possibility of someone inadvertently changing the account's password.

- **Patch Cache**   Specify the maximum size in MB for the Patch Cache, which works on the principle of First In First Out (FIFO). In other words, GC removes patches to keep the Patch Cache the specified size, beginning with the oldest automatically added patch.[37] Those patches you manually add must be manually removed.[38]

---

[37] When you run a patch job, Grid Control automatically downloads the patch and adds it to the Patch Cache, provided it has access to MetaLink.

[38] Manually adding a patch is useful when you currently have Internet access to reach MetaLink to stage a patch, but know you won't have Internet access later when running the patch job. To manually add a patch, on the Deployments page, click View Patch Cache, then the Upload Patch File button.

■ **Patch Validation for Critical Patch Advisories for Oracle Homes** The Patch Validation feature allows super administrators to filter patches and patch sets that appear on the Critical Patch Advisories page, although it does not prevent admins from staging or applying a patch. (A link to this page, called Patch Advisories, exists on both the home page and the Deployments page.) The patch validation feature is set to "No" by default. When enabling it by selecting Yes in the Validate Patches field, a link called Enter Patch Certification Setup appears as shown at the bottom of the above illustration. Clicking on this link brings up the following page.

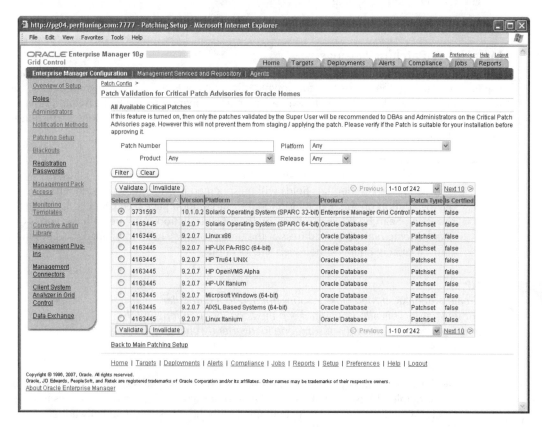

Validate or invalidate specific patches using the respective buttons. Click the Back to Main Patching Setup link at the bottom of the page, then click the Apply button to save your configuration.

Proxy & Connection Settings

If the OMS requires a proxy server to make an Internet connection to MetaLink, the Proxy & Connection Settings page allows you to enter and test proxy settings, which are stored in the Repository.

Chapter 10: General Console Configuration

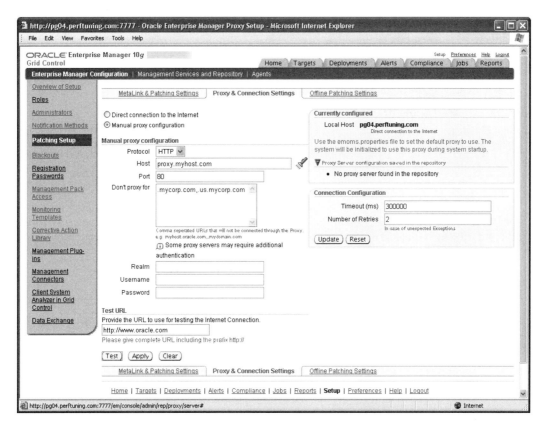

Proxy data can be set at two levels: in the Repository via this UI and/or in the *$OMS_HOME/sysman/config/emoms.properties* file. When the OMS starts, it reads from one or both places to get the proxy data, depending on where it is defined and its GC version.[39] Starting with GC 10.2.0.4, the OMS reads proxy data first from the Repository; if the proxy data is not found, the OMS reads it from the *emoms.properties* file. As soon as proxy data is set in the Console, the Repository is used as the single source, and proxy data set in *emoms.properties* is ignored.

This page is populated at GC installation time with your responses on the Specify Optional Configuration page (see Chapter 3, Figure 3-7) and has four sections:

- **Manual proxy configuration** This section only appears if you select "Manual proxy configuration." Following are the field names and descriptions in this section and their corresponding property names in *emoms.properties* (shown in parenthesis):

    - **Protocol** HTTP is the only protocol you can select.
    - **Host** (*proxyHost*) Enter the host name or IP address of the proxy server.
    - **Port** (*proxyPort*) Port number of the proxy service. Default value is 80.

---

[39] See Note 471842.1 for details on all proxy configurations and GC versions.

- **Don't proxy for** (*dontProxyFor*) Identifies URL domains for which the proxy will not be used.

- **Realm** (*proxyRealm*) Indicates the protected space that requires authentication. This shows up in a login window when you're using IE to access a web site that needs to go through a proxy.

- **Username** (*proxyUser*) Username for proxy authentication.

- **Password** (*proxyPwd*) Password for proxy authentication.

The input in the above illustration is for a proxy server requiring no authentication. To test your proxy server input, click Test, or to test and save it, click Apply. In either case, the OMS host must be able to contact the specified proxy server or it will not save your input.[40]

- **Test URL**   This section allows you to provide a URL for testing your Internet connection, which is required to reach the MetaLink web site. You must specify the *http://* part of the address, and you should use the default URL, *http://www.oracle.com*.

- **Currently configured**   This section reflects the current configuration in *emoms.properties* for the indicated host. Expand the "+" icon to see the current proxy server configuration stored in the OMR Database. Proxy data saved in the "Manual proxy configuration" section on this page should appear here.

- **Connection configuration**   Use this section to change the default properties for timeout (default 5 minutes) and number of retries (default 2) Grid Control uses to attempt a connection to MetaLink.

To test and save all input on this page, click Apply.

Offline Patching Settings

Grid Control allows you to patch Oracle software in Connected Mode (online) or Disconnect Mode (offline). Offline patching allows you to perform all patching tasks using the Patch Cache when no Internet connection exists for any OMS hosts to connect to MetaLink. Offline patching is applicable for GC sites with very tight security or when deploying new OMS servers that have not yet been configured on the company network.

The Offline Patching Settings page consists of three sections, as shown here.

---

[40] Grid Control will return an error that you are not connected to a DNS or that the provided host name is not correct.

Chapter 10: General Console Configuration 361

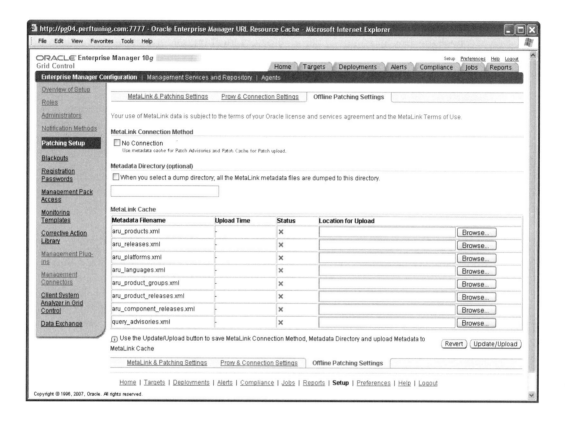

- **MetaLink Connection Method**   Select the No Connection box if you're not patching Oracle software in Connected Mode. This box is not checked by default. If you leave the box unchecked, no changes are required on this page. If you check the box, you must download the patch metadata in one of the following ways:[41]

    - Proceed to the next section "Metadata Directory (optional)" and follow the directions there.

    - Go directly to the metadata URLs through a browser and dump the XML files to a local directory on your PC. You then use the MetaLink Cache section below to specify the locations of each metadata XML file.

    - Download the metadata using the command-line utility.

    Whether you check the No Connection box or not, Oracle recommends running the RefreshFromMetalink job on a daily basis to enable Grid Control to collect all MetaLink metadata, analyze it, update Repository tables, and determine any new policy violations. If you are in Disconnected Mode, the RefreshFromMetalink job collects the metadata from the Grid Control MetaLink Cache rather than from MetaLink. See the following section titled "Schedule the RefreshFromMetalink Job" for instructions.

---

[41] From the Oracle Enterprise Manager Online Help for Offline Patching Settings.

- **Metadata Directory (optional)** If you checked the No Connection box, optionally check this box and enter a shared directory location (rather than using the default metadata cache location). This location must be accessible from the local OMS host and must map to a server with Internet access. Grid Control will use this server to download patch metadata from MetaLink.

- **MetaLink Cache** This section is for specifying a MetaLink Cache location (rather than the default metadata cache location in the Repository) where you have downloaded metadata files to the client machine from which you have launched the GC Console. For each named XML file, click the Browse button, select the corresponding metadata XML file location on your client machine, and click Update/Upload.

After completing this Offline Patching Settings page, click Update/Upload to save your changes to any of its three sections, or click Revert to roll back your changes.

Schedule the RefreshFromMetalink Job

The RefreshFromMetalink job downloads Critical Patch metadata from MetaLink, uploads it to the Repository, evaluates it with respect to current patch levels of GC components and other targets, and accordingly updates the Patch Advisories page in the Console. As such, this job is dependent on your having correctly set MetaLink credentials (and proxy server settings (if the OMS needs a proxy server for Internet access).

As of GC 10.2.0.4, the RefreshFromMetalink job is automatically scheduled on a daily basis. However, if you're running a GC release prior to GC 10.2.0.4, you must set up a recurring RefreshFromMetalink job as shown below. I don't provide any screen shots because it's a very simple process. Also, most of you will be running GC 10.2.0.4 and don't even need to perform this procedure. However, if you want to see the UI for the job system, see "RMAN Script Job" in Chapter 14.

1. Log in as a super administrator and select the Jobs tab. In the Create Job field, select RefreshFromMetalink from the pull-down menu and click Go. Note that this job does not require any input on the Parameters or Credentials pages.

2. On the General tab, enter a Name (REFRESH FROM METALINK) and Description (Refresh patch metadata from OracleMetaLink) for the job.

3. On the Schedule tab, enter a daily repeating schedule. The Frequency Type is "By Days", Repeat Every is "1 Days," and both Grace Period and Repeat Until are "Indefinite." Choose a time zone, start date and start time.

4. On the Access tab, choose to be notified only if the job status changes to Suspended or Problems.

5. Click Save to Library. Grid Control will confirm the job was created successfully.

6. To submit the job, click the Job Library subtab, select the RefreshFromMetalink job you just created, and click Submit. (Note that ".1" is appended to the job name to indicate the instance of the job. This number will increment each time the job runs, which should only take a few minutes.)

Summary

This chapter begins and propels us far into the reaches of general GC configuration in the Console. I demonstrate the UI input required for all configuration steps, and, along the way, cite practical examples and point out best practices. The configuration process deals with two menus: the Setup menu for global configuration and the Preferences menu for local administrator-specific settings.

The chapter has four major parts:

- Introduction
- Set up administrators
- Configure notifications
- Enable patching features

The rather lengthy introduction begins with a general sweep of all Setup and Preferences menu items. It then prepares you to configure notifications for administrators by offering a functional diagram that traces how an event trickles down to a notification. The diagram also crystallizes the interrelationship between two key menu items: notification rules and notification schedules. The intro concludes by deriving a viable order in which to accomplish GC configuration with minimal back-and-forth between screens. This brings you to the business of general setup of administrators, notifications, and patching.

Administrator setup begins by creating roles, which are useful for grouping GC system and target privileges that admins share. Roles and admins are defined by these same privileges, so it's easier to create admins with common privileges using roles than it is to create admins individually. Once you create administrator accounts, you can tell them to follow the remaining instructions in this section to configure their own accounts. This entails entering their individual preferences, including choosing e-mail accounts and customizing target subtabs, then entering and verifying their preferred credentials, either manually or automatically. They can programmatically set preferred credentials using EM CLI and verify their credentials using the GC features, Execute Host Command and Execute SQL.

Administrators can continue the general Console configuration process by setting up notifications. This setup entails three menu items: Notification Methods, Notification Schedules, and Notification Rules. The super administrator can define notification methods. E-mail is the principal method and is typically configured by specifying a mail server when installing Grid Control. On the Notification Methods page, you can identify additional mail servers for notification highly availability. The page also provides four other methods—OS scripts, PL/SQL procedures, SNMP traps, and Java callbacks—intended primarily to integrate GC notifications with third-party applications. The second aspect of notification setup is for each admin to define a schedule for receiving e-mail notifications. Notification schedules, which are modeled on the idea of a rotation frequency, are highly configurable, and can be suspended to reflect exceptions, such as all-too-needed DBA vacation times. The final step consists of notification rules, which are local in nature in that admins individually subscribe to them. However, admins can share rules they create by making them public. Each rule defines a set of targets, target availability states, metrics, policies, job status changes, and one or more notification methods for alerting the admin.

The final part of this chapter concerns ensuring that global GC patching features are enabled. A super administrator must check that the patching setup is to his or her liking. This setup consists of MetaLink account details (configurable when installing Grid Control), proxy and connection settings if the OMS needs a proxy server to contact MetaLink, and offline patching settings. This concludes patching setup for GC 10.2.0.4 or higher systems. For prior GC releases, however, you must schedule the RefreshFromMetalink job to run on a daily basis. (This job is automatically configured to run on GC 10.2.0.4.)

In the next few chapters, we discover and configure all remaining targets that Grid Control is to manage, set metrics and policies for these targets, and delve into backup and recovery, high availability and disaster recovery, security, and maintenance and tuning to ensure Grid Control is always available to help manage your IT environment.

CHAPTER 11

Configure Target Monitoring

While the last chapter dealt with general Console configuration, this current chapter focuses more on target monitoring. The sections are:

- Remove targets from monitoring
- Discover and configure targets for monitoring
- Grant management pack access
- Schedule blackouts for planned downtimes

The first two sections, forming the bulk of the chapter, deal with how to remove targets not requiring GC monitoring, then discover and configure certain commonly used GC target types: Agent, ASM, Database Instance, Cluster Database, Listener, Host, and Oracle Application Server. (In Chapter 12 we add Groups, Systems, and Services[1] to the fold.) The idea in this chapter is to bring the target list up to date for these target types. In other words, remove targets not requiring monitoring and discover (i.e., add) and set up any undiscovered targets that do require monitoring so that Grid Control collects metrics and sends notifications only on the targets that matter to administrators. You may need to add or remove quite a few targets in your environment, usually in direct proportion to the age of your GC installation. Grid Control is not like wine; it's like an engine—it needs to run or it begins to rust. For instance, if the OMS is down, collected Agent data cannot be uploaded, and Agent collections are eventually disabled.[2] If you don't patch GC, unresolved metric collection errors can cripple OMS loader throughput. If you're in the situation of taking over a dormant GC system, this chapter will come in handy because there will likely be lots of targets to add and remove.

After updating your target list, the next step is to grant management pack access to these targets. This step is well timed because it saves you from having to bounce back and forth between adding a target and granting it pack access. (Whenever you add a new target, you must grant management pack access specifically to that target.)

The final step is to schedule blackout jobs for targets brought down for routine maintenance or other reasons. When you intentionally stop targets, a blackout avoids false DOWN notifications, which increases the accuracy of target performance reporting. Metric collections are also disabled for blacked out targets, which eliminates spurious Agent uploads during maintenance periods. I go through the process of setting a blackout both in the Console and at the command line, and explain the benefits of each method.

Let's start with the best way to simplify anything, including Grid Control: remove what you don't need (i.e., targets that don't need to be monitored). Let's shrink our downstream work by getting rid of any dead weight upstream.

NOTE
Unless otherwise indicated, you can perform most of the steps in this chapter logged in as any super administrator. However, for all steps, I advise using your particular super administrator account created in Chapter 10 (there named GC_SUPER1) because some steps rely on preferred credentials.

---

[1] Services are built upon Systems, and a Web Application is one type of service covered in depth in Chapter 12.

[2] For details, see "How Agent Files Are Uploaded" in Appendix A (online at www.oraclepressbooks.com).

Remove Targets from Monitoring

Sometimes you may need to remove some or all targets (databases, listeners, application servers, etc.) from GC management, perhaps even the host and Agent itself. There are several reasons for needing to remove a target from GC monitoring:

- To remove nonproduction targets from a production GC environment or vice-versa.
- To remove targets not used, such as a dormant database RAC listener on a host where an ASM listener in a separate Oracle home services the databases on that node.
- To decommission or move a target to another server.
- To resolve a target status that is not shown correctly within the Console. This solution is employed only when bouncing or *resetting* the Agent does not fix the status mismatch. (Resetting an Agent is a method for resolving an OMS/Agent sync problem that causes the OMS to reject Agent uploads. For details, see "Resetting Agents" in Chapter 14.)
- To fix a target metric error that cannot be resolved any other way, even after resetting an Agent.

Follow the steps below to remove targets from Grid Control. I provide the methods to remove targets both in the Console and at the command line.

Ensure Agents Are Started

Make sure all Agents are running which are managing targets slated for removal, in one of the following ways:

- In the Console, check all Agent statuses at once by clicking Setup, then clicking the Agents subtab.
- In the Console, check all Agent statuses at once from the All Targets subtab under the Targets tab. Filter out all Agent targets by selecting Agent in the Search field and clicking Go.
- In the Console, check each Agent's status on each Agent's home page.
- Execute *emctl status agent* on each host that contains a target to be removed. If such an Agent is down, start it on the host by executing the OS command *$AGENT_HOME/bin/emctl start agent*.

Back Up targets.xml File

Back up the Agent inventory file, *$AGENT_HOME/sysman/emd/targets.xml*, on each Agent host where targets are to be removed. This is particularly important for the *targets.xml* file in the chain-installed Agent home located on OMS hosts, as rediscovering OMS targets that have been removed can be problematic.

Confirm No Targets Are in Blackout

Make sure no targets to be removed are in blackout, by using any one of the following methods:

- In the Console, on the Setup menu, click the Blackouts link. By default, the Blackouts page displays any active blackouts.

- Issue the command *emctl status blackout* on each Agent host. If no blackouts exist, the output of this command is "No Blackout registered."
- In the Console, check the home page of each target. The Status for a target will be reported as Under Blackout if the target is in blackout.

Either wait for the blackout to complete or stop the blackout. You can easily figure out how to stop blackouts by reading the section "Schedule Blackouts for Planned Downtimes" later in this chapter.

Remove Targets from Grid Control

Removing targets from Grid Control does not remove the targets themselves or change these targets in any way, but simply removes them from GC management. You can even rediscover these targets again in GC, if desired. The process described here removes target information on both the Agent host and in the Repository Database as required. On the Agent host, the OMS notifies the Agent monitoring the targets to remove target entries from the Agent inventory file (*targets.xml*) and to remove the special collection files and state files associated with the targets. In the Repository Database, the OMS removes the targets and their associated metadata and historical collection information from SYSMAN tables.

You can remove targets either from the Console or at the command line. Target removal in the UI is the recommended method. The command line method provides a way to script the removal of multiple targets, even through the GC job system. I've also found that the command line method usually works in the unlikely event that the UI method fails. Let's look at each of these methods.

Remove Targets in the Console

Remove targets in the Console as follows:[3]

1. Navigate to the home page for the host containing the targets to remove. (Click the Targets tab, the Hosts subtab, then click the link for the host under the Name column.)
2. On the host's home page, click the Targets page, which brings up only those targets located on the host.

---

[3] You can also remove targets one-by-one from the All Targets subtab by selecting the target and clicking Remove. However, the method I describe is safer because it better ensures you're on the right host, and is faster because you can remove multiple targets on a particular host in a single operation.

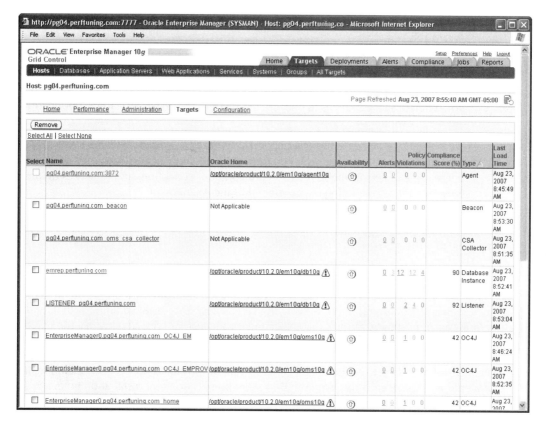

3. Choose the desired targets to remove in the Select column and click Remove. Reconfirm when prompted. (If you're removing a RAC cluster, you need to remove the Cluster Database target, then remove the Cluster in a separate operation.)

4. Repeat the procedure for each host containing a target to be removed.

Remove Targets at the Command Line

As already mentioned, the recommended method to remove targets is in the Console. However, you can also remove targets at the command line. I would only do this in two cases: if you have a lot of targets to remove and want to script it in the GC job system, or if the above UI process fails. Here is the procedure:

1. Query the Repository Database in SQL*Plus as SYSMAN to determine the exact target names and target types, both of which you must specify when removing these targets in step 2 below.

    SELECT HOST_NAME, TARGET_NAME, TARGET_TYPE FROM MGMT_TARGETS WHERE EMD_URL LIKE '%<Agent_hostname>%' ORDER BY 1,2,3;

2. Remove the targets by executing the following procedure in the Repository Database as SYSMAN:

   EXEC MGMT_ADMIN.DELETE_TARGET('<target_name>', '<target_type>');

   On rare occasions, I've observed that a database lock involving the DELETE_TARGET procedure causes it to hang. The solution allowing the DELETE_TARGET procedure to complete successfully is to execute the DELETE_TARGET_INTERNAL procedure in another session:

   EXEC MGMT_ADMIN.DELETE_TARGET_INTERNAL('<target_name>','<target_type>');

3. Having removed the targets directly from the Repository as just described, also remove these same targets from the Agent's configuration, collection, and state files on the Agent host by running the following command as the Agent owner:

   emctl config agent deletetarget <target_type> <target_name>

   where <target_type> and <target_name> are returned from the query in step 1 above.

Discover and Configure Targets for Monitoring

As mentioned in Chapter 1, as soon as you install an Agent, it starts and begins automatically discovering and monitoring itself, its host, any GC components installed along with it, and Oracle single-instance databases, clusters, listeners, and Application Servers already installed on the host. You must manually discover other target types. Auto-discovery occurs only the first time the Agent starts, not every time it starts. Therefore, after deploying an Agent on a host, you must manually discover *all* subsequent targets in Grid Control[4] installed on the host, including those that would have been auto-discovered if they were present at Agent installation time.

NOTE
Discovering a new target in Grid Control requires that the target already be installed on a discovered host that is running an Agent. By discovering a new target, you are not installing the target itself on the host, but rather are placing an existing target under GC management.

In this section, you learn how to discover and/or additionally configure six GC target types for monitoring (see Table 11-1).

Remember, you cannot discover an Agent or a host through the Console. When the Agent starts for the first time, it begins communicating with the OMS, the Agent and Host target are automatically discovered in the Console, and if you click the Configure button for them, you receive a kindly message that "there are no editable configuration properties" for these targets. The sections for Agent and Host target types cover additional configuration you may need to perform.

---

[4] If target discovery through the Console fails, you can also try using the Agent Configuration Assistant, *agentca*. See Oracle Enterprise Manager Grid Control Installation and Basic Configuration 10*g* Release 4 (10.2.0.4.0), Section 9.5.

Target Type	Auto-Discovered?	Must Configure?	Optional Configuration Steps?
Agent	Yes	No	Yes
ASM	Yes (when password file exists)	Yes	No
Database (Databse Instance, Cluster Database, and Cluster)	Yes (except Cluster Database)	Yes	Yes
Listener	Yes	No (unless secured)	No
Host	Yes	No	Yes
Oracle Application Server	Yes	No	No

TABLE 11-1. *Target Types Auto-Discovered, Requiring Configuration, or Additional Setup*

Although Grid Control can monitor over 100 target types, many sites manage little more than the six basic target types listed in Table 11-1. "Not that there's anything wrong with that."[5] These target types are the bread and butter of most Oracle shops.

In the next section, you learn three ways to navigate to the screen where you can discover or configure a target. These ways apply to all targets shown in Table 11-1, except Agent and Host. After this, subsequent sections explain how to discover and/or configure each of the six target types. Some of the configuration is required, and some is optional or conditional based on your environment.

Three Paths to Discover/Configure a Target

You can discover or configure targets from one of three pages:

- Agent home page
- All Targets subtab
- Subtab for the target type

I list these pages in order of desirability. For example, only the Agent home page can discover any type of target and is the easiest to use because it doesn't prompt you to select the host name on which to discover the target. I mention all three ways to illustrate that there are often several ways to navigate in the Console to achieve your goal and to give you a brief tour of frequently used Console pages. To demonstrate each of these three paths, take the example of discovering a new database instance on host name *ptc1*.

---

[5] Seinfeld.

372 Oracle Enterprise Manager 10g Grid Control Implementation Guide

Discover/Configure Targets from the Agent Home Page
To add or configure a target from the Agent home page, do the following:

1. As a super administrator, go to the Agent home page for the host (*ptc1*) where the installed target resides.

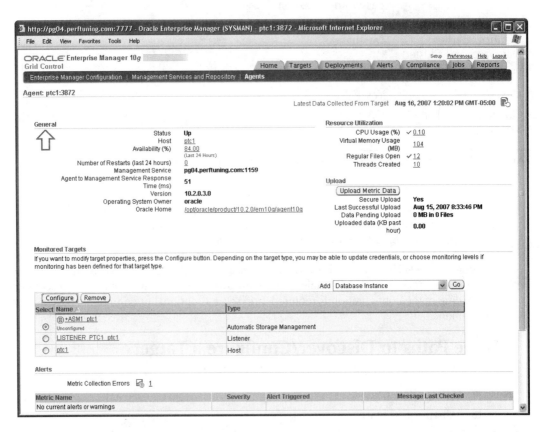

The Agent name in the Console is of the form *<Agent host>:<Agent port>*, where the default port is 3872. (In this case the Agent name is *ptc1:3872*.) You can reach an Agent home page in one of three ways:

- On the Console home page (the Home tab), in the Search field, select Agent and click Go. Click the desired Agent under the Name column.

- Click the Targets tab, All Targets subtab, and click the link for the desired Agent under the Name column.

- Click the Setup link located at the top right of the Console home page (and many Console screens). Click the Agents subtab and click the link for the desired Agent under the Name column.

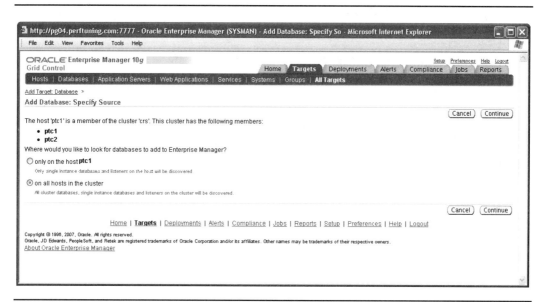

FIGURE 11-1. *The Add Database screen*

2. Once you're on the Agent home page, discover the target by selecting the target type (Database Instance in this example) in the Add field and clicking Go; configure the target by clicking the radio button in the Select column next to the target and clicking Configure.

3. The appropriate screen for the target type appears.[6] In the case of adding a database instance, the Add Database screen appears (see Figure 11-1). Complete this and subsequent screens for the target type you are discovering or configuring.

Discover/Configure Targets from the All Targets Subtab

In lieu of using the Agent home page, you can use the All Targets page to discover or configure many target types[7], predominantly aggregate ones, as described in Chapter 12.

1. Log on as a super administrator and navigate to the All Targets tab, as shown in the image on the following page.

---

[6] For more complicated targets (such as a Cluster Database), a wizard guides you through several screens to complete the target discovery process.

[7] You cannot add all target types here, such as Automatic Storage Management.

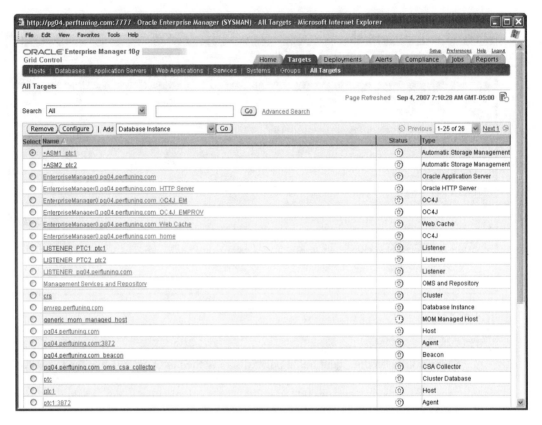

2. To discover a target, in the Add field, select the target type from the drop-down menu (Database Instance here) and click Go. To configure a target, select it and click Configure.

3. In the case of discovering a database instance, the screen shown in Figure 11-2 appears. Enter the host (*ptc1* in this case) where the target to be discovered resides, either directly or by clicking the flashlight icon and selecting it from a list of hosts, then click Continue. This additional step is required because you did not start out on the Agent home page for a particular host.

4. The Add Database screen appears as shown above in Figure 11-1. Complete this and subsequent screens for the target type you are discovering.

Discover/Configure Targets from the Target Type Subtab

The Targets tab contains subtabs for only certain target types as shown in Figure 11-3. As you can see, these subtabs are for Hosts (which need no configuration), Databases, Application Servers, Web Applications, Services, and Systems. (You must use one of the above two methods to add other target types.)

Chapter 11: Configure Target Monitoring **375**

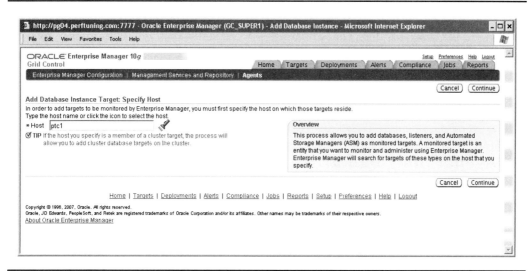

FIGURE 11-2. *The Add Database Instance Target: Specify Host screen*

To discover or configure one of these target types, do the following:

1. As a super administrator, click the Targets tab, then click the subtab for the target type you are discovering/configuring. This brings up the summary page for that target type (the Databases summary page [shown in Figure 11-3], in this case).

2. Click the Add or Configure button, as desired. For this example, click Add.

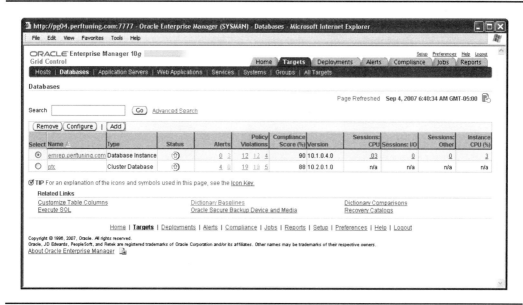

FIGURE 11-3. *The Databases summary page*

3. The screen shown in Figure 11-2 above appears. Enter the host where the target to be discovered resides, either directly or by clicking the flashlight icon and selecting it from a list of hosts, then click Continue.

4. The Add Database screen appears as shown above in Figure 11-1. Complete this and subsequent screens for the target type you are discovering.

The following sections explain how to discover (where applicable) and set up each of the six target types mentioned above. Do not forget to perform general Console configuration steps for newly discovered targets as covered in Chapter 10, such as to set preferred credentials.

Agent Configuration

As already mentioned, Grid Control automatically discovers and configures Agents to begin monitoring themselves and certain Oracle components installed on the host where they exist. However, you may need to perform two additional steps to set up Agents:

- Enable multi-inventory support (UNIX only)
- Enable notification if Grid Control goes down

Enable Multi-Inventory Support (UNIX Only)

Following is the procedure to ensure Grid Control has access to all inventories associated with discovered Oracle homes on target hosts. This is usually the case on Windows targets. However, on UNIX hosts, if you use a separate inventory for each product rather than a Central Inventory, you need to configure GC to discover each inventory located in a nondefault location. GC can display the contents of inventories it knows about and its patch apply feature will update these inventories accordingly. This section explains how to discover these separate inventories in Grid Control.

Refresher on Oracle Inventories First, let me provide a little refresher on inventories. As I explained in Chapter 2 in "Create Oracle Inventory Group (oinstall)," the Oracle Universal Installer (OUI) creates and uses a Central Inventory to store information about the Oracle software it installs on a host.[8] On Windows, the default location for the Oracle Central Inventory is *%SystemDrive%:Program Files\Oracle\Inventory* and cannot be changed. As such, a Windows target host has one Central Inventory. However, on UNIX hosts, the location of the Central Inventory is specified in the inventory pointer file, *oraInst.loc*, located in */etc/* on Linux and AIX, and in */var/opt/oracle/* on Solaris and HP-UX. The inventory pointer file consists of two lines following the example format here:

```
inventory_loc=/u01/app/oracle/oraInventory
inst_group=oinstall
```

The Central Inventory directory is usually *$ORACLE_BASE/OraInventory*, as shown above for the *inventory_loc* value.

I advised in Chapter 2 under the section, "Choose the Oracle Inventory Directory," that you have two choices for managing inventories on UNIX hosts: use a Central Inventory or a separate inventory for each product. I mentioned that the recommended practice was to use a Central Inventory, but that if you use a separate inventory, you need to configure GC to discover each inventory located in

---

[8] The OUI also creates a local *inventory* directory under each Oracle product home, which describes the particular product installed in that home; this is not the inventory I am addressing here.

Oracle Product Installed	Inventory Pointer Location File	inventory_loc[9] Value in File
10*g* Grid Control Agent	oraInst.loc.agent10*g*	/export/home/oracle/oraInventory.agent10*g*
9.2 Oracle Database	oraInst.loc.db92	/export/home/oracle/oraInventory.db92
10.1.3 Oracle Application Server	oraInst.loc.as1013	/export/home/oracle/oraInventory.as1013

TABLE 11-2. *Example Inventory Locations on Host with Multiple Inventories*

a nondefault location (other than that pointed to by the default *oraInst.loc* file). You may recently have installed such a separate inventory when deploying a standalone Agent, for example.

How to Find Nondefault Inventories Grid Control doesn't go searching for these nondefault inventory locations by itself; you must point GC to these locations. But you yourself may not know where these inventories are. If this is the case, use one of the means below to identify the Inventory location pointer file for each Oracle product's missing inventory and the missing *OraInventory* directory itself. Make note of these locations because you need them to enable multi-inventory support as described in the next section.

- You may know either where the pointer file or inventory is located because you installed the Oracle product in question or are familiar with the installation.

- Your company has a standard for where pointer files and inventories are located, or uses the default directory locations. Typically, DBAs store pointer files in the *default* directory for their platform. The default *oraInventory* directory location is under $ORACLE_BASE. This example follows such a practice.

- Check the Oracle home directory of the product installed for which Grid Control is missing the inventory. Many Oracle products, including the GC Agent itself, install an *oraInst.loc* file in this location as well as in the default location for your platform.

- Check the home directory of the owner of the Oracle product, as some products create both the *oraInventory* directory and *oraInst.loc* file in this location.

If you can't find the inventory pointer file for the Oracle product, but you can locate its *oraInventory* directory, don't worry. Just create a new two-line pointer file in the standard location for such files at your site to indicate this inventory directory.[10]

How to Enable Multi-Inventory Support Now let's get to the business at hand: how to configure Grid Control for separate inventories. Regardless of whether you employ a Central Inventory or separate inventories on target UNIX hosts, check that GC has accounted for all inventories associated with discovered Oracle homes using the procedure below. Let's examine the procedure using the case of a site with a Solaris host *ptcsun.perftuning.com* running the three Oracle products listed in Table 11-2.

---

[9] The *inventory_loc* parameter points to the location of the *OraInventory* directory for the Oracle product in question.
[10] One such product is a non-NFS Agent created via Agent Deploy (see Chapter 6, "Agent Deploy Post-Installation").

Each product uses a dedicated inventory pointer file named after that product and located in the */var/opt/oracle/* directory on the host (the default location for Solaris). Let's assume that the site has already followed the procedure below to account for the Agent and database inventories, but their GC environment is missing the inventory for the OracleAS target named *sunas1013.ptcsun. perftuning.com*.

Here is the procedure to add the missing AS10*g* inventory to Grid Control:

1. Click the Deployments tab. In the View field, note whether Software Targets Without Inventory appears when selecting Database Installations or Application Server installations. (Here, Grid Control is missing 1 of the 2 AS target inventories in the entire environment.[11])

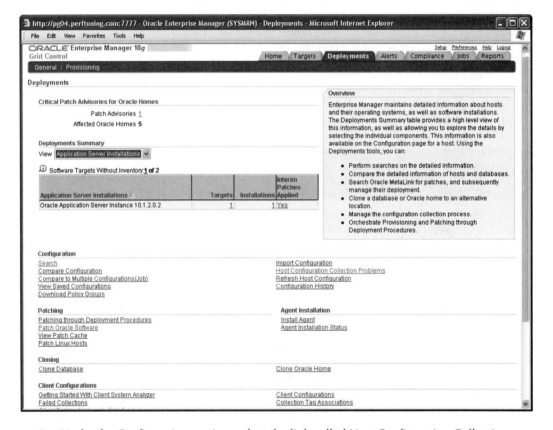

2. Under the Configuration section, select the link called Host Configuration Collection Problems, which brings you to this page.

---

[11] You can click the number of targets (1 in this case) to jump directly to the screen shown in step 2 (the Host Configuration Collection Problems screen).

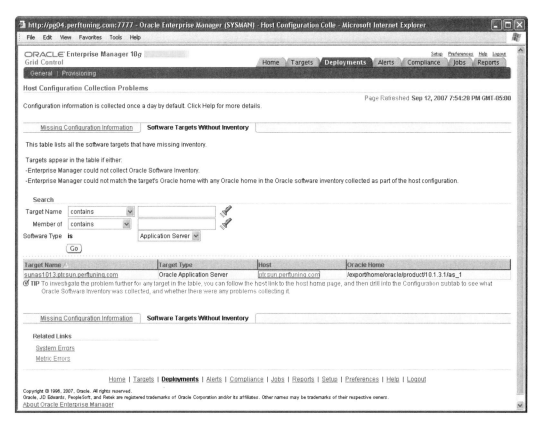

On the Software Targets Without Inventory page as shown above, view any missing inventories (note the missing inventory for the OracleAS target, named *sunas1013.ptcsun.perftuning.com*).

3. Use one of the means described above in the section, "How to Find Nondefault Inventories," to identify the Inventory location pointer file for each Oracle product's missing inventory and the missing *OraInventory* directory itself. In the example, assume we find that the pointer file for the missing OracleAS Inventory is */var/opt/oracle/oraInst.loc.as1013*, and is comprised of two lines:

   inventory_loc=/export/home/oracle/oraInventory.as1013
   inst_group=oinstall

4. Specify the path to the nondefault inventory pointer file found in step 3 in the Agent configuration file *$AGENT_HOME/sysman/config/OUIinventories.add* located on the host reported in step 2 with the missing inventory. Add a line of the following form:

   inventory: <inventory_pointer_location_file>

   In our example, the *OUIInventories.add* file already contains the first two lines (the site already added them), and you must add the third **bolded** line:

   inventory: /var/opt/oracle/oraInst.loc.agent10g
   inventory: /var/opt/oracle/oraInst.loc.db92
   inventory: /var/opt/oracle/oraInst.loc.as1013

5. To effect this change in the configuration file, reload the Agent on the target host:

   `emctl reload agent`

6. Manually refresh the host configuration so that the Console can immediately show the details of this OracleAS home with the nondefault Inventory. Go to the Targets tab, the Hosts subtab, and select the host link. Click the Configuration subtab, and click Refresh at the top right of the page, the results of which are shown here.[12] Otherwise, you must wait for Grid Control to refresh the host configuration information, which occurs every 24 hours.

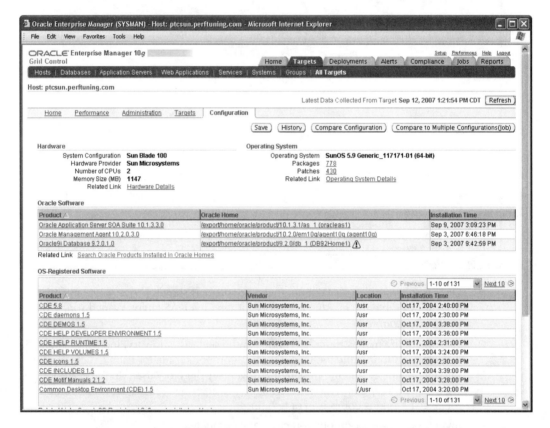

You receive a message that the host configuration is now being refreshed, along with a nifty clock with a sweeping hand.

7. Confirm that Grid Control now discovers the nondefault inventory. Click the Deployments tab. Under Deployments Summary, in the View field, select the appropriate deployment type for the formerly missing target: either Database Installations or Application Server Installations (the latter in this example).

---

[12] After you click Refresh, the Host page should list the previously missing Oracle product under the Oracle software section, as shown above for the Product "Oracle Application Server SOA Suite 10.1.3.3.0."

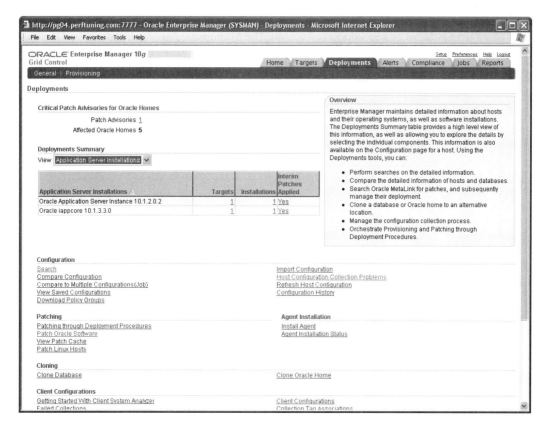

A new line should appear showing the target inventory, which for this example is an Oracle Application Server Installation called "Oracle iappcore 10.1.3.3.0." (This line did not appear in the illustration shown in step 1 above.)

Enable Notification if Grid Control Goes Down

If Grid Control is monitoring your entire IT environment and goes down unexpectedly, you need to know about it right away. Otherwise, you wouldn't receive alerts for any targets at your site. If Grid Control itself goes down, its self-monitoring capability can alert you of this all-important fact, but it requires a little configuration for some platforms. Self-monitoring is the responsibility of the OMS host containing the *Monitoring Agent*, which relies on the *mailx* utility to send you notification of a complete GC outage. Unfortunately, if your OMS platform does not supply *mailx*—which is certainly the case on Windows and is also the case for some UNIX flavors—then you need to configure another mail utility to send you notification. Following are details on what a Monitoring Agent is, how it determines that Grid Control is down, what built-in GC script calls *mailx*, how to fix this script if *mailx* doesn't exist for your OMS platform, and how to specify the e-mail address and mail gateway for receiving notification of a GC outage.

Grid Control monitors its own availability via scripts in conjunction with the Monitoring Agent. The Monitoring Agent is the Agent on the first OMS host installed, and is responsible for checking the availability of both the Management Repository Database and all Management Services.

You can identify the Monitoring Agent for your GC installation by clicking the Setup link, then the Management Services and Repository subtab. The Monitoring Agent is listed on the Overview page under the General section.

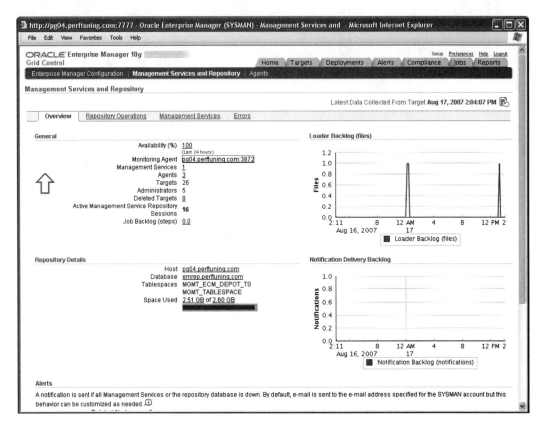

Here, the Monitoring Agent is *pg04.perftuning.com:3872*. The Management Repository and Management Services are logically modeled as a single target called the "Management Services and Repository" target.[13] The Monitoring Agent detects whether the Management Repository Database is down or whether all Management Services are down by running a script every five minutes called *$AGENT_HOME/sysman/admin/scripts/emrepresp.pl*. This script runs as an *out-of-band (OOB) notification* mechanism (i.e., it is not dependent on Grid Control notification methods), a requirement because this regular mechanism would be unavailable if the Management Services were down. If this script determines that the Repository Database is down or that all Management Services are unavailable, it calls a script named *$AGENT_HOME/bin/emrepdown.pl*. By default, the notification script will send e-mail via the *mailx* executable with the error message "Could not connect to Enterprise Manager Repository Database..." or "No active Management Services were found."

---

[13] During Repository creation, this target and the EM Website target, listed under the Web Applications tab, are created to monitor the overall health of your Grid Control environment.

The script runs without modification only on those platforms that use *mailx*, including Sun SPARC Solaris, Intel Linux, HP-UX, HP Tru64, and IBM AIX. If *mailx* is not installed on your OMS platform, change the following two bolded lines in *emrepdown.pl* to call another command line e-mail client or utility that is installed, such as *mailto.exe* or *Postie*, with the appropriate options.

```
if ("$return" eq "" )
{
    $command1 = "`mailx -s \"$subject\" $list \< $fn`";
    EMD_PERL_DEBUG("emrepdown: command1=$command1");
    system $command1;
}
else
{
    $command1 = "`mailx -s \"$subject\" -r $return $list < $fn`";
    EMD_PERL_DEBUG("emrepdown: command1=$command1");
    system $command1;
}
```

You should also specify the appropriate e-mail address and gateway information for these OOB notifications as follows in the *$AGENT_HOME/sysman/config/emd.properties* file[14] on the OMS host containing the Monitoring Agent:

- **emd_email_address**   The e-mail address to which the notification is sent. Specify one or more e-mail addresses, separated by commas, but not spaces. For example: *dba1@mycorp.com,dba2@mycorp.com*

- **emd_email_gateway**   The e-mail gateway or SMTP Server address, such as *smtp.mycorp.com*.[15]

- **emd_from_email_address**   The e-mail address that GC will identify as the sender, such as *<OMShost>@mycorp.com*. Specify any valid e-mail address for this property. When sending an OOB notification, Grid Control only validates that the SMTP server can send mail to the domain name part of this address. Send a test e-mail outside of Grid Control to independently verify that you receive e-mail at this address.

ASM Discovery and Configuration

Grid Control can manage ASM targets by way of alerting, performance monitoring, and administering of ASM disk groups. However, first you must discover (if not auto-discovered) and configure ASM targets. Grid Control typically auto-discovers an ASM instance target when you discover one of the databases using this target for storage, but only if an ASM password file already exists on the ASM node. Once discovered, configuring an ASM target simply requires supplying the ASM SYS password to Grid Control; otherwise, a metric collection error is reported

---

[14] *emd.properties* is the Agent properties file and *emoms.properties* is the OMS properties file. These files, which dictate Agent and OMS behavior, are located in the *sysman/config/* directory of their respective Oracle homes.

[15] I recommend setting this value explicitly. However, if you comment this line out, the value appears to default to the Outgoing Mail (SMTP) Server entered on the Setup page under the Notification Methods link, which corresponds to the value for *em_email_gateway* specified in the *$OMS_HOME/sysman/config/emoms.properties* file on the host containing the Monitoring Agent.

on the ASM target. Here then, are the overall steps to manually discover and configure an ASM instance for monitoring:

- Enable ASM password file authentication
- Discover and configure an ASM instance

To demonstrate the process, take the example of configuring the ASM instances named *+ASM1*[16] and *+ASM2* on hosts *ptc1* and *ptc2*, respectively. The steps shown below are for configuring *+ASM1*. Steps for *+ASM2* are not shown because they are identical.

Enable ASM Password File Authentication

Grid Control must be able to remotely access a target ASM instance with SYSDBA privilege. The reason is that, behind the scenes, Console functionality to monitor an ASM instance originates from the OMS host rendering the Console to an administrator. This equates to remote SYSDBA access, necessitating using password file authentication for an ASM managed instance. Complete the following steps to enable password file authentication:

1. Check that an ASM password file exists and create one if necessary for the ASM instance (*+ASM1*). To create a password file, issue the following OS command:

   On UNIX:

   ```
   cd $ORA_ASM_HOME/dbs
   ```

   On Windows:

   ```
   cd %ORA_ASM_HOME%/database
   ```

   On all platforms:

   orapwd file=orapw<*ASM_instance_name*> password=<*password*>

   You should use the same password for each ASM instance of a cluster.

2. If not already done, set the static initialization parameter REMOTE_LOGIN_PASSWORDFILE to EXCLUSIVE for the ASM instance, which requires bouncing the ASM instance. If you're using an SPFILE, enter the following command to set this parameter on *+ASM1*, assuming you're on UNIX (if on Windows, use *set* instead of *export*):

   ```
   export ORACLE_SID=+ASM1
   export ORACLE_HOME=$ORA_ASM_HOME
   sqlplus '/ as sysdba'
   ALTER SYSTEM SET REMOTE_LOGIN_PASSWORDFILE = EXCLUSIVE SCOPE=SPFILE;
   ```

   Then stop any databases that use this ASM instance and bounce the ASM instance.[17] Remember to do the same for all other ASM instances before restarting any dependent databases.

Discover and Configure an ASM Instance

If the ASM instance wasn't already auto-discovered during database discovery, discover it now. Then configure the ASM instance, (*+ASM1* in this case), which is simply a matter of supplying the

---

[16] The default ASM SID on RAC nodes is +ASM<*node#*>.

[17] For those unfamiliar with ASM, the procedure to stop ASM is as follows (reverse the procedure on startup): srvctl stop database -d <*db_name*>, srvctl stop asm -n <*node_name*>. (You don't need to stop the node apps to stop ASM. But for your information, the command to stop the node apps is srvctl stop nodeapps -n <*node_name*> and would be executed last.)

ASM SYS password to Grid Control. This password is stored encrypted in the Agent's *targets.xml* file[18] on the ASM host.

1. Navigate to the monitoring configuration page for the ASM instance. Do one of the following:

   - If discovering and configuring a new ASM instance, log in as a GC administrator with full host privileges on the ASM host. Navigate to the Agent home page (see "Three Paths to Discover/Configure a Target" above), select Automatic Storage Management from the pull-down menu in the Add field, and click Go.

   - If configuring an existing ASM instance, log in as a GC administrator with full target privileges on the ASM target. Navigate to the Agent home page for the host containing the ASM instance[19] (which I do here for host *ptc1*). Note that the word "Unconfigured" appears in red below the ASM target name. Select the radio button for the unconfigured ASM instance (+ASM1), as shown above in step 1 of the section "Discover/Configure Targets from the Agent Home Page," and click Configure.

2. If you're discovering and configuring a new ASM target, the Add Automatic Storage Management page appears. If you're just configuring an ASM target, the Monitoring Configuration page appears as shown here (both pages have the identical fields except that the former page has an additional Name field).

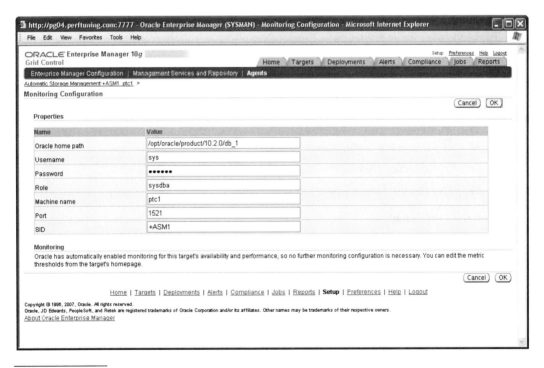

---

[18] For RAC Agents, this *targets.xml* file is located under <EMSTATE>/sysman/emd/, where <EMSTATE> contains the node name on which the Agent runs. An example EMSTATE directory for a RAC Agent is /u01/app/oracle/product/10.2.0/em10g/agent10g/<node_name>.

[19] You can also go to the All Targets page, select the radio button next to the ASM target, and click Configure.

3. Enter values for all empty fields (see example entries above), including the SYS Password for the ASM instance (*+ASM1*). If you're just configuring the ASM target, all fields except Password should already be completed, but double-check them and click OK when finished.

4. The Agent home page returns, and may still show Unconfigured, but this will disappear if you click the Refresh icon at the top right of the screen adjacent to the date for the Latest Data Collected from Target.

5. Check that Grid Control Shows the ASM Instance as UP. Check the ASM status on the Agent home page where the ASM instance is located. It shouldn't take more than 30 minutes for the ASM instance status to reflect correctly as UP. If the status does not change after this time, try resolving this by bouncing the Agent (i.e., execute *emctl stop agent* followed by *emctl start agent* on the Agent host).

Database Discovery and Configuration

At the time of Agent installation, Grid Control automatically discovers any existing Database Instance and Cluster targets but not Cluster Database targets.[20] This section shows you how to discover both single-instance databases (installed subsequent to the Agent on a host) and cluster databases in Grid Control. It also demonstrates how to run the Configure Database wizard for both single-instance and cluster database targets. Beyond this configuration, several additional steps are required or recommended as part of the database setup process. In sum, following are these steps, listed in a workable order:

- Prepare database
- Discover database
- Run Configure Database wizard
- Grant additional roles and privileges to DBSNMP
- Set up segment findings (optional)
- Create dictionary baseline (optional)
- Stop and disable Database Control (optional)

Prepare Database

There are three steps to prepare a potential target database for Grid Control management:

- Enable database password file authentication
- Start the database listener
- Address database management bugs

---

[20] Clusters and Cluster Database aggregate targets, unlike Database Instance, but set-up procedures for all of these target types are very similar.

Enable Database Password File Authentication   As for ASM targets, database targets must use exclusive password file authentication so Grid Control can monitor them. The reason is the same as for ASM targets—Console database features go through an OMS host, which is remote to a database target. To enable database password file authentication for a database instance, do the following.

1. Check that a database password file exists and create one if necessary for the database instance. To create a password file, issue the following OS command:

    On UNIX:

    cd $ORACLE_HOME/dbs

    On Windows:

    cd %ORACLE_HOME%/database

    On all platforms:

    orapwd file=orapw<i><instance_name></i> password=<i><password></i>

    You should use the same password for each RAC instance of a cluster.

2. If not already done, set the initialization parameter REMOTE_LOGIN_PASSWORDFILE to EXCLUSIVE for the database instance. This requires bouncing the database instance because it is a static parameter. If you're using an SPFILE, enter the following command to set this parameter on database instance *ptc1*:

    $ORACLE_HOME/bin/sqlplus system@ptc1
    ALTER SYSTEM SET REMOTE_LOGIN_PASSWORDFILE = EXCLUSIVE SCOPE=SPFILE;

    If you're on RAC, do the same for all other RAC instances, then bounce the RAC database (which includes all instances).[21]

Start Database Listener   If you don't start the listener for a target database, you will receive a network adapter error when Grid Control attempts to connect to the database on the Properties page of the Configure Database wizard. The wizard caches the downed state of the listener, so if you wait to start the listener until after receiving the error, you'd have to cancel out of the wizard, refresh the Agent home page, and restart the wizard. To avoid this unpleasantness, start the target database listener now from the OS command line:

- Start a single-instance database listener as follows:

    $ORACLE_HOME/bin/lsnrctl start *<listener_name>*

- Start a RAC database listener with the rest of the nodeapps, as follows:

    $ORACLE_HOME/bin/srvctl start nodeapps -n *<nodename>*

---

[21] For those unfamiliar with RAC, the procedure to stop a RAC database is: srvctl stop database -d *<db_name>*, srvctl start database -d *<db_name>*.

Address Database Management Bugs See the Database Management Issues section of the Enterprise Manager Release Notes for your OMS platform for database management problems. For the most up-to-date version of the release notes, see http://www.oracle.com/technology/documentation/oem.html.

One such database management bug not documented in the release notes (or anywhere else, apparently) is when the UTL_FILE_DIR initialization parameter is set to multiple values, each delimited by a single or double quote such as:

utl_file_dir='/usr/tmp','/tmp', ...

In this case, you receive the following repeated ERROR and WARNING in the Agent emagent.trc file:

```
2006-03-29 18:08:20 Thread-2015237 ERROR engine:
[oracle_database,GLDT_gldt1,utlFileDirSetting] : nmeegd_GetMetricData failed :
Result has
repeating key value : utl_file_dir_9i+,/usr/tmp,

2006-03-29 18:08:20 Thread-2015237 WARN   collector: <nmecmc.c> Error exit. Er-
ror message:
Result has repeating key value : utl_file_dir_9i+,/usr/tmp,
```

To avoid this error/warning, enclose all UTL_FILE_DIR directories in a single set of quotes as follows:

utl_file_dir='/usr/tmp,/tmp, ...'

This requires a database bounce to implement, as UTL_FILE_DIR is a static initialization parameter.

Discover Database

The procedures are similar for discovering single-instance and cluster database targets in Grid Control but are slightly more involved for cluster databases.

Discover a Database Instance Perform these steps to add a single-instance database target to Grid Control:

1. Log in to a super administrator account (such as the template account called GC_SUPER1 in Chapter 10) where preferred credentials are set. (The reason is that it's easier to implement the steps in "Create Dictionary Baseline" below if you can rely on database preferred credentials without having to manually override them.)
2. Navigate to the Add Database screen (see Figure 11-1 above). See "Three Paths to Discover/Configure a Target" for instructions.
3. The database will either be automatically or manually discovered:
   - If Grid Control automatically discovers the database, the following page appears.

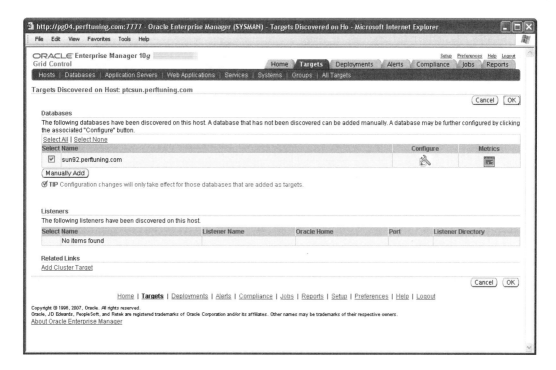

- If Grid Control does not automatically discover the database, click the Manually Add button. The Configure Database Instance wizard appears. For a complete description of this wizard, see the section entitled "Run Configure Database Wizard.")

Discover a Cluster Database   A properly installed RAC Agent should automatically discover an existing cluster, but will not automatically discover the cluster[22] databases on a cluster. Follow the procedure below to manually discover a cluster database and associated listeners that the cluster supports. I use the example of a two-node Oracle10g[23] Cluster Database named *ptc*, consisting of instances *ptc1* and *ptc2* located on nodes *ptc1* and *ptc2*, respectively, where the RAC Agent on *ptc2* has already discovered the cluster, *crs*.

1. As a GC administrator with full target privileges on the cluster database target, navigate to the home page of the Agent for the node where the cluster was discovered (node *ptc2* here[24]). This page is of the same form shown in step 1 of the section, "Discover/Configure Targets from the Agent Home Page."

2. Click the Add field, select the "Cluster Database" target type as shown in the illustration, and click Go. A screen appears asking you to choose whether to look for databases and their listeners only on the host where the Agent resides, or to look on all hosts in the cluster.

---

[22] This section also applies to discovering single-instance databases that use Clusterware.

[23] While this example is for an Oracle10g Cluster Database, the manual discovery process is the same for a 9i Cluster Database target.

[24] Usually the node that discovers the cluster is the node where the RAC Agent was installed. It may not be the "first" node, as this example demonstrates.

3. Here, choose "all hosts in the cluster" and click Continue. The following screen appears, showing which cluster databases were "automatically"[25] discovered.

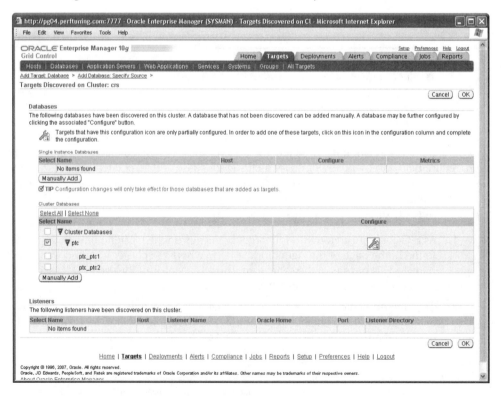

---

[25] I use quotes because I consider this a manual process. In my opinion, it would be automatic only if newly installed RAC Agents automatically discovered existing cluster databases. But I'll stick with Oracle's phraseology here to avoid confusion.

In this case, the Agent automatically finds the cluster database, named *ptc*. It is listed at the bottom half of the screen, and consists of two database instance targets, named *ptc_ptc1* and *ptc_ptc2*.[26] (The two listener targets are not yet discovered in GC.) If the Agent does not find the cluster database, you need to click the Manually Add button to specify the cluster database information explicitly.

4. At this point you can do one of the following:

   - Click OK, and on the confirmation page click OK again to complete the cluster database discovery process.[27]

   - Alternatively, click the Configure icon in the row that lists the cluster database (*ptc*) to jump directly to configuring monitoring for the cluster database, as covered in the next section.

5. For purposes of this discussion, click OK, as I cover the monitoring setup process separately in the next section. Only one Agent is allowed to manage a specific target. Therefore, while a Database Instance target will appear for each node, the Cluster Database target will be added on just one of the Agent nodes (*ptc1* in this example). This node may be different from the Agent node on which the cluster target was discovered (*ptc2* here). The cluster node on which a RAC target is discovered is immaterial to GC.

Run Configure Database Wizard

Single-instance and cluster database targets must be set up in Grid Control before an Agent can begin collecting monitoring data on them. Configuring target database monitoring in Grid Control is a wizard-driven process. The Configure Database Instance wizard consists of the following pages:

- Properties
- Install Packages (9*i* or lower only)
- Credentials (9*i* or lower only)
- Parameters (9*i* or lower only)
- Review

As of the time of this writing, Grid Control was supported to manage Oracle 11.1.0.6, 10.2, 10.1.0.4+, 9.2.0.6+, 9.0.1.5[28], and 8.1.7.4 single-instance and RAC database targets.[29] For Oracle10*g* target databases, the wizard jumps from the Properties page directly to the Review page and skips the three middle pages. Essentially, with a 10*g* database, you only need to set the DBSNMP password in the Properties page. The configuration process is so easy because Grid Control monitors an Oracle10*g* database using its built-in Automatic Workload Repository (AWR) feature.

---

[26] The Agent assigns the name <node>_<instance_name> to uniquely identify all database instances for this particular cluster database. This naming convention allows Grid Control to discover other cluster databases of the same name (i.e., *ptc*) on different cluster nodes.

[27] For Oracle9*i* databases, if the DBSNMP database password is not set to the default ("dbsnmp"), then the process succeeds with the result "Properties for instance … have been updated", but with an expected network adapter error that you can ignore. You will correct this error in the next section when you enter the DBSNMP password in Grid Control.

[28] 9.0.1.5 is supported only as the database release used for an OracleAS Metadata Repository, which is part of AS 9.0.4 with fips enabled only.

[29] See the Certify page on Oracle*MetaLink* for the latest Grid Control certification information.

By contrast, Grid Control must integrate with Statspack to monitor Oracle 9*i* and 8*i* targets, which requires completing all wizard pages. Statspack must be integrated with Grid Control to monitor certain types of 9*i* and 8*i* database performance metrics through the Console, and to enable certain reports under the Performance tab of the Database home page, such as Top SQL and Duplicate SQL. Part of that integration involves installing additional GC packages in the target 9*i* or 8*i* database, which the Configure Database Instance wizard accomplishes for you, as you shall see.

Following is the procedure to configure basic monitoring for any certified database target version. I use the examples of a single-instance Oracle9*i* Database named *sun92* and a RAC Oracle10*g* database named *ptc*. Along the way, I point out slight differences in the configuration between single-instance and cluster databases. Note that Grid Control does not commit configuration changes until you complete the wizard and click Submit—you can click Cancel at any time.

To bring up the Configure Database wizard, click the Targets tab and the All Targets subtab[30] as an administrator with full privileges on the database target. Select the radio button next to the desired Database Instance or Cluster Database target, and click Configure.

Configure Database Wizard: Properties Page   The Properties page is for entering basic database information. The page looks a little different for single-instance and cluster database targets.

For a single-instance database (*sun92*), the page looks like this.

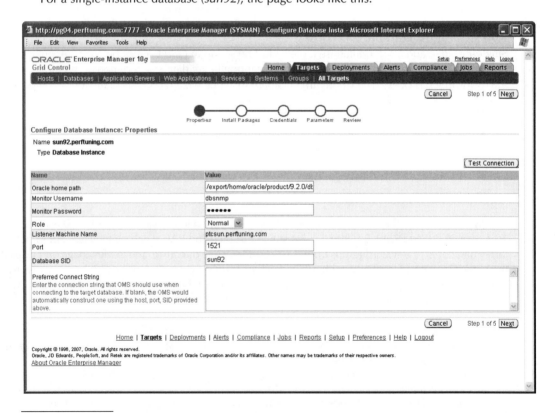

---

[30] You can also go to the Agent home page for the host containing the database instance. However, in the case of a cluster database, you need to know which Agent discovered it—it can be any RAC Agent node. Therefore, it is easier to use the All Targets subtab.

For a cluster database (*ptc*), this page looks a little different.

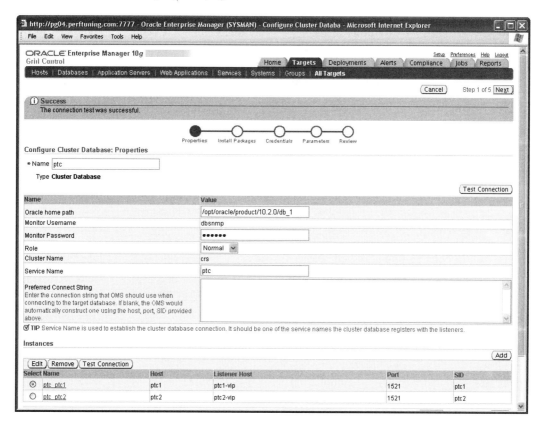

A brief comparison between the above two screen shots shows that both single-instance and cluster databases share the majority of the fields as described in Table 11-3 (see the last column for the fields that apply to each type of database).

1. Enter all input and click Test Connection to ensure it is correct, particularly the password. The information banner at the top indicates whether the connection test is successful. If successful, skip to step 3.

2. If the Test Connection above is not successful, following are typical errors, the reasons for them, and solutions. After implementing the solution, retry the Test Connection button.

   ■ **ORA-01017** The DBSNMP password entered does not match the actual password. If you don't know the DBSNMP database password, change it in SQL*Plus by executing the following SQL statement as a user with SYSDBA privilege:

   ALTER USER DBSNMP IDENTIFIED BY *<password>*;

Field	Description	Recommended Value	Applies to SI, Cluster Database, or Both
Name	Cluster Database name for GC to use. Defaults to value for Cluster Name field.	Use Cluster Name or a unique name if another cluster database by this name was already discovered.	Cluster Database
Oracle home path	The directory where the Oracle database software is located.	ORACLE_HOME. Do not specify a symbolic link or append a forward slash ("/").	Both
Monitor Username	Hard-coded as DBSNMP for primary databases.	DBSNMP (for primary) SYS (for standby)[31]	Both
Monitor Password	DBSNMP password. If changed later, must change it here as well.	<password>	Both
Role	The role for the Monitor Username to use.	Normal (for primary) SYSDBA (for standby)[32]	Both
Listener Machine Name	Hard-coded hostname on which listener runs.	<host name>	SI Database
Port	Port on which listener runs. Defaults to 1521.	<port>	SI Database
Database SID	System Identifier (SID), without domain, to uniquely identify database from any other on host.	<DB_NAME>	SI Database
Cluster Name	Hard-coded cluster name specified at installation time.	<cluster_name>	Cluster Database
Service Name	Service name registered with the listeners. Use DB_NAME.	<SERVICE_NAME> initialization parameter value	Cluster Database
Preferred Connect String	Connect string for OMS to access target database.	Leave blank. OMS will use host, port, and Service Name or SID.	Both

TABLE 11-3. *Properties Page for Cluster and Single-Instance Databases*

---

[31] See the section "Create an OMR Database Data Guard Configuration" in Chapter 15.
[32] Ibid.

- **ORA-28000** The DBSNMP user is locked, and perhaps also expired. If so, click the "Change dbsnmp Password" button. A screen appears where you must enter a username and password with SYSDBA privilege and the new DBSNMP password.[33] Click OK and Grid Control will unlock the user and change the password. You can also manually unlock the user and change the password (which also un-expires the account) by executing the following as a user with SYSDBA privilege:

  ALTER USER DBSNMP ACCOUNT UNLOCK;
  ALTER USER DBSNMP IDENTIFIED BY <password>;

3. Click Next to advance to the next page in the wizard, which varies by database version:
   - For Oracle10*g* databases, the wizard jumps from the Properties page directly to the Review page. Take this opportunity to review all entries. Click OK to accept all input and save the Oracle10*g* monitoring configuration.
   - For Oracle9*i* and Oracle8*i* Databases, the wizard advances to the Install Packages page as shown next.

Configure Database Wizard: Install Packages Page For Oracle 9*i* and 8*i* Databases, the Install Packages page appears to allow you to enable the following metrics.

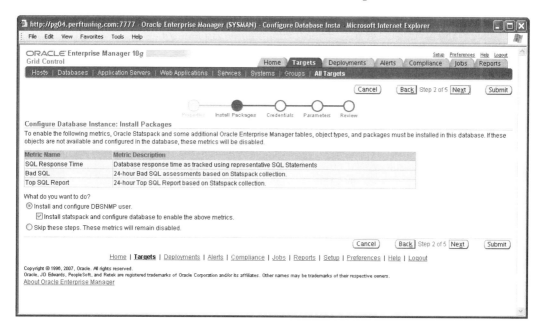

---

[33] Grid Control populates the Username field incorrectly with "dbsnmp"—you must change this username to a user with SYSDBA privilege. Also, the Connect As field offers a pull-down menu selection of Normal, SYSOPER, or SYSDBA, but only the SYSDBA privilege succeeds.

I suggest taking advantage of this metric functionality, but you may have a reason to enable these metrics manually rather than through the Console. Choose one of the following solutions:

- **Manually enable the metrics**   If you've previously installed Statspack in a 9*i* or 8*i* target database and have collected historical data you'd like to view in the GC Console,[34] or if you wish to script the process because there is a large number of target databases to configure, then select "Skip these steps. These metrics will remain disabled." Manually enable these metrics by following the instructions in Oracle Enterprise Manager Advanced Configuration, Section 18.3 entitled Manually Configuring a Database Target for Complete Monitoring. You can skip step 7, which creates the PERFSTAT user.

- **Let Grid Control automatically enable the metrics**   If the above does not apply, inform Grid Control to enable these metrics by selecting "Install and configure DBSNMP user" and by checking the box "install statspack and configure the database to enable the above metrics." This directs Grid Control to grant required roles and privileges to DBSNMP and to install Oracle Statspack and certain SYSMAN objects. If Statspack was previously installed but you are not using it, remove it now, before proceeding with the wizard. Grid Control's Statspack installation creates new SYSMAN tables and packages and grants certain privileges and roles to the DBSNMP user (all in the target database) that a standalone Statspack installation does not accomplish. To drop a pre-existing Statspack environment, log in to the target database with SYSDBA privilege in SQL*Plus and execute *$ORACLE_HOME/rdbms/admin/spdrop.sql*.

Make your selections and click Next.

Configure Database Wizard: Credentials Page   If you chose "Skip these steps..." on the previous page, the wizard skips to the Review page; you should accordingly skip now to the section below called "Configure Database Wizard: Review Page."

If you chose "Install and configure DBSNMP user" on the previous page, the Credentials page appears, which allows you to enter the SYS password and the privilege to connect as. Enter the SYS password and select SYSDBA in the "Connect As" field to allow the wizard to grant the DBSNMP user the appropriate privileges to monitor a 9*i* or 8*i* database. Click Next when finished.

Configure Database Wizard: Parameters Page   For the Parameters page, Grid Control requests information to create the Statspack user PERFSTAT.

---

[34] There's no point putting yourself through the paces to manually enable these metrics if there is no Statspack data to save. It makes more sense to remove Statspack and let Grid Control install it and automatically perform the other required integration steps—see next bullet.

Chapter 11: Configure Target Monitoring

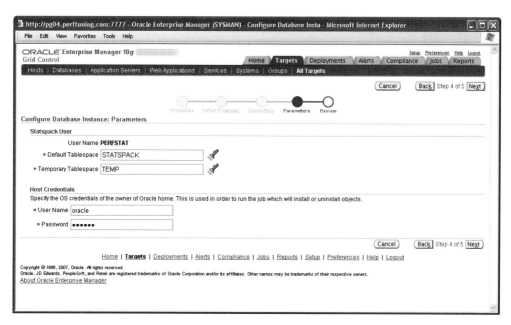

You must enter a DEFAULT and TEMPORARY tablespace and the OS credentials of the owner of the Oracle home to run a job through the job system to install the Statspack objects.

The DEFAULT tablespace is TOOLS by default, but best practices call for creating a separate tablespace for Statspack. The following statement creates a new tablespace called STATSPACK with the appropriate storage parameters[35]:

CREATE TABLESPACE STATSPACK
DATAFILE ['<file_name>']
SIZE 2048M AUTOEXTEND ON NEXT 32M MAXSIZE UNLIMITED
EXTENT MANAGEMENT LOCAL AUTOALLOCATE
SEGMENT SPACE MANAGEMENT AUTO;

Once you create this STATSPACK tablespace, select it as the DEFAULT TABLESPACE on this page. Enter the remaining fields as shown above and click Next.

Configure Database Wizard: Review Page   For the Review page, check that your entries are correct. If satisfied, click Submit to complete your configuration of basic monitoring for the database instance. If successful, a screen will appear reporting that properties for the particular database instance have been updated.

Unlock and Un-Expire PERFSTAT User

The PERFSTAT user account is expired and locked by default. Unlock the account and un-expire it on the target database by resetting the password as follows:

ALTER USER PERFSTAT ACCOUNT UNLOCK;
ALTER USER PERFSTAT IDENTIFIED BY <password>;

---

[35] ['<file_name>'] is in brackets because it does not need to be specified if using Oracle-Managed Files (OMF). You're using OMF if the initialization parameter DB_CREATE_FILE_DEST is set.

Role or Privilege	Valid for 9i?
ANALYZE ANY	Yes
ADVISOR	No
CREATE JOB	No
ADMINISTER SQL TUNING SET	No
EXECUTE ON SYS.DBMS_WORKLOAD_REPOSITORY	No
SELECT ANY DICTIONARY	Yes
OEM_ADVISOR	No
OEM_MONITOR	Yes

TABLE 11-4. *Additional Roles and Privileges to Grant DBSNMP on Target Databases*

Consider running these two commands across all target databases in one operation by using Grid Control's Execute SQL feature[36] or by running a SQL Script job in the job subsystem. You can combine these commands with those listed in the next section.

CAUTION
Grid Control configures Statspack at level 6 at a one-hour collection interval. Some Oracle customers report 9i database crashes at this level. If you experience this problem, change the snapshot level from 6 to 5 by entering the following command in SQL*Plus as the PERFSTAT user:[37]
`EXECUTE STATSPACK.MODIFY_STATSPACK_PARAMETER (I_SNAP_LEVEL=>5);`

Grant Additional Roles and Privileges to DBSNMP

Chapter 10 recommends that GC administrators specify DBSNMP as the Normal user in preferred credentials for target databases (refer back to the table in Chapter 10 in the section "Normal vs. Privileged Credentials"). Grid Control uses these Normal credentials by default when an administrator manages a target database in the UI. However, accessing some of the Related Links on a database's Performance page requires granting DBSNMP the additional roles, system and object privileges listed in Table 11-4, which are over and above the default DBSNMP privileges granted.[38]

---

[36] The Execute SQL link is located under Related Links under the Database tab. See the section "Automatically Verifying Default Credentials" in Chapter 10 for instructions on how to use this feature.

[37] Reducing the snapshot level to 5 will disable Grid Control functionality related to SQL plan usage, but must be done if using snapshot level 6 crashes the target database.

[38] Do not confuse these database privileges with GC administrator privileges discussed in Chapter 10, in the section "Create Administrators." These are target database, not GC administrator, privileges.

As Table 11-4 shows, some roles and privileges are new to Oracle 10g, and they are not valid for 9i targets. To grant these roles/privileges, enter the following command in SQL*Plus while logged in with SYSDBA privilege to the target database:

GRANT <*privilege*> to DBSNMP;

While you can wait to grant these roles/privileges to DBSNMP until you need the corresponding features in the UI, there are several advantages to being proactive about it. The most obvious nicety is that users won't receive an error when attempting to use the dependent functionality. For example, without some of these roles/privileges, clicking on the Related Link called Advisor Central on any database home page would generate a warning message that you have insufficient privileges to run advisors. Another benefit to pregranting these roles/privileges rather than doing so piecemeal is that you can use Grid Control's Execute SQL feature to implement this across all target databases in one operation. (This is why the step appears after running the Database Configuration wizard.)

There is nothing inherently wrong with granting these roles/privileges to DBSNMP, and I suggest doing so. However, if security or other reasons prevent this, you can do one of the following on an individual database target basis:

- **Grant DBSNMP select roles/privileges** Grant DBSNMP only those roles/privileges needed to use certain features that GC admins must have, such as AWR/ADDM, SQL Access Advisor, and SQL Tuning Advisor. See Note 420504.1 for the roles/privileges that enable these features for DBSNMP. This option is applicable when admins don't have SYSDBA access but need certain GC performance features for a target database.

- **Grant DBSNMP no additional roles/privileges** To use features that rely on any of these roles/privileges, administrators must click Logout on the Performance page for a target database, select "Display database login page after logout," and log in again as a user with SYSDBA privilege. This is more cumbersome for admins, so use this solution only when DBSNMP security concerns are very strict or when admins don't need such features.

Set Up Segment Findings (Optional)

Optionally, Grid Control can identify database space management issues and recommend how to address them to improve database performance. This reporting is available for segments with excessive reclaimable space and row chaining:

- **Excessive reclaimable space** Segments with excessive reclaimable space contain a lot of fragmentation (empty space). This fragmentation increases the cost of queries involving these segments because more blocks must be read from disk than necessary. Segments can be shrunk to a smaller size to release this unused space.

- **Row chaining** Row chaining is when a row spans two or more blocks. It occurs when a segment's row size exceeds the database block size, or when a record update increases the record length such that it cannot fit in the current block. Queries of chained rows may require fetching two or more blocks instead of one, which reduces query performance.

To set up fragmentation detection for a particular database target, do the following:

1. Go to the home page of the database for which you want to set up segment space monitoring.

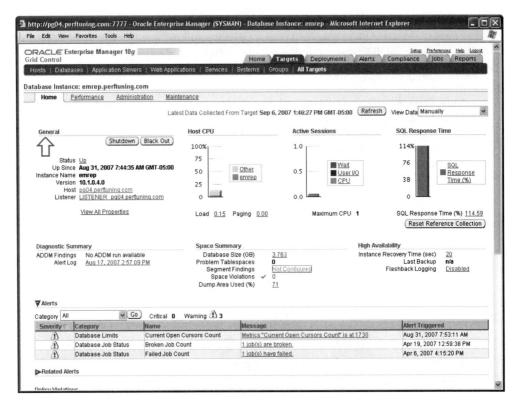

2. Under the Space Summary section, a Segment Findings link says "Not Configured", as shown above. Click this link.

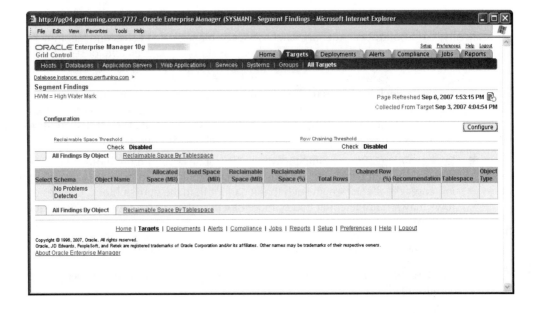

3. Click Configure, which brings you to this screen.

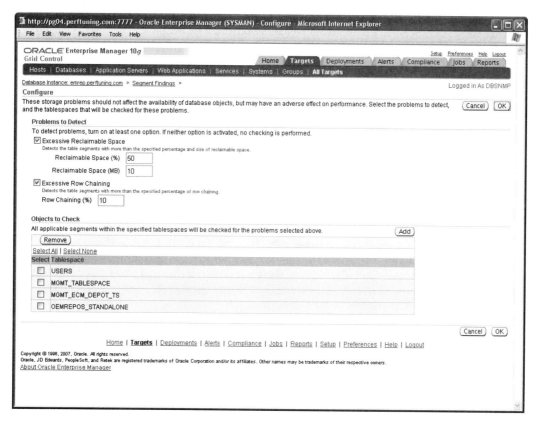

4. Choose the desired options and their values for detecting fragmentation. Above, I select both options—Excessive Reclaimable Space and Excessive Row Chaining—and leave the default values for each. These settings are appropriate for many databases. You can always adjust these values later using the process described here.

5. Click Add, select the tablespaces to check for problems, and click OK. I typically select all tablespaces, as the overhead to turn on this feature is minimal.[39]

6. Click OK to save your changes. If successful, you get a confirmation message that "the fragmentation findings configuration has been successfully updated."

Create Dictionary Baseline (Optional)

The *data dictionary* is a read-only set of tables in an Oracle database that contains metadata about database objects, such as object definitions, the space allocated to and used by these objects, user names, the privileges and roles granted to them, and auditing information. A dictionary baseline is

---

[39] No matter what tablespaces you choose, Grid Control skips checking certain segments, such as indexes, tables in SYSTEM, and temporary and UNDO tablespaces, to avoid unnecessarily degrading target database performance. Also, GC only performs excessive space checks for tables and table partition segments in tablespaces where SEGMENT SPACE MANAGEMENT is not set to AUTO.

a point-in-time capture of Oracle10g[40] or higher database object definitions contained in the data dictionary. Baseline functionality cannot be implemented for Oracle9*i* targets because it depends on Automatic Workload Repository (AWR) and on several sets of statistics that only exist in Oracle10g. Dictionary baselines are a feature provided by the Oracle Change Management Pack, which must be licensed separately. When creating a baseline, you choose which object types, schemas, or individual objects from any schema to capture, and whether to capture initialization parameters as well. The Repository stores the baseline information you capture for other Grid Control Change Management applications to tap.

As part of the initial database monitoring setup described in this chapter, it is prudent to identify those databases that require strict change management. For these select databases, and perhaps for only a subset of their schemas, you can set up a recurring job to capture multiple versions of a specific baseline to track the history of object changes. Baseline comparison features are rich—you can compare database to database, baseline to baseline, or database to baseline. But you need to capture baselines to do such dictionary comparisons.

Below is the procedure to create a dictionary baseline using the Create Dictionary Baseline wizard. The example used here is the Oracle 10.1 single-instance Repository Database called *emrep*, for which the example administrator GC_SUPER1 created in Chapter 10 collects all possible data for the most complete baseline. (I chose the OMR Database because every GC site has one; you probably don't need to capture baselines for it.) The baseline is submitted to run immediately and to re-run indefinitely at the same time once a week. You can only create dictionary baselines for database instances, not cluster databases. Therefore, to create a baseline for a cluster database, choose one of its instances.

To kick off the Create Dictionary Baseline wizard, follow these steps:

1. Log in to the Console using an administrator account with access to the database. It makes things easier if the admin's Privileged database credentials and Normal host credentials are already set. This does away with the need to enter SYSDBA database credentials when launching the wizard and on the Job Options page (see below).

2. Confirm you've already enabled the Change Management Pack for the database instance you are "baselining," including checking Pack Access Agreed.[41]

3. Navigate to the Dictionary Baselines page for the desired database instance in one of the following ways:

    ■ Click Dictionary Baselines under Related Links on the Databases subtab.

    ■ Click Dictionary Baselines in the Change Management section on the Administration page for a particular database instance.

4. On the Dictionary Baselines page, click Create to launch the Create Dictionary Baseline Wizard. The wizard consists of four pages:

    ■ Source

    ■ Objects

    ■ Job Options

    ■ Review

---

[40] This contradicts the instructions on the Create Dictionary Baseline Source Page and its OEM Online Help page, which incorrectly state that the source database must be 9*i* or later.

[41] For instructions on enabling packs, see the section called "Grant Management Pack Access" later in the chapter.

Create Dictionary Baseline Wizard: Source The Source page is the first wizard page that appears.

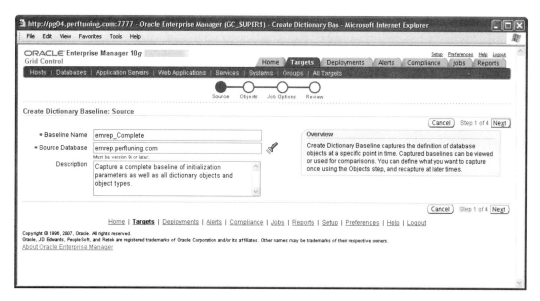

Complete the following fields (the first two fields require input, as indicated by asterisks on the Source page):

- **Baseline Name** Assign a unique name to the baseline. When choosing a name, take the following into consideration:
  - If you plan to submit a reusable baseline (schedule it to recur), as in this example, don't reference a date in the name, as it should describe the baseline for whenever it will run. An apt name would include the database name and, if applicable, any particular schemas and object types captured.
  - You may want to avoid using "baseline" in the name, as Grid Control constructs the *job* name (not the baseline name) that will create this baseline by prepending "BASELINE_" to the *baseline* name provided here. I wouldn't use the word "dictionary" in the name either.
- **Source Database** Select the source database instance for which you want to create the baseline. For cluster databases, choose one of its database instances. As already stated, contrary to the on-screen instruction "Must be version 9*i* or later," the source database must be version 10*g* or later.
- **Description** Enter an optional description for the baseline.

Click Next when finished entering your input on the Source page.

Create Dictionary Baseline Wizard: Objects On the Objects page, select which objects and object types to capture, and whether to capture initialization parameters too.

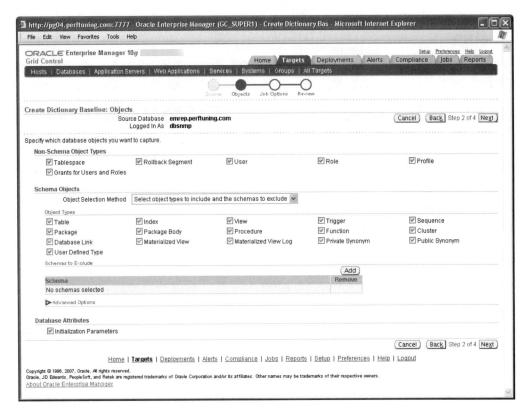

The Object Selection Method offers three ways to choose schemas:

- Select object types and the schemas to include
- Select the object types to include and the schemas to exclude
- Select individual objects from any schema

For this example, I select the second method, which is the easiest way to choose all schemas. The other two methods are more suitable for choosing a low percentage of all schemas or individual objects, respectively.[42] To choose schemas, click Add, make your selections, and click Next.

TIP
A session is considered inactive while on a particular wizard page. It may take longer than the 15-minute inactivity timeout period to select from a large list of schemas on the Objects page. To keep the session active while making your selections, click Back and Next to toggle between pages. You cannot change this timeout period, which differs from the configurable Console timeout described in Chapter 4 in the section "Modify the Default Console Timeout."

---

[42] On the Add page, Select All only selects schemas or objects on the page, and there is no choice in the pull-down menu to choose all schemas or objects.

Create Dictionary Baseline Wizard: Job Options   The Job Options page allows you to define the job schedule for executing this baseline. The schedule definition encompasses when to begin the job, whether and how often to repeat it, and what database credentials to use. For this example, the baseline is set to run immediately using preferred credentials as the Privileged user, and once a week thereafter, indefinitely. You must use Privileged rather than Normal database credentials to collect baseline information for schemas other than DBSNMP and PUBLIC. (Here I assume Normal credentials specify the DBSNMP user as recommended in Chapter 10.) Click Next to review all choices.

Create Dictionary Baseline Wizard: Review   Look at the job summary, and click Submit to send it to the job system for processing. The information banner will confirm the job was successfully submitted.

Monitoring, Viewing, and Comparing Dictionary Baseline Jobs

To monitor the baseline job run, click the View Job Details link provided. No refresh button exists on this page, but you can refresh your browser. (In general, however, use a GC refresh button when available on the page. On such pages, a browser refresh sometimes brings up cached data and does not reflect its current status.) Depending on the size of the database, the job can take from a few minutes to many hours. To provide a basis for comparison, creating a complete dictionary baseline of a Repository Database at a small, well-tuned Grid Control site should take about 30 minutes. To control the job run, click Stop, Suspend, or Resume as required.

For baseline jobs that take a long time to complete, consider creating a notification rule to alert you by e-mail when the status of the baseline job (or any job) changes. You can create a new database notification rule or edit an existing rule whose definition includes the target database. On the Jobs tab for the notification rule, choose Specific Jobs in the Add field, and click Go. (To receive notifications of status changes for all jobs, select "Jobs by Criteria" in the Add field.) The baseline job you just created will be selectable. Choose the statuses for which you want to receive notification. For this job, appropriate status choices are Suspended, Succeeded, and Problem. For details on creating notification rules, see "Create and Subscribe to Notification Rules" in Chapter 10.

If the job returns an error, Grid Control refers you to the "log file," but does not identify it. This log file is *$OMS_HOME/sysman/log/emoms.log*.

To view a completed dictionary baseline job or to generate DDL from it, return to the Dictionary Baselines page by clicking Dictionary Baselines under Related Links on the Databases subtab or by clicking Dictionary Baselines in the Change Management section on the Administration page for a particular database instance. Select the baseline of interest and click View. Then select the version of the baseline and click View or Generate DDL.

To compare dictionary baselines, click the Dictionary Comparisons link located on the Databases subtab, on the Administration page for a particular database instance, or as a Related Link on the Dictionary Baselines page. Click Create and follow an easy, wizard-driven process.

Stop and Disable Database Control (Optional)

For better or worse,[43] the same executable name, *emctl*, is used for the Central Management (i.e., Grid Control) Agent, Database Control Agent, and AS Control Agent. As mentioned in Chapter 1,

---

[43] My personal opinion is that this naming convention is for the worse, in that it confuses users. However, the code is shared to some extent between these "Controls," so I suspect it's easier for Oracle developers to use the same executable name.

a best practice when using Grid Control to manage your IT environment is to shut down Database (DB) Control if it's running on any target databases. You can go one step further and disable Database Control to preclude users from using it, either on purpose or mistakenly when meaning to use the Grid Control Agent.

When installing the Oracle Database software for a target database, DB Control is installed if, on the Select Installation Type screen, you select Enterprise Edition or if you select Custom followed by checking Oracle Enterprise Manager Console DB 10.x.x.x.x on the Available Product Components screen (it's checked by default, so is installed unless you explicitly de-select it).

The Repository Database won't be running DB Control, regardless of the GC installation type you chose. The GC New Database Option does not install DB Control, and if you're selecting the GC Existing Database Option, the installation will fail if DB Control is installed in the Database specified for the Repository (see "Install Database Software" in Chapter 3).

To shut down and, optionally, disable DB Control on target databases, do the following (ORACLE_HOME and ORACLE_SID below refer to the target Database software home and SID, respectively):

1. Check whether DB Control is installed by looking for the *$ORACLE_HOME/bin/emctl* executable.

2. If DB Control is installed, see whether it is running and, if so, shut it down by issuing the following commands (ORACLE_SID must be set in the *oracle* user environment):

   ```
   $ORACLE_HOME/bin/emctl status dbconsole
   $ORACLE_HOME/bin/emctl stop dbconsole
   ```

3. Optionally, disable DB Control to prevent accidental or intentional startup. It's easy to mistakenly call *emctl* for DB Control rather than for the central Agent because they have the same name. To disable DB Control, rename the DB Console *emctl* executable (the command below is shown in UNIX syntax):

   ```
   mv $ORACLE_HOME/bin/emctl $ORACLE_HOME/bin/emctl.bk
   ```

4. On Windows database hosts, prevent DB Control from automatically starting or attempting to start on boot. DB Control is implemented as a service called OracleDBConsole<*ORACLE_SID*>. Prevent auto startup in one of the following ways:

   - **Preclude DB Control auto startup using the Services Panel** Open the Services Panel from the Start menu or by using the *services.msc* command, right-click the OracleDBConsole<*ORACLE_SID*> Service, and select Properties. If the Startup type field is set to *Automatic*, select *Disabled* or *Manual* and click OK.

   - **Preclude DB Control auto startup using SC.EXE** Prevent auto startup using the Service Control utility *SC.EXE* at the command line. This utility, which is part of the Windows Resource Kit, achieves the same result as the Services Panel method, and is useful for scripting the process. The command is:

     SC.EXE CONFIG OracleDBConsole<*ORACLE_SID*> START= DISABLED |MANUAL

   - **Remove the DB Control Service** Use a Microsoft tool such as *Sc.exe* to remove the service.

The third option, removing the DB Control service, is the cleanest. If you choose one of the first two options, someone may inadvertently try to control the service in the Services Panel and it will fail (as desired) because you renamed the *emctl* executable in the previous step.

5. If you don't disable or remove DB Control, you can at least ensure that issuing an unqualified *emctl* command invokes the central Agent executable rather than that for DB Control. To do so, place *$AGENT_HOME/bin* before *$ORACLE_HOME/bin* in the PATH for the *oracle* user, as recommended in Chapter 7 under "Set Up Agent User Environment."

TIP
You can shut down and disable DB Control across all target databases in one Console operation by scripting the above steps and submitting the script as a job. You can also use the Execute Host Command feature to run the script (see "The Execute Host Command Feature" in Chapter 9).

Do *not* shut down the AS Control Agent or rename its executable. AS Control allows you to administer the OracleAS instance supplied with Grid Control through the AS Console. The *emctl* executable is located in the *$OMS_HOME/bin* directory on OMS hosts and in the *$AS_HOME/bin* directory for OracleAS standalone installations. You may recall from Chapter 9 that you control the AS Console and AS Control Agent together with the command *$OMS_HOME/bin/emctl start |stop iasconsole*.

Listener Discovery and Configuration

Grid Control should automatically discover and enable monitoring of a listener's availability and performance, unless the listener is secured with a password. In this case, you must enter the listener username and password in Preferred Credentials (see "Set Preferred Credentials" in Chapter 10). While no listener configuration is required under normal circumstances, you may need to remove and re-add a listener in Grid Control or manually change its configuration for a number of reasons. The most common reason for removing and re-adding a listener is to clear up a status mismatch (i.e., the listener is running, but Grid Control shows that it is down). Listener reconfiguration in the Console would also be required to match a change in the actual listener port.

To discover a listener or re-configure a listener's properties, do the following:

1. Log in to the Console as an administrator with full privileges on the host running the listener.

2. Remove any unused listeners from Grid Control (see "Remove Targets from Grid Control" earlier in this chapter).

TIP

As of Oracle Database 10gR2, the best practice when running a database on ASM is to install separate ASM and database homes. One consequence is that there are two potential listeners per node, one in each home, and GC usually discovers them both. Use the ASM listener or the database listener, but not both. The standard for a RAC database is to use the ASM listener.[44] The chosen listener then services both the ASM and database instance on the node. Listeners not configured to run (such as a database listener when using the ASM listener) will always show a DOWN status, and you should remove them from Grid Control.

3. If the *listener.ora* file contains an IFILE pointing to a file that doesn't exist, either create a real IFILE file or an empty file (using *touch <filename>*) in this location, or remove (i.e., comment out) the IFILE entry. You must bounce the listener for the change to take effect. This step is necessary because the file that the IFILE points to must exist or the listener status will show as DOWN in the UI, even if that listener is started. I've found this to be the case even if other valid entries exist in the *listener.ora* file.

4. Bring up the listener monitoring configuration page:

   - To discover a listener, click Add on the Agent home page of the host where the listener runs.

   - To configure a listener, go to the Agent home page of the host running the listener, select the listener, and click Configure. Alternatively, go to the Targets ->All Targets page, select the listener, and click Configure.

5. The following page appears when configuring a listener. When discovering a listener, an additional Name property appears on this page. It is best to abide by the naming convention Grid Control uses, which is *<listener_name>_<hostname.domainname>*.[45] Add or change the properties as required and click OK when you are finished.

---

[44] The advantage to using an ASM listener with RAC is that you don't have to start the listener separately with the *lsnrctl* command—in fact, you should not use *lsnrctl* with a RAC listener. Instead, the listener is a service of the Clusterware and as such is started along with the other services using the command "srvctl start nodeapps -n <node_name>".

[45] If two or more listeners with the same name exist on the same host, subsequent listeners are named *<listener_name>0_<hostname.domainname>*, *<listener_name>1_<hostname.domainname>*, etc.

Chapter 11: Configure Target Monitoring **409**

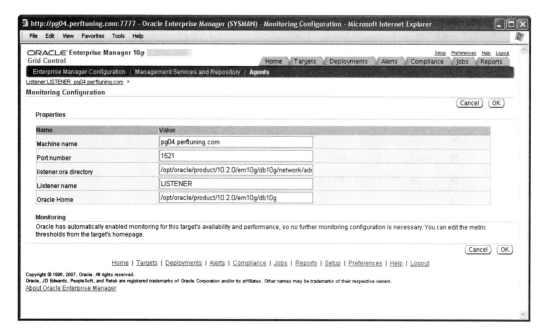

6. If the listener is password protected, enter the listener username (typically *oracle*, the OS user who created the listener) and password in the administrator's Preferred Credentials, if not already done (see "Set Preferred Credentials" in Chapter 10). For Oracle E-Business Suite middle tier listeners, specify the owner of the APPL_TOP (typically *applmgr*) as the Preferred Credential user.

7. Check that the status of the listener shows correctly in the Console. Go to the Targets tab and All Targets subtab or the Targets page for the host where the listener is running, and confirm the Status column shows a green arrow pointing up. The status may take up to 30 minutes to be reflected accurately in the Console.

Host Additional Monitoring

When you install an Agent, it automatically discovers the resident host target[46] and provides out-of-box metrics for monitoring the host's availability and performance. These default metrics are a good start, but you might be interested in adding some additional host monitoring and management capabilities, including:

- **Generic log file monitoring** Monitors OS log files for abnormal conditions by specifying Perl expression patterns to match and ignore. The metric is "Log File Pattern Matched Line Count." For instructions on setting up log file monitoring, see Configuring Generic Log File Monitoring Criteria in the Oracle Enterprise Manager Online Help. A best security practice is to place sensitive files that should not be monitored via the log facility in *$AGENT_HOME/sysman/config/lfm_efiles*.

---

[46] It is a little misleading that you can click Configure for a host target, as you can for any other target, but Grid Control just returns a message that there are no editable configuration properties for the host.

- **Program resource utilization monitoring** Monitors CPU and memory consumed by a particular program/owner combination on UNIX hosts. Following are the program resource metrics available:
  - Program's Max CPU Utilization
  - Program's Max Process Count
  - Program's Max Resident Memory (MB)
  - Program's Max Resident Memory PID
  - Program's Min Process Count
  - Program's Total CPU Time Accumulated (Minutes)
  - Program's Total CPU Utilization (%)

  For directions on how to configure these metrics, see Configuring Program Resource Utilization Monitoring Criteria in the Oracle Enterprise Manager Online Help.

- **Dell Linux hardware monitoring** Monitors Dell PowerEdge Linux hosts. This monitoring capability requires installing the Dell OpenManage Server Administrator (OMSA) software on the Linux hosts. Metrics are provided for the following hardware components and measurements:
  - Fans
  - Memory devices
  - PCI devices
  - Power upplies
  - Processors
  - Remote Access Card
  - System BIOS
  - Temperature

  For detailed steps on setting up Dell PowerEdge monitoring, see Enabling Hardware Monitoring for Dell PowerEdge Linux Hosts in the Oracle Enterprise Manager Online Help.[47]

- **Generic file and directory monitoring** Monitors files and directories on hosts running many flavors of UNIX. Following are the metrics available:
  - File or directory size (MB)
  - File or directory permissions
  - File or directory size change rate (KB/minute)

  For instructions on how to configure file and directory monitoring, see Configuring File and Directory Monitoring Criteria in the Oracle Enterprise Manager Online Help.

---

[47] Once step not noted in the OEM Online Help is to change the View Data field from Last 24 hours to Real Time:Manual Refresh to see the metrics for the Dell components.

Chapter 11: Configure Target Monitoring **411**

OracleAS Discovery and Configuration

Grid Control should automatically discover any Oracle Application Servers installed on a host at the time the Agent is installed. However, after the Agent is deployed, you must prompt Grid Control to discover subsequently installed OracleAS targets.

Discover OracleAS Instance

Grid Control automates the OracleAS instance discovery process, which is as simple as this:

1. Log in as an administrator with full privileges on the host where the OracleAS instance is running.

2. Navigate to a page where you can discover a target by following one of the procedures above in the section "Three Paths to Discover/Configure a Target." Select Oracle Application Server in the Add or Add New field (depending on which of the three ways you chose to add the target), and click Go.

3. The following page should appear. (The OracleAS target added here is SOA[48] Suite 10.1.3, installed by selecting the installation type "J2EE Server and Web Server.")

Click OK to add all discovered Application Server targets to Grid Control. You should get confirmation that all targets were added successfully. (If the above page does not appear, you receive the message, "All Application Server targets have been added to monitoring already. There are no targets to add." Only the OK button is available, indicating that you cannot add OracleAS targets that GC does not discover itself.)

4. Confirm the status of the OracleAS target is reflected correctly in the Console, either from the Targets tab and Application Servers subtab or from the Targets tab and All Targets subtab.

Configure OracleAS Instance

Occasionally, you may need to configure or reconfigure an OracleAS target instance in Grid Control if you change the OracleAS configuration in some way. For example, you may need

---

[48] SOA is an acronym for Service Oriented Architectures. This is the first SOA release and is not intended for production use.

to change the default Web Cache listen port from 7777[49] because of a port conflict with another product. To change the configuration of an OracleAS target in the Console accordingly, do the following:

1. Log in as a GC administrator with full host privileges on the OracleAS target instance.
2. Navigate to the OracleAS target Monitoring Configuration page by selecting the target and clicking Configure from any of the three pages described above in "Three Paths to Discover/Configure a Target."

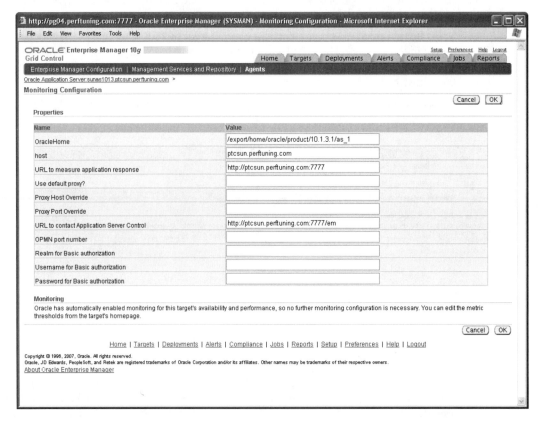

3. Change the applicable configuration properties and click OK when finished.

Address Application Server Bugs

See the Application Server Management Issues section of the Enterprise Manager Release Notes for your OMS platform for OracleAS management problems. For the most up-to-date version of the release notes, see http://www.oracle.com/technology/documentation/oem.html.

---

[49] This is the Web Cache HTTP Listen port listed in *$OMS_HOME/install/portlist.ini*.

Enter Target Properties

Once you add a new target of any type, you can ascribe target properties to it. Target properties are optional, user-defined attributes that administrators can employ to extend Grid Control's default ability to search for, classify, and filter targets. In addition, when setting up certain target types, you can assign properties to appear on select target pages. For instance, when configuring a System target type (see Chapter 12), you can add columns for target properties to be seen on the Components and Dashboard pages.

NOTE
GC 10.2.0.3 is missing an OK button to save values entered for target properties. Patch 5531715 fixes this bug.

To navigate to the Target Properties page, go to the home page for any target, such as the Host home page, and under Related Links, click Target Properties.

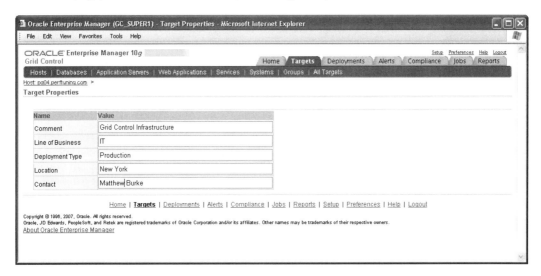

Administrators with OPERATOR target privileges can edit target properties. Here is a description of these properties as Oracle intended them to be used. You can, of course, use these properties in any way you see fit:

- **Comment**   Any description or information about the target
- **Line of Business**   The organization, business unit, or department the target supports
- **Deployment Type**   Can be anything, such as a production, test, or development
- **Location**   The physical location of the target
- **Contact**   Contact details of the administrator responsible for the target

When finished, apply your changes by clicking OK.

After you enter properties for targets, you can search for targets by entering one or more target property values. On the All Targets or Groups subtabs, or on the summary page for almost every target type such as for a Database Instance (see Figure 11-3 above), click Advanced Search,[50] enter the desired search criteria for one or more target properties, and click Go. Targets with matching properties are displayed.

Grant Management Pack Access

The list of targets you want Grid Control to manage should be complete, at least for now. (If your IT environment is growing, you'll doubtless need to add new targets later on.) You've removed any unnecessary targets and added/configured any previously undiscovered targets that require GC monitoring. Now is a good time to grant all *licensable targets* access to the management packs and management plug-ins installed by default. Licensable targets, also called parent targets, when granted access to the packs, propagate that access to their dependent targets. For instance, packs granted to a database propagate to the database host. A target must already be discovered in order to grant it management pack access, so by waiting until now, you've saved yourself a lot of back and forth between adding targets and granting pack access.

Management Options: Packs, Plug-ins, and Connectors

Management packs, System Monitoring Plug-ins, and management connectors are the three types of licensed *management options* (also known as extensions or add-ons) for Grid Control, and are distinguished as follows:

- **Management Packs** Management packs provide premium features related to specific areas of Grid Control, and "extend" the product's functionality considerably. (I quote "extend" because many Console links for a target are enabled only when you grant that target access to one or more management packs.) Most management packs are installed by default, but some require separate installation.

- **System Monitoring Plug-ins** System Monitoring Plug-ins (or Management Plug-ins) extend Grid Control's system coverage to non-Oracle components, including databases, middleware, storage, and network devices. A System Monitoring Plug-in is a target type that Oracle provides, or that a third-party vendor or user develops, to expand the set of predefined target types. Oracle offers many System Monitoring Plug-ins, such as for Microsoft SQL Server, F5 BIG-IP Local Traffic Manager, and EMC Symmetrix DMX. You download plug-ins individually from OTN and install and deploy them in the Console.

- **Management Connectors** Management Connectors enable you to integrate Grid Control bidirectionally with other management frameworks and trouble ticketing systems. As of EM10g Release 4, Oracle offers management connectors for three HP system management products, Microsoft Operations Manager (MOM), BMC Remedy Service Desk PeopleSoft Enterprise HelpDesk and Siebel HelpDesk.

---

[50] The Application Servers subtab is missing the Advanced Search link.

Unlike Oracle Enterprise Manager 9*i*, where management packs were installed separately, many management packs, System Monitoring Plug-ins, and management connectors are included with Enterprise Manager 10*g*R3 and later. You can download the remaining management option from MetaLink, OTN, and third-party vendors. For details on all System Monitoring Plug-ins and connectors, see http://www.oracle.com/technology/products/oem/extensions/index.html.

As for licensing, Oracle trusts its customers to license any management options they use. Licensing is by the number of hosts, transactions, or targets these options manage. Oracle Support Services does not provide support for unlicensed management options. Consult Oracle Enterprise Manager Licensing Information, which is part of the Oracle EM documentation set, for a description of all management options and their functionality, or speak with your Oracle sales representative.

Management Packs

Following is a list of the management packs installed and configured by default:

- Application Server Configuration Pack
- Application Server Diagnostics Pack
- Configuration Management Pack for Non-Oracle Systems
- Database Change Management Pack
- Database Configuration Pack
- Database Diagnostics Pack
- Database Tuning Pack
- Provisioning Pack
- Service Level Management Pack
- Linux Management Pack

See OTN for documentation on how to install and configure a few additional management packs that are not built-in. Examples include the Application Management Pack for Oracle E-Business Suite (EBS), available as patch 5489352, and the Oracle Application Management Pack for Siebel, whose installation involves certain UI tasks and running a script resident in the Management Agent software home.

Grant Access to Management Packs You must explicitly grant access to all built-in management packs for the current licensable targets in your GC environment as follows:

1. As a super administrator, select the Management Pack Access[51] link under Setup.

---

[51] The Management Pack Access link is a bit of a misnomer, as you use it to grant access to both management packs and System Monitoring Plug-ins.

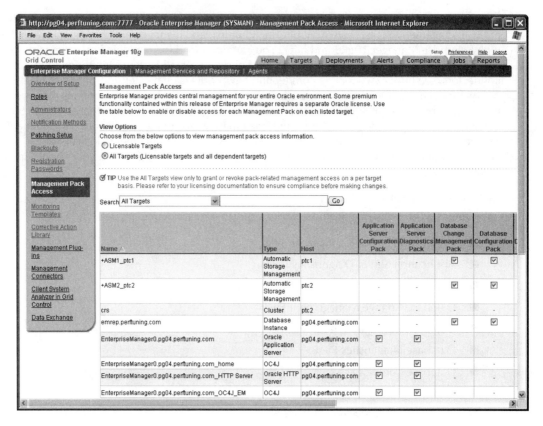

2. On this page, enable access to packs on a target-by-target basis by acknowledging that you've licensed a pack for a particular target. Under View Options select "All targets (Licensable targets and all dependent targets),"[52] as shown above, and in the Search field leave All Targets selected.

3. Check the box for each licensed pack/target combination, and check the Pack Access Agreed column on the far right for each licensed target.

4. Click Apply to save your changes.

You'll need to repeat this procedure later to grant management pack access to any targets you subsequently discover.

How to Display Management Pack Information in the Console Sometimes it's useful to know what management options provide certain GC functionality. To determine the packs used to display a particular screen or links on that screen, expand the About Oracle Enterprise Manager plus-sign ("+") icon at the footer of some Console screens. If no plus-sign exists, no

---

[52] Rather than tracking how licensed targets propagate pack access to their dependent targets, I prefer to explicitly choose pack access for each licensed target, whether dependent or not. Don't depend on dependencies.

Chapter 11: Configure Target Monitoring

management pack is required for that screen. For instance, when selecting the Compliance tab and Security At a Glance subtab, clicking on this icon displays the following information:

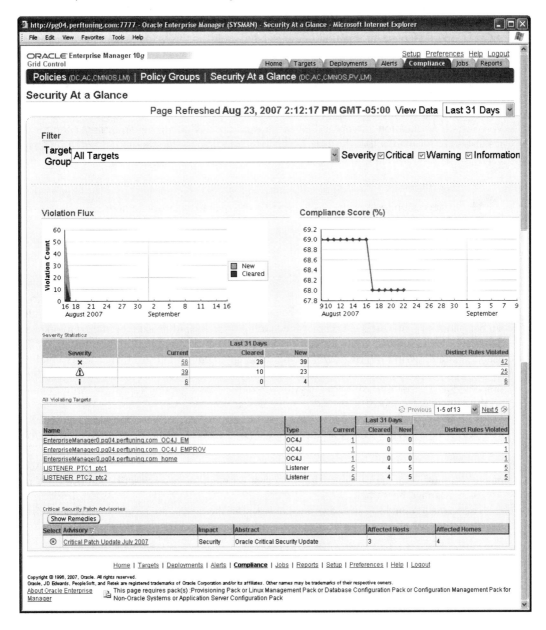

Both the Policies and Security At a Glance subtabs display acronyms for the Packs you must license to use the functionality of these subtabs, and the expanded icon lists the full names of each pack required to display the page. For a description of all management packs, including their acronyms and the target types they manage, click Oracle Enterprise Manager 10*g* Release 4 Grid Control License Information under Related Links on the Management Pack Access page.

System Monitoring Plug-Ins

A System Monitoring Plug-in (also known as a Management Plug-in) is a target type provided by a user or a third party to extend Grid Control's set of predefined target types. Grid Control comes with some built-in third-party target types, such as BEA WebLogic Managed Server, IBM WebSphere Application Server, and Network Appliance Filer. However, System Monitoring Plug-ins also exist for many other third-party products, offering additional metrics and monitoring capabilities. Plug-ins fall into five categories:

- System Monitoring Plug-ins for Network Management
- System Monitoring Plug-ins for non-Oracle Database Management
- System Monitoring Plug-ins for non-Oracle Middleware Management
- System Monitoring Plug-ins for Host Management[53]
- System Monitoring Plug-ins for Storage Management

The list of downloadable plug-ins is too long to include—38 at last count. Suffice it to say, plug-ins are available for the foremost products in the above categories from leading vendors such as Check Point, Cisco, Citrix Systems, Dell, EMC, IBM, Microsoft, Nortel Networks, and Symantec.[54] All plug-ins require additional installation and configuration. You install a Management Plug-in (MP) by downloading a Management Plug-in Archive (MPA), which is usually a jar file. An MPA is the container for an MP, and the form in which an MP is transported (imported and exported). An MP exists outside of the Management Repository only as an MPA. So the only way to transport an
MP from one Grid Control system to another is via an MPA. You can download these System Monitoring Plug-ins from OTN at http://www.oracle.com/technology/software/products/oem/index.html.

Once you install a System Monitoring Plug-in, it appears on the Management Pack Access page. You must disable or enable plug-in access for each target as described above in "Grant Access to Management Packs."

The Management Plug-ins page, available under the Setup menu on the vertical navigation bar, is the central place for administering plug-ins.

---

[53] The System Monitoring Plug-in for Hosts appears on the Management Pack Access page, and you must enable access to this plug-in for all licensed hosts.

[54] If you're looking to monitor Oracle's Hyperion products, a plug-in is not available. Instead, you model these products by creating systems and services using the service models called Generic Service and Web Application. for more information, see http://www.oracle.com/technology/products/bi/epm/pdf/sys9_oem_integration.pdf.

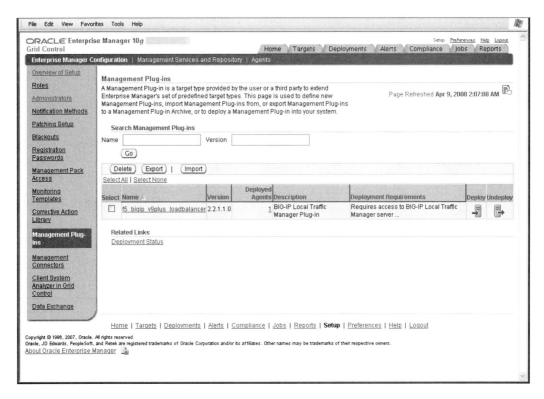

On this page, there are buttons for doing the following with Management Plug-ins:

- **Delete** Remove an MPA from the Repository
- **Export** Export an MP from the Repository to an MPA
- **Import** Import an MPA to the Repository
- **Deploy** Deploy an MP into your Grid Control system
- **Undeploy** Undeploy an MP from your Grid Control system

For more information on MPs and MPAs, see the EM Online Help and the Oracle Enterprise Manager Online Documentation Library, which contains a large System Monitoring Plug-ins section of installation and metric reference guides and a troubleshooting guide.

Management Connectors

Currently, Oracle has seven Management Connectors,[55] all of which require integration in Grid Control. Table 11-5 summarizes information about these connectors.

Documentation is available in the Management Connectors section of the Oracle Enterprise Manager Online Documentation Library. This section also offers an integration guide for interfacing the Enterprise Manager Connector Framework with any management framework system, either general purpose or specialized.

---

[55] Oracle is also planning to release many more connectors as Grid Control matures.

Management Connector Name	Product Compatible Versions	Bundled With
Microsoft Operations Manager (MOM)	MOM 2005	EM10gR3
BMC Remedy Service Desk	6.x	EM10gR3
PeopleSoft Enterprise HelpDesk	PeopleTools 8.48, 8.49	PeopleSoft CRM 9 Bundle #5
Siebel HelpDesk	8.x	Siebel 8.0.0.2
HP Service Manager	7	EM10gR4
HP ServiceCenter	6.1	EM10gR4
HP OpenView Operations	OVOU[56] 8.0	EM10gR4 (with patch 6884527)

TABLE 11-5. *Grid Control Management Connectors*

The Management Connectors page is on the vertical navigation bar under the Setup menu.

This page allows you to view the available connectors and configure, remove, and determine the status of connectors.

---

[56]OVOU is an acronym for HP OpenView Operations Unix.

Schedule Blackouts for Planned Downtimes

Blackouts disable notifications and target data collection during periods of exception, such as for scheduled maintenance. For example, you don't want to receive an alert that a particular database is down every time you run a scheduled cold backup of the database. Setting blackouts up-front prevents these types of false notifications and permits Grid Control to more accurately report on a target's performance. This, in turn, improves the accuracy of service level reporting in the Console (more about this in Chapter 12). Administrators are highly motivated to meet service level agreements (SLAs), and blackouts help maximize target availability calculations. For targets that belong to a service, you can configure a service level rule to consider the service and its targets as UP when under blackout.[57]

You can set blackouts either in the UI or at the command line, as discussed next. The Console provides more functionality than the command line. The main difference is that the Console allows you to specify targets for blackouts across multiple hosts, whereas the command line only allows you to set blackouts on a particular host using the Agent executable (*emctl*) command with the appropriate parameters. As such, the Console offers more options for choosing targets and for scheduling blackouts (such as duration, frequency, and repeatability), whereas the command line is better suited to letting administrators locally issue ad hoc blackouts.

Set Blackouts in the Console

When defining blackouts in the Console, you take advantage of three areas of GC functionality related to blackouts:

- **Manage the reasons for blackouts**   (Optional) Grid Control supplies nearly 60 out-of-box reasons from which administrators can select when implementing a blackout. You can remove reasons or add new ones. You can also require that administrators enter a reason for each blackout. As you will see below, these reasons are selectable from a pull-down list on the Properties page of the Blackout wizard.

- **Use the Blackout wizard**   Use the Blackout wizard as a straightforward way to create blackouts, as illustrated below.

- **Create retroactive blackouts**   Retroactive blackouts allow you to revise target availability data to show that the target was in blackout for a specified time period in the past. You can thus correct the negative impact on target availability of mistakenly not having set a blackout for the target. To enable this feature, click the Retroactive Blackout Configuration link at the bottom of the Blackouts page. The Create Retroactive Blackout link then appears under Related Links.

The following steps to create a blackout in the UI demonstrate these first two features. I offer a real-world case of having to place all targets under blackout from 2 AM to 6 AM to do a cold backup of the entire Grid Control environment.[58] For this purpose, we'll create a new reason for blackouts called EM: Cold Backup.

1. Log in as a super administrator, click Setup, and then click Blackouts on the vertical navigation bar.

---

[57] See "Edit Service Level Rule" in Chapter 12.
[58] A cold backup is recommended in Chapter 14 in the section "Backups Before and After a Configuration Change."

422 Oracle Enterprise Manager 10*g* Grid Control Implementation Guide

2. (Optional) Manage the reasons for implementing blackouts and whether you must specify a reason for each blackout.

   ■ Under Related Links, click Manage Reasons.

   ■ To enforce that an administrator must enter a reason, check "Reasons are required for a blackout" as shown.

   ■ Add or delete any reasons for blackouts by clicking on Add Another Row or Delete, respectively. Here we add another row named EM: Cold Backup. Click OK.

3. Click Create to kick off the Blackout wizard, which consists of the following five pages:

   ■ Properties
   ■ Member Targets
   ■ Targets on Host
   ■ Schedule
   ■ Review

These pages are fairly self-explanatory, but there are a few fine points worth mentioning.

Create Blackout: Properties Page
The Properties page looks like this:

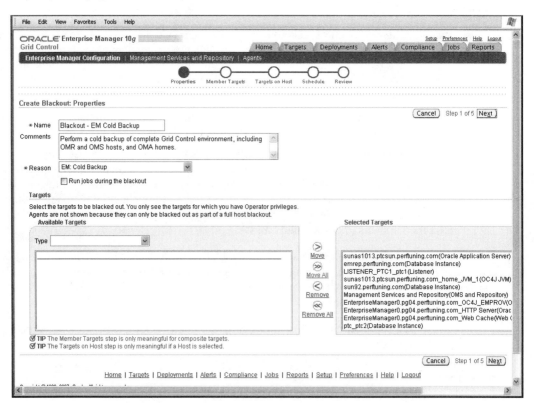

On this page, enter the following information (asterisks on the page indicate required fields):

- **Name** Use the automatically generated unique name containing the system timestamp, or change it to another unique name.
- **Comments** Optionally, enter comments that will appear on the page displayed when you later select the blackout and click View.
- **Reason** If you enforce that a reason for the blackout is required, then this is a required field. See step 2 above for details. In this example, I choose the custom reason added in that step.
- **Run jobs during the blackout** Choose whether jobs scheduled for execution during the blackout should run on targets defined in the blackout. You can't change this option after the blackout has begun.
- **Targets** Choose those targets you want to black out. Only targets for which you have Operator privileges appear under Available Targets. To select from all available targets, select "All target types."

Click Next to advance to the next page in the wizard.

Create Blackout: Member Targets Page

The Member Targets page looks like this:

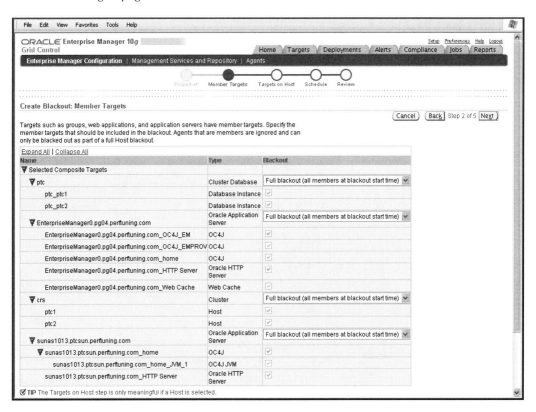

This page is meant for deciding which aggregate target members (or "member targets") to include in the blackout. Thus, it appears only if you select any aggregate targets or hosts on the Properties page; otherwise, the Blackout wizard skips over this page to one of the next two pages. Aggregate targets are top-level, composite targets composed of multiple members that can be modeled as a single entity.[59] Aggregate target types are application servers, Cluster, Cluster Database, Groups, Services, and Systems. The Blackout column on the Member Targets page offers three choices (for all choices, the top-level target is included in the blackout and you cannot deselect it):

- **All current member targets**   This choice propagates the blackout for an aggregate target or host to all members; you cannot unselect any member.

- **Full blackout (all members at blackout start time)**   This is the default option selected and blacks out all members belonging to the aggregate target (as chosen in this example). This is the only option for which an Agent can be blacked out; Agents are not otherwise blacked out, even if they are members of a group.

- **Selected member targets**   This option allows you to exclude one or more members of the aggregate target or host from blackout. Each member, including the aggregate target itself, is blacked out as an independent target.

Click Next after making your selections.

Create Blackout: Targets on Host Page

Here is what the Targets on Host page looks like:

---

[59] You can regard certain aggregate targets, particularly Services, as black boxes. This is useful when you are not concerned about the operational status of any single component in that box, only that the box as a whole is available to provide the intended function.

This page is intended for non-aggregate target selections. As such, it offers roughly the same three options for the Blackout column as on the previous page (sans "member" in the name):

- All current targets
- Full blackout (all targets at blackout start time)
- Selected targets

The page organizes targets by host rather than by aggregate target as for the Member Targets page. The appearance of and available choices on this page are relative to those made on the previous pages, as follows:

- When choosing a combination of aggregate and non-aggregate targets, the wizard jumps from the Properties to the Schedule page, skipping over both this page (Targets on Host) and the previous page (Member Targets).

- When choosing only non-aggregate targets on the Properties page, the wizard skips the Member Targets page, and the Targets on Host page allows for individual target selection on a host-by-host basis.

- When selecting aggregate targets on the Properties page, the Member Targets page appears, and regardless of your selections on that page, the Targets on Host page displays. However, this page is redundant, as you cannot use it to change any target selections made on the Member Targets page (as in this example).

A complete host blackout requires at least Operator privileges for all targets on the host. The blackout is still created, but only on those targets where the administrator has these privileges.

Make your selections and click Next.

Create Blackout: Schedule Page

Set the blackout schedule on this page (shown in the image on the following page).

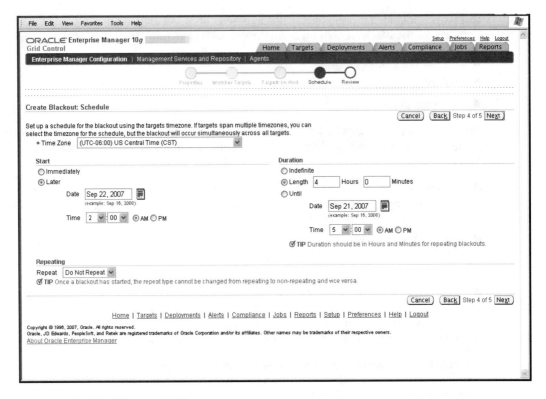

Complete the following fields:

- **Time Zone** Select the time zone to schedule the blackout. You can schedule a blackout on a target in another time zone than that selected without having to convert to the equivalent time at the managed target. For instance, say you're located in Boston (EST) and want to schedule a blackout at 4 AM EST for a target in San Diego (PST). Select EST for Time Zone and 4 AM for Time, and the target will be blacked out at 1 AM PST.
  If targets span multiple time zones, you can select one of the time zones for these targets and the blackout will begin across all targets simultaneously according to that time zone.

- **Start** Choose the start date and time. You can begin the blackout immediately or at some point in the future.

- **Repeating** Optionally, set a repeating schedule by choosing an appropriate interval, frequency, and end date. Once a blackout starts, you cannot change between repeating and nonrepeating selections.

  - **Repeat** Select from the following choices: Do Not Repeat, By Minutes, By Hours, By Days, By Weeks, Weekly, Monthly, or Yearly. If you select a repeating schedule, Frequency and End Date fields appear, which you need to complete; you must also select a length-based duration (i.e., it cannot be indefinite or end-dated).

  - **Frequency** Select an integer frequency greater than 0. The frequency (difference between two consecutive blackouts) should be at least five minutes greater than the duration.

- **End Date** Choose either no end date or a specified end date and time by which a repeating blackout will no longer be scheduled for the future. A blackout that is running when a specified end date is reached will not be stopped, but will be allowed to complete.

- **Duration** You can choose three types of durations: either an indefinite, length-based, or end dated blackout. Once started,[60] a blackout's duration cannot be changed between indefinite and length-based, but can be end dated. Once the blackout begins, you can't shorten the blackout period.

Make your scheduling selections and click Next to review all wizard choices.

Create Blackout: Review Page

Scan the Review page and click Finish if you're satisfied with your blackout definition.

Set Blackouts at the Command Line

Setting blackouts at the command line can be useful when you want to define blackouts on the target host itself. This is a matter of convenience for administrators who are already on a particular server running targets they need to black out, and who want to set ad hoc blackouts to stop data collection and prevent spurious alerts on these targets. Setting a blackout at the command line is also useful when you can't use the Console because the OMS or Repository is shut down, but will be started when a target to be blacked out will be down.

The command line syntax to start a blackout, as shown in the output from the command *emctl start blackout*, is as follows:

```
emctl start blackout <Blackoutname> [-nodeLevel]
[<Target_name>[:<Target_Type>]].... [-d
<Duration>]
```

The following are valid options for blackouts:

- *<Target_name>:<Target_type>* defaults to the local host target if not specified.

- If *-nodeLevel* is specified after *<blackoutname>*, the blackout will be applied to all targets on the node.[61] This is equivalent to setting a full blackout in the Console at the host level.

- Duration is specified in [days] hh:mm.

For example, to immediately start a four-hour blackout of all targets on the local node, enter the following command:

```
emctl start blackout blackout_adhoc -nodeLevel -d 04:00
```

To stop a blackout, issue:

```
emctl stop blackout <blackoutname>
```

To query whether a target is under blackout, execute:

```
emctl status blackout [<Target_name>[:<Target_Type>]]...
```

---

[60] The EM Online Help incorrectly states "once set," but should say "once started". You can switch between an indefinite and length-based blackout if the blackout is scheduled to run in the future.

[61] Any target list that follows *-nodeLevel* is ignored because this option places all targets on the node under blackout.

Summary

After gaining knowledge of general Console configuration in Chapter 10, in this chapter you learned how to remove, discover, and configure the most commonly used target types. The greater part of this chapter focused on discovering and/or configuring the following target types:

- Agent
- ASM
- Database (Database Instance, Cluster Database, and Cluster)
- Listener
- Host
- Oracle Application Server

The Grid Control Agent auto-discovers all of these target types, except Cluster Database. However, even though most target types are auto-discoverable at Agent installation time, they must be manually discovered if installed subsequent to the Agent. Therefore, I addressed how to discover all target types (except Agent and Host, which GC must discover).

It is not enough to simply discover a target. Most target types require configuration, and some benefit from additional setup to bring their monitoring to operational status (in addition to metric and policy tuning, which is covered in Chapter 13).

Once you discover and configure targets, you need to grant licensable targets access to built-in management packs, System Monitoring Plug-ins, and management connectors. By waiting until all targets are discovered before granting pack access, you save yourself a lot of repeat trips to the Management Pack Access page because the target must be discovered to grant it pack access.

Last, you began using blackouts to disable target data collection and notifications during planned downtimes. Not only do blackouts cut down on false alerts, but they also make Grid Control's reporting of a target's actual performance more accurate. I walked you through the process of setting blackouts both in the Console and at the command line, and I provided the rationale for choosing one method over the other.

In the next chapter, we move on to two important aggregate target types: Groups and Services. To set up a Service, we create a System, then build on it to configure one noteworthy service type in particular: the Web Application.

CHAPTER 12

Configure Group and Service Monitoring

This chapter picks up from where the last chapter left off in configuring two principal remaining target types: Group and Service. They garner recognition as the principal characters here, and each is accordingly bequeathed a section title in this chapter. However, the service takes the lion's share of the page count, for good reason. A service is conceptually more difficult to explain; it has its own vocabulary, objectives, and configuration intricacies. The complexity in configuring a service comes from having to first create the underlying System (yet another target type), then create and configure the service itself. We concentrate on one of a handful of service types, a specialized service called a Web Application, yet another target type in its own right. A service is like a Russian nested doll[1]: it has many layers, and just when you think you're finished configuring one layer, you find another one you need to address.

While some target types discussed in Chapter 11 are single targets, in this chapter all are *aggregate target types*. Aggregate target types have multiple components, and include application servers (such as Oracle Application Server), Cluster, Cluster Database, Group, System, Service, and Web Application. Aggregate targets are top-level, composite targets made up of multiple member targets. You can model certain aggregate targets—the System, Service, and Web Application in particular—as a single entity or "black box" in Grid Control. You may not always be overly concerned about the status of any single component in that box, only that the box remains available to provide the intended function.

Let's begin with how to configure a group before we move on to the more complicated process of setting up a service.

Group Configuration

Many Grid Control sites manage thousands of targets, and even the smallest sites typically contain hundreds of targets. The *group* target type helps you organize targets into logical sets to facilitate managing them as a unit. Following are some of the characteristics and advantages of groups:

- Once an administrator creates a group, he or she can refer to it as to an individual target. For example, an administrator can choose to grant access to, define notification rules for, or schedule blackouts or jobs for a group.

- Admins can customize their Consoles to directly access their groups on the Targets subtab (see the section "Create a Group" later in the chapter).

- Group charts, dashboards, out-of-box and custom reports help you visualize or summarize the overall status of member targets, and allow you to drill down to problem areas.

- At very large sites, you can add groups as members of other groups. This "roll-up" of groups allows for a hierarchical modeling of targets in your GC environment.

Members of groups can be of the same target type, such as Hosts, or of different types, like a System, Service, and Web Application. For instance, groups can cut across target types to contain

---

[1] For those who've never seen a Russian nested doll or matryoshka doll, you open the first doll, traditionally a woman dressed in a sarafan holding a rooster, only to find a smaller doll, and so on; eight dolls later, you get down to the smallest baby doll. I need not be so esoteric. It suffices to compare the complexity of a service to that of a woman.

all targets for a particular data center or application. One such application at every Grid Control site is Grid Control itself, which contains one or more database(s) hosts, Agents, listeners, Oracle Application Servers, Oracle HTTP Servers (OHS), etc.

NOTE
The System target type offers similar functionality to a group, but with many additional capabilities. A system defines components for a Service (yet another target type). For details, see the "Create a System" subsection below in the "Service Configuration" section.

Many administrators are not aware of a special type of group called a *redundancy group*. The reason for the obscurity of redundancy groups may be that you cannot create or manage them from the Groups home page, but must add them from the All Targets subtab. Redundancy groups provide all the features of regular groups, with the additional feature of allowing you to monitor the availability of target members as a single logical target. In other words, as long as at least one member of a redundancy group is UP, the group is considered UP. Such functionality requires members of a redundancy group to be of the same target type. The support that redundancy groups provide for the status (i.e., availability) metric is useful when configuring notification rules. (Example: You have two reports servers and want to be notified only if both reports servers go down, not just one. To do this, create a redundancy group comprised of the two reports servers. Then, on the rule's General page,[2] specify the Redundancy Group target type and select the reports server redundancy group. Choose the DOWN availability state on the Availability page for the rule, and a notification will be sent only if both reports servers go down.)

CAUTION
Do not use a redundancy group for the following types of targets: Cluster, Cluster Database, HTTP HA Group, or OC4J HA Group. These target types have specialized features applicable to the targets they manage.

Next, I describe how to create a group and finish the section by exploring some features of groups.

Create a Group

Defining a new group is easy. Here's how to define a normal group called GC_Targets, composed of all targets in your Grid Control system. I use this example only because everyone has these targets and can create such a group in their environment. However, you may want to remove this group after creating it because a prebuilt EM Website service and supporting EM Website System better models Grid Control's capabilities (as discussed below in "Service Configuration"). Consider waiting to remove the group until the end of the chapter (where I remind you) so that you can compare the features of groups and services.

1. Log in to the Console as an administrator with full privileges on all targets that comprise the group.

---

[2] See the section "Creating Notification Rules," General Page, in Chapter 10.

432 Oracle Enterprise Manager 10g Grid Control Implementation Guide

2. Choose to create either a regular group or redundancy group as follows:
   - To create a regular group: go to the Targets tab, the Groups subtab, and click Add.
   - To create a redundancy group: navigate to Targets, All Targets, select Redundancy Group in the Add field, and click Go. In the Member Type field, select the target type for the group and click Continue.
3. A set of pages appears, which functions similarly to a wizard.

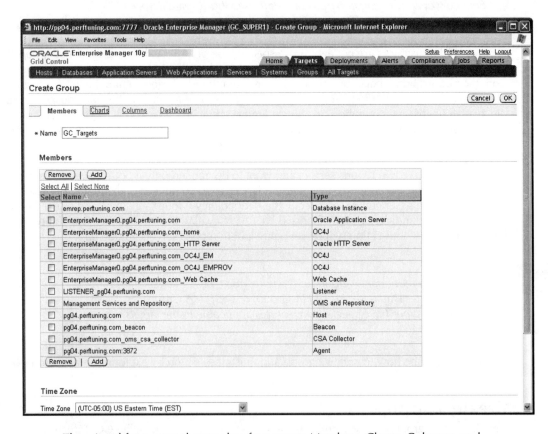

The wizard for a normal group has four pages: Members, Charts, Columns, and Dashboard; a redundancy group has all pages except the Dashboard page. On the Members page, enter the name for the group, click Add to select target members, and click Select.

Chapter 12: Configure Group and Service Monitoring **433**

NOTE
Carefully consider the time zone you choose for the group, as the statistics charts on the Charts page use this time zone and you cannot change it once the group is created. This time zone is also used as the default time zone for certain operations pertaining to the group, such as when setting blackouts.

4. Click the Charts page.

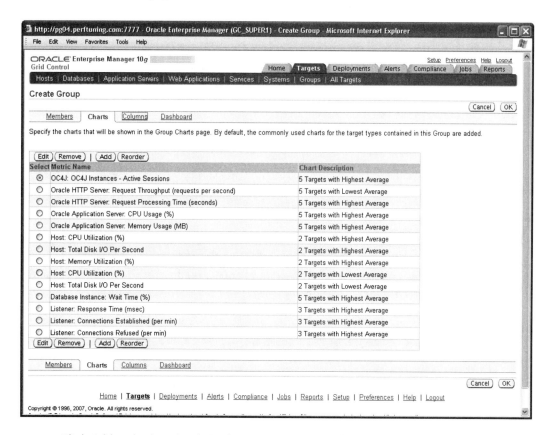

Click Add and select the desired Type, Metric, and Chart Type. For redundancy groups, also select any default metrics you don't want and click Remove.

5. Click the Columns page, shown in the next image.

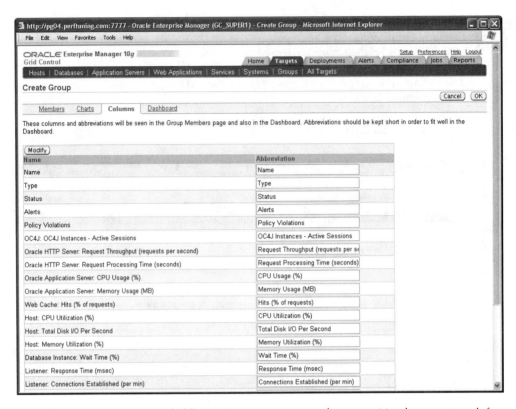

Choose the columns and abbreviations to appear on the group Members page and, for normal groups, on the Dashboard page.

6. For normal groups, click the Dashboard page.

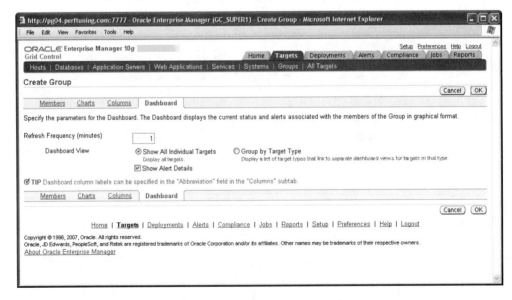

Chapter 12: Configure Group and Service Monitoring **435**

Choose the parameters you'd like for the Dashboard.

7. Click OK to save changes made on all pages.
8. The Groups summary table appears with the new group added to it. The columns of the table are group name, alerts, policy violations, and members. For normal groups, you can customize these columns to your liking. Click Customize Table Columns under Related Links, select from among the available columns, which include Comment, Contact, Deployment Type, Line of Business, and Location, and then click OK.
9. Add any desired normal groups to your target subtabs. On the Preferences menu, click Target Subtabs, select the group, and click Apply. Let other admins know that they can do the same.

Group Features

Once you create a normal or redundancy group, you can refer to it as to an individual target, such as to grant access to a group or define notification rules for a group. In addition to referencing groups, you can avail yourself of the following group features:

- Group pages
- System dashboard (normal groups only)
- Group out-of-box reports (normal groups only)

Group Pages

When you drill down into a particular group, it offers functionality on the following pages:

- **Home** The Home page shows a summary of the status of all targets in the group, including the percentage of components running, outstanding alerts and policy violations, recent configuration changes, and critical patch advisories.
- **Charts** The Charts page permits group performance monitoring via charts.
- **Administration** The Administration page allows for group management, such scheduling blackouts or jobs.
- **Members** The Members page customizes the Console to provide direct access to group management pages.

The next image shows a screen shot of the Home page. View the other three pages at your leisure.

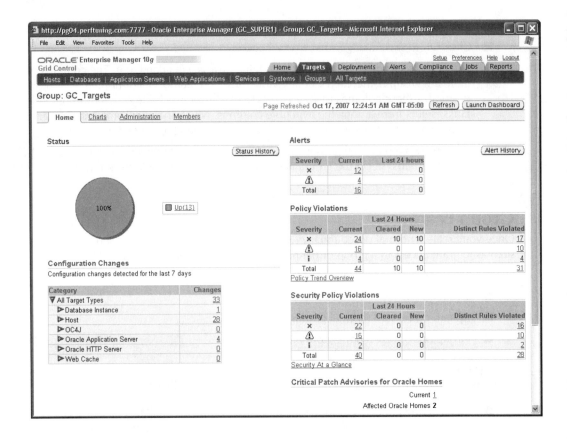

System Monitoring Dashboard

The System Monitoring Dashboard (also known as the System Dashboard or Dashboard), provides real-time monitoring of the overall health of targets in a group or system. The Dashboard presents information using graphics and icons that permit you to spot recent changes and quickly respond to problems. The Dashboard harks back to the days when operators pulled shifts staring at a dizzying array of monitors at server farms, just waiting for alert conditions to flash on the screens. Many data centers still have such a setup, only they now use large plasma display panels. Increasingly, operators are being replaced by "off-duty" DBAs with PDAs.

The Dashboard is accessible by clicking Launch Dashboard from any page for a specific group target. Here is the Dashboard for the GC_Targets group we just created.

Chapter 12: Configure Group and Service Monitoring **437**

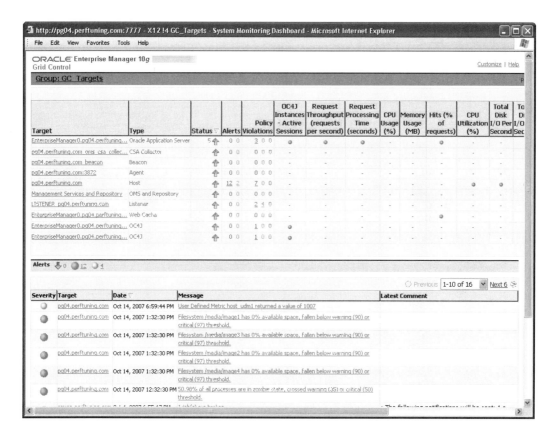

The UI displays UP and DOWN target states as green and red arrows, respectively, and metrics in critical, warning, and normal states as red, yellow, and green dots, respectively. These colors highlight problem areas and allow for drilldown as needed. An alert table lists all open alerts in the group. The Dashboard allows you to drill down for more details on a particular target, target status, alert, policy violation, metric, or alert message. You can also make the Dashboard publicly available on the Enterprise Manager Reports Web site.[3]

Out-of-Box Reports

There are many out-of-box reports for groups, accessible via the Reports link under the Related Links section of all Group pages. The image that follows shows a screen shot listing some of these reports.

---

[3] The EM Reports Web site is not discussed in this book because it requires no configuration. Many standard reports are available from the Reports tab.

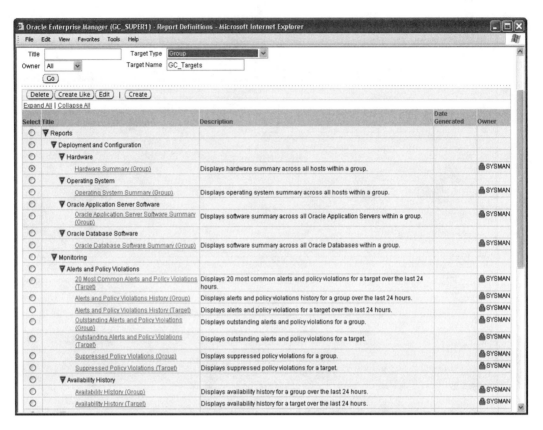

The reports, which fall under the categories of Deployment and Configuration, Monitoring, Security, and Storage, provide valuable summary information, particularly for IT managers.

Service Configuration

A *service* in Grid Control is a model of a business function or application that "services" end users. Example services are an e-mail service, an online store, an Enterprise Resource Planning (ERP) application, and a Customer Relationship Management (CRM) application. It is critical to business operations to monitor services so that they meet service level agreements (SLAs), which define availability, performance, and usage objectives for a service. When you model a service correctly in Grid Control, you can accurately measure service conformity with SLA specifications. To model a service in Grid Control, you must define a *system*. A system is the infrastructure used to host one or more services. This infrastructure is a logical grouping of targets, such as databases, hosts, listeners, and application servers. Like a service, a system is a target type in Grid Control, and is designed specifically for monitoring service components.

This section on services begins with a general discussion of the SLA objectives for services, including service availability, performance metrics, and usage metrics. Following this, I show you how to configure a service. There are no generic steps for setting up a service. There are just too many types of applications out there, and service functionality is such an all-encompassing subject. Still, the devil is in the details, so I decide to narrow the scope and present steps to configure a specific service that all Grid Control users already have at their disposal: the EM Website service. This service, defined and configured out-of-box, is a target in the Console and monitors the status, response time, and usage of the Grid Control application. I simply reverse engineer the steps required to create this service to demonstrate the configuration process and its end result. (Subsequently, I add more features to the EM Website service to fully leverage its capabilities.)

SLA Objectives for Services

A service level agreement (SLA) spells out service objectives for availability, performance, usage, capacity, security, and the like. You can configure a service in Grid Control to monitor adherence to these first three fundamental SLA objectives. An SLA must define service availability along with performance and usage metrics to further clarify that availability. A concise way to picture how performance and usage relate to availability is to imagine how a service slowly becomes more unavailable as performance degrades and usage increases. There will come a time when users consider the service unavailable, even if it remains "running."

What should an SLA look like for Grid Control? As for availability, the SLA for Grid Control should match or exceed the most rigorous SLA for a production target; if you're relying on Grid Control as the detection mechanism for a target outage, then your GC environment must be as available as any production target. (Grid Control monitors itself, so you don't need to play an endless chicken and egg game of "Who is monitoring Grid Control?".) A Grid Control SLA might include a 99.9% availability specification with a five-minute disaster Recovery Time Objective (RTO), a performance requirement of seven seconds or less for Console login, and a usage specification of no more than 95% memory utilization on hosts running the GC framework (OMS and OMR). Build a Grid Control SLA by taking the most stringent production target SLA; then add any specific GC requirements, such as the capacity to manage at least 10,000 targets.

Service Availability Definitions

The availability of a service is concerned with end-user accessibility to that service at any given point in time. You can define availability differently for each application. For instance, you may define availability for an e-mail application as the ability to access e-mail, and for an online banking application as the ability to perform an online transaction.

There are two ways to define availability:

- **Service test-based availability** This is the preferred option, and it defines a service as UP as long as critical functionality remains available. Such critical functionality may include generating a report or accessing a form. Administrators define critical functionality by creating *service tests* to run against a service to determine whether it is available. You must define locations for user communities from which to run service tests and deploy beacons to run the tests (see the "Beacons" section later in this chapter).

A service is considered available if at least one *key beacon* can execute either of the following, selectable when defining availability:

- At least one *key service test*
- All key service tests

■ **System-based availability** With this option, service availability is predicated on the availability of the underlying system's *key components* considered critical to running the service. The service is considered UP if either of the following is true, selectable when defining availability:

- At least one key component is UP
- All key components are UP

You should only choose the system-based availability option when you cannot find a suitable service test by which to define availability.

For both of these service availability models, the configuration process involves choosing performance and usage metrics (and setting their thresholds) to further qualify this availability. I will demonstrate this process when we set up a duplicate EM Website service. After creating a service, I will also show you how to proactively monitor your service by setting up alert notifications for metrics whose values cross the thresholds you chose.

Performance Metrics

Service performance is concerned with the response time a user experiences. For a service to be considered available, users must be able to access an application and perform critical functions in a set amount of time. In other words, service performance is closely tied to service availability in that, as performance degrades, a service slows to the point where it effectively becomes unavailable to users. Grid Control provides critical monitoring functionality by offering many ways to measure service performance. You can define *performance metrics* for both service test-based and system-based availability methods. However, the performance metrics each availability method employs are different, as described below:

■ **Metrics for system-based availability** When using system-based availability, administrators can define *performance metrics* to collect for system components. Performance metrics are simply the out-of-box metrics related to performance that you use with individual target types. You can choose metrics for each system component that vary by its target type. Example performance metrics for the host target type are Average Disk I/O Service Time (ms) and CPU Utilization (%). You can also define performance metrics for system components with service-based availability. However, it is customary to employ usage metrics for that purpose (see "Usage Metrics" further).

■ **Metrics for service test-based availability** For this availability definition, administrators can collect performance metrics and *response metrics* from service tests run by beacons located in particular user communities. Response metrics collected vary based on which model you choose for the service (discussed in "Types of Service Models"). For example, Connect Time (ms) and Transfer Rate (KB/second) are two of many response metrics the Web Application service model offers.

Irrespective of whether you choose a service test-based or system-based availability definition for your service, both response metrics and performance metrics provide a notable feature for a service over that for a single target: they allow you to statistically combine metric measurements for multiple system components of the same type. When adding metrics, you can use the metric results from one beacon or from an *aggregation function* of multiple beacons. An aggregation function is a combined statistical result of a metric taken from multiple beacons. There are four such aggregation functions:

- **Maximum** Useful for measuring the worst performance across all beacons so that you are notified if any location is suffering from performance problems.

- **Minimum** Not appropriate for exception reporting of problem beacons; it records the best performing beacon (invariably the local beacon) and ignores performance problems for the rest of the beacons. For this reason, don't rely solely on this aggregation function.

- **Average** Appropriate for getting a pulse on all user communities as a whole. It is advisable to use this function in conjunction with other metrics that use the Maximum aggregation function.

- **Sum** Provides a measure of the total response time for a metric across all beacons. The actual sum value is only significant when comparing it to sum values historically recorded.

Usage Metrics

Whereas service performance relates to user response time, usage performance is concerned with system load and user demand. Given that a system supports a service, regardless of how you model that service, it should not surprise you to learn that both service test-based and system-based availability methods offer *usage metrics* for system components. Usage metrics are identical to performance metrics for system-based availability (see "Performance Metrics" above). Usage metrics are a good way to introduce these system-based metrics into a service-based availability model. Administrators can define a usage metric from one system component or, as with performance metrics, they can aggregate a metric across multiple components using an aggregate function (e.g., average, minimum, maximum, sum). Metrics differ by target type, although many target types offer similar utilization metrics, such as for total and average utilization of CPU, memory, and I/O resources. To represent the usage of a web application service, for instance, you may define a host metric called Memory Utilization (%) and a Forms Listener metric called Total CPU Utilization (%).

Key Elements of Services

Before you can make sense of the service configuration process, you need to grasp some elemental concepts about services, which include the following:

- Beacons
- Transactions
- Types of service models

Beacons

A beacon is a target that allows an Agent to monitor services. Beacons are an integral part of monitoring service-based availability, and have the following characteristics and features:

- Beacons are offered only for testing service-based (not system-based) availability.
- Beacon functionality is built into the Agent, allowing service tests to execute at regularly specified intervals.
- Beacons monitor availability and performance of any service or web application.
- Beacons automatically "record" and "replay" *web transactions* of user actions at specified intervals from chosen user communities. A web transaction is a service test that allows a beacon to emulate a web client to test a web enabled service.
- Beacons offer basic network monitoring, including the ability to ping any network device or host and find network problems using interactive trace routing.
- Beacons support tests for various protocols, including database (TNS Ping, JDBC Script), web (SOAP, HTTP Ping), mail (POP, SMTP, IMAP, LDAP, NNTP), infrastructure (DNS, FTP), and network infrastructure (ICMP Ping, Port Checker, SQL Timing).
- The only beacon created by default is what I call the *OMS beacon* for the Agent that is chain-installed with the initial OMS.
- You must add beacons for other Agents as you would add any other Grid Control target, but you can only do so from the Agent home page (not from the Targets tab or All Targets subtab).
- When creating a web application target, you classify one beacon as the *local beacon* for the Agent host located geographically closest to the web application. This beacon should preferably be on a host in the same location and on the same network as the most underlying key system components.
- The local beacon is the benchmark against which you can compare the performance of service tests run from *remote beacons* (all other beacons created).
- When setting up a service, create a beacon in each geographical user community that uses the service.
- When adding beacons, you designate one or more of them as *key beacons* to determine the availability of the service. The service is considered available if at least one key beacon can execute one or more service tests.

As already mentioned, unless you have explicitly created beacons on other Agents, the only existing beacon will be the local beacon (or "OMS beacon") for the chain-installed Agent on the initial OMS installed. You will almost certainly want to add it as a key beacon to monitor any service you create, as this beacon is located in the "Grid Control user community" that must monitor the service, a community to which you as an administrator certainly belong!

Transactions

A transaction is a single URL or a series of user navigation steps that test whether a web application is available and performing adequately. Part of the service creation process is to use a transaction recorder to automatically capture critical business transactions for a web application. You designate one transaction as an *availability transaction*, which is the homepage URL for the application, and you can define additional transactions as well. Grid Control classifies a web application as "available" if at least one key beacon can execute this availability transaction.

Types of Service Models

There are six types of service models in Grid Control, and you must choose the one that most closely suits your application when creating a service. Here is a description of these service models:

- **Generic Service** The most elemental service model. You should use this service model only if no other service models are more appropriate.

- **Aggregate Service** A logical collection of one[4] or more *subservices*. Each subservice can in turn use any service model. Grid Control calculates availability, performance metrics, and usage metrics of an aggregate service based on the union sum of these metrics for its subservices. As an example, aggregate services are used for Business Process Execution Language (BPEL) target processes.

NOTE
Do not confuse this very specific service model called an Aggregate Service with "aggregate target types," which encompass all services, systems, groups, and other target types made up of multiple member targets.

- **Web Application** Software that provides information and functionality through the World Wide Web. The Web Application service model monitors web transactions and provides significant features over other service models. These include monitoring templates, transaction management resources, problem diagnostics (both interactive transaction tracing and request performance), and end-user performance monitoring.

- **Forms Application** A set of Forms related to one another in delivering a comprehensive service. An example of such a Forms application is Oracle Applications Release 11 or any other application built on Oracle Forms.[5]

- **Identity Management Service** A logical target configured for your Identity component instances, which supplies features such as a Single Sign-On (SSO) facility and an Internet Directory service. You must include at least one Oracle Internet Directory (OID) target and one SSO Server target as components of an Identity Management Service.

- **OCS Service** A specialized service model for representing the functionality that Oracle Collaboration Suite (OCS) provides. The OCS components include application servers, databases, and other processes, and must contain one and only one Identity Management Service.

To help you decide which service model is most fitting for your application, it's useful to know up-front what features these service models offer. Certain *test types* are available for the

---

[4] You may start out with just one service in an aggregate service if you are planning to add more services later.

[5] The Oracle E-Business Suite (EBS) R11*i* and R12 contain both a Forms and web-based user interface, but should be modeled using the Application Management Pack for Oracle E-Business Suite (EBS), which provides a target type called Oracle Applications. See "Can You Model the Oracle E-Business Suite as a Service" near the end of the chapter.

service models described above, except for Identity Management and OCS Service (each has its own test properties). Test types are categories of service tests offered for service models. Multiple test types are currently only available for a Generic Service. For this service model, you should choose the test type that best reflects how clients access the service. The test type you select determines the set of response metrics GC collects. If the test succeeds, the service is considered available. The test type chosen for a particular service test directs GC to collect the appropriate set of metrics for that test. Table 12-1 catalogs this information for easy reference. (It is not in the Oracle EM documentation, but collected metrics for each test type are listed at the bottom of the Service Tests pages and descriptions for each metric are given in the EM Online Help for these pages.)

As Table 12-1 clearly shows, for the Web Application service model, *Web Transaction* is the only test type. A Web transaction test type allows a beacon to emulate a browser to test a web enabled service. Similarly, the Forms Application service model hard codes the Forms Transaction test type. This table is not very original, but it's to the point. Part of its purpose is to highlight that only the service model, Generic Service, offers a plethora of test type choices, encompassing those for both Web Application and Forms Application (as noted in **bold**). If you have definite ideas about how you'd like to instrument performance metrics, and if these ideas go beyond the metric capabilities of Web Application or Forms Application, the Generic Service is likely your best choice.

Types of Service Models	Test Type	Metrics Offered for Test Type
Web Application	Web Transaction	Status, Perceived Total Time (ms), DNS Time (ms), Connect Time (ms), Redirect time (ms), First Byte Time (ms), HTML Time (ms), Non-HTML Time (ms), Total Time (ms), Transfer Rate (KB per second), Perceived Slowest Page Time (ms), Perceived Time per Page (ms), Time per Connection (ms), First Byte Time per Page (ms)
Forms Application	Forms Transaction	Status, Total Time (ms), Login Time (ms), Forms Time (ms), Average Time per Message (ms), Slowest Time (ms), Network Latency (ms), Database Time (ms), Server Time (ms), Status Description
Generic Service	Custom Script, DNS, **Forms Transaction**, FTP, Host Ping, HTTP Ping, IMAP, JDBC SQL Timing, LDAP, NNTP, Oracle SQL Timing, POP, Port Checker, SMTP, SOAP, TNS Ping, **Web Transaction**	(Vary by Test Type—too many to mention)

TABLE 12-1. *Service Models and Their Test Types and Metrics*

I would be remiss if I did not temper this good news about Generic Services with the fact that only the Web Application service type allows you to set up alert notifications on metrics you configure for the service. It's a trade-off, like most things.

Configure a Service

Armed with this basic knowledge of Grid Control services, you are now in a better position to configure the services upon which your enterprise depends. There are many types of applications you may have to place under Grid Control service management. As already mentioned, our example is to create a duplicate EM Website service (called EM Website 1) of the one provided out-of-box, then improve upon it with additional service features. The configuration process is representative of that for a typical service. It is also useful to know how to re-create the EM Website service when it is not created during OMS installation due to a slow OMS machine,[6] when it's unintentionally removed, intentionally removed along with the initial OMS,[7] or when it is forcibly removed because of configuration problems, such as being stuck in a STATUS PENDING state.[8]

There are three main steps to configuring a service, which I demonstrate in this section for the duplicate, then the improved EM Website service. *Specifics of this EM Website 1 service example are in italics.*

- **Create a system**   A system is the set of targets (hosts, databases, etc.) forming the backbone of your data center. You must create at least one underlying system in Grid Control to host one or more services. You construct a system in the UI using the Create System pages, which function similarly to a wizard. *To host a service called EM Website 1, we create EM Website System 1, which is a little more robust than the built-in EM Website System (differences are indicated).*

- **Create the service**   Creating the service is a wizard-driven process where you specify the system created above on which it will run, the availability method (system-based or service-based), and an initial service test. You also select one or more beacons to run the service tests, and choose performance and usage metrics, either here or in the next step. I defer nonessential wizard input until the next step for two reasons. First, you can only save your input after finishing the wizard, so I aim to get through it ASAP. *Second, this minimum input re-creates the default EM Website service (which you may need to do for the reasons stated above), and the next step customizes the service suitable for a typical production environment.*

- **Perform monitoring configuration tasks**   Here you specify additional service tests—only one can be defined in the wizard. You can also add beacons and performance/usage metrics configurable but not required when initially creating the service, but which are appropriate customizations for a production GC site. Finally, you perform monitoring configuration tasks not available when creating the service in the wizard, including managing Web server data collection, Watch Lists, service level rules, and Root Cause Analysis (see the respective sections below for details). *All tasks are modifications beyond the built-in EM Website service.*

---

[6] See Note 419886.1.

[7] I highly discourage you from removing the initial OMS. There is no documentation on the procedure to follow to reassign its unique characteristics to another OMS.

[8] See Note 334180.1.

TIP
Although you can easily remove the new EM Website System 1 and EM Website 1 service targets later, consider keeping them as improvements over the out-of-box targets. It's also preferable to customize these new targets rather than the built-in ones so other admins can reference them untouched.

As always, required fields shown in the UI screen shots below are notated with an asterisk ("*").

Create a System

A system is the set of infrastructure components—for example, hosts, databases, and application servers—that work together to host your applications. When creating a service later, you will specify the system used to host the service and mark the key components critical to operating that service. These key components are used to determine service availability; they factor in to Grid Control's *Root Cause Analysis (RCA)* feature, which helps identify possible causes of service failure (discussed later in the "Perform Root Cause Analysis Configuration" section). *We're adding a duplicate of the underlying EM Website System (with additional features as noted) for the EM Website Service.*

Following is the process to add a system on which the service is to be hosted. There are five pages to the Create System UI:

- Components
- Topology
- Charts
- Columns
- Dashboard

To create a system, log in as an administrator with ADD ANY TARGET system privilege. *In this example, log in as a super administrator you may have created in Chapter 10 (such as GC_SUPER1) or as SYSMAN.* Click the Targets tab and the Systems subtab. The Systems summary table appears, showing each system's name, alerts, policy violations, and members. On this page, you can add, configure, or remove a system. Leave System selected in the Add field, then click Go.

Components Page The Components page of the Create System page is where you name the new system, and select its components and time zone.

Chapter 12: Configure Group and Service Monitoring **447**

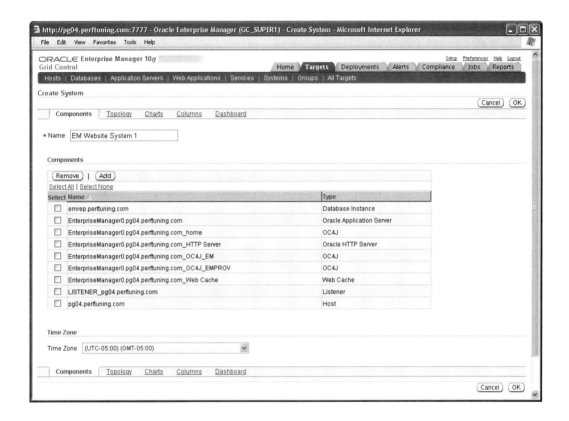

- **Name** Enter a unique name for the system. *Use a name such as EM Website System 1 to distinguish it from the original name, EM Website System.*

- **Components** Check the targets to comprise the system by clicking Add. *To match the built-in EM Website System, check the GC Oracle Application Server, Web Cache, and Host targets. Also select the remaining GC components shown above for additional RCA functionality as described below under "Perform Root Cause Analysis Configuration."*

- **Time zone** Choose a time zone for the system, bearing in mind that you cannot change it once the system is created. If you have OMSs in two different time zones, I suggest choosing the time zone where the OMR Database node(s) are located.

Topology Page The Topology page, shown next, graphically displays the relationships, which are not necessarily dependencies, between system components.

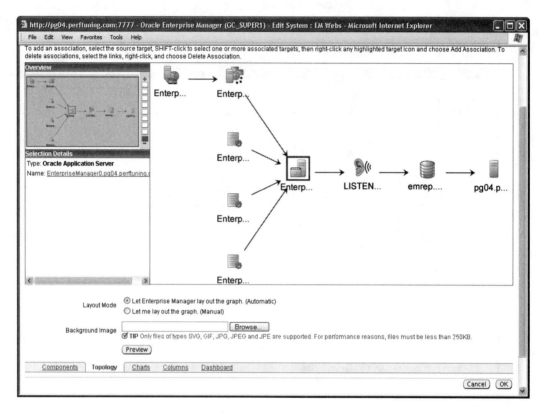

Arrows indicate associations between components, which are represented as icons. Status indicators over each icon let you quickly determine which components are down or have alerts, and allow you to drill down to these alerts. This page also highlights the causes of service failure as identified by RCA. A third-party plug-in called Adobe SVG Viewer is required to use this page, as described in the section "Install Adobe SVG Viewer" in Chapter 8. If not already done, install this plug-in now by clicking the link "Adobe SVG Plugin," which will appear at the top of the page. No topology associations exist for the out-of-box EM Website System. *Configure the topology associations shown above to take advantage of additional RCA functionality as explained below under "Perform Root Cause Analysis Configuration."*

Charts Page   Select the Charts page; this page shows which component metrics are displayed to depict the overall performance of the system.

Chapter 12: Configure Group and Service Monitoring 449

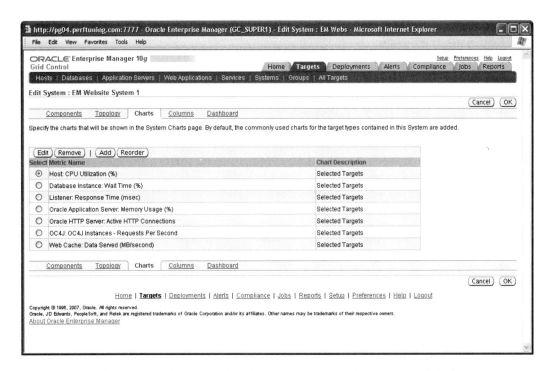

You can add, remove, edit, or reorder charts appearing on this page. By default, GC selects typical charts for the system's target types. The out-of-box EM Website System contains no charts. *Select some appropriate charts for the GC system, such as those shown above, to tap this graphical feature. When you edit a chart for a particular component, click Selected Targets and choose the specific GC target name. This is reflected on the above screen as Selected Targets under the Chart Description column.*

Columns Page The Columns page sets the columns and names that appear in the Components and Dashboard pages. Click Modify to change and reorder these columns. The Name column is not selectable because it is always included. In addition to Name, Type, Status, Alerts, and Policy Violations, I suggest choosing the same metrics you selected on the Charts page. *For the columns on this page, select the same metrics as the ones you chose on the Charts page.*

Dashboard Page The System Dashboard provides a real-time view of the health of managed targets in a system (or group). On this page, shown in the following image, specify the few parameters that affect how the Dashboard is displayed. See the section "System Monitoring Dashboard" earlier in this chapter for a description of Dashboard features.

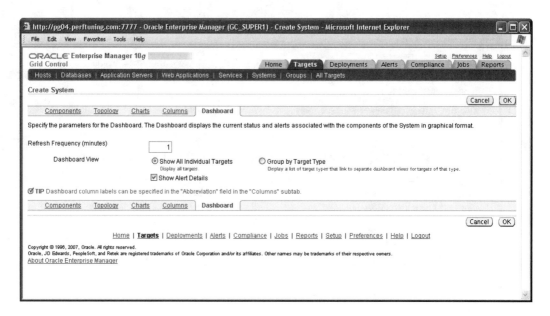

Choose the options you want. *Leave the default options shown above, which are those used in the original EM Website System.* Click OK to save the system.

Create the Service

Now that you have created the system, it's time to create the service that runs on it. As I already mentioned, you must complete all pages of the Create Service wizard to save your input. I suggest you enter only that information the wizard requires; otherwise, a network glitch or other failure while you're in the wizard might force you to have to start the wizard from scratch.

To get started in the UI, log in as an admin with ADD ANY TARGET system privilege (typically the same admin used to create the system). Click the Targets tab and Services subtab. On the Services summary page, the first decision you need to make when you're creating a service is to choose the service model that most closely emulates your application. The six service models already discussed (see the earlier section "Types of Service Models") are selectable from the Add drop-down menu:

- Aggregate Service
- Forms Application
- Generic Service
- Identity Management Service
- OCS Service
- Web Application

Make your selection and click Go, which launches the wizard. This wizard is specific to your choice of service model and directs you through the creation process for that model. *For the duplicate EM Website service we're creating, select Web Application. This is the only service model covered here, as it is used for the built-in EM Website service and also happens to be the most frequently used model.*

General Page The General page permits you to enter general information for the Web application.

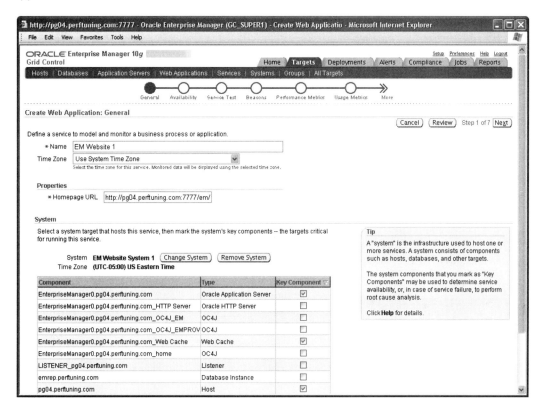

On this page, select the following:

- **Name** Choose a meaningful, unique name for the Web Application. *Enter EM Website 1.*
- **Time Zone** For consistency, I advise selecting Use System Time Zone to select the same time zone for the service as for the underlying system. Choose carefully because, as with a system, you cannot change the time zone of a service once you create it. *Choose Use System Time Zone.*
- **Homepage URL** Enter the homepage URL, which is the login page for your web application and the availability transaction mentioned earlier. *To match the EM Website, specify HTTP access via Web Cache at http://<OMS_host>:7777/em/ if your initial OMS host was UNIX, and if it was Windows, specify OHS access on port 80. Note the trailing slash to indicate that "em" is a virtual directory and not a filename (GC adds this slash anyway when you save a new system target).*

452 Oracle Enterprise Manager 10g Grid Control Implementation Guide

NOTE
If you later lock down HTTP Console access via Web Cache and/ or OHS (as explained in Chapter 16, section "Secure Console Connections"), you need to accordingly change the Homepage URL and any service test URLs defined below from HTTP to HTTPS. You may also need to configure beacons to monitor your web application over HTTPS (see OEM Advanced Configuration 10g Release 4 (10.2.0.4.0), Section 5.7.2).

- **System** Click Select System to choose an underlying system for the service, as well as the key components critical for running that service. Grid Control defines the service as available if all key components remain available. Key components also factor into Grid Control's Root Cause Analysis (RCA) feature based on the relationship you define between components on the Topology page when creating the system. *Select the system just created, EM Website System 1. For now, to exactly duplicate this service, leave the Key Components box checked only for Oracle Application Server, Web Cache, and Host, and uncheck the remaining components. In my view, all components except the OC4J home should also be key components, as they must be running to provide all available GC features. You will correct this later in the section "Perform Monitoring Configuration Tasks."*

Availability Page Perhaps because of the weightiness of the decision, Oracle dedicated the entire page to deciding whether to use service-based or system-based availability.

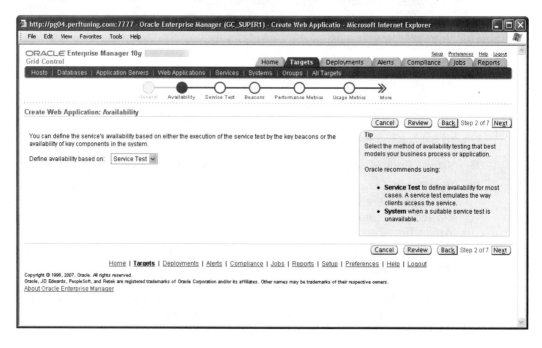

Chapter 12: Configure Group and Service Monitoring

Select either Service Test or System. (See the "Service Availability Definitions" earlier in the chapter for more information.) If you choose Service Test, service availability is based on one or more key beacons executing a service test, as defined next on the Beacons Page of the wizard. If you select System, service availability is based on the status of one or more key system components. As the Tip on this Availability page suggests, it is best to choose Service Test, unless such a test is not available. You should be able to choose a suitable service test for a web application, as the next page shows. *Choose Service Test.*

NOTE
If you define availability based on System, which you should not do for the Web Application service model, but which may be a valid choice for other service models, the wizard skips the Service Test and Beacons pages and goes straight from the Availability page to the Performance Metrics page.

Service Test Page The Service Test page is where you specify the first service test to confirm availability.

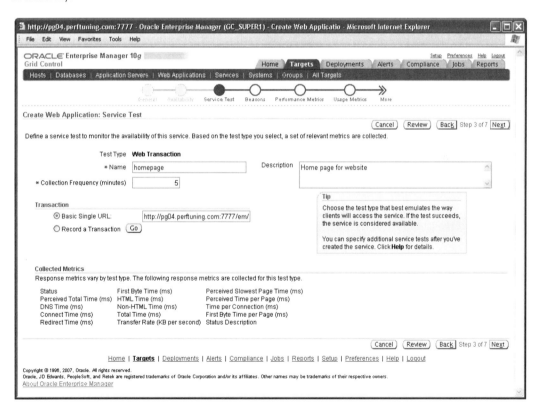

This should be the main test for how clients access the service, which for a web application is typically the application login URL. This is why Grid Control auto-populates the Basic Single URL field with the Homepage URL you entered on the General page. After you create the service, you can stipulate additional service tests, or this first test can be the only test, as is the case for the default EM Website service. Enter a name for the service test, an optional description, and a collection frequency, which should not be less than 5 minutes. Then choose the transaction type, which can be either Basic Single URL or a web transaction containing multiple steps (selected by clicking Record a Transaction, then Go).

The Collected Metrics listed at the bottom of the page are those gathered when executing beacons added or created on the next page. You can also set thresholds on these metrics (see "Performance Metrics Page" below), which when crossed will generate notifications. After entering your input on this page, click Next. *For our example, enter/make the selections shown above.*

Beacons Page   On the Beacons page, you add or create one or more beacons on hosts located in your key user communities. These beacons can execute the service test just created, and others you may later create, to monitor service availability.

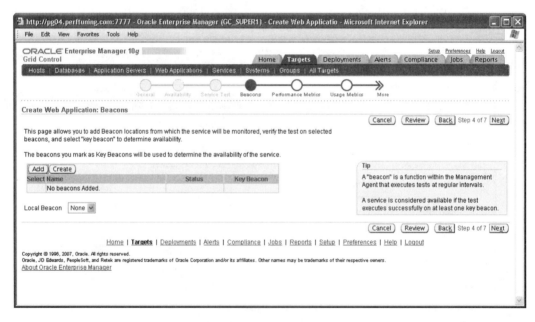

As I explained in the "Beacons" section above, unless you've explicitly created beacons on other Agents, the only existing beacon will be that for the Agent chain-installed with the initial OMS. This OMS beacon may not be the local beacon for custom services you create; however, you should add this beacon for all services created because it represents the Grid Control administrator user community. *In the case of the EM Website, this beacon is also the local*

beacon as it is the location of your Grid Control infrastructure. Add just the OMS beacon here to match the built-in EM Website service and to finish the wizard ASAP. You can create any additional beacons later (see the "Add Beacons" section later in this chapter).

Add the beacon on the initial OMS host as follows:

1. On the Beacons page, click Add.

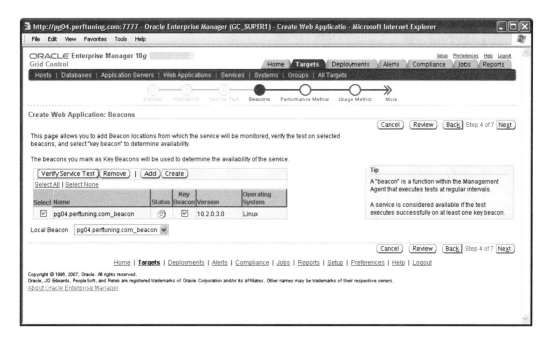

2. Choose the desired beacon. *Choose the OMS beacon (the only beacon offered by default), and click Select. As shown in the above illustration, check the Select box for the beacon and leave the Key Beacon box checked, as this is the beacon located on the server hosting the Grid Control infrastructure.*

3. If this is also the local beacon for the service, select this beacon in the Local Beacon field pull-down menu. *For the EM Website, select the OMS beacon in the Local Beacon field.*

4. Confirm the service test succeeds by clicking the Verify Service Test button. A test automatically runs for the beacon just added, and statistical results are displayed for those metrics listed on the Service Test page. *Click Verify Service Test.*

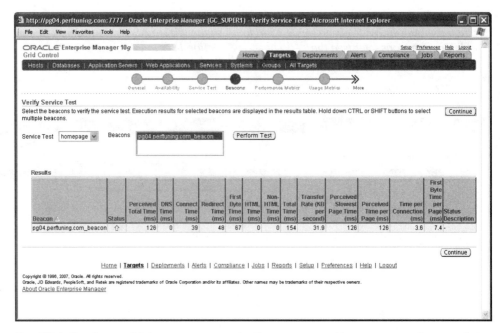

5. Click Continue, which returns you to the Beacons page. You can create beacons by clicking Create, but as I already recommended, I would wait to do this until after you've created the service. *Do not create any additional beacons because the default EM Website only employs a local beacon.*

Performance Metrics Page   The Performance Metrics page allows you to select metrics to measure the performance of the service.

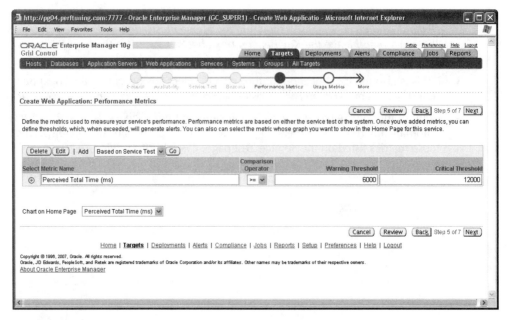

Chapter 12: Configure Group and Service Monitoring

For service-based availability, you can base these performance metrics either on service test(s) or on the system, whereas for system-based availability (where service tests are not offered), you can only base performance metrics on the system. This is a little confusing, because you may have thought that by choosing service-based availability, you've dispensed with the idea of basing anything on the system; however, remember from the "Performance Metrics" section above that you can base performance on that of service tests, but also on that of individual system components:

- **Based on Service Test** You can use response times for defined service tests. Test one or more metrics in the test types applicable to the service type. This option is only applicable when selecting Service Test on the Availability page (see above).

- **Based on System** If you select availability based on System, you can define performance metrics relevant to system components. This selection is available for both choices on the Availability page (i.e., for both system and service based availability).

TIP

For sake of clarity, I prefer to define system-based performance metrics on the Usage Metrics page, as the menus for both are identical. That is, the Performance Metrics menu when selecting Based on System is the same as the Usage Metrics menu. This way you are not mixing system and service-based metrics under the umbrella of "performance metrics." It makes it easier to remember where metrics are defined if performance metrics are always service-based and usage metrics are always system-based. The only exception to this is if you want the value for a usage metric to impact the service level calculation, then you should define the usage metric on the Performance Metrics page (see "Edit Service Level Rule" later in the chapter).

While I encourage you to define usage metrics on the Usage Metrics page and not on the Performance Metrics page, it is important to understand that, for both service test-based and system-based availability, usage metrics defined on the Usage Metrics page *cannot* be chosen to factor into the Actual Service Level (%) calculation displayed on the service home page. You must define usage metrics on the Performance Metrics page for them to be eligible in determining service level availability (you can select from these metrics on the Edit Service Level Rule page—see below). Even for system-based availability, usage metrics chosen on the Usage Metrics page are not considered in service level calculations.

The default EM Website service only defines the performance metric, Perceived Total Time (ms), which is automatically selected, as shown in the above screen shot. See the section below, "Add Performance Metrics," for instructions on selecting additional performance metrics.

Usage Metrics Page On this page you can measure user load on the underlying system by adding usage metrics relating to the performance of one or more system components. You can define usage metrics for both service- and system-based availability models. Here is the main Usage Metrics page.

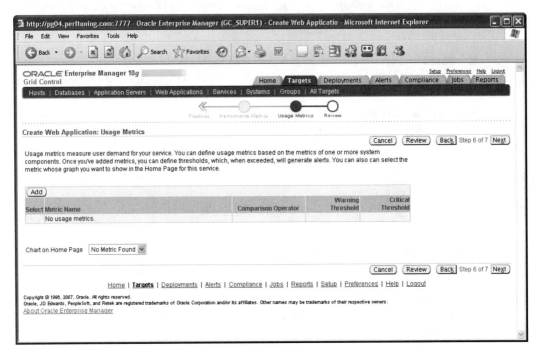

The default EM Website service does not employ any usage metrics. See the section later in the chapter, "Add Usage Metrics," for instructions on adding usage metrics.

Chapter 12: Configure Group and Service Monitoring

Review Page Review your entries as shown here.

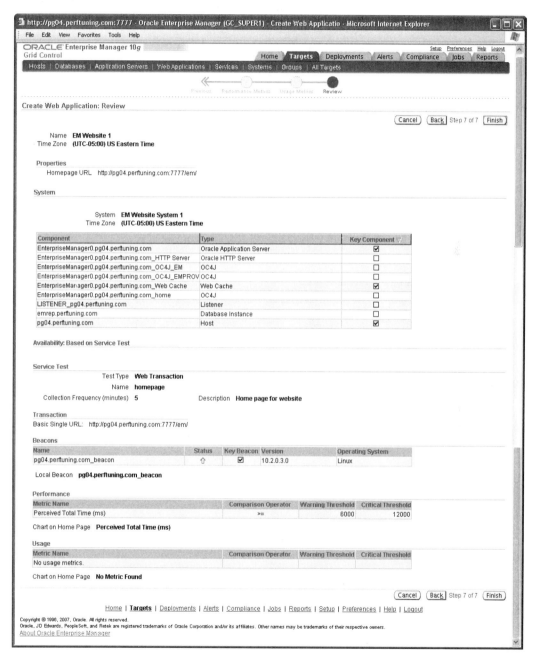

If you're satisfied with your input, click Finish to create the service.

This concludes the basic creation process for a service, but you're not finished configuring it yet. Read on for the remaining steps to beef up and tweak a service to maximize its utility to monitor the application it models.

Perform Monitoring Configuration Tasks

The Create Service wizard is the main throughway for defining a service, but the wizard is not comprehensive. You need to exit the highway now and take the roads less traveled to complete the service configuration to tap additional functionality. It's easy to find the roads because, with one exception, there are links to all configuration steps on the service's Monitoring Configuration page. To navigate to this page, select a service from the Services main page (*EM Website 1 in our case*) and click Configure (see Figure 12-1 below).

As was the case in Chapter 10 for navigating the Preferences and Setup menus, the most streamlined way to configure a service is to bounce back and forth between the sections of this page in the order below (link names are underlined):

- Enable management pack access for Web application
- Add Service Tests (<u>Service Tests and Beacons</u>)
- Add Beacons (<u>Service Tests and Beacons</u>)
- Change <u>System Configuration</u>

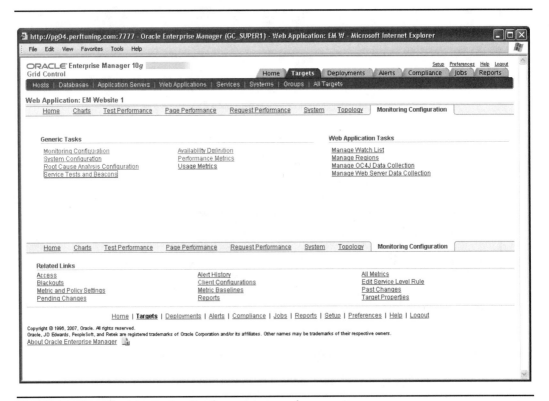

FIGURE 12-1. *The Monitoring Configuration page for a service*

Chapter 12: Configure Group and Service Monitoring

- Manage Web Server Data Collection
- Change the Availability Definition
- Add Performance Metrics
- Add Usage Metrics
- Manage Watch List
- Edit Service Level Rule
- Perform Root Cause Analysis Configuration

All tasks except the first have corresponding links on the Monitoring Configuration page. Furthermore, all tasks are in the Generic Tasks and Web Application Tasks sections, except Edit Service Level Rule, which is under Related Links. As I already noted, the Create Service wizard provides the interface to perform some of these tasks. However, I suggested dividing the service creation and monitoring configuration steps this way, given that the wizard is long and does not allow you to save your work as you step through it. *This served the additional purpose of isolating all steps to re-create the default EM Website Service to the "Create a Service" section above. Only in this current section do I add sample customizations suitable for a production environment. As above, I use italics to list steps applicable to our example service, EM Website 1.*

Enable Management Pack Access for Web Application If you create a Web application, as in the above example, grant it access to all licensed management packs to avail yourself of all licensed functionality. For instructions, see "Grant Access to Management Packs" in Chapter 11. The quickest way to find the Web application on the Management Pack Access page is to select Web Application in the Search field and click Go.

Add Service Tests The first link to visit on the Monitoring Configuration page is Service Tests and Beacons to create any additional service tests and beacons needed. This brings up the page shown in Figure 12-2.

As already cited, the service creation wizard only allows you to create one service test. Let's create an additional service test now for our EM Website 1 service. We'll make the service test a little more complicated than the initial single URL to test the Console login page.

Perform the following steps to add a new service test:

1. If you're planning to use the Web Transaction Recorder (see step 4 below), turn off all pop-up blockers on your client, as they cause the recorder to fail.

2. If you're planning to use the Web Transaction Recorder for a Console service test, log in to the Console via the protocol (HTTP or HTTPS) you want the service test to use to access the Console. This is required because the recorder, which opens in a spawned browser window, must inherit the credentials used to log in to avoid being prompted to supply them (which it can't).

3. Click Service Tests and Beacons on the Monitoring Configuration page (refer back to Figure 12-1) for the service.

4. Click Add next to the Web Transaction Test Type.

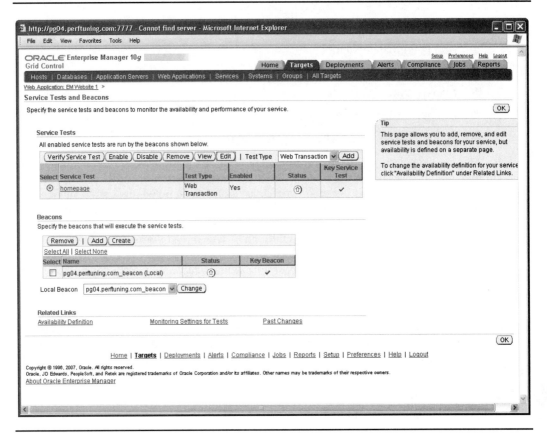

FIGURE 12-2. *The Service Tests and Beacons page for a service*

5. This brings up the Create Service Test page, which is shown in step 10 below, except that the steps you are about to configure are not shown yet at the bottom of the page. You can click Go on this page to use a Web Transaction Recorder to automatically record a transaction or click Create to manually specify one or more transaction steps. It is much easier to reproduce the steps in a recording session, as Grid Control automatically constructs the URLs for each step. *To demonstrate, let's record the steps to display Top Activity for the Repository Database.* To create a service test by recording a transaction, click Go on this page.

6. The Record Web Transaction page is displayed.

Chapter 12: Configure Group and Service Monitoring   **463**

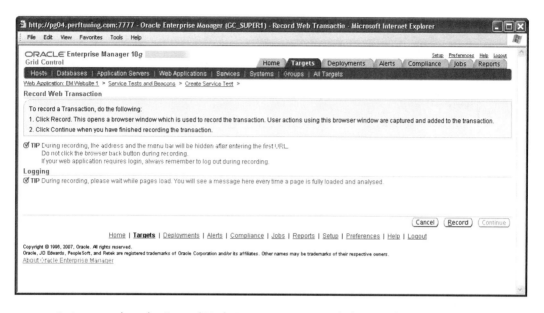

As instructed on the Record Web Transaction page, click Record, which opens a browser window to record a transaction.

7. In the address field of the new browser window, access the target web site and click through the desired actions. *Pull up the Top Activity page as follows: click the Targets tab, Databases subtab, choose the emrep Database target, click the Performance page, and the Top Activity link under the Additional Monitoring Links section. Wait for the Top Activity page to appear.* Observe the following points during recording in the new browser window:[9]

   ■ For an EM Website service test, enter the Console login URL using the same protocol (HTTP or HTTPS) over which you're currently logged in.

   ■ After entering the first URL, the address and menu bar are hidden.

   ■ Do not click the browser back button.

   ■ If your web application requires login, log out completely.

   ■ Wait to perform the next action until the previous action is recorded as reflected by a "Page Loaded" message for that action under the Logging section of the Record Web Transaction page.

   ■ Do not close the window manually. The Continue button does this in the next step.

8. Return to the above page and click Continue. This ends the transaction and closes the browser window in which you performed the actions.

9. The following page appears containing the steps you just recorded.

---

[9] Note 554824.1.

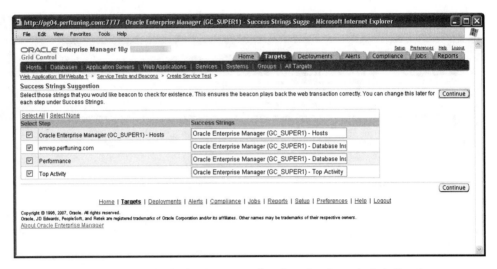

Select all strings to verify the beacon can play them back, and click Continue.

10. The Create Service Test reappears with the service test information. For GC 10.2.0.3 or earlier, go to the next step. For GC 10.2.0.4, a new Transaction section contains two additional selections to consider: Playback Mode[10] and Collection Granularity.

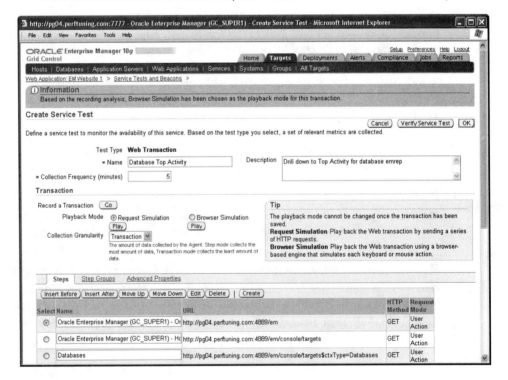

---

[10] Some of the material on Playback Mode is from the OEM Online Help for this page.

Chapter 12: Configure Group and Service Monitoring

- **Playback Mode** GC automatically picks one of the following playback modes based on a simple heuristic of the actions performed:

Request Simulation	This default selection is equivalent to Web transaction monitoring in GC 10.2.0.3. The OEM Online Help for this page advises this mode as more platform independent and scalable, but warns it is not appropriate for applications with an Asynchronous JavaScript and XML (AJAX)[11] based client request system. GC 10g does not use such an AJAX system.[12] Therefore, I would agree with the Online Help and select this playback mode for EM Website service tests. The shortcoming of this mode is that it only measures server response time, not browser rendering time.
Browser Simulation	In this mode, playback is via a browser-based engine simulating keyboard and mouse actions. This mode is appropriate for AJAX applications and allows for capturing both server response time and browser side rendering time. This mode requires IE 6.0 or higher and a 10.2.0.4 or later beacon on a Windows OS. Additional configuration steps are required, but, contrary to the Online Help, they are not documented in the OEM 10.2.0.4 Advanced Configuration Guide.

Manually modify the playback mode if necessary and click the corresponding Play button to verify the chosen mode can successfully play back the transaction. *Keep the default selection of Request Simulation.*

- **Collection Granularity** The granularity setting influences the amount of data the beacon collects. A web transaction consists of a series of steps, and each step contains one or more HTTP requests. You can either record a transaction as shown above or manually create one or more steps or step groups.

Transaction	Collects data for the overall transaction only. This is the default granularity level.
Step Group	Collects data for the overall transaction and the step groups in the transaction. To manually combine a set of steps into a Step Group, select this level, click the Step Group tab, and click Create.
Step	Collects data for the overall transaction, the step groups, and each step in the transaction. To manually create a step, select this level, click the Step page, and click Create.

For this example, use Transaction, the default Collection Granularity.

11. Optionally, for Request Simulation mode only, I recommend clicking Verify Service Test to confirm the transaction runs correctly. *Click Verify Service Test here.* On the results page, make note of the value for Perceived Total Time (ms). This metric determines how long it would take a web browser to play the transaction, which simulates the user experience. As such, this is an excellent metric on which to set thresholds for the service test, as done below. Click Continue on the Verify Service Test page.

---

[11] With AJAX, a client can retrieve/post any type of data from/to a server from a browser without refreshing the web page. Thus, AJAX can load the static portion of the page just once, and can load/update the dynamic portion on the fly. Source: Note 445852.1.

[12] Request Simulation is not an apt choice for GC 11g because it employs a new UI, the Oracle Application Development Framework (ADF), which leverages AJAX.

12. Click OK on the Create Service Test page to save the web transaction.

TIP
An alternative to the Verify Service Test button is to replay the J2EE transaction. To do so, click the service test name on the Service Tests and Beacons page (Figure 12-2). Click Play to return performance results on the client-side or Play with Trace for results on the server-side. This feature is only available when using Microsoft Internet Explorer 5.0 or later.

13. Select the service test and click Enable to activate it. *Do so for this example.* The Status might still not show UP, which you can rectify by going through the Verify Service Test steps again.

14. To set thresholds for service tests at the transaction, step, or step group level, under Related Links at the bottom of the Service Tests and Beacons page (refer back Figure 12-2), click Monitoring Settings for Tests.

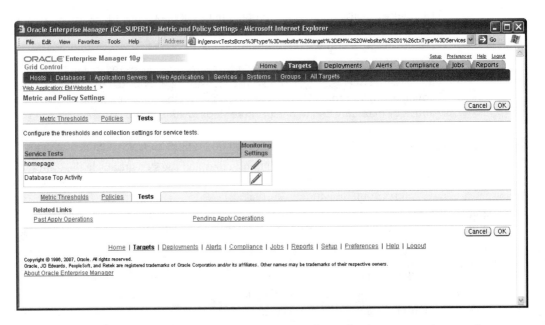

15. Click the pencil under the Monitoring Settings column for the new service test, *here called Database Top Activity.*

Chapter 12: Configure Group and Service Monitoring 467

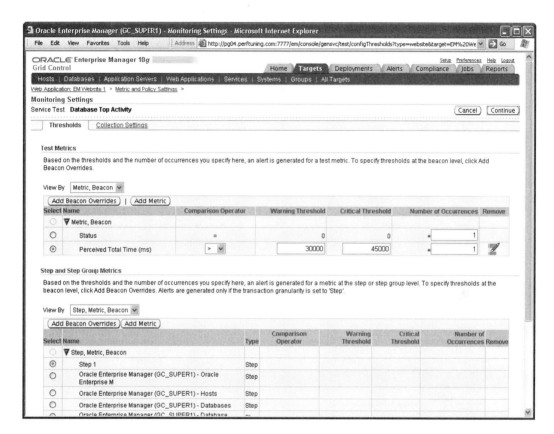

16. See the online help for this page and Note 433520.1, which provide details on how to define thresholds at the various granularity levels to generate alerts for service tests. The process is similar to that possible on the Performance Metrics page when first creating the service. You can even specify different metrics at the beacon level by clicking Add Beacon Overrides. Repeat for as many metrics as you want to receive alerts. *The above page contains no changes.* Click Continue when ready, then click OK to update the service test with your changes.

Add Beacons As you may recall, beacons are located in different user communities to monitor the availability and performance of services and web applications. At defined intervals, results of service tests, such as web transactions, are recorded from these beacons. You can add multiple beacons while creating the service. However, as suggested above, I prefer to add just the local OMS beacon when creating a service, and defer the creation of additional beacons for the service until now. There are many reasons for using multiple beacons at select geographical user communities; following are some of the more common ones:

- To compare service test performance against benchmark results from the local beacon.
- To more accurately reflect the user communities represented. For example, you may need to add beacons that run unique service tests in certain user communities, thereby broadening the list of representative user actions.

- To bring new hardware online, for which you need to designate new key beacons to determine service availability. (Remember, a service test is considered successful if at least one key beacon can execute it.)
- To run ping tests to network devices and hosts to gauge network latency.

To create a beacon for an existing service, follow the procedure below. *In our example, we create a beacon called "Dallas" for the Agent on host ptc3-vl3.perftuning.com.*

1. Click Create under the Beacons section on the Service Tests and Beacons page (Figure 12-2).
2. The Create Beacon page is displayed for you to enter the beacon information, including any proxy settings required for it to run a service test.

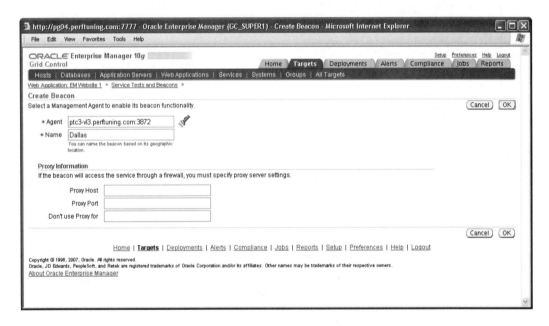

- **Agent** Select the Agent for a host on which you want to run a beacon to monitor the service.

    Name Choose a Name for the Agent. The default name for the OMS beacon created by default is the host name with "*_beacon*" appended. Enter a name using this convention (which would be *ptc3-vl3.perftuning.com_beacon* for the Agent in our case) or a name based on the geographical location of the Agent host *(such as "Dallas" as used in our example)*. Choose the name carefully as it will not only appear on the Service Tests and Beacons page, but also on pages that list target names, such as the Agent home page, the All Targets subtab, and the Targets page for a given host target.

Chapter 12: Configure Group and Service Monitoring   **469**

- **Proxy Information**   If the beacon and the service are separated by a firewall, enter the Proxy server settings in this section.

Click OK when finished to create the beacon.

3. Wait five minutes for the beacon to execute and confirm the status changes to UP. On the Service Tests and Beacons page, the status of the beacon will initially appear as STATUS PENDING. This is expected for a newly added target whose availability state is being calculated.

4. Response time metrics for service tests are collected for this beacon, as for all beacons. You can also designate this as a key beacon for determining service availability (a service test is successful if it executes on at least one key beacon). To mark the newly created beacon as a key beacon:

   - Click OK on the Service Tests and Beacons page.

   - On the Monitoring Configuration page (see Figure 12-1), click the Availability Definition link under Generic Tasks.

   - On the Availability Definition page (see Figure 12-5 later under the "Change Availability Definition" section), check the Key Beacon box for the beacon you created and click OK.

Change System Configuration   Use the System Configuration page (see Figure 12-3) to change the system on which the service is chosen to run, or to change your definition of which components to classify as key components.

You may need to elevate a new or existing system component to key component status. Check or un-check the Key Component column for components of the system. *For our example, we select the following GC target types as additional key components: OC4J_EM and OC4J_EMPROV, Oracle HTTP Server, Database, and Listener. Elevating these to key component status accurately reflects that these components must be running for GC to be fully functional, and permits Root Cause Analysis to automatically test the availability of these components (see the "Perform Root Cause Analysis Configuration" section below).* Click OK when finished and then click Yes when prompted to save your changes.

Manage Web Server Data Collection   Grid Control taps the *end-user performance monitoring* capability of the Apache-based web server of a web application to help improve its performance. This monitoring provides a way to examine response times on pages as end users navigate a web site. A GC administrator can configure an OMS Web server to collect monitoring data in a log file for their GC Web site. The OMS uploads this data to the Repository, and admins can then view and analyze this data on the Page Performance page for the EM Website service.

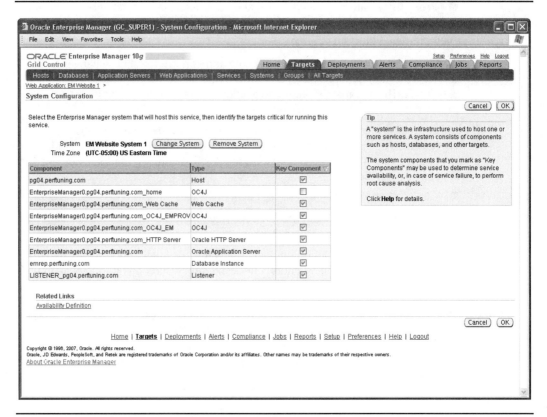

FIGURE 12-3. *The System Configuration page for a service*

CAUTION
It is important to understand that you do not use the GC Web Cache to collect data for any of your Web applications—only for the EM Website itself, the subject of our example. We are turning Grid Control on itself, so to speak, and using the GC Web Cache to collect performance data on the EM Web application. Similarly, to monitor another web application, use the web server associated with that application.

To configure a web application's web server to collect performance data, navigate to the Monitoring Settings page (Figure 12-1) of the web application and click Manage Web Server Data Collection. This brings up the page shown in Figure 12-4.

Chapter 12: Configure Group and Service Monitoring

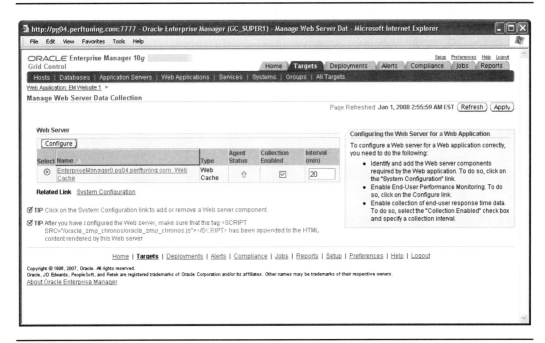

FIGURE 12-4. *The Manage Web Server Data Collection page for a service*

The write-ups on the right sides of this and other configuration screens are very descriptive, and should suffice in explaining the rationale behind the end-user performance monitoring configuration process. As the write-up on Figure 12-4 makes clear, configuring the web server for a web application is a three-step process:

1. Add the Web Cache associated with the application as a component to the underlying system. *In our case, we already completed this step when configuring the EM Website 1 system by adding the Grid Control Web Cache as a component.*

2. Configure the End-User Performance Monitoring feature with Web Cache.[13] This is the lengthiest of the three steps, but a straightforward process nonetheless:

    a. Click Configure, which launches Application Server Control. You are prompted to enter the *ias_admin* username and password.[14]

    b. The following screen appears to allow setup of end-user performance monitoring.

---

[13] Oracle Enterprise Manager Concepts 10g Release 4 (10.2.0.4) states in Chapter 12 under Monitoring Web Application Services that you can also configure this feature with Oracle HTTP Server. However, this component is not listed on the Manage Web Server Data Collection page, even when chosen as a key component.

[14] The *ias_admin* password by default is the same as the SYSMAN password supplied at GC installation time.

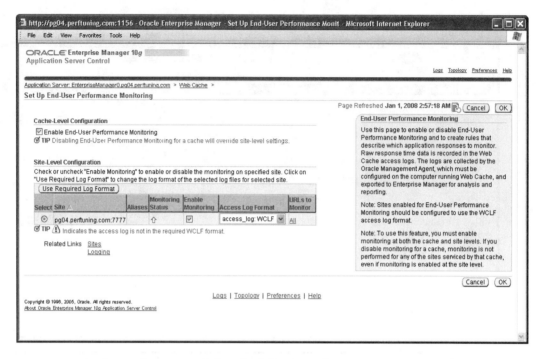

As shown above, enter/select the following information:

i. Under the Cache-Level Configuration section, check **Enable End-User Performance Monitoring**.

ii. Under the Site-Level Configuration section:

1. Ensure **Enable Monitoring** is checked.

2. Leave the default (and only) setting in the Access Log Format column. This column lists all available access log file names and file formats, separated by a colon, for storing monitoring data. The log format must be in end-user performance Monitoring format, which is WCLF (Web Cache Log Format) by default. If set differently, you would click Use Required Log Format, then click Yes to confirm this choice.

3. In the URLs to Monitor column, leave All selected so as not to filter any URLs from end-user performance monitoring as configured from this page. If you want to limit the log file from getting too large, you can click All to configure rules to filter certain URLs for the site. However, I provide alternate instructions to both filter and rotate the *access_log* in Chapter 4, "Reduce Grid Control Logging."

Click OK to save your changes.

c. You're prompted to restart Web Cache, but you can make any necessary properties changes first.

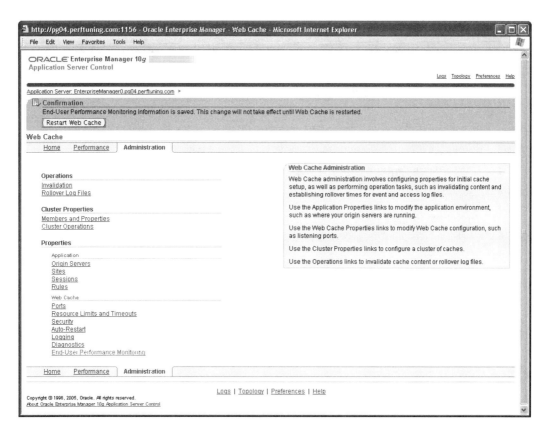

As the Web Cache Administration page describes, you can make many such properties changes. *For our example, the default properties are the correct ones for monitoring the EM Website.* However, monitoring other web applications may require using the Application and Web Cache Properties links to change their respective configurations. Click Restart Web Cache to put end-user performance monitoring into action.

d. The next page warns that you will lose all cached objects when restarting, and prompts you to confirm. Click Yes to restart Web Cache. You should receive confirmation that it restarted.

3. The last step is to enable collection of end-user response time data. Return to the Manage Web Server Data Collection page in Grid Control (Figure 12-4). Make sure the Collection Enabled box is checked, specify a suitable collection interval (20 minutes is sufficient), and click Apply to save all changes.

This concludes the process to set up web server data collection for your GC web application.

Change Availability Definition On the Availability Definition page (shown in Figure 12-5) you can modify the basis of the availability definition for the selected service (i.e., service test-based

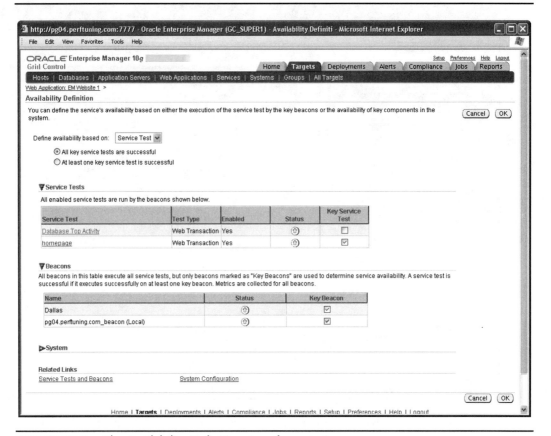

FIGURE 12-5. *The Availability Definition page for a service*

or system-based) and allocate key service tests, key beacons, and key system components. *For our example, EM Website 1 service, no changes are needed on this page.*

1. **Define availability based on**   Specify whether to define availability based on Service Test or System. If you choose Service Test, choose one of the following for determining when the service is deemed to be available:

   - All key service tests are successful (default).
   - At least one key service test is successful.

2. **Service Tests**   Mark one or more service tests as key tests by enabling the Key Service Test checkbox. A service is considered available if at least one or all key service tests (depending on your choice above) can execute from at least one key beacon. To use service-based availability, you must designate at least one service test as a key service test.

3. **Beacons**   Choose at least one beacon as a key beacon (typically the local beacon at a minimum). If there are no key beacons, the service test will have an UNKNOWN status.

4. **System** Expand the System heading to see all system components currently chosen as key components. You can change key component choices here as well as on the System Configuration page, as done for the EM Website 1 service. Key system components are used to determine availability and the possible root cause of a service failure (refer to "Perform Root Cause Analysis Configuration" below for more information).

Add Performance Metrics Use the Performance Metrics link on the Monitoring Configuration page to add or change performance metrics used to determine service performance, or to select the graph shown on the home page for the service. The Performance Metrics page is the same page that appears in the Create Service wizard.

Following is the procedure for adding a metric on the Performance Metrics page. *For our example, I add a performance metric called Total Time (ms) for the Database Top Activity service test to be run by the initial OMS beacon.*

1. Click the Performance Metrics link on the Monitoring Configuration page (see Figure 12-1).

2. Select either Based on System or Based on Service Test in the Add drop-down menu. If you're using service-based availability, I recommend choosing Based on Service Test, as already discussed. Regardless of your choice, metrics selected will be eligible for notification after creating the service, but only for the Web Application service type, as mentioned above. *Select Based on Service Test.* Click Go.

3. If you choose Based on Service Test, the following page appears.

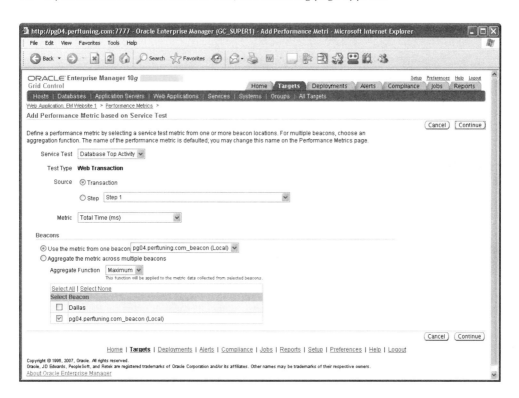

Make choices on the following:

- **Source**   Select the Source, either Transaction or Step.
- **Metric**   Select a metric on which you want to base your service's performance or for which you want to receive notifications.
- **Beacons**   Choose to use the metric from one beacon, or to aggregate the metric across multiple beacons along with a corresponding aggregate function. If you choose to aggregate, select the beacons for which you want to evaluate the metric.

For our example, make the choices shown in the screen shot. Click Continue.

4. This returns you to the Performance Metrics page (identical to that shown above in the "Performance Metrics Page" section), where you can optionally enter a comparison operator and warning and/or critical threshold values if you want to receive a notification of metric out-of-bounds conditions. This notification requires a corresponding notification rule for the web application. You must enter a Critical threshold value for the performance metric to be considered when computing the service level (see "Edit Service Level Rule" below). Click OK.

CAUTION
Based on my testing, it appears that a bug in GC 10.2 does not allow you to choose the same metric name for two different service tests.

5. GC warns you that this performance metric modification will affect the way the service level assessment is calculated. Click Yes to proceed.
6. Repeat steps 1 through 5 for any additional desired metrics.
7. In the Chart on the Home Page field, select the metric critical to your site or that best represents the overall performance of the service. The service home page will display this metric's historical values from the past 8 hours or so. *Select Perceived Total Time (ms).*
8. When you've finished with the Performance Metrics page, click OK.

Add Usage Metrics   Use the Usage Metrics link on the Monitoring Configuration page to add or change usage metrics that establish the workload for a service. The Usage Metrics link takes you to a page offering the same options as the Usage Metrics page in the Create Service wizard (see the "Usage Metrics Page" section above).

TIP
When you click Metric and Policy Settings for a service, the metrics listed on the Metric Thresholds page are a compilation of the Performance Metrics and Usage Metrics currently configured. However, nothing on the Metric Thresholds page indicates which metrics are for performance and which are for usage. This underscores how seamless it is to use both service-based and system-based testing by selecting service-based availability, then adding Usage Metrics. Such a combination is not possible when selecting system-based availability.

Chapter 12: Configure Group and Service Monitoring **477**

Following is the procedure to add usage metrics. *For our example, I add a usage metric called Request Processing Time (seconds) for the OC4J_EM module.*

1. On the Monitoring Configuration page for a service (see Figure 12-1), click Usage Metrics.
2. Click Add on the page.
3. This brings up the Add Usage Metric based on System page (see Figure 12-6).

    This is the same page that appears if you select Based on System on the Performance Metrics page. Make selections for the following:

    - **Target Type**   Choose any target type. All target types are selectable, even those for components not designated as key components.

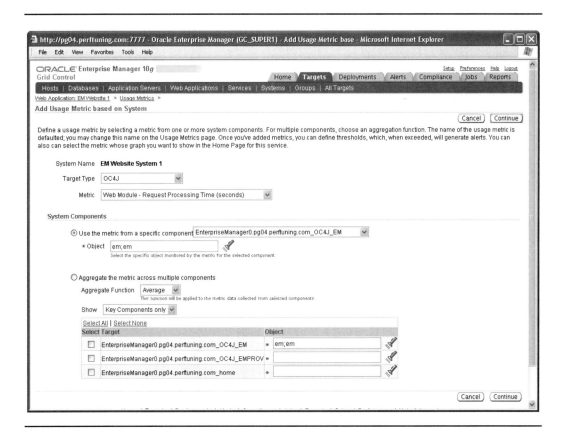

FIGURE 12-6.   *The Add Usage Metric based on System page*

- **Metric** Choose a metric of interest for the target type selected. Avoid selecting usage metrics for a system component whose metric thresholds are already set for the component under its Metric and Policy Settings page, unless you're choosing an aggregate function (see next bullet), which would differentiate the usage metric calculation from the metric calculation for the component.

- **System Components** All components of the chosen target type appear in the fields for each radio button. Choose to use the metric from a specific component or to aggregate across multiple components. If you choose the latter, select the aggregation function and those components you wish to aggregate.

Make the selections shown above. (Under the System Components section, the "em;em" selection in the Object field corresponds to selecting the Web Module Name "em.") Click Continue after making your selections.

4. This returns you to the Usage Metrics page, already shown above in the "Usage Metrics Page" section, but which now reflects the usage metric just added. As with performance metrics, optionally specify a comparison operator, and critical and/or warning thresholds if you want to receive a notification of metric out-of-bounds conditions, and enter a critical threshold value if you want the metric to be evaluated when calculating the service level (see "Edit Service Level Rule" below). *Enter/choose the values shown above for the EM Website 1 service.* Click OK.

5. Repeat steps 1 through 4 above for additional usage metrics you'd like to see.

6. For the Chart on Home Page field, specify the metric to display graphically on the service home page which best represents the web application. *Select Web Module—Request Processing Time (seconds).*

7. Click OK to save all changes.

Manage Watch List After configuring end-user performance monitoring, Grid Control begins collecting response metrics for every application URL that users visit. The Watch List is a place to keep tabs on crucial user pages the web application administrator identifies. For example, every user must go through the login page, which is a user's first impression of your site. Well, you know how important first impressions are.

Whether it is your corporate intranet portal, or Grid Control itself, it's important not only for job efficiency purposes, but for employee morale, to provide well-performing Web sites for them to do their jobs. One of these Web sites that DBAs and other IT professionals have come to rely on is their Grid Control site.

A fair percentage of my consulting engagements are to do Grid Control performance tuning. It's immediately clear at many of these sites that such tuning is needed when I try to log in to their GC environment and it just hangs. Amazingly, these companies still rely on the GC monitoring and notification mechanisms initially set up and which continue to work in a relatively timely manner, but the system is coasting downhill. When it's absolutely necessary to make a configuration change, an administrator, like a reluctant mechanic on an "ol' Jalopy," rolls up his or her sleeves and logs in. It helps to be good at multitasking—between clicks, that is.

When arriving at one of these Grid Control sites, after providing acute care and tuning (see Chapter 17), I work with administrators to set up a Watch List to give them immediate feedback on just how well their site is performing. It's easy to set up a Watch List; here's how:

1. Identify those pages that are important to users or that have a history of not performing well. The login page (i.e., home page) is certainly one of these URLs. You can identify nonperforming pages using several methods. Aside from your personal knowledge of slow URLs from navigating in the Console, there are at least three other ways to identify slow pages:

   ■ Evaluate the results of the Request Performance page for the service, which shows the slowest requests broken down by server tier. On the Services main page, click the link for the service in the Name column, then click the Request Performance page link.

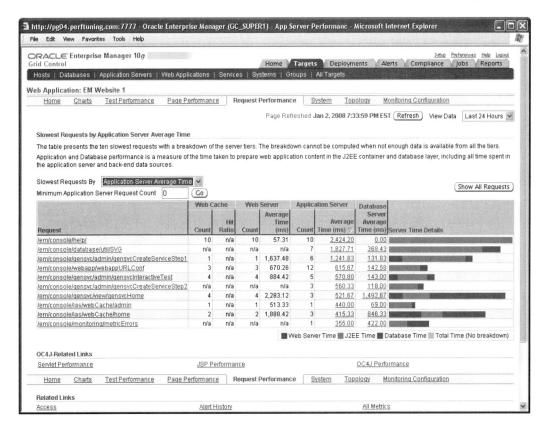

Note the worst performing URL requests. Choose representative samples of the worst performing URLs for each tier by selecting Web Server Average Time, Application Server Average Time, and Database Server Average Time from the Slowest Requests By pull-down menu.

   ■ Add URLs for metric tests you've configured that repeatedly generate alert notifications.

- Manually play service tests, either using Play or Play with Trace, and from the Trace Results, select the URLs for steps that contributed the largest share to the overall response time of the service test. *Here are the results of clicking Play for the service test, called Database Top Activity, we created earlier.*

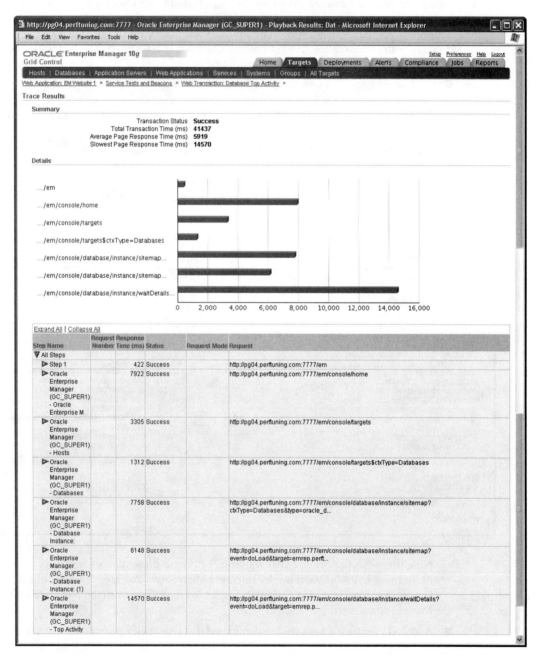

The graph makes it easy to spot the slowest page, but the Request column may cut off some of the URL if it's too long, so you may need to manually step through the test in the Console to get the complete URL. *Above, the slowest URL was the very last step: clicking Database Top Activity. Click manually through to the Database Top Activity page and copy the URL from the browser Address field to your clipboard to paste in the next step.*

2. Add the URLs you identified above to the Watch List. Click Manage Watch List on the Monitoring Configuration page (Figure 12-1), which brings up a summary page showing any existing Watch List items, their URLs, and descriptions. Click Add.

Complete the fields shown above. *In this example, I add the URL for the step that takes the longest when executing the Database Top Activity service test created earlier, namely, selecting the Top Activity link from the emrep database Performance page.*

Once you have a Watch List for your web application, you can get a first glimpse at how well your site is performing. You should return to this Watch List periodically, perhaps daily—and weekly at the outset—to get response times for critical user pages, beginning with the login page, which many visitors use as a litmus test for each repeat visit to the site. On a weekly basis, you should review your Watch List, pruning it for URLs no longer needing monitoring, and adding to it those application URLs that creep up to the top of your worst performing pages list.

Edit Service Level Rule A *service level* in Grid Control is a measure of service quality, defined as the minimum percentage of time during business hours in which a service is expected to meet certain availability and performance criteria. A service level agreement (SLA) for services at a company should specify such criteria. Grid Control wants to know your service level expectations to evaluate how well the service conforms to SLA objectives. Indeed, the home page for the service displays the current Actual Service Level (%) next to the Expected Service Level you define.

The Edit Service Level Rule page is where you define service level related parameters. Here are instructions for this rather straightforward interface to configure the service level for a service:

1. Under Related Links, click Edit Service Level Rule.

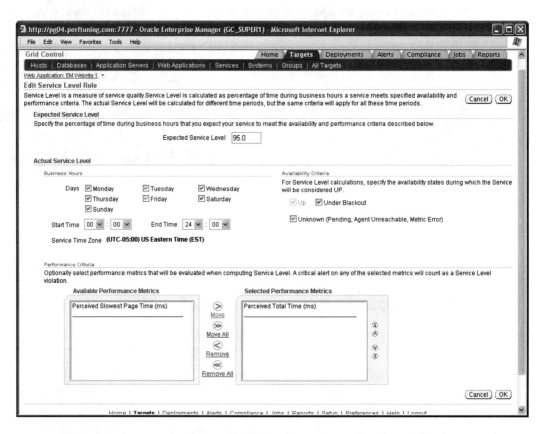

2. This page has two main sections:[15] Expected Service Level and Actual Service Level.

   - **Expected Service Level** Choose the service level percentage, or the minimum period of time (default 85%) during business hours for which a service must meet service levels.

   - **Actual Service Level** This is where you define business hours and service level criteria for availability and performance.

---

[15] The names for these sections—Actual Service Level and Expected Service Level—are a bit misleading, in my opinion, given that the whole page is for entering expected service level values.

Business Hours — Choose the days of the week and times during which the service must perform, accounting for availability criteria, without triggering a critical performance metric alert (see Performance Criteria below).

Availability Criteria — Choose the availability states for which a service is evaluated as UP. Check Under Blackout (not checked by default) or Unknown (checked by default) to relax service availability calculations not to include such statuses. Make your selections in accordance with your SLA, or if it does not account for such factors, amend your SLA to coincide with these selections.

Performance Criteria — Select from the available metrics already chosen as performance metrics (see the section "Performance Metrics" earlier in this chapter) to consider when you're calculating the service level percentage. As mentioned above, for both service test-based and system-based availability, the only usage metrics under Available Performance Metrics on this page are those chosen under the Performance Metrics page, not the Usage Metrics page (see respective sections for these pages above).

Make the selections for the EM Website 1 Service as shown in the screen shot above. Click OK after entering the desired SLA criteria for the service.

3. You can click the Home page for the service to see the impact of your service level choices. Click the Availability % and Actual Service Level (%) values and drill down to other links to see what functionality Grid Control offers in this regard.

Perform Root Cause Analysis Configuration Grid Control offers a Root Cause Analysis (RCA) feature to help you narrow down the cause of service failure. When multiple alerts fire and application users begin reporting problems, it is no small comfort to be able to rely on RCA to filter out all alerts and user symptoms downstream of the underlying problem. This analysis can allow you to more quickly resolve the issue and return your application to operational status. RCA results consist of graphical flags of the system component(s) responsible for the failure, with the topology page serving as a backdrop to illuminate component interrelationships. In this way, multiple alerts are interpreted for you. Before that can happen, however, you need to configure RCA by choosing the analysis mode (automatic or manual) and by defining any relevant metrics and thresholds for component tests. In this way, you program the artificial intelligence that RCA can later exhibit. In either mode, the service must currently be down to see the results.

The example system and service we've been using is from an installation using the GC New Database Option. Since all key components must be running to access the Console, the easiest way to demonstrate RCA is to create a component test with a threshold artificially set such that the condition is not met. For this example, we add a condition that the metric Load > Run Queue Length (5 minute average) on the GC host must not exceed the set critical threshold of 0. This, in turn, requires setting a critical metric threshold for the Host target on its Metrics and Policy Settings page.

To define how RCA is run, navigate to the Monitoring Configuration page (Figure 12-1 above) and under Generic Tasks, click Root Cause Analysis Configuration.

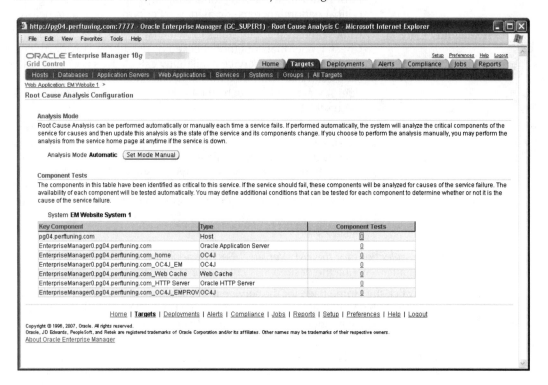

This page has two sections: Analysis Mode and Component Tests.

- **Analysis Mode**   Choose either Automatic (the default) or Manual mode:
    - **Automatic**   This mode automatically runs RCA whenever a service fails. Grid Control analyzes key components for the cause of the failure and updates the analysis in real-time. Choose this mode unless you do not wish to take advantage of this functionality. *Choose Automatic Mode.*
    - **Manual**   This mode disables RCA but allows you to perform RCA from the service home page by choosing Perform Analysis. One drawback is that RCA does not store historical results in Manual mode.
- **Component Tests**   This section lists the components and component tests defined for the service. By default, only the unavailability of a key component is considered a cause of service failure. However, you can add component tests as additional conditions for flagging a component as a cause of failure. To manage (add, remove, or edit) component tests, do the following:
    1. Click the link in the Component Tests column for the key component you want to test. *For our example, EM Website 1 service, click the number zero "0" link for the Host type.*
    2. Click Add on the Component Tests page for the chosen target.

Chapter 12: Configure Group and Service Monitoring

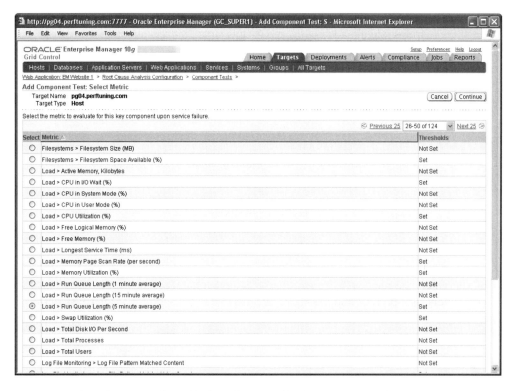

3. Select the Metric from the list of metrics for the target type chosen. *In our example, choose Load > Run Queue Length (5 minute average), as shown above.* Click Continue.

4. Set the fault threshold for a component test.

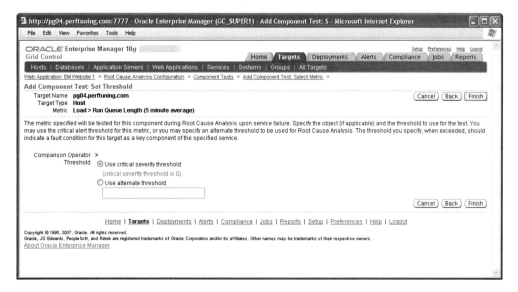

Choose either "Use critical severity threshold" (the default) to use the Critical metric threshold already configured for this target or "Use alternative threshold" to enter a custom fault threshold against which to compare the last collected value. It is customary to choose the default (the current critical threshold). If this threshold is not set or is not correct, change it at the target level (in Metric and Policy Settings) so that the RCA can inherit this default value on the above page. *Select "Use critical severity threshold" as it is set artificially low (to 0) to purposefully cause the component test to fail.* Click Finish.

5. Repeat steps 2 through 4 to add any more metrics for this component. *No other metrics are needed to demonstrate this feature, but you may add any other metrics desired.*

6. Click the Root Cause Analysis Configuration link at the top and repeat steps 1 through 4, selecting any other components for which you want to add component tests. *Our example does not require any other component tests, but you may add any other tests deemed appropriate.*

To view the RCA results, go to the home page or topology page of the service; the service must currently be down to see the results. The service topology page shows the system component relationships and highlights target(s) that RCA identifies as the cause of the service failure. In this example, the icon for the GC host, pg04, would contain a down arrow on the topology page as shown above in the section, "Create a System."

TIP
You may elect to remove the duplicate EM Website 1 System and service now (see "Remove Targets From Monitoring" in Chapter 11). However, consider keeping this service, as its configuration is a good starting point for customizing your EM Website. You may also want to contrast the functionality that services afford over groups by examining the EM Website 1 service alongside the GC_Targets group we created. You may then want to remove this group if it's not needed.

The business is likely counting on you to keep its services available and performing well. Grid Control allows you to constantly monitor services so you can address problems before they take irreversible hold. This monitoring allows you to intervene before services succumb to the second

---

Can You Monitor the Oracle E-Business Suite as a Service?
Oracle offers the Oracle Application Management Pack for Oracle E-Business Suite (EBS), which considerably extends Grid Control's ability to govern Oracle EBS Release 11*i* and Release 12, known as the Oracle Applications in releases prior to 11*i*. This pack adds a new target and Targets subtab called Oracle Applications. The pack is available as patch 5489352, and the latest version (2.0.2) is certified for Linux x86, Linux x86-64, Solaris SPARC (64-bit), AIX5L-based (64-bit), HP-UX (PA-RISC and Itanium), and Windows. For links to all documentation, including an installation guide that provides detailed configuration steps, see Oracle*MetaLink* Note 394448.1.

law of thermodynamics. The only part of a service that defies the second law is the clock displayed in Grid Control when restarting a component of that service. If you stare at it long enough, you'll see that it actually reverses at times.

Summary

As you have seen, setting up groups is a relatively easy process, as contrasted with configuring services. However, as the configuration process itself reveals, features for services are more far-reaching than those for groups.

As for groups, I began by showing you how to create a group consisting of all Grid Control targets, called GC_Targets. Following this was a tour of group features, including all group pages, the System Monitoring Dashboard, and out-of-box reports related to groups.

I introduced the concept of services by explaining service level agreement (SLA) objectives in Grid Control terms (service availability definitions—service test or system-based, performance metrics, and usage metrics). I then provided some background on three key service elements—beacons, transactions, and service types. Having dispensed with the science of services, I illustrated the configuration process for a service, which takes up the bulk of the chapter. As with groups, I used Grid Control as a paradigm service by building a duplicate EM Website system and service. The main steps are to create a system, create the service using a wizard, and perform monitoring configuration tasks for the service. Creating a system is easy enough. Creating the service requires a bit more work, but I explained how to get through the wizard ASAP (because you must complete it to save your entries) and how to also reproduce the default EM Website at the same time. The final section on monitoring configuration tasks customized the EM Website service as is appropriate for most sites. The section covered configuring features such as additional service tests and beacons, web server data collection, availability definitions, performance/usage metrics, Watch Lists, service level rules, and Root Cause Analysis (RCA).

Now that we have set up the basic target types (Chapter 11) and additional aggregate targets (this present chapter), we can turn our attention in the next chapter to tuning metrics and policies for all target types.

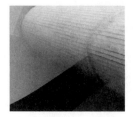

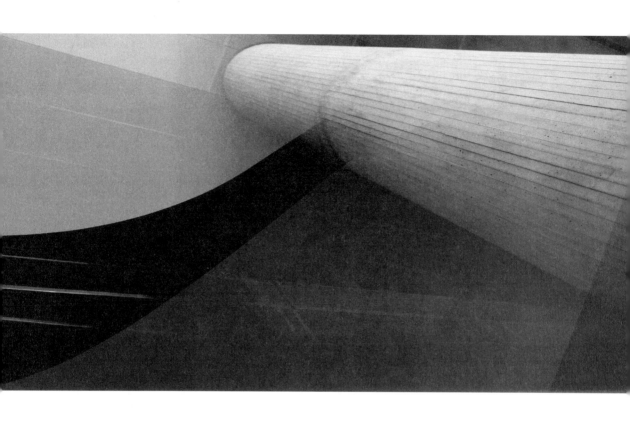

CHAPTER 13

Tune Metrics and Policies

he number, variation, and complexity of targets in corporate IT environments increase every year. Grid Control offers a way to manage these resources by automating its approach to target problem detection and resolution. Metric and policy alerts are configured out-of-box to notify administrators of such problems.

This chapter addresses how to set up and tune metrics and policies for targets so that you receive notifications that are meaningful for your IT environment. Which notifications are considered "meaningful" can differ widely by site. However, almost no GC site escapes the initial spray of notifications from out-of-box metrics and policies. This is not a faulty product design, in my opinion, but rather a strategic decision by Oracle that it's better to send too many notifications than too few. Still, the majority of these notifications will be false, due, by and large, to thresholds that have not been properly adjusted yet for your specific environment. Notifications usually lagging a close second will be for uninteresting metric and policy conditions—uninteresting to you, that is. Some notifications will be duplicates for the same condition, due primarily to rapid-cycle threshold crossings, but also due to multiple metrics reporting the same condition. Finally, there will be legitimate notifications, but they'll be hiding in plain site among all the notifications crying wolf.

Perception is everything. If you install Grid Control and set up multiple administrators for notifications before tuning metrics and policies, they will receive waves of initial "raw" notifications. These folks will probably walk away with a bad first impression of Grid Control. For some, it may be their last impression. The equivalent of this share-all approach in automotive terms would be if a car manufacturer in the early stages of developing a new model, asked would-be buyers to take it out for a spin—it wouldn't exactly instill confidence that the car would be reliable when it went into production. In the same manner, nobody wants to be a GC "crash test dummy." Yet one administrator at your site needs to volunteer for the position. The good news is that it's a short-term position: this admin just needs to accomplish the steps in this chapter so that other admin can rely on notifications to alert them of *meaningful* out-of-bounds conditions for their site.

Here's what the job entails. Monitor Grid Control for at least one week. During this time, let it collect metric data on all managed targets, let notifications fire at will while you begin chiseling away at them as directed in this chapter, and don't enable e-mail notifications during this time for anyone except perhaps yourself.[1] Use the super administrator account discussed in Chapter 10 (which I called GC_SUPER1) for the task, for the reasons explained in that chapter. Ideally, this one administrator can take on the responsibility of tuning notifications for all targets, because this makes it easier to standardize the tuning process. However, if there are too many targets to make this approach feasible, create accounts for other administrators and let them tune notifications for targets they will oversee. During this week of mayhem, the responsible super administrator(s) must perform the arduous task of attempting to rein in notifications.

Many companies with existing, languishing installations also need a GC crash test dummy. You may be "volunteered" for the position (you know—everyone else took a step backward). Many admin take over these GC installations with "preexisting conditions." The scenario is all too familiar: an Oracle site installs Grid Control, including Agents on target hosts. Then, when administrators are overwhelmed by notifications, they shut down the entire GC environment.

---

[1] You don't even have to enable notifications for yourself unless you need to immediately begin monitoring targets. You can get all the alert information you need from the Console.

Later on, someone at the site wants to give GC another try, and asks the volunteer to take it into the server room (where no one can hear) and restart it.

Sites that call on outside help to implement Grid Control may not fare much better when it comes to the flood of notifications. Clients usually underestimate how long it takes to do the job right, forcing consultants to cut corners. Most of my Grid Control consulting assignments, and those on which I've advised fellow consultants, have been for one week—perhaps two weeks if the client required a highly available infrastructure. The inherent problem with such a short-term engagement is that by the time a consultant installs and completes basic setup of Grid Control, little time remains to tune notifications. The consultant implements "high availability" all right—Grid Control showers administrators with alerts 24×7. Not long after the consultant leaves, someone left holding the bag understandably disables notifications or shuts down Grid Control altogether. You then come whistling along, and are asked to dust off and fire up the old GC engine. If you're in this situation, and must tune frequent and copious alerts, this chapter in particular will speak to you.

The following table lists the overall tuning objectives of this chapter for metrics and policies and the corresponding GC features to accomplish those objectives. The objectives are listed in a logical order of attack, from general to specific, relegating to last position the advanced functionality of automating responses to events.

Tuning Objective For Metrics And Policies	Feature In Grid Control to Meet Objective
Coarse-tune metrics and policies	Metrics and Policies
Fine-tune metrics	Metric Baselines and Metric Snapshots
Customize metrics	User-Defined Metrics
Apply standardized monitoring settings across targets	Monitoring Templates
Automate responses to certain metric alerts and policy violations	Corrective Action Library

Regrettably, dependencies in the Console between GC features make it inefficient to implement the tuning goals in the preceding functional order. Rather than bore you with these dependencies, let me simply present a derived tuning order that streamlines the tuning process for metric alerts and policy violations:

Change Default Metrics and Policies → Enable Metric Baselines → Add Corrective Actions → Implement User-Defined Metrics → Use Metric Snapshots → Leverage Monitoring Templates

Given that tuning is an iterative, circular process, the more tuning iterations (or circles) you complete, the less critical the tuning order becomes. However, when first beginning to tune metrics and policies, as we are about to do, the above order will maximize your productivity and results.

TIP

For this chapter in particular, I highly recommend navigating in the Console and performing the steps as you read along. There's nothing like "face time" in the user interface to get the concepts of this chapter out of your head and on to your fingertips.

Change Default Metrics and Policies

Grid Control comes with a set of predefined out-of-box metrics (with default threshold values) and policies for each target type, to provide immediate target monitoring. These default metrics and policies are just a subset of those available to an administrator, but they still usually manage to flood your inbox with notifications.

I've found that recommendations detailed below for changing default metrics hold true, regardless of the size or complexity of a GC installation. Most of the suggested metrics changes reduce the frequency of alerts, but a few changes, mainly in additional host metrics, may increase the rate of alerts—such is the price of better monitoring. Overall, the net result of tuning metric alerts should be significant to vast reduction in their frequency. To help you set appropriate expectations, I'll go out on a limb and say that a "normal" rate of metric alerts for a tuned environment would be no more than a few per week per target.

My general suggestion for choosing policies differs greatly from the specific metrics recommendations found below. Your company should endeavor to adhere to all available policies, as they constitute IT best practices. However, initially, you need to add or remove policies in accordance with company IT standards. Given that most policies are security-related, changes needed to out-of-box policies will vary between sites as widely as their IT security standards.

Let's first examine how you might go about changing metrics, and then look at changing policies.

TIP

If standardizing monitoring across targets, make metric and policy changes to one "template" target for each target type, then use monitoring templates (discussed later) to propagate these changes to other targets of that type. It's best to use Grid Control infrastructure targets templates because these targets will never be removed.

Change Default Metrics

By and large, default metric settings provide the right amount of raw material "to chisel away at" for the myriad of GC environments out there. However, I've found that for the majority of sites, out-of-box metrics benefit from certain changes, ranging from removing some metrics, to adding others, to adjusting thresholds.

The reasons for these metric changes vary. Here are some of these reasons:

- Thresholds are too tight (i.e., not tolerant enough), causing an inordinate number of alerts to display.

- Thresholds are not tight enough, or are arbitrary, such as the warning for Owner's Invalid Object Count if more than two objects are invalid.

- A metric monitors a condition not of interest to most administrators, such as Audited User notifying every time the SYS user connects to the database.

- Unset metrics, such as Wait Time (%), could benefit target monitoring and help detect problems before they become too serious.

- Multiple metrics monitor the same condition, such as Total Invalid Object Count and Owner's Invalid Object Count.
- Thresholds are not standardized across similar metrics, such as those monitoring free space percentages available.
- Some metric thresholds are difficult to standardize across a particular target type, such as Current Open Cursors Count, and should be unset, then added back with adjusted values on a target-by-target basis.
- A metric does not apply to a particular operating system, such as Memory Utilization (%) for Linux hosts.
- Metric thresholds for a properly tuned GC component differ from those appropriate for most targets of that target type, such as CPU Utilization (%) for an OMS or OMR host.

Table 13-1 contains recommended metric changes from the default thresholds, along with reasons for the suggested changes. New threshold values of Unset/Unset effectively mean that you are removing the metric from collection and monitoring. Keep in mind that your site's alert history should inform these recommendations. (Empirical data always trumps general advice.)

Target Type	Metric	Comparison Operator	Original Thresholds: Warning/Critical	New Thresholds: Warning/Critical	Reason for Change
Agent	Virtual Memory Utilization Growth (%)	>	0.5/2	5/20	To allow for normal Agent memory spikes
Automatic Storage Management	Used (%)	>	75/90	90 / 97	To use a standardized space threshold
	Used % of Safely Usable	>=	90/100	<Tune>/<Tune>	Adjust values based on redundancy[2]
Cluster Database	Open Instance Count	<	Unset/Unset	Unset/<n>, where <n> is the number of RAC instances	To receive notification of an instance going down
	Owner's Invalid Object Count	>	2/Unset	Unset/Unset	Is a duplicate of Total Invalid Object Count
	Total Invalid Object Count	>	Unset/Unset	0/Unset	Compile invalid objects or drop them
Database Instance	Archive Area Used (%)	>	80/Unset	Unset/Unset	Is already covered under Filesystem Space Available (%)

TABLE 13-1. *Recommended Metric Changes from Default Settings*

---

[2] This metric indicates the amount of free space that can be safely used, taking mirroring into account. See Oracle Enterprise Manager Oracle Database and Database-Related Metric Reference Manual 10g Release 4 (10.2.0.4), Section 1.7.7.

Target Type	Metric	Comparison Operator	Original Thresholds: Warning/Critical	New Thresholds: Warning/Critical	Reason for Change
	Audited User	=	SYS/Unset	Unset/Unset	Remove unless you want notification every time SYS connects
	Current Open Cursors Count	>	1200/Unset	Unset/Unset	Does not necessarily indicate a problem
	Database Time Spent Waiting (%) - Other	>	30/Unset	Unset/Unset	Will cause recurring alerts
	Dump Area Used (%)	>	95/Unset	Unset/Unset	Is already covered by Filesystem Space Available (%)
	Free Dump Area (KB)	<	2000/Unset	Unset/Unset	Is already covered by Filesystem Space Available (%)
	Generic Alert Log Error	Matches	<Default>/Unset	<Default>/ ORA-19815||ORA-16014||ORA-16038||ORA-19809||ORA-00312	To alert when Flash Recovery Area (FRA) fills up, which hangs a database and halts GC notifications[3]
	Interconnect Transfer Rate (MB/s)[4]	>	Unset/Unset	107[5]/Unset	Set to 90% of cluster interconnect rate
	Owner's Invalid Object Count	>	2/Unset	Unset/Unset	Is duplicate of Total Invalid Object Count
	Process Limit Usage (%)	>	Unset/Unset	90/97	Reaching max processes would prohibit new database connections
	Total Invalid Object Count	>	Unset/Unset	0/Unset	Compile invalid objects or drop them

TABLE 13-1. *Recommended Metric Changes from Default Settings (continued)*

---

[3] You must set these FRA space-related thresholds, despite the fact that Oracle Enterprise Manager Oracle Database and Database-Related Metric Reference 10g Release 2 (10.2) lists Cluster Database metrics called Usable Flash Recovery Area (%) and Recovery Area Free Space (%). Apparently, this is a documentation errata, as these metrics are not provided in GC 10gR2.

[4] Oracle Enterprise Manager Oracle Database and Database-Related Metric Reference Manual 10g Release 4 (10.2.0.4), Section 4.23.1, states this metric is available, but it is only available in the Console (see "RAC OMR Interconnect Rate" in Chapter 17). I include this metric because it will be available in later GC releases.

[5] This threshold assumes a GigE Interconnect (1 Gbps = 119MB/sec).

Target Type	Metric	Comparison Operator	Original Thresholds: Warning/Critical	New Thresholds: Warning/Critical	Reason for Change
	Wait Time (%)	>	Unset/Unset	90/97	To detect wait events responsible for the bulk of wait time for resources
Host	CPU Utilization (%)	>	80/95	40/80 (OMR host) 20/40 (OMS host) 80/95 (all other hosts)	To adhere to a properly tuned OMR/OMS host, which should consume < Warning threshold
	Filesystem Space Available (%)	<	20/5	10/3	Set as required for each mount point[6]
	Memory Page Scan Rate (per second)	>	Unset/Unset	300/Unset	To advise when paging/swapping is excessive
	Memory Utilization (%)	>	99/Unset	95/Unset (all except Linux) Unset/Unset (Linux[7])	To alert when host becomes memory-bound
	Network Interface Collisions (%)[8]	>	Unset/Unset	150/Unset	To provide early indication of bandwidth problems[9]
	Network Interface Combined Utilization (%)	>	Unset/Unset	90/Unset	To indicate overall bandwidth based on total reads/writes from/to the network
	Network Interface Total Error Rate (%)	>	Unset/Unset	25/Unset	To check whether host is dropping packets[10]
	Network Interface Total I/O Rate (MB/sec)	>	Unset/Unset	107[11]/Unset	Is measured as sum of reads and writes on interface

TABLE 13-1. *Recommended Metric Changes from Default Settings (continued)*

---

[6] Do not set this metric below 3 percent for the Agent file system, because the Agent disables metric data collections when this percentage falls below 2 percent.

[7] Don't set on Linux to account for Linux memory management. Based on Oracle*MetaLink* Note 297627.1, which is no longer available.

[8] Though the units in the metric name are "%," the metric measures the number of collisions per second.

[9] Network collisions greater than 120 per second generally indicates a network bottleneck.

[10] To compensate for bandwidth-limited network hardware, reduce the packet size (for example, set *rsize/wsize* to 2048).

[11] This threshold represents 90 percent of available network bandwidth, assuming a GigE corporate LAN (1 Gbps = 119MB/sec).

Target Type	Metric	Comparison Operator	Original Thresholds: Warning/Critical	New Thresholds: Warning/Critical	Reason for Change
	File or Directory Permissions	!=	Unset/Unset	<as desired>	To monitor certain files/dirs
	File or Directory Size (MB)	>	Unset/Unset	<as desired>	To monitor certain files/dirs
	File or Directory Size Change Rate (KB/minute)	>	Unset/Unset	<as desired>	To monitor certain files/dirs
	Log File Pattern Matched Line Count	>	0/Unset	<as desired>	To monitor desired log files
Management Services and Repository	Loader Throughput (rows per second)	>	2700/3000	Unset/Unset	To avoid incorrect comparison operator, which should be <
Oracle HTTP Server (OMS only)	Active HTTP Connections	>	7500/8000	100000/Unset	Must tune empirically; default thresholds are usually too tight
	Active Requests for a Virtual Host	>	7500/8000	100000/150000	Tune empirically; default thresholds are usually too tight
	CPU Usage (%)	>	Unset/Unset	80/95	To show CPU % web server is using
Web Cache	Web Cache CPU Usage (%)	>	Unset/Unset	80/95	To show CPU % Web Cache is using
	Hits (% of requests)	<	30/20	Unset/Unset (for OMS Web Cache) 30/20 (all other Web Cache targets)	Unset for OMS because it gets most data dynamically

TABLE 13-1. *Recommended Metric Changes from Default Settings (continued)*

To add, edit, or remove metric settings for a target, including making the metric threshold changes indicated in Table 13-1, do the following:

1. Log in as an administrator with OPERATOR privilege on the target. Without this privilege, you can only view metric settings, not edit them.
2. Navigate to the target home page and, under Related Links, click Metric and Policy Settings.

3. The Metric Thresholds page appears, shown here for an OMS host target.

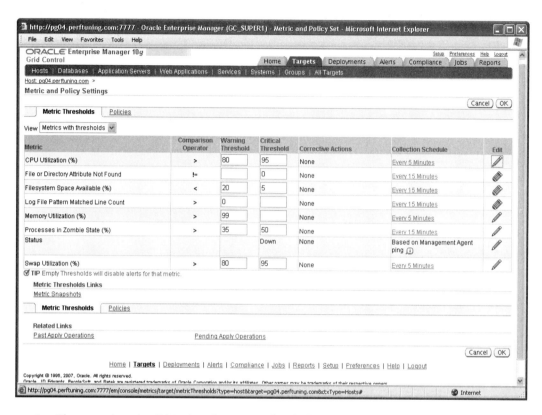

4. Change values for Critical and Warning thresholds as desired.
5. If a metric is enabled, you can change its default collection settings by clicking the value in the Collection Schedule column. This brings up the Edit Collection Settings page, shown in Figure 13-1 for the CPU Utilization (%) metric.

Oracle recommends that you not change the default collection settings, if for no other reason than that it's difficult to anticipate how changes will affect dependent metrics. However, there are a few legitimate reasons for changing settings. For example, you may need to change the collection frequency if, irrespective of threshold settings, it samples metric changes too quickly (as evidenced by multiple false alerts) or too slowly (in that an alert doesn't fire when it should). In any case, try not to reduce the collection frequency to less than 5 minutes, as this can place excessive load on the Agent. Another justifiable change is to set the use of metric data to Alerting Only if you're not interested in metric data, but only in receiving metric alerts and data about alerts. Make any needed changes and click Continue.

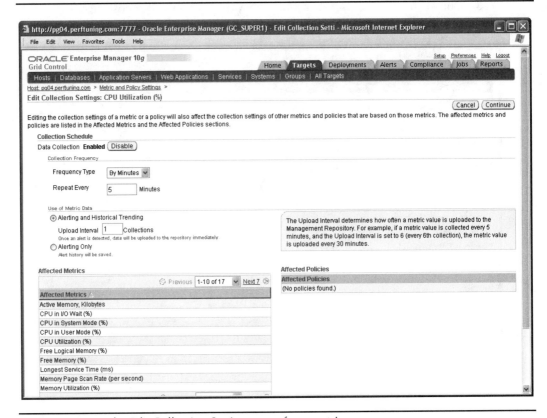

FIGURE 13-1. *The Edit Collection Settings page for a metric*

CAUTION
The Healthcheck metric may not work, especially on hybrid systems, which are 64-bit platforms running 32-bit databases (Bug 5872000). Healthcheck is a composite metric that includes seven metrics (see Note 469227.1 for background and solutions on resolving related metric collection errors). Oracle Support may recommend disabling this metric to make the Agent more stable. To disable the Healthcheck metric, choose one of the composite metrics, such as Instance Status, navigate to the screen shown in Figure 13-1, and click Disable (see Note 379423.1 for details). You can also disable it at the command line (see Note 431330.1). Healthcheck is a standalone metric, so disabling it does not affect any other metrics.

Chapter 13: Tune Metrics and Policies 499

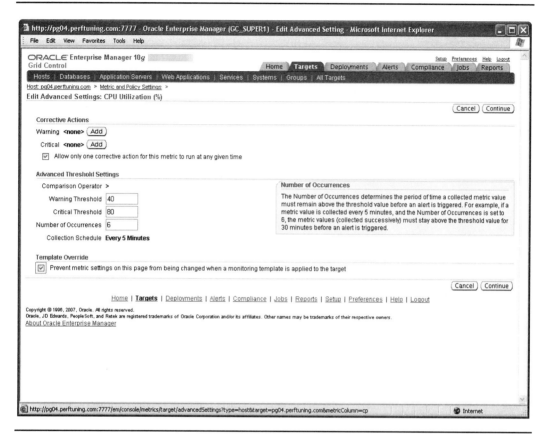

FIGURE 13-2. *The Edit Advanced Settings page for a metric*

6. To edit advanced settings for the metric, click the pencil icon in the Edit column for that metric. The pencil icon signifies whether you can enter multiple thresholds for the metric.

   ■ A single pencil means that the metric monitors a single object at a single Warning and Critical threshold level. In the case of the CPU Utilization (%) metric, the advanced settings shown in Figure 13-2 are offered.

   On this page, you can define corrective actions for the alert conditions and specify (by checking the box under Template Override) that metric settings not be overridden when applying a monitoring template (checked here because it is an OMS target that has specific threshold values as defined in Table 13-1). You can also change the *number of occurrences,* or the consecutive number of times a metric value, sampled according to the collection schedule, must go beyond the threshold before it will trigger an alert. Make any necessary changes and click Continue to modify the settings.

■ A group of pencils denotes that the metric monitors multiple objects, which can each have different thresholds. For instance, the Filesystem Space Available (%) metric allows you to set different thresholds for each mount point object.

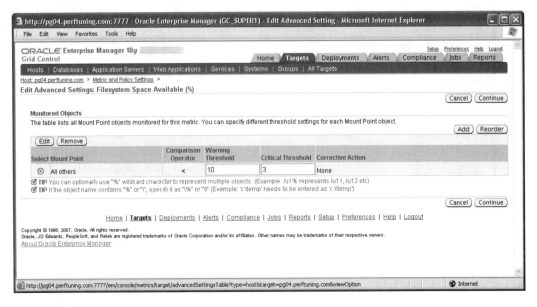

7. Make any changes and click Continue to modify the settings.
8. Make necessary changes to other metric settings and click OK to save all changes to the Repository.

Perl Syntax for Alert Log Metric Thresholds

Grid Control uses Perl syntax for threshold values of database metrics with "Alert Log Error" in their name. The comparison operator for these metrics is hard-coded as Contains. Given the importance of alert log errors, the following list shows the Perl syntax available for these threshold settings:

Perl Syntax	Meaning
*	Joins each individual expression that follows, which is enclosed in parentheses.
|	Bitwise Or.
||	A binary || performs a C-style short-circuit logical OR operation. If the left operand is true, the right operand is not even evaluated.
!	Logical Not. Returns the logical negation of the expression to its right.
?	To match one character.
[0-9]	To match the range of numbers in brackets.
^	Exclusive Or.
&	Bitwise And.

Here are some examples of Perl syntax to specify thresholds for these Alert Log Error metrics:

Perl Syntax	Description
ORA-0*(600?|7445|4[0-9][0-9][0-9])[^0-9]	Alerts on ORA-600 (Critical database errors), ORA-7445 (Critical database background errors), and ORA-4xxx (Operating specific errors).[12]
ORA-0*(?!(16166|16040|00002))	Excludes noncritical Data Guard errors ORA-16166 (LGWR network server failed to send remote message), ORA-16040 (standby destination archive log file locked), and ORA-00002[13] (primary to standby network outage). Useful when network is slow between primary and standby site.
ORA-19815||ORA-16014||ORA-16038||ORA-19809||ORA-00312	Alerts when Flash Recovery Area (FRA) fills up due to any of these errors, which would hang the database. Recommended in Table 13-1 as Critical threshold for Generic Alert Log Error.

Change Default Policies

Policies define how you want your systems to behave, to remain in compliance with company security, configuration, and storage standards. Policy monitoring enables automated checking of target adherence to *policy rules* (expressions that test values from a target against a condition), notification if targets deviate, and compliance scoring to measure their compliance status. (A target with a 100% compliance score for a given policy rule fully conforms with that policy rule.) Comparing target compliance scores permits you to focus on the worst-offending targets. For an enabled policy, you can customize its weight on a target basis by changing the policy's default importance rating of High, Normal, or Low. This change, in turn, affects the overall compliance rating for that target.

The above policy features elevate Grid Control beyond many other System Management Products (SMPs) in this area. Such functionality is only rivaled by the few remaining vendors in the IT security space[14], such as Application Security, Inc. The security products that these vendors offer are typically very expensive and best-of-breed. The base GC product and accompanying policies, by contrast, is free, so you are getting a lot of bang for your buck in the policy arena. Grid Control offers 222 predefined policies (also known as "policy rules") to help keep systems in compliance. To give you a feel for the relative number of default policy rules for each target type, Table 13-2 lists the number of policies for the most common target types.

---

[12] This is the Generic Alert Log Error default Warning threshold.
[13] This is a nonvalid ORA error caused by Bug 2327312.
[14] Most of this attrition is due to mergers and acquisitions (M&A).

Target Type	Number of Policy Rules
Database Instance	83
Cluster Database	36
Listener	31
Oracle HTTP Server	8
Host	4
Web Cache	4
ASM	4
OC4J	1

TABLE 13-2. *Policy Rule Count by Target Type*

Because there are so many default policies for databases and listeners, your policy tuning will probably focus on clearing these types of policy violations and tuning them to model the policy dictates of your GC environment.

Policy rules fall into three categories:

- **Security** Most policy rules are security related. Security requirements at Oracle shops differ widely, making it difficult to suggest which of the default security policies to change. These policy rules are in accordance with the Oracle security concepts identified by the SANS Institute.[15] An example database policy rule in the Security category is Audit File Destination, which ensures that access to the directory containing audit files is restricted to the Oracle software owner and DBA group.

- **Configuration** Configuration policies are the second most plentiful category, and consist primarily of Database Instance policies. Configuration specifications can also be quite disparate between sites, so your company needs to assess which out-of-box configuration policy rules to change. One such policy in the Configuration category is Insufficient Number of Control Files, which flags when only one control file is being used. This exposes you to a single point of failure if a media error were to occur on this file. (It's a best practice to mirror control files at the Oracle level.)

- **Storage** There are less than 30 storage policies, and they are predominantly for Database Instance and Cluster Database targets, with a small number of ASM rules. An example of a database policy rule in the Storage category is Users with Permanent Tablespace as Temporary Tablespace. Such users will likely experience poor performing SQL due to space management inefficiencies for sort operations and the like.

---

[15] The SANS (SysAdmin, Auditing, Networking, and Security) Institute, founded in 1989, is a cooperative education and research organization with more than 165,000 members. Pete Finnigan, a noted Oracle security expert, has written an excellent workbook on Oracle security, based on SANS Institute research, called *Oracle Security Step-by-Step*, Sans Press, 2004. Also see his excellent web site, http://www.petefinnigan.com, for additional security resources.

Each target type has policy rules that fall into one or more of these three categories. Furthermore, each policy rule is associated with a particular target type. You need to evaluate the policy rules in each category to determine their suitability to your environment. While it is difficult to offer general advice on policy rules, for your particular GC environment you should settle on certain standard changes to default policies for each target type to make them compliant with corporate IT policies. Ideally, your company will adhere to most of the policy rules (not just the default ones), as they constitute best practices for the target types that Grid Control manages. However, it is not realistic to expect firms to adopt new rules en masse. It is a slow process, because some of these rules are difficult to implement and enforce, particularly for international companies with data centers in many countries. Aim to align out-of-box policies with your firm's IT standards. As you improve these standards, add more policies to match and enforce these standards. The GC policy rules can actually drive your company standards, as these rules define a well-regarded and proven system management framework.

Following is the procedure to remove, set, or change policies for a given target.

1. Log in to the Console as an administrator with OPERATOR privilege on the target. Without this privilege, you can only view metric settings, not edit them.

2. Navigate to the target home page and, under Related Links, click Metric and Policy Settings.

3. Click the Policies page, shown here for a Repository Database target:

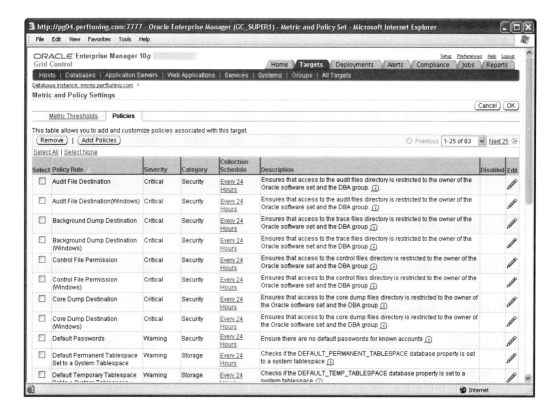

4. Remove any policies by selecting the policy rule(s) and clicking Remove. Add policies by clicking Add Policies and following the prompts.

5. Edit settings for a policy rule by clicking the pencil icon in the Edit column; following is the page that opens after clicking the pencil icon for the Audit File Destination policy:

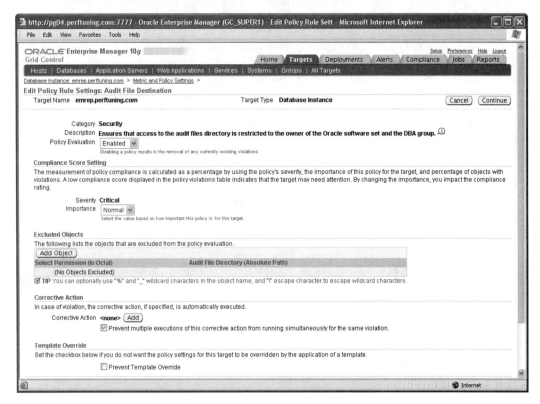

Here you can change the Policy Evaluation to Enabled or Disabled, and set the Compliance Score Setting to High, Normal, or Low Importance. You can also exclude objects from policy evaluation, add corrective actions, and prevent a template from overriding the policy settings made here. See the EM Online Help for more details on this page. Click Continue after making any changes.

6. Make necessary changes to other policy rules and click OK to save all changes to the Repository.

Enable Metric Baselines

A metric baseline is a chosen time period for a target during which Grid Control automatically evaluates its metric performance. A baseline computes statistics over this time period for target performance and workload-related metrics that an administrator chooses to include. When an administrator enables a baseline for a target, Grid Control uses the statistical history of the metrics collected to automatically set and dynamically adjust metric thresholds based on the target workload pattern.

CAUTION
Avoid creating a metric baseline for the built-in EM Website service, as this causes loss of monitoring data.

Metric baselines compare current system performance with past system performance, so they are appropriate only for relatively stable systems. A stable system is one with predictable performance that can vary but that you can model according to a certain workload pattern. Performance of these systems can slowly evolve (i.e., drift or change gradually over the long term) in proportion to user or transaction volume, but the workload pattern should not wildly fluctuate. If it does, you should set fixed target metrics using the Metric and Policy Settings page.

When you enable a metric baseline for a 10*g* Database target, a history of the database instance performance metrics you select for the baseline are collected and retained in the Automatic Workload Repository (AWR) for a certain *retention period* (or number of days stored in the AWR). By default, a 10*g* Database collects certain AWR statistics, which allow the database to diagnose and tune itself. These statistics are operational in nature and reflect both the state of the whole database (such as the Program Global Area and System Global Area structures) and various areas within the database (such as wait events, enqueues, latches, and expensive SQL). The default AWR retention period is the trailing (previous) 7 days, which is suitable for baselines of most OLTP systems, but you can set the retention period to as long as 91 days. The initialization parameter STATISTICS_LEVEL determines the level of detail of the statistics collected. TYPICAL is the default value, which is sufficient in most cases. Metric baselines extend the number of AWR statistics normally gathered and control the length of time these statistics are retained. (The Cost-Based Optimizer also collects 10*g* statistics, but they are related to database objects, such as the number of rows in a table or index and the number of distinct index keys.)

Because of their dependence on the AWR, metric baselines are supported only for Oracle Database 10.2 or higher and for Service and Web Application target types. For other target types, metric snapshots offer similar functionality by allowing you to reference collected metric data to set metric thresholds (see "Use Metric Snapshots" later in this chapter). However, with metric snapshots, you must set thresholds to fixed values. By contrast, metric baselines automatically set thresholds, and moving window baselines (discussed in a minute) adjust thresholds dynamically to accommodate slowly evolving metric patterns.

Consider using a metric baseline for a qualified target instead of setting fixed metric values (either manually on the Metric and Policy Settings page or using with the advice of metric snapshots) under any of the following conditions:

- To track the performance of targets whose transaction size or rate changes predictably, thereby causing movement in performance metrics, such as physical reads or writes per second. Moving window metric baselines will adapt to these slowly moving creeping averages.[16]

- To let Grid Control set performance metric thresholds when you don't know what values to use or when values in Metric and Policy Settings are not causing alerts to fire when they should.

- To repetitively capture baselines at certain times of interest (e.g., month end) to provide a record for historical comparison of system performance over time.

---

[16] This reminds me of the time my brother Jon put on a few pounds and explained that he reached his "new" equilibrium weight.

- To normalize spikes (up or down) in a performance metric's value that, without baselines, would produce numerous false alerts. These are situations where you cannot prevent false alerts by relaxing fixed metric thresholds; as you do relax them, false alerts continue to occur until, at some point, they stop. However, at that point, thresholds are so relaxed that legitimate alerts don't occur either.[17] Metric baselines normalize these occasional metric spikes so that they do not spur an alert unless they remain elevated and start "making their home there."

Metric Baseline Concepts and Terms

You must learn to speak the language of metric baselines to understand how to create them and fully tap their usefulness. Here are the critical terms of metric baselines:[18]

- Baseline period—two types: moving window and static
- Time groups
- Normalized metric values
- Adaptive thresholds—two types: significance level and percentage of maximum

I'll define and explain each of these terms before I get into how to create and use metric baselines.

Baseline Period

Metric baselines use a target's captured performance metrics over a certain time interval, called a *baseline period*, to dynamically set thresholds for these metrics to detect unusual performance conditions. The baseline period is the sampling period over which metrics are statistically summarized. The baseline period must be at least 7 days and can be as long as 91 days. Two types of baseline periods exist:

- **Moving window baseline** Period between midnight of the current day (i.e., does not include the current day's metric data) and the prior *n* days, as selected by the administrator creating the baseline. Moving window baselines dynamically set metric thresholds based on captured metric data over a trailing number of days—7 by default, but adjustable to 21, 35, or 91 days. A moving window rather than a static baseline time period is recommended for ongoing alerting, provided the performance metric patterns evolve slowly enough over the chosen trailing period that the baseline can "make sense of" the change in system performance. Moving window baselines are suitable for evolving systems (i.e. those that increase in size or number of users) with predictable workload cycles.

- **Static baseline** Captures a past workload period with fixed beginning and ending dates to compare against a current occurrence of this workload (e.g., month end from one month to the next). The baseline must be at least seven days long. Grid Control calculates static baseline statistics and derives thresholds *only once*—at the time the baseline is created. When you enable the static baseline, captured performance metrics going forward are compared with these derived thresholds, and alerts fire accordingly. You can enable any previously captured static baseline at any time. Static baselines are fitting for times of particular interest to you, such as during month- or year-end processing.

---

[17] You might try adjusting advanced metric settings, such as the number of occurrences or the collection schedule, but unnecessary alerts might still be generated.

[18] A partial source of these definitions is "About Metric Baselines" in the EM Online Help.

Only one type of baseline can be active at a time, and this active baseline automatically sets metric thresholds.

Time Groups

A *time group* is the workload pattern an administrator chooses to divide time over the baseline period, whether moving window or static. If a time group can accurately model a target's workload pattern, the resultant statistics grouping by time allows adaptive thresholds (see the upcoming "Adaptive Thresholds" section) to adjust threshold values throughout the day in concert with workload patterns. There are seven such time groups, and each aggregates the data collected over the baseline period in a different way. Choose a time group from Table 13-3, some of which are only available for moving window baselines, as noted. Day hours are from 7 A.M. to 7 P.M. and Night hours are from 7 P.M. to 7 A.M. and cannot be changed as of GC10gR2.

Baseline Type (Moving Window, or Both Static and Moving Window)	Time Group	Time Group (Explained)	Number of Groups	Minimum AWR Retention Period Required for Time Group	System Availability	System Type	Workload
Moving window	None	Over Trailing period	1	7	24×7	Online or Batch	24×7 online access or constant batch feeds
Both	By Day and Night	Day hours, Night hours	2	7	24×7	Online/Batch	Online days and batch nights
Both	By Weekdays and Weekend	Weekdays together, Weekends together	2	7	24×5	Online or Batch	Online weekdays and batch weekends
Moving window	By Day and Night, over Weekdays and Weekend	Day hours, Night hours, Weekdays together, Weekends together	4	7	24×5	Online/Batch	Online weekdays and batch weeknights; batch weekends
Both	By Day of Week	Daily	7	7	24×7	Batch	Weekly recurring jobs
Both	By Hour of Day	Hourly	24	35	24×7	Batch	Hourly recurring jobs
Moving window	By Day and Night per Day of Week	Day hours for 7 days, Night hours for 7 days	14	35	24×7	Online/Batch	Online days one week, batch nights the next week, with weekly recurring jobs

TABLE 13-3. *Time Groups Available for a Moving Window Baseline*

Keep in mind these points about time groups:

- The number of time groups spans the entire AWR retention period. For example, a time group By Day and Night contains two time groups regardless of the AWR retention period (if it's 91 days, the Day time group would contain 91 days of data, and the Night time group would contain 91 nights of data).

- For static baselines, you can combine time groups (e.g., selecting both By Day and Night and By Weekdays and Weekend yields four groups).

- Activating a moving window baseline causes the system to compute new time group statistics every night for the next day's thresholds.

Normalized Metric Values

The *normalization* of metric values is a concept that those with a background in statistics will better understand. A baseline normalizes metric values over time by converting raw metric baseline data to well-defined statistical significance percentile levels (0.95, 0.99, 0.999, 0.9999). These levels are buckets for categorizing the raw data and indicate the positioning of these raw values in the statistical profile of the baseline. For example, a baseline for a particular metric may calculate a raw value of 2000 at the 95th percentile (0.95) and 3000 at the 99th percentile (0.99). In this case, a normalized value of 2040 might equate to a percentile level of 0.95, which is the largest baseline percentile for values less than 3000.

Adaptive Thresholds

Both moving and static window baselines normalize a target's performance by implementing *adaptive thresholds* for performance metrics included in the baseline. Adaptive thresholds automatically adjust Critical and Warning threshold values statistically to account for their normal fluctuation within a time group. The goal of adaptive thresholds is to detect a performance problem signified by a statistically significant spike in a metric's value.

For each registered metric in a baseline, you must select one of two adaptive threshold types:

- **Significance level (default type)** Computes a metric's thresholds based on well-defined, selectable statistical significance levels. When using this type, Warning and Critical thresholds need to occur as frequently as the chosen level to trigger an alert: Very High (0.99) is 1-in-100, Severe (0.999) is 1-in-1000, and Very Severe (.9999) is 1-in-10000. Use significance level thresholds for relatively stable metrics that intermittently spike during times of poor system performance.

- **Percentage of maximum** Computes a metric's Warning and Critical thresholds based on respective specified percentages of the *trimmed maximum* observed over the selected time group. The trimmed maximum is defined as the 99th percentile, rather than the true maximum, to avoid distortion by highly atypical measurements. Example percentages for a metric are a 110% warning level and a 120% critical level. You also choose the consecutive number of occurrences where the metric value must exceed the thresholds before an alert triggers. Use percentage of maximum thresholds to detect when metric values approach or surpass previously observed peaks. This threshold type suits targets on machines whose resource use is near capacity and requires closer scrutiny.

As just explained, these two adaptive threshold types "typify" targets with two different performance characteristics. Therefore, I suggest choosing the same threshold type for all metrics in a given target's baseline. Irrespective of the threshold type that you choose, try also to standardize on the same threshold type settings for all metrics, unless certain metrics require otherwise. This means that if you choose the significance level threshold type, use the same significance level (Very High, Severe, or Very Severe) for all metrics if feasible. Likewise, if you select the percentage of maximum threshold type, enter the same Warning and Critical levels and same number of occurrences for all metrics if possible. At least start with these settings and see how well the baseline performs in terms of alerting you of out-of-bounds conditions.

NOTE
You cannot use metric baselines to tune your system using the Oracle Wait Interface (OWI) methodology. OWI tuning requires a much smaller time scope and action scope than metric baselines currently offer.[19] OWI time scopes are on the order of minutes or hours, whereas both moving window and static baselines require data collection for at least seven days. OWI action scopes are for a particular user or transaction, whereas metric baselines capture statistics at the database level. Finally, metric baselines are not instrumented with all the wait-based statistics that are integral to OWI tuning.

Create Metric Baselines

Now that we've defined the terms that represent the building blocks of metric baselines, we can move on to creating a metric baseline of an Oracle 10.2 Database to capture its performance characteristics. The procedure to create a metric baseline for Services and Web Applications is similar.

Following is the specification for the example 10.2 Database metric baseline:

Administrator: GC_SUPER1

Baseline period: Moving window

AWR retention values: Default

Baseline retention period: Trailing 7 day

Time group: By Day and Night, over Weekdays and Weekend

Adaptive threshold type: Percentage of maximum

Activate the baseline: Yes

Metrics to register: All

Warning/Critical thresholds: 130%/150% for all metrics

Occurrences: Three for all metrics

Threshold action for insufficient data: Suppress alerts

---

[19] I enthusiastically credit Cary Millsap for the terms *time scope* and *action scope*. See page 48 of *Optimizing Oracle Performance*, by Cary Millsap with Jeff Holt, O'Reilly & Associates, Inc., 2003. I cannot praise this book highly enough.

Following are the steps to create a baseline, using the above example:

1. Log in to the Console using your administrator account (e.g., GC_SUPER1 rather than SYSMAN) so that the metric baseline can rely on your preferred credentials without having to override them with manual entry.
2. Go to a 10.2 Database Instance (not a Cluster Database)[20] home page, and click the Performance page.
3. Log out of the database in Grid Control as DBSNMP and log back in as SYS with SYSDBA privilege. For instructions, see "Web Console Login As a User Other Than DBSNMP" in Chapter 9.
4. Scroll down to the bottom of the Database Instance home page. Under Related Links, click Metric Baselines.
5. A screen appears with the Enable Metric Baselines button if you have not previously enabled metric baselines for this target. (If you've already enabled metric baselines, the button on this screen will toggle to Disable Metric Baselines.) Click Enable Metric Baselines to initiate instance baseline performance metrics to be stored in the AWR.
6. On the Confirmation page, click Yes to continue.[21]
7. The Metric Baselines page appears (see Figure 13-3). Following is a description of the fields (listed in the order in which you should select them) and guidelines to help you choose the appropriate settings for your system:

    - **Accumulated trailing data**   This field is applicable only for moving window baselines. Click Change AWR Retention if you are not satisfied with the default values, which are a retention period of 7 days, a snapshot interval of 1 hour, and a collection level of TYPICAL (appropriate for most database systems).[22] You can change both the AWR snapshot interval and the retention period on multiple databases by clicking Execute on Multiple Databases. You can capture metric data over any specified AWR interval, but keep in mind that the only selectable baseline retention periods are 7, 21, 35, and 91 days. Therefore, it makes sense to select only these values for the AWR retention period as well.
    - **Select Active Baseline**   Here you can stop a baseline or select the type of baseline (either moving window or static).
        - **No active baseline**   Select this option to stop a baseline or switch between a moving window and a static baseline. Grid Control stops computing statistics for moving window baselines, stops automatically setting alert thresholds, and clears existing thresholds.
        - **Moving window baseline**   This type of baseline is recommended for systems with predictable workload cycles.

---

[20] You can get to a Database Instance home page from the Targets tab and All Targets subtab or by clicking the instance name under the Instances section of a Cluster Database home page.

[21] If you are not logged in to the Console as a database user with SYSDBA privilege, you will receive an "insufficient privileges" error after clicking Yes.

[22] You can also change the retention time in SQL*Plus as follows:
begin
DBMS_WORKLOAD_REPOSITORY.MODIFY_SNAPSHOT_SETTINGS(<system_snapshot_interval>, <retention_period>); end;/
where <system_snapshot_interval> is in minutes and <retention_period> is in days.

Chapter 13: Tune Metrics and Policies **511**

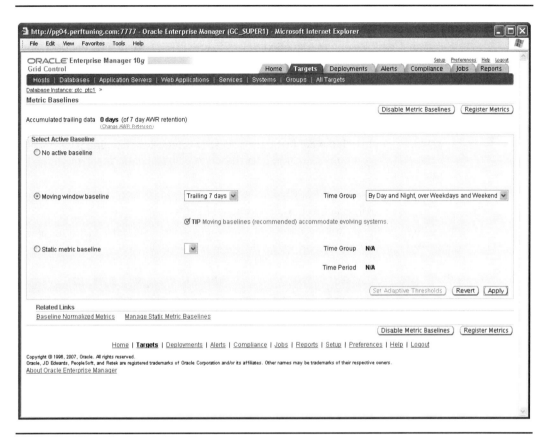

FIGURE 13-3. *The Metric Baselines page*

- **Static metric baseline**   This type of baseline is appropriate for time periods of particular interest, such as end of fiscal year.
- **Retention period**[23]   The only baseline retention period available is Trailing 7 days if you use the default AWR retention period, which is sufficient for most OLTP systems. Baseline retention periods of Trailing 21 days, Trailing 35 days, and Trailing 91 days also become selectable as you increase the AWR retention period to at least these trailing values. I recommend using the same AWR and retention periods. Use higher retention periods only when your system's performance is better modeled by a longer sampling period. Such an example is a Decision Support System (DSS) with a batch job cycle extending beyond 7 days (i.e., where repeating batch jobs run every 7+ days). You must increase the retention period to 35 days or more to choose time group By Hour of Day or By Day and Night per Day of Week (see Table 13-3).

---

[23] These two fields on the Metric Baselines page are not labeled. They are the fields with the pull-down list of values located adjacent to the Moving window baseline and Static metric baseline radio buttons.

- **Time Group** Choose a time group from Table 13-3, some of which are only available for moving window baselines.

  Change the AWR Retention if needed, select to create either a moving window or static baseline, choose the retention period and time group, and click Apply.

8. For static window baselines, click Compute Statistics. Metrics appear based on the time period and time group selected so that you can determine whether these selections are valid. If not, an alert message is displayed in the Sufficient Data column indicating either that the data is not adequate to compute reliable statistics or that the baseline model chosen does not match the data characteristics. This requires changing your selection of static baseline beginning or ending dates, baseline type (static or moving window), and/or time group.

9. On the Confirmation page, click Yes to activate the baseline. When you activate a baseline, any previous active baseline for that target is deactivated.

10. Choose which metrics to register by clicking Register Metrics on the Metric Baselines page (see Figure 13-3). This brings up the Register Metrics page.

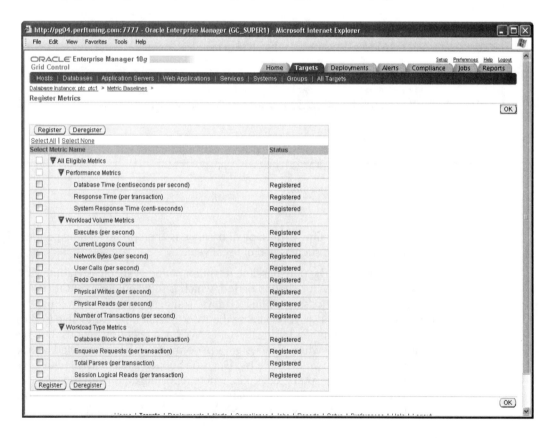

Chapter 13: Tune Metrics and Policies **513**

All eligible metrics are registered by default. Metric categories for databases are all performance related and include:

- Performance Metrics
- Workload Volume Metrics
- Workload Type Metrics

Each category has specific metrics. Select any metrics you want to deregister and click Deregister. Click OK to save your changes.

11. Determine the adaptive threshold type to use for each registered metric. Return to the Metric Baselines page (see Figure 13-3) and click the Set Adaptive Thresholds button, which will now be active. This brings you to the Manage Adaptive Thresholds page.

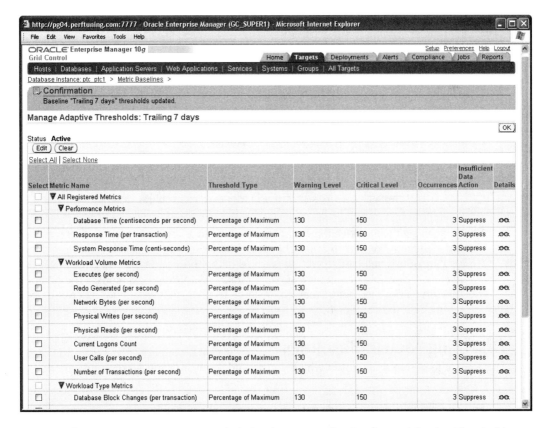

12. Select one or more metrics, and click Edit to open the Configure Adaptive Thresholds page.

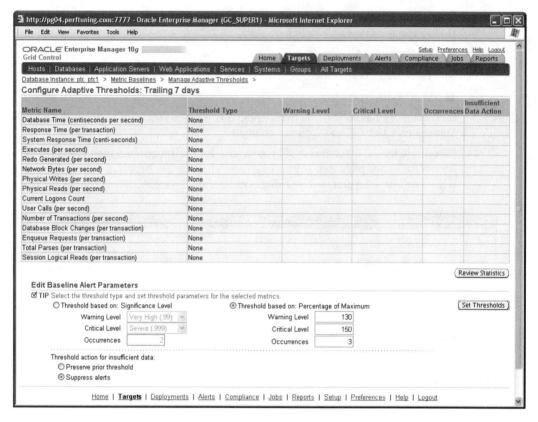

13. You can set the adaptive threshold type (significance level or percentage of maximum) independently for each metric. Choose one of these types for the one or more metrics you previously selected to edit. Regardless of the threshold type you select, I suggest leaving "Threshold action for insufficient data" set to "Suppress alerts." The other option, "Preserve prior threshold," may cause spurious alerts that would not otherwise occur if metric data were sufficient.

Disabling Metric Baselines

There are at least two situations in which you should seriously consider disabling metric baselines:

- **The system's performance pattern becomes unpredictable**   There is no sense in continuing to use a metric baseline if it no longer models a system's performance pattern. If your system does not adhere to the currently selected time group, or to any repeatable time groups that Grid Control offers, then no metric baseline for that system will be capable of setting accurate alert thresholds. At some point, a system's performance may become better suited to using metric snapshots, where you set thresholds to a certain percentage worse than actual metric values captured when performance was acceptable.

- **The costs outweigh the benefits** There is overhead associated with using metric baselines. Agents monitoring targets with enabled baselines require incrementally more CPU and memory resources on the target host. (Baselines will not require any additional storage on targets hosts because the Agent limits its upload directory, which contains the baseline data, to a certain size—50MB by default.) In addition, the number of baselines you enable has a cumulative effect on CPU, memory, and storage consumption on the Repository Database and OMS hosts. The computation involved in using baselines primarily takes its toll on the CPU of the Repository Database host. For OMS hosts, loader or notification backlog may occur due to spurious alerts, such as from selecting the wrong time group. If any of these resources on the target, Repository, or OMS hosts are already overtaxed, you may need to disable one or more metric baselines.

To disable the currently enabled metric baseline for a target, click Metric Baselines under Related Links at the bottom of the target's home page. The page will only contain the button Disable Metric Baselines. Click this button and confirm the choice. You can also disable a metric baseline from the Metric Baselines page (refer to Figure 13-3).

Add Corrective Actions

Corrective actions automate responses to target alerts or policy violations. A corrective action can run on the target where an alert or policy violation triggers or can execute multiple tasks across different targets. You can configure separate corrective actions for Critical and Warning alerts and set notification rules to tell you when corrective actions succeed or fail.

Following are some characteristics of corrective actions:

- A single corrective action can be used with multiple metrics or policy rules, either for a target or for a monitoring template.
- You can copy a corrective action from the Corrective Action Library[24] to a template and vice versa.
- When you apply a monitoring template containing a corrective action to a target, the corrective action is associated with the target.

CAUTION
A corrective action is added to a target metric only if it is defined in the monitoring template when that template is first applied. In other words, you cannot edit a template to add a corrective action, and reapply the template to a target to propagate that corrective action. You only get one chance to mass-deploy corrective actions, so be sure to follow this order (which is reflected in the chapter order): define all corrective actions, add them to monitoring templates, and apply these templates.

---

[24] All super administrators share and have access to the Corrective Action Library.

One way to leverage corrective actions is in reducing a target's Mean Time To Recovery (MTTR). The MTTR for an outage is defined as the detection time plus the repair time to bring a target back online. Grid Control can address both detection time and repair time. Administrators can minimize detection time by monitoring a target's availability with metrics, and minimize repair time by predefining a corrective action to restart the target if it goes down. Reducing the overall MTTR can be crucial to meeting a target's service level agreement (SLA). For more background on MTTR, see "Use Fast-Start Fault Recovery to Control Instance Recovery Time" in Appendix C (online at www.oraclepressbooks.com).

Another SLA specification that a corrective action can address is to guarantee that a specified *time delay*[25] is sufficient for a standby database in a Data Guard environment. A time delay is the interval between when primary redo data is received by and applied to the standby database. This time delay can help prevent database corruption or user error from propagating to the standby database, which can compromise a site's disaster recovery setup. A Data Guard configuration best practice of Oracle Maximum Availability Architecture (MAA) is to set the time delay to 30 minutes or less regardless of the Data Guard protection mode chosen.[26] Achieving such a small delay is only possible if you can detect a problem with the primary database and stop the standby database managed recovery process within the time delay period. For example, you can define a corrective action to cancel managed recovery for a standby database after detecting a non-zero value for the metric Corrupt Data Block Count in the primary database.

Create Corrective Actions

There are three different ways to create a corrective action in the Console:

- From the Corrective Action Library sidebar on the Setup menu
- From the Edit Advanced Settings page for a specific target metric
- From the Edit Policy Rule Settings page for a given target policy rule

To illustrate how to add corrective actions, let's use the Corrective Action Library sidebar to create a corrective action to start up a single-instance database. Here are the steps:

1. Log in to your super administrator account and set preferred credentials for all targets on which you plan to use the corrective action.

2. Click Setup and then click Corrective Action Library on the sidebar.[27]

---

[25] The recovery delay is the time lag between archiving a redo log at the production site and applying that archived redo log file on the standby database.

[26] From Maximum Availability Architecture Overview - An Oracle White Paper, June 2005, Section 5.2.3, "Establish a Recovery Delay," pages 5–16. For more information on MAA, see http://www.oracle.com/technology/deploy/availability/htdocs/maa.htm.

[27] Note the two built-in corrective actions relating to failed login attempts.

Chapter 13: Tune Metrics and Policies **517**

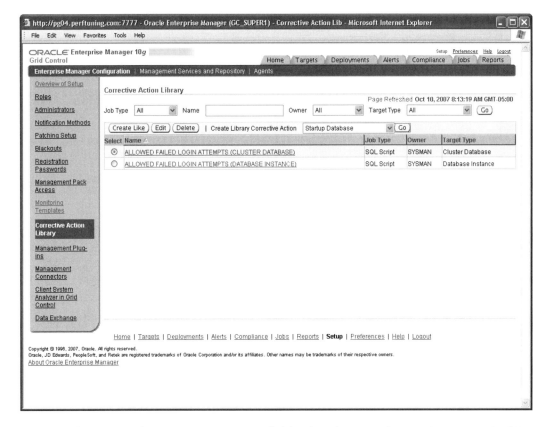

3. In the Create Library Corrective Action field, select the type of corrective action—in this case, Startup Database. There are 11 types of corrective actions—for running OS and SQL commands, security policy responses, RMAN scripts, and multitask[28] jobs, among others. Choose the correction action type and click Go.

4. The following three pages exist for all but the Multi-Task type of corrective action, which has only the first two pages:

   - **General**   Enter a name and optional description for the corrective action.
   - **Parameters**   The Parameters page differs depending on the type of corrective action selected. The image on the following page shows what it looks like for the Startup Database type.

---

[28] The Multi-Task type allows you to specify different targets for different tasks. This functionality would be required to define a corrective action to cancel managed recovery for a standby database if the primary database were to experience corruption errors, as discussed in the introduction to the section.

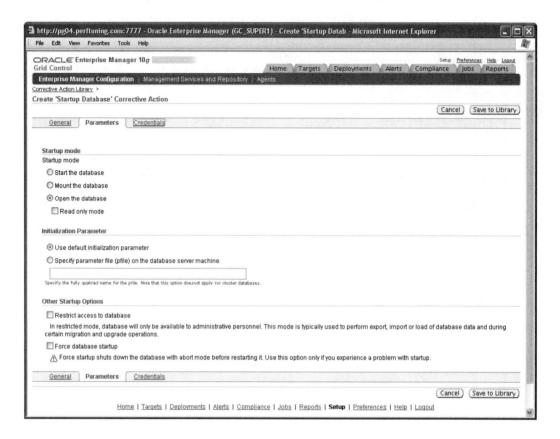

Enter the desired values—in this case to open the database in read/write mode using the default initialization parameter file.

- **Credentials** Choose either to use the preferred credentials set at the time the corrective action runs or to override these credentials, in which case the page displays fields for you to enter the appropriate credentials for the corrective action type selected. Database corrective actions such as Startup Database require both database credentials and host credentials to run the corrective action job.

5. Click Save to Library. A banner message confirms that the corrective action was created successfully.

Add Corrective Actions to Metrics or Policies

Once a corrective action is stored in the Corrective Action Library, the next step is to add it to a metric or policy. You can add corrective actions to metrics in one of two ways:

- **Add a corrective action directly on a target** To do this, edit the metric on the Metric and Policy Settings page. This method is useful when a corrective action is meant for a small number of targets.

■ **Add a corrective action to a monitoring template and apply the template** This method enables you to add a corrective action to multiple targets in one operation. You must first define one or more corrective actions in a monitoring template. Then, in one fell swoop, you can apply the template to many targets of the same type.

For our Startup Database corrective action, let's take the second route to demonstrate how to keep Console keystrokes to a minimum. Here's the procedure to add this corrective action (and any others):

1. Log in to your super administrator account and set preferred credentials for all targets to which you plan to apply the monitoring template containing the corrective action(s).[29]

2. Navigate to the Edit Monitoring Template page by clicking Setup, selecting Monitoring Templates on the sidebar, selecting the template, and then clicking Edit (see "Leverage Monitoring Templates" later in the chapter for details).

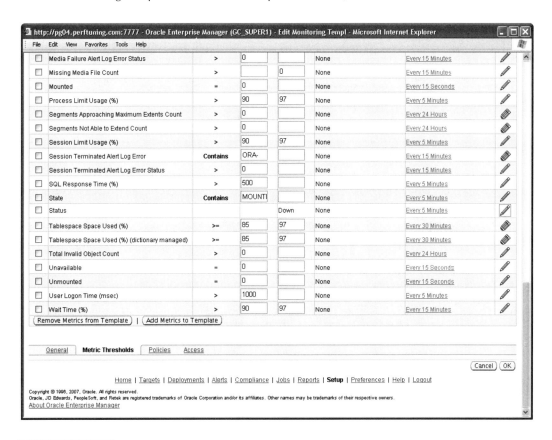

---

[29] When the corrective action runs, GC uses the target preferred credentials of the administrator who applied the template to that target.

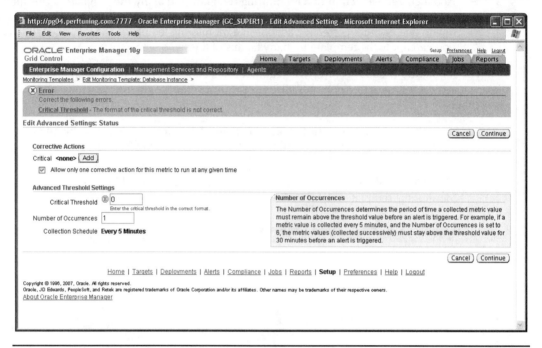

FIGURE 13-4. *The Edit Advanced Settings page*

3. Click the Edit (pencil) icon next to the metric for which you want to add a corrective action (the Status metric in this case).

4. The Edit Advanced Settings page appears. Figure 13-4 shows Bug 5724886, which you'll encounter after clicking Add when editing advanced Status metrics settings for many target types. The workaround until the bug is fixed in GC11*g* is to change the default Critical threshold setting from Down to 0, as already done in Figure 13-4. I suggest leaving the "Allow only one corrective action for this metric to run at any given time" box checked (a default setting).[30]

5. The Add Corrective Action page appears. Here you can create a new corrective action on-the-fly, select one from the library (as for our example), or reuse one already defined on the template for another metric (not applicable here). Make your selection (From Library in this example) and click Continue.

6. Corrective actions already in the library appear.

---

[30] The setting prevents a metric's warning and critical corrective actions (if defined) from overlapping due to the metric value quickly crossing both warning and critical thresholds. Checking the box also prevents multiple executions of the same corrective action due to the metric value crossing back and forth over just one of the thresholds. Either behavior would be caused by a corrective action that takes longer to run than the metric's collection interval. Unless you want to get tricky and manually manage this quagmire, I'd leave this box checked.

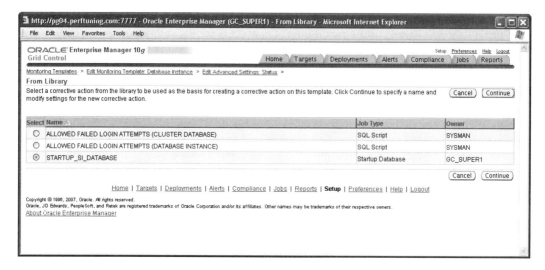

7. Select the desired corrective action and click Continue.

8. The corrective action will now be listed on the Edit Advanced Settings page (not shown but almost identical to Figure 13-4). Click Continue.

9. A banner informational message states that the settings were modified but not saved. The Corrective Actions column now reflects that a Critical Only corrective action is defined for the Status metric.

10. Repeat steps 3 through 9 for each metric to which you want to add a corrective action. Click OK to save all changes to the Repository.

11. A banner confirmation message says you edited the template successfully. Reselect the monitoring template just edited and click Apply. Walk through applying the edited monitoring template to all targets for which you want the corrective action(s) to run when the metric triggers. (For complete instructions, see "Leverage Monitoring Templates" later in the chapter.)

12. Change notification rules as desired that include the target(s) to which you added corrective actions. You can be notified if the corrective action had problems or succeeded by checking Problem or Succeeded under Corrective Actions on Target Down on the Availability Page for the rule. (See Availability Page under "Creating Notification Rules" in Chapter 10.)

13. An optional but highly recommended step is to test that the corrective action(s) works (by artificially triggering the metric threshold, for instance) and that you receive notification to this effect. Again, this notification is dictated by your administrator account's notification rule governing corrective action states for the target.

Implement User-Defined Metrics

User-Defined Metrics (also known as UDMs) are custom monitoring scripts that you *register* in the UI to extend built-in GC monitoring. (I use the term "register" to describe just one part of the UDM creation process, in which you supply the UDM script and other parameters in the UI. There are other tasks to creating a UDM, which I delineate in this section, including composing the script itself and configuring notifications for it.) Many sites that implement Grid Control have previously developed custom monitoring SQL and OS scripts to monitor certain target conditions or metric values. These sites can usually replace a good portion of these scripts with built-in GC metrics that check for these same conditions, such as for databases that go down. However, some custom scripts may have no equivalent built-in GC metrics. GC sites can co-opt or plug these custom scripts into the GC monitoring framework by registering them as UDMs. Grid Control then evaluates UDMs as it does built-in metrics, but using the custom scripts specified to return the metric value of the condition checked. Registered UDMs are imbued with all monitoring features available to built-in metrics—they are evaluated at a specified frequency against specified Warning/Critical thresholds, they store custom metric data in the Repository for historical purposes, they trigger alerts and notifications, and you can define corrective actions for them.

The following two types of UDMs are for operating system scripts and SQL queries:

- **OS UDMs**   Custom OS scripts registered in the UI to monitor conditions not covered by built-in GC OS metrics. An example of an OS script is to check that a system backup job completes successfully. You register the OS script as a UDM by creating the script and pointing to its location on the target machine.

- **SQL UDMs**   Custom SQL queries or function calls registered in the UI that perform database monitoring not supplied by built-in GC database metrics. Unlike OS UDMs, SQL UDMs do not use external scripts. Rather, you can enter the SQL script, query, or function call directly into the Console when creating the UDM.

The method for registering these two types of UDMs is similar, but different enough to warrant providing separate instructions for each type, as described next. What both types have in common is that you must register them for particular targets, as instructed below. However, after registering either UDM type for a single target, you can use the target as a monitoring template to register the UDM against other targets of that type (see "Leverage Monitoring Templates" later in the chapter for details).

Create an OS User-Defined Metric

Creating an OS UDM consists of five steps:

1. Compose the OS monitoring script.
2. Register the script as an OS UDM.
3. Configure notification for the OS UDM.
4. Verify the OS UDM is working.
5. Define the OS UDM on other hosts.

To illustrate these steps, let's use the simple example of creating an OS UDM on a Repository Database node using a UNIX shell script to check every 15 minutes for the total number of processes on the host. The UDM will send Warning and Critical alerts if the total number of processes exceeds 1000 and 1500, respectively. Please define the UDM on an OMR node so you can see how monitoring templates handle UDMs later in the section, "Leverage Monitoring Templates."

Compose the Monitoring Script

The first step in creating an OS UDM is to compose the monitoring script. You can employ any scripting language your host supports to check for a particular condition of interest.

1. Log in as the Agent user (typically *oracle*) to one of the hosts on which you want to run the script. If you are planning to run it against other hosts, choose a host you want to use as the monitoring template for the host target type (see "Leverage Monitoring Templates" later in the chapter).

2. Create a directory for the script. I suggest creating a subdirectory below *$AGENT_HOME/sysman/emd/*,[31] because this is the Agent directory in most need of mirroring for high availability (see "Back Up the Agent" in Chapter 14).[32] As such, UDM scripts you create will be backed up continuously. In this example, create a subdirectory named *custom*[33] as follows:

   ```
   mkdir -p $AGENT_HOME/sysman/emd/custom
   ```

   The Agent user must have full permissions on whatever directory you choose for storing the UDM scripts.

3. Make sure the script's run-time environment is ready to execute the script by installing any missing interpreters, such as Perl, that the script requires on the Agent host.

4. Compose the script itself, in this example called *numProcs.sh*, in the directory you just created for OS UDMs. The script must contain code to do the following:

   - Define logic to check the condition being monitored. In our example, the condition is the total number of processes on the host, and the logic to check it is the following OS command:

     ```
     ps -ef | wc -l
     ```

   - Return the script results to Grid Control via standard output (stdout) using the syntax of the scripting language. In our example of a shell script, use the *echo* command. (For a Perl script, use the *print* command.) The results sent to stdout must contain a tag that Grid Control recognizes: *em_result* and, optionally, the tag *em_message*. If using just the *em_result* tag, the format sent to stdout must be:

     ```
     em_result=<em_result_value>
     ```

   - The value for the *em_message* tag defaults to "The value is <em_result_value>." If you are using both tags, send a string format delimited by newline characters (\n) to stdout as follows:

     ```
     em_result=<em_result_value>\nem_message=<em_message_text>\n
     ```

---

[31] For RAC Agents, the Agent home is the EM STATE directory, which includes the node name. See "What Is an EMSTATE Directory?" in Chapter 5.

[32] GC stores its built-in scripts in the *$AGENT_HOME/sysman/admin/scripts/* directory by default, as defined by the *scriptsDir* property in the *$AGENT_HOME/sysman/config/emd.properties* file.

[33] I suggest assigning a generic name to this directory, as it can serve as the central location for all types of GC custom scripts, not just those for UDMs. For instance, you can place RMAN Script jobs in this custom directory (see Chapter 14).

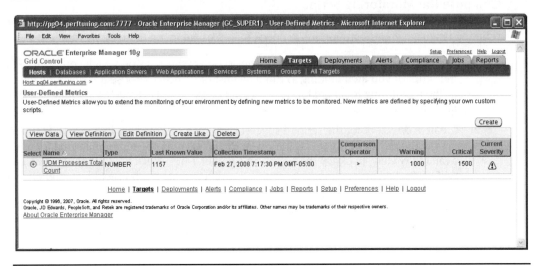

FIGURE 13-5. *The OS User-Defined Metrics summary page*

- Both tags must be lowercase, but the result value or message can be mixed case. For our shell script, we send the result to stdout and use the standard message. So the script syntax is:

  ```
  echo "em_result=`ps -ef | wc -l`"
  ```

5. Grant the Agent user (typically *oracle*) permissions to execute the monitoring script. In this case:

   ```
   cd $AGENT_HOME/sysman/emd/custom
   chmod u+x numProcs.sh
   ```

6. Confirm the script can run in the OS, independently of Grid Control, as the user whose credentials you intend to specify below when registering the UDM (typically the Agent user, *oracle*). Following is the command execution and sample output for testing purposes:

   ```
   ./numProcs.sh
   em_result=1157
   ```

Register the Script as an OS UDM

Now that you've created the script for the OS UDM, you need to register the UDM in Grid Control, as follows:

1. Log in to the Console as an administrator with privileges to monitor the OS target.

2. Navigate to the host target's home page and click User-Defined Metrics under Related Links. The Agent on the target host must be running—if not, no Related Links will appear.

3. A summary page appears, as shown in Figure 13-5, with a list of any previously registered UDMs. (The UDM we're registering will not appear as shown in the figure until after completing this registration process.) You can edit, view, and delete UDMs, or use the Create Like functionality to model new metrics using existing ones. To register a new UDM, click Create to open the Create User-Defined Metric page, shown next.

Chapter 13: Tune Metrics and Policies **525**

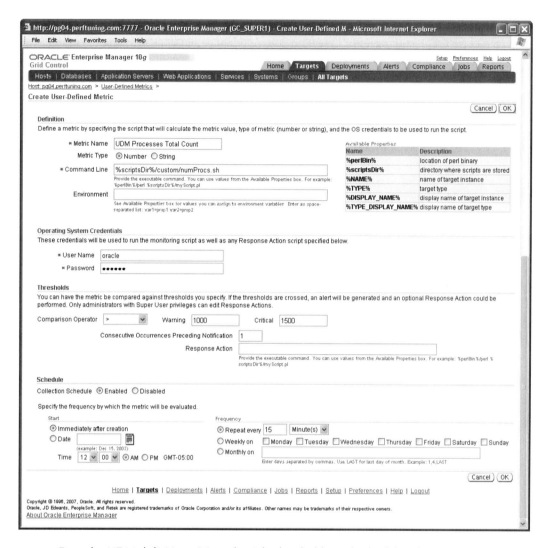

Enter the UDM definition, OS credentials, thresholds, and schedule information. Here are some rambling thoughts and suggestions on what to input. I recommend prefacing the name of User-Defined Metrics with "UDM" so you can easily spot them among the list of all target metrics. Note that I use the *%scriptsDir%* property in the Command Line field under the Definition section. This field name and the instructions below it beginning with "Provide the executable command" are misleading. You must enter a script name, not a command, in the Command Line field (and Response Action field, which is an additional feature on top of any corrective actions you may later define for the UDM). Schedule the UDM to start immediately after creation so that you can test it right away. Click OK when finished.

4. A summary page appears listing the new OS UDM and a confirmation message that the metric was created. The OS script doesn't even have to exist for the UDM to be created. However, after running the UDM according to its defined schedule, GC will report any execution problems (as you'll see shortly).

Configure Notification for the OS UDM

To receive notifications on this OS UDM and, optionally, all OS UDMs, edit the notification rules for your administrator account. See "Create and Subscribe to Notification Rules" in Chapter 10 for detailed instructions. Here is a short synopsis of what to do:

1. Edit the appropriate notification rule. Click Preferences, then Notification Rules on the top left. Select the desired host-based notification rule, and click Edit. Take the example of a rule already created using the Create Like feature, based on the built-in rule, Host Availability and Critical States. (I am following my own advice from Chapter 10 to leave the built-in rules alone and use Create Like functionality to copy and change them.)

2. Add the OS UDM to the notification rule. Click the Metrics page for the rule, then click Add to open the Add Metrics page, shown next. In the pull-down field just below the Cancel button, click Show All or the last set of metrics. Check the metric called User Defined Numeric Metric (which would be called User Defined String Metric if you had defined a String metric type). In the Objects column, you can add notifications for all OS UDMs by leaving All Objects selected, or you can choose specific OS UDMs by using the Select button and then the Flashlight. In this case, let's choose to be notified on all UDMs. Select the desired Severity States and Corrective Action States (here we choose all such states according to best practices), and click Continue.

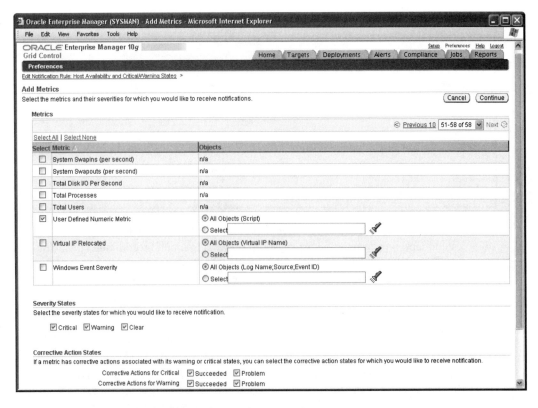

3. Click OK to save changes to the notification rule.

Verify the OS UDM Is Working

After you create the UDM, the last step is to confirm whether the UDM is functioning properly, as follows:

1. Assuming you adopted my suggestion to schedule the OS UDM to start immediately after creation, wait a few minutes. Then refresh[34] the summary page to confirm that Grid Control populates the Last Known Value and Collection Timestamp columns, as shown in Figure 13-5.

2. Check that no metric errors occur when the OS UDM executes by clicking the Alerts tab and Errors subtab and confirming the metric is not listed for the target that registered it.

3. For proof-positive that the OS UDM works, artificially reduce the Warning or Critical threshold so that the metric triggers an alert. Confirm that you receive the alert notification. Then reset the threshold, confirm that the alert clears on the corresponding subtab (Warning or Critical) under the Alerts tab, and check that you receive a clear notification if defined in the notification rule.

Define OS UDM on Other Hosts

Once you test that the OS UDM works, you can add this metric to any other target hosts you like. The easiest way to accomplish this is with monitoring templates. For instructions, see "Leverage Monitoring Templates" later in the chapter. Remember, keep this OS UDM defined on an OMR node for now, as later it will help to demonstrate how monitoring templates handle UDMs.

Create a SQL User-Defined Metric

The procedure to create a SQL UDM is a little less involved than that for creating an OS UDM. This is because you enter the SQL directly in the UI, whereas for an OS UDM, you must create an OS script on the target host that outputs the result in a specific format that Grid Control can parse. In this section I use the example of creating a SQL UDM against the Repository Database that alerts you when SYSMAN tables need to be reorganized due to a high number of inserts and deletes (covered as a maintenance task in Chapter 17 under the section "Rebuild Tables and Indexes as Required").

1. Log in to the Console as an administrator with privileges to monitor one of the database targets for which you want to create the SQL UDM. If you are planning to run the SQL UDM against multiple databases, choose a database you want to use as the monitoring template for the database target type. In this case, the UDM is solely for the Repository Database.

2. Navigate to the Database Instance target's home page and click the link called User-Defined Metrics under Related Links. This link appears for Database Instance targets, but not for Cluster Database targets, so for a cluster database, choose one of its instances.

3. A summary page appears with a list of any previously defined UDMs. You can edit, view, and delete UDMs, or use the Create Like button to model new metrics using existing ones. To register a new UDM, click Create.

---

[34] There is no Refresh button on the UDM summary page. Refresh your browser using its menu or shortcut key, such as by pressing F5 in Internet Explorer.

4. Complete the Create User-Defined Metric page (see Figure 13-6). Enter the UDM definition, database credentials, thresholds, and schedule information. Note the following characteristics of the SQL Create User-Defined Metric page that are not present in the equivalent OS UDM page:

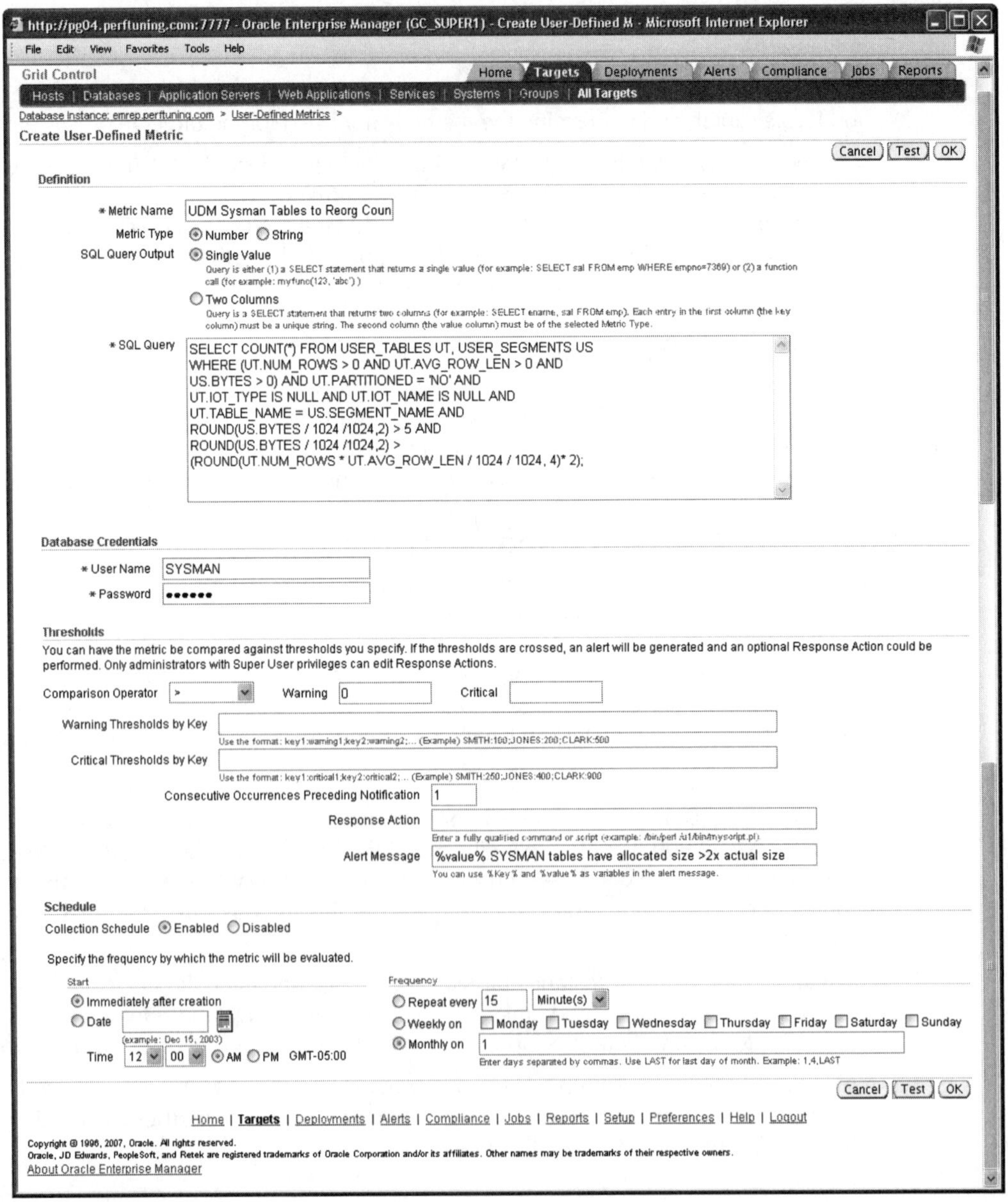

FIGURE 13-6. *The Create User-Defined Metric page for a SQL UDM*

- You enter the SQL query or function call directly rather than entering a script name, as with OS UDMs. You cannot enter a SQL script name.
- In the Definition section, select whether the SQL query returns either a single value or two columns. You can also specify a function call, but it must return a single values. In this example, the query returns a single value.
- If the query returns only one column, it must return only one row. To return multiple rows, the query must return two columns.
- When the query returns two columns, the first is the key column, which must be a unique string. The second column is the value column, and must be of the selected Metric Type.
- For a single-value SQL query, the %Key% variable is not applicable, so remove "key = %Key%, " (including the space at the end as noted) from the Alert Message field in the Thresholds section.

Enter all input to this screen and click Test to confirm that the SQL UDM can execute successfully (a feature the OS UDM does not offer).

5. Grid Control evaluates the current value of the SQL UDM and returns the results at the top of the page (not shown here). Click OK if you're satisfied with the SQL UDM definition and result. If you're not satisfied, edit the query or function and retest it.
6. The UDM summary page appears with an informational message that the SQL UDM was created. If you scheduled the SQL UDM to start immediately after creation, wait a few minutes. Then refresh[35] the summary page to confirm GC populates the Last Known Value and Collection Timestamp columns. The SQL UDM summary page contains the same columns as that for an OS UDM, as shown earlier in Figure 13-5.
7. Configure notification for the SQL UDM as for an OS UDM (see "Configure Notification for the OS UDM" earlier). Choose a database rather than a host notification rule. The rest of the procedure is identical.
8. Ensure that no metric error occurs when the SQL UDM executes by clicking the Alerts tab and Errors subtab and confirming the metric is not listed for the target that registered it.
9. For proof-positive that the SQL UDM works, artificially reduce either the Warning or Critical threshold so that the metric triggers an alert. Confirm that you receive the alert notification. Then reset the threshold, confirm the alert clears on the corresponding subtab (Warning or Critical) under the Alerts tab, and check that you receive a clear notification if defined in the notification rule.
10. Define the SQL UDM for other databases using the Monitoring Templates functionality (see "Leverage Monitoring Templates" later in the chapter).

Use Metric Snapshots

While 10.2 or higher databases, services, and web applications can use the robust metric baseline feature, metric snapshots are supported for all other targets types. Metric snapshots guide you in setting fixed metric threshold values based on a target's past performance. Like static metric baselines, metric snapshots are collections of a target's performance metrics from a past time period that you specify.

---

[35] Ibid.

However, unlike static metric baselines, which define thresholds statistically and allow no user input, metric snapshots calculate and display fixed threshold values that you can then manually edit.

The procedure for using metric snapshots is twofold. First, you create one or more metric snapshots for a specified target, where a snapshot contains chosen metrics and Warning/Critical threshold values. Then you can copy any snapshot to that target, which sets metrics and thresholds for the target as defined in the snapshot.

Let's start with the procedure to create metric snapshots. It is a conceptually simple procedure. The tricky part is in selecting which metrics to include—and, for metrics with obviously problematic calculated thresholds, what threshold values to set instead.

Create Metric Snapshots

Follow the procedure detailed below to create a metric snapshot for a target. The example target is an Oracle 10.1 Repository Database, *emrep*.

1. Go to the home page for the target (the database *emrep* in this example) and, under Related Links, click Metric and Policy Settings. On the bottom of this page, under Metric Threshold Links, click Metric Snapshots.

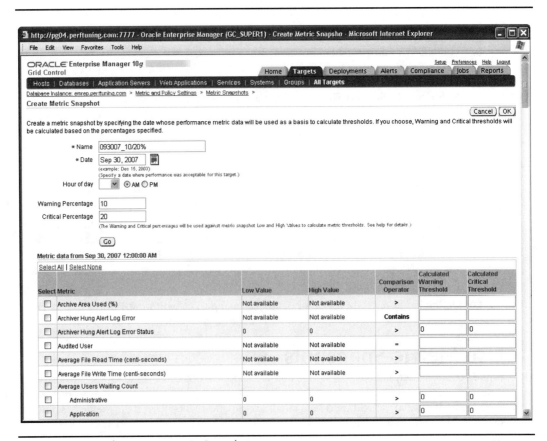

FIGURE 13-7. *The Create Metric Snapshot page*

☑ Buffer Cache Hit (%)	97	100	>	110	120
☑ CPU Usage (per second)	1	15	>	16	18
☑ CPU Usage (per transaction)	1	15	>	16	18
☑ Consistent Read Blocks Created (per second)	0	0	>	0	0
☑ Consistent Read Blocks Created (per transaction)	0	0	>	0	0
☑ Consistent Read Changes (per second)	0	0	>	0	0
☑ Consistent Read Changes (per transaction)	0	1	>	1	1
☑ Consistent Read Gets (per second)	30	12695	>	13964	15234
☑ Consistent Read Gets (per transaction)	33	19703	>	21673	23643
☑ Consistent Read Undo Records Applied (per second)	0	0	>	0	0
☑ Consistent Read Undo Records Applied (per transaction)	0	0	>	0	0
☑ Corrupt Data Block Count	0	0	>	0	0
☑ Cumulative Logons (per second)	0	0	>=		
☑ Cumulative Logons (per transaction)	0	0	>	0	0
☑ Current Logons Count	29	35	>	38	42
☑ Current Open Cursors Count	711206	919348	>	1011282	1103217
☑ Cursor Cache Hit (%)	14	96	>	105	115
☑ DBWR Checkpoints (per second)	0	0	>	0	0
☑ Data Block Corruption Alert Log Error	Not available	Not available	Contains		
☑ Data Block Corruption Alert Log Error Status	0	0	>	0	0
☑ Data Dictionary Hit (%)	91	100	>	110	120
☑ Database Block Changes (per second)	5	860	>	946	1032
☑ Database Block Changes (per transaction)	9	861	>	947	1033
☑ Database Block Gets (per second)	4	2367	>	2603	2840
☑ Database Block Gets (per transaction)	7	2369	>	2605	2842
☑ Database CPU Time (%)	53	96	>	105	115

FIGURE 13-7. *The Create Metric Snapshot page (continued)*

2. Click Create, the only button available until you create at least one metric snapshot.
3. The Create Metric Snapshot page appears (see Figure 13-7). Create the first iteration of the snapshot by entering information for the following fields:

   ■ **Name** Enter a name, usually containing the target name. If you're going to create multiple snapshots, enter a date or description that characterizes this particular snapshot, such as "Year-End." (For a descriptive name, I suggest using the Warning/Critical percentages and any hour of day chosen below.)

   ■ **Date** Choose a recent date of characteristic workload for the target when performance was acceptable. Optionally enter a time (see next field). If you only specify a date, the metric snapshot is the set of average daily values of the target's performance metrics for that date.

   ■ **Hour of day** (Optional) If you also indicate an hour within the date, then the metric snapshot is the set of average values for the preceding hour. I would only advise using this short of a sampling period for capturing periods of particular interest, such as a slow month-end process.

- **Warning and Critical Percentages** Specify Warning and Critical percentages as integers greater than zero (10% and 20%, respectively, are good starting values) for the snapshot to use in calculating Warning and Critical thresholds that are the specified percentages worse than the actual metric snapshot values captured.

4. After you provide the above information, click Go.
5. The initial calculated data is returned in the metric data table shown in Figure 13-7 (shown on previous page). The Low Value, High Value, Calculated Warning Threshold, and Calculated Critical Threshold are based on your input in step 3.
6. After this first result set, you can experiment with different Date, Hour of day, and Warning/Critical Percentage settings to derive more acceptable calculated thresholds. After you change any of these inputs, click Go to recalculate the snapshot.
7. When you're satisfied that raw calculations of Warning and Critical thresholds are as close as they will get before you edit them as instructed below, change the name of the snapshot if needed to reflect the final Warning/Critical percentages (and hour of day, if chosen), as you cannot change the name once you create the snapshot.
8. Select those metrics in the metric data table in Figure 13-7 to include in the metric snapshot, and deselect those metrics not desired. You must choose at least one metric to create a snapshot, and all metrics are selected by default.[36] Be aware that once you click OK to save your changes as instructed below, you cannot add back any metrics you've removed; you'd need to create a new snapshot to use these metrics. Also note that metrics selected for the snapshot are combined with those already set for the target (see "Apply Metric Snapshots" below for how the two are combined).
9. For metrics you choose to include in the snapshot, edit the values as you see fit in the Calculated Warning Threshold and Calculated Critical Threshold columns. You will almost certainly need to change some thresholds, which will not be usable for the reasons cited later in "Why Snapshot Computed Thresholds Require Editing."
10. When you're finished selecting and editing metrics, click OK to save your changes.
11. The Metric Snapshots summary page appears, where you can review or edit metric snapshots you've created by clicking the View or Edit buttons, respectively. On this page, you can also click Copy Thresholds From Metric Snapshot to apply snapshot thresholds to a target, as described next.

Apply Metric Snapshots

Once you create a metric snapshot for a target, you must apply it for its metrics to go into effect for the target. Here's how:

1. On the Metric Snapshots summary page, select the snapshot you want to apply to the target and click the button, Copy Thresholds From Metric Snapshot.

---

[36] To select metrics for a snapshot, thresholds for these metrics do not have to be previously set in Metric and Policy Settings for the target.

Why Snapshot Computed Thresholds Require Editing

Metric snapshots compute thresholds based on user-supplied Warning and Critical percentages only when (a) metrics have numeric values (e.g., Archiver Hung Alert Log Error is not eligible because it's a string metric with the default Critical threshold "ORA-"), and when (b) the snapshot populates both Low and High Values for the metric as shown in the metric data table in Figure 13-7.[37] Even when a snapshot computes thresholds for a metric, threshold values are far from infallible. There are several reasons why a snapshot may not calculate usable threshold values for a metric. Below I refer to *x/y thresholds* as a Calculated Warning Threshold value of *x* and a Calculated Critical Threshold value of *y*:

- A 0/0 Low/High Value for a chosen metric results in 0/0 thresholds. These are not acceptable thresholds, because, if used, any nonzero metric value would result in an alert. Your options in this case are to:
    - Change the Date and/or Hour of Day fields of the snapshot until it computes nonzero threshold values.
    - Change thresholds to nonzero values based on some other criteria, such as best practices. For example, use 200 as a Warning threshold for Memory Page Scan Rate (per second).
    - Deselect the metric so that it is removed from the snapshot definition.

- A small High Value where Warning and Critical Percentages are in a normal range (such as 10/20%) results in identical or nearly identical Calculated Warning/Critical Thresholds. For example, Consistent Read Changes (per transaction) thresholds of 0/1 produce Calculated Warning/Critical Thresholds of 1/1. You generally need to adjust these values, particularly the Critical threshold, which is too close to the Warning threshold and should be raised.

- Threshold percentages sometimes increase above the theoretical absolute of 100%, which you may rightly regard as a bug. For example, as shown in the metric data table in Figure 13-7, a High Value of 96 calculated for the metric Database CPU Time (%) produces Calculated Warning/Critical Thresholds of 105/115% when using 10/20% Warning/Critical Percentages. You need to reduce these calculated thresholds below 100%, though GC will not prevent you from saving and copying a snapshot with incorrect thresholds. In this case, the metric would never produce an alert because Critical Database CPU Time (%) cannot be higher than 100%.

- Snapshots incorrectly calculate database metrics with "Hit (%)" in their names[38], so you cannot use these metrics in snapshots (or for alerting). The problem is that these metrics use a > (greater than) Comparison Operator in both metric snapshots and Metric and Policy Settings, but should use a < (less than) Comparison Operator. This is a bug for which no patch is available, as far as I know.

Look for these conditions to flag problematic threshold values that would trigger spurious alerts or no alerts at all (even when they should fire).

---

[37] If Low/High Values are not available for a particular metric in a snapshot, you can still manually edit thresholds for the metric after getting the result set, provided the metric has a numeric value.

[38] These metrics are Buffer Cache Hit (%), Cursor Cache Hit (%), Data Dictionary Hit (%), Library Cache Hit (%), PGA Cache Hit (%), and Redo Log Allocation Hit (%).

2. The Metric and Policy Settings page appears for the target.

The snapshot settings are combined with current Metric and Policy settings. That is, metric thresholds in the snapshot override any corresponding metric thresholds previously set for the target. Also, any metrics not in common between the snapshot and previous metric settings are combined (i.e., a logical AND is performed). Edit the metrics further if desired, then click OK to save your changes to the Repository.

Leverage Monitoring Templates

Monitoring templates provide a method to apply standardized monitoring settings across targets in your GC environment. Rather than specifying monitoring settings for each target, with monitoring templates you choose a *source target*, which is a representative target of a particular target type, whose monitoring settings (or subset thereof) you want copied to the template for that target type. You can modify the copied settings for the template before saving it. You then apply the template to selected *destination targets* of that type. For example, you might create two database monitoring templates: one for production databases and the other for nonproduction databases, each with different metric and policy settings.

Monitoring templates are defined by target type, and contain all the monitoring settings of the source target, including metrics (UDMs included), policy settings, and metric/policy properties (thresholds, comparison operators, metric collection schedules, and corrective actions). If you change a template, you must reapply it to propagate the changes; but you can change and reapply templates as often as you like.

Following are some general guidelines and considerations for using monitoring templates:

- Once you create a template, make changes to the template (not the source target), and reapply the template to all targets, including the source target.

- Create separate monitoring templates for single-instance and RAC databases, because a RAC template can contain additional metrics unique to RAC, such as Open Instance Count and Global Cache statistics.

- Remember to adjust notification rules for targets to which you apply templates so that you receive e-mail notification on any new metrics in the template.

- Some template metrics will not apply to destination targets due to target versioning support for that metric, such as the Processes in Zombie State (%), which is available for Linux but not for Solaris.

- Administrators who apply templates that contain corrective actions must have the correct preferred credentials set for destination targets. The reason is that, when executing the corrective action, Grid Control uses the target preferred credentials of the administrator who applied the template.

- When applying templates containing UDMs, Grid Control prompts you to provide credentials for the UDM. The administrator applying the template does not have to be the same user as the one who created the UDM.

- When applying templates with OS-based UDMs, make sure the OS script exists in the same location on each destination target.

- When applying a monitoring template, UDMs are always included, regardless of which option you choose for metrics with multiple threshold settings (as shown later in Figure 13-9).

- If all UDMs in a template have the Mark for Delete option set (under Edit Advanced Settings for the metric), the User-Defined Metrics step will not appear in the Apply wizard.

- An apparently undocumented bug is that Cluster Database Apply operations don't appear on the Past Apply Operations page.

The next four sections discuss how to create and use monitoring templates:

- Create monitoring templates
- Maintain custom metrics when applying monitoring templates
- Apply monitoring templates
- Compare settings between targets and template

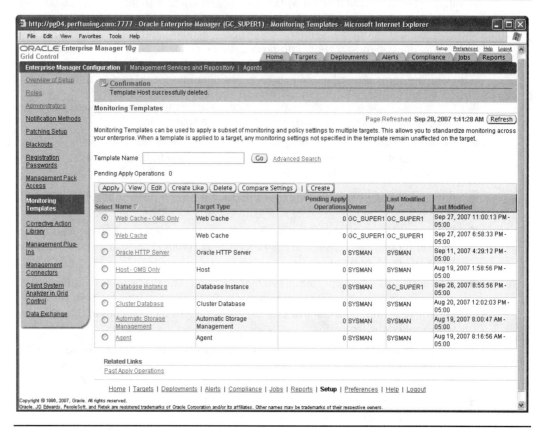

FIGURE 13-8. *The Monitoring Templates page*

Create Monitoring Templates

In this section, we take a real-world example and create a monitoring template based on the current OMR host settings in your environment. You'll apply this template to all hosts (see "Apply Monitoring Templates" next). I throw two curve balls. One is that for the host *pg04.perftuning.com*, which is the source target, you want to maintain custom Warning/Critical thresholds of 40/80 for the CPU Utilization (%) metric,[39] as a properly tuned OMR host should not consume more than 40% CPU.[40] The second curve ball is to use the OS UDM that you created earlier on an OMR Database node to demonstrate an additional step required (and shown below) to apply templates containing UDMs.

Following, then, is the procedure to create a monitoring template with the above specifications:

1. Select the OMR node on which you earlier created the OS UDM as the source target for the monitoring template and log in to GC as an administrator with FULL privilege on this target.

---

[39] As you will see, you can maintain custom metric thresholds for source targets as readily as for destination targets.
[40] This is one of the metrics listed in Table 13-1, located in the earlier section "Change Default Metrics."

Chapter 13: Tune Metrics and Policies **537**

2. Go to the source target home page (host *pg04.perftuning.com* here), click Metric and Policy Settings, and set all metrics, including UDMs, policy settings, and metric/policy properties that you want the monitoring template to possess.

3. For the source target and each destination target to which you are planning to apply the template, re-enable data collection for any metrics you may have previously disabled. Even if you apply a template where data collection is enabled for a metric on the source target, you will not receive an alert on the metric for destination targets where data collection is disabled.[41]

4. Maintain any desired custom metric settings for destination targets, as shown in the next section, "Maintain Custom Metrics When Applying Monitoring Templates." In this example, for the host *pg04.perftuning.com*, we want to set the CPU Utilization (%) metric as shown in that section to the custom Warning/Critical threshold settings of 40/80, and to prevent the monitoring template we're creating from overriding these settings.

5. Click Setup, then click Monitoring Templates in the sidebar (see Figure 13-8).[42] Click Create.

6. Select the source target you chose in step 1 above to represent the target type for which you are creating the monitoring template (the host *pg04.perftuning.com* here). Click Continue.

7. The Create Monitoring Template page appears, itself composed of three pages (a fourth, Access, page appears after you create the template).

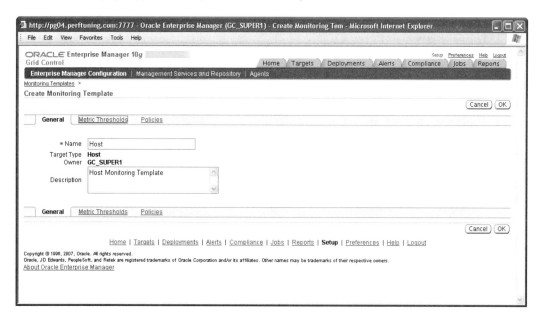

---

[41] Data collection is enabled by default, but if disabled can be re-enabled on the Edit Collection Settings page (see Figure 13-1).

[42] Note that, as this figure reflects, I've already created other monitoring templates for other target types.

8. On the General page, enter a name and description for the monitoring template, which must be unique among all templates, regardless of target type.
9. Click the Metric Thresholds page.

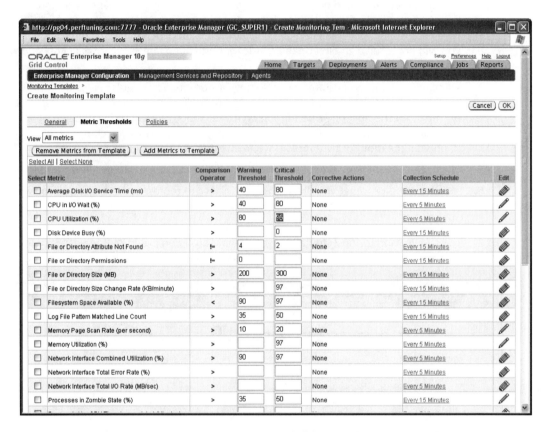

10. Change the View setting "Metrics with thresholds" to "All metrics." Add or remove metrics from the template and change metric thresholds as required. (For this example, reset the CPU Utilization (%) metric thresholds back to the defaults of 80/95 as illustrated above.)

- Only if warranted (see step 5 in "Change Default Metrics" above), click the Collection Schedule link for a metric to change its collection settings. (I don't change anything for this example). Beware that if you disable a metric in the template, this propagates to destination targets, even if you subsequently set thresholds for the metric on the destination target.[43] Therefore, do *not* disable collection of a metric

---

[43] Disabling metrics complicates GC configuration in general, not just that for monitoring templates. You don't need to disable a metric. Just don't set its thresholds or include it in any notification rules.

for a template unless you want to disable alerts for that metric across all destination targets. Click Cancel here if you made no changes, or click Continue if you changed any settings.

- Click the Edit icon for any metrics to which you want to add corrective actions. Click Add to include a corrective action (not needed here). See Figure 13-4 and its associated text in the "Add Corrective Actions to Metrics or Policies" section for instructions on how to add a corrective action for a metric.

11. Click the Policies page. Select those policies to add or remove from the template. When you click Add Policies to Template, you can select policies from the library or from any target of the same type as the source target. Click OK when finished.

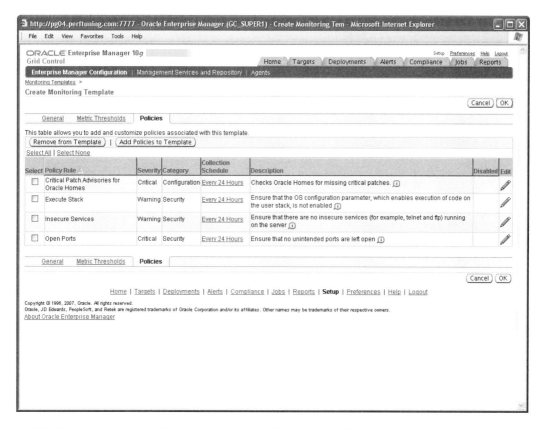

12. You are returned to the Monitoring Templates page with a confirmation message in the banner stating that the monitoring template was successfully created.

Maintain Custom Metrics when Applying Monitoring Templates

Targets can maintain certain custom metric settings even after a template with different metric settings has been applied to these targets. Take the following example of the host target *pg04.perftuning.com*:

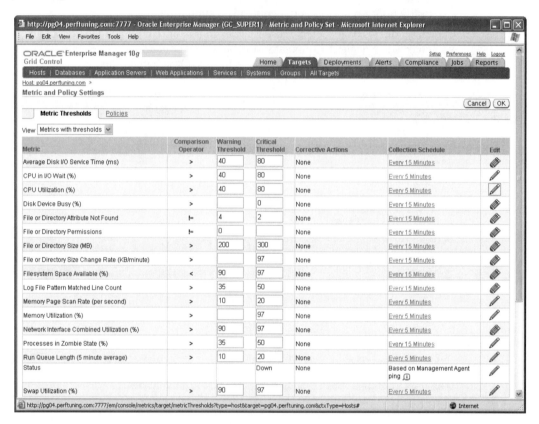

Here, I want to apply a monitoring template to the host to change all metrics except CPU Utilization (%), for which I need to retain custom Warning/Critical thresholds of 40/80 instead of using the template values of 80/95. To do this, click the Edit (pencil) icon for this metric, which brings up the screen shown in Figure 13-2 earlier in the chapter. As shown there, check the box under the Template Override section to prevent any monitoring template from overriding this metric setting, and click Continue. Repeat this procedure for any other custom metric settings you want to retain. Then, click OK to save all changes to the Repository.

Apply Monitoring Templates

Once you create a monitoring template and prepare destination targets to retain custom metric settings, you are ready to apply the template to the destination targets. When you apply a monitoring template, settings for all metrics except those employing Template Override are copied to the destination target. A destination target can be an aggregate target such as a group or system, and the template is designed to apply only to those members of an aggregate target of the same type. If, after applying a template, you add new members to a group (or simply add new targets), you must reapply the template directly to those new members (or new targets). Grid Control applies templates in the background asynchronously using the job system. After you apply a template, click the link under the Pending Apply Operations column on the Monitoring Templates page to view any pending Apply operations.

How Templates Apply Metrics

The way templates are applied differs for metrics with *single threshold settings* and metrics with *multiple threshold settings*. A metric with a single threshold setting monitors a single object. For instance, the Failed Job Count can only monitor for failed RDBMS jobs. However, a metric with multiple threshold settings can monitor multiple objects at different thresholds. An example of such a metric is Filesystem Space Available (%), which can monitor different mount points using different Warning and Critical thresholds for each mount point. Following is a description of how templates are applied to destination targets for metrics with single versus multiple threshold settings:

- **Metrics with single threshold settings**   Template settings are paramount for metrics with single threshold settings in two ways. First, settings for metrics with single thresholds in templates override corresponding settings on destination targets. Second, any metrics set on destination targets are removed if not defined in the applied template.

- **Metrics with multiple threshold settings**   When a template or destination target contains a metric with multiple threshold settings, GC offers the following two options for applying a template to a target. To illustrate, see Table 13-4,[44] which takes the example of the Filesystem Space Available (%) metric and summarizes the three permutations: where sample thresholds are set for objects—mount points */u01*, */u02*, and */u03*—on both template and target, on just the template, or on just the target.

    - **Apply threshold settings for monitored objects common to both template and destination target**   GC performs a union sum operation on metrics with multiple thresholds, where template settings prevail for common objects (*/u01* in this case), template object settings that are not set on the destination target are not copied (*/u02*), and destination target object settings remain if not defined in the applied template (*/u03*),

    - **Duplicate threshold settings on destination target**   The template settings for metrics with multiple thresholds, whether defined or not, are cloned to the destination target. That is, as with the first option above, template settings override target settings for common objects (*/u01*). However, the two differences are that a template adds any object-specific thresholds existing only in the template (*/u02*), and removes any object-specific thresholds existing only on the destination target (*/u03*).

---

[44] Source: EM Online Help section entitled "Apply Monitoring Template." I just improve on the presentation by using a single table to summarize both options.

Object Threshold Set on ...	Template Object (Thresholds)	Target Object (Thresholds)	Resultant Thresholds if Select Common Option	Resultant Thresholds if Select Duplicate Option
Both template and target	/u01 (20,10)	/u01 (15,5)	/u01 (20,10)	/u01 (20,10)
Just template	/u02 (25,15)			/u02 (25,15)
Just target		/u03 (30,20)	/u03 (30,20)	

TABLE 13-4. *How Templates Apply for Metrics with Multiple Threshold Settings*

This conceptual understanding of how templates are applied is the difficult part of applying templates. Below are the simple steps to apply a template.

How to Apply Monitoring Templates

Applying a monitoring template involves selecting destination targets for the template and, if the template contains metrics with multiple thresholds, specifying one of two options (just described) for how to copy metrics to the destination targets. Let's apply the Host template just created back onto the source host itself and to the host *sun.perftuning.com*:

1. Log in as an administrator with VIEW privilege on the template (assigned on the Monitoring Templates Access page) and OPERATOR privilege on all targets to which the template will be applied.

2. Click Setup, then click Monitoring Templates on the sidebar.

3. Select the monitoring template to apply and click Apply (see Figure 13-8).

4. Choose the options for applying the monitoring template:

    - Select the template's destination targets by clicking Add. This brings up the page shown in Figure 13-9. Choose the target(s) to which you want to apply the template, and click Select.

    - Choose one of the two options for applying metrics with multiple thresholds, as described in the previous section. These options appear only if the source target has any metrics with multiple threshold settings. (Here I chose the first option because, in this case, I am applying the template to hosts running two different operating systems, and there may be differences in the settings of metrics with multiple thresholds).

Chapter 13: Tune Metrics and Policies **543**

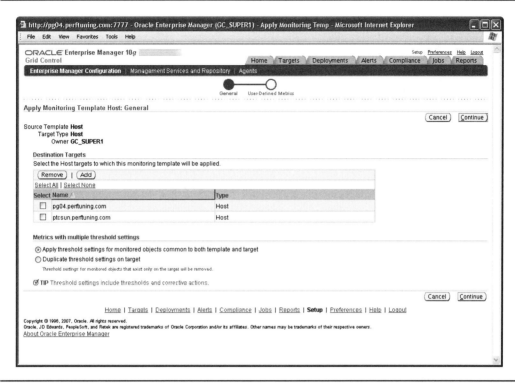

FIGURE 13-9. *The Apply Monitoring Template Host page*

■ If the template does not contain any UDMs, click Finish. If the template does contain any UDMs (as in this example), you are prompted to click Continue.

5. You can choose one of three options related to credentials for UDMs:
   - All User-Defined Metrics use the same credentials
   - Each User-Defined Metric uses its own credentials
   - Each User-Defined Metric uses different credentials for different targets

   Select the appropriate option (here I chose the first option, because OS credentials are the same for both targets) and click Finish.

6. Check the status of the Apply operation in the Pending Apply Operations column (refer to Figure 13-8). Click Refresh to update this status. This operation usually completes within a minute. If necessary, click the link for the number of Pending Apply Operations to get details. If the operation is no longer pending, you can also click Past Apply Operations under Related Links on the Monitoring Templates page to get the details of this operation.

Compare Settings Between Targets and the Template

The Compare Settings feature makes it easy to see how a target's metric settings differ from those defined in a template. This feature is useful for verifying whether a monitoring template that you applied to multiple targets successfully changes all target monitoring settings (except those on template override) to those of the template. Let's compare the Host monitoring template we just created to the two destination hosts where we applied it.

To use the Compare Settings feature, do the following:

1. If you are a super administrator or the owner of the template, log in to your account (owners have full privileges on their own templates). Otherwise, ask the template owner, shown in the Owner column on the Monitoring Templates page, to log on and grant this access to you. (To grant access, the owner must edit the template and then, on the Access page that appears after creating the template, click Add, select the administrator(s) to add, then click OK.) Then log in using your administrator account.

2. Click Setup, then click Monitoring Templates on the sidebar (see Figure 13-8).

3. Select the template you want to compare to one or more targets (the Host template in this case) and click Compare Settings to open the Compare Monitoring Template Host page.

Chapter 13: Tune Metrics and Policies **545**

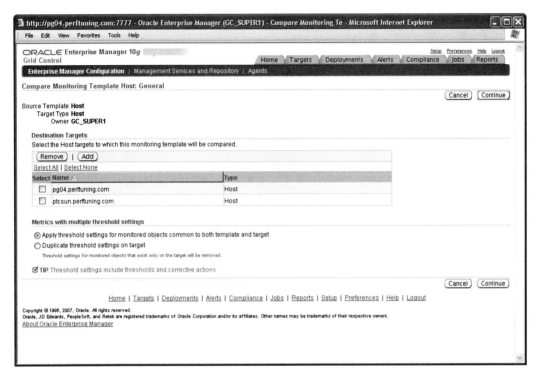

4. Click Add. In the pop-up that appears, check the targets to compare against the template, and click Select. A later option in this procedure allows you to apply the template to the targets being compared. If you plan to do this, choose the option you prefer of the two options listed on the page in the section on the page entitled "Metrics with multiple threshold settings." (Otherwise, it doesn't matter which option you select.) Click Continue.

5. If you choose one target to compare against, the results are displayed immediately. However, if you compare the monitoring template against more than one target (as in the illustration on the following page), you receive a confirmation page that a job is submitted. Click OK to go to the Job Run page.

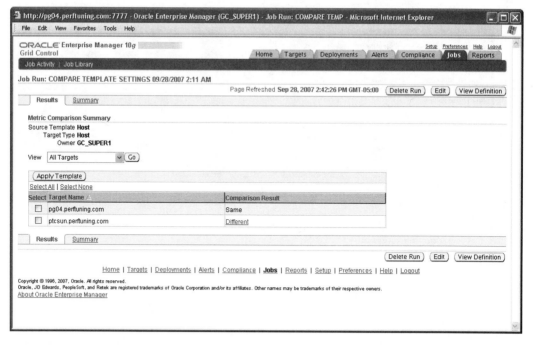

6. Refresh the page until the above page appears showing the results. In the View field, choose View Targets with Difference or select All Targets (as shown above).

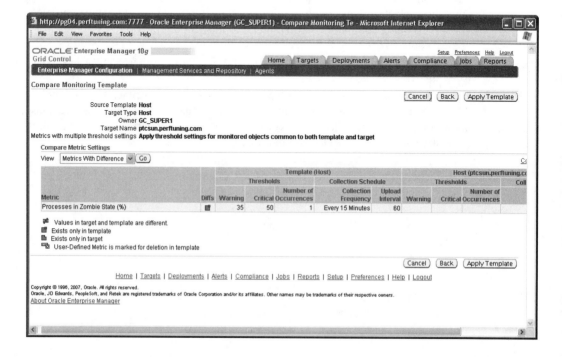

7. Each target name that is different has a Comparison Result column. Click this column to view the results. The screen shot above shows a single difference between metrics on *sun92.perftuning.com* and those of the Host template. This difference is attributable to a metric available for Linux hosts (the source of the template) but not for Solaris hosts.[45]
8. (Optional) Select any targets where you want to apply the template and click Apply Template (not applicable in this case).

Summary

Well, you've made it through a tough chapter. I hope that while you were reading, you were able at least to navigate the Console along with me. The best-case scenario is that you actually tuned metrics and policies for some of your targets using the tools presented here.

What did you learn? I organize this summary according to the logical order of our tuning goals rather than by the sequence of tuning steps we had to follow.[46] These two things actually align for the first few steps, but never mind that. Here are the chapter steps we completed, in a nutshell:

- Perform high-level tuning of both *metrics* and *policies*:
    - For metrics, I recommendeded some metrics changes, including adjusting thresholds, adding desired metrics, and removing troublesome metrics. With any luck, this cranks down your alert volume from a deafening "11"[47] to perhaps 6 or 7.
    - For policies, each company must tune default policies to align with its IT standards. Database policies demand most of your time, given the sheer number of default policy rules compared with other target types. Evaluate all three categories of policy rules: security, configuration, and storage.
- Complete fine-tuning of metrics, which has two prongs:
    - *Metric baselines* set adaptive thresholds for performance metrics on 10.2 or higher Oracle Databases, systems, and web applications.
    - *Metric snapshots*, for all other target types, calculate thresholds that we edit with extreme prejudice (changing or discarding many of them), then apply.
- Customize alerts with two kinds of User-Defined Metrics:
    - *OS UDMs*, where you create and point to a local script that returns a tag that Grid Control parses for the metric value.
    - *SQL UDMs*, where you enter a SQL query or function call directly in the UI, which returns either a single value or two columns for comparison with thresholds you define.

---

[45] There is a documentation bug on page 3-62 of the Oracle Enterprise Manager Framework, Host, and Third-Party Metric Reference Manual 10g Release 2 (10.2), which states that this metric's Data Source for Solaris hosts is the *ps* command. While it is true that the Solaris *ps* command provides zombie state information, for some reason GC does not offer this metric for Solaris hosts.

[46] Frankly, if I were designing the next release of the product, I would align the two.

[47] From the movie *This is Spinal Tap*. If you haven't seen it, consider it homework for this chapter.

- Both kinds of UDMs plug into Grid Control's monitoring framework, which confers on them all the bells and whistles that built-in metrics possess.
  - Use *monitoring templates* to apply standardized monitoring settings across targets of the same target type. The monitoring settings include built-in metrics and UDMs, policy settings, and metric/policy properties (thresholds, comparison operators, metric collection schedules, and corrective actions).
  - Create *corrective actions* to automate responses to target metric alerts and policy violations. Be on the lookout to leverage corrective action scripts in UDMs.

Here ends yet another long, intricate chapter. I hear the ghosts of a million DBAs of yore grumbling that they prefer the command line to the Console any day. There was a time when all "real" DBAs spoke pejoratively about any type of GUI work. That's a tough stance to take these days, given the complexity and power of Grid Control and other GUI DBA tools that stand tall in their own right. For those who prefer infrastructure work to more functional target-based efforts, the next chapter on backup and recovery of Grid Control will be a welcome relief. If you're happier in GUI land, the next chapter also features many GUI islands you can hop by puddle-jumper.

CHAPTER 14

Backup and Recovery of Grid Control

A voiding Grid Control monitoring outages requires adopting best practices in the following areas, covered in the chapters indicated:

- Backup and recovery (this chapter)
- High availability and disaster recovery (Chapter 15)
- Security (Chapter 16)
- Maintenance and tuning (Chapter 17)

This chapter covers how best to set up backup and recovery (B&R) for your Grid Control environment. The focus is on implementing a sound strategy to back up and recover all GC components (including Agents on all target hosts), *not* on using Grid Control to back up and recover managed targets.[1] (That said, I use Grid Control whenever possible to complete B&R tasks for the OMR Database so that you may learn how to leverage GC to perform B&R for target databases.) The purpose of Grid Control B&R is to provide a restore point to the current time if possible, should one or more GC components fail. The B&R strategy presented here is the first line of defense in meeting your GC Service Level Agreement (SLA) requirements. High availability (HA) and disaster recovery (DR) measures must take over where B&R measures leave off in meeting your SLA *Recovery Time Objective (RTO)*. The RTO is the duration of time within which a company must restore business processes to a specified service level after an outage to avoid unacceptable consequences associated with a break in continuity.[2] In Chapter 15, you can extend HA for the GC framework to meet your RTO, as needed. However, as covered in earlier chapters, you may have had to initially build the GC framework (OMR, OMS, and OMAs) with some HA characteristics, such as using fault-tolerant storage (see "Hardware Installation Requirements" in Chapter 2). Fault-tolerant storage, in particular, is a foundation stone for the B&R strategy presented here and for the HA recommendations given in Chapter 15. For instance, in the section "OMR Database Backup and Recovery," later in this chapter, the methods discussed would likely not allow for recovery to the point of failure if archive logs that accrued between backups were stored solely on an unprotected disk that failed.

Following are the Grid Control B&R topics covered in this chapter:

- Introduction to Grid Control B&R
    - Grid Control B&R concepts
    - Resetting Agents
    - Recommended Grid Control backups
- OMR Database backup and recovery
    - Back up the OMR Database in Grid Control
    - Recover the OMR Database
    - Manage OMR Database backups in Grid Control

---

[1] Technically, the OMR Database, OMS, and OMA are also "targets" in Grid Control. However, in this chapter I refer to GC components explicitly and use "targets" to signify all remaining targets

[2] Source: Wikipedia.

- Grid Control software backup and restoration
    - OMR Database software backup and restoration
    - OMS software backup and restoration
    - Agent software backup and restoration

I need to spend a minute on the terms *restore, recover,* and *B&R* used above and throughout the chapter. "Restore" and "recover" are applicable to a database. To restore a database is to locate and return backup files to online media (e.g., disk). To recover a database is to use the restored files to bring the database back online to just before the point of failure or to a previous point in time. By contrast, GC software contains OS files, not database structures (except as noted later for two Oracle Clusterware components when using a RAC OMR Database), so the restoration process is almost entirely OS-based. I use the term "B&R" freely to mean "backup and recovery" or "backup and restoration," depending on the context.

Given that the GC software is almost all file-based, B&R of Grid Control software is relatively uncomplicated when compared with that of the OMR Database. If you thumb through the chapter, you'll find that the OMR Database B&R section accordingly comprises the bulk of the page count, although GC software B&R is given equal hierarchical billing.[3]

Introduction to Grid Control B&R

Let's ease into the subject of Grid Control B&R by beginning with an overview. I've divided this introduction into three sections:

- Grid Control B&R concepts
- Resetting Agents
- Recommended Grid Control backups

Grid Control B&R Concepts

To formulate a strategy for backing up your entire Grid Control environment, you need to distinguish those GC *elements* that change during normal operations from those that don't change. I use the term "elements" rather than "components" because these elements exist across all components of the GC framework. The GC elements that change call for a different approach to B&R from that taken for those elements that don't change. For example, most of the directories in the software homes for the OMR, OMS, and Agent don't change (except Agent management files being uploaded, log and trace files) unless you patch the software or edit related configuration files. Therefore, you need to back up the directories in the software homes only if you make changes. On the flip side, the OMR Database is changing constantly, and thus should be backed up daily. The OMR Database should be in ARCHIVELOG mode, so daily backups will allow you to recover it completely to the current point in time just before a failure without having to rely on more than a day's worth of archive logs. The task at hand is to identify those elements that vary and those that don't, and to determine how to back up everything so as to be able to recover your GC environment completely to just before the time of failure.

---

[3] The coverage here of Grid Control B&R should prove useful, as the Oracle EM documentation set devotes only three pages to the subject in Sections 11.3.1 and 11.3.2 of Oracle Enterprise Manager Advanced Configuration 10*g* Release 4 (10.2.0.4.0).

Grid Control Elements That Continually Change

There is a common thread among all the elements that change on a continual basis across the GC framework—Agents, OMS, and OMR. You can trace this thread back to its source: Agent files containing management data that continually accrue on target hosts, are uploaded to the OMS, and then loaded into the Repository. Below are all elements that change continually, broken out by component:

- **Agent files in /sysman/emd/ directory**   The files that change on Agent hosts are located in the *$AGENT_HOME/sysman/emd/* directory. You can place them in two categories according to whether they are uploaded to the OMS or not (see "Agent Files and Directories" in Appendix A, online at www.oraclepressbooks.com, for details):

    - *Agent management files* contain target management information such as metric data, alerts, and policy violations. These files are continually created on the Agent host and uploaded to the OMS, which in turn loads them into the Repository.

    - *Agent configuration files* remain resident on the Agent host and are not uploaded to the OMS. Some of these files change continually (such as *agntstmp.txt*, which contains a timestamp); others change only when metric thresholds or target statuses change (such as collection or state files); and still others don't change unless you perform some Console action (such as *targets.xml*, which changes only when discovering, reconfiguring, or removing a target).

- **OMS files in receive directory**   The OMS receive directory is where the OMS stages files received from Agents before it loads the data from these files into the Repository. This directory either resides locally on each OMS host in the default location *$OMS_HOME/sysman/recv/* or is configured as a *shared receive directory* on storage accessible by all OMS hosts (see "Set Up the Shared Filesystem Loader" in Chapter 15).

- **OMR Database datafiles**   Data in the SYSMAN schema located in the OMR Database is updated when the OMS uploads Agent files, when GC application background processing occurs, and when administrators use certain features in the Console.

    - The SYSMAN schema contains data from Agent files, initially loaded into the MGMT_METRICS_RAW table.

    - Other SYSMAN objects change in response to the DBMS background job processing that keeps the GC engine running. The Repository DBMS Job Status summary table lists these jobs in the Console (click Setup, the Management Services and Repository subtab, then the Repository Operations page). One such job, called Rollup, aggregates data from the MGMT_METRICS_RAW table over one hour and one day into the MGMT_METRICS_1HOUR and MGMT_METRICS_1DAY tables, respectively.

    - Many administrator operations in the Console alter data in the SYSMAN schema, such as submitting jobs, patching Oracle homes, and saving custom reports.

Grid Control Elements That Don't Change

So what doesn't change in your Grid Control environment? Well, everything else, almost. Unless you count new log and trace file entries,[4] the GC software itself doesn't change, except when

---

[4] All software homes contain log and trace files. Although these files are important, particularly for troubleshooting purposes, it's not essential to restore them to the current point in time. It should suffice to back them up weekly when performing server-level backups (see "Daily or Weekly Server-Level Backups.")

patching or upgrading, or editing an associated configuration file. This GC software is found both inside and outside GC component homes:

- **Agent software home** Software in the Agent home does not change, except for files in the *$AGENT_HOME/sysman/emd/* directory, as just cited.

- **OMS software home** OMS software homes do not vary with day-to-day operations, except files in the OMS receive directory, as already mentioned.

- **OMR Database software home** The OMR Database software includes the database home, but may also contain a separate ASM home and Clusterware home, depending on your storage selection and whether you're running a RAC OMR Database. You can back up all software homes at the OS level, except two Clusterware components that must reside on shared raw devices or on a shared file system, and which therefore require special backup procedures, covered later.

- **GC files located outside GC software homes** These Grid Control files include those in the *oraInventory* directory, *oraInst.loc* and *oratab* files, initialization files for automatic GC startup, GC patches, DBA scripts (for user-defined metrics, RMAN backups, etc.), and any GC documentation stored on GC hosts.

How to Back Up All Grid Control Elements

Now that you know which GC elements do and don't change, how do you approach backing up both types of elements? Of course, a concurrent, GC site-wide cold backup, including Agents on all target hosts, guarantees that you'll capture all components in a perfectly synchronized state with regard to the changing elements previously described. At the time of such a cold backup, with all component operations frozen, some management files just created would be backed up on the Agent host, other Agent files already uploaded to the receive directory on the OMS host would be backed up there, and still other Agent files processed by the Repository would be caught in a cold database backup.

Though a cold backup captures the changing elements in their respective component locations, performing regular GC site-wide, cold backups would not be an effective strategy for a number of reasons:

- For GC sites that must remain highly available, regular cold backups and time-costly site-wide restorations are simply not acceptable.

- No matter how small or localized the failure, to recover/restore all GC components from backup to maintain synchronicity, including Agents on all target hosts, is overkill and too costly a way to have to recover from any type of failure.

- You'd have to settle for incomplete database recovery and GC software restoration to the time of the last cold backup rather than being able to perform complete recovery and restoration to just before the time of failure.

- Target states and metric severities after recovering from GC component failure would almost certainly not match those captured during an earlier cold backup, creating a mismatch between actual and reported states and severities after restoring/recovering from backup. This would require resetting Agents (see the next section), which is the only additional step necessary when recovering/restoring from hot backup. Therefore, recovery from cold backup is no easier than from hot backup.

If a regularly scheduled, site-wide cold backup is not the solution to backing up these changing elements, what is? You may be surprised to learn that you can perform hot backups of all GC components. Integral to the success of such a backup strategy is the ability to reconcile changing elements across components by resetting Agents as part of the recovery and restoration process.

Resetting Agents

Many GC component recovery and restoration scenarios require you to *reset Agents*, so it seems logical to look at this topic to round off general discussions on Grid Control B&R. As detailed in the "Procedure to Reset an Agent" section below, resetting an Agent is the simple, manual process of removing certain Agent timestamp, state, and upload files to force it to re-evaluate the metric severity, status, and metadata changes for each target on the Agent host and resend this information to the OMS. Often, recovering the OMR Database or restoring the OMS or Agent returns the GC component to an earlier state, which prevents Agents on their own from resuming normal communication with the OMS and Repository. Such loss of normal communication manifests itself in various ways:

- Agents cannot upload data to the OMS.
- Agents are out of sync with the OMS in their state information about the severity and status of targets.
- The OMS rejects uploaded Agent files because it does not have the corresponding metadata for them or severities appear out of sequence.

My approach to the subject of resetting Agents is as follows:

- When to reset an Agent
- Procedure for resetting an Agent
- Rationale behind the procedure for resetting an Agent

CAUTION
Resetting an Agent is not the same as clean starting an Agent, which causes more problems than it solves and should be avoided except when pointing an Agent to another OMS and Repository (see Note 413228.1). The Agent clean-start process involves removing additional Agent files and requires issuing the emctl clearstate agent command.

When to Reset an Agent

Resetting Agents is an inevitable part of most recovery scenarios for GC components. Therefore, it's much easier to first list the recovery situations in which you *don't* need to reset an Agent (in all situations, I assume you have a backup of the affected GC component home):

- When recovering/restoring all GC components from cold backup (though not a suggested backup method), including Agents on all target hosts
- When only the OMR Database must be recovered and you can perform a complete recovery

Chapter 14: Backup and Recovery of Grid Control

- When the device containing the OMS receive directory (whether local or a shared receive directory) fails but is located on fault-tolerant storage
- When the device containing Agent management files (in the *$AGENT_HOME/sysman/emd/* directory) fails but is located on fault-tolerant storage

So when *must* you reset Agents? Essentially, whenever an Agent cannot resume uploading to the OMS after recovery or restoration of a GC component due to loss of Agent data, such as:

- After performing OMR Database incomplete recovery. Restoring the OMR Database to a previous time causes it to lose synchronization with associated Agents, which maintain current state data. An Agent will almost certainly be out of sync with the OMR Database if it uploaded management data to the OMS since the time of OMR Database failure.
- After restoring an OMS to an earlier state whose local receive directory contains Agent files at failure time and is not located on fault-tolerant storage. In such a case, Agent files are lost that are located in the local receive directory at the time of failure. The reason is that Agents remove their local copy of files after uploading them to the OMS, but the OMS fails before uploading these files to the Repository.
- After restoring an Agent to an earlier state whose *$AGENT_HOME/sysman/emd/* directory contains management files at failure time and is not on fault-tolerant storage. In this scenario, collected Agent information at the time of failure that is not yet uploaded to the OMS is lost.

Procedure to Reset an Agent

Before I explain what resetting an Agent accomplishes, I'd like to dispense with the procedure so that you see what it does and how easy it is to complete. To reset an Agent, execute the following steps at the OS command line as the Agent user on the Agent host:

1. Determine whether the Agent can upload data to the Repository. During GC component failure, you cannot rely on the Console to accurately report this, so check from the command line on the Agent host:

   `$AGENT_HOME/bin/emctl upload`

   If the Agent can upload data, the message "EMD upload completed successfully" is returned, and you can skip the rest of this procedure. If an error message is returned, the Agent cannot upload data to the Repository, and you must carry out the remaining steps below.

2. Shut down the Agent:

   `$AGENT_HOME/bin/emctl stop agent`

3. Remove the following files (for RAC and NFS Agents, AGENT_HOME contains the node name, i.e., is the EMSTATE directory[5]):

   ```
   rm $AGENT_HOME/sysman/emd/agntstmp.txt
   rm $AGENT_HOME/sysman/emd/lastupld.xml
   rm $AGENT_HOME/sysman/emd/state/*.*
   rm $AGENT_HOME/sysman/emd/upload/*.*
   ```

---

[5] For a description of the EMSTATE directory, see "What Is an EMSTATE Directory" in Chapter 5.

4. Restart the Agent:

   `$AGENT_HOME/bin/emctl start agent`

5. Confirm the Agent can now upload to the Repository, beginning with pending files:

   `$AGENT_HOME/bin/emctl upload`

After resetting the Agent, it should resume normal uploads thereafter.

Rationale Behind Resetting an Agent

Following is an explanation of the files and directories that are removed when resetting an Agent. I describe their function and the consequence of removal when resetting an Agent.

agntstmp.txt File

- **Function**
    - Tracks the timestamp and the EMD_URL used in the most recent heartbeat the Agent sends to the OMS.
    - The Agent re-creates or updates this file after each heartbeat.
- **Consequence of removing** The Agent re-creates the file after the next heartbeat to accurately reflect the current timestamp. If the EMD_URL did not change (i.e., the Agent was not relocated to a different host name), the OMS will accept subsequent uploads from the Agent.

lastupld.xml File

- **Function**
    - Contains a list of already uploaded metadata settings and the timestamp of all relevant Agent monitoring files.
    - Used to determine on startup what information has changed so that the Agent can send only changes in metadata to the OMS. This optimizes Agent performance on startup in that the Agent needs to send only changes and not all metadata.
    - The Agent re-creates or updates this file immediately after startup.
- **Consequence of removing** Forces the Agent to regenerate currently timestamped, accurate metadata settings and to upload them to the Repository for processing.

Files in upload Directory

- **Function**
    - The *upload* directory holds the files pending upload to the OMS.
    - The Agent generates three forms of data: metadata (A files), state information (B files), and metric data (C, D, and E files). See "How Agent Files Are Uploaded" in Appendix A (online) for a description of these files.

- **Consequence of removing** If you remove from the *upload* directory data that the Agent collected but did not relay to the OMS to push to the Repository, and if you also remove the *lastupld.xml* file and files in the *state* directory, all files the Agent subsequently uploads will be accepted by the OMS. (However, if you don't also remove the *lastupld.xml* file and files in the *state* directory, the Agent may later send files that the OMS will reject, either because it does not have the corresponding metadata or because severities appear out of sequence).

Files in state Directory

- **Function**
    - The Agent maintains state information about target statuses and metric severities but, to prevent unnecessary network traffic, does not keep resending this information to the OMS unless a change to the target occurs that warrants the upload.
    - The files in this directory generate the appropriate "clears" for metric severities or errors when they occur.
- **Consequence of removing** Removing these files while also removing the *lastupld.xml* file and files in the *upload* directory will keep the Agent's memory of previous severities and statuses in sync with that of the OMS. (However, simply removing files in the *state* directory might remove the Agent's memory of previous severities and statuses, which would prevent the Agent from sending clears and cause the Agent and OMS to be out of sync).

Recommended Grid Control Backups

Now that you have an idea of the elements that change and those that don't change during normal GC operations, and understand how resetting Agents allows you to synchronize relatively concurrent, hot backups of GC components, you can appreciate how a GC backup plan is derived.

Regular backups of GC software and GC hosts allow you to recover or restore rather than having to reinstall, patch, and reconfigure GC if media failure occurs or user error corrupts or accidentally removes GC files. Table 14-1 summarizes overall recommendations for GC backups, detailing what to back up, the backup schedule, and the media to use, and whether to perform a hot or cold backup. (The latter is no mystery, as all suggested backups are hot.) This suggested backup strategy is appropriate for most Grid Control sites, but you may need to adapt it to suit your current IT backup schedule or the specific availability needs of your GC environment.

What to Back Up	Schedule	Media to Use	Backup Type
OMR Database	Daily	Tape (and disk if possible)	Hot
GC components affected	Before and after a configuration change	Tape (and disk if desired)	Hot (before and after)
GC servers	Daily or weekly	Tape	Hot

TABLE 14-1. *Suggested Grid Control Backups*

Following is a narrative description of each of these types of OMS backups.

Daily Backups of the OMR Database

Take daily hot backups of the OMR Database to capture Repository changes. You may choose between full, incremental, or incrementally updated[6] backups to best suit your RTO. Tape backups will suffice, but if disk space is sufficient, perform disk backups to separate devices as well for quicker media recovery.

Refer to the upcoming section "Back Up the OMR Database in Grid Control" for instructions on configuring daily OMR Database backups. See the particular subsection, "Oracle-Suggested Backup," for example combinations of daily disk and tape backups.

Backups Before and After a Configuration Change

Just before making any significant change to your GC configuration, such as when patching or upgrading, take a hot backup to tape of all GC components that will be affected. If the configuration change fails, you can roll it back by restoring from backup. Performing a hot backup of all components to be changed, including standalone and chain-installed Agents, immediately before making the change allows you to return these GC components to their former state. You should set a server-wide blackout on all GC hosts just before they undergo a configuration change to suppress notifications due to the change. If you must later restore from backup, you will need to reset Agents.

If you don't encounter problems from the configuration change, bring GC back online and take a new concurrent, hot backup to tape of all GC components affected by the change. You can restore from this new hot backup if a subsequent failure occurs, thereby avoiding having to perform the configuration changes again. If desired, before or after a configuration change, substitute server-level hot backups for those of affected GC components to kill two birds with one stone.

For directions on backing up a specific GC component before or after a configuration change, refer to the appropriate section below:

GC Component	Section in Chapter to Consult
OMR Database	Back Up the OMR Database in Grid Control
OMR Database software	OMR Database Software Backup and Restoration
OMS software	Back Up the OMS
Agent software	Back Up the Agent

Daily or Weekly Server-Level Backups

You should already have squirreled away a hot server-level tape backup of the Grid Control OMS and OMR Database hosts from the time of the initial build, as recommended at the end of Chapter 4 (see "Begin Regular Grid Control Backups"). You should also have performed a hot tape backup of each Agent host or, at a minimum, backed up the Agent software home (see "Back Up the Agent" in Chapter 7). You need to continue these server-level backups going forward.

System administrators (SAs) typically take online tape backups, both daily and weekly, of all hosts in their IT environment. These recurring backups will include software for OMR, OMS, and OMA homes that post-date one-time backups taken after making a configuration change, providing yet another layer of protection in case the post-configuration backup tape is unreadable or you

---

[6] Incrementally updated backups are a good way to minimize recovery time and are discussed in the section "Oracle-Suggested Backup" later in the chapter.

Suggestions for Working with Backup Media

Following are some general suggestions on how to best use disk and tape media for GC B&R. Many of these suggestions apply to any type of software backups, not just those for Grid Control. Note that "backups" in the following list refer to those of all GC components.

- Locate disk backups on separate file systems, RAID devices, controllers, and logical volumes from active[7] GC software.

- While you should aim to back up to tape, you may also elect to back up to disk, as determined by the number of components to be backed up, available disk space, and likelihood of having to restore.

- Disk-to-disk B&R is much faster than that from either disk to tape or tape to disk. Therefore, if a particular backup exists on both disk and tape, only restore from tape as a secondary means of recovery if disk restoration fails.

- If you need to minimize the time it takes to perform a GC tape backup, you may elect to back up to disk, then copy the disk backup to tape. The trade-off is that the backup may be compromised if a backup disk experiences media failure before the tape backup completes.

- Test the extent to which backups drain resources (I/O, CPU, and RAM) on GC hosts, particularly when performing backups concurrently to disk and tape, in parallel to disk, or in parallel to tape.

made minor subsequent changes, such as to a configuration file, that didn't warrant a complete GC environment backup. All in all, daily or weekly online server-level tape backups capture the following files that have been changed but not yet backed up:

- GC software homes, including configuration files
- Related GC files located outside GC software homes, such as the *oraInventory* directory, *oraInst.loc* and *oratab* files, initialization files for automatic GC startup, GC patches, DBA scripts (for user-defined metrics, RMAN backups, etc.), log and trace files, and documentation
- Non-GC operating system files, such as binaries, configuration files, and logs
- Target software that GC monitors

Notify the SA which directories to exclude in server-level tape backups, such as excluding old backups of GC software to disk and active database files.

---

[7] By "active" I mean the Grid Control software currently installed; it need not be running to be considered the active software, such as during a cold backup. Active OMR Database files are known specifically as the *working set* and all backup files needed to perform full or point-in-time database recovery after failure are known as the *redundancy set*. These backup files include datafiles, control files, redo logs, archive logs, the SPFILE, password files, and flashback logs.

As mentioned in the previous section, after making a configuration change, you can kill two birds with one stone by performing a server-level backup (perhaps earlier than scheduled) instead of a one-time backup of affected GC components.

OMR Database Backup and Recovery

It's time to shift gears now from the general to the specific, beginning with OMR Database B&R. I heartily advocate using Grid Control as the principal backup and management method for both OMR Database and target database backups, and as the main recovery method for target databases. The techniques for backing up, managing, and preparing to recover the OMR Database in Grid Control are no different from those for any target database. The RMAN[8] interface in the Console allows you to back up and manage both the OMR Database and target databases, and lets you choose their recovery settings. The only distinction between the OMR Database and target databases with respect to Grid Control B&R is that the Console would be available to recover target databases, but would itself be down if the OMR Database required recovery. Given the above, you can see that this section on using Grid Control for implementing OMR Database backups, managing them, and choosing recovery settings applies equally well to target databases.

There are several key advantages to using Grid Control's capabilities to administer OMR Database B&R:

- Grid Control was designed to integrate seamlessly with built-in features of the *Oracle database system*[9] (the database and the software that runs it). These features—the Recovery Manager (RMAN) client[10], SQL*Plus, ARCHIVELOG mode, Flashback Technology, automatic undo management, restore points, and block change tracking—form the building blocks of an Oracle database B&R solution.[11]

- On top of the native Oracle server B&R technology, the Grid Control product possesses unique B&R attributes not found anywhere else, such as the Oracle-Suggested Backup and Customized Backup strategies, as well as the RMAN Script job, as discussed later in the chapter.

- Beyond its B&R abilities, Grid Control offers many administrative features that add to the page count (presented in the section "Manage OMR Database Backups in Grid Control").

---

[8] While the material in this section on RMAN B&R for a database (the OMR Database in this case) strives to be complete, you should consult the Oracle Database documentation set for comprehensive coverage. Start with Oracle Database Backup and Recovery Basics 10g Release 2 (10.2), which explains many of the concepts presented here. For extended coverage of the more complex B&R scenarios and techniques, see Oracle Database Backup and Recovery Advanced User's Guide 10g Release 2 (10.2). Also see the excellent Oracle Press book, *Oracle Database 10g RMAN Backup & Recovery*, Matthew Hart and Robert G. Freeman, The McGraw-Hill Companies, Inc., 2007.

[9] This is not surprising, given that the Oracle Database is one of the most common Grid Control targets among users.

[10] You launch the RMAN client by executing *$ORACLE_HOME/bin/rman* at the OS command prompt.

[11] If you followed my advice in Chapters 3 and 4 for installing Grid Control, then you've already implemented many of these B&R features for the OMR Database.

This section on OMR Database B&R is broken up into three subsections:

- Back Up the OMR Database in Grid Control
- Recover the OMR Database
- Manage OMR Database backups in Grid Control

Let's start with how Grid Control itself can take care of performing regular OMR Database backups.

NOTE
In this section, I use the example of an OMR RAC Database named emrep and perform all Console steps logged in as the built-in SYSMAN user. If you don't log in as SYSMAN, log in using your own administrator account, as many steps rely on having set your preferred credentials.

Back Up the OMR Database in Grid Control

It's understandable if you hesitate to entrust your OMR Database backups to GC itself. It may seem like circular reasoning to use an application such as Grid Control to regularly back up part of that application (the OMR Database). Perhaps your main objection is that regularly scheduled OMR Database backups in Grid Control won't run when Grid Control is down. Well, if any GC components are down—the OMS (which dispatch jobs), the Agents on OMR nodes (which run jobs), or the OMR Database (which stores job information)—you can always use the RMAN client to directly execute the backup script that Grid Control creates. (Instructions below for setting up backups in GC include copying the RMAN script to the OS on all OMR Database nodes.)

If you implement GC high availability and disaster recovery practices to maximize uptime (see Chapter 15), you should need to resort to OMR Database backups outside of GC only on rare occasions. One such occasion might be to take a cold backup of the OMR Database immediately after a failure so that you can roll back any unsuccessful recovery attempts that do not succeed or that further damage database integrity.

Setting up OMR Database backups in Grid Control involves four steps:

- Direct the OMR Database to use a recovery catalog in Grid Control
- Configure OMR Database backup settings in Grid Control
- Implement a Grid Control backup method for the OMR Database
- Set up OMR Database exports

Direct the OMR Database to Use a Recovery Catalog in Grid Control

The goal of this section is to "attach" an existing *recovery catalog* to Grid Control so that backups of the OMR Database in the Console use this recovery catalog. The procedure applies equally to other target databases. To make sense of the process, let me first provide some background on a recovery catalog and its benefits.

Oracle stores all metadata about a database's RMAN operations in an *RMAN repository*. The primary storage for RMAN repository information is in the control file of the *RMAN target database*. This is the database that the RMAN client is connected to (via the CONNECT TARGET command) and that RMAN commands will act upon. (In the context of Grid Control, the RMAN target database is also the GC target database, for all intents and purposes, which in this chapter is the OMR Database itself.) A recovery catalog is an optional, supplemental RMAN component that can house copies of RMAN repositories for multiple RMAN target databases in a centralized location—an RMAN schema in a database. Grid Control licensing[12] allows you to place a recovery catalog in the OMR Database; however, RMAN best practices call for housing a recovery catalog in a dedicated database isolated from all RMAN target databases. (No separate database license is required for an RMAN repository, provided that all Oracle databases managed in it are licensed.)[13] This protects the recovery catalog in the event an RMAN target database fails. Other RMAN best practices are to configure the recovery catalog database for high availability, to physically isolate its host from all hosts containing RMAN target databases, and to use one centralized recovery catalog to manage all RMAN target databases in your enterprise.

Since primary storage for an RMAN repository is in the control file of the RMAN target database, a recovery catalog is not required to use RMAN. However, a recovery catalog offers many advantages, which apply when you run the RMAN client either directly or in GC. (The Console "proxies" many of these RMAN features when you set up a recovery catalog in GC, as described later in the chapter.) Following are the principal benefits of using a recovery catalog in GC to store the RMAN repository for the OMR Database and repositories for other target databases:

- Allows you to utilize all Grid Control B&R functionality.
- Centralizes your RMAN repository information for the OMR Database and other target databases that you back up.
- Maintains historical data as far back as you like, without needing to purge it.
- Stores RMAN scripts in the recovery catalog.
- Persistently maintains default *channel* configuration information between backups. RMAN channels are connections to server sessions on RMAN target databases that perform all RMAN tasks.
- Improves backup reporting functionality, such as to report on a target database at a time other than the present.

A recovery catalog requires extra effort to maintain, but the above benefits are worth the effort, particularly if using RMAN to back up multiple databases in GC. An example of a UI feature that uses a database's recovery catalog configuration in GC is the RMAN Script job, which implicitly connects the database you're backing up to the recovery catalog (i.e., performs a CONNECT CATALOG… command) when submitting the RMAN script to the job system.

You must complete two simple steps to direct Grid Control to use a recovery catalog when backing up the OMR Database (or any target database) in Grid Control:

- **Configure a recovery catalog in Grid Control**   This procedure involves entering recovery catalog information in GC so it can grant the recovery catalog owner the necessary roles and privileges to perform B&R operations in the Console.
- **Register the OMR Database in the recovery catalog**   Once a recovery catalog is configured in GC, you must register the OMR Database (or other target database being backed up) with the recovery catalog.

---

[12] See Special-Use Licensing in Chapter 1 of Oracle Database Licensing 10g Release 2 (10.2). Also see Oracle*MetaLink* Note 394626.1.

[13] Ibid.

Chapter 14: Backup and Recovery of Grid Control 563

Configure a Recovery Catalog in Grid Control You must configure at least one recovery catalog in Grid Control so that you can register the OMR Database (and other target databases, if desired) in this recovery catalog. Most companies abide by best practices and use just one recovery catalog for all RMAN target databases. Below are the steps to configure a recovery catalog for Grid Control to use. To illustrate, assume you want to configure a single recovery catalog located in an existing GC target database called *rman* with recovery catalog schema owner also called *RMAN*.

1. Create a recovery catalog outside of Grid Control.[14] You cannot use Grid Control itself to create a recovery catalog: it must already exist.
2. Click on Targets, Databases. Under Related Links, select the Recovery Catalogs link.
3. Click Add, which takes you through the Add Recovery Catalog wizard, beginning with the following screen:

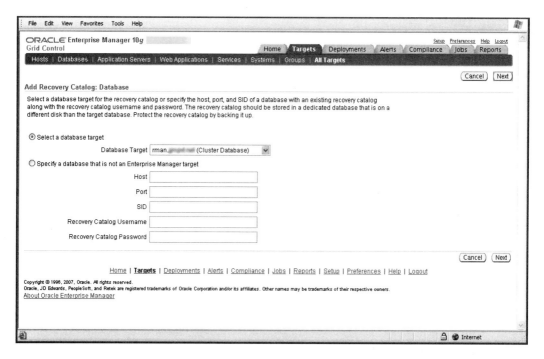

4. Select a database housing an existing recovery catalog. You can select an already discovered database target or a new database by specifying its host, port, and SID. (In this example, I selected an existing *rman* database target.) Click Next.

---

[14] For instructions on creating a recovery catalog, see Chapter 10 of Oracle Database Backup and Recovery Advanced User's Guide 10*g* Release 2 (10.2).

5. The Credentials page appears.

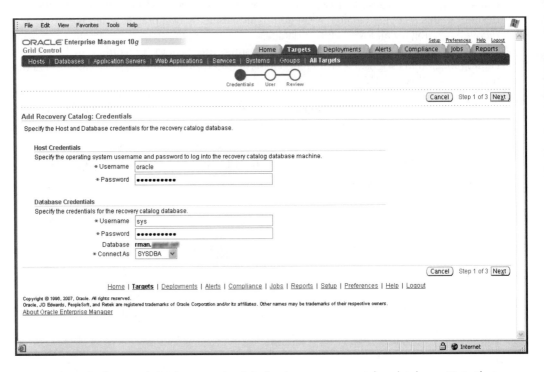

Enter the host and database credentials for the recovery catalog database. Note that you should enter a database user with SYSDBA privilege, not the owner of the recovery catalog schema. Click Next.

6. On the User page, enter the existing recovery catalog owner (*RMAN* in this example) and password. GC grants this user the necessary roles and privileges to perform B&R operations in the Console using the recovery catalog you're pointing to. Click Next.

7. On the Review page, click Finish to complete the recovery catalog configuration in GC. A banner message should confirm that the configuration succeeded and the recovery catalog should appear in the list of recovery catalogs.

Register the OMR Database in the Recovery Catalog Having prepared a recovery catalog for use, you must now register the OMR Database in the recovery catalog before GC can begin performing RMAN backups. You can do this either in the Console or directly in the RMAN client, so I provide instructions for both methods below. (In the example, the OMR Database is called *emrep* and the recovery catalog is called *rman*.)

Chapter 14: Backup and Recovery of Grid Control 565

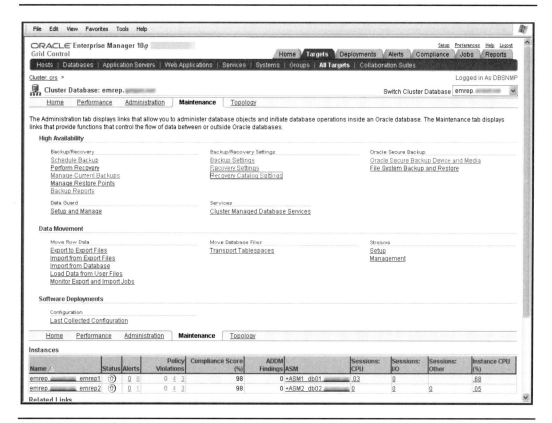

FIGURE 14-1. *The Maintenance page for a database target*

In Grid Control, register the OMR Database as follows:

1. Drill down to the OMR Database target and click the Maintenance page (see Figure 14-1), from which you access all database B&R functionality in GC.

2. Under the Backup/Recovery Settings section, select Recovery Catalog Settings, which brings you to the page shown in the following image.

3. This page displays the database's current registration status in a recovery catalog as indicated by which radio button is selected: Use Control File or Use Recovery Catalog. If not already the case, select Use Recovery Catalog, choose the desired recovery catalog in which to register the database, enter the OS username and password for the owner of the recovery catalog database, and click OK.

Using the RMAN client, register the OMR Database as follows:

1. First, confirm the database is registered in the desired recovery catalog. From an operating system prompt, start the RMAN executable, connecting to the target database and to the recovery catalog:

    $ORACLE_HOME/bin/rman TARGET sys/<password>@emrep CATALOG rman/<password>@rman

    connected to target database: EMREP (**DBID=3839157144**)

    connected to recovery catalog database

List the incarnations[15] of the OMR Database to confirm that the DBID[16] for the current incarnation in RMAN matches that of the target database you connected to above.

```
RMAN> LIST INCARNATION OF DATABASE emrep;
List of Database Incarnations
DB Key  Inc Key  DB Name  DB ID         STATUS   Reset SCN  Reset Time
-------  -------  -------  -------------  ---  ----------  ----------
1        2        EMREP    3839157144     CURRENT 1          02-MAR-07
```

The DBIDs match, as shown in bold. This means that the current OMR Database *emrep* is registered in the recovery catalog.

2. If the *emrep* database is not registered in the recovery catalog, register it using the RMAN client by executing the following command:

   `RMAN> REGISTER DATABASE;`

At this time, you may want to register other target databases not already registered in the recovery catalog. You can do so either in GC or with the RMAN client, as shown above.

Configure OMR Database Backup Settings in Grid Control

All B&R operations for a database begin on its Maintenance page (see Figure 14-1), and the operation to configure databse backup settings is no exception. Navigate to the Maintenance page for the OMR Database and click the Backup Settings link. This brings up the first of three Backup Settings pages, the Device page, shown in Figure 14-2.

Use these pages to override RMAN configuration settings for the target, whether set by default or previously user-entered. (Default settings apply unless you explicitly change a setting.)[17]

NOTE
Going forward, I use "Backup Settings" in title capitalization to indicate the backup configuration settings you can modify on these Backup Settings pages.

The input on the Backup Settings pages is database-specific, stored persistently (i.e., it doesn't change, even with a database bounce) in the control file of the database and also in the recovery catalog, if the target database is registered there. The input does not change settings for any other databases. If you change Backup Settings for a database, they will be in effect for any RMAN-based backups of that database, whether in GC or using the RMAN client, even if you don't configure a recovery catalog in GC or register the database with the recovery catalog.

While the Backup Settings you input here are persistent for the database you are configuring, two of the three GC backup methods presented later permit you to override most of these settings (see the section "Implement a Grid Control Backup Method for the OMR Database").

---

[15] A new database incarnation is created in RMAN each time an RMAN target database is opened with RESETLOGS.

[16] The recovery catalog enforces that all databases have a unique DBID (the DBID shown above is just an example).

[17] To view default configuration settings for a target, connect to the target in RMAN and execute SHOW ALL. Default settings are suffixed with *#default*.

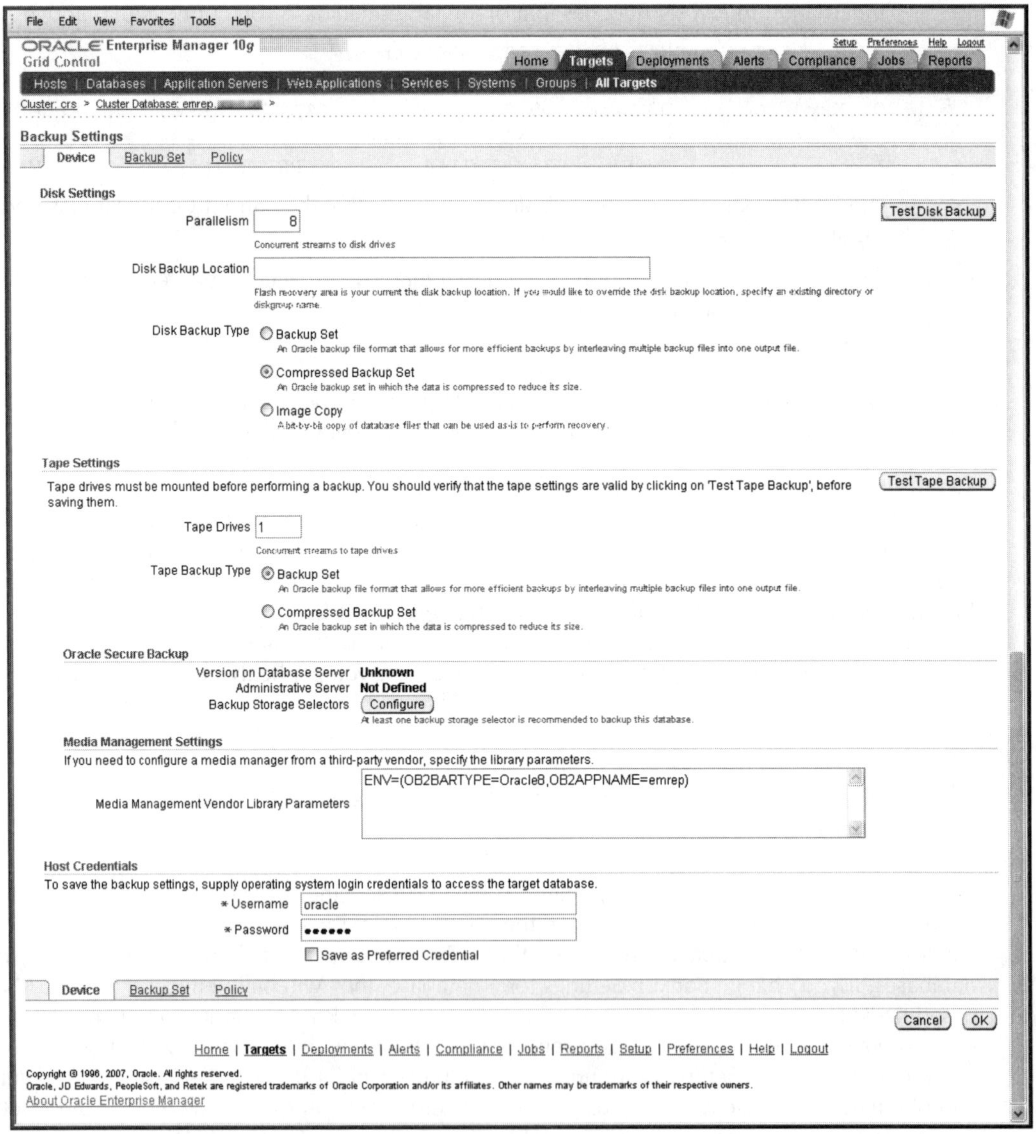

FIGURE 14-2. *The Backup Settings Device page*

It's time now to tour the Backups Settings pages, which are:

- Device
- Backup Set
- Policy

Chapter 14: Backup and Recovery of Grid Control 569

Figure 14-2 above and screen shots of the two remaining pages capture representative settings for an OMR Database. For each page, I provide a corresponding table[18] that lists its settings, their default and recommended values, and the equivalent RMAN commands that GC runs when you change settings.

Device The Device page (see Figure 14-2) is where you enter and test disk and tape configuration settings. These settings include disk backup location, the number of tape drives, media manager library parameters, and Oracle Secure Backup (OSB) setup, if configured in GC. A *media manager* is software that loads, labels, and unloads sequential media such as tape drives used to back up and recover data.

Table 14-2 distills the fields found on the Device page, detailing default and recommended values for the OMR Database and the underlying RMAN commands that Grid Control executes when inputting values on the page. (The Media Management Vendor Library Parameters specified above are for a specific vendor, HP Data Protector, and vary by vendor.)

After entering your preferred disk and tape device settings, confirm they are valid by clicking Test Disk Backup and Test Tape Backup, respectively.

Backup Set Click the Backup Set page to instruct RMAN on how to generate backup sets.

Table 14-3 reflects the default and recommended values and the equivalent RMAN commands that Grid Control runs.

---

[18] Grid Control requires Host Credentials to submit a job to implement the configuration changes made in the Backup Settings pages. Host Credentials fields are located at the bottom of each page. Because of this redundancy, these fields are not represented in the three tables describing these pages.

Section	Field	Default Value	Recommended Value	Equivalent RMAN Command
Disk Settings	Parallelism	1	2 × <total CPUs on all database nodes>	CONFIGURE DEVICE TYPE DISK PARALLELISM <n>;
	Disk Backup Location	Flash Recovery Area location	<leave blank> (defaults to FRA location)	None required
	Disk Backup Type	Backup Set	Compressed Backup Set	BACKUP AS COMPRESSED BACKUPSET …
Tape Settings	Tape Drives	None	Number of tape drives (allocates same number of channels)	The number of parallel ALLOCATE CHANNEL commands issued (see Media Management Settings in last row of table)[19]
	Tape Backup Type	Backup Set	Backup Set (for local device), Compressed Backup Set (for remote device)[20]	BACKUP AS [COMPRESSED] BACKUPSET…
Oracle Secure Backup	Configure Backup Storage Selectors	None	See Oracle Secure Backup Installation Guide Release 10.1	None
Media Management Settings	Media Management Vendor Library Parameters	None	Variables and values that vendor recommends	ALLOCATE CHANNEL OEM_SBT_BACKUP<n>[21] TYPE 'SBT_TAPE' FORMAT '%U' PARMS 'ENV=(var1=val1,var2=val2)';

TABLE 14-2. *Recommended Values for the Device Page*

---

[19] This is equivalent functionally to the command CONFIGURE DEVICE TYPE 'SBT_TAPE' PARALLELISM <n>.

[20] For backups to locally attached tape devices, compression that the media management vendor provides is preferable to the binary compression that the RMAN command BACKUP AS COMPRESSED BACKUPSET affords. Oracle Database Backup and Recovery Reference 10g Release 2 (10.2), pages. 2–41.

[21] Here, <n> is the Tape Drives setting specified on the Device page. Multiple commands of this form are generated.

Section	Field	Default Value	Recommended Value	Equivalent RMAN Command
	Maximum Backup Piece (File) Size	None	None	CONFIGURE MAXSETSIZE TO <n>M;
Tape Settings	Copies of Datafile Backups	1	1	CONFIGURE DATAFILE BACKUP COPIES FOR DEVICE TYPE SBT TO <n>;
	Copies of Archivelog Backups	1	1	CONFIGURE ARCHIVELOG BACKUP COPIES FOR DEVICE TYPE SBT TO <n>;

TABLE 14-3. *Recommended Values for the Backup Set Page*

While disk backups can be either image copies or backup sets, tape backups must be backup sets. Therefore, Maximum Backup Piece (File) Size pertains to both tape and disk backups using backup sets.

Policy Use the Policy page to define backup and retention policies.

Section	Field	Default Value	Recommended Value	Equivalent RMAN Command
Backup Policy	Automatically backup the control file and SPFILE	OFF	ON	CONFIGURE CONTROLFILE AUTOBACKUP ON;
	Optimize the whole database backup by skipping unchanged files	OFF	ON (provided meets Recovery Time Objective)	CONFIGURE BACKUP OPTIMIZATION ON;
	Enable block change tracking	Not enabled	Enabled (tracking file is auto-populated when enabled)	ALTER DATABASE ENABLE BLOCK CHANGE TRACKING --SQL command;
	Tablespaces excluded	None excluded	None excluded	CONFIGURE EXCLUDE FOR TABLESPACE '<tbsp_name>'
Retention Policy	None, recovery window or redundancy	REDUNDANCY 1	Disk: RECOVERY WINDOW OF 2 DAYS (or more) Tape: REDUNDANCY 2[22]	CONFIGURE RETENTION POLICY TO RECOVERY WINDOW OF <n> DAYS	REDUNDANCY <n>;

TABLE 14-4. *Recommended Values for the Policy Page*

As done for the other two pages, I provide a table (Table 14-4) of all fields on the Policy page, with default and recommended policy settings for the OMR Database, and equivalent RMAN commands corresponding to each field.

By way of background, let me cover these backup and retention policy settings and the ramifications of your selections:

- **Control file and SPFILE autobackups: Enable** (10g only) RMAN can automatically back up control files and server parameter files (SPFILEs) after every backup and after database structural changes. (The CONFIGURE CONTROLFILE AUTOBACKUP ON command backs up the control file and SPFILE.) The control file autobackup contains metadata about the previous backup, which is critical for disaster recovery. The SPFILE backed up is the one the instance is currently using. If you start the instance with an initialization parameter file, RMAN does not back up anything when autobackups are configured.

---

[22] Try to keep at least two copies of the redundancy set to protect against the possibility of tape failure.

- **Backup optimization: May increase recovery time**   Make certain that enabling backup optimization does not slow down *media recovery* such that you cannot meet a database's RTO. Media recovery means to restore datafiles or control files from backup and recover them to the last transaction before the failure or to a consistent past point in time. The reason a slowdown is possible is that backup optimization skips unchanged files, which sometimes necessitates going back to dated tapes for the last concurrent backups taken of these files.

- **Block change tracking: Enable**   (10*g* only) Enable block change tracking to improve incremental backup performance. Block change tracking permits Oracle to track the physical location of all database changes for incremental backups. RMAN automatically uses a change tracking file to determine which blocks it needs to read during an incremental backup, and directly accesses these blocks to back them up. When block change tracking is not enabled, Oracle must read the entire datafile during each incremental backup to find and back up only the changed blocks, even if just a very small part of that file changed since the previous backup.

- **Retention policy: Retain enough disk space to meet recovery window**   A retention policy determines how many backups are retained to meet a point-in-time recovery goal. There are two mutually exclusive retention policies for a given media (disk or tape): *recovery window* and *redundancy*. A recovery window policy ensures that RMAN will keep the backups required for database recovery to a specified past number of days or *point of recoverability*. A redundancy policy guarantees that RMAN will keep a *specified* number of backups of each database file, whenever they are taken (typically daily). You can specify different retention policies for disk and tape, as recommended in Table 14-4. These policies are well advised for the Repository Database, and in fact are good policy settings for any target database, with the following caveat applicable to when disk space is low.

As reflected in Table 14-4, I recommend a recovery window–based retention policy for disk backups, as it guarantees you can recover back to the point of recoverability. (By contrast, I recommend a redundancy-based policy for tape backups). However, it's difficult to maintain this policy when available backup disk space is low. The reason is that RMAN must usually retain backups older than the point of recoverability, restore these backups, then roll forward using archived redo logs to the point of recoverability. If disk space is low, you may prefer the predictability of a redundancy retention policy for disk backups, which maintains the same number of backups of each datafile.

Implement a Grid Control Backup Method for the OMR Database

Having dispensed with the preliminary step of configuring OMR Database Backup Settings in Grid Control (device, backup set, and policy settings), you are now ready to perform full database backups. You can use one of three methods to back up the OMR Database in GC, which I list here in order of least to most configurable (I capitalize method titles throughout because they are rather generic):

- **Oracle-Suggested Backup strategy**   The least robust of the three methods, the Oracle-Suggested Backup uses a proven out-of-box strategy to perform daily and, optionally, weekly backups of the entire database to disk, tape, or both. A wizard offers a few simple options, such as media type, scheduling, and tape retention policy (which is the only one of the Backup Settings you can override), guarantees recovery to within the past day (or more, depending on the retention policy), and integrates with third-party media management software for tape backups.

- **Customized Backup strategy** A good middle ground of the three backup methods, the Customized Backup also provides a wizard that constructs an RMAN backup script, but with more features than those of the Oracle-Suggested Backup. There are more backup options (including hot or cold, full or incremental, and whole or partial database backups), and you can override almost all Backup Settings for the current backup, customize the generated RMAN script, and submit it with more scheduling options. Customized Backup integrates with Oracle Secure Backup and third-party media management software for tape backups.

- **RMAN Script job** The most configurable method, the RMAN Script job allows you to supply your own RMAN script and runs it through the job system using the recovery catalog configuration you set up in GC. You can enter the script in the UI or call a fully qualified script name that must exist on the database host(s) you are backing up. This method uses the GC Backup Settings, unless overridden. A unique advantage over the other methods is that you can configure notification of backup job status (such as Suspended or Problems).

TIP
If you set up a standby OMR Database in a Data Guard configuration with the primary database, as discussed in Chapter 15, you can back up the standby rather than the primary to offload the resource drain from backup processing (CPU, RAM, and disk) onto the standby. In this case, you need to evaluate the feasibility of using these backup methods for the standby OMR Database.

These three backup methods in GC provide certain advantages over the RMAN client to perform regular backups of the Repository Database (and target databases):

- **Connect automatically to target and to recovery catalog** All methods have built-in intelligence to connect implicitly to both the target database and the recovery catalog without the need for CONNECT TARGET or CONNECT CATALOG commands, respectively.

- **Can set it and forget it** The Oracle-Suggested Backup is an automated but configurable solution that is tested and proven to work. This method runs with minimal user input and purges older backups according to the retention policy you set. The customized backup strategy is also proven to back up the database or portion thereof, as you direct.

- **Use the Grid Control job system** All methods of GC backup take advantage of its job system to some extent. The RMAN Script job does so completely, thereby offering a rich set of features for job scheduling, exception handling, and reporting.

- **Provide backup management capabilities** Management tasks available for all methods include upkeep (crosscheck, validate) and reporting (view history, status, and details of backup jobs). Backup reporting and listing in GC is much easier to use and decipher than the equivalent RMAN command-line options and message logs.

Following are several drawbacks to using these GC backup methods and the ramifications for performing recovery of the Repository Database:

- **Cannot recover OMR Database through Grid Control**   While you can use GC to back up the OMR Database, the Console would not be available to perform OMR Database recovery if the OMR Database itself were out of commission.[23] However, you can recover the OMR Database in RMAN using backups created in GC.

- **Cannot centralize B&R if you also use other backup tools**   If you use GC for regular backups of only certain databases, and rely on the RMAN client or third-party backup tools to back up other databases and Oracle homes, your backup configuration will not be centralized. This approach adds administrative overhead and can cause mix-ups or oversights in your backup coverage. If you cannot set up GC to manage all your database backups, perhaps use it only for nonproduction or ad hoc database backups.

- **Some backup methods don't provide job notification**   The Oracle-Suggested Backup and Customized Backup methods don't provide for administrator e-mail notification of job status (such as Problems), either through notification rules or through the job itself in the Schedule Backup page. This is a shortcoming that alone may dissuade you from using either of these backup strategies. The RMAN Script job allows for such job notification, but if it doesn't suit your backup needs, and you require backup job notification (which excludes the other two methods), then none of these methods may appeal to you.

Due to one or more of these drawbacks, you may decide to set up RMAN backups completely outside of GC, using third-party backup tools or job file schedulers such as Unix *cron*. In this case, you can skip to the section "Recover the OMR Database." The scope of this section is to explain how to back up the OMR Database in GC; I could not begin to discuss the large variety of other database backup methods.

How do the three GC backup methods compare and contrast with each other? As already mentioned, they offer varying degrees of customization. Both Oracle-Suggested Backup and Customized Backup offer a nice user wizard interface to RMAN to accomplish backups, particularly for administrators not well versed with RMAN syntax. You may find their wizard-generated scripts useful as a starting point for building your own RMAN Script job. The Oracle-Suggested Backup is not nearly as versatile as the other methods, opting instead to automate the process of scripting a sensible backup solution by asking for minimal user input. The Customized Backup provides heftier UI "hooks" into RMAN to generate a more advanced script, which it then lets you customize, and can also temporarily override Backup Settings. The RMAN Script job is the most configurable in that you can enter RMAN commands or the name of an RMAN script located on the database host and can fully leverage all job system features. Both the Customized Backup and RMAN Script job can set up multiple backup jobs to meet any scheduling requirements, whereas the Oracle-Suggested Backup only provides for daily and weekly backups.

It's time now to go through the machinations of each backup method, which should help cement your choice of backup method for the OMR Database. The rationale for your decision should apply equally to backing up all target databases at your site.

Oracle-Suggested Backup   The primary purpose of the Oracle-Suggested Backup is to make backups easy, so it is intentionally less flexible when compared with the other two methods. The Oracle-Suggested Backup can only back up the entire database, not a portion of it, such as a

---

[23] You could, however, recover all target databases through the Console, provided the OMR Database is running.

datafile, tablespace, or archive log. This method does not change RMAN configuration settings (that is, the wizard-generated scripts don't contain the CONFIGURE command), except for being able to override the tape retention policy in the UI. Furthermore, the wizard doesn't present the choice of online ("hot") or offline ("cold") backups, but backs up the database in its current state, provided it's at least mounted. (For RMAN to back up a database, it must be mounted so RMAN can record backup information in its control file.) That said, I would only recommend Oracle-Suggested Backup for hot backups (which are recommended anyway). Using this method for cold backups is a bit dodgy because you'd have to independently shut down and mount the OMR Database before backups execute, and then, after they complete, reopen the database.

Despite its relative inflexibility, Oracle-Suggested Backup offers three variations of backups, the last of which has four types, which a wizard guides you through:

- **Disk** Perform daily *incrementally updated* disk backups. These backups avoid the overhead of taking a daily full image copy backup, but provide the same recovery advantages. They work by creating an image copy backup, followed by a daily level 1 incremental backup, which is applied to the image copy backup to roll it forward to the next day. Media recovery time is minimized in that you never need to apply more than one day of archive logs.

- **Tape** Perform daily cumulative incremental tape backups and weekly full tape backups. Remove obsolete backups, copies, and archive logs from tape daily according to the tape retention policy.

- **Both disk and tape** Use disk for daily backups so you can recover quickly, and use tape for weekly backups to extend the recovery window. Perform weekly tape backups of *recovery files* on disk (all disk backups and archive logs not already backed up to tape), and daily incrementally updated disk backups combined with one of the following types of daily tape backups:

    - None
    - Archivelogs
    - Archivelogs and the Incremental Backup
    - Archivelogs and the Full Database Copy

I'll expound on all Oracle-Suggested Backup variations and types while navigating the wizard; all of them guarantee full database recovery or point-in-time recovery to any time within the past day (or more, depending on your retention policy).

To ply the wizard's wares, consider the following real-world assignment. Implement an Oracle-Suggested Backup variation and type to perform OMR Database hot backups to both disk and tape for a small GC site (with fewer than 1,000 targets[24]). Perform one full database image copy disk backup, and then daily incrementally updated disk backups. Also set up daily full database tape backups, including archive logs. Finally, back up all recovery files to tape once a week. All backups should run at 2 A.M., with weekly backups to occur on Saturday. Specify a 31-day recovery window for tape backups and remove obsolete tape backups. Assume you already specified a two-day recovery window for disk backups in Backup Settings.

---

[24] This is based on the definition of small, medium, and large Grid Control sites listed in Table 2-5 of Chapter 2.

Chapter 14: Backup and Recovery of Grid Control 577

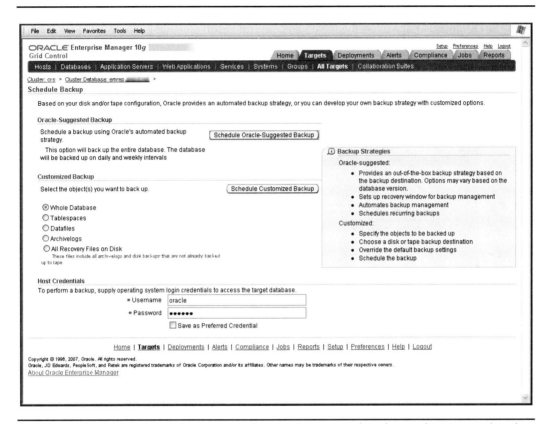

FIGURE 14-3. *The Schedule Backup page for Oracle-Suggested Backup and Customized Backup*

You may already see that, to implement this example, you need to select the third variation, "Both disk and tape," and the type "Archivelogs and the Full Database Copy." To accomplish this variation and type of Oracle-Suggested Backup, log in as an administrator with VIEW ANY TARGET system privilege, FULL target privilege on the OMR Database, and whose OMR Database SYSDBA preferred credentials are set. Navigate to the Maintenance page (see Figure 14-1 earlier in the chapter) for the database to be backed up—the OMR Database in our case. Under the Backup/Recovery section, click the Schedule Backup link to open the Schedule Backup page (see Figure 14-3).

Click Schedule Oracle-Suggested Backup. Grid Control launches a wizard with the following pages on which to make your backup selections:

- Destination
- Setup
- Schedule
- Review

Let's cover each page in turn.

Destination

The Destination page offers the three variations for destination media to store the backups:

- Disk
- Tape
- Both Disk and Tape

If the database is registered in GC, this page also supplies the recovery catalog username and database. Choose the destination media desired, which for this example is Both Disk and Tape, and click Next.

Setup

The Setup page specifies the location and parameters of the media chosen on the Destination page, and varies according to the media chosen. Below are screenshots and descriptions of the three Setup page variations.

VARIATION 1: DISK

When you select Disk on the Destination page, the following Setup page appears:

For 9*i* or earlier databases, you must specify a disk backup location. For 10*g* databases that use a FRA, the DB_RECOVERY_FILE_DEST location is used as the disk backup location. The Disk option takes one full database backup, then performs incrementally updated backups to disk every day. This allows for full recovery or point-in-time recovery to any time within the past day.

VARIATION 2: TAPE

The second variation of the Setup page is when you choose Tape on the Destination page.

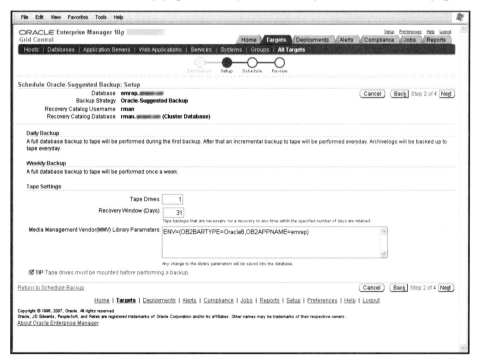

When using tape as the only storage for backups, Oracle-Suggested Backup carries out both daily and weekly tape backups:

- **Tape-only daily backup** Take one full tape backup, followed by daily incremental tape backups of datafiles and archive logs. (You cannot take incrementally updated backups to tape because it is sequential media, which cannot store image copies.)
- **Tape-only weekly backup** Perform a weekly full tape backup.

This strategy ensures that you won't have to apply more than one day of archive logs from tape.

VARIATION 3: BOTH DISK AND TAPE
The last variation of the Setup page is when you select Both Disk and Tape on the Destination page.

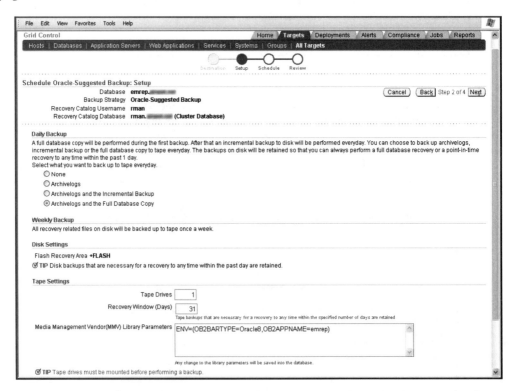

This is the most robust Oracle-Suggested Backup variation and type for the OMR Database—or any production target database, for that matter. It offers the best of both worlds: backup to both disk and tape. The daily disk backups let you recover quickly, and weekly tape backups allow you to maintain a longer recovery window than for the disk backups (31 versus 2 days, in this example). A nightly tape backup also gives you a "Plan B" if your daily disk backup becomes unusable due to backup disk failure.

Here is a functional breakdown of the selections on this page (only the main two bullet points, "Daily Backup" and "Weekly Backup," correspond with field names on the page):

- **Daily Backup**   Perform daily disk backups along with one of the four types of daily tape backups, which are combined in one script:
    - **Daily disk backup**   Perform a full database image copy followed by daily incrementally updated[25] backups to disk. Allows full or point-in-time recovery to any time within the past day (or more, depending on your retention policy). This backup is the same as Variation 1 above.
    - **Daily tape backup**   Provide four types of daily tape backups, three of which can fill in where disk recovery is not possible, such as in a disaster recovery scenario. The four types are listed in decreasing order of recovery time (i.e., the highest recovery time is for "None"):
        - **None**   Back up nothing to tape on a daily basis, only weekly. Do not select this option if backup disks aren't redundant, because failure to restore from disk could result in losing as much as one week of data.
        - **Archivelogs**   Perform daily archive log tape backups. Purge obsolete backups, copies, and archive logs from tape daily according to the tape retention policy. Before choosing this type of tape backup, you should test restoring and applying up to a week's worth of archive logs from tape to guarantee your RTO if disk recovery is not possible.
        - **Archivelogs and the Incremental Backup**   Perform daily archive log and incremental tape backups. Purge obsolete backups, copies, and archive logs from tape daily according to the tape retention policy. This option is suitable for medium and large GC environments[26] because it consumes less tape than "Archivelogs and the Full Database Copy."
        - **Archivelogs and the Full Database Copy**   Perform daily archive log and full database backups. Purge obsolete backups, copies, and archive logs from tape daily according to the tape retention policy. This option is appropriate for small GC sites, as a relatively small database size should keep tape usage to a reasonable level. Recovery is less involved than for incremental backups.
- **Weekly Backup**   Back up all recovery files on disk to tape once a week.

For the example, choose "Archivelogs and the Full Database Copy" and click Next.

---

[25] The text on the Setup page for Both Disk and Tape incorrectly states that the daily disk backup is incremental, but the generated RMAN script reveals that it is an incrementally updated backup.

[26] Ibid. Footnote 24.

Schedule

The Schedule page varies a little based on your selection on the Destination page. However, it's basically the same, providing the same schedule that the job system offers, but stripped of its recurrence-related fields. Such fields aren't needed because all Oracle-Suggested Backup variations and types are daily or weekly. Following is the Schedule page for our example, when choosing Both Disk and Tape:

As a general rule, schedule daily recurring backups of the OMR Database during periods of low update transaction activity. For Grid Control, this equates to running backups when other jobs aren't running. This prevents you from becoming too dependent on an excessive number of archive logs to perform recovery, which would increase recovery time and the risk of recovery failure due to a missing or corrupted archive log.

Select the schedule as shown above to match the backup objectives in this example, and click Next.

Review

The Review page summarizes the recurring backups that will be put in motion based on all user input and also lists the automatically generated RMAN script to accomplish these backups. Here's the Review page for the selections made above:

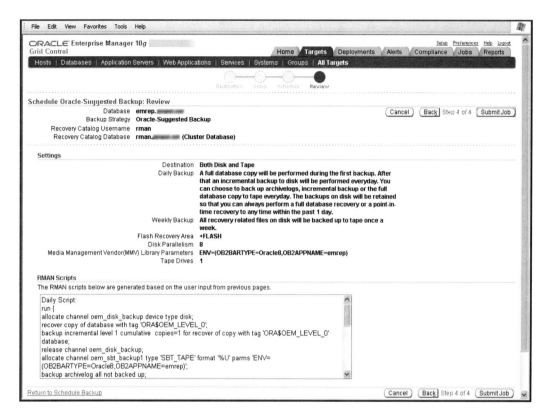

Appendix F (online) lists the RMAN scripts that the wizard generates for all Oracle-Suggested Backup variations and types and summarizes which commands are used in each variation and type. This appendix is helpful to review even if you don't opt for the Oracle-Suggested Backup, as it provides command syntax you may use either in a Customized Backup or RMAN Script job.

Check that your input will result in the backup you wanted, and click Submit to send it off to the GC job system. You should receive confirmation that the job was submitted successfully, and can click View Job to track its status.

Leave a back door open to take OMR Database backups outside of GC using the RMAN client. Prepend the following commands to the wizard-generated RMAN script and copy it to all OMR Database nodes:

CONNECT TARGET sys/<password>@<OMR_db_name>;
CONNECT CATALOG <rman_user>/<password>@<recovery_catalog_db>;

Customized Backup   A Customized Backup is more flexible than an Oracle-Suggested Backup, though not as capable as a user-defined RMAN Script job. Like an Oracle-Suggested Backup, a Customized Backup also leads you through a wizard to collect input that it uses to generate an RMAN script. However, a Customized Backup is more versatile than an Oracle-Suggested Backup for three primary reasons:

- Offers more backup options in the UI, including:
  - Whole or partial backups, such as tablespaces, datafiles, archive logs, or disk-based recovery files
  - Hot or cold backups, where the database is shut down and mounted, backed up, then opened afterward
  - Full, incremental, or incrementally updated backups
  - Removing archive logs after successful backup
- Generates an RMAN script that you can edit before submitting
- Allows you to override any Backup Settings for the current backup, except media management parameters
- Provides more job scheduling options when submitting the RMAN script

To demonstrate the Customized Backup method, I will set up the same daily disk backups as used for the Oracle-Suggested Backup: a daily incrementally updated online disk backup of the OMR Database, including archive logs, and which deletes obsolete backups. I will not override any Backup Settings, but will edit the generated RMAN script in the wizard and add some additional commands before submitting the backup.

Here are the steps to accomplish a Customized Backup, illustrated using the above backup specifications:

1. Log in as an administrator with VIEW ANY TARGET system privilege, FULL target privilege on the OMR Database, and whose OMR Database SYSDBA preferred credentials are set.

2. Go to the Maintenance page (see Figure 14-1) for the database you want to back up (the OMR Database) and, under the Backup/Recovery section, click the Schedule Backup link.

3. On the Schedule Backup page that appears (see Figure 14-3), in the Customized Backup section, select Whole Database, and click Schedule Customized Backup.

4. This launches a wizard, composed of four pages, starting with the Options page.

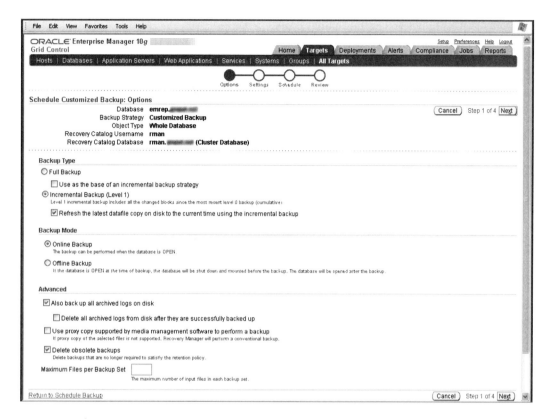

On the Options page, make your selections under Backup Type (full or incremental), Backup Mode (Online or Offline), and Advanced sections (various choices). The only ambiguously named feature, located under Backup Type, is called "Refresh the latest datafile copy on disk to the current time using the incremental backup." This feature is the "updated" part of incrementally updated backups.[27] For the example, make the selections shown above and click Next.

---

[27] This feature applies the changes in the incremental backup to the image copy of the database, so that the image copy will contain the latest changes.

5. This brings you to the Settings page.

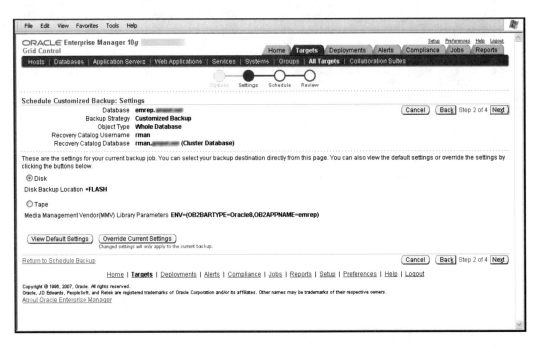

You can click View Default Settings or Override Current Settings to take you to the Backup Settings Devices page (see Figure 14-2) in read-only mode or read-write mode, respectively. Changed settings apply only to the current backup, which is fine for ad hoc backups. However, to schedule recurring backups, rather than overriding the current settings here, it's better to change the Backup Settings directly before entering the Customized Backup wizard. In this example, you don't need to do anything because no Backup Settings changes are necessary. Click Next.

6. The Schedule page is displayed, where you enter a job name and description, and the typical job system choices of time zone, start date and time, repeating interval, and end date. Make your scheduling selections and click Next.

7. This brings you to the Review page, a misnomer, really, given that you can edit the RMAN script generated from your input.

Chapter 14: Backup and Recovery of Grid Control **587**

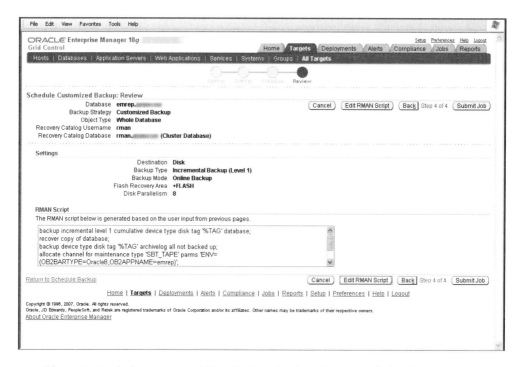

The script includes any overridden Backup Settings. You can click Edit RMAN Script, as I do here, to add or edit any RMAN commands you like.

8. The Edit RMAN Script page appears.

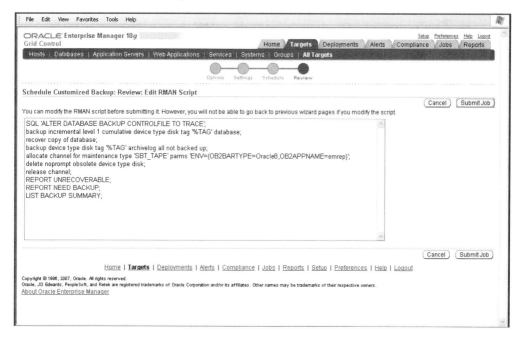

9. Edit or add the commands as needed (for this example, I added those shown in CAPITAL letters above).

10. To leave a back door open to take OMR Database backups outside of GC using the RMAN client, copy the backup script to your clipboard. Paste it to a file on an OMR Database node—it's typical to name the file with a *.rcv* extension. Prepend the following RMAN commands to the script:

    CONNECT TARGET SYS/<password>@<OMR_db_name>;

    CONNECT CATALOG <rman_user>/<password>@<recovery_catalog_db>;

    Save the script and copy it to any remaining OMR Database nodes.

11. Return to the wizard and click Submit to hand off the backup to the job system. When receiving confirmation that the job was submitted successfully, you can track its status by clicking View Job.

RMAN Script Job If neither of the other two methods suffices for your backup needs, consider the advantages of a user-defined RMAN Script job. This method takes full advantage of the GC job system, whereas the other two methods have more limited interfaces to the job system. The RMAN Script job provides the following additional functionality over the other two methods:

- Is able to submit an RMAN Script job against multiple databases in one operation
- Accepts either an RMAN script name or RMAN commands directly in the UI
- Allows for a multitask job; for instance, a cold backup job that consists of three tasks: Shutdown Database, RMAN Script, and Startup Database
- Provides for e-mail notification based on any selected job status values (Scheduled, Running, Suspended, Completed, and Problems)
- Offers advanced scheduling options and storage in the job library

Let's create and save in the job library a recurring RMAN Script job that performs the same daily incrementally updated disk backups of the *emrep* Repository Database configured using the Customized Backup. Set up the RMAN Script job to call a backup script located on the OMR Database host. You can use the script that the Customized Backup wizard generated. Set up notification of job status exceptions only—that is, Suspended and Problems.

1. Log in as an administrator with VIEW ANY TARGET system privilege and FULL target privilege on the OMR Database. Ensure that all OMR Database preferred credentials are set for the administrator.

2. Click the Jobs tab, then the Job Library subtab, as shown below. (If you don't need to save the job, click the Job Activity tab.)

Chapter 14: Backup and Recovery of Grid Control **589**

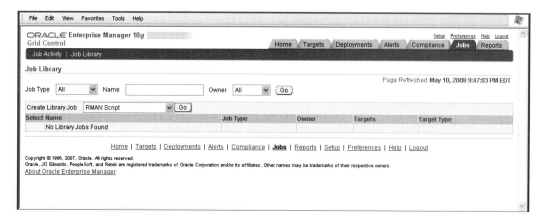

3. Make one of the following selections in the Create Library Job field and click Go:
   - **RMAN Script**   Choose when you only need a single RMAN script, as in this example.
   - **Multi-Task**   Choose when you require one or more of the following job tasks to run in a specified sequence with the RMAN Script: OS Command, Security Policy Configuration, Shutdown Database, SQL Script, Startup Database, and Statspack Purge. For more details, see "Working with Multi-Task Jobs" in the EM Online Help.
4. A page appears for entering job characteristics.

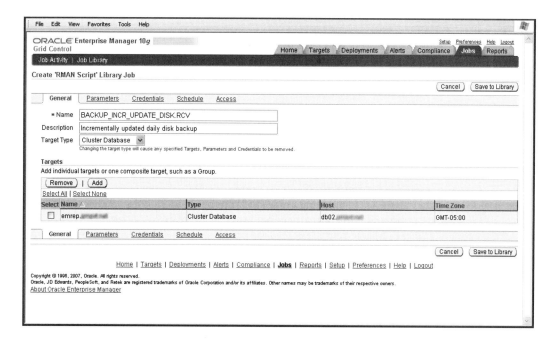

- General
- Parameters
- Credentials
- Schedule
- Access

5. On the General page, click Add to select one or more targets to run the job against. (In this example, the target is the RAC OMR Database *emrep*.)

6. Click the Parameters page. Enter commands directly in the RMAN Script field or enter an RMAN script name (as in this case).

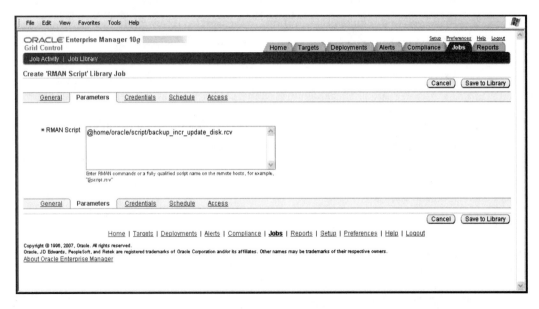

7. If you're entering RMAN commands in the UI, leave a back door to run them directly in the RMAN client if GC is not available. Copy the backup script to your clipboard and paste it to a file on an OMR Database node—it's typical to name the file with a *.rcv* extension. Ensure the following commands are prepended to the RMAN script and then save the script and copy it to any remaining OMR Database nodes:

CONNECT TARGET SYS/<*password*>@<*OMR_db_name*>;

CONNECT CATALOG <*rman_user*>/<*password*>@<*recovery_catalog_db*>;

8. If you're entering an RMAN script name, note the following points about the script you specify:

   a. Create the script as the host username entered in the administrator's preferred credentials for the OMR Database. Place the script on all OMR Database nodes, as the RMAN Script job may attempt to run the script from any node.

   b. You must use a hard-coded path when calling the script, and you cannot pass any parameters to the script. Furthermore, you cannot employ target properties (e.g., %emd_root%, %SID%) in the path, as with user-defined metrics.

   c. The script does not require nor should it contain the CONNECT CATALOG command to connect to the recovery catalog, as an RMAN Script job implicitly connects any registered database to the recovery catalog[28], provided you've configured it in Grid Control (see "Direct the OMR Database to Use a Recovery Catalog in Grid Control" earlier in the chapter).

   d. The script also does not need the CONNECT TARGET command because the job implicitly connects to the specified target databases selected.

9. Click the Credentials page and then select SYSDBA Database Credentials. RMAN requires them.

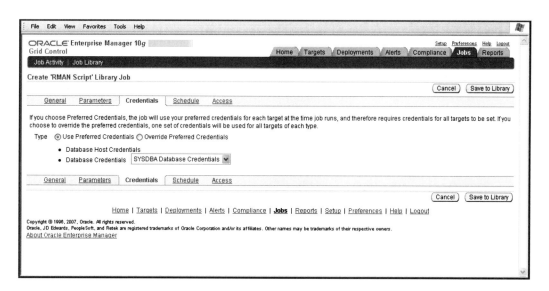

10. Click the Schedule page (shown in the following image).

---

[28] This is a very useful feature, not having to worry about connecting to the recovery catalog anymore. "One less thing," as Forest Gump would say.

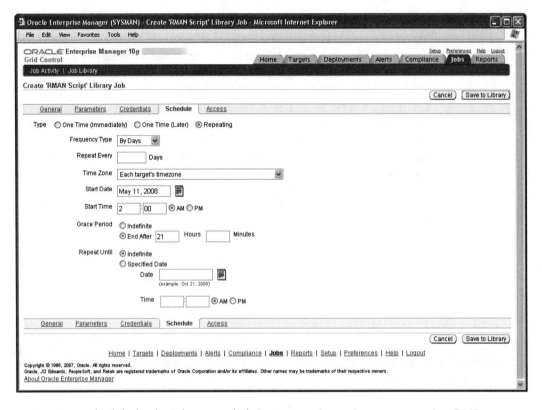

11. Set a schedule for the job, as needed. Because we're setting up a recurring OMR Database backup, choose the following selections:

   - **Type** Choose the Repeating type to set up a recurring job to back up the OMR Database.

   - **Frequency Type** Select By Days so the job will run every day as stipulated next.

   - **Repeat Every** In the preceding illustration, the job is scheduled to run every day.

   - **Grace Period** Select a grace period[29] with a cushion that allows for the backup job to complete within a reasonable time. In the preceding illustration, I select an indefinite grace period, but you may want to choose a 21-hour grace period, as the job usually takes two hours, and I give it an extra one-hour buffer to finish. If the job doesn't start within 21 hours, it is skipped and the job system waits until the next scheduled backup.

   - **Repeat Until** Choose Indefinite, as this is a regular backup job of the OMR Database.

---

[29] The grace period is the maximum permissible delay when attempting to execute the job. If the job system cannot start the job within the scheduled time plus the grace period, it will set the job status to Skipped. Source: Oracle Enterprise Manager Online Help, Job Schedule Page.

Chapter 14: Backup and Recovery of Grid Control   **593**

12. Click the Access page. Grant administrators and roles access as desired to edit this job.

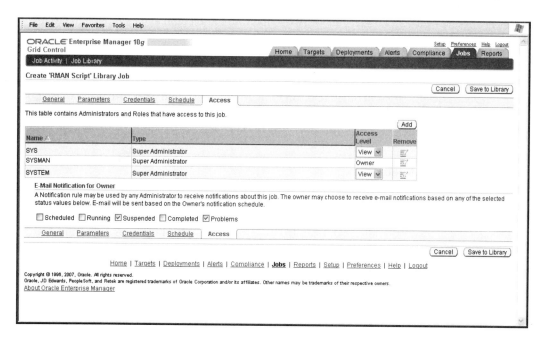

To enable notifications of certain backup job execution statuses, do the following:

- In the section E-Mail Notification for Owner, choose those job status values you want to be notified about. (For this example, select Suspended and Problems to notify you of exceptions.)

- Set up a database notification rule to which administrators can subscribe to receive e-mails of the RMAN Script job status. See "Create and Subscribe to Notification Rules" in Chapter 10 for instructions on how to set up such a notification.

13. When finished, click Save to Library, as this is a repeating job that you'll want to be able to view or change. The Job Library page returns with a confirmation message that the job was created successfully and with an entry reflecting the job you created.

    - Ensure the job is selected and click Submit.

    - A page appears with details of the job to be run, including a unique job name that GC forms by appending ".1" to the job name you supplied. (The appended number is incremented by 1 for each future execution of the job.) You can edit the job, including the name assigned. Then click Submit again to run the job.

    - You should receive another confirmation message that "the job was created successfully," which is code for "the job was submitted successfully." Click the job name link to track its progress.

14. Leave a back door open to take OMR Database backups outside of Grid Control using the RMAN client. If you configured an RMAN Script job by entering commands rather than calling a script, create an RMAN script from these commands. Prepend the following commands to the script and copy it to all OMR Database nodes:

    CONNECT TARGET sys/<password>@<OMR_db_name>;
    CONNECT CATALOG <rman_user>/<password>@<recovery_catalog_db>;

Once backups are scheduled and run through the job system, you can avail yourself of all job system editing, viewing, and reporting mechanisms integrated into the GC product to maintain these backups. See "Manage OMR Database backups in Grid Control" below for coverage of these features.

Set Up OMR Database Exports

In addition to taking daily full physical OMR Database backups as just described, use an Oracle export utility to take daily *logical backups* as well.[30] Logical backups contain definitions of database objects (tables, procedures, etc.) exported using an Oracle export utility and stored in a binary file that you can reimport if needed for recovery. Physical backups, by contrast, are backups of physical database files—datafiles, control files, redo logs, and archive logs. An OMR Database export leaves open the option of using an export dump to help recover from logical corruption. (See the upcoming section "Recover the OMR Database" for an example export scenario.) For an Oracle 9.2 OMR Database, use the original Oracle Export and Import utilities. (See Oracle 9*i* Database Utilities Release 9.2, Chapters 1 and 2, for more information.) For an Oracle 10*g* OMR Database, use the more sophisticated Oracle Data Pump Export and Import utilities, which support all Oracle Database 10*g* features, perform better, and are easier to use. Grid Control offers a UI to both the Original Export and Data Pump Export utilities.

Below are the procedures to take a full Data Pump Export of an OMR 10*g* Database, both at the OS level and in Grid Control.

Set Up Exports at the OS Level   Following are the steps to set up Data Pump Exports of the entire OMR Database at the OS level:

1. Create the directory at the OS level:

   ```
   mkdir -p <data_pump_dir>
   ```

2. As a user with CREATE DIRECTORY privilege, create a directory in SQL*Plus:

   ```
   CREATE DIRECTORY DPUMP_DIR AS '<data_pump_dir>';
   ```

3. Call a script containing the following three lines (when executing the script, pass the single argument of the OMR DB_NAME):

   ```
   export ORACLE_SID=${1}
   rm -f <data_pump_dir>/expdp_${1}.dmp
   expdp system/<password> DUMPFILE=dpump_dir:expdp_${1}.dmp \
   LOGFILE=dpump_dir:expdp_${1}.log JOB_NAME=expdp_${1}
   ```

---

[30] Exports are also a secondary backup method for the recovery catalog, in addition to doing a NOCATALOG backup in RMAN.

Chapter 14: Backup and Recovery of Grid Control   595

Set Up Exports in Grid Control   Below is the procedure for setting up Data Pump Exports of the entire OMR Database in Grid Control:

1. In the Console, log in as the SYSTEM[31] user with Normal role to the OMR Database. For instructions, see "Web Console Login As a User Other Than DBSNMP" in Chapter 9.
2. Navigate to the Maintenance page for the OMR Database (see Figure 14-1 earlier in the chapter) and click Export to Export Files under the Data Movement section.
3. The Export Type page is displayed for choosing whether to export the Database, Schemas, Tables, or Tablespace.

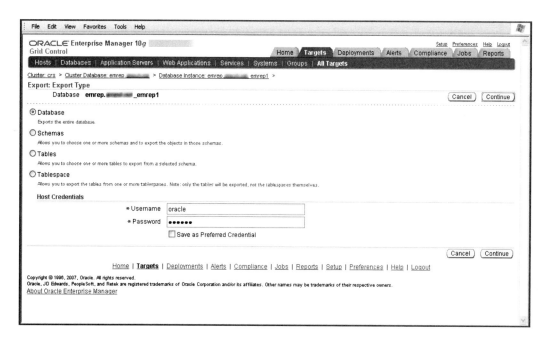

4. For this example, select Database, enter host credentials, and click Continue.
5. The Options page appears, shown next, which is the first page of a wizard to guide you through the export process.

---

[31] You cannot execute Data Pump as the SYS user.

All settings on this page are optional. Following are the possible selections:

- Enter the maximum number of threads (or parallelism) for the export job. The default value is 1. A good value for parallelism is twice the number of CPUs on all OMR Database nodes combined.

- Estimate the disk space required for the export dump file by clicking Estimate Disk Space Now.

- Select whether to generate a log file and its path. Enter a log file name. For the directory object, you have two choices:

  - Select an existing directory object in the Directory Object field.

  - Click Create Directory Object to create a new directory object in the database. A page is displayed for entering the object name, path, and host login credentials. You must also create this directory at the OS level as follows:

    mkdir -p <data_pump_dir>

Chapter 14: Backup and Recovery of Grid Control **597**

6. Select any advanced options desired to limit the export content to data only, metadata only, or both (the default), and the objects. When finished with your input, click Next.

7. The Files page appears. Choose a directory object, filename, and, optionally, a maximum file size (unlimited by default). For the filename, leave the *%U* wildcard, which will be replaced by 01, 02, 03, etc. for each dump file name in the set. You can also add a date wildcard by specifying *%D*. Click Next.

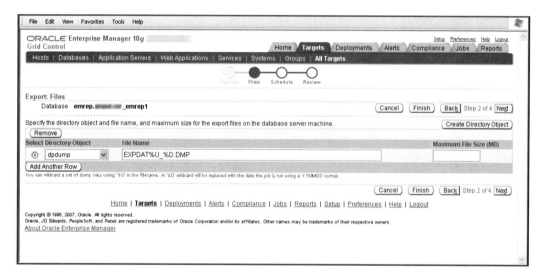

8. On the Schedule page, choose the time zone, start time, and whether to repeat the job. To schedule a recurring job, you cannot start the job immediately but must select a future start time. Click Next.

9. On the Review page, look over your choices, and click the Show PL/SQL link to see the script generated. If satisfied, click Submit Job.

10. You will receive confirmation that the job was created and can click a link to view its status and log file.

Recover the OMR Database

This section provides instructions for recovering the OMR Database. The structure of this section is as follows:

- Use Grid Control to choose OMR Database recovery settings
- Restore Grid Control operations after OMR Database failure
- Leverage Flashback Technology for OMR Database recovery

> ### Should You Use Grid Control to Recover Target Databases?
> If you're a DBA, you're likely unphased by my warning to steer clear of Grid Control for OMR Database recovery. DBAs tend not to rely on tools when they need to recover *any* database. The reason is that, in stressful recovery situations, DBAs (myself included) want one thing above all others: control. RMAN and SQL*Plus provide that sense of control, whereas a GUI tool like Grid Control adds what many DBAs perceive as an unwanted layer between them and the database. Also, GC cannot perform every type of recovery available in RMAN.
>
> While I advise against attempting any type of OMR Database recovery with Grid Control, I do encourage you to consider using GC to recover target databases, particularly those backed up using Oracle-Suggested Backup. To recover target databases backed up in other ways, I still suggest using Grid Control, depending on your comfort level with and testing of its recovery features, and on its capability to address the type of recovery needed. For an account of all GC recovery features, which are not in scope for this chapter on OMR Database B&R, see "Perform Recovery" in the Oracle EM Online Help.

As you've learned, setup and execution of OMR Database backups in GC is easy enough, but you can't expect to rely on the Console to recover a downed OMR Database because the Console itself requires this database to be running. Remember that analogy from Chapter 4 equating Grid Control to a doctor who can self-treat for minor medical problems, but not operate on himself or herself under general anesthesia? A downed OMR Database is definitely Grid Control under general anesthesia.[32]

Technically, you may be able to use GC to recover from certain "localized" OMR Database failures, provided the database is running and healthy enough to provide the Console functionality needed to perform the recovery. However, using the Console even to recover from object-level logical corruption of the OMR Database is a risky proposition. You'd need to be positive that the failure of the corrupted objects would not affect GC recovery functionality itself, ideally by simulating both the failure and the recovery in a GC test environment. If you're considering using Grid Control to recover Grid Control, remember that an attorney who represents himself has a fool for a client. Forgive me for "mixing my adages" here (first doctors, now lawyers), but they're both appropriate.

If you can't use GC as a management interface to RMAN for OMR Database recovery, on what tool should you rely? I recommend RMAN[33] of course, though you could also employ

---

[32] Even if the OMR Database were hobbling along enough to somehow render the Perform Recovery UI for the OMR Database, selecting Perform Whole Database Recovery would return a big error that "Whole database restore is NOT supported for the Repository Database."

[33] You should certainly consider using RMAN, either directly or through Grid Control, to recover from target database backups taken using the Oracle-Suggested or Customized Backup strategies, which both employ RMAN.

user-managed recovery to directly manage database files using OS and SQL*Plus commands. RMAN is easier to use and more powerful than user-managed recovery. Salient RMAN recovery features not possible with user-managed recovery include the advantages of a recovery catalog, quicker and easier block media recovery, and recovery integration with RMAN backup features such as incremental, encrypted, binary, and unused block compression backups.

Use Grid Control to Choose OMR Database Recovery Settings

While you are well advised to use the RMAN client, not Grid Control, for *actual* recovery of the OMR Database, there is no danger in using Grid Control to *prepare* for recovery. This preparation involves establishing certain settings related to instance, media, and flash recovery. These settings, shown in the two images on the next page, are accessible by navigating to the Maintenance page for the OMR Database (see Figure 14-1 earlier in the chapter) and clicking the Recovery Settings link.

As you see, this page is organized into three sections: Instance Recovery, Media Recovery, and Flash Recovery. Table 14-5 lists the fields in these three sections and their equivalent initialization parameters and SQL commands, just as I provided the equivalent RMAN commands earlier to explain Backup Settings. The table should demystify what GC would do behind the scenes if you were to change any settings for these fields.

The settings on the page shown in the two screen shots are consistent with those desired for the OMR Database, though you need to tune the "Desired Mean Time to Recover" and "Flash Recovery Area Size" for your particular site. If you followed the directions in Chapter 3 and in its referenced Appendix C (online) for configuring the OMR Database, then its recovery settings are already modified as I recommended. Below are references to these earlier directions, embellished to make this chapter on OMR Database B&R comprehensive:

- **Instance Recovery** Tune the Mean Time To Recover (MTTR). Also known as fast-start fault recovery, you can set FAST_START_MTTR_TARGET to a desired RTO from instance or system failure. (Recommended in the online Appendix C section "Use Fast-Start Fault Recovery to Control Instance Recovery Time.")

- **Media Recovery** Operate the OMR Database in ARCHIVELOG mode and send all archived redo logs to the FRA so you can recover the database to the current point in time. Regularly monitor the FRA to ensure there is adequate disk space to store archived redo log files. A message is sent to the alert log when the FRA becomes 83% and 93% full. Monitor this message via Grid Control by implementing the Generic Alert Log Error critical threshold specified in Table 13-1 of Chapter 13. You should never need to manually delete archived redo logs. The RMAN retention policy, when running backups, and the FRA mechanism should handle removing archive logs and managing all FRA space issues. (Recommended in the online Appendix C section "Enable ARCHIVELOG Mode".)

Oracle Enterprise Manager 10g Grid Control Implementation Guide

Screenshot of Oracle Enterprise Manager 10g Grid Control — Recovery Settings page for a cluster database, showing Instance Recovery (FAST_START_MTTR_TARGET settings with MTTR Advice chart), Media Recovery (ARCHIVELOG mode, Log Archive Filename Format, and a table of 10 Archive Log Destinations), and Flash Recovery (Flash Recovery Area location +FLASH, size 2302 GB, Reclaimable 41.96 GB, Free 2,155.74 GB, Flashback Database enabled with 48 Hours retention, current flashback logs 6.969 GB, Lowest SCN 238336140, Flashback Time May 10, 2008 12:32:46 AM, and a Flash Recovery Area Usage pie chart: Archive Log 204.42GB (8.9%), Control File 0GB (0%), Online Log 0.69GB (0%), Backup Piece 0.69GB (0%), Image Copy 5.76GB (0.2%), Flashback Log 0GB (0%)).

Chapter 14: Backup and Recovery of Grid Control   **601**

Section	Field Name	Corresponding Initialization Parameter(s) or SQL Command	Description	Dynamic Parameter?
Instance Recovery	Desired Mean Time to Recover	FAST_START_MTTR_TARGET	Tunes database instance crash recovery time	Yes
Media Recovery	ARCHIVELOG Mode	No init parameter in 10g. SQL is ALTER DATABASE ARCHIVELOG in mount mode.[34]	Places database in ARCHIVELOG mode	No
	Log Archive Filename Format	LOG_ARCHIVE_FORMAT	The default filename format for archive logs; ignored if use FRA	No
	Archive Log Destination	LOG_ARCHIVE_DEST_<n> where <n> = 1 to 10	Sets attributes to direct where and how to archive redo	No[35]
Flash Recovery (10g only)	Flash Recovery Area Location	DB_RECOVERY_FILE_DEST	Default path for FRA	Yes
	Flash Recovery Area Size	DB_RECOVERY_FILE_DEST_SIZE	Hard limit on total space available to recovery files in FRA	Yes
	Enable Flashback Database	SQL is ALTER DATABASE FLASHBACK ON in mount mode. There are multiple parameters that influence flashback behavior.[36]	See "Enable Flashback Database" in Appendix C (online)	No
	Flashback Retention Time	DB_FLASHBACK_RETENTION_TARGET	Upper limit for how far back in time you can flash back a database	Yes

TABLE 14-5.   *Fields on the Recovery Settings Page*

---

[34] In Oracle 9i, you must manually enable automatic archiving by setting the initialization parameter LOG_ARCHIVE_START=TRUE. In Oracle 10g, automatic archiving is enabled by default.

[35] You can actually modify most of this parameter's attributes dynamically, and thus can change them here for the OMR Database, but it's easier to make all media recovery changes together in SQL*Plus.

[36] Initialization parameters related to Flashback Database are DB_FLASHBACK_RETENTION_TARGET, DB_RECOVERY_FILE_DEST, DB_RECOVERY_FILE_DEST_SIZE, and UNDO_RETENTION.

- **Flash Recovery** All FRA features are specific to Oracle 10g.

    - **Use a Flash Recovery Area** Use an FRA to store and automate the management of all B&R-related files; RMAN backups, multiplexed copies of control files and online redo logs, archived redo logs, and flashback logs (covered a bit later in the section "Leverage Flashback Technology for OMR Database Recovery") are automatically written to a specified filesystem or ASM disk group. An FRA is required to use both Flashback Database and guaranteed restore points, which are useful recovery features for any database, including the OMR Database. (I recommend an FRA when installing the OMR Database—see "Recovery Configuration" in Chapter 3.)

    - **Enable Flashback Database** Flashback Database allows you to "rewind" the database to a previous point in time without having to restore a backup of the database. It is a more efficient alternative to database point-in-time recovery (DBPITR) for logical data corruptions or user errors because you don't need to restore backups from media. For more details, see the upcoming section "Leverage Flashback Technology for OMR Database Recovery." (Recommended in the online Appendix C section "Enable Flashback Database.")

For those of you who have not adjusted recovery settings as needed for the OMR Database or want to change such settings for target databases, it's useful to know about this Console functionality. Note the following about the Recovery Settings page:

- Notice that all options are grayed out except the sole Instance Recovery setting, "Desired Mean Time to Recover." The reason is that you are logged in to the Console as the default DBSNMP user (as shown at the top right of the page), who does not have the SYSDBA privilege required to alter any media or flash recovery settings.

- To change any grayed-out settings, log out of the Console and log in again as SYS.[37]

- Grayed-out fields correspond to both *static* and *dynamic initialization parameters* and, in some cases, to SQL commands. (Static initialization parameters require a database restart to implement, whereas the dynamic parameters on this page are modifiable globally[38] while the instance is running.) The rightmost column of the table indicates which parameters are dynamic.

- GC asks for confirmation to restart the database if it needs to alter any static parameters. Before confirming, you can click Show SQL to see what SQL Grid Control will execute on your behalf. GC will inform you that the startup command will use a temporary file as the pfile, containing the following single parameter:

    spfile='<spfile_path>'

---

[37] To log out of the Console and log in again with SYSDBA privilege, click Logout at the top right of the Recovery Settings page, check "Display database login page after logout," then log in as the SYS user, choosing Connect As SYSDBA.

[38] In general, dynamic parameters are modifiable either globally for all sessions in the instance with the ALTER SYSTEM SET *<parameter_name>* = *<value>* statement, or locally for the duration of the current session with ALTER SESSION SET *<parameter_name>* = *<value>*.

- GC can change all settings (both static and dynamic) for both target databases and the OMR Database. It may surprise you that GC can change static parameters in the UI for the OMR Database. GC bounces the OMR Database, which "bounces" you out of the Console[39]—like Tigger bounces Eeyore into the river in the Hundred Acre Wood.[40] The OMS "goes funny for a wee bit."[41] But like Eeyore, who is soon fished out, the OMR Database soon returns, and with it, the Console.[42] Mr. Milne could have undoubtedly woven the tale into three chapters, perhaps named: *In Which GC Bounces OMR*, *In Which OMR Bounces GC*, and *In Which OMR and GC are Unbounced*.

- If you want to avoid the drama of a funny Console error and an OMS that temporarily goes "nutty," use SQL*Plus instead to make any needed static parameter changes for the OMR Database. You can, however, log in to the Console as SYS and change all dynamic parameters for the OMR Database without affecting Console functionality one bit.

Restore Grid Control Operations After OMR Database Failure

This section tackles the subject of bringing Grid Control back online from an OMR Database failure, which mainly concerns recovering the OMR Database—no surprise there. What you may not know is that the procedure for resuscitating GC hinges on whether the OMR Database requires complete or incomplete recovery (I define this resuscitation procedure in a minute).

Take one paragraph to count your DBA blessings before considering the pain and suffering that a recovery may bring. Let us remember that Oracle handles some forms of recovery automatically. First, an instance may fail, or perhaps multiple instances in the case of a RAC database. However, instance or crash recovery[43] automatically occurs on startup to bring the datafiles to a transaction-consistent state, preserving all committed changes up to the point of instance failure. Second, let us give thanks to the Oracle trinity[44] that data loss does not occur, making manual recovery unnecessary, for failures of database background processes, network failures, or statement execution failures due to resource problems. Lastly, we praise the Oracle RAC Pack[45] that a RAC database automatically addresses node failures by "failing over" connections to instances on surviving nodes. These instances recover any in-progress transactions by rolling them back if needed, and by applying changes to committed transactions. Amen.

---

[39] The Console displays "Internal Server Error" and "The server encountered an internal error or misconfiguration and was unable to complete your request."

[40] *House at Pooh Corner* by A.A. Milne, 1928.

[41] As you might expect, when GC shuts down the OMR Database to change the recovery settings you indicate, the OC4J EM application goes down too, as evidenced by the OC4J_EM process-type temporarily reporting an Init status in the output of *opmnctl status*.

[42] When the OMR Database restarts, the OMS reestablishes a connection to it and the OC4J_EM process restarts.

[43] Crash recovery is when a single instance or all instances in a cluster fail. The related process of instance recovery occurs when some, but not all, instances of a RAC database fail.

[44] In 1977, Larry Ellison, Bob Miner, and Ed Oates founded Software Development Laboratories, later to be renamed Oracle.

[45] Oracle's RAC Pack is a team of RAC implementation specialists that is part of the RAC Development organization.

OK, back to the tribulations. While Oracle performs automatic recovery for the above failures, most remaining forms of recovery require manual intervention. An Oracle Database can fail, and therefore must be recovered, in many different ways. However, for all recovery variations, the process of returning your GC environment to service varies only slightly, depending on whether you perform *complete recovery* or *incomplete recovery* on the OMR Database:

- **Complete recovery** A complete (point-of-failure) recovery brings the database back to the most recent point in time, without the loss of any committed transactions. You can completely recover from various types of failure, including *media failure* at the database, database file, or block levels, and recovery from node failure of a single-instance database by relocating it to another database node at a disaster recovery (DR) site. (Media failure is the failure of the read or write of a disk file required to run the database due to a physical problem with the disk, such as a head crash. Media recovery is the process of restoring database files to the same point-in-time image prior to the media failure.) You can even classify certain Flashback Technology features and techniques in this category.

- **Incomplete recovery** If complete recovery is not possible because of user error or logical corruption, for example, you must resort to incomplete recovery. Also known as database point-in-time recovery (DBPITR), incomplete recovery returns the database to a past point in time. DBPITR does not completely recover all database changes, such as those stored in online redo logs; DBPITR usually results in some data loss. If the error in data affects only one tablespace, you have the option of performing tablespace point-in-time recovery (TPITR) rather than DBPITR. Flashback Technology is an alternative to DBPITR, and is usually faster.

I cannot provide RMAN commands and procedures for all forms of recovery, as they are too varied and expansive to present in a book about implementing Grid Control.[46] Oracle does a thorough job of documenting all these recovery scenarios in Oracle Database Backup and Recovery Basics 10g Release 2 (10.2), which you can consult for the specific steps to recover the OMR Database. See Chapter 6 for complete recovery topics and Chapter 7 for incomplete recovery coverage. What I am examining here is how the steps differ for returning GC to normal operations based on the type of OMR Database recovery performed.

Outside of the difference in the recovery process itself for complete and incomplete recovery, the overall steps to return GC to service are nearly the same, with one significant distinction. If incomplete recovery is required, Agents will likely lose contact with the Repository, requiring that you remove certain Agent files. (See "Resetting Agents" earlier in the chapter). This is a minor procedure, but can be labor-intensive for sites with a large number of Agents. Thankfully, a complete recovery scenario is not fraught with this Agent problem, so it does not require any Agent changes.

Below are the specific procedures to bring Grid Control back online after complete or incomplete recovery, taking into account procedures for the OMS and Agent components.

Restore Grid Control After OMR Database Complete Recovery Complete OMR Database recovery, unlike incomplete recovery, recovers the database to the current time, resulting in no loss of committed data. The recovery may involve online redo logs, the control file, or the SPFILE,

---

[46] Below in "Restore Grid Control After OMR Database Complete Recovery," I provide the few RMAN commands needed to perform a straightforward complete database recovery.

and may be at the database, tablespace, or datafile level. Once you use the RMAN client to completely recover the OMR Database, Agents should resume normal communication with the Repository without the need for any manual intervention. The steps to bring Grid Control online in a complete OMR Database recovery situation are simple: stop the OMS, restore and recover the database, and restart the OMS. Here is the procedure, including the RMAN commands, to recover the entire OMR Database:

1. The OMS will likely be in an unusable state from the time the OMR Database goes down. Therefore, shut down all OMS processes on all OMS hosts:

   `$OMS_HOME/opmn/bin/opmnctl stopall`

2. Restore and recover the OMR Database (in this example, the entire database) to the last saved transaction, using backups created with RMAN either in GC or at the RMAN command-line interface. To perform complete recovery, log in to an OMR Database node as the OS owner of the OMR Database (typically, *oracle*) and execute the following OS command:

   `$ORACLE_HOME/bin/rman`

3. At the RMAN command prompt, enter the following RMAN commands:

   CONNECT TARGET sys/*<password>*@*<OMR_db_name>*;

   CONNECT CATALOG *<rman_user>*/*<password>*@*<recovery_catalog_db>*;

   SQL 'STARTUP MOUNT';

   RESTORE DATABASE;

   RECOVER DATABASE;

   SQL 'ALTER DATABASE OPEN';

4. After recovering and restarting the OMR Database, restart the OMS processes on all OMS hosts:

   `$OMS_HOME/opmn/bin/opmnctl startall`

5. Test that Agents resume normal uploads via the OMS to the Repository Database, beginning with uploads of any pending files to the OMS:

   `$AGENT_HOME/bin/emctl upload`

Restore Grid Control After OMR Database Incomplete Recovery If complete recovery is not possible, incomplete recovery is your next best bet. Given that DBPITR returns the OMR Database to a previous point in time, this form of recovery usually causes the OMR Database and Agents uploading to it to lose synchronization, requiring you to reset Agents (see "Resetting Agents" earlier in the chapter).

Perform Incomplete Recovery of OMR Database Incomplete recovery is so named because it does not completely recover all changes to your database, so it usually results in some OMR Database management data loss. You can perform incomplete recovery either using DBPITR or Flashback Database. With DBPITR, you must restore the database from backup, then use archived redo logs to apply all changes up to a point in time just prior to the error. Flashback Database is preferable to DBPITR because it's faster and does not require media recovery (see the next section, "Leverage Flashback Technology for OMR Database Recovery"). However, DBPITR is your only choice if the flashback window does not extend far enough back in the past to reach the desired recovery target time.

To return GC to operational status after incomplete recovery of the OMR Database, you will likely need to perform an additional step: manually reset those Agents that cannot reestablish contact with the Repository. (See "Resetting Agents" earlier in the chapter for the background and steps to reset an Agent.)

Following is the procedure to get Grid Control back on its feet using incomplete recovery of the OMR Database:

1. Shut down all OMS processes on all OMS hosts:

   $OMS_HOME/opmn/bin/opmnctl stopall

2. Perform incomplete recovery and restart the OMR Database.

3. Restart all OMS processes on all OMS hosts:

   $OMS_HOME/opmn/bin/opmnctl startall

4. Reset all Agents so they can communicate with the OMR by completing the procedure documented in "Resetting Agents," earlier in the chapter.

Leverage Flashback Technology for OMR Database Recovery

I would be remiss if I did not at least touch on Flashback Technology, given that flashback features can figure so prominently in an OMR Database complete or incomplete recovery scenario. Flashback Technology is a set of recovery features available only in Oracle 10*g* that allows you to view and rewind to past states of data and wind data back and forth in time.

Most flashback features operate at the logical level, allowing for object-level recovery, sometimes while a database remains open.[47] These features include:

- **Flashback Drop**   Reverses the effects of a DROP TABLE statement
- **Flashback Table**   Restores a table to its state at a previous point in time while the database remains online
- **Flashback Query**   Retrieves the contents of lost rows by viewing results of database queries as they would have appeared at a previous point in time
- **Flashback Version Query**   Allows you to see how updates affect the versions of rows in one or more tables during a specified time frame
- **Flashback Transaction Query**   Lets you view changes that one or more transactions make

As you've already read in this chapter, Flashback Technology also includes a recovery feature that works at the physical level, called Flashback Database. This feature employs *flashback logs* to revert current datafiles to a previous point in time. Flashback logs record past versions of changed data blocks, and are stored in the FRA. Flashback Database is a substitute for DBPITR, and is faster because it avoids the need for media recovery.

---

[47] You should consider using flashback features while GC remains running only if all of the following are true: it's a complete recovery scenario, GC must remain available (you can't shut it down or switch to a DR site), you're sure the failure won't affect GC functionality, and there's no chance the problem would cause OMR Database corruption if you left the OMS running.

Flashback Technology is powerful because it is versatile. It provides another means to achieve both complete and incomplete recovery, either using one flashback feature or combining it with other backup technologies. You can use Flashback Database alone as a substitute for DBPITR recovery or combine Flashback Database with export utilities (Oracle Export in 9*i* or Oracle Data Pump Export in 10*g*) for complete recovery.

Take the example of logical corruption, whose extent may dictate performing either incomplete or complete recovery. Both forms of recovery start with Flashback Database to rewind to a point just before the earliest logical corruption. The recovery scenario then branches off in two directions, one equivalent to DBPITR and the other to complete recovery:

- **Incomplete recovery**   If many objects are logically corrupted after the rewind point, you may be forced to open the OMR Database to this earlier point in time with the RESETLOGS option, producing the same results as DBPITR. The consequence is that all database changes made after the flashback point would be abandoned, including any valid changes to non-corrupted objects. You stand to lose many desired changes if the logical corruption is not discovered right away.

- **Complete recovery**   If only a few objects are logically corrupted, another option is to surgically reverse this corruption, yet preserve subsequent, valid changes to other objects, which is akin to complete recovery. Here's how. While the database is flashed back to a point before the logical corruption, export the objects in their uncorrupted state. Then perform complete recovery of the database to the present time to undo the effect of the Flashback Database, open the database read-write, and import the exported objects.[48] Of course, for this technique to work, you must be certain that no valid changes occurred with these imported objects from the point of logical corruption to the present time.

As for backups, you have the choice of RMAN or user-managed recovery (i.e., SQL*Plus) when implementing flashback features for OMR Database recovery. To mirror my earlier recommendation for backups, I also favor RMAN over user-managed recovery when using Flashback Technology. One of many RMAN advantages is that it integrates with Flashback Database, so it automatically retrieves archive logs needed during a flashback operation. In SQL*Plus, this same procedure would entail restoring these archive logs yourself.

TIP
I do not recommend using the Flashback Technology UI in Grid Control to operate on the OMR Database; flashback actions may disrupt the EM application or upset synchronization between OMR and OMS, causing unexpected results. However, I heartily endorse using the Console to run flashback operations against target databases. Grid Control can greatly simplify execution of flashback features, particularly Flashback Query, Flashback Transaction Query, and Flashback Versions Query. Flashback features are available on the Perform Recovery page in the UI under the Object Level Recovery section of the page.

---

[48] Oracle Database Backup and Recovery Basics 10*g* Release 2 (10.2), Chapter 7, pages 7–15.

For a thorough treatment of Flashback Technology recovery, see Chapter 7 of Oracle Database Backup and Recovery Basics 10*g* Release 2 (10.2).

Manage OMR Database Backups in Grid Control

As stated earlier, Grid Control can *back up* the OMR Database, but you can't expect Grid Control to *recover* part of itself (i.e., the OMR Database) when that part needs to be running to do the recovery. What about using Grid Control to *manage* OMR Database backups? This is actually a good idea. Here's why:

- All GC backup management pages query the recovery catalog, if configured in GC, or query the database control file otherwise. All backup data is available on these pages, not just data for backups created in GC.

- On many of the backup management screens, after making your selections in the Console, you can click Show RMAN Script to view the RMAN commands that will execute. This makes Grid Control a good tool to learn RMAN syntax for LIST, REPORT, and other commands.

- You can drill down to a history of OMR Database backups scheduled through the job system. The OMR Database home page even provides a timestamped link to the Last Backup.

Let's examine Grid Control's backup management features by briefly touring the three applicable links on the Maintenance page (see Figure 14-1 earlier in the chapter) for the OMR Database:

- **Manage Current Backups**   Perform common management tasks on existing backups, with a UI that constructs and executes RMAN commands on your behalf, such as CROSSCHECK, DELETE, and VALIDATE.

- **Manage Restore Points**   Create, delete, and display restore points, a database feature used to bookmark System Change Numbers (SCNs). You can flash back a database to these bookmarks without having to query the database for the SCN corresponding to the flashback time.

- **View Backup Report**   Offers a more user-friendly interface than the RMAN client for viewing output from LIST and REPORT commands.

TIP
Many operations on these three pages require the SYSDBA privilege. Therefore, before accessing any of these pages, it is best to log out of the Console as the default DBSNMP user, and log in again as SYS.

A high-level tour of each page should be enough to familiarize you with Grid Control's main management and reporting features, and hopefully motivate you to investigate them further on your own.

Manage Current Backups

Following is the Manage Current Backups page:

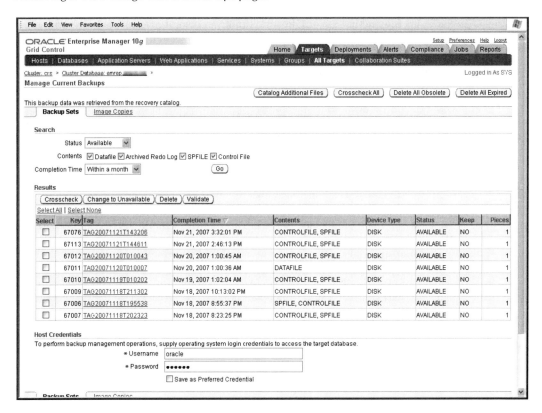

Here you can perform typical RMAN management tasks. As you work down the page, you'll find the following features:

- Buttons at the top provide the following global-level RMAN commands affecting the recovery catalog: Catalog Additional Files, Crosscheck All, Delete All Obsolete, and Delete All Expired.

- Two subtabs, Backup Sets and Image Copies, allow you to easily filter backups.

- The Search section lets you quickly find particular backups of interest.

- Backup-specific commands are available under the Results section for selected backups, namely Crosscheck, Change to Unavailable, Delete, and Validate.

- Once you select backup-specific commands, a job page appears, allowing you to schedule it to run in future[49] (which the RMAN command-line interface does not offer).

---

[49] You can also submit such jobs on a recurring basis, but this is not very useful, as the generated scripts usually run RMAN commands for specific backup files, and you cannot change these scripts in GC10gR2.

Manage Restore Points

Grid Control offers a straightforward interface to administer *restore points*. A restore point is a designated point in time to which you can restore a database using Flashback Database (which relies on flashback logging if enabled). Restore points are useful bookmarks for returning the OMR Database to a previous point in time. You should create a guaranteed restore point before any OMR Database upgrade or patch, to make certain you can roll back in case of failure without having to resort to media recovery.

There are two kinds of restore points:

- **Normal restore point** An alias corresponding to an SCN recorded to simplify flashback operations in that you won't have to determine this SCN later if you need to flash back to it. Eventually, normal restore points age out of the control file if they are not manually deleted.

- **Guaranteed restore point** An alias corresponding to an SCN to which you can always flash back the database. Flashback logs required to enforce the restore point are not deleted in response to space pressure in the FRA, but, if space is unavailable, the database will hang until space in the FRA is freed. A guaranteed restore point does not age out of the control file unless it is manually deleted, and uses considerably less disk space than a normal restore point when flashback logging is disabled.

The Manage Restore Points page displays existing restore point information and allows you to create and delete both kinds of restore points, all of which you can do for the OMR Database.

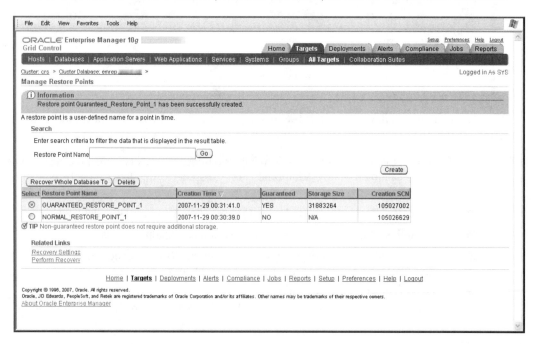

This page also offers the ability to flash back a database to an existing restore point, but don't try this in Grid Control with the OMR Database—use RMAN or SQL*Plus instead. The information displayed shows the restore point name, creation time, whether it is guaranteed or normal, the storage size required to maintain it (which is very useful to know), and, of course, the creation SCN.

View Backup Reporting

The View Backup Report page shows all backup jobs recorded in the recovery catalog if configured in Grid Control (or in the control file if not configured in GC).

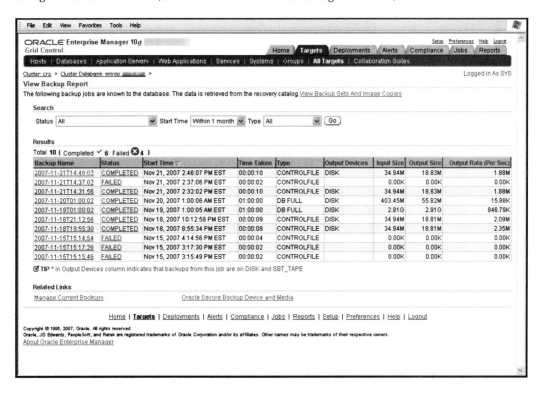

The HTML report formatting in the Console makes backup reports "easy on the eyes," offering the following benefits:

- Backup summary information not available in the RMAN command-line interface
- Search capabilities based on backup job status, start time, and type of object backed up
- Drill-down to the RMAN output log by clicking the Status field for a backup job
- Drill-down to detailed input and output information by clicking the Backup Name field for a job, as shown next.

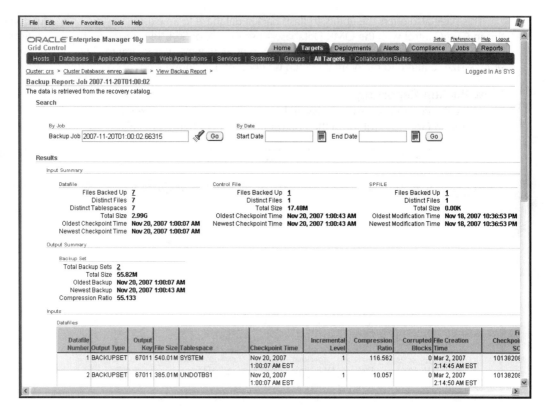

In this drill-down, the input shows statistics on the files backed up, and the output lists the backups created and their associated statistics. It's not exactly riveting reading but is useful information on OMR backups (and when recovering target databases).

Grid Control Software Backup and Restoration

The previous section of this chapter was devoted to laying out how to integrate Grid Control with Oracle database features (such as RMAN and Flashback Technology) to back up the OMR Database, prepare it for recovery, and manage its backups. However, neither GC nor any Oracle features supports B&R of GC's three major software homes:

- OMR Database software (except Oracle Cluster Registry and Voting Disk as noted below)
- OMS software
- Agent software

What follows are suggested B&R strategies for each of these GC software components. You can call upon these strategies to back up one or more GC components before patching or upgrading them. (An earlier section, "Backups Before and After a Configuration Change," refers you here.) This section does not cover server-level backups, which are addressed in "Recommended Grid Control Backups" earlier in the chapter. A system administrator (SA) is usually tasked with such backups as part of daily server-level backups across the enterprise.

It is wise to back up the software for the OMR Database, OMS, and Agents. Restoration of these GC homes becomes less trivial as your site matures and as you make more configuration changes to GC components and to target configurations in GC. Component changes include GC and Database software upgrades and patches. Example target configuration changes in GC are metric and policy settings, groups created with the targets as members, and administrator rules referencing targets. It's true that this information is safe in the Repository, but sadly, you must remove it when reinstalling an Agent from scratch. By contrast, if you back up Agents, you can simply restore the latest backup of the Agent home and reset the Agent to return it to a consistent state with that known to the OMS and Repository.

NOTE
If an OMR, OMS, or Agent host fails and cannot be returned to service, you cannot restore a backup of any of the corresponding software for these GC components to another host, even to the same directory path. Moving an OMS (or OMR Database, for that matter) to another host is considered a migration,[50] not a restoration, and is too time consuming to consider as a B&R solution. If you do migrate an OMS and OMR to new hosts, however, you can reconfigure Agents to point to them (see Note 413228.1).

OMR Database Software Backup and Restoration

The OMR Database[51] software comprises the Database home and, optionally, an ASM home and Clusterware home, depending on your GC configuration. The files in the three Oracle Database software homes are all OS files, no different from non-Oracle OS files, so they can be backed up at the OS level, except for two Clusterware components discussed below in "Clusterware Components Backup and Restoration." Files in these Oracle homes include executables, initialization parameter files, password files, networking files (*tnsnames.ora*, *listener.ora*, and *sqlnet.ora*), configuration files, log files, and the like. Following is a summary of these three OMR Database software homes.

- **Database home** The Oracle database home must always exist to support either a single-instance or RAC[52] OMR Database. This is the home that the GC New Database Option installs in the *db10g* directory path below the GC parent directory. If using the GC Existing Database Option, you preinstalled the Database home, hopefully using the Oracle 10.2 software distribution.

---

[50] If you need to migrate the OMS or the Repository to another host, there is a detailed, involved procedure for doing so. Open an SR requesting Oracle internal *MetaLink* Note 388090.1, which Support is allowed to distribute to customers provided they guide you through the process.

[51] In this section, I assume the OMR is running on Oracle Database 10g Release 2 (10.2). If using Oracle 10g Release 1 (10.1) or Oracle 9i, the B&R techniques presented still apply, with a few differences. One is that Oracle Clusterware in Oracle 10.2 is known as Cluster Ready Services (CRS) in Oracle 10.1 and earlier releases. Another is that ASM is an Oracle 10g new feature.

[52] As explained in Chapter 1, there is only one active "primary" OMR Database for a given GC site, though you may also build a standby OMR Database for HA/DR reasons, as discussed in Chapter 15. A RAC OMR Database is a single database with multiple instances, each located on a separate RAC node.

- **ASM home** An ASM home may exist if chosen to manage an OMR Database (single-instance or RAC) and if installed in a separate home from the Database home, which is a best practice for an Oracle 10.2 Database. You install the software home for ASM identically to that for the Database (provided they're the same version, which is recommended), so you back up both homes the same way as well. ASM storage, on the other hand, is clustered and contains the OMR Database files, which you back up in GC using RMAN (covered earlier in the chapter in the section "OMR Database Backup and Recovery").

- **Clusterware home** You usually install a Clusterware home only when building a RAC OMR Database, although you can also install a single-instance OMR Database on a cluster node. The Oracle Clusterware consists of both nonclustered and clustered file types. B&R of the clustered files requires a different approach from that taken for the nonclustered files, as presented next.

To back up these homes in UNIX, use the "cp -Rp" command rather than the "cp -rp" command. The capital "R" recursively copies the symbolic links to the same location that the source did, whereas the lowercase "r" replaces these links with actual destination files to which the links point.

Clusterware Components Backup and Restoration

If you are running a RAC OMR Database, a separate Oracle Clusterware home exists. The Clusterware not only has a software tree that you back up at the OS level, but also contains two components that an OS-level backup of the Oracle Clusterware home will not address. OS backups are not applicable because both components must reside on either shared raw partitions or a supported shared file system, such as the Oracle Cluster File System 2 (OCFS2)[53] or OCP-Certified NAS Network File System (NFS). The two components, to which all RAC nodes must have access, are as follows:

- **Oracle Cluster Registry (OCR)** A configuration repository that manages cluster and RAC database information, including instance-to-node mapping, and manages Clusterware processes.

- **Voting disk** A file that manages information about node membership via a health check. RAC uses the voting disk to ascertain which instances are members of the cluster and arbitrates cluster ownership among the instances in case of network failure.

I speak of these components in the singular, but, as the Oracle documentation points out, they need to be mirrored externally and/or in the Clusterware itself to guarantee high availability.[54]

Oracle Cluster Registry Backup and Restoration The Oracle Clusterware backs up the OCR automatically and also allows you to manually "export" or back up the OCR online when making configuration changes to it.

The Cluster Ready Services Daemon (CRSD) process automatically backs up the OCR every four hours to the $ORA_CRS_HOME/cdata/<*cluster_name*> directory, retains the last three OCR backups, and retains one backup for each day and one at the end of each week. OS backups of the Clusterware home therefore include these OCR backups. (In other words, once backup files are automatically created, you can back them up at the OS level.)

---

[53] OCFS2 is Oracle's proprietary cluster file system software that is available for Linux and Windows platforms.

[54] Directions for multiplexing these components are provided in Chapter 3 of the Oracle Clusterware and Oracle Real Application Clusters Administration and Deployment Guide 10g Release 2 (10.2), which is part of the Oracle Database 10g Release 2 documentation library.

The procedure to restore the automatic OCR backups involves stopping the Clusterware and using the *ocrconfig -restore* command; for complete steps, see Chapter 3 of the Oracle Clusterware and Oracle Real Application Clusters Administration and Deployment Guide 10*g* Release 2 (10.2). You can find the separate procedures for UNIX and Windows in the section "Restoring the Oracle Cluster Registry from Automatically Generated OCR Backups."

To supplement these automatic backups, also export the OCR before and after changing the Clusterware configuration, such as when adding or removing a node or creating a database. To manually back up the OCR on Windows or UNIX, log in as a user who is a member of the local Administrators group (on Windows) or as *root* (on UNIX) and execute the following command:

$ORA_CRS_HOME/bin/ocrconfig -export<*file_name*>

When doing the export, you should concurrently back up the Oracle Clusterware home and all databases running on it, as most OCR configuration modifications also change Clusterware home files and database objects.

If you make OCR changes that you want to back out, restoring your previous configuration involves using the *ocrconfig -import* command in a procedure that varies for UNIX and Windows platforms. See "Administering the Oracle Cluster Registry with OCR Exports" in Chapter 3 of the Oracle Clusterware and Oracle Real Application Clusters Administration and Deployment Guide 10*g* Release 2 (10.2).

Voting Disk Backup and Restoration Perform a new backup of the voting disk on a regular basis; at a minimum, back up the voting disk both before and after adding a new OMR Database node or removing an existing node. As with the OCR, a hot backup of the voting disk is valid, meaning you don't need to stop the CRS daemons to take this backup.

To back up the voting disk, use the *dd* command on UNIX or the *ocopy* command on Windows. Execute the following *dd* commands on UNIX to back up and restore the voting disk, respectively:

dd if=<*voting_disk_name*> of=<*backup_file_name*> bs=4k
dd if=<*backup_file_name*> of=<*voting_disk_name*> bs=4k

where *if* is the input file, *of* is the output file, and *bs* is the block size.

The format of the *ocopy* commands on Windows to back up and restore the voting disk, respectively, are:

ocopy <*voting_disk_name*> <*backup_file_name*>
ocopy <*backup_file_name*> <*voting_disk_name*>

For more details on voting disk backup and restoration, see "Administering Voting Disks in Real Application Clusters" in Chapter 3 of the Oracle Clusterware and Oracle Real Application Clusters Administration and Deployment Guide 10*g* Release 2 (10.2). Oracle*MetaLink* Note 279793.1 also shows how to restore a lost voting disk.

OMS Software Backup and Restoration

OMS software consists entirely of plain OS files (located in the *oms10g* directory under the GC parent directory), which makes OMS B&R inherently less complex than that for the OMR Database or OMR Database software. Restoring an OMS from backup is far quicker and less painful than having to remove and reinstall an OMS from scratch, then patch and reconfigure it. This is particularly true of the original OMS installed, which has a few unique characteristics over additional OMSs. For instance, the original OMS host is where the monitoring Agent is located and configured (see "Enable Notification if Grid Control Goes Down" in Chapter 11), it is by default the only host member of the EM Website System (see "Create a System" in Chapter 12 for steps to re-create this system), and it contains the logical target called Management Services and Repository. You would need to manually reproduce these characteristics if you had to reinstall the original OMS, because you'd be installing it as an additional OMS. Backing up and restoring the original OMS is far preferable, particularly given the fact that there's documentation on how to reinstate these characteristics if you lose the initial OMS. I've done it for a few customers, and it's not pretty, I can tell you that.

Back up each OMS in your environment as instructed below, regardless of your GC configuration. As mentioned earlier, each OMS is unique because configuration files contain hard-coded references to the host name and OMS home. Below are some generic reasons to back up each OMS, irrespective of your particular GC configuration or whether it's the initial OMS:

- When only one OMS exists, it's critical to back it up because a single OMS means a single point of failure. Running just one OMS puts pressure on OMS B&R to supply availability in an outage. If the one OMS goes down, the integrity and speed of the OMS backup and restoration process factor heavily into availability.

- When running multiple OMSs without a server load balancer (SLB), Agents are hard-coded to upload to a specific OMS. If that OMS goes down, its Agents cannot upload until it is restored or you reconfigure the Agents to point to a working OMS.

- Even when multiple OMSs are set up behind an SLB, backing up all OMSs allows you to quickly restore any that may fail so that you can resume the increased performance levels from running multiple OMSs.

CAUTION
Don't try to back up or recover an OMS using the OracleAS Backup and Recovery Tool in the AS Console. This tool is intended for standalone OAS installations, not the OMS. Also, in an OMS recovery situation, the AS Console associated with the OMS, and thus the tool itself, may be offline as part of the failure. This is akin to GC's inability to recover the OMR Database. You cannot rely on a component for recovery when that component may be affected by the failure.

Back Up the OMS

According to the section "Recommended Grid Control Backups," earlier in the chapter, you should expect to take hot OMS backups to maintain your GC environment. A hot backup is sufficient for both server-wide backups and for backups taken befor/after a significant configuration change, such as applying an OMS patch. These hot backups are as easy as backing up any other software tree and do not require any special consideration. Note that when you back up an OMS, you should also back up its chain-installed Agent. This is because an OMS failure sometimes extends to its chain-installed Agent.

Following are the steps to execute a hot backup for a given OMS, placed in the context of performing a configuration change afterward:

1. Back up the Inventory for the OMS and chain-installed Agent if they will be affected by the configuration change.

2. Create an indefinite, server-wide blackout of the OMS host, Agent hosts uploading only to this OMS, and OMR nodes if changing its configuration too. Blackouts suspend data collections, uploads, and alerts. In this way, blackouts avoid DOWN alerts caused by the configuration change.

NOTE
When you must shut down Grid Control for situations other than a configuration change, such as to reboot GC servers, you don't have to set blackouts to avoid triggering alerts if you shut down and restart GC components in the correct order, as listed in "EM Startup/Shutdown Order" in Chapter 9.

3. Take a hot backup to tape (and disk, if desired) of the OMS and chain-installed Agent software homes. See "Suggestions for Working with Backup Media" earlier in the chapter for ideas to streamline this backup.

4. Make the OMS configuration change (patch, upgrade, etc.), then test. If the configuration change does not require you to stop the OMS, try bouncing the OMS to test that it restarts without error.

5. Stop the blackout created in Step 2.

Restore the OMS

The steps to restore an OMS depend on whether you backed it up. I start with the procedure for restoring the OMS from backup. For those without an OMS backup, I also give broad instructions for reinstalling and reconfiguring an additional OMS.

Restore the OMS from Backup To restore an OMS to the same directory path and host that you've backed up via hot backup as described above in "Back Up the OMS," perform the following steps:

1. Create a server-wide blackout encompassing all Agent hosts configured to upload specifically to the failed OMS. This will suppress Agent upload attempts and alert notifications due to the downed OMS.

2. Shut down or kill any OMS processes that remain running. (If you must also restore the chain-installed Agent, stop or kill any running Agent processes.)

3. Restore the backup of the OMS home to its original directory path. Restore from the latest disk backup or, if not accessible, from tape. (Do the same for the chain-installed Agent, if required.)

4. If not using a shared receive directory, remove all old XML files just restored to the local $OMS_HOME/sysman/recv directory on the OMS. (These files should already be uploaded to the Repository, unless the OMS failure occurred immediately after the OMS backup and these files did not get a chance to upload to the OMR.)

5. Reset those Agents not able to upload to the restored OMS by following the instructions in "Resetting Agents," earlier in the chapter.

6. Start the OMS and perform manual Agent uploads to confirm that the chain-installed Agent on the OMS host and other Agents can communicate with the OMS:

   ```
   $OMS_HOME/opmn/bin/opmnctl startall
   $AGENT_HOME/emctl upload
   ```

7. Stop the server-wide blackout created in step 1 that included all Agent hosts uploading to the specific OMS that failed.

Reinstall an Additional OMS If you don't back up an additional OMS, you need to reinstall it from scratch. In this case, perform the following steps, which only apply to reinstalling an additional OMS, not the initial OMS:

1. Shut down or kill any OMS processes that remain running on the affected host.

2. If other active OMS hosts are running, thus allowing Console access, then remove all failed OMS targets from Grid Control, including the chain-installed Agent:

   a. Log in to the Console via a working OMS as an administrator with full privileges to manage the OMS.

   b. Cancel any blackouts involving the failed OMS and chain-installed Agent.

   c. Click the Targets tab, All Targets subtab, select all OMS components one by one, and click Remove.

   d. Select the host target and click Remove.

   e. Shut down the chain-installed Agent if still running, or kill orphaned Agent processes. Return to the All Targets page, select this Agent, and click Remove.

3. If the Oracle Inventory is still accessible and was not lost as part of the failure, remove the OMS and chain-installed Agent installation from the Inventory by running the OUI from the GC installation media, as when first installing GC:

   a. On UNIX, execute *runInstaller*; on Windows, click *setup.exe*.

   b. Click Deinstall Products… .

   c. Under Oracle Homes, select the components *oms10g* and *agent10g*, click Remove…, and confirm.

   d. Leave the installer open, because you need to reinstall the OMS (see Step 5).

4. Manually remove the original OMS and chain-installed Agent software installations.

5. Reinstall the additional OMS (and chain-installed) Agent from scratch. Click Next in the OUI session already open, and select the GC Additional OMS Option.

6. Upgrade the OMS and chain-installed Agent and reapply any patches formerly applied.

7. Reconfigure OMS and chain-installed Agent targets as before (e.g., set Metric and Policy settings, add targets to groups, etc.). See Chapters 11 through 13.

8. Consider beginning backups of all OMSs and chain-installed Agents, now that you know what a chore it is to reinstall the OMS.

Agent Software Backup and Restoration

Agent backup and restoration choices are very similar to those for the OMS. Here I am speaking of both standalone Agents and chain-installed Agents that fail but whose corresponding OMS remains running. Regularly back up Agents as shown below.

Let's start with Agent backup procedures, then move on to Agent restoration steps, both when restoring from backup and when no backup is available.

Back Up the Agent

According to the section above, "Recommended Grid Control Backups," you should expect to take both hot and cold Agent backups to maintain your GC environment. A hot backup is sufficient both for server-wide backups and for backups taken before/after a sizeable configuration change, such as applying an Agent patch. These hot backups are as easy as backing up any other software tree and do not require any special consideration.

Following are the steps to execute a hot backup for a given standalone or chain-installed Agent, placed in the context of performing a configuration change afterward:

1. Back up the Agent Inventory if it will be affected by the configuration change.
2. Create an indefinite, host-wide blackout of the Agent and any other GC components whose configuration you are changing. Blackouts suspend data collections, uploads, and alerts, thereby avoiding Agent Unreachable alerts.
3. Take a hot backup to tape (and to disk, if desired) of the Agent software home. See "Suggestions for Working with Backup Media" earlier in the chapter for ideas to streamline this backup. (Also take a hot backup of the other GC components that will be affected by the configuration change.)
4. Stop the host-wide blackout of Agent and any other GC components blacked out.
5. Make the Agent configuration change (patch, upgrade, etc.), then test. If the configuration change does not require you to stop the Agent, try bouncing the Agent to test that it restarts and uploads to the OMS.

Restore the Agent

Again, as with the OMS, the procedure to restore a standalone or chain-installed Agent depends on whether you backed it up.

Restore the Agent from Backup   To restore a standalone or chain-installed Agent that you've backed up, as shown in the previous section, to its original host and directory path, perform the following steps:

1. Shut down or kill any Agent processes that remain running.
2. Restore a hot backup of the Agent home to its original directory path. Restore from the latest disk backup or, if none is available, from tape.
3. Reset the Agent according to the procedure in the section "Resetting the Agent" earlier in the chapter. This procedure includes restarting the Agent and testing that it can upload to the OMS.

Reinstall the Agent If you don't back up a standalone or chain-installed Agent, you'll need to reinstall the Agent from scratch. An Agent is easy enough to reinstall, but you will lose all data about that Agent's targets and must reconfigure the targets as needed.

To perform a fresh Agent installation on the same host and in the same home directory, perform the following steps:

1. Shut down or kill any Agent processes that remain running.
2. Remove all information about the Agent targets from Grid Control:
    a. Log in to the Console as an administrator with full privileges to manage the Agent.
    b. Cancel any blackouts of Agent targets.
    c. Click Targets, All Targets, select all Agent components one by one, and click Remove. Confirm these targets are removed by clicking Setup, and then the Management Services and Repository subtab. In the General section, click Deleted Targets. The Time Delete Completed must have a date. If it does not within 10 minutes, place a check next to the target and click Force Delete.
    d. Select the host target and click Remove. See step c above for how to force delete the host if it is not removed.
    e. Select the Agent target and click Remove. See step c above for how to force delete the Agent if it is not removed.
3. If the Oracle Inventory is still accessible and was not lost as part of the failure, remove the standalone Agent installation from the Inventory by running the OUI from the GC installation media, as when first installing GC:
    a. On UNIX, execute *runInstaller*; on Windows, click *setup.exe*.
    b. Click Deinstall Products... .
    c. Under Oracle Homes, select the component *agent10g*, click Remove..., and confirm to remove.
    d. Leave the installer open if you plan to reinstall the Agent using the OUI (see Step 5).
4. Manually remove any remaining part of the original Agent software installation.
5. Reinstall the Agent from scratch. Click Next in the OUI session already open, and select the GC Additional OMA Option, or use an alternative Agent installation method.
6. Upgrade the Agent and reapply any Agent patches formerly applied.
7. Reconfigure Agent targets as before (e.g., set Metric and Policy settings, add targets to groups, etc). See Chapters 11 through 13.
8. Back up the Agent (see "Back Up the Agent" above).

Summary

I began this chapter with an overall look at Grid Control B&R. A key concept was the division of GC elements into two groups: those that change and those that don't. Agent management files change and are found across the entire GC framework, uploaded from Agent to OMS to Repository.

However, you don't need to regularly back up these files because you can reset Agents during recovery so that they synchronize with the OMS and Repository. I presented the procedure to reset an Agent up front, as you must rely on it for recovery of all components in many recovery scenarios. Rounding off my introduction is a recommendation for three types of hot backups: daily hot OMR Database backups, hot backups of GC components before and after making configuration changes, and daily or weekly GC server-level hot backups.

I then launched into B&R of the OMR Database, which a page count will confirm spans more than half the chapter. This should come as no surprise. Just as all roads lead to Rome, all target data leads to the Repository. I presented three daily OMR Database backup methods in Grid Control: Oracle-Suggested Backup, Customized Backup, and RMAN Script job. I explained how to set up for OMR Database recovery in GC and how to actually recover outside of Grid Control. Recovery techniques presented were for both complete recovery from media errors and incomplete recovery from logical corruption, using RMAN and Flashback Technology. Finally, I pointed out Grid Control's backup management features, which offer a nice UI alternative to equivalent RMAN command-line functionality.

The last section on B&R of the Grid Control software was comparatively uncomplicated. Software for the OMR Database (which optionally includes ASM and Clusterware), for the OMS, and for the Agent is file-based (except two Clusterware components for those using a RAC OMR Database), so it's easy to back up in the OS. I also explained how to restore OMS and Agent software homes, both from backup and from scratch.

I'd like to wrap up this chapter on Grid Control backup and recovery by urging you to test your B&R mechanisms regularly. I cannot overly stress the importance of making certain all necessary OMR Database and Grid Control software files are backed up and easily accessible. The best way to verify B&R processes and your readiness to tackle them is to perform recoveries in a similar test environment throughout the GC lifecycle. At a minimum, for the OMR Database, avail yourself of the RMAN commands BACKUP VALIDATE DATABASE, VALIDATE BACKUPSET..., CROSSCHECK..., and RESTORE DATABASE VALIDATE to test the integrity of your B&R processes in place without disrupting the production GC environment. These commands validate all online and backup database files needed for backup or recovery without actually performing a backup or recovery. For example, to check the entire active OMR database and archive logs for physical and logical corruption and to confirm these files can be backed up, use the command BACKUP VALIDATE CHECK LOGICAL DATABASE ARCHIVELOG ALL.

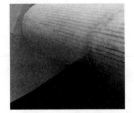

CHAPTER 15

Configure Grid Control for High Availability and Disaster Recovery

Grid Control high availability (HA) measures protect against local disruptions such as failure of an OMR or OMS server or disk, Agent malfunction, or GC application outage. Grid Control disaster recovery (DR) measures provide geographic isolation to guard against a catastrophic outage of the data center that's housing your primary GC environment. Outages can be due to events such as natural disaster, electrical failure, fire, war, sabotage, or internal data corruption. A type of outage often overlooked in DR planning is a *localized catastrophic event*, a combination of events or faults that affects system availability. HA systems guard against individual faults, but an amalgamation or a particular series of faults can prove fatal to system availability and require switching to a DR site.

A Grid Control environment is, by definition, a production system if it monitors production targets. GC requirements for high availability and disaster recovery should exceed those for its production targets, because if GC monitoring were to fail, you'd lose all visibility to your production environment. It is therefore essential to harden your GC infrastructure by implementing an HA/DR architecture to ensure that the failure of one or more GC components does not disrupt the ability to monitor your environment.

Recall from the last chapter that implementing HA/DR best practices for Grid Control is one of four ways to avoid monitoring outages, presented in the following chapter order:

- Backup and recovery (Chapter 14)
- High availability and disaster recovery (this chapter)
- Security (Chapter 16)
- Maintenance and tuning (Chapter 17)

To determine which HA/DR solutions to implement for your GC installation, you must place them on the same board, so to speak, as the backup and recovery (B&R) solutions established in Chapter 14. For instance, evaluate whether B&R alone can satisfy the Recovery Time Objective (RTO) for Grid Control specified in its service level agreement (SLA), which must match or surpass the most stringent target SLA. If the Grid Control RTO is tighter than B&R capabilities alone can provide, then you need to consider adopting one or more HA/DR strategies presented in this chapter.

Grid Control has been architected to allow you to meet the most stringent HA/DR requirements. You can design your Grid Control lattice in a wide variety of ways, which run the HA/DR gamut. The default configuration, based on the GC New Database Option, provides very little in the way of HA/DR, as it installs all GC components on a single host. This default choice may be appropriate for a nonproduction GC system, or for initially exploring the product's capabilities and features. Contrast this with an advanced HA/DR configuration, as described in this chapter, in which server load balancers act as a front end for both Agent and Console connections to multiple Active/Active OMSs, Agent management files on redundant storage upload to a shared, redundant OMS location, both Oracle Real Application Clusters (RAC) and Oracle Data Guard protect the Repository, and a GC standby site is ready and waiting for any disaster recovery situation. Other levels of HA/DR architectures fall between the default and most advanced configurations.[1]

An Oracle DBA, perhaps with the moniker "architect," is usually responsible for the design of a Grid Control HA/DR environment. The architect's objective is to ensure the fault tolerance of all GC components and to protect the flow of management data between them in case of performance or

---

[1] Chapter 3 of Oracle Enterprise Manager Advanced Configuration 10g Release 4 (10.2.0.4.0) presents some of these intermediate options.

Chapter 15: Configure Grid Control for High Availability and Disaster Recovery

availability problems, such as failure of an OMS host. The designer must take advantage of operational best practices, Grid Control features, hardware technologies, and both Oracle and non-Oracle software technologies. Most of these best practices have been formalized under Oracle's Maximum Availability Architecture (MAA) blueprint for Enterprise Manager, and documented in the Oracle white paper, "MAA Best Practices: Enterprise Manager 10gR2 and 10gR3," January 2007.

This chapter is not intended to replace the MAA material on EM, but to flesh it out in places, particularly in the areas of OMS and Agent DR. The chapter is organized into two main sections: one for Grid Control HA recommendations, and the other for Grid Control DR recommendations. Each section is further broken into subsections for each GC component, as follows:

- Grid Control high availability recommendations
    - OMS HA recommendations
    - OMR Database HA recommendations
    - Agent HA recommendations
- Grid Control disaster recovery recommendations
    - OMR Database DR recommendations
    - OMS DR recommendations
    - Agent DR recommendations

The beginning of each HA and DR subsection contains an outline of recommendations for that area. OMS HA/DR recommendations include the chain-installed Agent, when applicable, and Agent HA/DR recommendations pertain to standalone Agents.

NOTE
Unless otherwise noted, all MAA white papers cited in this chapter are available from OTN at http://www.oracle.com/technology/deploy/availability/htdocs/maa.htm.

Grid Control High Availability Recommendations

Backup and recovery (B&R) and high availability (HA) are related topics, but they differ significantly. B&R addresses the integrity of the system and how to recover it in the event of system failure. System failure has many causes, including hardware malfunction, operating system defects, application bugs, and operator or user-induced error (e.g., a DBA deleting a GC table). After any one of these events, the primary goal is to restore the system to the current state or, if logical corruption occurs, to the moment immediately prior to the fault.[2]

High availability is premised on a reliable B&R system. HA emphasizes the latter (i.e., Recovery) portion of the B&R equation. HA assumes an appropriate backup method exists, and its goal is to reduce both downtime (through redundancy, for example) and recovery time. The degree to which HA must reduce downtime and recovery time depends on user needs. The cost of availability rises substantially as the availability requirement approaches 100%. Therefore, it is important to identify the acceptable window of unavailability and to "differentiate between the needs at critical

---

[2] I credit Matthew Burke, my mentor, for much of the material in this "Grid Control High Availability Recommendations" section, taken from his white paper entitled "Oracle High Availability Guidelines and Recommendations."

times and the needs for other periods."[3] Although a system probably needs to be available during business hours, greater flexibility may be permissible at other times.

You attain high availability both by decreasing the length and frequency of downtime and by reducing the recovery time after failure. In the HA subsections below for OMS, OMR Database, and Agent, I cover hardware and software options that improve availability for these components. The cost of these different options has a wide range. The expense increases dramatically as the requirements move from 98% to 99.9% availability. You can meet availability in the 99.99+% range with fault-tolerant hardware used in conjunction with Oracle and non-Oracle technologies.

Not all high availability enhancements require expensive hardware and software improvements or modifications. You can implement the following general measures to improve and enhance availability at minimal expense. (I provide database-specific HA recommendations in the section "OMR Database HA Recommendations" later in the chapter.)

- **Enforce security**   Enforcing both physical and system security provides the most cost-efficient increase in availability. Limit access to Grid Control hardware, allocate GC application privileges only as needed, and secure data that is transmitted between components (see Chapter 16).

- **Maintain software configuration management (SCM)**   Develop, install, maintain, and document your GC environment in conformance with a standard software release procedure. Apply these practices to all software, including the operating system, network, database, and GC application. A key objective of any GC maintenance life cycle is to test any changes on a nonproduction system before implementing those changes on a production system. GC environments, particularly large ones, warrant at least two separate environments: production and nonproduction (which was one of the decisions you made in Chapter 2 in the section "Decide How Many Grid Control Environments to Build"). Here is a short description of these environments:

    - **Production environment**   A GC site that only monitors production servers. A mission-critical GC environment calls for instituting the B&R, HA, disaster recovery, security, and maintenance and tuning measures covered in Chapters 14 through 17.

    - **Nonproduction environment**   Monitors all nonproduction targets. Provides a smaller-scale hardware environment (compared with the production site) in which the DBA can test patches, upgrades, configuration changes, and B&R solutions intended for the production GC framework or OS. You should execute a well-defined suite of tests every time you patch or upgrade. The SLA for this environment need not be as strict as for production, but the nonproduction environment should ideally be configured as closely to production as possible.

- **Provide fault-tolerant network devices**   Fault-tolerant network devices in each GC host allow the application to use dual-ported NICs (which are both active) or to seamlessly (or after restart) fail over to a standby network card in the event of primary network card failure. Work with your system administrator or network specialist to build fault-tolerant network interfaces into each OMS, OMR, and OMA server.

---

[3] "Designing VLDBs for High Availability," Gupta, Deepak and Erik Peterson. *Oracle Magazine*. November/December, 1996. pp. 65–72.

Now that we have touched on these general HA recommendations, applicable to Grid Control environments as well as to any other type of Oracle environment, let's take on high availability at the GC component level. As I will mention when applicable, some of these HA suggestions also improve the scalability of your GC site.

OMS HA Recommendations

Out-of-box, the OMS has HA capabilities. A built-in restart mechanism called the Oracle Process Management and Notification Service (OPMN) attempts to restart any OMS processes that terminate unexpectedly. Also, built-in metrics monitor the OMS processes and functional performance, such as loader throughput.

Still, you can take significant additional measures to increase OMS high availability. It's difficult to discuss each OMS HA measure outside the context of the others, so let me briefly define them now before examining them individually:

- **Install multiple Oracle Management Services**   Running multiple OMSs provides redundancy for an OMS host failure. You can install multiple OMSs in either an Active/Active or Active/Passive configuration. Active/Active is preferred because all OMSs operate concurrently, whereas only one OMS is active at a given time in an Active/Passive environment.

- **Install the OMS on fault-tolerant storage**   Placing the OMS host on fault-tolerant storage obviates the need to restore the OMS from backup due to media failure, thus reducing downtime and increasing HA. It also protects against loss of Agent data uploaded to an OMS using a local receive directory.

- **Set up the Shared Filesystem Loader**   A GC feature called the Shared Filesystem Loader allows you to store uploaded Agent files on fault-tolerant storage in a shared central location that all Active/Active OMS hosts can access. All OMSs process uploaded Agent files, and none of these files is lost if a particular OMS fails.

- **Use a server load balancer**   Front-ending all Active/Active OMSs with a server load balancer (SLB) distributes Agent uploads and Console requests evenly between OMSs. Without an SLB, Agents configured to point to a particular OMS cannot upload data if that OMS goes down; also, administrators accessing the Console only through that OMS lose connectivity and must log in through an active OMS by specifying its unique URL.

- **Configure automatic OMS startup on boot**   When an OMS server reboots, the OMS should be configured to automatic restart. Otherwise, GC functionality provided by that OMS host would not resume.

Figure 15-1 depicts this HA architecture of two OMS hosts in an Active/Active configuration with a Shared Filesystem Loader and SLB.[4] (It's difficult to pictorialize the remaining HA measures, so they are not represented in the figure.) A two-node Active/Active RAC Database is also shown, which supplies Repository HA (see "Install the Repository in a RAC Database" later in the chapter).

Let's examine each of the above-listed measures that you can put into operation to make the OMS more highly available.

---

[4] From the Oracle white paper, "MAA Best Practices Enterprise Manager 10gR2 & 10gR3," January 2007, Figure 3.

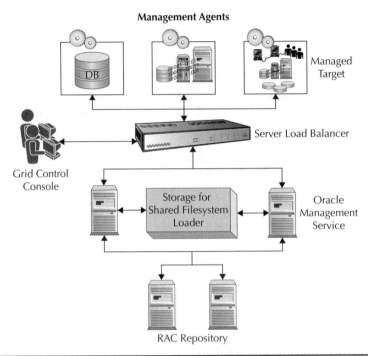

FIGURE 15-1. *An Active/Active OMS and OMR architecture with an SLB and Shared Filesystem Loader*

Install Multiple Oracle Management Services

In Chapter 2, the section "Selecting the GC Additional OMS Option" offered advice on the number of additional active OMS hosts to deploy for scalability purposes, based solely on the size of your GC environment. As explained there, medium or larger GC sites (defined as containing 1,000 or more targets) should use at least two active OMSs to disperse the load and improve the efficiency of data flow. Chapter 2 also touched on guaranteeing high availability for the OMS, even in a small GC environment, by installing at least two OMSs, either in an Active/Active or Active/Passive configuration. Any HA solution for your GC site must employ multiple OMSs for redundancy in case of an outage. If no OMS were available, management data uploads, Console access, and notifications would all come to a halt.

Whether for reasons of scalability or high availability, you may have already installed an additional OMS. The two possible configurations, Active/Active and Active/Passive, are described next. If you've waited until now to add an Active/Active OMS for either reason, now is the time to install one (directions below refer you back to the appropriate sections in Chapters 3 and 4). As noted at the outset of Chapter 2 and again in Chapter 3, you needed to refer to this present chapter if you were considering an OMS Active/Passive configuration, because it requires that you follow a special procedure when first installing Grid Control. As you will see, an Active/Passive OMS doesn't offer the scalability and performance gains of an Active/Active configuration and requires using Oracle or third-party CFC software and shared storage. On the positive side, an Active/Passive OMS architecture doesn't require an SLB.

Install the OMS in an Active/Active Environment I recommend an Active/Active OMS over an Active/Passive architecture, because the former allows all installed OMS hosts to operate concurrently. In an Active/Active environment, multiple OMSs, each located on a separate host, receive Agent uploads and Console requests. This allows the GC environment to scale horizontally and maximizes your hardware utilization, resulting in better performance than when using the same number of OMSs in an Active/Passive configuration. You can add more OMS servers, unlike with an Active/Passive setup, which prevents scaling beyond a single OMS. An Active/Active OMS configuration also provides for uninterrupted service, whereas an Active/Passive configuration requires failover.

You can install an additional OMS in an Active/Active configuration at any time in the GC life cycle, as follows:

1. If you removed the emkey from the Repository for security purposes after the last OMS installation (as recommended in Chapter 4 once all OMSs are deployed), you must run the following command from the first OMS home to copy the emkey back to the Repository:

   ```
   $OMS_HOME/bin/emctl config emkey -copy_to_repos
   ```

   If you attempt to install an additional OMS with the emkey missing from the Repository, the GC installer will prompt you to run the above command.

2. Launch the GC installer and choose the GC Additional OMS Option. See "Install an Additional OMS" at the end of Chapter 3.

3. After installing the additional OMS, configure it according to the post-installation steps listed in Chapter 4:

   a. Complete the steps in the "Patch Grid Control" section for the additional OMS.

   b. Finish all steps in the "Oracle Management Service Configuration" section for the additional OMS.

   c. If desired, follow the procedure in the subsection "Configure *i*SQL*Plus Access in Grid Control" under the "Oracle Management Repository Configuration" section for the additional OMS.

   d. Back up the additional OMS as reminded in "Begin Regular Grid Control Backups" at the end of Chapter 4.

Install the OMS in an Active/Passive Environment While an OMS Active/Active configuration allows you to run multiple OMSs concurrently, an Active/Passive configuration runs only one active OMS at a time in a CFC environment and limits you to just one Active OMS. Your IT department's policy may be to use CFC technology for application HA, which may weigh heavily in deciding to adopt an Active/Passive OMS architecture. However, consider that an Active/Active OMS configuration is a superior solution to an Active/Passive OMS for the reasons stated in the previous section. Of course, to adopt an Active/Active OMS solution, a server must be available to run additional OMSs and an SLB must be accessible to all OMSs.

A CFC consists of two or more servers (or nodes) placed in close physical proximity so as to share the same logical hostname, IP address, and storage subsystem. The *active node* (also called the primary node) runs the OMS application, and one or more *passive nodes* (also called secondary nodes) remain available to host the OMS should the active node fail. You may also perform a planned failover to bring down the active node for maintenance, and fail back when finished.

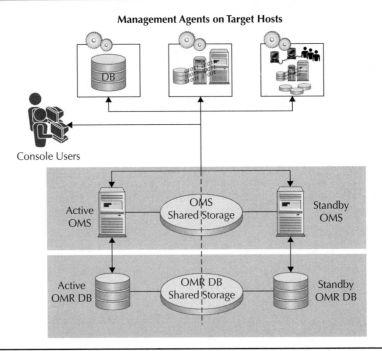

FIGURE 15-2. *An Active/Passive OMS (and OMR) architecture*

Figure 15-2 illustrates an Active/Passive OMS (and OMR) configuration.[5] The OMR need not be Active/Passive as well, unless adopting an Active/Passive OMS configuration on a single tier (i.e., shared by the OMS and OMR). A single tier GC solution is the norm for CFC environments. Unless a RAC OMR is an option, companies want an Active/Passive OMR because HA for the OMR is as important as for the OMS. Furthermore, CFC hardware is scalable enough to host both the OMS and OMR on the same cluster.

CFC sends constant heartbeats to monitor the OMS running on the active node; if it detects an application or resource outage, such as with processors, memory, or the node itself, the CFC fails over the OMS file systems to a passive node and restarts the OMS on that node. Grid Control clients (both Console users and Agents) are locked out during failover, but are then connected to the new active node after the OMS begins running there.

CFC provides application high availability by offering redundant modules, including power supplies, controllers, cabling, and cooling systems. The storage is most commonly Network File System (NFS), on which you install all OMS software, including binaries, Oracle inventory, configuration data, and run-time data located in the OMS receive directory. Oracle provides a CFC solution for Windows systems called Oracle Fail Safe. UNIX operating system vendors offer native CFC solutions, including:

- HP Serviceguard
- IBM High Availability Cluster Multi-Processing (HACMP)

---

[5] From the Oracle white paper, "MAA Best Practices Enterprise Manager 10gR2 & 10gR3," January 2007, Figure 2.

Chapter 15: Configure Grid Control for High Availability and Disaster Recovery

- Sun Solaris Cluster
- Veritas Cluster Server

There is a special procedure described in Oracle*MetaLink* Note 405642.1 for installing the OMS in an Active/Passive environment, beginning with the initial OMS host. (The procedure given in the note has not been tested on Windows platforms.) You perform the installation at the same time for all existing cluster members by selecting the GC Existing Database Option. After the installation completes, you can immediately log in to Grid Control via Oracle HTTP Server, but you must make a minor change to log in via Web Cache (see Note 305412.1).

Install the OMS on Fault-Tolerant Storage

You can enhance OMS availability in either an Active/Active or Active/Passive configuration by using fault-tolerant storage for the OMS host or part of it. The desired scope for extending fault-tolerant storage is as follows (listed in order of most to least desirable):

- The entire OMS host
- The OMS home (and chain-installed Agent home)[6]
- The local *$OMS_HOME/sysman/recv* directory (if not using the Shared Filesystem Loader)

NOTE
See "Hardware Installation Requirements" in Chapter 2, which points you to the MAA documentation specifying the desired characteristics of such fault-tolerant storage.

Using fault-tolerant storage in the first two cases avoids the need to restore the OMS host or OMS home, respectively, from backup if an OMS disk fails, thereby decreasing downtime.

The third case applies only if the OMS receive directory is local and you are not employing the Shared Filesystem Loader as discussed in the next section. Placing a local OMS receive directory on fault-tolerant storage prevents Agent data loss, which you can't avoid by restoring from backup because it would not contain any current files located in the local OMS receive directory at the time of failure.

The OMS receive directory is typically empty when the OMS is able to load data as quickly as it is received. However, small loader backlogs do occur, even in well-tuned GC systems, and XML files in an OMS receive directory located on non-redundant storage are susceptible to loss. The data in these XML files would be lost if not placed on fault-tolerant storage because, once the OMS receives a file, the Agent considers the file committed and removes its local copy. The data loss is resolvable, but necessitates synchronizing all Agents reporting to the affected OMS (see "Procedure to Reset an Agent" in Chapter 14). Using fault-tolerant storage averts the need to reset Agents, which can be a real time saver for sites with many target hosts.

Set Up the Shared Filesystem Loader

While fault-tolerant storage protects a local *$OMS_HOME/sysman/recv* directory, using it to protect shared storage accessed by the Shared Filesystem Loader in an Active/Active OMS environment provides the most HA and scalability benefits. A GC 10.2 feature, the Shared Filesystem Loader

---

[6] The GC installer always locates the chain-installed Agent home on the same mount point as the OMS home. Therefore, media failure would almost certainly affect both OMS and chain-installed Agent homes.

temporarily stores Agent files at a shared location between OMS hosts, called the *shared receive directory*, before loading them into the Repository. The OMSs coordinate among themselves to distribute the task of uploading files into the Repository. If an OMS fails in an Active/Active configuration, the surviving OMSs assume its workload. This improves processing efficiency of Agent data for given OMS resources, and consolidates the required fault-tolerant storage in one place rather than requiring each OMS host to possess such storage for its local receive directory.

The shared receive directory can be on a Network File System (NFS) or Common Internet File System (CIFS)[7] mounted disk. (For instructions on how to create an NFS shared directory, see "Configure Shared Storage" under the "nfsagentinstall Installation" section in Chapter 7.) However, the most fault-tolerant solution for achieving true high availability is to use two NetApp filers (which employ a network-attached storage [NAS] architecture) in an Active/Active cluster via a private, high-speed link. Irrespective of the storage solution that you choose for the shared receive directory, replicate it to a standby OMS if you're running one for disaster recovery purposes (covered later in the chapter in the section "Replicate the Shared Receive Directory at the Standby Site").

To configure the OMS to use the Shared Filesystem Loader, execute the steps below, one by one, on each OMS host in parallel. You must configure the Shared Filesystem Loader on all Active/Active OMS hosts, or the OMSs will fail to start.

1. Configure a shared receive directory on fault-tolerant storage (see the section, "Install the OMS on Fault-Tolerant Storage") that the *oracle* user and *dba* group on all Agent and OMShosts can write to and read from (set directory permissions to 770 on UNIX, for example). The shared receive directory disk space requirement is based on the size of your GC deployment, as follows:

Deployment Size	**Receive Directory Space Required**
Small (<1,000 targets)	2GB
Medium (1,000 to 10,000 targets)	(Number of OMSs) × 5GB
Large (>10,000 targets)	(Number of OMSs) × 10GB

2. Make sure all XML files are uploaded to the Repository from the local *$OMS_HOME/sysman/recv* directory on each OMS host.

3. Stop the Management Service on each OMS host:

   $OMS_HOME/bin/emctl stop oms

4. Configure each OMS host to use the shared receive directory:

   $OMS_HOME/bin/emctl config oms loader -shared yes -dir *<shared receive directory>*

5. Restart the OMS on each OMS host:

   $OMS_HOME/bin/emctl start oms

6. Confirm that all Agents begin uploading files containing their management data to the shared receive directory. Choose a few representative Agents and manually upload data.

   $AGENT_HOME/bin/emctl upload

   Then, check that the shared receive directory contains new Agent upload files.

---

[7] The CIFS protocol is the successor to the SMB protocol and is supported by most Windows and Linux servers, Network Attached Storage appliances, and the Open Source server Samba.

Use a Server Load Balancer

As already mentioned in the introduction to this section and shown earlier in Figure 15-1, HA best practices call for using a clustered hardware server load balancer (SLB) to virtualize both Console and Agent connections to multiple Active/Active OMS installations. SLB vendors include F5 BIG-IP Local Traffic Manager, Radware WSD/CT100c, Nortel Alteon, Foundry ServerIron, NetScaler, and Cisco ACE/CSM.[8] A hardware SLB such as one of these is recommended, but you can use a software load balancer instead if desired. Using a clustered SLB provides for high availability of the SLBs themselves. The SLB is the front end for OMSs serving the same application workload, and is capable of redirecting both Agent uploads and Console requests to a specific OMS. The SLB thereby supplies load balancing for both Console and Agent data flow to multiple OMSs, and failover for OMS malfunction. You set up the SLB such that, if an OMS were to fail, the remaining OMSs would take over the workload.

Your SLB likely provides an advanced cache of its own, so you should configure it to bypass the OMS Web Cache and route traffic directly to the Oracle HTTP Server subcomponent of the OMS. Set up the load balancer to distribute requests according to a least-loaded rather than a round-robin policy to more evenly distribute workload.

NOTE
See OracleMetaLink Note 549270.1 for complete instructions on configuring 10.2.0.4 OMSs for high availability behind an SLB. With an SLB, Agents bypass direct uploads to a particular OMS. Therefore, the SLB configuration instructions include setting up the Shared Filesystem Loader (see the previous section).

If you use multiple Active/Active OMSs without an SLB, you must manually assign Agents to active OMSs. In this case, consider setting up each Agent to upload to the particular OMS most physically proximate to it.

Configure Automatic OMS Startup on Boot

When a host containing one or more GC components reboots, those components should automatically restart to ensure continued monitoring of targets in your environment. The OMS is a critical component that, if not started, renders most GC functionality inoperable. For example, Agents still collect data, but can't upload it, and alert notifications cease. See Appendix E (online at www.oraclepressbooks.com) for instructions on setting up all OMS processes to start up automatically on system boot.

OMR Database HA Recommendations

Grid Control monitors the OMR Database for availability and performance with out-of-box metrics at both the database level (e.g., database status and broken jobs), as for any database target, and at the OMR functional level (such as DBMS job processing time).

This section presents additional HA configuration best practices for the Management Repository, which are as follows:

- Implement any Database HA best practices not already adopted
- Use clustering for the OMR Database
- Configure automatic OMR Database startup on boot

---
[8] Source: "MAA Best Practices, Enterprise Manager 10gR2 and 10gR3," an Oracle white paper, January 2007.

Note that while HA best practices for Oracle databases actually encompass using both Oracle RAC (for HA) and Oracle Data Guard (for both DR and HA), Data Guard is primarily a DR solution, so I defer Oracle Data Guard treatment until a later section of this chapter, "OMR Database DR Recommendations."

Following is a description of each of the above HA best practices for the OMR Database.

Implement Database HA Best Practices

Chapter 2 and Appendix C (online; referenced in Chapters 3 and 4) contain some recommendations for the Repository Database that accord with Oracle Database High Availability Best Practices 10g Release 2 (10.2), part of the Oracle Database 10.2 documentation set. It is out of scope here to restate all database HA recommendations from this Oracle guide. Comb through it now for any remaining HA best practices needed for the Repository Database to meet Grid Control SLA objectives.

Use Clustering for the OMR Database

As you may recall from the section "Selecting the GC New Database Option" in Chapter 2, true HA for the OMR requires creating it either in an Active/Passive CFC environment or in an Active/Active database (Oracle RAC). Oracle RAC is a separately licensed database product, formerly known as Oracle Parallel Server, OPS. A RAC Repository is preferred over CFC for both HA and scalability reasons. However, a fair percentage of IT shops have standardized on CFC technology as the HA solution for their databases. If this describes your IT department, you may be swayed by these standards to use CFC for the OMR Database.

These two distinct forms of Oracle clustering, CFC and RAC, use a Shared-Nothing and Shared-All approach, respectively, to accessing their storage devices. In a Shared-Nothing architecture (CFC), multiple nodes can mount and access disk partitions, but can only logically share them. That is, only one node can access a given partition at a given time, though the clustering software does provide a global view of disks on the other nodes. In the second form of clustering, Shared-All (RAC), one or more nodes can access the same storage partition at the same time. This is a more efficient technique where disks (managed by ASM or raw devices) are physically rather than logically shared between nodes.

Due to the throughput requirements of the disk subsystems, the devices in both CFC and RAC configurations must be in close proximity to each other, generally within the same machine room. Neither architecture offers geographic isolation of its nodes, so both are susceptible to localized catastrophes. For this reason, with a clustered Repository at the primary site, it is an MAA best practice to also install a standby database (which can be non-clustered, CFC or RAC, depending on DR requirements) at a remote site in an Oracle Data Guard configuration (see the section "OMR Database DR Recommendations" later in the chapter).

I characterize Data Guard as a DR solution, but it may also satisfy your HA requirements for the OMR in lieu of RAC or CFC. For example, you can automate Data Guard failover to occur within one minute, which meets most HA specifications for Grid Control. In that time frame, you can make the corresponding change to point active OMSs to the OMR Repository at the standby (rather than primary) site. Even so, Data Guard, like CFC, is technically a cold failover solution from an HA perspective because the primary Repository is shut down at failover time. RAC, on the other hand, is a hot failover solution, requiring no downtime whatsoever.

Install the Repository in a RAC Database   As already stated above, RAC provides better high availability and scalability an CFC for the Repository Multiple cluster nodes increase availability by avoiding single points of failure. If a node were to crash, the Repository would continue to run on the surviving node(s), and the OMS would automatically direct all communication to these nodes—that is, provided you followed the instructions in Chapter 4 under the section "Configure OMS for Failover and Load Balancing." As for scalability, a RAC system can support a large number of users and transactions. If you need more processing power, you can add another node to the cluster without interrupting GC operations.

RAC software allows multiple nodes to run separate Oracle instances[9] that reference a shared collection of data files. This concept of sharing disks among a cluster of machines differs somewhat from distributed file systems such as Network File System (NFS) and Andrew File System (AFS). While distributed file systems have the advantage of allowing file sharing between heterogeneous machines, they lack two important features required by the RAC architecture. The first is a guaranteed write mechanism,[10] which asserts to the database that data has been securely written to disk. The default NFS/AFS specifications do not provide this type of assurance; rather, these utilities buffer write requests and assume that other background processes will complete the operation later. The second feature is a distributed lock manager that coordinates access to pieces of data for the purpose of updating them. Without this coordination, it is possible to lose data or experience data corruption. Therefore, distributed lock management is a crucial element in the RAC architecture.

> **NOTE**
> *A RAC Repository requires a RAC Database license, whereas a single-instance Repository does not need licensing. Managed targets must always be licensed.*

For instructions on installing a RAC Repository, see the relevant parts of the section "Install Grid Control Using an Existing Database" in Chapter 3.

Install the Repository in an Active/Passive Environment   Like RAC, CFC is a method for improving availability of the OMR Database. However, the benefits discussed above for using RAC far outweigh those for using CFC. CFC works the same for the OMR Database as for the OMS. The drawbacks on CFC are the same as well. For example, as with an OMS in a CFC, you can only run one active OMR database instance in a CFC. See "Install the OMS in an Active/Passive Environment" earlier in the chapter for a description of CFC. That section includes Figure 15-2, which reflects an Active/Passive architecture for both the Repository and OMS. Most sites that elect to use an Active/Passive GC architecture run both the OMS and OMR Database on the same CFC.

As with CFC for the OMS, you must follow a special procedure when initially installing Grid Control to implement CFC for the Repository. See Note 405979.1 for this procedure.

---

[9] An Oracle instance, technically, is the collection of system processes that facilitate access, control, and perform the requisite functions of the DBMS, such as managing Oracle shared memory and archiving redo log data. An Oracle database is the collection of data files residing on disks that actually contains the data. In a RAC configuration, multiple instances share access to a single database.

[10] In addition to using Oracle ASM, Oracle has certified the use of specific NFS vendor products for use with Oracle RAC Databases. Rigorous testing has validated their ability to perform secure, guaranteed writes.

Configure Automatic OMR Database Startup on Boot

When a system reboots, GC components on the system should restart automatically so that they can resume their role in monitoring targets as soon as possible. The Repository Database and listener are critical components that must be running not only to save target monitoring data that Agents upload to the OMS but also to provide access to the Console. See Appendix E (online) for instructions on ensuring that the OMR Database and listener are configured to automatically start on boot.

Agent HA Recommendations

Following are our Agent HA steps to guarantee that your targets are monitored around the clock:

- Configure the Agent in an Active/Passive environment
- Install the Agent on fault-tolerant storage
- Provide a redundant notification method
- Test Agent restart on failure
- Configure automatic Agent startup on boot

Configure the Agent in Active/Passive Environment

Many clients today use virtual host environments to achieve high availability and scalability for their applications. Third-party vendors providing such virtual host environments include Sun (Solaris Cluster), Veritas (Cluster Server), HP (Serviceguard), and IBM (HACMP). If you are running virtual hosts, you'll want to monitor them. The procedure to install and configure Agents to monitor a virtual hostname in an Active/Passive environment is detailed in Oracle*MetaLink* Note 406014.1. This procedure has been tested on all UNIX platforms, but not on Windows systems.

NOTE
Oracle Fail Safe is Oracle's solution for running an Oracle database on a Windows CFC. For instructions on installing an Agent to manage Oracle Fail Safe databases from the Console, see "Oracle 10g Grid Control with Oracle Fail Safe: Highly Available Management Agent," an Oracle white paper, November 2004.[11]

Install the Agent on Fault-Tolerant Storage

The Agent collects management data and alerts in local files under the *$AGENT_HOME/sysman/ emd/* subtree before uploading them to the OMS. If these files were lost or corrupted before being uploaded, you would lose monitoring data and any pending alerts contained in these files. At a minimum, for both standalone and chain-installed Agents, configure this Agent subdirectory on fault-tolerant storage. You can further enhance availability by placing the entire Agent home on fault-tolerant storage. See "Hardware Installation Requirements" in Chapter 2, which in turn references the relevant Oracle MAA documentation describing the characteristics of such fault-tolerant storage.

Provide Redundant Notification Method

It is crucial that Agent alerts reach administrators, and a potential single point of failure in this regard actually resides outside of Grid Control. I am speaking of the SMTP server(s) that the OMS

---

[11] This white paper is available on OTN at http://www.oracle.com/technology/products/oem/deployments/fs_with_gc.pdf.

Chapter 15: Configure Grid Control for High Availability and Disaster Recovery   **637**

relies on to relay Agent alerts. To establish e-mail notification HA, configure e-mail notifications in the Console to notify administrators using multiple SMTP servers, if available. This is required only if the currently specified SMTP server is not already in an HA cluster. Click Setup in the UI and modify the Outgoing Mail SMTP Server setting under Notifications Methods to add a second SMTP server, separated by a comma or space. (This HA measure could justifiably be classified under the OMS rather than the Agent. However, the way I see it, the OMS is just the intermediary in notifying administrators on behalf of the Agent.)

Test Agent Restart on Failure

Once the Agent is started, the watchdog process monitors and attempts to restart it on failure. Before starting the Agent, you can set two variables in the Agent environment to change the default behavior of the watchdog, though default settings are usually sufficient for HA purposes. These two variables are EM_MAX_RETRIES and EM_RETRY_WINDOW. The watchdog will stop trying to restart the Agent after more than EM_MAX_RETRIES attempts within a time period of EM_RETRY_WINDOW. These two variables are defined as follows:

- **EM_MAX_RETRIES**   (3 times by default) The maximum number of times the watchdog will attempt to restart the Agent (including the first attempt) within the EM_RETRY_WINDOW time period.

- **EM_RETRY_WINDOW**   (600 seconds by default) The time period in seconds during which the watchdog will attempt to restart the Agent EM_MAX_RETRIES times.

The attempts that the watchdog makes to restart the Agent are evenly spaced over the time period EM_RETRY_WINDOW. In other words, the watchdog attempts to restart the Agent every (EM_RETRY_WINDOW / EM_MAX_RETRIES) seconds.

You can test that this automatic Agent restart feature works (in a nonproduction GC system) by killing the *$AGENT_HOME/bin/emagent* process at the OS level and confirming that the Agent restarts. Before killing the Agent, tail the *$AGENT_HOME/sysman/log/emagent.nohup* log file to verify that the message "Agent launched with PID…" appears on restart.

Configure Automatic Agent Startup on Boot

It is important that the Agent automatically start when a target host is rebooted. The Agent must be running for Grid Control to monitor targets on a host, though the Console would still be available, as well as target historical data, in the Repository. To ensure continued target monitoring on a host, the GC installation process creates mechanisms to automatically start the Agent and other installed GC components on server boot. See Appendix E (online) for instructions on how to make certain the Agent starts on boot.

Grid Control Disaster Recovery Recommendations

The high availability steps outlined in the previous section safeguard against the local failure of a GC component. GC disaster recovery should dovetail with your company's overall DR plan to survive a catastrophic failure of your entire production data center so as to guarantee business continuity. These days, most companies approach the subject of disaster recovery very seriously, as the loss of business continuity can cost them dearly in lost revenue per hour. IT departments at most firms maintain a DR data center that stands at-the-ready to assume a production role in the event of a disaster.

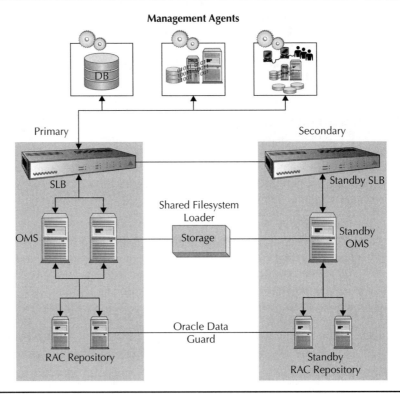

FIGURE 15-3. *Disaster recovery architecture for Grid Control*

Production GC components are typically in the same location as the production targets they monitor, and similarly, standby GC components are located at the standby site. The DR plan for Grid Control hinges on establishing secondary GC components to monitor standby targets when they assume production roles. IT departments routinely perform such role reversals for reasons other than disaster response, such as to test their DR strategy or to remain operational while performing primary site offline maintenance.

Setting up a DR site for Grid Control involves installing a standby OMR Database (single-instance, RAC, or CFC), a standby OMS, and a standby Shared Filesystem Loader (and, ideally, SLB), if used at the primary site. To tie it all together, an MAA best practice is to configure the standby OMR Database and OMS to automatically start if the primary site goes down.

To protect standalone Agents on production servers, you have a range of choices, depending on your DR configuration and needs (which hopefully align). On the do-nothing side, you can wait until failover to install new Agents on replacement servers. The polar opposite is to preinstall and configure standalone Agents on standby target servers —when circumstances permit—before any role reversal. (We'll explore all possibilities for standalone Agent DR at the end of the chapter.)

Figure 15-3 depicts the above described architecture for a standby OMR and OMS in relation to the primary site.[12] (The DR architecture for the Agent is not depicted because of its numerous variations.)

---

[12] From the Oracle white paper, "MAA Best Practices Enterprise Manager 10gR2 & 10gR3," January 2007, Figure 6.

Chapter 15: Configure Grid Control for High Availability and Disaster Recovery **639**

The following sections on OMR Database and OMS disaster recovery detail how to set up these standby GC components.

OMR Database DR Recommendations

The approach to disaster recovery of the Repository Database is Data Guard, Oracle's robust solution for both DR and HA protection of any Oracle database. Following is some background on how Data Guard can protect the OMR Database, followed by suggestions for setting up and executing OMR Database DR with Data Guard. I divide this discussion into four parts:

- Overview of Oracle Data Guard
- Create an OMR Database Data Guard configuration
- Configure the Data Guard observer for Fast-Start Failover
- Data Guard role reversal procedures

Overview of Oracle Data Guard

Maintaining a standby database at a geographically distant site in sync with the production Repository Database provides data protection against production disasters, errors, data corruption, and planned or unplanned outages. By switching or failing over to the standby database (and standby OMS and Agents), you can begin monitoring standby targets.

Only physical standby database is supported for the OMR Database; a logical standby database is not supported.[13] The database structure on disk of a physical standby database is physically identical, block-for-block, to that of the primary database. The primary and standby hardware platforms do not need to be identical—RAM and processor speed can be different, and the standby can be RAC or single-instance even if the primary is RAC. However, it's better if the systems are identical so that Repository performance can be the same in a switchover (planned downtime) or failover (unplanned downtime) scenario. The two machines must be from the same vendor, must run the same version of operating system and database software, and must be binary compatible in order to process the transferred redo log data. The worst-case risk window with respect to data loss, when a standby database is in Maximum Performance mode, is on average equal to the data contained in one-half of the current, active redo log. If the primary platform crashes before forwarding the contents of this log, the data contained in it may be lost.

NOTE
If you install a physical standby database for the Repository, you must purchase an Oracle Data Guard license for the standby.

Standby databases work by transporting redo log data from the primary to the standby database, then using the Oracle recovery mechanism to apply this redo data to the standby database, with an optional apply delay. This delay is an MAA best practice because it can prevent a user error or logical corruption from propagating to the standby. An administrator must detect the error or corruption and stop the standby managed recovery process within this lag period.

---

[13] For a description of logical standby databases, see Chapter 2 of Oracle Data Guard Concepts and Administration 10g Release 2 (10.2).

640   Oracle Enterprise Manager 10*g* Grid Control Implementation Guide

You can take a standby out of recovery mode, and open it in read-only mode for reporting or backup operations. You can even open a physical standby in read-write mode for testing or development or to create a clone database. While opening a physical standby in read-write mode invalidates the standby, when finished with your task, you can flash back to revert to a time when the standby was valid, provided you've enabled Flashback Database.

Maintaining an Oracle 10*g* standby database is an easy task, compared to maintaining standby databases in earlier Oracle Server releases. Oracle's 10*g* Data Guard facility assumes the burden of keeping the control, log, and data files consistent between the two databases. Grid Control can use the Data Guard broker to create, administer, and monitor the Data Guard configuration for the OMR Database and any other target databases with standbys. (You can use Grid Control to manage the OMR Database Data Guard configuration, except to perform role reversals, as the Console only works when the OMR Database is running.) The Data Guard broker logically groups primary and standby databases as an integrated unit. The broker ensures that data is transferred and applied to the standby database, and controls the various facets of switchover, failover, and failback processing to ease administration and minimize outages.

Create an OMR Database Data Guard Configuration

To install an OMR standby database in a Data Guard configuration, placing it under control of the Data Guard broker, perform the following steps:

1. Prepare the primary database for Data Guard management according to best practices by performing actions that require a database restart. This allows you to use Grid Control to create the Data Guard configuration (see Step 3 below).

    a. Shut down the GC middle tier with *opmnctl stopall*.

    b. Shut down the OMR Database as a user with SYSDBA privilege.

    c. Set the Data Guard broker to start on instance startup by setting the following initialization parameter:

    ALTER SYSTEM SET DG_BROKER_START=TRUE;

    d. Mount the database and place it in FORCE LOGGING mode:

    ALTER DATABASE MOUNT;

    ALTER DATABASE FORCE LOGGING;

    e. If you're planning to use Fast-Start Failover, establish maximum availability mode by running the following command on the primary database, and then open it:

    ALTER DATABASE SET STANDBY DATABASE TO MAXIMIZE AVAILABILITY;

    ALTER DATABASE OPEN;

    f. Configure OMSs to point to the new standby OMR Database in addition to the primary OMR Database to which they already point. Add load balancing and connect-time failover connect strings[14] for the standby database node(s) to the

---

[14] Oracle Enterprise Manager Advanced Configuration 10*g* Release 4 (10.2.0.4.0), Section 11.3.3.1 incorrectly refers to the connect string with the FAILOVER parameter as Transparent Application Failover (TAF). TAF, however, is already in place by use of the parameter FAILOVER_MODE.

Chapter 15: Configure Grid Control for High Availability and Disaster Recovery **641**

following property in the *$OMS_HOME/sysman/config/emoms.properties* file on each currently active OMS host:

oracle.sysman.eml.mntr.emdRepConnectDescriptor=<*OMR Database connection string*>

For example, assuming the current connect string is that from Chapter 4 in the section "Configure OMS for Failover and Load Balancing," add the new connect string shown in bold for a two-node standby RAC Database on nodes *OMR_node3* and *OMR_node4* (enter the entire property as a single line):

oracle.sysman.eml.mntr.emdRepConnectDescriptor=(DESCRIPTION\=(LOAD_ BALANCE\=on)(FAILOVER\=on)(ADDRESS_LIST\=(ADDRESS\=(PROTOCOL\=TCP) (HOST\=<*OMR_node1*>)(PORT\=1521))(ADDRESS\=(PROTOCOL\=TCP) (HOST\=<*OMR_node2*>)(PORT\=1521))**(LOAD_BALANCE\=on)(FAILOVER\=on) (ADDRESS_LIST\=(ADDRESS\=(PROTOCOL\=TCP)(HOST\=<*OMR_node3*>) (PORT\=1521))(ADDRESS\=(PROTOCOL\=TCP)(HOST\=<*OMR_node4*>) (PORT\=1521))**(CONNECT_DATA\=(SERVICE_NAME\=<*SERVICE_NAMES*>) (FAILOVER_MODE\=(TYPE\=select)(METHOD\=basic)))))

Then restart the GC middle tier:

opmnctl startall.

2. Install the same version of Oracle Database software on the standby as on the primary host, choosing the same installation type. Do not select the option to create a new database; Grid Control will create the standby from a backup of the primary database. You can stop the Oracle Net Configuration Assistant after it kicks off, as GC will copy the Oracle network configuration files from the primary if they don't exist on the standby.

3. Create either a single-instance or RAC standby database at a DR site and bring it under Data Guard broker control, using either the SQL*Plus command-line interface, the Data Guard command-line interface, or the GC Console.

   If using the SQL*Plus command-line interface:

   - For a single-instance primary database, create a single-instance standby according to Chapter 3 of Oracle Data Guard Concepts and Administration 10*g* Release 2 (10.2).
   - For a RAC primary database, see the following white papers, depending on whether you're creating a single-instance or RAC standby:
     - To create a single-instance standby, see "MAA/Data Guard 10*g* Setup Guide— Creating a Single Instance Physical Standby for a RAC Primary."
     - To create a RAC standby, see "MAA/Data Guard 10*g* Setup Guide—Creating a RAC Physical Standby for a RAC Primary."

   If using the Data Guard command-line interface (DGMGRL), see Chapter 7 of Oracle Data Guard Broker 10*g* Release 2 (10.2).

If using the Grid Control UI itself,[15] the configuration is automatically placed under broker control. See the instructions in Section 6.1 and 6.2 of Oracle Data Guard Broker 10g Release 2 (10.2), which are to select Setup and Manage from the Data Guard section on the Maintenance page for the OMR Database (Figure 14-1 in Chapter 14 provides a screen shot of this page) and complete the wizard. Note the following UI limitations and behavior:

- Neither the primary nor the standby database can be RAC. In either of these cases, use the command line (SQL*Plus or DGMGRL) to build the standby.
- If using ASM, first manually install and start the ASM instance on the standby node.
- To follow best practices in keeping the standby filenames and locations the same as the primary filenames and locations, apply patch 5162803 or follow the workaround in Oracle*MetaLink* Note 368693.1.
- After completing the wizard, create standby redo logs on both primary and standby by clicking Verify on the main Data Guard page under Additional Administration and checking the box "Create standby redo logs for the following database(s)."

4. If you did not create the standby database in the Console, add it as an existing standby database to GC and bring the primary and standby database under Data Guard broker control, as explained in Section 6.3 of Oracle Data Guard Broker 10g Release 2 (10.2). Essentially, the procedure is to click Setup and Manage under Data Guard on the Maintenance page for the primary OMR Database, select "Manage an existing standby database with Data Guard broker," and complete the wizard.

5. If you haven't already done so, discover the standby OMR Database in the Console (see "Database Discovery and Configuration" in Chapter 11). Specify SYS as the Monitor Username and SYSDBA as the role. These credentials allow the status of the database to show as UP when in managed recovery mode, where only SYSDBA connections are allowed.[16]

6. Complete any general Console configuration steps for the standby database target as described in Chapters 10, 11, and 13 of this book. When entering preferred credentials, ignore the error "Connection to <db_name> as user dbsnmp failed." This is expected behavior for a standby database in managed recovery mode.

7. As an MAA best practice, change the log apply delay by adding the parameter DELAY=<number_of_minutes> to LOG_ARCHIVE_DEST_2 on the primary database. Also do this on the standby database to prepare for failover/switchover. You can change this parameter dynamically using the ALTER SYSTEM SET LOG_ARCHIVE_DEST_2 command.

8. Configure monitoring for the standby database. See Chapter 6 of Oracle Data Guard Broker 10g Release 2 (10.2) for directions on using the UI to monitor and manage a Data Guard configuration.

---

[15] "MAA Best Practices Enterprise Manager 10gR2 & 10gR3," January 2007, page 13 states that you should use the command line for setting up the standby because you need to bounce the primary database when creating the standby. This is true, but you can use the Console if you first perform the changes requiring the bounce, as indicated in Step 1.

[16] In GC 10.1, the Monitor Username is hard-coded as DBSNMP. The workaround is to grant SYSDBA to DBSNMP.

Configure the Data Guard Observer for Fast-Start Failover

Fast-Start Failover (FSFO) is a Database 10.2 feature that automatically fails over to a standby database upon primary database failure and reconfigures the primary database as a new standby database. With FSFO, Data Guard can restore disaster protection quickly and without user intervention. FSFO relies on the Data Guard *observer* to trigger a failover from primary to standby database in the event of primary database outage. The observer is a separate process incorporated into the DGMGRL client that continuously monitors the health of the primary and standby databases. Ideally, place the observer on an OMS standby host, other standby host, or, least preferably, on the standby OMR host itself. Setting up the observer involves specifying its host and Oracle Database home. You also indicate the number of seconds you want the observer to wait after detecting that the primary database is down before the observer initiates a failover. The FSFO process is automatic and fast, so you must carefully manage your Data Guard environment to avoid accidentally triggering an FSFO. For instance, manually stopping the primary OMR Database for maintenance without first stopping the observer process can cause a failover to occur. For this reason, many GC sites prefer not to enable FSFO, opting for more manual control.

You can enable FSFO in one of two ways:

- In the Grid Control UI. See Section 6.4 of Oracle Data Guard Broker 10*g* Release 2 (10.2).
- In the Data Guard command-line interface (DGMGRL). See Section 7.6 of Oracle Data Guard Broker 10*g* Release 2 (10.2).

Note that FSFO requires using the maximum availability mode (which sets the log transport mode to LGWR SYNC AFFIRM) to guarantee no loss of data if an FSFO occurs. Maximum availability mode carries with it the chance that the standby database may adversely affect the primary if network bandwidth and latency are not adequate between primary and standby sites or if disk I/O rates on the standby server cannot keep up with the primary redo stream.

NOTE
For details on Data Guard network issues, see Data Guard Redo Transport & Network Best Practices, Oracle Database 10g Release 2," Oracle Maximum Availability Architecture White Paper, February 2007. For best practices related to FSFO, see "Fast-Start Failover Best Practices: Oracle Data Guard 10g Release 2," Oracle Maximum Availability Architecture White Paper, May 2007.

Data Guard Role Reversal Procedures

Following is an overview of how to perform a role reversal of the primary and standby OMR Database—that is, a switchover, switchback, failover, failback, or FSFO.

1. Shut down the middle tier on each OMS host with *opmnctl stopall*.
2. Manually perform all database role change operations using one of the following methods:
   - Use standard SQL*Plus commands.
   - Use the Data Guard command-line interface (DGMGRL).

For instructions on both methods, see Chapter 5 of Oracle Data Guard Broker 10*g* Release 2 (10.2), and for DGMGRL, see Sections 7.8 and 7.9 of this manual. Do not use Grid Control to perform role reversals, as they involve the primary database being shut down.

3. Configure the OMS to point to the new primary (former standby) database node. Change the following property in the *$OMS_HOME/sysman/config/emoms.properties* file on each currently active OMS host to point to one of the new primary OMR node name(s):

   oracle.sysman.eml.mntr.emdRepServer=<OMRnode>

4. Start the middle tier on each OMS host with *opmnctl startall*.

NOTE
For best practices on switchover and failover, see "Switchover and Failover Best Practices: Oracle Data Guard 10g Release 2," Oracle Maximum Availability Architecture White Paper, January 2007.

OMS DR Recommendations

Disaster recovery for the OMS involves preparing one or more OMS hosts at a DR site to take over for an active OMS that fails. Following are the OMS disaster recovery steps you can take:

- Install additional OMSs at the standby site
- Add a server load balancer at the standby site
- Replicate the shared receive directory at the standby site
- Set Up the Shared Filesystem Loader at the Standby Site
- Trigger standby OMS startup from standby OMR startup

Install Additional OMSs at the Standby Site

If a disaster occurs, you can't just restore a backup of an OMS home (and chain-installed Agent) to a new host and modify OMS and Agent configuration files to reflect the new host name. Host name dependencies exist in OMS (and chain-installed Agent) configuration files, so you should always do a fresh OMS installation.

Irrespective of your primary OMS configuration—whether it is a single primary OMS or multiple OMSs in an Active/Active or Active/Passive configuration—it is a GC best practice to install at least one OMS at the DR site. (In discussions, I will refer to only one DR OMS, but you may need more, depending on DR scalability requirements.) I call this the "DR OMS" rather than a standby OMS to signify that you can configure it either to remain online or offline when the primary GC site is running, depending on whether the DR OMS meets the minimum network performance requirements for active communications with the primary OMR Database.[17] (As shown in Table 2-1 of Chapter 2, the minimum bandwidth requirement is 1Gbps and maximum latency is 30ms.) I use the words "online" and "offline" to avoid confusion with Active/Active and

---

[17] A DR OMS cannot report to a DR OMR Database. It is not open, but is kept in managed recovery mode to synchronize with the primary OMR Database.

Chapter 15: Configure Grid Control for High Availability and Disaster Recovery 645

Active/Passive OMS configurations. "Online" simply indicates that the OMS remains running, and "offline" means the OMS is kept shut down until needed. While an "online" OMS is equivalent to an "Active" OMS, an "offline" OMS is not passive (i.e., in a CFC Active/Passive configuration). The DR OMS, whether online or offline, is pre-configured to communicate with the former standby (new primary) OMR Database in the event of failover. Let me elaborate a little on these two configurations for the DR OMS:

- **The DR OMS remains online with the primary OMSs**   This is the preferred configuration because you can continuously monitor the DR OMS host and actively leverage its hardware resources. The DR OMS can receive data from local standby targets that run alongside their primary counterparts such as standby databases, thereby avoiding the longer network hop to the primary OMS. During primary operations, the DR OMS points to the primary OMR Database, but also provides disaster recovery simply by virtue of its location. The only requirement is that the DR OMS meet the minimum network performance requirements to the primary OMR.

- **The DR OMS remains offline while the primary OMSs are online**   The only reason to leave the DR OMS offline is if it doesn't meet the above-mentioned OMS/OMR network requirements. For example, the DR OMS host may be located too geographically distant from the primary OMR nodes, or the network connection between them may not be fast enough.

Regardless of which configuration you choose, the OMS and OMR Database at the DR site can share a single tier or run on different nodes (as at the primary site) to suit Grid Control DR performance needs.

The procedure to install either configuration is nearly the same (differences are noted below):

1. Install the DR OMS using the GC Additional OMS Option to point to the Repository running on the primary database (see "Install an Additional OMS" in Chapter 3).

2. Perform post-installation configuration tasks in Chapter 4 for the OMS at the DR site.

3. If using a Shared Filesystem Loader at the primary site, configure it at the standby site as well, even if using only one standby OMS, so that the primary and standby sites are identically configured. This makes the process of activating the DR site as smooth as possible and minimizes Agent management data loss. Follow the directions in the section "Set Up the Shared Filesystem Loader" earlier in the chapter.

4. Shut down the DR OMS, its chain-installed Agent, and AS Control (if started) at the DR site.

    ```
    $OMS_HOME/opmn/bin/opmnctl stopall
    $AGENT_HOME/bin/emctl stop agent
    $OMS_HOME/bin/emctl stop iasconsole
    ```

    For an offline DR OMS, you will not restart these components again unless you need to activate the DR site after a role reversal.

5. If configuring an offline OMS at the DR site, edit the *$OMS_HOME/sysman/config/emoms.properties* file and modify the following property, changing its value to the standby OMR node (or one of them if using a standby RAC OMR Database):

    oracle.sysman.eml.mntr.emdRepServer=<*standby OMR_node*>

6. Change the following additional property in the *emoms.properties* file according to whether the DR OMS will be online or offline when the primary site is running:

   oracle.sysman.eml.mntr.emdRepConnectDescriptor=<*database connect string*>

   ■ If the DR OMS will be online, append the standby OMR Database connect string to that for the primary OMR Database to automatically redirect communications to the standby OMR Database after a role reversal (as shown in step 1f of "Create an OMR Database Data Guard Configuration" above).

   ■ If the DR OMS will be offline, change the existing primary OMR Database connect string to point only to the standby OMR Database. (For example, when both primary and standby OMR Databases are RAC or when both are single-instance, Just replace the primary OMR host name(s) with the standby OMR host name(s), as the primary and standby database names should be the same.)

7. If the DR OMS will be online, restart it, the chain-installed Agent, and AS Control:

   ```
   $OMS_HOME/opmn/bin/opmnctl startall
   $AGENT_HOME/bin/emctl start agent
   $OMS_HOME/bin/emctl start iasconsole
   ```

Add a Server Load Balancer at the Standby Site

To provide a DR environment for a primary SLB, install a second hardware SLB at the standby site, if at all possible. (This can be cost-prohibitive for some companies.) Configure the standby SLB in the same way as the primary, with the same IP address, within the standby DMZ, and using the same virtual pools, even if only one OMS exists at the standby site. (For instructions, see "Use a Server Load Balancer" above.) This allows Agents to retain their configuration in pointing to the same SLB virtual host name.

In addition to a standby SLB, you can also use global rather than local SLBs at both primary and standby data centers. Networking vendors offer such global SLB products, which you assign the same Big IP address to and place outside the firewall. Instead of manually manipulating DNS to point Grid Control traffic to a primary or standby SLB, you can use global SLBs to automatically direct traffic to the standby site without users even knowing of the switch. You can program global SLBs with complex business rules so that they automatically negotiate when to pass control of the virtual IP address from the primary to the standby SLB. One such global SLB product is the F5 BIG-IP Global Traffic Manager (GTM).[18]

Local hardware SLBs are expensive, on the order of $20,000 apiece, and you should buy a pair and cluster them for HA. Global SLBs are considerably more expensive (50 to 100 percent more for some vendors) than local SLBs. IT departments usually place all applications requiring HA or DR, including e-mail, behind whichever type of SLB they use. Their cost demands a cost-benefit analysis, but their benefits extend far beyond load balancing and failover; they also provide SSL acceleration and network and application security, for example.

Replicate the Shared Receive Directory at the Standby Site

As explained in Chapter 14 in the section "Resetting Agents," at the time of OMS host failure, those Agent management files trapped in the OMS local receive directory are lost. (Agents remove their local copy of these files after uploading them to the OMS, and the OMS had failed before it could load these files into the Repository.) In the same way, in the event of entire primary site failure, Agent files located in the shared receive directory at the time can be lost, even if this directory is protected by fault-tolerant storage (as this storage is also at the primary site). This loss of Agent data

---

[18] For details, see http://www.f5.com/products/big-ip/product-modules/global-traffic-manager.html.

Chapter 15: Configure Grid Control for High Availability and Disaster Recovery   **647**

requires resetting Agents after failover, unless you replicate the shared receive directory from primary to DR site. Keeping the primary and standby shared receive directories in sync allows the standby OMS to process all Agent files up to the last transaction.

To replicate the shared receive directory to a standby OMS site, use one of the following methods:

- Preferably, use third-party remote data replication software such as EMC Symmetrix Remote Data Facility (SRDF), Sun Availability Suite, Hitachi TrueCopy, or HP Data Protector. These products use either a host-based or a storage-based replication facility and can be configured for synchronous or asynchronous replication. Synchronous replication guarantees that no Agent files will be lost, but requires sufficient network throughput to the DR site to keep up with the rate of Agent file accrual in the shared receive directory at the primary site. Note that some of these replication products, such as SRDF, work only over a limited geographical distance of 100 miles or so.

- Alternatively, schedule a frequently recurring cron job (perhaps run every minute) that executes an operating system utility such as *rdist* or *rsync* to synchronize files from the primary to the DR shared receive directory.

Set Up the Shared Filesystem Loader at the Standby Site

Irrespective of the number of standby OMS hosts (i.e., even if using only one OMS), if using the Shared Filesystem Loader at the primary site, set it up at the DR site as well (see "Set Up the Shared Filesystem Loader" above). Agents can then upload to the shared receive directory on the DR site after primary site failover. For consistency, use the same shared receive directory name at both the primary and standby sites.

Trigger Standby OMS Startup from Standby OMR Startup

You can manually perform a failover or switchover of your Grid Control environment to the DR site. However, you can also strive to automate the startup of standby GC hardware and processes, which is the aim of two of the DR measures previously described. The first measure, addressed in the section "Configure the Data Guard Observer for Fast-Start Failover," automates the failover of the OMR Database. The second measure, brought up in the section "Add a Server Load Balancer at the Standby Site," can use global SLBs to automatically pass control of Agent and Console traffic from a primary to a standby SLB on failover.

A third measure for mechanizing role reversal between primary and standby sites is to configure a standby GC middle tier—OMS, AS Control, and chain-installed Agent—to automatically start on standby OMR Database startup. (This solution applies only to DR OMSs that remain offline while the primary OMSs are online, as described above in "Install Additional OMSs at the Standby Site.") An Oracle Database trigger can accomplish this for you. The following trigger calls a script, here called */home/oracle/script/start_oms.sh*, which starts the OMS and Agent:[19]

```
CREATE OR REPLACE TRIGGER manage_service after startup on database
DECLARE
role VARCHAR(30);
BEGIN
SELECT DATABASE_ROLE INTO role FROM V$DATABASE;
IF role = 'PRIMARY' THEN
DBMS_SERVICE.START_SERVICE('gcha_oms');
begin
dbms_scheduler.create_job(
job_name=>'oms_start',
job_type=>'executable',
```

---
[19] Source: "MAA Best Practices, Enterprise Manager 10gR2 & 10gR3," an Oracle white paper, January 2007.

```
job_action=>'/home/oracle/script/start_oms.sh',
enabled=>TRUE
);
end;
ELSE
DBMS_SERVICE.STOP_SERVICE('gcha_oms');
END IF;
END;
/
```

The script to start the OMS, AS Control, and Agent, called *start_oms.sh*, is as follows (substitute the hard path for *<OMS_HOME>* and *<AGENT_HOME>*):

#!/bin/sh

<OMS_HOME>/opmn/bin/opmnctl startall

<OMS_HOME>/bin/emctl start iasconsole

<AGENT_HOME>/bin/emctl start agent

Agent DR Recommendations

Unfortunately, the Oracle documentation on standalone Agent disaster recovery is almost nonexistent. Oracle Enterprise Manager Advanced Configuration 10*g* Release 4 (10.2.0.4.0) covers Agent DR in one sentence: "In the event of a true disaster recovery, it is easier to reinstall the Management Agent and allow it to do a clean discovery of all targets running on the new host." This sentence is vague, but clearly recommends installing the Agent from scratch as the only DR solution, despite the wide variety of Grid Control DR implementations. If you're tasked with overseeing Grid Control DR readiness for a site with lots of Agents, you'd probably like post-disaster alternatives to having to install Agents from scratch, then discover and configure targets. You have two more DR options for standalone Agents, which accompany the online and offline DR OMS solutions described in the above-entitled section, "Install Additional OMSs at the Standby Site." Following are the three options. (When I refer below to Agents uploading to a DR OMS, this can be via a global or local standby SLB):

1. **No Agent DR**   Oracle's suggestion may suffice if monitoring can wait after failover until you install Agents and discover/configure targets on the new production (former DR) hosts.

2. **Preinstall a DR Agent against an online DR OMS**   Some DR sites run an Active/Active production OMS configuration with one of the OMSs located at the standby site doubling in a DR capacity. You can preinstall standalone Agents on DR hosts, pointing them to the online DR OMS. Three permutations exist to this solution, depending on the characteristics of the DR host and its targets:

    ■ **Start the DR Agent and discover unique targets**   This approach works for a DR server with a unique hostname and IP address (i.e., that doesn't exist at the primary site) whose DR targets (at least those auto-discovered) are also unique. This scheme is appropriate at a smaller company where a DR host contains both nonproduction databases and standby databases, which are all unique targets.[20]

---

[20] It is a best practice to use the same DB_NAME for a primary database and its standby(s), but each has a different DB_UNIQUE_NAME. This is a database initialization parameter that distinguishes them and allows GC to manage Data Guard configurations consisting of primary and standby databases.

- **Do not start the DR Agent** You can take this tact for a DR host configured identically to a primary host (including host name, IP address, and target configurations). Install the DR Agent with Agent Deploy, agentDownload, or silent methods, which allow you to suppress startup at installation time (see Table 7-2 in Chapter 7 for details).[21] After failover, start GC at the standby site, remove the failed primary targets from GC, then start and discover all DR targets, including the DR host. This is a workable Agent DR policy when primary targets are synchronized at the OS level to their DR targets, and when targets have no database component, such as a web server or host.

- **Start the DR Agent but don't discover targets** This variation is applicable to DR Agents with a unique host name and IP address, but with duplicate targets to their primary counterparts. Install the DR Agent using Agent Deploy, agentDownload, silent, or Agent Cloning methods, which can block target discovery at installation time. (Again, see Table 7-2 in Chapter 7.) Start the DR Agent but do not discover targets until failover. At that time, remove the failed primary targets from GC, and in their place, manually discover the duplicate DR targets. I do not offer any examples because this configuration is not typical. A DR host and its targets are usually an exact replica of a primary host and its targets, or the DR host and targets are all uniquely named.

3. **Preinstall a DR Agent against an offline DR OMS** If you chose to keep the OMS down at the DR site until such time as it needs to assume a primary role, you can still preinstall DR Agents on all DR target hosts. To do so, temporarily start the DR OMS and install DR Agents against it. Use Agent Deploy, agentDownload, or silent methods with the appropriate option to prevent Agent startup at installation (see Table 7-2 in Chapter 7). Then shut down the DR OMS and restart it only if needed for failover. At that time, start DR Agents and discover DR targets to begin monitoring them.

You can evaluate these Agent DR options on a DR host-by-host basis, but this is an administrative bad dream, and with many DR targets, a nightmare or more aptly put, a full-blown migraine (since you'll be awake). It's probably best to adopt the same "least common denominator" strategy across all hosts at your DR site.

Summary

Oracle designed Grid Control to integrate with both Oracle and non-Oracle HA and DR solutions. Grid Control's HA architecture is based on two critical factors: redundancy and component monitoring,[22] which span all GC components:

- For the OMS, redundancy includes installing multiple OMSs, preferably on fault-tolerant storage, front-ending those OMSs with an SLB, and setting up the Shared Filesystem Loader (also on fault-tolerant storage), all to allow multiple OMSs to process Agent management files and Console requests. As for OMS component monitoring, Grid Control provides built-in metrics for self-monitoring, the OPMN mechanism restarts failed OMS processes, and the OMS can restart on system boot.

---

[21] If you tried to start such a DR Agent, it might fail or cause duplicate targets to appear in the Console.
[22] Source: "Configuring Oracle Enterprise Manager for High Availability," an Oracle white paper, April 2004.

- For the OMR Database, you implement redundancy by using RAC or CFC to cluster the database. As for OMR component monitoring, GC provides out-of-box metrics to monitor the database and OMR processes, and you can configure automatic Repository startup on boot. Finally, you can choose from among any remaining database HA best practices to meet SLA objectives for the Repository.

- For the Agent, redundancy in Active/Passive environments is made possible by Agents installed on each CFC node to monitor targets irrespective of which node is active. You can enhance redundancy by installing Agents (or at least the *$AGENT_HOME/sysman/emd/* directory for Agent management files) on redundant storage. You achieve e-mail notification redundancy by specifying multiple SMTP servers. As regards component monitoring the Agent (like the OMS) has a built-in restart mechanism, called the watchdog process. Also, as with the other GC components, built-in metrics perform Agent monitoring and a mechanism exists for automatic startup on bootup of a target server.

Grid Control's DR architecture for the OMR Database relies on standard Oracle Database DR facilities, and for the OMS and Agent capitalizes on the same technologies as used to protect the primary OMS:

- For the OMR Database, Oracle Data Guard is the principal DR solution. You can automate and quicken role reversals between the primary and standby OMR Database by configuring the observer for FSFO.

- For the OMS and chain-installed Agent, it is important for DR purposes to deploy an additional OMS at the standby site. The OMS can remain online or can stand by offline. Either way, the DR OMS will take over and report to a standby OMR Database in case of primary OMS failure. You can add a local or global SLB at the standby site, replicate the primary shared receive directory and set up the Shared Filesystem Loader there, and create an OMR Database trigger to start the standby OMS on database startup.

- For standalone Agents, DR ranges from the nonexistent to the sublime; you can wait to do a fresh installation on a standby host in response to a disaster, However, a more proactive approach is warranted in proportion to the sophistication of your DR site. You can preinstall DR Agents against an offline DR OMS and wait to start these Agents until failover. Better still, you can preinstall Agents against an online DR OMS. Depending on DR host and target uniqueness, you may have to leave a DR Agent down, start it but suppress DR target discovery, or start it and allow target discovery. The approach to Agent DR is as varied as the DR configurations out there.

By bolstering Grid Control with respect to both HA and DR, you can rely on it to monitor your entire data center. However, your particular site may not require realizing all the best practices specified in this chapter. This chapter is intended to serve as a guideline for configuring a fault-tolerant, highly available management solution. You must decide which recommendations are required in order to meet the GC availability and business continuity requirements for your particular environment. Remember that Grid Control's HA requirements must match or surpass those of all your managed targets. As with any other application, you should test those Grid Control HA and DR measures you put in place, both before implementing Grid Control to monitor production targets and throughout its life cycle as a system management product.

CHAPTER 16

Securing Grid Control Data Transfer

Oracle designed Grid Control to integrate into and manage the most security-hardened data centers. Security functionality is built into the product, both at the application and underlying component levels. The Grid Control application extends built-in features to tighten its own security. These features are available both through the Console and in executables (such as *emctl* under the OMS and Agent homes). The underlying components, the Oracle Database 10*g* and Oracle Application Server 10*g* instances (Oracle HTTP Server in particular), also provide their own built-in security services that Grid Control co-opts to protect itself.

Consider Table 16-1 the "big board"[1] on Grid Control security. It lists the primary GC security goals (those in **boldface** are the subject of this chapter), how you achieve them, and where they are documented. References include previous chapters, this chapter, and OEM Advanced Configuration 10*g*, which you can consult for the last two goals listed.

Let's run through the big board and break it down this way:

- **Previous chapters** The first four security goals receive attention in earlier chapters for several reasons—to meet time-sensitive security objectives (e.g., protecting passwords and auditing GC operations from the start) and as part of GC setup (e.g., choosing privileges when creating administrators/roles and tuning security policies for GC targets).

- **This chapter** This chapter addresses a particular aspect of Grid Control security: how to secure data transmitted between GC components. (I also briefly cover how to restrict Console access to certain clients.) Mission critical systems containing sensitive data should implement secure connections between all points of communication. For Grid Control, the communication points are between its components—Repository, OMS, Agents, and Console—and such communication is secured by strong encryption.

- **SSO and EUS** If you use either of these products, evaluate leveraging them in tandem with GC for authentication.

- **Firewalls** Configuring GC with firewalls is an important security matter, but a relatively straightforward process, which receives excellent treatment in Chapter 6 of OEM Advanced Configuration 10*g*. As explained there, before installing GC, open the firewalls between GC hosts to all traffic on GC communication ports. After completing the steps in this current chapter, you can lock down the firewalls.

Let's focus now on the main subject of this chapter: securing data transfer between GC components. The major contributor to preserving the privacy and integrity of GC traffic is *EM Framework Security* ("Framework Security"), so named because it protects the flow of data within the GC framework (i.e., between Agents, OMS, and Repository). Bringing up the rear is the Oracle HTTP Server (OHS) bundled with GC, responsible for protecting browser/OHS or browser/Web Cache exchanges and restricting Console access based on client network characteristics. Let me introduce this dynamic duo by drawing the big picture, which you can refer back to as we delve in. Figure 16-1 shows how Framework Security and OHS together protect all GC communications.

---

[1] A reference to the big board in the War Room in the film *Dr. Strangelove*, directed by Stanley Kubrick.

GC Security Goal	Method to Achieve Goal	Documentation Coverage
Protect sensitive data such as passwords	Secure the emkey	Chapter 4, section "Secure the emkey"
Provide auditing capability for sites requiring it	Turn on the GC audit function	Chapter 4, section "Set Up Grid Control Auditing"
Deny unauthorized users access to the GC Console; grant privileges to authorized users	Enforce administrator access and privileges	Chapter 10, sections "Create Roles" and "Create Administrators"
Comply with company/industry standards for securing all GC components and hosts	Set security policies for GC components and hosts	Chapter 13, section "Change Default Policies"
Secure data transfer between Agents and OMS	**Enable EM Framework Security**	**Chapter 16, section "Enable Framework Security Between Agents and OMS"**
Transmit Repository data in encrypted form	**Use Oracle Advanced Security (OAS)[2]**	**Chapter 16, section "Secure Repository Data Transmissions"**
Safeguard data transferred between browser and OHS (or Web Cache) and restrict Console clients	**Leverage Oracle HTTP Server security**	**Chapter 16, section "Secure Console Connections"**
Integrate GC with other Oracle technologies used to authenticate users	Integrate GC with Single Sign-On (SSO) or Enterprise User Security (EUS)	OEM Advanced Configuration 10g, Sections 5.3 (SSO) and 5.4 (EUS)[3]
Deploy GC when firewalls separate components	Set properties in GC configuration files, leverage proxy servers, and configure firewalls	OEM Advanced Configuration 10g, Chapter 6

TABLE 16-1. *Grid Control Security Goals and How to Accomplish Them*

As you can see, Framework Security and OHS are tightly integrated and complement one another to secure all data channels between GC components. Table 16-2 presents this same information in tabular form for those of you who find tables more palatable than figures.

Both Figure 16-1 and Table 16-2 detail the security method employed to protect data flow, whether the method is considered part of Framework Security or OHS security, and the ports where communications take place. HTTPS signifies that *Secure Sockets Layer (SSL)* is used, which is an industry-standard technology for encrypting data.

---

[2] OAS is an acronym for Oracle Advanced Security, a separately licensed component of Oracle Database Server.
[3] I briefly address this at the end of this chapter. See "A Peek at the Extras in the Oracle EM Advanced Configuration Guide."

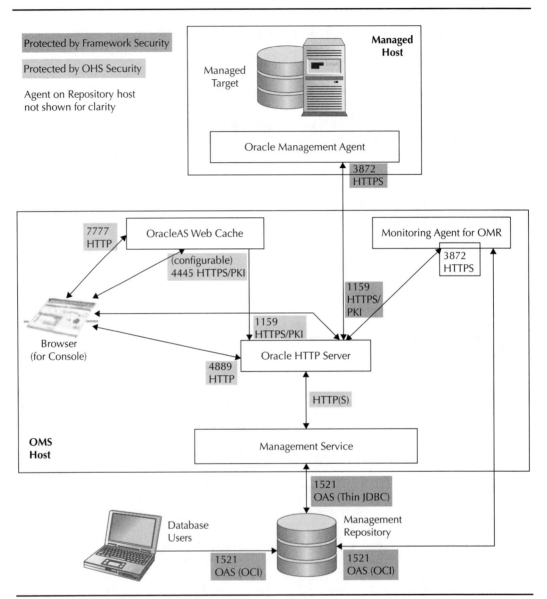

FIGURE 16-1. *Framework Security and OHS together secure GC communications.*

Component 1	Port 1 (Default)	Component 2	Port 2 (Default)	Security Method Used	Framework Security or OHS
Database users	Various	OMR	1521	OAS (OCI[4])	Framework Security
OMS	Various	OMR	1521	OAS (thin JDBC)	Framework Security
Monitoring Agent,[5] Local OMR Agent(s)	3872	OMR	1521	OAS (OCI)	Framework Security
Agents (except those noted above)	3872	OMS	1159	HTTPS and PKI[6]	Framework Security
Browser	Various	OHS	1159	HTTPS and PKI	OHS security
Browser	Various	Web Cache	Configurable	HTTPS and PKI	OHS security

TABLE 16-2. *Methods and Ports Involved in Securing Grid Control Traffic*

What Is Secure Sockets Layer?

Throughout this chapter, I refer to SSL as the encryption method used to secure the data transmitted between GC components. Therefore, let me give you an overview[7] of SSL, for those not completely familiar with it. SSL is an encrypted communications protocol that securely sends messages across the Internet. The mod_ossl plug-in to OHS enables SSL capability. SSL resides between OHS on the application layer and the TCP/IP layer and transparently handles data encryption and decryption between client and server.

A typical application of SSL is to secure HTTP traffic (i.e., to use HTTPS) between a browser and a web server. Indeed, as described later in this chapter, the OHS supplied with Grid Control employs SSL to secure browser/Console OHS connections. However, as you also learn further, configuring SSL does not restrict unsecured (i.e., HTTP) access. It simply provides an additional secure channel of HTTP access over SSL (called HTTPS). (Throughout this chapter, I use the terms "SSL" and "HTTPS" interchangeably.) SSL thereby allows URLs beginning with *https://* without disturbing working URLs starting with *http://*. In addition to the http/https difference in URL specification, you must dedicate different ports for HTTP and HTTPS traffic for a given web application. The default ports used vary by web application. Out-of-box, the GC Console provides both non-secure and secure access via OHS on http://<*OMShost*>:4889 and https://<*OMShost*>:1159, respectively.

---

[4] OCI is the Oracle Call Interface, used for Oracle Net traffic between Database clients and Oracle Database Server.
[5] The Monitoring Agent is the Agent on the first OMS host installed. For background, see "Enable Notification if Grid Control Goes Down" in Chapter 11.
[6] PKI stands for Public Key Infrastructure, which includes use of signed digital certificates.
[7] Source: Oracle HTTP Server Administrator's Guide 10g Release 2 (10.1.2), Chapter 11, Overview.

Enable EM Framework Security

What does Framework Security encompass, as defined by the Oracle EM documentation? A quick scan of Figure 16-1 or Table 16-2 shows that Framework Security safeguards data sent between Agents, OMS, and Repository, as well as between database users and the OMR Database. (OHS protects browser/OHS (and browser/Web Cache) connections.)

Framework Security for Agent/OMS traffic is enabled when you install Grid Control. This piece of Framework Security also affords default HTTPS browser access to the Console via OHS over the same port used for secure Agent uploads. But such access is not considered part of Framework Security, as reflected by Figure 16-1 and Table 16-2. As the EM documentation puts it, "Framework Security works in concert with—but does not replace—the security features you should enable for your Oracle HTTP Server."[8] This classification of what is and isn't part of Framework Security is somewhat arbitrary, in my opinion. To wit, the security model called "Framework Security" secures Agent/OMS traffic and browser/OHS traffic using the same OHS security mechanism (HTTPS and PKI), in the same installation process, and over the same port to boot. Yet, Framework Security is credited with protecting Agent/OMS traffic, and OHS is credited with safeguarding browser/OHS traffic. I don't get it. I'm tempted to recharacterize the whole ball of wax as Framework Security, but never mind. Let's not muddy the waters any more than they already are.

In deference to the EM documentation, I also characterize the Repository security that Oracle Advanced Security (OAS) provides as part of "Framework Security." In actuality, the OAS configuration is no different from a generic OAS implementation between database client and server. The only crossover into Grid Control is that you set up OAS for the OMS and the Agents indicated above using their respective Grid Control configuration files, *emoms.properties*, and *emd.properties*. Again, I'm shaking my head, but I'm not going to fight city hall on this one either.

In the end, it doesn't matter which elements of GC security you draw a box around and call "Framework Security." We'll adopt Oracle's definition of the term. Forgive my ranting. I just wanted to debunk "Framework Security," then in the same hypocritical breath, turn around and say that we'll define it exactly as the Oracle EM documentation does. My rationale for not redefining Framework Security is twofold: to explain exactly what Oracle means by the term so that you can decipher the Oracle EM documentation relating to security, and to avoid confusing you further over such a frivolously logical construct, in my humble view.

NOTE
OMS hosts at different sites use distinct Framework Security. Therefore, in this section, when I refer to the OMS, it signifies that you should perform these steps on each OMS host at you're particular site. If running multiple GC sites, such as for production and non-production, you must enable Framework Security independently at each site.

Enable Framework Security Between Agents and OMS

Enabling Framework Security between Agents and OMS hosts is a three-step, ordered process. The good news is that, whether you know it or not, this part of Framework Security was enabled by default when you installed Grid Control. However, for various reasons, some referenced in

---

[8] OEM Advanced Configuration 10*g* Release 4 (10.2.0.4.0), Section 5.2.1.

this section, you may have partially or totally disabled Agent/OMS Framework Security. So it doesn't hurt to double check that it's completely enabled. Here is a summary of the three steps, including the basic command to accomplish each step (explicit directions for these steps follow):

- **Step 1: Enable Framework Security for OMS hosts** You must secure the first OMS host at installation time by specifying an *Agent Registration password*, then enter the same password when installing any additional OMSs. (This password is stored in the Repository and must be specified when configuring Agents to securely communicate with the OMS.) Securing an OMS modifies its OHS to *enable* a secure Agent/OMS connection, but this process does not restrict unsecure Agent uploads. (To truly secure an OMS host against HTTP Agent uploads, you must lock that OMS host—see step 3.) If you need to manually resecure an OMS host, the command is *$OMS_HOME/bin/emctl secure oms*.

- **Step 2: Enable Framework Security for Agents** Chain-installed Agents are automatically secured at installation time. When deploying standalone Agents against a secure OMS host, the Agent is secured depending on the installation method employed (see Table 5-1 in Chapter 5). While it is recommended to secure Agents, you don't have to do so unless the associated OMS host is locked. The command to secure an Agent is *$AGENT_HOME/bin/emctl secure agent*.

- **Step 3: Lock OMS hosts to prevent Agent HTTP uploads** Locking an OMS host *requires* Agents to communicate with it via HTTPS, prohibiting further HTTP access. In other words, by locking an OMS host, you require that its associated Agents be secure. I recommend locking all OMS hosts at installation time by checking the box "Require Secure Communication for all agents." You can also lock an OMS host at any time by executing the command *$OMS_HOME/bin/emctl secure lock*.

The following three subsections provide all the details on completing these three steps.

Step 1: Enable Framework Security for OMS Hosts

GC Release 10.2 enforces Framework Security at installation time for OMS hosts. This enforcement takes the form of requiring you to enter an Agent Registration password when installing the initial OMS (see Figure 3-8 in Chapter 3), to specify the same password when installing any additional OMSs at a given site. This password is stored in the Repository. After installation, you may need to resecure an OMS host so as to deploy a server load balancer (SLB) to virtualize multiple OMS hosts, or to change the incoming OMS port (default 4888) that Agents use to communicate over HTTPS due to a port conflict.

You can resecure an OMS host with the *$OMS_HOME/bin/emctl secure oms* command, which performs the following actions:

- Generates a *Root Key* in the Repository. A Root Key is the public/private key pair generated by the certificate authority (CA), which is an entity that issues digital certificates for use by other parties. A Root Key is used during distribution of Oracle *wallets* that contain unique digital certificates for Agents. A wallet is a data structure that stores and manages security credentials for an entity.

- Modifies the OHS subcomponent of an OMS host to allow HTTPS traffic from Agents enabled for Framework Security (done in Step 2 below) on default port 1159.

- Permits administrators to access the Console over HTTPS using the same port (default 1159) that Agents use for uploading to the OMS.

Enabling Framework Security for the OMS authorizes Agent uploads and Console logins over HTTPS, but OMS Framework Security does *not* prohibit the following:

- Agent access over HTTP (default port 4889). You must lock the OMS, as explained in Step 3 below, to prevent Agents from uploading to the OMS over HTTP.
- HTTP access to the Console via OHS (default port 4889 for Unix/Windows) or via Web Cache (default port 7777 for Unix and 80 for Windows). See the section "Secure Console Connections" later in the chapter for information on how to secure Console access.

To test whether Framework Security is still enabled for a particular OMS host (as it was at installation time), and enable it if necessary, do the following:

1. Test whether OMS Framework Security is enabled either in a browser session on your client or at the command line on the OMS host:

    Open a browser and enter the following URL:

    https://<OMShost>:<secure_port>

    OMS Framework Security is enabled if your browser displays the Oracle Application Server Welcome page. The default <secure_port> is 1159 for all platforms, but may be different if this port was unavailable during GC installation or if you specified another port using the static ports feature.[9]

    On the OMS host, enter the following command:[10]

    `$OMS_HOME/bin/emctl status oms -secure`

    If Framework Security is enabled, the command will return "OMS is secure on HTTPS Port 1159."

2. If OMS Framework Security is not enabled, continue with the rest of this procedure. Stop the GC middle tier on the OMS host (which includes the OMS, OHS, and other OracleAS components):

    `$OMS_HOME/opmn/bin/opmnctl stopall`

3. Back up the *$OMS_HOME/sysman/config/emoms.properties* file.

4. Secure the OMS host. You can do so with or without prompting, as follows:

    To be prompted for input, enter the following command:

    $OMS_HOME/bin/emctl secure oms [-host <SLB_hostname>] [-secure_port <SLB_secure_port>]

    You are asked for the SYSMAN password (the prompt calls it the "Enterprise Manager Root Password") and the Agent Registration password. The *-host* and *-secure_port* parameters are only required to enable Framework Security for an OMS host configured behind an SLB. In this case, enter the name and port of the SLB, respectively.

---

[9] If you are not sure of the port number, it is the "Enterprise Manager Central Console Secure Port" in the *$OMS_HOME/install/portlist.ini* file. (Note that this file is not updated if the port was changed after installing Grid Control.) You can also confirm the port number by executing *$AGENT_HOME/bin/emctl status agent* for a working Agent that has already been secured. The https port is listed in the value for Repository URL.

[10] The command is incorrectly documented as *emctl secure status oms* in OEM Advanced Configuration 10g, Section 5.2.3.1, Example 5-3.

To include all required and optional input at the command line, which is useful for scripting purposes, enter the following command:

$OMS_HOME/bin/emctl secure oms -sysman_pwd <sysman_password>

-reg_pwd <agent_registration_password>[-host <SLB_hostname>][-reset]

[-secure_port <SLB_secure_port>][-root_dc <root_dc>]

[-root_country <root_country>][-root_state <root_state>][-root_loc <root_loc>]

[-root_org <root_org>][-root_unit <root_unit>][-root_email <root_email>]

The parameters in brackets are all optional. For a description of them, see Example 5-2 in Section 5.2.3 of OEM Advanced Configuration 10g.

5. Restart the entire OMS middle tier:

   ```
   $OMS_HOME/opmn/bin/opmnctl startall
   ```

6. Test that the OMS is now secure and that Agents will be able to connect via HTTPS to the OMS host you just configured by repeating Step 1 of this procedure.

Now that Framework Security is enabled (or you've confirmed it's enabled) for an OMS host, the next step is to enable Framework Security for Agents uploading to that OMS host, if not already enabled.

Step 2: Enable Framework Security for Agents

Enabling Framework Security for an Agent is a fancy way of saying that you're "securing" the Agent. The following actions are performed behind the scenes to secure an Agent:

- The Agent obtains from the OMS an Oracle wallet containing a unique digital certificate[11], which is required for the Agent to communicate via SSL with a secure OMS host (i.e., with an OMS enabled for Framework Security).

- The Agent obtains an *Agent Key* from the OMS, which is also registered with the OMS. (An Agent Key is similar to a Root Key, described in the preceding section, "Step 1: Enable Framework Security for OMS Hosts.")

- The Agent is configured to use HTTPS for all future communication with the OMS.

A chain-installed Agent is automatically secured when installed with its OMS. A standalone Agent can be secured at install time for some installation methods (as listed in Table 5-1), provided Framework Security is already enabled for the OMS. You can also manually secure an Agent at any time using the *emctl secure agent* command. Here are the possible scenarios:

- If Framework Security is enabled for the OMS and the chosen Agent installation method secures the Agent, you must supply the Agent Registration password at install time to enable Framework Security for the Agent. (If the associated OMS host is locked, you must enter the password; if the OMS is not locked, you don't technically have to enter the password, but it's easier to do so rather than having to manually secure the Agent later if the OMS is locked.)

---

[11] The EM secure framework is based on a closed CA model with one CA that signs all certificates deployed in the infrastructure. This model is acceptable for most customers, but some require that OMS/Agent communications use their own trusted CA. This has been logged as Bug 5732607, dubbed an "extremely desirable feature," and is currently in Internal Review status.

- If Framework Security is enabled for an OMS host and the chosen Agent installation method does not automatically secure the Agent, you must manually secure the Agent after installation.

- If an OMS host is not configured with Framework Security, the Agent can't be configured with Framework Security either. You can't have a secure Agent without a secure OMS. In this case, after Agent installation, you must first enable Framework Security for the OMS (step 1 above) and then enable it manually for the Agent.

The basic procedure to secure an Agent, as already described in earlier chapters dealing with Agent installation, involves stopping, securing, and restarting the Agent. Following is a thorough method for securing an Agent (all commands below assume *$AGENT_HOME/bin* is in your PATH, which contains the *emctl* executable used here):

1. If not already known, determine whether Framework Security is already enabled for the Agent, either in the Console or at the command line:

    - In the UI, you can view the upload status (i.e., secure or unsecure) of all Agents at once on the Agents summary page:

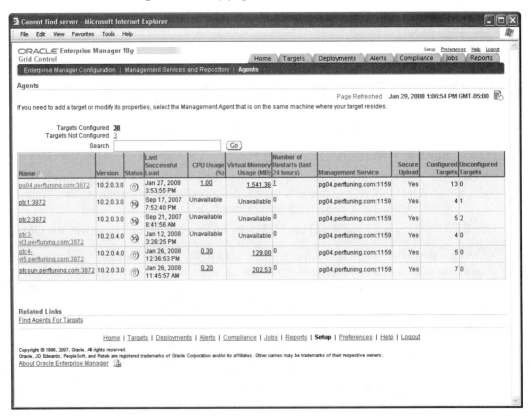

Each Agent home page also has a Secure Upload field showing whether Framework Security has been configured for the Agent.

Chapter 16: Securing Grid Control Data Transfer 661

- Do the following at the command line on the Agent host:
    - Execute the following command to report whether Framework Security is enabled for the Agent:[12]

    ```
    emctl status agent -secure
    ```

    If the Agent is secure, The command will report "Agent is secure at HTTPS port 3872," and will also verify that the OMS is secure.

    - Force an Agent upload to confirm that it's working, and determine the Agent's status (for specifics, see "Confirm Agent Is Working" in Chapter 7):

    ```
    emctl upload agent
    emctl status agent
    ```

    If "Last successful upload" contains the current timestamp and if Agent URL contains "https" rather than "http," then the Agent is already secure and you can skip the rest of this procedure.

2. Ensure that the Agent and associated processes are shut down:
    a. Stop the Agent:

    ```
    emctl stop agent
    ```

    b. Verify that the Agent is not listening on the Agent port, which would render the port unavailable (this may occur if the Agent crashes):

    netstat -an | grep <Agent_port>

    The default unsecure Agent listening port is 4889 and the default secure port is 3872. You can look up the Agent listening port in the $AGENT_HOME/sysman/config/emd.properties file. The port is in the value for the EMD_URL parameter. If the port is in use, the above command will return a line of the following form:

    ```
    tcp        0      0 0.0.0.0:3872            0.0.0.0:*               LISTEN
    ```

    c. If the Agent port is in use, determine which Agent processes are running so that you can kill them:

    ```
    ps -ef | grep emagent | grep -v grep
    ```

    No lines should be returned. If they are, kill these rogue emagent processes. The following is example output showing the three Agent processes that could still be running:

    ```
    oracle    3065  3050  0 Feb23 ?        00:18:20 /opt/oracle/product/10.2.0/em10g/oms10g/bin/emagent
    oracle    3313     1  0 Feb23 ?        00:00:58 /opt/oracle/product/10.2.0/em10g/agent10g/perl/bin/perl /opt/oracle/product/10.2.0/em10g/agent10g/bin/emwd.pl agent /opt/oracle/product/10.2.0/em10g/agent10g/sysman/log/emagent.nohup
    oracle    3333  3313  1 Feb23 ?        00:39:21 /opt/oracle/product/10.2.0/em10g/agent10g/bin/emagent
    ```

    In the above case, you would kill the following three processes:

    ```
    kill -9 3065 3313 3333
    ```

---

[12] The command is incorrectly documented as *emctl secure status agent* in OEM Advanced Configuration 10g, Section 5.2.4, Example 5-5.

3. Back up the Agent configuration file, *$AGENT_HOME/sysman/config/emd.properties*.

4. Secure the Agent, which prompts for and authenticates the Agent Registration password, and reconfigures the Agent to use Framework Security:

   emctl secure agent [-reg_passwd <*agent_registration_password*>]

   If you don't specify the Agent Registration password on the command line, you are prompted for it.

5. Restart the Agent:

   `emctl start agent`

6. Confirm that the Agent is working using HTTPS by repeating Step 1 of this procedure.

Step 3: Lock OMS Hosts to Prevent Agent HTTP Uploads

As previously mentioned, even after enabling Framework Security for an OMS host and for associated Agents, Agents can still communicate with that OMS host via HTTP and HTTPS. OMS Framework Security simply provides the ability for Agents to upload over a separate HTTPS channel. To prevent an Agent from uploading over the HTTP channel, you must lock each OMS host with the *emctl secure lock* command. The OMS host is already locked if, when installing Grid Control or an additional OMS, you checked the box "Require Secure Communication for all agents" as suggested in Chapter 3 (see Figure 3-8, which is the Specify Security Options screen). Once you lock an OMS host, Agents will no longer be able to upload to that OMS host at http://<*OMShost*>:4889/em/upload.

CAUTION
Locking an OMS host restricts HTTP Agent uploads to that OMS host, but does not prevent HTTP Console access to that OMS host. To secure Console connections so that administrators can only log in via HTTPS, see "Secure Console Connections" later in the chapter.

If you did not lock an OMS host when first installing Grid Control or an additional OMS host, you can lock the OMS host at any subsequent time as follows:

1. Determine whether the OMS host is already locked by opening a browser session and attempting to connect to the following HTTP Agent upload URL on the OMS host:

   http://<*OMShost*>:4889/em/upload

   An OMS host is already locked if you get the following error message:

   ```
   Forbidden
   You don't have permission to access /em/upload on this server.
   ```

   An OMS host is not locked if the following is displayed:

   ```
   Http XML File receiver
   Http Receiver Servlet active!
   ```

2. If not locked, stop the OMS middle tier (which includes the OMS, Oracle HTTP Server, and other OracleAS components):

   `$OMS_HOME/opmn/bin/opmnctl stopall`

3. Back up the *$OMS_HOME/sysman/config/emoms.properties* file.

4. Lock the OMS host to block it from unsecured Agent uploads over HTTP:

   `$OMS_HOME/bin/emctl secure lock`

5. Restart the OMS host middle tier in question:

   `$OMS_HOME/opmn/bin/opmnctl startall`

6. Confirm that you can no longer access the HTTP Agent upload URL on the OMS host by repeating Step 1 of this procedure.

7. Also verify that you can now access an OMS host's secure Agent upload URL:

   https://<*OMShost*>:1159/em/upload

   You should receive this confirmation:

   ```
   Http XML File receiver
   Http Receiver Servlet active!
   ```

You can secure Agents against a locked OMS host, either when deploying Agents or afterward. Furthermore, you don't have to unlock an OMS host, secure the Agent, and then relock that OMS host. To unlock an OMS host to re-enable Agent uploads via HTTP, repeat the preceding steps, replacing the *emctl secure lock* command with *emctl secure unlock*.

Secure Repository Data Transmissions

Having dealt with enabling Framework Security between Agents and the OMS, let's turn our attention now to the second half of Framework Security: protecting the transfer of data to and from the Repository Database. As depicted earlier in Figure 16-1, three clients exchange data directly with the Repository Database: the OMS, database users, such as DBAs, and certain Agents (the Monitoring Agent located on the initial OMS installed, and local Agents on OMR Database nodes. (This describes only one Agent for single node GC installations.) Only these Agent(s) directly communicate with the OMR Database using OCI. Like all Agents, they also upload management files to the OMS, but those uploads are protected by HTTPS/PKI Framework Security, as detailed earlier.

To protect Repository data in transit with these clients, you must tap two capabilities of Oracle Advanced Security (OAS): data encryption and data integrity (referred to in short as "data encryption and integrity"). OAS is a comprehensive, industry-standard solution to Oracle security problems that also supplies data privacy, authentication, access authorization, and Single Sign-On. I begin with a general description of the data encryption and data integrity OAS features, and follow with specific instructions on how to set up and test OAS communications between the above-mentioned clients and the OMR Database. Expressed in bulleted form, the structure of this section is as follows:

- Overview of OAS data encryption and integrity
- Installing OAS
- Configuring OAS for all OMR data transmissions
- Testing that OAS secures OMR data transmissions

Overview of OAS Data Encryption and Integrity

This overview of OAS data encryption and integrity is more than just an introduction. It contains conceptual material, but a good portion of it details the OAS parameters and properties to be configured. Such content applies to remaining sections on installing, configuring, and testing OAS, which are more procedural. This front loaded structure, which offers a rationale for the procedures that follow, allows me to simply list the steps in each procedure without explanation.

The OAS conceptual content related to securing the Repository data stream is divided into three bite-size chunks:

- Description of OAS data encryption and integrity
- Applicable OAS data encryption and integrity parameters/properties
- Generating random characters for an encryption seed

Description of OAS Data Encryption and Integrity   To secure Oracle Net traffic between the OMR Database and the three clients mentioned above, you must leverage OAS. As already cited, Grid Control uses two Oracle Net Services components of OAS: *data encryption* and *data integrity*.[13]

The goal of Oracle Net data encryption is to provide data privacy so that unauthorized parties cannot view the plaintext data when transmitted over the network. Secure cryptosystems convert plaintext into unintelligible ciphertext based on a secret key. After the data is sent from client to server, the same key (in the case of symmetric cryptosystems such as employed here for Grid Control data) converts the ciphertext back to its original plaintext. Without the key, it is extremely difficult to decode the ciphertext. OAS supports all the standard data encryption algorithms, including DES, 3DES, RC4, and Advanced Encryption Standard (AES).

The purpose of Oracle Net data integrity in the context of OAS is to ensure the validity of transmitted data. Data integrity protects data from both malicious altering (interception or repeated retransmission) and accidental altering (transmission or hardware errors) using one of two industry-standard hash algorithms, MD5 or SHA-1. These algorithms create a checksum that changes if the data is altered in any way.

Data encryption and integrity processes operate independently, allowing you to enable one, the other, or both. In the case of protecting Repository data, we want to enable both data encryption and integrity for maximum protection. OAS chooses one data encryption algorithm and one data integrity algorithm for each session connection.

Applicable OAS Data Encryption and Integrity Parameters/Properties   The Repository Database uses the Oracle Call Interface (OCI) layer to provide OAS data encryption and integrity at the database server side for Agents and database users. By contrast, each OMS relies on the capabilities of thin JDBC to connect to the OMR Database using OAS. The configuration file in which you set OAS parameters for OCI connections is *$ORACLE_HOME/network/admin/sqlnet.ora*; for thin JDBC connections, the configuration file is *$OMS_HOME/sysman/config/emoms.properties*. To achieve the goal of securing the transfer of OMR data, we must assign the values shown in Table 16-3 to the data encryption and integrity parameters for OCI connections and to the equivalent properties for OMS thin JDBC connections. (Individual parameters and properties must be displayed on separate lines to fit in the table.) Don't do anything yet: we'll accomplish this in the upcoming section "Configuring OAS for All OMR Data Transmissions."

---

[13] See Chapter 4 of Oracle Database Advanced Security Administrator's Guide 10g Release 2 (10.2) for more information. Note that we are not implementing OAS authentication using SSL, but rather are using native Oracle Server password file authentication.

Parameter in sqlnet.ora (for Both Clients and Server)	Property in emoms.properties for OMS Client	Value to Use	Description
(No equivalent)	oracle.sysman.emRep.dbConn.enableEncryption	TRUE	Defines whether encryption is used between OMS and OMR.
SQLNET.CRYPTO_CHECKSUM_CLIENT	oracle.net.crypto_checksum_client	REQUESTED	Defines data integrity requirement. REQUESTED enables checksumming only if server supports it.
SQLNET.CRYPTO_CHECKSUM_TYPES_CLIENT	oracle.net.crypto_checksum_types_client	(Unset)	Defines different types of checksum algorithms supported.
SQLNET.CRYPTO_CHECKSUM_TYPES_SERVER	(Not applicable)	(Unset)	Specifies ordered list of data integrity algorithms for server to use if available on client.
SQLNET.CRYPTO_SEED	(Not available)	<unique user-defined value of 10–68 random characters>	Seeds random number generator used to generate cryptographics on client or server.
SQLNET.ENCRYPTION_CLIENT	oracle.net.encryption_client	REQUESTED	Defines encryption level. REQUESTED enables secure connection only if server supports it.
SQLNET.ENCRYPTION_TYPES_CLIENT	oracle.net.encryption_types_client	(Unset)	Specifies ordered list of encryption algorithms for client to use to negotiate with server.
SQLNET.ENCRYPTION_TYPES_SERVER	(Not applicable)	(Unset)	Specifies ordered list of encryption algorithms for server to use if available on client.

TABLE 16-3. *Data Encryption and Integrity OCI Parameters and Equivalent Thin JDBC Properties*

Note the following about Table 16-3:

- Parameter names all contain CLIENT or SERVER, making it clear whether the parameter is applicable to a client or server, except SQLNET.CRYPTO_SEED, which is valid for both, as indicated.

- Parameters for both data encryption and integrity algorithm types (those with TYPES in the name) are not set. The reason is that, in the absence of specified algorithms, the OMR Database server automatically chooses mutually acceptable algorithms for data encryption and integrity from among those installed on both client and server.

- No particular server or client algorithms are supplied. (If you want to optimize algorithm negotiation speeds, you can dictate that a particular algorithm be used on both server and client. However, this introduces a single point of failure and does not buy you much speed, so is not recommended.) For OCI connections:

    - The encryption algorithms that both database server and client accept by default, in decreasing order of preference, are AES256, RC4_256, AES192, 3DES168, AES128, RC4_128, 3DES112, RC4_56, DES, RC4_40, and DES40.

    - The default data integrity algorithms that database server and client accept, in decreasing order of preference, are SHA1 and MD5.

You must manually edit *emoms.properties*, but you can either manually edit *sqlnet.ora* or use the GUI tool, Oracle Net Manager. The OAS configuration procedures below call for modifying the *sqlnet.ora* file manually. However, some Oracle shops prefer to alter configuration files using administration tools such as Net Manager because these tools are more impervious to user error. See Appendix G (online at www.oraclepressbooks.com) for instructions on using Net Manager to configure OAS data encryption and integrity.

Generating Random Characters for an Encryption Seed   In an upcoming section "Configuring OAS for All OMR Data Transmissions," which covers configuring OAS for certain clients and for the OMR Database, you need to enter an encryption seed between 10 and 68 random characters[14] for the parameter SQLNET.CRYPTO_SEED. For security purposes, encryption seeds should be unique for the OMR Database server and each client. The encryption seed value does not become part of the cryptographic key itself, but rather seeds the random number generator used to generate cryptographic key material. The longer the random string, the more random (and therefore stronger) is the resultant key, so I would suggest using a 68-character random string. To generate ten unique 68-character random strings on Unix, enter the following shell command:[15]

```
cat /dev/urandom | tr -dc "a-zA-Z0-9" | fold -w 68 | head
```

You can then use these unique random strings generated as encryption seeds to configure OAS for each client and for the OMR Database.

---

[14] The Oracle Database Advanced Security Administrator's Guide 10g Release 2 (10.2) says the encryption seed can be up to 70 characters, but my testing with Oracle Net Manager shows that the maximum allowed encryption seed you can enter is 68 characters. This length is amply sufficient to securely seed the Oracle random number generator.

[15] This command is care of the UNIX and Linux Forums at http://www.unix.com/shell-programming-scripting/42663-generate-random-password.html.

Installing OAS

To use OAS data encryption and integrity, OAS must be installed on both database server and client sides. I briefly cover the following points relating to OAS installation:

- When OAS is installed by default
- Checking whether OAS is installed
- Installing OAS in an existing Oracle home

When OAS Is Installed by Default Is OAS installed by default in the software homes for OMR Database and clients? Well, it depends:

- **OMR Database server** OAS is installed in the OMR Database server home, depending on how you installed Grid Control:

  - When selecting the GC New Database Option, OAS is installed in the new OMR Database home that the GC installer creates.
  - When choosing the GC Existing Database Option, OAS may be installed, depending on whether you chose the OAS component when preinstalling the Database intended for the Repository (as recommended in the Chapter 3 section "Install Database Software").

- **Clients** OAS is installed in GC component homes, but may not be for database users:

  - **OMS** OAS is installed in the OMS home as an OracleAS component. Both the initial and additional OMS homes contain OAS.
  - **Agent(s) monitoring and local to the OMR Database** OAS is installed in the Agent home for all Agent installation methods.
  - **Database users** OAS is installed depending on whether you selected it when installing the database "client" software. (Client software can be a database server acting as a client to the OMR Database server.)

NOTE
Oracle Advanced Security is an optional, add-on product bundled with Oracle Net Server and Oracle Net Client that must be licensed separately for both the OMR Database server and its clients—database users, OMS, and Agent(s) monitoring the OMR Database.

Checking Whether OAS Is Installed As just stated, OAS is always installed in OMS and Agent homes. The method to confirm that OAS is installed in OMR Database server or client homes depends on the platform:

- On Unix, enter the following command:

  $ORACLE_HOME/bin/adapters

  The output from this command should include the following lines:

  Installed Oracle Advanced Security options are:
  RC4 40-bit encryption
  RC4 56-bit encryption
  etc.[16]

---

[16] You should see approximately 17 lines of ASO options.

- On Windows, the *adapters* command does not exist. To verify whether OAS is installed, you need to run the database installer and confirm that Oracle Advanced Security is listed as a component. You can run the OUI either from the command line or from the Start menu:

    - From the command line, execute the database server or client OUI as follows:

        `%ORACLE_HOME%\oui\bin\setup.exe`

    - From the Start menu, select All Programs, Oracle - *<Oracle_Home_Name>*, Oracle Installation Products, and then select Universal Installer. Click Installed Products, expand the options, and check whether Oracle Advanced Security (10.x.x.x) is listed.

Installing OAS in an Existing Oracle Home   Again, OAS is always installed in OMS and Agent homes. You can install OAS after the fact in the existing OMR Database server home or database client home only if all the following are true:

- You originally selected the Custom Installation Type.
- You have not since patched the originally installed Database software.
- You have available or can download the original software distribution release.

To install OAS, run the Database OUI from the software distribution, point to the *products.xml* file, reselect the Custom Installation Type, choose OAS from the Available Product Components screen, and complete the installation process. Afterward, confirm that OAS is installed, as covered above.

Configuring OAS for All OMR Data Transmissions

Having ensured that OAS is installed on both client and server sides, you're now ready to configure OAS so that it secures all OMR data transfers with clients. Here are the steps:

1. Configure OAS for the OMS
2. Configure OAS for Monitoring/local OMR Agents
3. Configure OAS for OMR Database users
4. Configure OAS for the OMR Database
5. Test all OAS client connections with the OMR Database

Generally speaking, the only real dependency in the above order is to set up OAS for the OMR Database only after doing so for its three clients. This allows the clients to maintain their existing unsecured connections with the OMR Database until it is enabled for OAS. At that time, all clients will switch to encrypted, checksum-enabled communication when they next establish a connection with the OMR Database. Here's a more detailed look at the order and the settings to use when configuring OAS on both client and database server sides:

- **Client encryption setting**   Configure all clients first to use the default setting of *REQUESTED* for encryption parameters (*oracle.net.encryption_client* for the OMS and *SQLNET.ENCRYPTION_CLIENT* for other clients). An encryption requirement of *REQUESTED* means that clients use encryption if the OMR Database supports secure connections, and otherwise connect unsecured.

- **Server encryption setting** Only after configuring all clients for OAS do you set it up for the OMR Database server. On the server side, specify *SQLNET.ENCRYPTION_SERVER = REQUIRED* in the *sqlnet.ora* file,[17] dictating that client connections must be encrypted. (If a client specified *REJECTED* or did not possess a compatible encryption algorithm, the connection would fail.)

Configure OAS for the OMS Thin JDBC does not offer all the OAS options and encryptions provided by the OCI layer used at the OMR Database server side. However, thin JDBC does offer the necessary properties, listed in Step 3 of the procedure below, to enable the OMS to connect to an OMR Database configured for OAS.

To enable OAS for OMS hosts, complete the following steps on each OMS host in turn:

1. Stop the OMS:

    ```
    $OMS_HOME/bin/emctl stop oms
    ```

2. Back up the OMS configuration file *$OMS_HOME/sysman/config/emoms.properties*.

3. Add the following properties to *emoms.properties*:

    ```
    oracle.sysman.emRep.dbConn.enableEncryption=TRUE
    oracle.net.encryption_client=REQUESTED
    oracle.net.crypto_checksum_client=REQUESTED
    ```

    Contrary to Section 5.2.9.2 of OEM Advanced Configuration 10g, I advise the following:

    - Don't specify *oracle.net.encryption_types_client*, which is supposed to define the different types of encryption algorithms the client supports. Table 5-1 of OEM Advanced Configuration 10g recommends setting this to DES40C, but this appears to be an invalid setting according to permitted values listed in Table 5-2 of Oracle Database Advanced Security Administrator's Guide 10g Release 2 (10.2). My testing shows that the RC4_256 encryption algorithm is chosen for the OMS client regardless of the value assigned to this property.[18]

    - It's not necessary to set *oracle.net.crypto_checksum_types_client=(MD5)*, as shown in Table 5-4 of OEM Advanced Configuration 10g, because MD5 is the only supported data integrity algorithm for thin JDBC. (Setting this parameter does no harm, except to my suggestion above not to set parameters for any data encryption or integrity algorithm types so that the OMS and OMR Database can automatically choose mutually acceptable algorithms.)

4. Restart the OMS to implement the new properties:

    ```
    $OMS_HOME/bin/emctl start oms
    ```

As already mentioned, the above property settings allow the OMS to continue its unsecured communication with the OMR, even before the OMR Database is configured with OAS. When

---

[17] You can also specify REQUESTED for this parameter if you don't want to enforce encryption from the server. This is what the Oracle EM documentation recommends, but I'd rather err on the side of more security than less security.

[18] According to server-side tracing, the encryption algorithms that the OMS thin JDBC client accepts are (in decreasing order of preference) RC4_40, DES, DES40, RC4_256, RC4_56, RC4_128, 3DES112, and 3DES168.

the OMR Database is enabled for OAS, the OMS will switch to a secured, checksum-enabled connection, as will be required by server-side OAS settings.

Configure OAS for Monitoring/Local OMR Agents   Before you enable OAS for the OMR Database, you should enable OAS for the Monitoring Agent and for local Agent(s) on OMR node(s):

- The Monitoring Agent is the Agent on the first OMS host installed, and is responsible for checking the Management Services and Repository logical target (see "Enable Notification if Grid Control Goes Down" in Chapter 11). The Monitoring Agent is listed on the Overview page of the Management Services and Repository tab under Setup.

- You need to configure OAS for the local Agent on each OMR node. For a RAC Repository, an Agent runs on each RAC node. A local Agent must be able to connect via Oracle Net to the OMR Database instance on the host in order to display database instance pages.

Remember that the OAS configuration performed here is for direct communications between these Agents and the OMR Database; OAS does not protect Agent management files uploaded to the OMS; Framework Security protects the transfer of these files. The procedure is simple for setting up OAS for the monitoring Agent and for local Agents on OMR Database servers. Perform this procedure for each of these Agents:

1. Create the file *$AGENT_HOME/network/admin/sqlnet.ora* and add the following entries:

    SQLNET.ENCRYPTION_CLIENT = REQUESTED

    SQLNET.CRYPTO_CHECKSUM_CLIENT = REQUESTED

    SQLNET.CRYPTO_SEED = '<*encryption_seed*>'

    Enter a unique encryption seed. (See "Generating Random Characters for an Encryption Seed" earlier in the chapter.)

2. Restart the Agent to put this *sqlnet.ora* change into effect:

    ```
    $AGENT_HOME/bin/emctl stop agent
    $AGENT_HOME/bin/emctl start agent
    ```

Configure OAS for OMR Database Users   Database users, such as DBAs, who require Oracle Net connectivity with the OMR Database must configure OAS on their computer workstations. They can do so by manually configuring their local *sqlnet.ora* file or by using Oracle Net Manager (see online Appendix G). Below are directions for manual configuration on a database user's machine. To avoid having to reconfigure all database user machines in the unlikely event that you misconfigure OAS the same way on all of them, I suggest configuring your own workstation first, and then, if it works, configuring other workstations the same way. Following is the configuration procedure for a database user:

1. On the workstation, back up the client-side *$ORACLE_HOME/network/admin/sqlnet.ora* file, if it exists.

2. Add the following entries to an existing client-side *sqlnet.ora* file or create a new file if none exists:

   SQLNET.ENCRYPTION_CLIENT = REQUESTED

   SQLNET.CRYPTO_SEED = '<encryption_seed>'

   SQLNET.CRYPTO_CHECKSUM_CLIENT = REQUESTED

   Enter a unique encryption seed. (See "Generating Random Characters for an Encryption Seed" earlier in the chapter.)

3. Add the following trace parameters to the client-side *sqlnet.ora* file, which are used to confirm in the test section below that OAS is working properly:

   TRACE_LEVEL_CLIENT = SUPPORT

   TRACE_DIRECTORY_CLIENT = <Client_Oracle_Home>\network\trace

   TRACE_FILE_CLIENT = cli

   TRACE_UNIQUE_CLIENT = on

   After testing is complete, I'll remind you to comment out these parameters. The example value for TRACE_DIRECTORY_CLIENT is shown in Windows path format, as most OMR Database users connect from PC workstations.

Configure OAS for the OMR Database   Now that all clients are enabled to communicate with the Repository using OAS (except database clients other than yourself, for the moment), we can proceed with configuring OAS on the server side—for the OMR Database itself.

NOTE
If you installed separate OMR Database and Automatic Storage Management (ASM) software homes, I suggest configuring OAS in both homes. Only one of these homes (typically ASM) runs the listener that services the OMR Database and from which OAS runs. But a GC system change, such as with the listener configuration or TNS_ADMIN environment variable, might cause the listener to run out of the other home.

Configure OAS on the server side for each OMR Database node as follows (*ORACLE_HOME* signifies the OMR Database software home):

1. Configure server-side data encryption and integrity. Add the entries below to the *$ORACLE_HOME/network/admin/sqlnet.ora* file on each OMR node:

   SQLNET.ENCRYPTION_SERVER=REQUIRED

   SQLNET.CRYPTO_SEED= '<encryption_seed>'

Enter a unique encryption seed. (See "Generating Random Characters for an Encryption Seed" earlier in the chapter.)

2. Add server-side tracing parameters to the above *sqlnet.ora* file for use in testing OAS for OMS/OMR communications, as described next (it's easier to trace from the OMR Database node than from the OMS because it uses thin JDBC):

TRACE_LEVEL_SERVER = SUPPORT

TRACE_DIRECTORY_SERVER = <ORACLE_HOME>/network/trace

TRACE_FILE_SERVER = srv

The example value for TRACE_DIRECTORY_SERVER is in a Unix path format because most clients run the OMR Database on Unix.

Testing All OAS Client Connections with the OMR Database

Now that data encryption and integrity are enforced on the OMR Database side, it is important to test that all clients can now use these OAS security services when communicating with the OMR Database. Without such testing, you may find that you've locked out all your OMR Database clients...and we all need clients. In this worst-case scenario, the OMR Database, and therefore Grid Control, would play dead. The Oracle EM documentation doesn't provide OAS testing instructions, expecting you, perhaps justifiably, to dig it out of the Oracle Database documentation set. My expectations aren't so lofty. The steps to test OAS for all three client connections with the OMR Database are contained in the following subsections:

- Test OAS for the OMS
- Test OAS for OMR Database Users
- Test OAS for Monitoring/Local OMR Agents

For the latter two client connections, I employ separate denial of service (DoS) tests for data encryption and integrity, which is a kind of proof by transposition.[19] We expect the server to reject connection requests in both cases, thereby confirming that the server is enforcing each security service. A server data encryption setting of REQUIRED means that the client cannot reject encryption or the connection will fail. The DoS test is intended to do just that: set the client encryption parameter to REJECTED and make sure the connection is not allowed. The data integrity DoS test is a little different because there's no server-side parameter to reject data integrity. However, you can set the client data integrity parameter to REJECTED and validate that the connect request is denied.

Let's begin testing, shall we? If you're running a RAC OMR Database, it's best to run the tests on all OMR nodes to be sure each is configured correctly. Finally, please forgive some duplication between tests. I wanted the test procedures to be bullet-proof and stand on their own. When OAS fires, you don't want GC to die.

---

[19] Proof by transposition establishes the conclusion "if p then q" by proving the equivalent contrapositive statement "if not q then not p." Source: http://en.wikipedia.org/wiki/Mathematical_proof.

CAUTION
After completing testing, remove all trace parameters (as I will remind you to do along the way), especially on the OMR Database server side. Trace files can quickly consume gigabytes of filesystem space and bring down your GC system. If trace files completely fill up the OMS filesystem, the emoms.properties file is reduced to a zero-byte file the next time you try to start the OMS. That's why it's also important to back up the emoms.properties file (see "Back Up Critical OMS Files" in Chapter 4).

Test OAS for the OMS   To confirm that the OMS uses data encryption and integrity over its thin JDBC connection with the OMR Database, use the server-side tracing already configured on each OMR node, as follows:

1. On the OMR Database node, identify just one server-side trace file generated in *$ORACLE_HOME/network/trace/* that relates to OMS/OMR traffic by searching for "jdbc" in all trace files:

    ```
    cd $ORACLE_HOME/network/trace
    grep jdbc *.trc
    ```

    The output will contain the trace filename in the following format:

    ```
    srv_2289.trc:[06-FEB-2008 03:58:55:693] nsprecv: 6A 64 62 63 5F 5F 29 28
    |jdbc__)(|
    ```

2. Open the indicated trace file and search for "encryption is active" and "crypto-checksumming is active" to verify that data encryption and integrity are in force (and which algorithms the server chooses). You should find two lines that look like this:

    ```
    [06-FEB-2008 03:58:55:742] na_tns:      encryption is active, using
    RC4_256
    [06-FEB-2008 03:58:55:742] na_tns:      crypto-checksumming is active,
    using MD5
    ```

3. When finished testing, comment out all trace parameters in the *sqlnet.ora* file on the OMR Database node. Also remove all generated trace files.

Test OAS for OMR Database Users   Test that a database user (or "client") configured for OAS can connect using data encryption and integrity. Then, run separate DoS tests, one to prohibit data encryption, and the other to prohibit data integrity. All tests leverage client-side tracing:

1. The client-side trace parameters are already in place to capture OAS details of a database user connection to a Repository Database. Connect via SQL*Plus using the alias for any OMR Database instance you choose (e.g., to the *<OMR_SID>* called *emrep1*), and then disconnect as follows:

    sqlplus system/<password>@<OMR_SID>

    quit

Check that the two lines below appear in the larger of two client-side trace files generated on the database client machine in the location defined in *sqlnet.ora* by TRACE_DIRECTORY_CLIENT. These lines tell you what data encryption and integrity algorithms the server chooses:

```
[06-FEB-2008 04:24:54:536] na_tns:     encryption is active, using
AES256
[06-FEB-2008 04:24:54:536] na_tns:     crypto-checksumming is active,
using SHA1
```

2. Run a DoS test on the database client side to confirm that encryption is enforced. Change the value for the encryption parameter in the client side *sqlnet.ora* file as follows:

   SQLNET.ENCRYPTION_CLIENT = REJECTED

   When you try to establish a new connection via SQL*Plus, as in Step 1 above, from client to OMR Database, you should receive the following error:

   ```
   ERROR:
   ORA-12660: Encryption or crypto-checksumming parameters incompatible
   Exit SQL*Plus, then change this parameter value back to REQUESTED.
   ```

3. Run a DoS test on the database client side to verify that data integrity is enforced. Change the value for the data integrity parameter in the client side *sqlnet.ora* file as follows:

   SQLNET.CRYPTO_CHECKSUM_CLIENT = REJECTED

   You should receive the same ORA-12660 error as in the previous step. Reset the value of the parameters in Steps 2 and 3 to REQUESTED in the *sqlnet.ora* file.

4. Your testing is complete, so comment out all trace parameters in the database client-side *sqlnet.ora* file to disable client-side tracing. You can also remove client-side trace files that were generated.

Test OAS for Monitoring/Local OMR Agents   Database clients can connect via SQL*Plus to the OMR Database. With Agents, a logical way to test is in Grid Control. You can test that the Monitoring Agent and local Agents on OMR nodes are using data encryption and integrity by running separate DoS tests similar to the second and third tests earlier for OMR Database users. Then after temporarily changing the data encryption or integrity parameter value in the client-side *sqlnet.ora* file to REJECTED, you can rely on Grid Control rather than SQL*Plus for DoS proof. The following is the procedure to perform for each Agent:

CAUTION
Don't set a server-wide blackout on Agent hosts that you are testing, because results depend on the Agent remaining up.

1. Notify administrators who are set up to receive notifications on the Agent host that you are expecting "false" e-mail notifications for DoS tests.

2. Run a DoS test on the Agent side to prohibit encryption. That is, test that encryption is enforced on the server side by temporarily changing the value for the encryption parameter in the Agent side *sqlnet.ora* file as follows:

SQLNET.ENCRYPTION_CLIENT = REJECTED

The way to corroborate test results for a Monitoring Agent is different from that for local OMR Agents.

- To validate results for a Monitoring Agent (and your Monitoring Agent notification process at the same time), wait 5 minutes, by which point the administrator(s) listed for emd_email_address in *$AGENT_HOME/sysman/config/emd.properties* should receive the following e-mail alert (for background, see "Enable Notification if Grid Control Goes Down" in Chapter 11):

    Subject: Severe Enterprise Manager problem
    Error message: Could not connect to Enterprise Manager Repository database: 12660

- To confirm test results for a local Agent on an OMR node, log in to Grid Control and try to go to the home page for the OMR Database instance local to the Agent. You should receive the same ORA-12660 error under the Agent Connection to Instance section.

    Reset the value for SQLNET.ENCRYPTION_CLIENT to REQUESTED.

3. Perform a similar DoS test for data integrity by changing the value for the data integrity parameter in the Agent side *sqlnet.ora* file as follows:

    SQLNET.CRYPTO_CHECKSUM_CLIENT = REJECTED

    You should receive the same ORA-12660 error as in the previous step. Reset the value for SQLNET.CRYPTO_CHECKSUM_CLIENT to REQUESTED.

4. Your Agent testing is now completed, so you can comment out all trace parameters in the *sqlnet.ora* file in the Agent home.

Note that Agent uploads from both the Monitoring Agent and local Agents on OMR nodes will continue to work even if OAS is misconfigured (i.e., even if Agent/OMR Database Oracle Net connectivity is lost). This is because Agents upload via HTTP(S) rather than Oracle Net, and these uploads are to the OMS, not to the Repository.

Secure Console Connections

As you've seen, Framework Security secures the Agent/OMS data stream out-of-box and allows you to configure OAS for Repository communication with its clients: the OMS, Monitoring/local Agents, and database users. However, Framework Security does not secure browser/OHS connections for the GC and AS Consoles. Instead, for such communication, Grid Control and AS Control both leverage the security features of the Oracle HTTP Server (OHS) bundled with OAS 10.1.2. Like other web applications deployed with OAS, the OMS and AS web applications rely on OHS to encrypt browser connections.

There are two ways to secure these Console communications: to encrypt *traffic* between client browsers and OHS (or Web Cache) using SSL, and to restrict client *access* to the Console using *host-based access control*. This is a native OHS security feature that limits access based on characteristics of the client request, such as host name, IP address, or domain name.[20] To comply

---

[20] The Oracle EM documentation is conspicuously silent about securing browser/OHS communication, and just references the voluminous Oracle HTTP Server Administrator's Guide for such efforts. In my opinion, the current section provides much-needed direction on this critical leg of intercomponent security.

with security objectives for client access to GC and AS Control, SSL and host-based access control can safeguard or place restrictions on the following traffic channels:

- **Browser/OHS traffic for AS Console** GC administrators log in to the AS Console via OHS over HTTP, but you can configure OracleAS for HTTPS access only so that admins must log in securely.

- **Browser/OHS traffic for GC Console** Out-of-box, admins can access the GC Console via OHS over either HTTP or HTTPS. However, you can restrict HTTP access and force admins to use HTTPS.

- **Browser/Web Cache traffic for GC Console** Post-installation, admins can access the GC Console via Web Cache or OHS. While OHS connections can be over HTTP or HTTPS, Web Cache access is only available over HTTP. If you must enforce security for all GC Console connections, you need to disable HTTP access via Web Cache (and, optionally, spend considerable effort configuring Web Cache with SSL, but only if you really need such access).

- **Browser/GC Console client restrictions** You can restrict GC Console access to clients by using one or more host-based access control schemes.

Your company's security requirements will dictate which measures may be warranted. To encrypt all browser traffic with GC, you must take into account the first three communications channels. Consider the fourth channel only if your company requires additional security to restrict Console access to certain GC clients.

CAUTION
Enforcing SSL for all browser/OHS connections breaks the iSQL*Plus link in Grid Control unless you set up iSQL*Plus to run in secure mode. For instructions, see "Enabling SSL with iSQL*Plus" in the "Configuring SQL*Plus" chapter of the SQL*Plus User's Guide and Reference.

Enforce HTTPS Between the Browser and AS Console

Administrators can normally connect to the Oracle Application Server Control Console via HTTP, but not via HTTPS. (See "AS Console Login" in Chapter 9 for login details.) To secure browser connections to the AS Console so that admins must connect over SSL, do the following:

1. Test that you can access the AS Console via HTTP. The login URL differs by platform:
   - On UNIX: http://<*OMShost*>:1156
   - On Windows: http://<*OMShost*>:18100

2. Stop AS Control:

   `$OMS_HOME/bin/emctl stop iasconsole`

3. Back up the OMS files, listed below, that will be affected by securing AS Control, so that you can restore the configuration if the process fails or if you need to unsecure AS Control later:

   `$OMS_HOME/sysman/j2ee/config/emd-web-site.xml`
   `$OMS_HOME/sysman/config/emd.properties`
   `$OMS_HOME/sysman/emd/targets.xml`

4. Secure AS Control:

   ```
   $OMS_HOME/bin/emctl secure em
   ```

5. Restart AS Control:

   ```
   $OMS_HOME/bin/emctl start iasconsole
   ```

6. Confirm that you can not access the AS Console via HTTP (see Step 1 above).

7. Verify that you can now connect to the AS Console via HTTPS via the platform-specific URL:

   - On UNIX: https://<*OMShost*>:1156
   - On Windows: https://<*OMShost*>:18100

If you need to subsequently unsecure the AS Control Console, stop AS Control and restore the files backed up in Step 3 above. Alternatively, see the procedure in Oracle*MetaLink* Note 280034.1. This note requires manually editing the affected files because there is no *emctl unsecure em* command.

Enforce HTTPS Between Browser and OHS

Framework Security, such as Oracle defines it, does not prohibit HTTP communication between a browser running the GC Console and OHS (or between browser and Web Cache, for that matter). To harden security for GC Console connections via OHS, you must restrict browser access to OHS on default web server HTTP port 4889. If all Agents are inside the firewall,[21] you can block incoming HTTP traffic on this port at the firewall to all outside HTTP Console requests via OHS. However, this does not prevent such Console access from within the firewall or via VPN.

To disallow all HTTP access (both externally and internally) to the Console via Oracle HTTP Server, you must use host-based access control. (For background, see "Limit GC Console Access to Certain Clients" below.) Following are the steps, which you must perform on each OMS host:

1. Confirm that the out-of-box OHS configuration with SSL in OracleAS 10*g* works by logging in to the Console with this URL (1159 is the default port): https://<*OMShost*>:1159/em. If HTTPS Console access were not available for some reason, you wouldn't want to restrict HTTP access as well; doing so would probably lock you out of the Console. If the Console login page does not display, stop here and fix the problem before continuing.

2. Determine whether you can log in to the Console over HTTP via OHS by going to the URL http://<*OMShost*>:4889/em. If the Console login page appears, continue to Step 3 to restrict HTTP access. If you receive an error message such as "Forbidden, You don't have permission to access */em/console/home* on this server," HTTP access via OHS is disabled already, at least to some degree. However, it's prudent to continue this procedure to make certain HTTP access is completely disabled.

3. Confirm that you are not forbidden browser access to the wallet location */em/wallets/emd* by entering the URL http://<*OMShost*>:4889/em/wallets/emd. The emdWalletSrcUrl property in the Agent *emd.properties* file defines this wallet location. When an Agent tries to secure against a secure OMS, the Agent needs to exchange wallet information

---

[21] It is best if Agents are located inside the firewall because they must be able to contact OHS in an nonsecure fashion over HTTP on port 4889 to resecure if needed. See Step 3 for more detail.

with the OMS at this nonsecure URL. (Otherwise it becomes a chicken and egg issue where the Agent can never secure.) You should not get an error like "Forbidden, You don't have permission to access /em/wallets/emd on this server." Instead, you should get the expected, "good" servlet error, which in Internet Explorer will be "500 Internal Server Error, Servlet Error: Wallet Dispatch operation failed." (After all, you're not an Agent and don't know the secret handshake.)

4. Define a <Location> directive with an Allow/Deny rule, placing it anywhere in the <VirtualHost *:4889> block in *$OMS_HOME/sysman/config/httpd_em.conf*. This rule excludes all Console access, but accepts wallet requests from Agents located on a specific subnet and domain (define a separate rule for each Agent subnet and domain):

   <Location /em/wallets/emd*>

   Order deny,allow

   Deny from all

   Allow from <agent_subnet> <agent_domain>

   </Location>

   #

   <Location /em/console/*>

   Order allow,deny

   <agent_subnet> is the agent subnet of the form *a.b.c.**, *a.b.*.**, or *a.*.*.**, the first 3 to 1 bytes of an IP address, respectively. Note that keywords may only be separated by a comma and no whitespace is allowed between them, such as in the *Order deny,allow* directive.

5. Add the same lines as in Step 4 to *$OMS_HOME/sysman/config/httpd_em.conf.template*, because any subsequent *$OMS_HOME/bin/emctl secure* operation would use this template file to regenerate *http_em.conf*, thereby overwriting the manual change.[22]

6. Register this change in *httpd_em.conf* with the Distributed Configuration Management (DCM) repository and bounce OHS to put the change into effect:

   $OMS_HOME/dcm/bin/dcmctl updateconfig -ct ohs -v -d
   $OMS_HOME/opmn/bin/opmnctl restartproc process-type=HTTP_Server

7. Repeat Step 1 above to check that you can still log in to the Console via OHS over HTTPS.

8. Repeat Step 2 above, but check that HTTP access is now forbidden. (If the login page is cached, you may need to clear your browser cache and open a new browser session.) You should receive an error message such as "Forbidden, You don't have permission to access */em/console/home* on this server."

9. Repeat Step 3 above to verify that you are still not forbidden browser access to the virtual directory */em/wallets/emd*.

10. Optionally, as a secondary layer of protection, if all Agents are within the firewall, block incoming HTTP traffic on port 4889 on the firewall so that no outside GC clients could reach the Console via this port, even if the above host-based access control rule were not in place.

---

[22] Source: Oracle*MetaLink* Note 339819.1

Chapter 16: Securing Grid Control Data Transfer **679**

Security Considerations for Console Web Cache Access

As mentioned in Chapter 3 (see the section "Log in to the Web Console"), you can log in to the Console via Web Cache or bypass Web Cache and log in directly via OHS.[23] Administrators can connect out-of-box using OHS over both encrypted and unencrypted channels. To provide encrypted communications, the OHS subcomponent of OracleAS is preconfigured with SSL. However, to secure logins to the Console via Web Cache, you would need to manually configure the Grid Control Web Cache with SSL. As far as I know, there is no documentation on how to do this specifically for the Grid Control Web Cache.[24] The EM documentation set certainly does not cover setting up Web Cache SSL Console access, so de facto it's not standard practice.

If you must encrypt all Console traffic, my first suggestion is to keep Web Cache shut down, lock down OHS access (see "Enforce HTTPS Between Browser and OHS" above), and use the out-of-box OHS SSL access.[25]

In my opinion, the only customers who need to configure Web Cache with SSL are those who must encrypt all Console communications *and* who require the End-User Performance Monitoring feature for the built-in EM Website web application. (The EM documentation states that you can configure this feature with Web Cache or OHS, but my testing shows that, currently, only Web Cache supplies this feature. See "Manage Web Server Data Collection" in Chapter 12 for details.) Given that most GC sites don't need to secure Web Cache access to the Console, I don't supply such a procedure here; feel free to contact my consulting firm for help with this, as the configuration process is nontrivial (to put it mildly).

For those who must secure GC Web Cache access, I will point out that you'll want to set up Web Cache with SSL for all hops between browser, Web Cache, and OHS, rather than just for the first hop. In other words, opt for the first of these two configurations:

Browser <- HTTPS -> **Web Cache** <- HTTPS -> **OHS**

Browser <- HTTPS -> **Web Cache** <- HTTP -> **OHS**

The latter configuration is not a realistic one for GC sites. Outside a GC context, if a Web Cache were in a demilitarized zone (DMZ)[26] and exposed to the Internet, and OHS was behind your firewall, you might want to set up HTTPS for browser/Web Cache (i.e., traffic), and only use HTTP for Web Cache/OHS. However, in Grid Control both OHS and Web Cache are on the same OMS server, which is either in the DMZ or not. If the OMS server were in the DMZ, which

---

[23] In Chapter 3, I suggest logging in via Web Cache because page performance is generally slightly better, but advise leaving OHS access open to bypass Web Cache (in case it crashes or performance degrades significantly, for example).

[24] Opening a Service Request with Oracle Support to configure Web Cache for SSL with GC will likely not get you far, either, because it is not a standard configuration and tends to falls in the gap between the EM and Application Server analysts.

[25] My experience, and that of a good number of Oracle Support analysts who shall remain anonymous, shows that, aside from its use with the EM Website, Web Cache doesn't add much value in the Grid Control world. While Web Cache promises a slight Console performance improvement, it can complicate the diagnosis of GC problems.

[26] A DMZ is a subnetwork that contains and exposes an organization's external services to a larger, untrusted network, usually the Internet. A DMZ adds a layer of security to an organization's LAN; an external attacker only has access to equipment in the DMZ, rather than to the whole network.

is not a sound security practice,[27] you'd want to enforce SSL for both browser/Web Cache and Web Cache/OHS. If the OMS server is not in the DMZ, but is behind a corporate firewall (as is usually the case), then whatever reason you have for insisting on HTTPS for browser/Web Cache also holds for Web Cache/OHS.

If you want additional levels of security to limit Console login by client characteristics, read on. Otherwise, you can skip to the Summary.

Limit GC Console Access to Certain Clients

As I pointed out at the beginning, this chapter's primary focus is how to protect the *transfer* of data between GC components; however, I also alluded to restricting Console access to certain clients. I'd be leaving you in the lurch if I didn't provide a little detail on this Oracle HTTP Server security feature that OEM Advanced Configuration 10g simply waves at.

Your GC site may have security concerns related to Console access that go beyond just authenticating administrators and encrypting the data transmitted between their browsers and the Console. You may want to limit Console login based on characteristics of the client machine. For this you can rely on host-based access control, the native OHS feature mentioned earlier, to authenticate and authorize access based on details of the incoming HTTP request. OHS supplies four host-based access control *schemes* (explanations are tailored to Grid Control):

- **Control access by IP address** This is the preferred method of host-based access control. No DNS lookups are necessary, which cost time and OMS resources, and which open your system to DNS spoofing attacks. An example is to specify a valid IP address range from which administrators can log in, such as 192.168.50.*.

- **Control access by domain name** This scheme is not as robust as IP address-based access control because domain name spoofing is easier to achieve than IP address spoofing. However, you can combine both schemes by authorizing originating requests from an IP address range or particular domain name by specifying "192.168.50.* mycorp.com," for example.

- **Control access by network/netmask** This controls access based on a network/netmask pair. It provides more fine-grained access control than just an IP address range, as you can distinguish which part is the network address and which part is the host address. For instance, the network/netmask pair 192.168.50.0/255.255.255.0 allows access for class C network addresses and host addresses 192.168.50.*, where the last octet can range from 0 to 254.

- **Control access with environment variables** This scheme is not typically used to implement security policies, but provides request handling based on browser characteristics such as type and version. Therefore, it is not very applicable to Grid Control, unless you want to enforce that Console users employ a particular browser or version.

---

[27] It is not recommended to place the OMS in the DMZ. From a general networking standpoint, it is not a good idea to allow a web server (OHS) in the DMZ to communicate directly with an internal database server (the OMR). Instead, an application server (the OMS) within the firewall should act as a medium of communication between the web server and database server. However, this is not possible to do with GC because both web server and application server must be co-located on the same host.

These schemes let you control Console access based on different client machine characteristics. You can use one or more of these schemes to grant or deny GC clients this access. You implement schemes using *directives*, which are configuration instructions for OHS placed in the *$OMS_HOME/Apache/Apache/conf/httpd.conf* file.

Directives have a *scope*; that is, they apply to the entire Oracle HTTP Server or to only part of a server (i.e., at the virtual host level). Scopes that apply to the entire OHS are best suited for controlling GC Console access. That is, you want to configure access control the same way across all virtual hosts, regardless of how someone logs in to the Console—whether via HTTP Server or Web Cache. To accomplish this, add directives to the main *httpd.conf* file and specify HTTPS directives in an <IfDefine SSL> block. There are two *types* of directives:

- **Container directives**   Place access restrictions on particular directories, files, URL locations, or virtual host levels
- **Block directives**   Detail a condition that must be true, such as for a module or definition, for directives contained within to be enforced

The <Location> container directive is very useful for Grid Control purposes. To exclude access to all GC Console pages, you can specify a <Location> directive prohibiting all */em/** URLs, as done in Step 4 of the earlier section "Enforce HTTPS Between Browser and OHS." Additional OHS directives that are of particular interest to Grid Control and allowed in the context of a <Location> directive are *order deny, allow, deny from,* and *allow from* directives.

Let's put this all together now with an example of a host-based access control scheme for a Grid Control site with HTTPS-only Console access (i.e., you've followed the directions in this chapter to enforce SSL for OHS logins and you've shut down Web Cache). Let's say we want to limit Console access to clients with a 192.168.50.* IP address or with the domain name *mycorp.com*. To achieve this, add the following <Location> container directive within an <IfDefine SSL> block in *httpd.conf* to combine IP address–based and domain name–based schemes:

```
<Location /em/*>
order deny,allow
deny from all
allow from 192.168.50.* mycorp.com
</Location>
```

Placing the directive in an <IfDefine SSL> block means that the directive applies to HTTPS connections. The */em/** stipulates that the directive applies to Console access on the port specified in the Virtual Server name or any port, if none are specified. The *order deny, allow* server directive determines the order in which Oracle HTTP Server reads the conditions of the *deny* and *allow* directives that follow. The *deny* directive first denies all requests. The *allow* directive then permits requests originating from any IP address in the 192.168.50.* range or with the domain name *mycorp.com* to access the URL. It is common practice to deny all access in this way, and then open up specific host-based authentication to explicitly define the access policy.

NOTE
For nonproxied requests, you cannot specify an absolute URL for a Location directive such as https://<OMShost>:1159/em. You can only use that portion of the URL that appears after the port (i.e., which you can define as part of the virtual host name).

For more details on OHS access control, including additional examples, see "Host-based Access Control" in Chapter 10 of Oracle HTTP Server Administrator's Guide 10g Release 2 (10.1.2). For an excellent writeup on using directives, see the Apache document at http://download.oracle.com/docs/cd/B14099_19/web.1012/q20201/mod/mod_access.html. For a more theoretical description of directives, see "Configuration Files Syntax," Chapter 2, Oracle HTTP Server Administrator's Guide 10g Release 2 (10.1.2).

A Peek at the Extras in the Oracle EM Advanced Configuration Guide

Enterprise Manager security has many facets, to which Chapter 5 of OEM Advanced Configuration 10g devotes 40 pages. The security material you just read in this chapter (and in earlier chapters) overlaps most of the material in the Oracle guide, and adds considerably more. However, some of the material in OEM Advanced Configuration 10g strays from the intent of this book, which is on implementing, not using Grid Control. The subset of material I speak of in the Oracle guide also veers from the focus of this chapter, which is to secure communications between GC components. For evidence of the auxiliary nature of some of the Oracle material, look no further than a few section titles, notably "Section 5.7: Additional Security Considerations" and "Section 5.8: Other Security Features." These are not the strongest section titles I've ever seen—understandably, even Oracle didn't know how to classify them.

While some of the material in Chapter 5 of OEM Advanced Configuration 10g does not serve the purposes of this book, in all fairness, the features discussed there may interest you, and I'm thankful to the EM documentation team for writing them up. These security features fall into two categories:

- How to use Grid Control with secure targets or users:
    - To configure beacons to monitor HTTPS web applications
    - To patch an Oracle home when the OS owner (typically *oracle*) is locked
    - To patch an Oracle home for which you've set credentials (a new GC 10.2 feature)
    - To use *sudo* to submit a job in Grid Control
- How to use alternative methods for Grid Control authentication

I refer you to OEM Advanced Configuration 10g for complete coverage of the first set of ancillary security matters. However, I would like to touch on the second subject: alternative methods to authenticate in Grid Control. These methods are not a core part of GC security, but are worthy of brief mention here because they involve integrating other Oracle security technologies you may be using into the GC security "paradigm."

As we know, Grid Control has built-in mechanisms for authentication. Administrators access the Console via Oracle HTTP Server or Web Cache and are challenged to enter their administrator credentials. Once logged in, they can drill down and log in to target databases, either manually (such as from a database Performance page) or using preferred credentials already entered and saved. In addition to this traditional GC authentication for Console and target database access, the EM security model offers two respective but mutually exclusive ways to authenticate in Grid Control: Console login via Oracle Application Server Single Sign-On (SSO), and target database login via Enterprise User Security (EUS). Below is a description of each authentication method:

- **OracleAS Single Sign-On** SSO login is an alternative to GC Console login where administrators are instead challenged to enter SSO credentials on the standard SSO login page. You can leverage SSO for administrator *authentication* (including creating and administering users). However, once logged in, Grid Control takes over and manages *authorization* (e.g., granting administrators privileges and roles available in the Console), along with everything else. See Section 5.3 of OEM Advanced Configuration 10g for details.

- **Enterprise User Security** EUS centralizes the administration of users and roles across multiple target databases using Oracle Internet Directory, Oracle Virtual Directory, or Microsoft Active Directory. You configure GC to authenticate database users with EUS rather than using preferred credentials pre-entered in the Console and stored in the Repository. The result is that you are allowed to bypass the login page for managed databases accessed in the Console. See Section 5.4 of OEM Advanced Configuration 10g for more information.

Again, you can set up Grid Control to use SSO or EUS, but not both at the same time—they are mutually exclusive options when used with Grid Control.

Summary

There are many facets to securing Grid Control. You must protect sensitive data, audit GC operations, set up administrators and grant them system and target privileges, secure all GC components and hosts in accordance with industry standards, evaluate integrating GC with existing Oracle authentication products used, and configure GC with firewalls. For all these worthy security endeavors, I provide a roadmap to other chapters in this book and to Oracle EM Advanced Configuration 10g.

One security aspect not in the above list that is key to Grid Control security takes most of this entire chapter to convey: how to safeguard communications between GC components. (A small portion of the chapter, as covered in the previous section, also explores how to restrict client access to the GC Console.) Protecting these data transfers between components involves teaming the security features of EM Framework Security and the Oracle HTTP Server bundled with Grid Control:

- Framework Security safeguards all data exchanged between GC components, namely:
    - Between Agents and OMS hosts using the *emctl* executable in OMS and Agent homes. The procedure is to enable Framework Security for the OMS and for Agents and, finally, to lock the OMS.

- Between the OMR Database server and its clients by using Oracle Advanced Security. Clients for the OMR Database are the OMS, Agents monitoring and local to OMR Database nodes, and database users. The process is to make sure OAS is installed on server and clients, then to configure and test OAS for each client/server connection.

■ Oracle HTTP Server security protects AS and GC Console browser traffic flowing through the Oracle Application Server, whether via OHS or Web Cache. This protection takes the form of:

- Securing browser communication with OHS to render the AS Console.

- Enforcing out-of-box SSL browser access to the GC Console and disabling HTTP access.

- Disabling browser/Web Cache GC Console login and, optionally, configuring Web Cache SSL login.

- Limiting browser/OHS access to the GC Console based on a client machine's characteristics. The most relevant access control schemes for the Console are based on a client's IP address, domain name, network/netmask pair, or a combination of these.

In the next and final chapter, we pull into the pit to learn how to maintain and tune your Grid Control engine, mostly while it's running. Then I'll get out of the passenger's seat, because you'll be qualified for solo driving. I only hope your first solo run is better than mine. At Sebring Raceway, a Carrera GT filled the rear view mirror of my humble 98' Boxster off a long straight, entering the famous hairpin "spin." I panicked, did a 180, and ended up face-to-face with him, remarkably without slamming into him or the wall. I'm sure you'll fare much better.

CHAPTER 17

Maintain and Tune Grid Control

In this last chapter, we examine how to maintain and tune your Grid Control environment. As mentioned in Chapter 1, Grid Control can scale for hundreds of users and thousands of targets, but only if you're diligent about maintenance for the life of the deployment. Such maintenance is necessary regardless of the workload or size of your GC site. Without it, Grid Control, like your car, would start knocking, and at some point would stop working altogether.[1] The cornerstone of software maintenance is periodic housekeeping. In addition, you need to monitor Grid Control, which involves collecting an initial baseline of metric performance, with the help of both built-in and User-Defined Metrics (UDMs). Using these baseline values as a guide, you can adjust thresholds over time, thereby allowing GC to monitor its own performance. Finally, you periodically evaluate metric data for trends, to pinpoint the tuning needed to remove GC application bottlenecks. Such tuning improves GC performance, which lets you accurately forecast GC capacity planning requirements without overstating them.

Stated in summary form, below are the iterative steps covered in this chapter, which form the basis for maintaining and tuning Grid Control:

- Perform routine GC maintenance tasks
- Gather and document GC metrics
- Tune GC to reduce bottlenecks

Let's take on each of these steps in turn.

Perform Routine Grid Control Maintenance Tasks

Grid Control routine housekeeping is perhaps the most overlooked type of DBA maintenance. This is understandable, because administrators tend to focus on maintaining production targets rather than the infrastructure that monitors them. However, without consistent care, GC performance will suffer, which will reduce its ability to monitor production systems.

The tasks to maintain Grid Control fall into two categories:

- Weekly online tasks
- Monthly offline tasks

A good number of the weekly and monthly tasks covered in this chapter are also mentioned in Oracle Enterprise Manager (OEM) Advanced Configuration 10g Release 4 (10.2.0.4.0), Chapter 11. However, the Oracle guide lacks many of this chapter's features, such as explaining how to tap the Metrics Browser to fix target metric collection errors, how to resolve certain OMR Database alert log errors using an Oracle lookup tool, how to identify the patches required for your particular GC version to ensure that OMR Database partition self-maintenance occurs, how to automate the rebuilding of OMR Database tables and indexes, and how to leverage user-defined metrics (UDMs) as an integral part of an overall plan to automate metric collection, evaluation, and tuning.

---

[1] GC operations would likely cease first at the Agent level, where upload problems would disable Agents after management files consumed their maximum allotted disk space (see "How Agent Files Are Uploaded" in Appendix A, online at www.oraclepressbooks.com).

Weekly Online Maintenance Tasks

Following are the tasks that should go on every administrator's calendar with a one-week recurring schedule:

- Clear OMS and OMR system errors
- Fix target metric collection errors
- Start or remove targets in DOWN status
- Clear Critical and Warning alerts
- Address OMR Database alert log errors
- Analyze metric rollup tables (9*i* OMR Database only)

Clear OMS and OMR System Errors

Every week, you should review the Errors page for the Management Services and Repository target. Click Setup, Management Services and Repository, and then the Errors page.

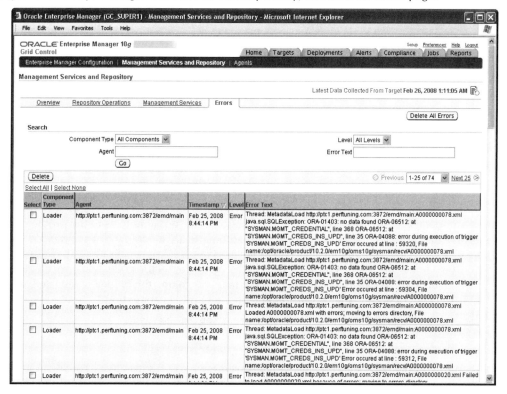

This page lists the errors for each OMS and Repository subsystem (such as the OMS Loader displayed here), with error text gleaned from OMS log files. Some errors may be due to unconfigured or misconfigured targets, such as those shown above for unset target credentials. Other errors may result from bugs. Resolve as many errors as possible by configuring targets correctly or applying applicable patches. Then clear the errors you resolved by selecting them in the error table and clicking Delete. These system errors do not clear by themselves, once resolved.

Fix Target Metric Collection Errors

In addition to clearing OMS and OMR system errors, fix all metric collection errors that accrued during the last week. These errors are associated with target installation, configuration, or status issues. To display these errors, click the Alerts tab, then the Errors subtab:

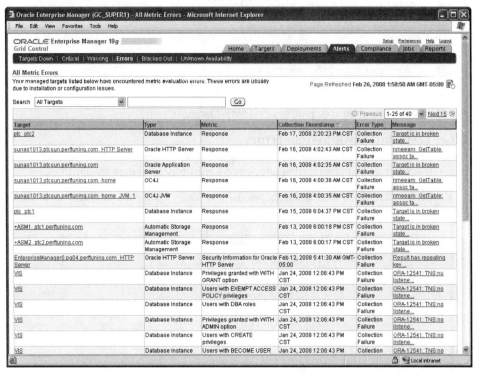

For each error, click the Message column to view the complete error message. Consult Oracle*MetaLink* and other sources, such as Google, for possible fixes to the problem. If you resolve the problem, the target error should automatically clear. Work your way down the error table, fixing as many errors as possible.

You can use the Metrics Browser to help resolve target metric collection errors. I did so for a client a few years back on a GC10gR1 system that was getting the following repeated metric error in *$AGENT_HOME/sysman/log/emagent.trc* on the OMR Database host:

```
[oracle_database,database_service_name,problemSegTbsp]
: nmeegd_GetMetricData failed:
Result has repeating key value : tablespace_name
```

After consulting *OracleMetaLink*, I found that this error was due to Bug 3306591. A workaround for this bug at the time required identifying the offending tablespace name. To do so I enabld the Metrics Browser for the Agent by uncommenting the following line in the *$AGENT_HOME/sysman/config/emd.properties* file on the Agent host (as covered in "Metrics Browser Login" in Chapter 9):

enableMetricBrowser=true

This is a reloadable parameter, meaning that you can reload the Agent rather than having to bounce it in order to put this property change into effect:

```
emctl reload agent
```

Chapter 17: Maintain and Tune Grid Control **689**

I launched the Metrics Browser at the URL: https://<Agent_host>:3872/emd/browser/main.

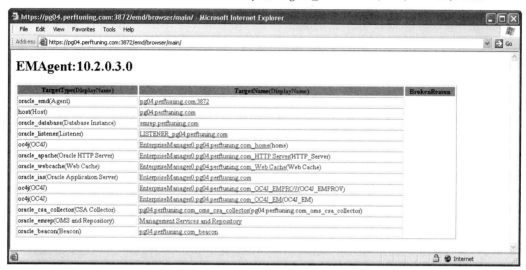

I clicked the Target Name, which in this case was *emrep.perftuning.com* of Target Type *oracle_database(Database Instance)* to pull up the list of database metrics that were being collected:

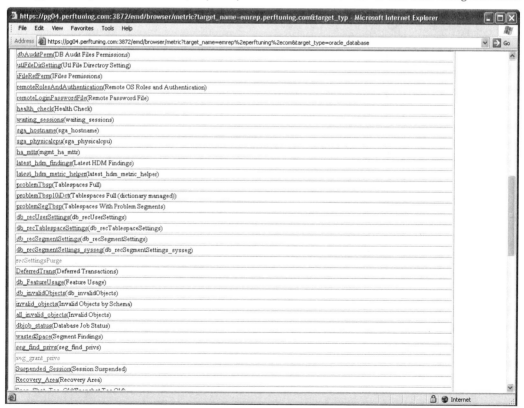

From the available database metrics, I selected the metric reported above in the *emagent.trc* file, *problemSegTbsp(Tablespaces With Problem Segments)*. Doing so on the GC10gR1 system under the conditions of the bug bought up a Metrics Browser page displaying duplicate rows containing the offending tablespace name.

In situations like this, the Metrics Browser is a good tool for resolving target metric collection errors. With it, you can drill down for metric data in real time rather than having to mine the Agent trace file for such data.

Start or Remove Targets in DOWN Status

Targets in DOWN status are shown in the Alerts tab as well as the Targets Down subtab. These targets may actually not be running, or may be running with mismatched status. Every week, review this subtab for any targets displaying DOWN status. Either start these targets, resolve why they are showing as DOWN, or drop them from Grid Control as described in Chapter 11, "Remove Targets from Monitoring." A standard first approach to clear mismatches in target status is to bounce the Agent on the target host.

Clear Critical and Warning Alerts

Also on the Alerts tab, Critical and Warning subtabs display recent target alerts that have occurred (such as the Critical alerts below):

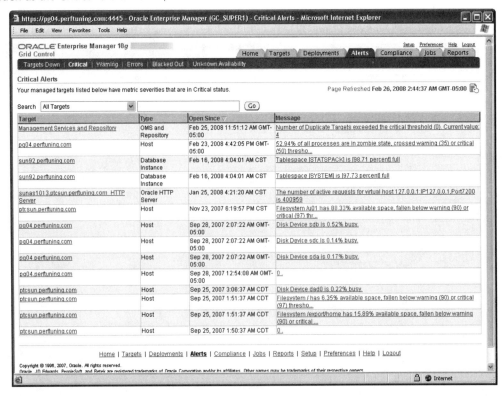

Most alerts that you address are cleared automatically from these subtabs. What remains are alerts that do not automatically clear or that you have not responded to. Some alerts do not clear automatically because Grid Control has no way of verifying whether the underlying problem has been taken care of. For these alerts, a Clear button appears on the alerts page (for either Critical or Warning) along with check boxes next to the alerts that need to be manually cleared. Once you resolve these types of alerts, select them, click Clear, enter an optional message, and click OK to confirm.

For those alerts that you have not yet resolved, some are legitimate alerts that require your attention. (These alerts may have triggered notifications that you haven't responded to yet; others may not have associated notification rules, and you are learning about them in the Console for the first time.) For these alerts, you must remedy the underlying condition. Other alerts are false alerts that fire, sometimes repeatedly, due to inappropriate Critical and Warning metric threshold values. Repeated alerts for the same metric occur when the metric's value cycles across Critical and Warning thresholds within the collection interval, which is typically five minutes. Repeated Warning, Critical, and Clear alert notifications ensue, sometimes for hours on end. These false alert notifications annoy administrators and can easily damage the "monitoring credibility" of a new GC site. This type of rapid-fire alerting wastes OMS resources and can cause a backlog in notification delivery or metric data uploads. When this happens, relax threshold tolerances for the offending metric. Increase the split between Warning and Critical threshold values, use a null value for one or the other severity, raise or lower metric values (as appropriate) for both severities, or use a combination of these methods. Adjusting thresholds is the standard way to keep an alert from constantly firing, and is the appropriate response for such situations.

Address OMR Database Alert Log Errors

Alerts that deserve special mention for a GC system are alert log errors for the Management Repository Database. You should promptly investigate both Warning and Critical errors. For example, the default Warning threshold for the Database Instance metric called Generic Alert Log Error includes serious errors such as ORA-600 (internal errors for Oracle program exceptions) and ORA-7445 (OS exceptions that generate a trace or core file). Below are details you may find useful as a starting point for researching these two alert log errors for the OMR Database (or any target database for that matter). Again, Oracle*MetaLink* is the principal source for examining such errors for patches or workarounds.

ORA-600 Errors   The Oracle RDMBS kernel code raises ORA-600 errors, which indicate some unexpected condition or internal inconsistency. Most DBAs associate these errors with bugs. However, they can also be due to OS or hardware resource problems tracing back to OS upgrades, hardware modifications, database recovery, power failures, and the like. A sample ORA-600 error found in one OMR Database alert log is as follows:

Errors in file /u03/app/oracle/admin/emrep/bdump/emrep1_j004_25620.trc:
ORA-00600: internal error code, arguments: [kjpsod1], [], [], [], [], [], []

Other ORA- errors may follow the ORA-600 error, which you can use in your *MetaLink* search criteria. Every ORA-600 error generates a trace file, whose location is listed in the alert log above the error itself. The trace file contains information about what caused the error condition, and is located either in USER_DUMP_DEST or BACKGROUND_DUMP_DEST, depending on whether the error is caught in a user or background process, respectively. The ORA-600 error itself has one or more arguments listed in square brackets. The first argument ("kjpsod1" in the above example) is the most important, as it pinpoints where the error was trapped in the kernel code. Oracle*MetaLink* Note 175982.1 provides a table on how to interpret ORA-600 error categories, supplying high-level background information for Oracle Server internal error codes.

ORA-7445 Errors ORA-7445 errors are raised by an Oracle Server foreground or background process that receives a fatal signal from the operating system. An ORA-7445 error usually generates a trace file, which is identified in the alert log. If a trace file is not generated, a core dump is created, which contains an image copy of a process's state at the time it aborted. The core dump is located in the CORE_DUMP_DEST directory, in the directory where you issued the offending command, or in the $ORACLE_HOME/dbs/ directory. You can use an OS debugger on the core dump to get a stack trace (see Oracle*MetaLink* Note 1812.1), which you can upload to Oracle Support when opening a Service Request for the ORA-7445 error.

ORA-600/ORA-7445 Error Lookup Tool Oracle*MetaLink* Note 153788.1 contains an ORA-600 and ORA-7445 Error Lookup Tool that parses the particular argument and call stack that the OMR Database is reporting in an attempt to produce relevant hits, such as bugs:

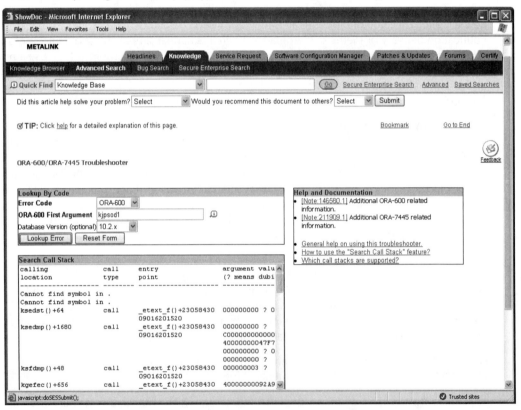

As shown here, choose the Error Code (either ORA-600 or ORA-7445), for ORA-600 errors enter the first argument in brackets, and for either type of error, paste the contents of the stack trace in the Search Call Stack field. The stack trace is found in the trace file starting below the line:

----- Call Stack Trace -----

and ending above the line:

--------------------- Binary Stack Dump ---------------------

As already stated, for ORA-7445 errors that produce no trace file but only dump core, you must run the core dump file through an OS debugger to get a stack trace.

Analyze Metric Rollup Tables

NOTE
This section applies only to Oracle9i Repository Databases. You can skip to the next section if you're running Oracle Database 10g or later.

Grid Control aggregates management data by hour and by day to minimize the size of the Repository Database. Management data is stored initially in a raw data table called MGMT_METRICS_RAW. Grid Control rolls up this raw data into a one-hour metric table called MGMT_METRICS_1HOUR. One-hour records, in turn, aggregate to a one-day table called MGMT_METRICS_1DAY.

One of the myriad reasons for running an Oracle 10g rather than an Oracle 9i Repository Database is that you don't need to analyze these three major metric rollup tables. Oracle 10g databases automatically analyze these tables weekly. However, for an Oracle 9i OMR Database, you must gather table statistics by executing the following commands on a weekly basis:

```
EXEC DBMS_STATS.GATHER_TABLE_STATS('SYSMAN', 'MGMT_METRICS_RAW', NULL, .000001,
    FALSE, 'for all indexed columns', NULL, 'global', TRUE, NULL, NULL, NULL);
EXEC DBMS_STATS.GATHER_TABLE_STATS('SYSMAN', 'MGMT_METRICS_1HOUR', NULL, .000001,
    FALSE, 'for all indexed columns', NULL, 'global', TRUE, NULL, NULL, NULL);
EXEC DBMS_STATS.GATHER_TABLE_STATS('SYSMAN', 'MGMT_METRICS_1DAY', NULL, .000001,
    FALSE, 'for all indexed columns', NULL, 'global', TRUE, NULL, NULL, NULL);
```

As this is an online task, you can schedule it as a recurring SQL Script job in Grid Control.

Monthly Offline Maintenance Tasks

Whereas you can complete weekly tasks while Grid Control is running, you must briefly shut down Grid Control to execute the following monthly tasks, the first of which depends on the release versions of GC and the OMR Database that you're running:

- Perform OMR partition maintenance
- Rebuild tables and indexes as required

Perform OMR Partition Maintenance

NOTE
You can skip the steps described in this section if you're running Grid Control 10.2.0.4 or later in conjunction with Repository Database 10.2.0.4 or later.

Grid Control's default retention policy defines how long all types of management data is retained. These policies are stated in Chapter 4 under "Modify the Data Retention Policy." Management data includes the three metric tables described above in "Analyze Metric Rollup Tables," namely MGMT_METRICS_RAW, MGMT_METRICS_1HOUR, and MGMT_METRICS_1DAY. Grid Control uses partitioning to enforce data retention policies, truncating data in these three tables and reclaiming space used by partitions older than the default retention times.

There's a twist to this seamless self-maintenance, however. You may need to apply patches to fix two bugs, depending on your Grid Control and OMR Database versions:

- Grid Control Bug 5357916, fixed in GC 10.2.0.4, causes the EMD_MAINTENANCE. PARTITION_MAINTENANCE routine that Grid Control internally calls to hang.

- Database Bug 5618049, introduced in Database 10.2.0.2 and fixed in 10.2.0.4, prevents Grid Control from dropping partitions while the OMS is running. The bug invalidates SQL cursors, causing unexplained errors in the Console.

After fixing these two bugs, Grid Control runs monthly partition maintenance as designed, while online, but only for GC 10.2.0.4+ *and* OMR Database 10.2.0.4+ or 10.2.0.1. (The plus sign indicates "or later"; e.g., 10.2.0.4+ signifies 10.2.0.4 or later.) For earlier release combinations of GC and OMR Database, you must run monthly partition maintenance manually with the OMS down, as explained below.

Following is the convoluted procedure to fix the bugs and run partition maintenance manually, if required, based on the releases of Grid Control and OMR Database in use.[2]

1. Do the following with regard to Grid Control Bug 5357916:

    - If running GC 10.2.0.4+, do nothing, as it contains the fix for the bug.

    - If running GC 10.2.0.2 or 10.2.0.3, apply patch 5357916 (which differs for each release) to fix the bug.

    - If running GC 10.2.0.1 or earlier, patch up to the latest GC patchset, which includes the fix for the bug.[3]

2. Do the following with regard to OMR Database Bug 5618049:

    - If running OMR Database 10.2.0.4+, do nothing, as it contains the fix for the bug.

    - If running OMR Database 10.2.0.1 or earlier, Bug 5618049 was not yet introduced, so you do not need to address it.

    - If running OMR Database 10.2.0.2 or 10.2.0.3, do the following:

        a. Apply patch 5618049 to all OMR Database homes.

        b. To activate the fix for this patch, set the following event in the OMR Database SPFILE:

            ```
            ALTER SYSTEM SET EVENT='14532 trace name context forever, level 1'
            SCOPE=SPFILE;
            ```

            If using a PFILE rather than an SPFILE, add the following line to the PFILE:

            ```
            event='14532 trace name context forever, level 1'
            ```

        c. Shut down all OMSs:

            ```
            $OMS_HOME/bin/emctl stop oms
            ```

        d. Bounce (i.e., stop and start the OMR Database) to put this event into effect.

        e. Restart all OMSs:

            ```
            $OMS_HOME/bin/emctl start oms
            ```

---

[2] The procedure is based on Note 456101.1.

[3] If you don't want to apply the latest GC patch set, execute Step 4a under Section B.4 in Note 456101.1.

Finally, if running a RAC OMR Database Release 10.2.0.1 or 10.2.0.2, apply the patch for bug 4151363 or upgrade to DB 10.2.0.3+ (see Note 4151363.8). At this point, if you're running GC 10.2.0.3+ *and* OMR Database 10.2.0.2+, GC begins monthly automatic partition maintenance while online, so you can skip the rest of this section.

However, if you're running GC 10.2.0.2 or lower, or OMR Database 10.2.0.1 or lower, you must manually perform monthly partition maintenance while the OMS is down. Here is the procedure:

1. Shut down all OMSs:

    ```
    $OMS_HOME/bin/emctl stop oms
    ```

2. Connect as SYSMAN to the OMR Database and run the following two procedures.[4] The second procedure may take an hour or more to complete, depending on how much data is being purged.

    ```
    EXEC EMD_MAINTENANCE.ANALYZE_EMD_SCHEMA('SYSMAN');
    EXEC EMD_MAINT_UTIL.PARTITION_MAINTENANCE;
    ```

3. Confirm that partition maintenance was successful using the following query. It should return 0 rows, indicating that no raw data exists that is older than seven days, the default retention period.[5]

    ```
    SELECT COUNT(*) FROM SYSMAN.MGMT_METRICS_RAW
    WHERE COLLECTION_TIMESTAMP < SYSDATE - 9;
    ```

4. Restart all OMSs:

    ```
    $OMS_HOME/bin/emctl start oms
    ```

Rebuild Tables and Indexes as Required

As with any Oracle Database, the Repository Database requires periodic upkeep of its tables and indexes. Every month you should pinpoint those tables and indexes that have undergone high inserts and deletes, causing their allocated size to balloon and actual size to become a fraction thereof. You must then shut down all OMSs and rebuild these tables and indexes offline. Rebuilding an object re-creates its physical structure, reduces its allocated size to "actual" size (based on the number of rows multiplied by average row length), and resets its high water mark (HWM). OMR tables and indexes, by default, are located in tablespaces set to AUTOEXTEND ON, so thereafter they can increase in size as needed.

Rebuilding all OMR Database tables and indexes is overkill and places an unnecessary burden on the Repository Database. I'm sure many DBAs recall defragmenting entire databases by doing a full export with COMPRESS=Y, dropping all objects, then importing them into one extent per segment. The advent of Oracle 9*i* and 10*g* Database self-management features, including locally managed tablespaces, automatic segment space management, and automatic extent management, renders this approach unnecessary. Instead, for the OMR Database, you should rebuild only those tables and indexes that require reorganization.

A number of tables and associated indexes are known to require more frequent rebuilding than others because they experience a high level of inserts and deletes. Such Repository tables tend to grow unevenly and to contain large areas of unused, allocated space due to occasional

---

[4] OEM Advanced Configuration 10*g* Release 4 (10.2.0.4.0) incorrectly advises to execute the procedure EMD_MAINTENANCE.PARTITION_MAINTENANCE.

[5] If you increased the default retention period value for raw data in Chapter 4, use a value greater than 9. For example, use a value that is two days higher than the retention period you're using.

Table Name	Can Reorganize With DBMS_REDEFINITION?
MGMT_CURRENT_METRIC_ERRORS	Yes
MGMT_CURRENT_VIOLATION	Yes
MGMT_ECM_GEN_SNAPSHOT	Yes
MGMT_JOB_HISTORY	No
MGMT_JOB_OUTPUT	Yes
MGMT_JOB_PARAMETER	No
MGMT_METRIC_ERRORS	Yes
MGMT_STRING_METRIC_HISTORY	Yes
MGMT_SYSTEM_ERROR_LOG	No
MGMT_SYSTEM_PERFORMANCE_LOG	No
MGMT_VIOLATION_CONTEXT	Yes
MGMT_VIOLATIONS	Yes

TABLE 17-1. *SYSMAN Tables that Require Frequent Rebuilding*

spikes in data volume. Table 17-1 lists these tables and indicates which of them are candidates for online reorganization using the DBMS_REDEFINITION package.[6] This package allows you to rebuild certain tables online (i.e., while the database is running). Those tables that cannot be rebuilt with DBMS_REDEFINITION must be rebuilt using Oracle export/import utilities. However, as mentioned below in "How to Rebuild Tables and Indexes," if your site can tolerate a little downtime, it's best to rebuild all tables offline, even those eligible for online rebuild.

Next, I provide two ways to identify the tables and indexes that need rebuilding, followed by instructions on how to rebuild these objects.

Manually Identify Tables to Rebuild To manually identify the tables with a lot of allocated space and a relatively small number of rows, execute the query below as SYSMAN. The query looks for tables that are greater than 5MB in allocated size and at least twice as large as their actual size. (These values, which work well for GC, are shown in boldface, but you can adjust them as desired.) Grid Control can remain running while you execute the query.

```
SELECT UT.TABLE_NAME,
  ROUND(UT.NUM_ROWS * UT.AVG_ROW_LEN / 1024 / 1024, 4) "CALCULATED SIZE MB",
  ROUND(US.BYTES / 1024 /1024,2) "ALLOCATED SIZE MB",
  ROUND(US.BYTES / (UT.NUM_ROWS * UT.AVG_ROW_LEN),4)
   "TIMES LARGER" FROM USER_TABLES UT,
USER_SEGMENTS US WHERE (UT.NUM_ROWS > 0 AND UT.AVG_ROW_LEN > 0 AND
US.BYTES > 0) AND UT.PARTITIONED = 'NO' AND
UT.IOT_TYPE IS NULL AND UT.IOT_NAME IS NULL AND
UT.TABLE_NAME = US.SEGMENT_NAME AND
ROUND(US.BYTES / 1024 /1024,2) > **5** AND
ROUND(US.BYTES / 1024 /1024,2) >
  (ROUND(UT.NUM_ROWS * UT.AVG_ROW_LEN / 1024 / 1024, 4)* **2**)
ORDER BY 4 DESC;
```

---

[6] See Oracle Database PL/SQL Packages and Types Reference 10g Release 2 (10.2) for a description of the DBMS_REDEFINITION package. See Note 177407.1 and Chapter 15 of Oracle Database Administrator's Guide 10g Release 2 (10.2) for package features, restrictions, and examples.

Following is sample output from the above query:

```
TABLE_NAME                CALCULATED SIZE MB ALLOCATED SIZE MB TIMES LARGER
------------------------- ------------------ ------------------ ------------
MGMT_JOB_PARAMETER                      .0002                  9    41031.2348
MGMT_JOB_HISTORY                        .0099                  8      809.0864
MGMT_VIOLATION_CONTEXT                  .3756                  6       15.9753
MGMT_ECM_GEN_SNAPSHOT                  1.2211                  6        4.9136
```

Identify Tables to Rebuild with a UDM   To automate the process of identifying tables in need of reorganization, create a SQL UDM to alert you when any tables meet the above criteria (greater than 5MB in allocated size and at least twice as large as their actual size). You may recall from Chapter 13 the example of creating just such a UDM. (See Figure 13-6 in the section "Create a SQL User-Defined Metric.") As mentioned then, you enter the following SQL query on the Create User-Defined Metric page and evaluate it on the first of every month (again, criteria are shown in boldface):

```
SELECT COUNT(*) FROM USER_TABLES UT, USER_SEGMENTS US
WHERE (UT.NUM_ROWS > 0 AND UT.AVG_ROW_LEN > 0 AND
US.BYTES > 0) AND UT.PARTITIONED = 'NO' AND
UT.IOT_TYPE IS NULL AND UT.IOT_NAME IS NULL AND
UT.TABLE_NAME = US.SEGMENT_NAME AND
ROUND(US.BYTES / 1024 /1024,2) > 5 AND
ROUND(US.BYTES / 1024 /1024,2) >
(ROUND(UT.NUM_ROWS * UT.AVG_ROW_LEN / 1024 / 1024, 4)* 2);
```

How to Rebuild Tables and Indexes   Once you identify the offending tables, rebuild them at least monthly using the DBMS_REDEFINITION package and rebuild all indexes on these tables too. To improve the speed of the "reorg" process and to keep things simple,[7] I strongly suggest shutting down Grid Control for this monthly maintenance, at least for the table reorg. Index rebuilds while GC is running are less likely to cause GC operational problems. Following is the procedure for reorganizing a table and its indexes, using the MGMT_VIOLATION_CONTEXT table as an example:

1. Determine whether the table(s) can be reorganized using the DBMS_REDEFINITION package. As SYSMAN, execute the following procedure:

   `EXEC DBMS_REDEFINITION.CAN_REDEF_TABLE('SYSMAN','MGMT_VIOLATION_CONTEXT');`

   - If an error message is returned, you will need to rebuild the table using another method, such as with Data Pump, the Export/Import utilities for 10*g* OMR Databases, or with the original Export/Import utilities for 9*i* Databases. (For an example Data Pump script, see "Set Up OMR Database Exports" in Chapter 14.)

   - If "PL/SQL procedure successfully completed" is returned, you can reorganize the table with DBMS_REDEFINITION.

2. Shut down GC so it can't access the tables during its organization:

   a. Shut down the OMS on all OMS hosts:

      `$OMS_HOME/bin/emctl stop oms`

   b. Stop the Agents on all OMR Database nodes:

      `$AGENT_HOME/bin/emctl stop agent`

---

[7] The table reorg process is complicated when GC concurrently executes DML on the table. For example, you may need to perform intermediate synchronization, abort the process due to problems, or run a procedure to finish the redefinition. See the references in the previous footnote for details.

3. Rebuild the tables using DBMS_REDEFINITION or an Export/Import utility.
   - For tables you can reorg with DBMS_REDEFINITION, log in to the OMR Database as SYSMAN and execute the commands below for each table.[8] These commands create an empty interim table, reorganize the source table using the interim table, and then drop the interim table.[9]

   ```
   CREATE TABLE TEMP_MGMT_VIOLATION_CONTEXT AS
   SELECT * FROM MGMT_VIOLATION_CONTEXT WHERE 1=2;
   EXEC DBMS_REDEFINITION.START_REDEF_TABLE('SYSMAN', 'MGMT_VIOLATION_CONTEXT',
   'TEMP_MGMT_VIOLATION_CONTEXT');
   EXEC DBMS_REDEFINITION.START_REDEF_TABLE('SYSMAN', 'MGMT_VIOLATION_CONTEXT',
   'TEMP_MGMT_VIOLATION_CONTEXT');
   EXEC DBMS_REDEFINITION.FINISH_REDEF_TABLE('SYSMAN', 'MGMT_VIOLATION_CONTEXT',
   'TEMP_MGMT_VIOLATION_CONTEXT');
   DROP MATERIALIZED VIEW TEMP_MGMT_VIOLATION_CONTEXT;
   DROP TABLE TEMP_MGMT_VIOLATION_CONTEXT CASCADE CONSTRAINTS;
   ```

   - For tables you cannot rebuild with DBMS_REDEFINITION, use an Export/Import utility. (For an example Data Pump script, see "Set Up OMR Database Exports" in Chapter 14.)

4. For the tables you reorganized, identify all corresponding indexes so that you can rebuild them to make index range scans more efficient.

   ```
   SELECT INDEX_NAME FROM DBA_INDEXES
   WHERE TABLE_NAME = 'MGMT_VIOLATION_CONTEXT';
   INDEX_NAME

   ------------------------------
   MGMT_VIOLATION_CONTEXT_PK
   ```

5. If rebuilding these indexes offline, do so with the following command syntax (or you can rebuild them online in step 7):

   ```
   ALTER INDEX MGMT_VIOLATION_CONTEXT_PK REBUILD;
   ```

6. Restart Management Servers on all OMS hosts:

   ```
   $OMS_HOME/bin/emctl start oms
   ```

7. Restart Agents on all OMR Database nodes:

   ```
   $AGENT_HOME/bin/emctl start agent
   ```

8. If rebuilding indexes online, do so using the following command syntax:

   ```
   ALTER INDEX MGMT_VIOLATION_CONTEXT_PK REBUILD ONLINE;
   ```

If you're on Oracle Repository Database 10.2 or later, you can also shrink tables using Segment Advisor in Grid Control, as OEM Advanced Configuration 10g points out. (To navigate to Segment Advisor, click Advisor Central under Related Links on a Database home page and choose it from among the many advisors.) I mention this in passing only because advisors such as this are generally useful for managing targets. However, the UI is not suitable for narrowing down which SYSMAN tables to rebuild. Selecting any sizeable portion of SYSMAN tables in the UI would require hours of

---

[8] The equivalent commands in OEM Advanced Configuration 10g Release 4 (10.2.0.4.0), Section 11.2.3.2, contain syntax errors.

[9] If you did not shut down the OMS and OMR Database Agents, the commands shown would also drop the materialized view created to handle any table DML that might occur during reorg.

click-through on page after page. The only viable solution would be to start with a select set of SYSMAN tables, such as those given in Table 17-1, and then add any newly discovered "trouble tables" every month. Also, with Segment Advisor you cannot use your own criteria to select which tables to redefine, as you can with UDMs.

Gather and Document Grid Control Metrics

Your site might now make the cover of "Grid Housekeeping," but don't stop there. You must not only perform routine maintenance, but gather and document Grid Control metrics, then tune accordingly. All three are critical activities in the service of good GC health.

You and other users experience GC performance firsthand, but it's important to quantify that performance with empirical metric data.

What metrics do you gather and how do you evaluate them? It's useful first to identify the types of metrics to call upon. Then, of course, you need to choose which metrics to gather, including built-in metrics and UDMs, and specify their thresholds to receive alerts for out-of-bounds conditions. Finally, you need to develop a procedure for periodically evaluating and addressing metric results. Here, then, are the three tasks to accomplish in this section:

- Types of GC metrics to gather
- Specify GC metrics to gather
- Procedure to evaluate GC metrics

Types of Grid Control Metrics to Gather

Grid Control offers many built-in metrics. You are familiar with regular metrics for which you can set thresholds under the Metric and Policy Settings link, as discussed in Chapter 13. However, you may not be aware that Grid Control has *informational metrics* for which historical data is available on the All Metrics page, but for which you cannot set thresholds. The All Metrics page, available under Related Links on most target pages, contains a comprehensive list of both regular and informational metrics for the target.

You would do well to gather three types of regular and informational metrics for trending purposes:

- **Management Services and Repository metrics** Grid Control has instrumented many self-monitoring metrics, both regular and informational, under the umbrella of this logical target.
- **OMS and OMR host metrics** As for any host target, regular and informational metrics, such as the normal metric, CPU Utilization (%), exist for OMS and OMR hosts. Target pages for these hosts also supply "slow-moving" GC characteristics, such as the amount of RAM.
- **UI page performance metrics** You can capture metrics related to Console page performance by creating service tests and beacons, as described in Chapter 12. UI metrics measure the response time in bringing up certain Console pages by creating service tests and executing them from beacons representing your user community locations.

Management Services and Repository Metrics

Management Services and Repository is a logical target that groups OMS and OMR operational metrics under one roof. These are metrics describing the workings and performance of the GC engine and do not include host metrics where the OMS and OMR reside. Figure 17-1 shows the All Metrics page for this target.

FIGURE 17-1. *The All Metrics page for the Management Services and Repository target*

Chapter 17: Maintain and Tune Grid Control **701**

You cannot set thresholds for more than half of the metrics (i.e., the informational metrics) listed here—those marked "Not Set" in the Thresholds column.[10] However, historical metric values are available for all listed metrics on this page—both regular and informational—by drilling down, as shown here when clicking the Steps Per Second informational metric in the Name column:

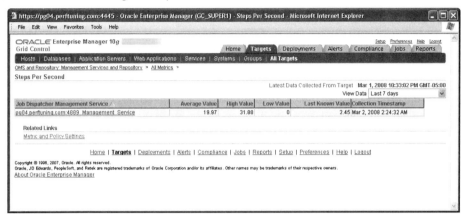

To display an average metric value for the last week, select "Last 7 days" in the View Data field (as shown above). By clicking the link pg04.perftuning.com:4889_Management_Service, a graph of this metric's history over a chosen period is displayed:

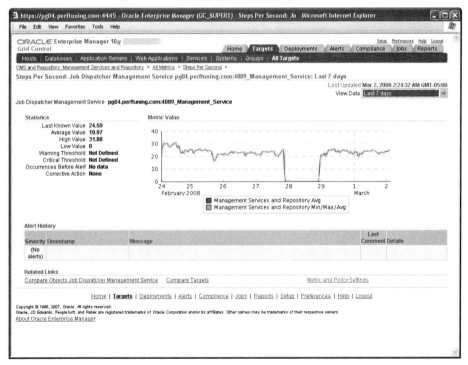

---

[10] Regular metrics with default thresholds would also display as "Not Set" if you were to remove their thresholds. This is not recommended for the Management Services and Repository target.

To get a longer-term view of metric values, choose "Last 31 days," or step back even further by selecting Custom, which allows you to view data as far back as the retention policy provides. This is how we will be recording weekly and monthly metric trends, as explained below in "Specific Grid Control Metrics to Gather."

OMS and OMR Host Metrics

Average weekly values for key metrics on OMS and Repository hosts are good indicators of GC resource consumption. Most host metrics of interest are regular metrics such as Memory Utilization (%) that appear on the Metric and Policy Settings page and allow for thresholds to be set. You will also find configuration and resource-related properties on OMS/OMR target pages, such as the number of OMS/OMR nodes, number and speed of CPUs, amount of RAM, number, speed and configuration of network adaptors, and RAC Repository interconnect information. You should keep track of such properties over the long term (with the help of UDMs, discussed below in "Specific Grid Control Metrics to Gather"), because they can play a large role in GC performance. It is easy to overlook the tracking of host configuration changes alongside other metrics. Without such records whose datelines can be compared against other metric values, you may be left with unexplained performance changes and the inability to accurately forecast the effect of proposed changes. Say you increase the database buffer cache on an OMR host, which causes a memory bottleneck. (The buffer cache is an SGA memory structure that holds copies of data blocks read from data files.) Months later you add RAM to this host, thereby improving performance. A well-documented timeline of these two changes may reveal that a particular ratio of buffer cache to RAM optimized memory usage. Such documentation, which GC can contribute to via drilldown into UDM history, allows a retroactive evaluation of the effect of all GC host changes.

UI Page Performance Metrics

Creating service tests and the beacons to execute them from various locations quantifies and standardizes how quickly administrators can access chosen Console pages. GC service tests rely on the built-in GC web application target, the EM Website. Without service tests, administrators must manually time Console access response, which is labor intensive and inaccurate. Service tests automate the periodic collection of metrics for a sequence of page hits. You can create multiple service tests to build a representative sample of Console URL responses, set thresholds for desired UI metrics, and review historical values.

As an illustration of the power of service tests, consider the built-in service test for homepage response. Let's look at an historical graph of the Perceived Total Time (ms) taken to display the EM Website home page in the past 31 days. You can view this graph by navigating to Targets: Web Applications and choosing EM Website. Click the Monitoring Configuration page, and then click the Service Tests and Beacons link. Click the homepage Service Test, go to the Performance page, and select "Last 31 days" in the View Data field. The Perceived Total Time (ms) metric is graphed as shown in the following image.

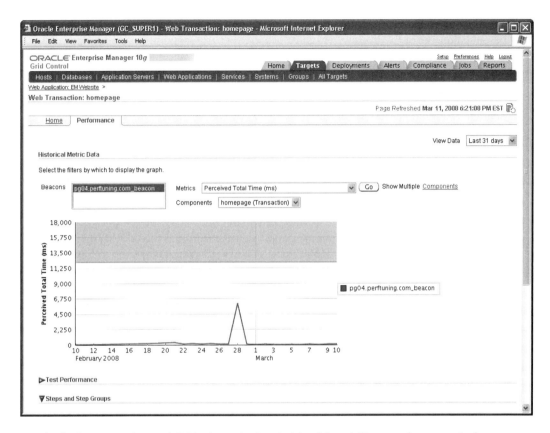

The built-in Warning and Critical metric thresholds of 6 and 12 seconds, respectively, are indicated with horizontal yellow and pink lines. In this example, the spike in homepage login time on February 28 might warrant investigating what changed on that day to cause such a performance slowdown. In this way, you may be able to prevent recurring Console performance problems.

Adding beacons at various other network locations to execute service tests can help pinpoint any WAN latency issues impacting UI response. Take the homepage service test again as an example. If you create beacons at different geographical locations for this service test along with individual metric thresholds for each location, you can compare homepage performance across network locations for a particular timeframe. You might find that response time is not degraded across the board on a particular day, but is isolated to a certain network subnet, as indicated by an alert sent from just one beacon location. You could then ask your network administrator to investigate performance on that subnet. See "Service Tests and Beacons" in Chapter 12 for instructions on how to use this powerful web application functionality to monitor Grid Control UI pages and other web applications as well. In the next section, I recommend five service tests and UDMs to monitor them.

Specific Grid Control Metrics to Gather

Now that you know the types of metrics that relate to GC health, let's get more specific and identify exactly what metrics you need to track to establish a baseline of what is considered acceptable performance at your site. Baselines are a fundamental precept of tuning. Here, a baseline consists of GC average metric values for the past week used as a basis for comparison with future values as your site ages. Such trending often reveals emerging bottlenecks that you can proactively address before they get a chance to take hold and choke GC performance.

Grid Control comes with many metrics to monitor its own health. To supplement these GC-supplied metrics, you can also create UDMs (covered in Chapter 13) as directed below to query the Repository directly for metric data. UDMs can fill gaps in the built-in GC self-monitoring in at least two ways:

First, UDMs can duplicate (or mirror) informational metrics, with the advantage of allowing you to set appropriate thresholds. As such, Grid Control comes closer to full self-monitoring, in that it alerts you about metric boundary conditions, without needing to manually check informational metrics on a weekly basis. One such UDM, referenced below in Table 17-2, is the equivalent of the informational metric *DBMS Job Status: Throughput Per Second: Rollup*, indicating the number of rows per second of raw data processed into the OMR rollup table MGMT_METRICS_RAW. This is one of several UDMs that mines the OMR Database for valuable GC vital signs, and which, in particular, may be key in detecting emerging rollup performance problems, as discussed later in the section "Minimize Rollup Delays."

Second, you can define UDMs to supplement the built-in regular and informational metrics offered. For instance, as suggested in Table 17-2 below, a UDM, called Alerts Per Hour calculates the number of metric severities generated per hour, which is a metric not offered in the Console, and an excellent way to track whether metric thresholds remain well tuned as your site ages. By defining these UDMs, you tap into Grid Control's ability to record and graph historical changes in metric values.

Table 17-2 contains a list of Grid Control metrics to collect, which cut across all metric types described above.[11] Following is a description of each column:

- **Metric Type** Groups metric by type as defined above in the section "Types of Grid Control Metrics to Gather."

- **Metric Subtype** Provides further detail of metric type. The metric subtypes for the Management Services and Repository metric type are found on the All Metrics pages for this target.

- **Description** Lists the metrics themselves, with regular and informational metrics as found in the UI identified in *italics*.

- **UI Navigation for Historical Data** Provides the UI navigation for these metrics, which is predominately on the pages (including the All Metrics pages) for the Management Services and Repository or OMS/OMR host targets.

- **Create UDM or Service Test?** Suggests whether you should create a UDM or service test for the particular metric. (UDMs are defined in Appendix H, online at www.oraclepressbooks.com.) Their names are composed of "UDM" followed by the moniker given in the Description column, such as "UDM Number of Targets".

- **Recommended Thresholds—Warning/Critical** Lists metric thresholds to set, some of which are nondefault values as recommended in Chapter 13, Table 13-1.

---

[11] Many of the metrics for the Management Services and Repository target are identified in a table in Section 11.2.2 of OEM Advanced Configuration 10g Release 4 (10.2.0.4.0), but this table does not provide the metric names, UI navigation, or UDMs offered here.

Chapter 17: Maintain and Tune Grid Control **705**

Metric Type	Metric Subtype	Description (Metric Name in Italics)	UI Navigation for Historical Data	Create UDM or Service Test?[12]	Recommended Thresholds— Warning/ Critical
Management Services and Repository	Configuration	Number of Targets	Management Services and Repository: All Metrics	UDM	Site dependent[13]/ Unset
	Loader Statistics	Loader Threads	Management Services and Repository: All Metrics: count threads in *Loader Throughput (rows per hour)*	UDM	<1[14]/Unset
		Loader Throughput (rows per hour)	Management Services and Repository: All Metrics	No	N/A
		Loader Throughput (rows per second)	Management Services and Repository: All Metrics[15]	UDM	< 500/Unset
		Total Loader Runtime in the Last Hour (%)[16]	Management Services and Repository: All Metrics	UDM	> 90/Unset
	Rollup Statistics	DBMS Job Status: Throughput Per Second: Rollup	Management Services and Repository: All Metrics	UDM	< 1000/Unset
		DBMS Job Processing Time (% of Last Hour): Rollup	Management Services and Repository: All Metrics	No	> 50/75
	Job Statistics	Job Dispatchers	Equal to the number of OMSs	UDM	<1/Unset
		Job Dispatcher Performance: Steps Per Second	Management Services and Repository: All Metrics	UDM	Site dependent/ Unset
	Notification Statistics	Notification Performance: Throughput Per Second	Management Services and Repository: All Metrics	UDM	Site dependent/ Unset

TABLE 17-2. *Grid Control Metrics, UI Navigation, and Recommended Thresholds*

---

[12] See "User-Defined Metrics on Grid Control Performance" in Appendix H (online) for how to define these UDMs in the Console, including the SQL query, thresholds, and frequency. Many of these metrics use queries from the script *repohealth.sql*, also listed in that appendix.

[13] Use thresholds of > 1000 for a small site and > 10,000 for a medium site. This will send an alert indicating your site size has grown to the next size, which requires changing OMR Database initialization parameters accordingly per Table C-2 in the online Appendix C.

[14] The number of threads you should be using is at least 2 × (total #CPUs on all OMR nodes) / #OMSs. For details, see "Tune OMS Thread Pool Size" in Chapter 4. I set a minimum threshold of 1 because there should be at least be one thread running.

[15] The default metric Warning/Threshold values of > 2700/3000 are incorrect because the comparison operator should be less than (<) rather than greater than (>). Therefore, I suggest unsetting both default thresholds, creating an equivalent UDM, and using a warning threshold of < 500.

[16] You can calculate this from [*Total Loader Runtime in the Last Hour (seconds)*] × 100 / [3600 × (Loader Threads)].

Metric Type	Metric Subtype	Description (Metric Name in Italics)	UI Navigation for Historical Data	Create UDM or Service Test?	Recommended Thresholds— Warning/ Critical
		Notification Processing Time (% of Last Hour)	Management Services and Repository: All Metrics	UDM	> 50/75
	Alert Statistics	*Alerts Per Hour*	None	UDM	Site dependent/ Unset
OMS and OMR Host Metrics	OMS Host Statistics (per host)	*Number of OMS hosts*	Management Services and Repository: Management Services page	No	N/A
		OMS *CPU Utilization (%)*	Host: All Metrics	No	> 20/40
		OMS Number of CPUs	Host: CPUs[17]	No	N/A
		OMS Memory (GB)	Host: Memory Size (MB)	No	N/A
		Network Interface Collisions Per Sec[18]	Host: All Metrics	UDM	> 150/Unset
		Network Interface Combined Utilization (%)	Host: All Metrics	No	> 90/Unset
		Network Interface Total Error Rate (%)	Host: All Metrics	No	> 25/Unset
	Repository Host Statistics (per node)	*OMR Number of RAC nodes*	Cluster Database: Home page: Instances	No	N/A
		OMR *CPU Utilization (%)*	Host: All Metrics	No	> 40/80
		OMR Number of CPUs	Host: CPUs[19]	No	N/A
		OMR Buffer Cache Size (MB)	Database Instance: Administration: Initialization Parameters[20]	UDM	Site dependent[21]/ Unset

TABLE 17-2. *Grid Control Metrics, UI Navigation, and Recommended Thresholds (Continued)*

---

[17] Speed of CPUs is available on the Host: Configuration page, Hardware Details link, CPU Implementation column.

[18] In Grid Control, this metric is called *Network Interface Collisions (%)*, but the units in that name are wrong: the metric measures the number of collisions per second, not the percentage of collisions. I chose to reflect the proper units when naming the UDM.

[19] Ibid. Footnote 17.

[20] When using automatic shared memory management, the DB_CACHE_SIZE initialization parameter specifies a minimal value. You can see the actual buffer cache size automatically allocated by navigating to Database Instance: Advisor Central: Memory Advisor.

[21] Thresholds depend on the environment size. Use thresholds of < 512 for a small site, < 2048 for a medium site, and < 4096 for a large site per Table C-2 in Appendix C online.

Metric Type	Metric Subtype	Description (Metric Name in Italics)	UI Navigation for Historical Data	Create UDM or Service Test?	Recommended Thresholds—Warning/Critical
		OMR Memory (GB)	Host: Memory Size (MB)	No	N/A
		Repository Tablespace Used (GB)	Management Services and Repository: All Metrics	UDM	Site dependent/Unset
		OMR *Private Interconnect Transfer Rate (MB/Sec)*	OMR Cluster: Interconnects page	No	Site dependent
		OMR *Network Interface Total I/O Rate (MB/Sec)*[22]	Host: All Metrics	No	> 107[23]/Unset
		OMR *Network Interface Collisions (%)*[24]	Host: All Metrics	UDM	> 150/Unset
		OMR *Network Interface Combined Utilization (%)*	Host: All Metrics	No	> 90/Unset
		OMR *Network Interface Total Error Rate (%)*	Host: All Metrics	No	> 25/Unset
		Network Latency OMR/OMS (ms)	None	UDM	> 30/Unset
		ASM: Cluster Disk Performance: I/O per second or *Disk Activity: Transfers (per second)*	ASM: All Metrics or Host: All Metrics	No	Site dependent
		Network Latency OMR/Datafiles (ms)	None	UDM	> 1/Unset
UI Page Performance Metrics[25]		Home Page Response (seconds)	Targets: Web Applications: EM Website: Monitoring Configuration: Service Tests and Beacons: homepage: Performance page: Perceived Total Time (ms)	No	> 6/12

TABLE 17-2. *Grid Control Metrics, UI Navigation, and Recommended Thresholds (Continued)*

---

[22] This metric represents total OMR Host I/O only if the host is dedicated to the Repository Database.

[23] Represents 90% of available network bandwidth, assuming a GigE corporate LAN (1Gbps = 119MB/sec). This is a requirement for OMS to OMR host communication (see Table 2-1).

[24] Although the units in the metric name are "%," the metric measures the number of collisions per second.

[25] Use the most common user access method: HTTP(S) via OHS or Web Cache. To use HTTPS, follow instructions in OEM Advanced Configuration 10g Release 4 (10.2.0.4.0), Section 5.7.2, "Configuring Beacons to Monitor Web Applications Over HTTPS."

Metric Type	Metric Subtype	Description (Metric Name in Italics)	UI Navigation for Historical Data	Create UDM or Service Test?	Recommended Thresholds—Warning/Critical
		Hosts Page Response (seconds)	Targets: Web Applications: EM Website: Monitoring Configuration: Service Tests and Beacons: <service_test_name>: Performance page: Perceived Total Time (ms)	Service Test	> 3/Unset
		Databases Page Response (seconds)	Targets: Web Applications: EM Website: Monitoring Configuration: Service Tests and Beacons: <service_test_name>: Performance page: Perceived Total Time (ms)	Service Test	> 3/Unset
		OMR Database Instance Home Page Response (seconds)	Targets: Web Applications: EM Website: Monitoring Configuration: Service Tests and Beacons: <service_test_name>: Performance page: Perceived Total Time (ms)	Service Test	> 5/Unset
		OMS Host Home Page Response (seconds)	Targets: Web Applications: EM Website: Monitoring Configuration: Service Tests and Beacons: <service_test_name>: Performance page: Perceived Total Time (ms)	Service Test	> 5/Unset

TABLE 17-2.   *Grid Control Metrics, UI Navigation, and Recommended Thresholds (Continued)*

Procedure to Evaluate Grid Control Metrics

Now that you know what metrics to define and how, it's time to get to work implementing and trending these metrics. The general idea is to add the GC metrics with thresholds suggested above to alert you of impending problems in your GC environment. Then, gather a baseline of these GC metrics (or vital signs), and periodically review trends in metric values. These trends help you gauge your site's relative state of health compared with the baseline and provide a basis for tuning, which is the third and last phase of this chapter, discussed in the next section.

Following is an outline of this basic procedure to establish a baseline of GC performance and monitor trends in the vital signs of your site on a weekly and monthly basis:

- One-time: Instrument the metrics identified in Table 17-2, including setting initial thresholds not based on empirical historical values. At the end of the week, add empirically based thresholds.

- **Weekly:**
  - Review 7-day graphs of metric history in the UI to spot trends. Record the week's average metric values outside of GC in a spreadsheet or in the table provided at the end of the online Appendix H in the section "Table to Record Grid Control Metric History."[26]
  - Fix any underlying problems that trends reveal. Common problems and tuning solutions are presented in the next section, "Tune Grid Control to Reduce Bottlenecks."
  - Adjust thresholds and notification rules based on the previous week's alerts and fixes.
- Monthly: Review 31-day graph of monthly metric values to trend overall performance and plan future resource requirements. Record monthly average values in a separate copy of the table listed at the end of Appendix H (online).

Remember, as described above, to view historical average values over a 7- or 31-day period for regular or informational metrics, go to the All Metrics page for the target, click the metric of interest, and select "Last 7 days" or "Last 31 days" in the View Data field. To view this history for UDMs, click the User-Defined Metrics link under Related Links on the home page for the target, click the link for the UDM name, and select "Last 7 days" or "Last 31 days" in the View Data field. For all metrics, use the values in the Average Value column.

TIP
Rather than viewing graphs of each metric in the Console each week, you can mine metric data directly from the Repository using the script repohealth.sql, shown in the online Appendix H in the section "Script to Report on Grid Control Performance Metrics." Spooled output from this script contains average values from the past week for many metrics in Table 17-2. Some metrics appear in both the UI and in repohealth.sql output, while other metrics appear exclusively in one or the other.

Tune Grid Control to Reduce Bottlenecks

What do you do when metric values consistently veer outside established thresholds, as evidenced by repeated notifications? The primary cause of GC bottlenecks is overlooked housekeeping. However, if you're diligent about carrying out the maintenance already delineated in the first section, "Perform Routine Grid Control Maintenance Tasks," then you need to look elsewhere for the culprit. Likely causes include incorrectly configured resources (hardware or software) and hardware resource constraints. These and other causes, some identified in this section, will reveal themselves during tuning.

Many of the vital signs listed in Table 17-2 measure resource utilization and throughput in Grid Control. Consult the Oracle Enterprise Manager Framework, Host, and Services Metric Reference Manual 10g Release 4 (10.2.0.4) for all GC-supplied metrics referenced in this table. Each metric has a User Action that may apply to your site, depending on the metric's trend.

---

[26] It is prudent to record this metric data outside of GC because it ages out of the Repository according to the default or custom retention periods you define (see "Modify the Data Retention Policy" in Chapter 4).

For example, according to the User Action for *Total Loader Runtime in the Last Hour (seconds)*, if the value for this metric increases along with that for *Loader Throughput (rows per hour)*, this indicates the need for another OMS or for more loader threads for existing OMSs. If *Total Loader Runtime in the Last Hour (seconds)* increases but *Loader Throughput (rows per hour)* does not increase, the user action is to check for OMS resource constraints, such as high CPU utilization, insufficient RAM, or OMR Database deadlocks.

Following are the most common Grid Control tuning tasks required:

- Reduce high CPU utilization
- Resolve loader backlog
- Minimize rollup delays
- Evaluate job, notification, and alert metrics
- Tune I/O bottlenecks
- Improve poor Console performance

Let's look at each of these tasks in detail to furnish you with a basic game plan for dealing with them should they crop up at your Grid Control site.

Reduce High CPU Utilization

High CPU utilization can occur on both OMR and OMS servers, but is usually a problem on OMR nodes. OMR node consume more CPU than OMS hosts because the Repository Database performs most of the heavy lifting in GC. While the OMS does its own share of processing by uploading Agent files to the Repository, the other major OMS function, rendering the Console, is more of a memory-draining than a CPU-intensive operation. By contrast, the Repository must use processing power to load and roll up Agent data and execute many other tasks. The DBMS jobs listed on the Repository Operations page for the Management Services and Repository target reveal the number of these processing tasks (see Figure 17-2).

What constitutes high CPU for OMR and OMS hosts? You may recall that we already set such expectations when suggesting thresholds for CPU utilization in Chapter 13. Table 13-1 recommends setting a CPU Utilization Warning threshold of 40% for OMR nodes so that spikes do not saturate processing resources. This threshold aligns with the principle that a properly tuned OMR host should consume an average of less than 40% CPU. Table 13-1 suggests a 20% Warning threshold for OMS hosts, which should not see more than this amount of CPU loading on average.

The ailment of high CPU utilization can plague any application—not just Grid Control—and is one of those catch-all symptoms with many possible underlying causes. The most common causes of CPU bottlenecks for Grid Control, as for other applications, are as follows:

- Bad application SQL, which is by far the principal contributor
- Locking problems between multiple SQL transactions vying for the same user objects, such as tables and rows
- Contention or waits for OMR Database System Global Area (SGA) memory structures, such as the library cache and shared pool

Chapter 17: Maintain and Tune Grid Control   **711**

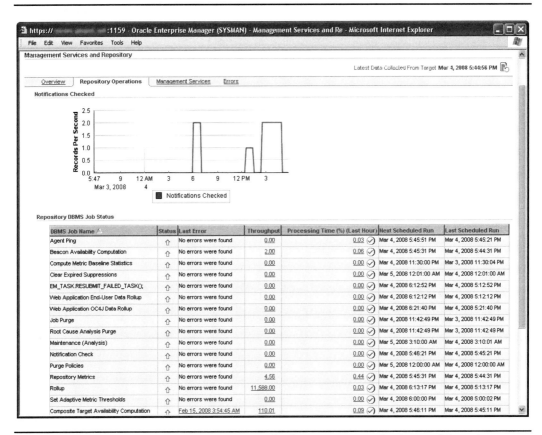

FIGURE 17-2.   *The Repository Operations page for the Management Services and Repository target*

You can use Grid Control itself to evaluate the cause of high CPU usage on either OMR or OMS servers. While it is beyond the scope of this book to delve deeply into database or host performance tuning using Grid Control, my intent here is to point you to the UI pages that can reveal details on the above causes for CPU resource issues. Do not just throw more CPU at the problem. In the majority of cases, CPU bottlenecks are not attributable to undersized CPUs for the current workload (provided your system was initially sized correctly), but to bad application SQL code.

Below are some of the UI pages that can help isolate the cause of CPU bottlenecks:

- To determine the processes taking up most of the CPU on the affected OMR or OMS host, go to the Host Performance page. It shows a graph of recent CPU Utilization and you can select a listing of the current top 10 processes by CPU Utilization.

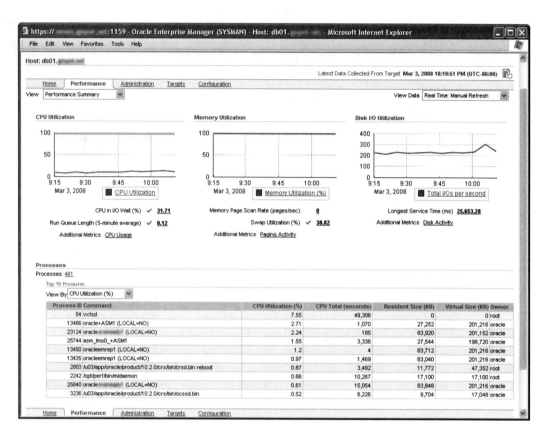

- If you confirm that a GC process is hogging CPU, click the relevant link, under Related Links, which may be Top Consumers, Top Activity, Duplicate SQL, Blocking Sessions, Hang Analysis, or Instance Locks.[27] As an example, click Top Activity (see next image).

---

[27] If you drilldown to one of these links from the OMR Cluster Database page, you may get a message that you need to indicate you've licensed the Diagnostics Pack. If you've already done this on the Management Pack Access page, then drilldown via one of the OMR Database Instance targets instead. This appears to be a bug.

Chapter 17: Maintain and Tune Grid Control 713

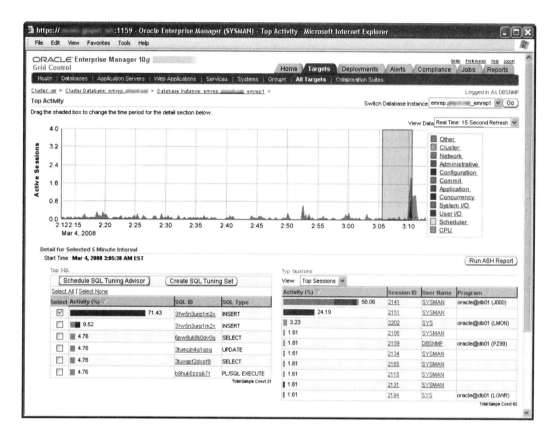

Here you can select a particular SQL ID and click Schedule SQL Tuning Advisor to display the SQL statement and get help in tuning it. After identifying the offending SQL, check Oracle*MetaLink* for any patches relating to it. Oracle does not support changing the GC application code, but if no patch is available, you may need to tune a particular SQL statement or add an index to address a critical performance problem. Just realize that you may need to manually maintain this customization when upgrading to the next GC release, unless a patch is subsequently supplied to redress the problem.

- Clicking a particular Session ID under the Top Sessions view on the bright shows you a breakdown of resources comprising the session. Clicking Session ID 2141, shown above, which comprises 58 percent of the sessions during the chosen time period (selected by dragging the shaded box over the graph), reveals the following, as shown in the next image.

Oracle Enterprise Manager 10g Grid Control Implementation Guide

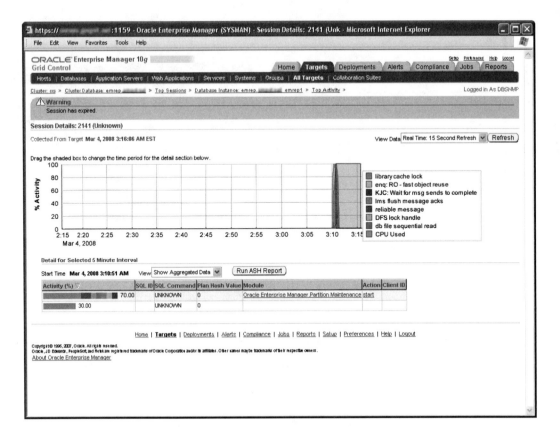

The above graph is color coded to indicate contributions to percentage activity from all wait events. From the above breakdown, you would see, if the screen shot was printed in color, that almost all wait time is due to waits from *lms flush message acks* when running the partition maintenance module, as the Module column reports. Searching Oracle*MetaLink* for this wait event reveals that this system was hitting Bug 4151363, described in the earlier section "Perform OMR Partition Maintenance," and for which a patch is available.

This is the extent of general suggestions I can provide on using Grid Control to tune excessive CPU Utilization (%). While I cannot present specific solutions, I can pass on the general recommendation to enable hyperthreading (HT) for Intel OMR and OMS hosts in order to increase processing power for existing CPUs. (See the section "Enable Hyperthreading for Intel OMR/OMS Hosts" later in the chapter.)

Resolve Loader Backlog

The OMS loader processes Agent data that is continuously uploaded as HTTP requests, and in turn uploads this data to the Repository. Files uploaded to the OMS by default are temporarily stored in the $OMS_HOME/sysman/recv/ directory on the OMS host. However, if you configure a Shared Filesystem Loader, as described in Chapter 15, all Agent XML files are uploaded to shared OMS storage. Loader backlog occurs when the loader is not keeping pace with uploaded Agent data volumes. You can get a high-level look at recent loader performance by going to the Overview page for the Management Services and Repository target.

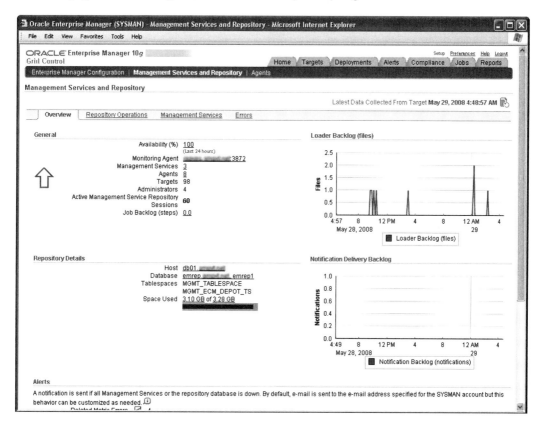

This page contains a graph of the number of XML files in the past 24 hours that queued up on OMS hosts for upload to the Repository. (The screen shot above does not depict any backlog.) The Management Services page, shown next, breaks down this loader backlog by OMS host.

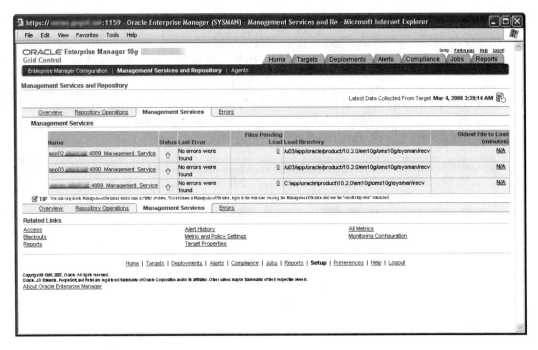

(Here, there are three OMS hosts.) This page allows you to pinpoint which OMS hosts are experiencing the backlog.

To accurately gauge loader backlog, you must examine three related Management Services and Repository metrics given in Table 17-2. Drill-down on these metrics to review their historical metric values by loader thread for the last 7 days:

- **Loader Throughput (rows per second)** This regular metric (which nevertheless requires a UDM)[28] measures the number of lines of XML text processed by each loader thread per second averaged over the past hour. The higher the value, the better. Threads should process at least 500 rows per second, but this can climb to 1200 rows/second or more for smaller, well-tuned sites, as shown next.

---

[28] Ibid. Footnote 15.

Chapter 17: Maintain and Tune Grid Control 717

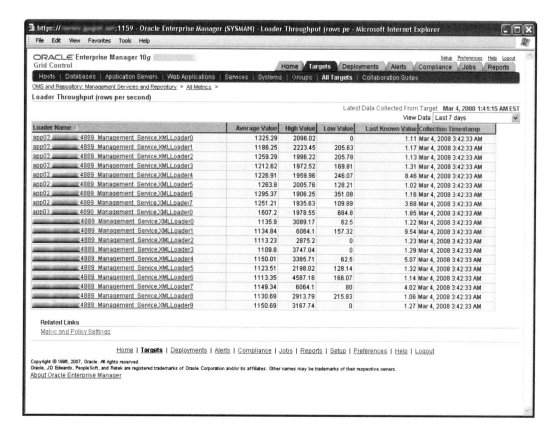

Here, loader throughput is over 1000 for all loader threads, which indicates a healthy loader configuration.

- **Total Loader Runtime in the Last Hour (%)** This UDM is calculated from the regular metric called *Total Loader Runtime in the Last Hour (seconds)* and signifies the average percentage of the hour during which loader threads process Agent XML files. It indicates whether the existing loader threads are keeping pace with incoming Agent data volumes. The lower the value, the better. I suggest a Warning threshold of 90% to alert you if loader bandwidth is becoming saturated.

- **Loader Throughput (rows per hour)** This informational metric measures the total number of lines of XML text on average that each loader thread processes per hour. The higher the values for each loader thread, the better. Example data for this metric over the last 7 days is shown in the following image.

[Screenshot of Oracle Enterprise Manager showing Loader Throughput (rows per hour) report with a table of loader names, average values, high values, low values, last known values, and collection timestamps.]

I do not suggest a UDM for this metric because I recommend creating a similar UDM for Loader Throughput (rows per second). However, when loader throughput is diminishing, summing the values in the Average Value column for all threads indicates the cumulative effect of all threads on loader throughput.

When values for either of the first two metrics cross defined thresholds, the solution is to add more loader threads (which should come as no big surprise). The loader for a particular OMS host consists of one to ten threads, as configured in *$OMS_HOME/sysman/config/emoms.properties* (see the Chapter 4 section "Tune OMS Thread Pool Size"). You may need to revisit the number of loader threads initially configured at installation time to keep pace with a higher volume of incoming Agent data. This higher volume is likely due to additional monitored targets, metric thresholds, policies and jobs you defined, and the like. For OMS hosts that are CPU bound, be careful not to add too many loader threads, as each thread increases overall host CPU utilization by 2 to 5 percent. As you add loader threads to reduce loader backlog, the following should occur

- *Loader Throughput (rows per second)* should decrease for each loader thread because the workload is distributed among more loader threads.

- *Total Loader Runtime in the Last Hour (%)* may increase, but this is fine as long as it does not cross the 90% threshold.

- The sum of *Loader Throughput (rows per hour)* for all loader threads should increase appreciably, as this is the indicator of overall throughput contributed by all loader threads.

There comes a point when adding more loader threads does not reduce loader backlog significantly. This is the point to stop adding loader threads. If adding more threads does not prevent backlog from increasing, look for a more pressing bottleneck elsewhere in the application. Examine the other vital signs, as described in the earlier section, "Specific Grid Control Metrics to Gather," for clues to what else could be causing the backlog.

Minimize Rollup Delays

After the OMS loader process uploads XML files containing raw Agent data, the Repository inserts this data into the MGMT_METRICS_RAW table. The rollup process then runs once an hour to aggregate all new raw data by calculating averages and storing them in the tables MGMT_METRICS_1HOUR and MGMT_METRICS_1DAY. The rollup process is just one of many DBMS jobs that Grid Control runs in the background to maintain the Repository (see Figure 17-2 earlier in the chapter). The rollup process is by far the largest consumer of OMR Database buffer cache space, which must hold the considerable volume of raw and aggregate data from these three tables.

The rollup process has two metrics that are equivalent to those pointed out above for the loader process, and are of equal importance here. These metrics, which appear under DBMS Job Status on the All Metrics page for the Management Services and Repository target (see Figure 17-1), are:

- **DBMS Job Status: Throughput Per Second** This informational metric reports the number of rows per second processed, or aggregated and stored in the rollup tables. The higher the value, the better. A smoothly running GC site should maintain 2000 +/– 500 rows per second. My earlier suggestion was to creating a UDM with a Warning threshold of 1000 to allow for occasional performance lags. The following page shows an exceptionally well-tuned Repository with a 7-day average rollup throughput of over 8000 rows per second.

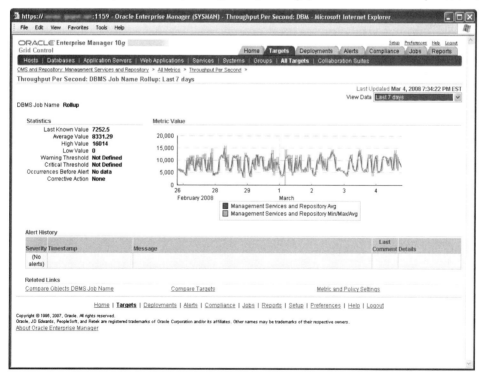

- **DBMS Job Processing Time (% of Last Hour)** This regular metric exhibits the average percentage of the hour during which the rollup process is active and thus indicates how hard the rollup process is working to keep up with incoming loader volume. The lower the value, the better. The default Warning/Critical thresholds of 50/75% are low enough to give you plenty of advance notice that rollup is decreasing.

If either of these rollup metrics grows worse over time, a larger OMR buffer cache[29] may halt the decline in rollup efficiency. Assuming you are using automatic shared memory management (implemented when choosing the GC New Database Option and suggested in Chapter 3 when using the GC Existing Database Option), you can increase the buffer cache in one of two ways, or both. You can increase the SGA size, which increases the buffer cache along with other memory structures, and/or set DB_CACHE_SIZE explicitly, which sets a minimum value for the buffer cache. If you choose to increase the entire SGA, the database Memory Advisor in Grid Control can help you decide how much to increase it. To use the Memory Advisor, click Advisor Central under Related Links on the home page for the OMR Database instance (or one of them for a RAC Repository), then click Memory Advisor.

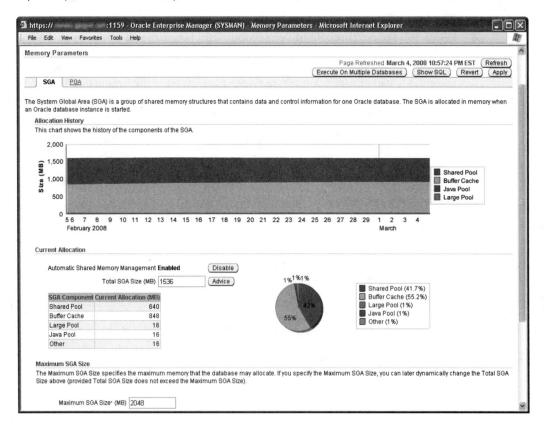

---

[29] As already mentioned, the buffer cache is an SGA memory structure that holds copies of data blocks read from data files.

This page displays the current SGA size, SGA max size, and component sizes, including that of the buffer cache. To see the benefit of increasing the SGA, click Advice.

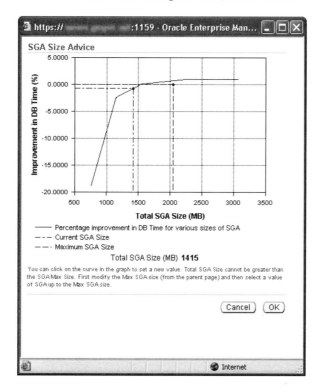

The above graph predicts that there would be very little improvement (about 2 percent) in database performance from increasing the SGA any further, but the above-mentioned rollup statistics in particular might get better. You can try increasing DB_CACHE_SIZE so that it takes up a larger percentage of the total SGA. You may also want to increase the SGA (provided there is sufficient RAM), then empirically measure whether it improves both rollup metrics.[30] Rising rollup statistics should be accompanied by improved loader statistics on OMS hosts, and reduced CPU utilization and I/O on OMR nodes. Do not pay attention to the Buffer Cache Hit Ratio metric, as it is not a good measure of database performance.[31]

If rollup vital signs do not noticeably improve, you can return the buffer cache to its previous size, as the increase would consume memory unnecessarily. Look for resource constraints elsewhere in the application that are causing rollup performance problems. For example, high CPU utilization on OMR nodes may be the cause of rollup slowdowns (see "Reduce High CPU Utilization" earlier in the chapter). Another culprit could be an I/O subsystem bottleneck. Given that the buffer cache reads from and writes to datafiles on the disks that contain the rollup tables, I/O subsystem speeds must be

---

[30] Increasing the SGA is preferable, free RAM permitting, because you can increase both SHARED_POOL_SIZE and DB_CACHE_SIZE to maintain the same ratio (320/512) suggested in "Set Variable Initialization Parameters" in Appendix C. This ratio seems to produce good overall performance for the OMR Database.

[31] Connor McDonald demonstrated this with an application that lets you increase the database buffer cache hit ratio as much as you like by adding unnecessary workload, which of course degrades system performance. *Optimizing Oracle Performance* (O'Reilly & Associates, Cary Millsap, 2003), page 35.

sufficiently high that they do not throttle the higher rollup throughput from an increased buffer cache. If I/O throughput cannot keep up with rollup processing in memory for a given buffer cache size, then no further increase in buffer cache size will improve rollup throughput.

Finally, as a general recommendation for increasing rollup performance, consider the Oracle Very Large Memory (VLM) option on 32-bit OMR nodes to extend the SGA (and therefore the buffer cache as well) beyond 1.7GB. VLM allows you to get more horsepower from existing RAM on 32-bit platforms, just as hyperthreading leverages existing CPUs on both OMS and OMR hosts, as already mentioned. For details on implementing VLM, see the section "Use VLM Option for OMR Hosts on 32-bit Platforms" later in the chapter.

Evaluate Job, Notification, and Alert Metrics

Job, notification, and alert metrics are best suited for measuring the relative improvement, after the fact, of adding another OMS. These metrics, unlike those for other vital signs listed in Table 17-2, are used more reactively than proactively. While positive trends in job, notification, and alert statistics provide a good indication of overall OMS health, negative trends indicate contention elsewhere in the GC system. Following are descriptions of the four related metrics for these OMS functions:

- **Alerts Per Hour** This UDM indicates how many severities per hour are triggered. While values may grow nominally over the long term as you add targets to GC monitoring, a sudden, uncharacteristic increase signifies a problem with the monitoring setup. For example, rapid-cycle threshold crossings due to tight thresholds may be artificially increasing this metric.

- **Job Dispatcher Performance: Steps Per Second** This informational metric is the number of job steps per second that the job dispatcher processes, averaged over the past hour. Once you set thresholds for the UDM based on an initial baseline, a decrease in this metric throughout the week alerts you to a burgeoning job dispatcher performance problem.

- **Notification Performance: Throughput Per Second** This informational metric calculates the number of notifications delivered per second, which is averaged over the past hour. After establishing a baseline and setting a suitable threshold for the UDM, investigate the reason for any sudden increases in notifications. One reason could be rapid-cycle threshold crossings, which also cause *Alerts Per Hour* to increase. In addition, you may need to remove extraneous or overburdening notification schedules, methods, or rules.

- **Notification Processing Time (% of Last Hour)** This regular metric represents the average percentage during the past hour in which notification delivery has been running. If this percentage continues to increase, there's usually a corresponding increase in *Notification Performance: Throughput Per Second*, and for the same reasons (cited for that metric).

If these metrics all experience a downward trend in performance, decide whether your GC system must be scaled up by adding an OMS, or whether the trend is symptomatic of OMS contention in other areas (such as memory shortages or poor I/O bandwidth). Only one job dispatcher exists for each OMS, and this is not configurable. Therefore, an additional OMS should improve job dispatcher performance in particular, and should also reduce *Notification Processing Time (% of Last Hour)*. In general, adding OMS hosts improves overall throughput for all GC metrics when there are no resource contention issues. Trends in job, notification, and alert vital signs as a group are a good gauge for measuring that improvement.

Tune I/O Bottlenecks

I/O bottlenecks in Grid Control typically cause many of its performance metrics to plummet because these metrics depend on I/O performance to one degree or another. Why is this? Well, I/O includes both network I/O and file I/O, and many metrics depend on disk or network channels. Most statistics on GC hosts, loader or rollup processes, and job, notification, and alert throughput have an I/O component to them. Many of these statistics involve disk access to the OMR Database, communication between OMR and OMS, or RAC Repository intercommunication.

I/O Channels

Following, then, are three I/O channels on which this disk access takes place:[32]

- Network I/O between OMR and OMS
- Disk I/O between OMR instance and datafiles
- RAC OMR interconnect rate

Certain metrics in Table 17-2 monitor whether your site is meeting I/O channel requirements. You may recall from Chapter 2 (see Table 2-1) that I provided bandwidth and latency requirements for the first two of these I/O channels. These requirements came up when hashing out how many regional GC sites were needed. In answering that question, you learned that the determining factor was the stringent network performance requirements between Repository and OMS hosts (the first channel). I have not yet brought up the third channel, the RAC OMR interconnect rate, but I will do so below. Let's start with the first I/O channel.

Network I/O Between OMR and OMS   As Table 2-1 indicates, OMR/OMS network bandwidth requirements are 1Gbps with 30ms maximum latency. To measure I/O speeds between OMR and OMS, you set the OMR metric *Network Interface Total I/O Rate (MB/Sec)*. Note that thresholds are defined for the OMR nodes, not for OMS hosts. You should measure I/O from the OMR side because it includes only OMS traffic (for OMR nodes dedicated to an OMR Database), whereas I/O from the OMS side includes both OMR and Agent traffic. The recommended Warning threshold of 107MB/sec represents 90 percent of the 1Gbps minimum bandwidth required between OMR and OMS hosts. This means that you are alerted if the I/O rate approaches 90 percent of the available bandwidth, assuming your intranet network is GigE.

Another metric, *Network Interface Combined Utilization (%)*, also measures bandwidth consumption, but in percentage terms. For both OMR and OMS hosts, additional network metrics—*Network Interface Collisions (%)* and *Network Interface Total Error Rate (%)*—are designed to sniff out whether network integrity is compromised.

No built-in GC metric is instrumented for measuring network latency, so we must create a host UDM called *UDM Network Latency OMR/OMS (ms)*. This is easy enough to do using the *ping, grep,* and *awk* commands (see Appendix H online for the definition).

Disk I/O Between OMR Instance and Datafiles   Table 2-1 also specifies I/O requirements between the OMR/OMR storage of 1Gbps bandwidth and 10ms latency.

---

[32] This term, *channel*, is an apt description, and the three network channels identified here are cited in Section 11.2.4.5 of OEM Advanced Configuration 10*g* Release 4 (10.2.0.4.0).

If using Automatic Storage Management (ASM), you can measure bandwidth using the informational metric *Cluster Disk Performance: I/O per second* for the disk group containing the datafiles (here named DATA).

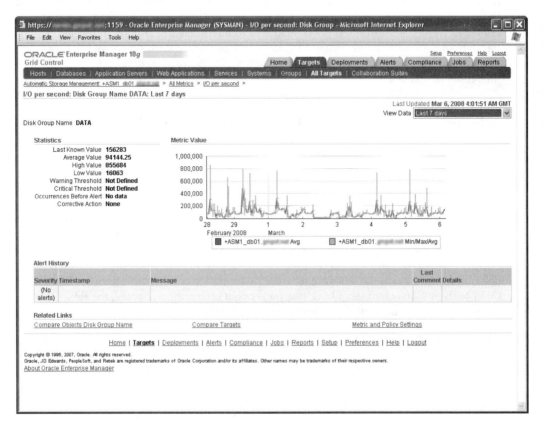

If using storage other than ASM, you can use the host metric called *Disk Activity: Transfers (per second)*.

As with the previous network channel, a UDM is required to set thresholds for network latency between the OMR instance and its datafiles, and is named *UDM Network Latency OMR/ Datafiles (ms)*.

RAC OMR Interconnect Rate The metric called *OMR Private Interconnect Transfer Rate (MB/ Sec)* covers the last of the three I/O channels listed above, namely, RAC Repository interconnect I/O. The threshold to set for this metric is site dependent, as it varies according to the interconnect speed, the number of databases that the cluster supports, etc. This metric is actually not a regular or informational metric, but a value you must obtain manually from the Interconnects page for the Cluster target servicing the Repository Database. Following is a screen shot of this page for a cluster supporting multiple cluster databases, one of which is the Repository Cluster Database, named *emrep*.

Chapter 17: Maintain and Tune Grid Control 725

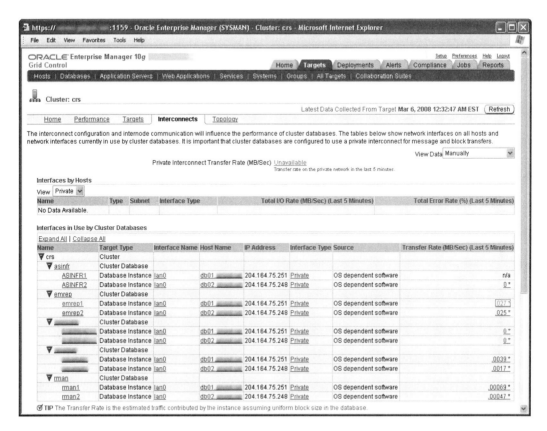

Such a configuration, with one cluster supporting multiple databases, is very common. In fact, it is the grid computing architecture introduced in Chapter 1 and recommended for Grid Control to integrate into. The history of the private interconnect transfer rate for just OMR Database traffic is accessible in the section Interfaces in Use by Cluster Databases. Click the rightmost column, Transfer Rate (MB/Sec) (Last 5 Minutes), for the rows showing the *emrep* Cluster Database interfaces. In the above case, the transfer rates are .027 and .025 MB/Sec. Clicking the .027 link brings up a page, shown next, from which we can select a 7-day history.

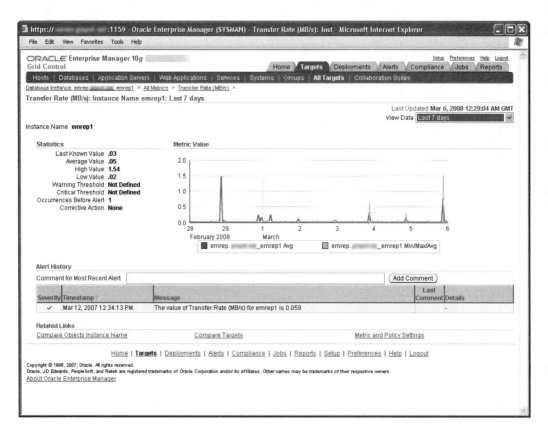

We see from this history that the average value reported for the most recent 7-day period is 0.05. The RAC interconnect bandwidth taken up by other databases that the cluster supports illustrates that the observer (Grid Control) is affected by the observed (target databases) and visa-versa when adopting a grid computing strategy. You should take this into account as one of the few drawbacks when integrating Grid Control into a grid computing environment.

Determine the Cause of I/O Bottlenecks

As for tracking down the cause of I/O bottlenecks, Grid Control can once again be of service. Following are the likely causes of I/O throughput problems:

- Bad housekeeping, causing OMR Database fragmentation, inflated high water marks, etc.
- Poorly tuned SQL code
- I/O from non-EM sources, such as backups, another application, or user direct OMR queries
- Asynchronous I/O not enabled on both OMR and OMS hosts

As with high CPU utilization (for which a troubleshooting example is supplied above in the section "Reduce High CPU Utilization"), use the OMR Database Performance page to find the source of the bottleneck.

Chapter 17: Maintain and Tune Grid Control **727**

Improve Poor Console Performance

To improve Console Performance, you must first pinpoint where it is deficient. Many larger sites find that the entry point to the UI, the login page, takes an inordinate time to load. Other sites discover certain Console pages are afflicted with slow response times. I admittedly divide Console performance problems into two very unequal parts: the login page and everything else.

Improve Console Login Performance

Grid Control offers a way to quickly display a stripped-down version of the Console home page when the Repository is very large. By default, the home page contains such data as target status, the number of alerts and policy violations, errors, jobs, a deployments summary, and patch advisories. When your site is sufficiently large, aggregating this summary data can significantly prolong rendering of the home page. In such cases, Grid Control offers a shortcut to work around this problem. It's not an elegant solution, because it removes almost all home page functionality, but it gets you in the door, so to speak. Once in, you can drill down to specific UI pages on any element that a full-function home page summarizes (alerts, policies, etc).

Following is the "before" look of a normal home page.

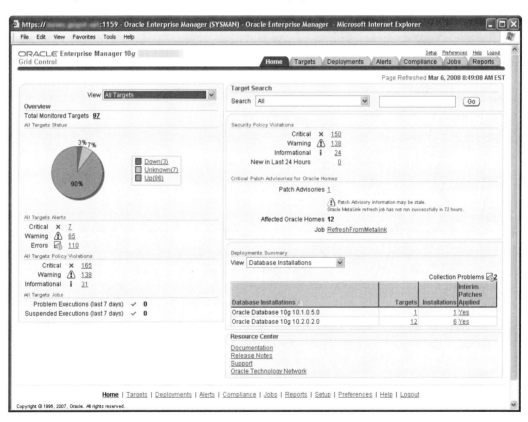

To implement the reduction home page, simply add the following property to *$OMS_HOME/sysman/config/emoms.properties* on all OMS hosts:

```
LargeRepository=true
```

Then bounce the OMS on each OMS host as follows:

```
$OMS_HOME/bin/emctl stop oms
$OMS_HOME/bin/emctl start oms
```

Following is the "after" look at the same home page.

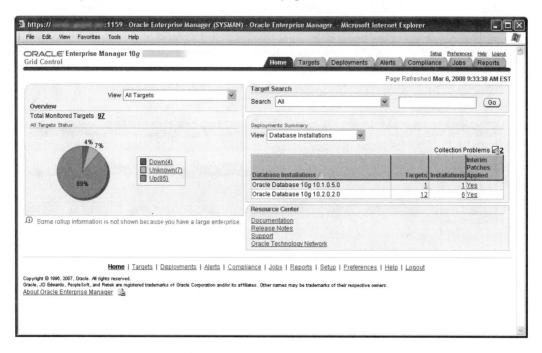

As you can see, setting the above parameter removes all page elements except the All Target Status and Deployments Summary. As I said, it's not elegant, but it beats waiting minutes for a home page that you may bypass anyway in favor of more specific pages needed for the task at hand.

Increase UI Page Performance

If it looks like Renoir is rendering his impression of your home page, don't be surprised if other Console pages are slow as well. To improve the performance of Console pages other than the login page, look to the built-in EM Website web application.

If you want to test a specific URL reported as being slow, create a service test and run it from beacons defined in key user communities. See "Add Service Tests" and "Add Beacons" in Chapter 12 for instructions.

However, a more common scenario is that users report general UI sluggishness. Below is a reasonable approach to efficiently investigate and resolve such generalized reports of slowdowns:

1. Before looking into UI performance beyond the home page, first eliminate any of the causes of performance degradation cited in the above section, "Perform Routine Grid Control Maintenance Tasks". As with so many other metrics, UI response time metrics suffer the slings and arrows of outrageous housekeeping. Shirking GC housekeeping in favor of reactive performance tuning is like sweeping during the Dust Bowl of the 1930s. In this analogy, good housekeeping would have been maintaining the top soil in the 1920s before the winds stripped it.

2. Assuming housekeeping is up-to-date, first review UI page performance trends for those service tests already created (such as those in Table 17-2), taking beacon locations into account. The best scenario is that you find troubling trends in UI response metrics for which service tests already exist, as you can immediately fall back on historical data for analysis.

    a. If service test results are anomalous for only select beacon locations, determine whether WAN latency is causing a UI response problem along just part of your network. Create beacons in new locations if you need more data points. You may also find that the problem is unrelated to the network, but is instead a configuration issue with one or more client machines (such as resource problems, unsupported browser, etc.).

    b. If service test results are slow from all beacon locations, eliminate the possibility of a site-wide network problem. Look at network metrics in Table 17-2 for evidence of a problem. Many DBA administrators are accustomed to users blaming all performance problems on the database and wisely rule out network problems before delving into database tuning.

3. If UI response is sluggish for pages lacking service tests:

    a. Compare these pages to those that already claim service tests.

    b. Compare complaining user locations to beacon locations from which service tests are run.

    You may find a correlation or pattern in either case that illuminates the problem, or some other boon. For example, a similar UI page to one already used in a service test might provide relevant historical data not otherwise available for slow pages not yet instrumented with service tests.

4. Use *end-to-end (E2E)* performance monitoring functionality in Grid Control to trace the worst-performing business transactions and drill down all URL invocation paths through to the JSP/servlet, EJB, and SQL statement levels. (Note that you must alter the EM

Website web application in order to capture J2EE and database performance data in the OC4J layer. See "Configure a Service" in Chapter 12 for details on the *EM Website 1* web application created with the additional functionality required.) The focus below is at the SQL statement level because you can address these problems more readily than those on the middle tier:[33]

a. Click Targets: Web Applications, select *EM Website Service 1*, then click the Request Performance page.

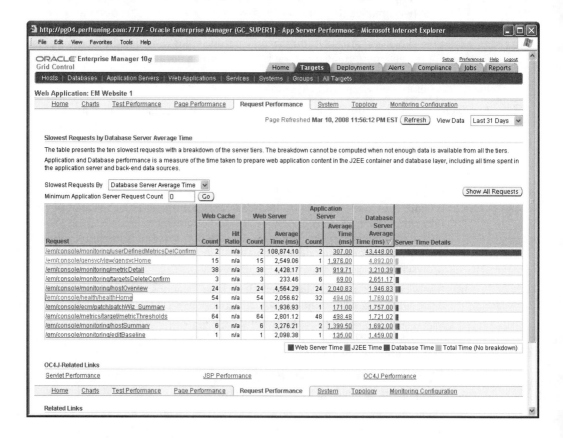

---

[33] You can diagnose performance problems in Java applications such as Grid Control using a new feature in GC 10.2.0.4 called the Java Applications Diagnostic Expert (JADE).

b. As shown above, mine data over a longer period by selecting Last 31 Days in the View Data field. Query the slowest requests by database time by selecting Database Server Average Time in the field, Slowest Requests By. Choose a popular request (i.e., with a high Web Server count). In this example, I clicked the request /em/console/health/healthHome with 54 hits.

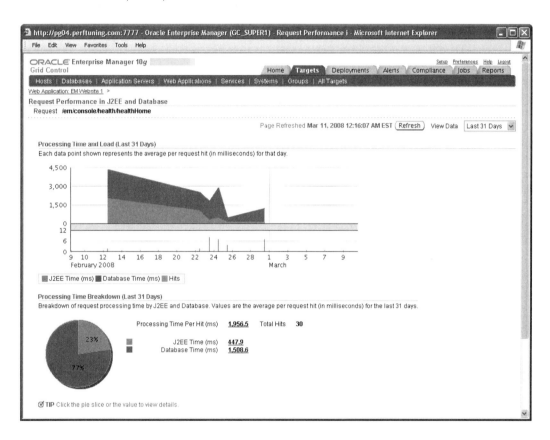

c. Click the largest slice of the Processing Time Breakdown pie chart, which is Database Time. This provides the SQL component details of this UI page.

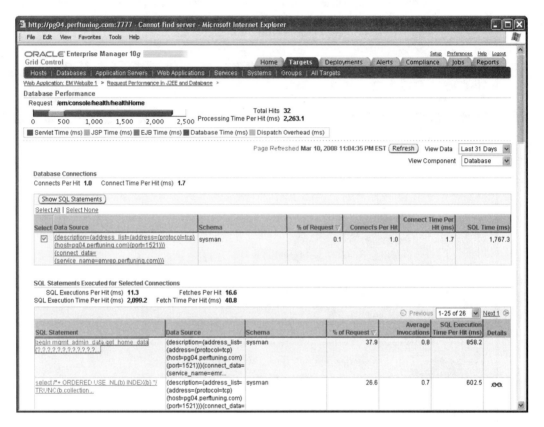

d. Above, the multiple SQL statements constituting this database time are broken out. In our case, the first SQL is a PL/SQL procedure and the second is an individual SQL statement. Click the PL/SQL statement, which takes up the largest % of Request (37.9%).

Chapter 17: Maintain and Tune Grid Control **733**

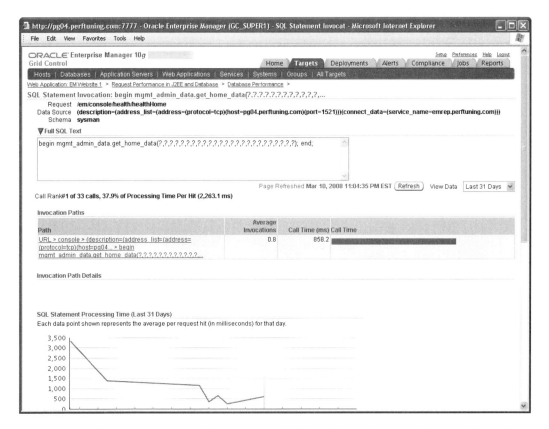

e. Select Last 31 Days in the View Data field (as shown above) and expand the Full SQL Text label at the top. This reveals a package and procedure called MGMT_ADMIN_DATA.GET_HOME_DATA. For PL/SQL, you must look at the procedure's source code to determine the tables and indexes called. Examine these segments for housekeeping problems (fragmentation, stale statistics, invalid objects, etc.).

f. The second SQL statement (see next image), beginning "select /* + ORDERED USE_NL (m)…," reveals the tables MGMT_TARGETS and MGMT_METRICS are involved (see screen shot below). Here, inspect these tables and their indexes for housekeeping problems as well.

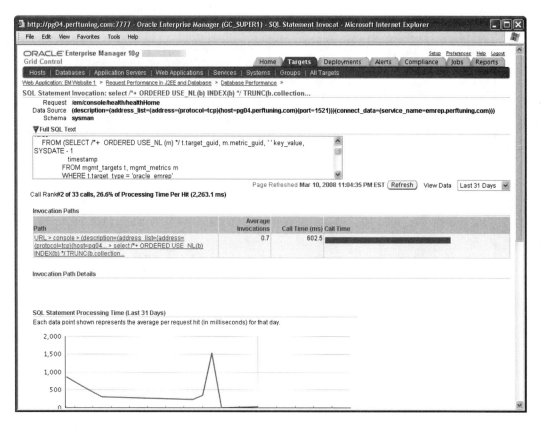

g. Repeat this procedure, beginning with step b, for all requests suffering from performance slowdowns.

5. Bring down the OMS, rebuild and reorganize the segments identified as problematic, and then restart the OMS. (See "How to Rebuild Tables and Indexes" earlier in the chapter). This should improve page performance involving these segments.

6. When finished looking into the most frequently reported UI pages, consider creating new service tests for these pages along with beacons in key GC user network locations. Set thresholds based on the now acceptable response levels so that you will be the first to know if these pages start slowing down again.

Platform-Specific Tuning Recommendations

Following are OMS and OMR platform-specific performance recommendations to consider implementing for Grid Control. While you may not need to call upon these solutions when first installing Grid Control, you may need to do so later to scale your environment. These tuning recommendations, which do not appear in the Oracle EM documentation library, suggest OS-related steps that will improve GC performance.

OMR Platform Requiring VLM Option	Reference Documentation for Implementing VLM
Asianux 2.0	Oracle*MetaLink* Note 361323.1
RHEL3	Oracle*MetaLink* Note 317055.1 Oracle Database 10*g* Linux Administration, Edward Whalen (Oracle Press, 2005), page 260
RHEL4	Oracle*MetaLink* Note 317141.1
SUSE9	Oracle*MetaLink* Note 317139.1
Windows 2003	Oracle*MetaLink* Note 225349.1

TABLE 17-3. *OMR Platforms Supporting VLM, and Reference Documentation*

Tune Windows to Optimize OMR Database Performance

Aside from the minimum requirements for Windows already covered, you should also tune Windows to optimize the performance of your OMR Database, as you would any Oracle database on Windows. For example, you should adjust processor scheduling for best performance of background services rather than programs. For all database tuning recommendations, see the tuning section in Chapter 7 of the Oracle Database Platform Guide for your particular Windows platform.

Enable Hyperthreading for Intel OMR/OMS Hosts

If running Grid Control on Intel-based hosts that do not employ dual- or quad-core processors, you can improve performance of the OMS and Repository hosts by enabling hyperthreading (HT). Many of the Intel processors currently produced offer HT, which allows CPU instructions to execute in parallel. This can increase CPU processing power by roughly 50 percent and double the appearance of the number of processors on the system (called *logical processors*) to applications such as Grid Control. Given that both OMS and OMR hosts routinely execute multiple processes simultaneously, both of these GC components can benefit from HT.[34]

Use VLM Option for OMR Hosts on 32-bit Platforms

In Chapter 2, under "Hardware Operating Requirements," I recommend running the OMR host(s) on a 64-bit platform. Nevertheless, if you must run the OMR host(s) on a 32-bit platform, one operating system mechanism to improve SGA performance is the Oracle Very Large Memory (VLM) option, which can create a large database buffer cache. Using the VLM option on 32-bit platforms can significantly improve overall GC application performance. On Windows, the VLM option is also known as Address Windowing Extensions (AWE) or Physical Address Extensions (PAE) to allow an SGA greater than 4GB. Table 17-3 provides reference documentation for implementing the VLM option (or equivalent) on most 32-bit-certified platforms for the OMR, and on Linux x86 platforms.

---

[34] Source: "Enterprise Manager Grid Control Performance Best Practices," an Oracle white paper, September 2004, page 16.

Summary

The maintenance and tuning in this last chapter is essential to your GC system's health. Without it, your GC engine would eventually stop running, but with it, you'll be well poised to take the product into the next release. The maintenance and tuning described here is the humdrum stuff you do behind the scenes to let Grid Control shine in its moments of glory. When an alert notifies you in plenty of time to avoid a critical production system outage, you can thank your diligent maintenance routine—not your lucky stars—for such reliable service. Maintenance and tuning is as easy as one, two, three:

- Perform routine GC maintenance tasks
- Gather and document GC metrics
- Tune GC to reduce bottlenecks

The routine housekeeping is straightforward. Instrumenting the metrics takes a little doing, especially to create the UDMs, but it's well worth it and I provided all UDM metric definitions. Tuning, compared with housekeeping and metric gathering, is more of an art in that each GC site can act out in its own idiosyncratic way. For this reason, it's impossible to offer specific advice for every possible performance problem you may encounter. Instead, I offered advice for three general areas of tuning: resource tuning (high CPU utilization, I/O bottlenecks), GC process tuning (for loader, rollup, job, notification, and alert processes), and Console performance tuning. The upside is that good housekeeping should keep the need for such reactive tuning to a minimum. At the end of the tuning section, I offered general performance recommendations, though they are for specific platforms.

Hopefully, you've implemented Grid Control using this book. You've installed, configured, and maintained it and are now a Grid Control aficionado. If you read the book and played with Grid Control in the lab, but haven't gotten your hands on a production GC system yet, you can implement and maintain Grid Control with confidence and know what to expect along the way from this exemplary, unrivaled Oracle management tool.

Index

:: (double colon), 169
= (equal sign), 139
\ (backslash), 139
; (semicolon), 169

A

adaptive thresholds, 508–509
Additional Management Agents, 158
  See also standalone Agents
additional_agent.rsp, 225–227
administrators, 299, 306
  creating, 313–317
  creating roles, 307–313
  setting preferences, 317–320
Adobe SVG Viewer, 13, 244
  installation, 262–263
Agent. *See* Oracle Management Agent
Agent Cloning, 158, 160, 162
  installation, 223–225
Agent configuration files, 552
Agent Configuration Manager, 291
Agent Deploy, 156, 158, 159, 161, 162
  backing up the current SSH
    configuration, 179
  choosing inventory location, 187–188
  configuring SSH user equivalence,
    178–184
  Fresh Install, 192–195
  including additional files, 189
  installation types, 190
  installing required packages, 178
  modifying properties file for SLB, 187
  NFS Agents, 195–196, 198
  overview, 176–178
  post-installation steps, 197–198
  preparing for cross-platform Agent
    push, 189
  running, 190–197
  running sshConnectivity.sh script,
    180–184
  setting ORACLE_HOME on OMS host,
    179–180
  setting up SSH Server on Windows, 179
  setting up time zone for SSH Server,
    184–185
  troubleshooting, 198
  Upgrade Agent, 196–197
  validating all command locations,
    185–187
  verifying Agent user permissions,
    188–189
  verifying SSH user equivalence is
    configured, 184
Agent management files, 552
Agent ping, 21
Agent Registration passwords, 657

agentDownload, 158, 160
    confirming WSH version on target
        host, 217
    copying script from OMS to target host,
        217–219
    download sites, 218
    downloading script and response file,
        214–217
    executing on target host, 219–223
    installation, 213–214
    installing a cluster Agent on a RAC
        cluster, 222
    installing an Agent on a single
        node, 221
    installing an Agent on each RAC cluster
        node individually, 221
    parameters available for, 220–221
Agents
    Additional Management Agents, 158
    backups, 241
    chain-installed, 33, 657
    cluster Agents, 161, 163–164
    configuring, 376–383
    confirming Agent is working, 238–240
    confirming Agent restart on reboot is
        configured, 240–241
    CPU requirements, 167
    cross-platform agent installation,
        171–173
    database control Agents, 164
    disaster recovery recommendations,
        648–649
    disk space requirements, 166–167
    enabling Framework Security between
        Agents and OMS, 656–663
    for Grid Control, Database Control, and
        AS Control, 26
    high availability recommendations,
        636–637
    local Agents, 161, 163–164
    master Agent, 203
    master Agent host, 203
    NFS Agents, 160, 164, 195–196, 198
    RAC Agents, 164
    RAM requirements, 168
    refreshing host configuration, 240
    resetting, 554–557
    running agentca for a cluster Agent, 240
    setting up an Agent user environment,
        237–238
    software backup and restoration,
        618–620
    software home, 553
    software requirements, 168–173
    standalone, 33
    upgrading, 196–197
    See also Agent Cloning;
        agentDownload; Interactive Agent;
        nfsagentinstall; Silent Agent;
        standalone Agents
Aggregate service model, 443
alerts
    addressing OMR Database alert log
        errors, 691–693
    clearing, 690–691
    per hour, 722
aliases, fully qualifying the first alias in a
    hosts file, 51
allow directive, 681
allow from directive, 681
Application Server tier (Tier 2), 7
architectural design, 30
    key preinstallation decisions, 33–47
archiving, 109
AS Console, 281, 285
    controlling, 279–280
    enforcing HTTPS between the browser
        and, 676–677
    login, 273
    OracleAS Backup and Recovery
        Tool, 616
    security, 675–676
AS Control, 290
    agents for, 26
    and Grid Control, 22, 24–27
ASM, discovery and configuration, 383–386
ASM home, 614
ASM instances, 291
    discovering and configuring, 384–386
auditing, setting up, 151
Automatic Storage Management (ASM), 7, 109
    installing software, 105–106
availability transactions, 442–443

Index

B

B&R. *See* backup and recovery (B&R)
backslash (\), 139
backup and recovery (B&R), 612–613
    Agents, 618–620
    concepts, 551–554
    OMS software, 615–618
    recommended GC backups, 557–560
    resetting Agents, 554–557
    See also disaster recovery; OMR Database backup and recovery
backup media, 559
backup optimization, 573
backup reports, 608, 611–612
backups
    Agents, 241
    backing up critical OMS files, 140–141
    beginning regular GC backups, 152
    Customized Backup, 574, 575, 583–588
    how to back up all GC elements, 553–554
    logical backups, 594
    Oracle-Suggested Backup, 573, 575–583
    recommended, 557–560
    RMAN Script job, 574, 575, 588–594
    targets.xml, 367
    See also backup and recovery (B&R)
bandwidth
    LAN bandwidths for different LAN devices, 39
    specifications between GC tiers, 37, 38
baselines. *See* metric baselines
beacons, 442, 703
    adding, 467–469
    See also services
blackouts, 300, 421
    Blackout wizard, 421
    confirming no targets are in blackout, 367–368
    creating, 422–427
    reasons for, 421
    retroactive, 421
    setting up at the command line, 427
    setting up in the Console, 421–422

block change tracking, 573
broadband transmission rates, 39
bugs, 77–83
    root.sh, 212
    See also patches
built-in Java callbacks, 332

C

Central Inventory, 66–67
certificates, installing the Grid Control security certificate in your browser, 115–116
certification, verifying certification information, 60–61
chain-installed Agents, 33, 657
Change Manager, 245–246, 249
    creating a database user for, 254–260
    creating a standalone repository for, 260–262
    creating a tablespace for, 251–254
    See also EM Java Console
Client System Analyzer (CSA), 302
Client tier (Tier 1), 6
Cluster Disk Performance, 724
Clusterware home, 614
Clusterware software, 291
    installation, 105
configuration assistants, 101–103
Configure Database Instance wizard, 391–397
connectivity checks, 53–55
Console. *See* Grid Control Console
Corrective Action Library, 302
corrective actions, 515–516
    adding to metrics or policies, 518–521
    creating, 516–518
CPU utilization, reducing, 710–714
Create Dictionary Baseline wizard, 402–405
Create Service wizard, 450–451
    Availability page, 452–453
    Beacons page, 454–456
    General page, 451–452
    Performance Metrics page, 456–457
    Review page, 459

Create Service wizard (*cont.*)
    Service Test page, 453–454
    Usage Metrics page, 458
    See also services
Create System user interface, 446
    Charts page, 448–449
    Columns page, 449
    Components page, 446–447
    Dashboard page, 449–450
    Topology page, 447–448
    See also services
creation options, 110
Critical Patch Update (CPU), 124–125
    See also patches
Customized Backup, 574, 575, 583–588
Cygwin Suite, 179

D

Dashboard, 436–437
data encryption, 663–666
Data Exchange, 302
data flow between components, 19–22
Data Guard, 639–644
data integrity, 663–666
data retention policy, modifying, 148–150
database clusters, 7
Database Control
    agents for, 26
    and Grid Control, 22, 23–24, 26–27
    stopping and disabling, 405–407
database discovery and configuration, 386
    Configure Database Instance wizard, 391–397
    creating a dictionary baseline, 401–405
    database management bugs, 388
    discovering a cluster database, 389
    discovering a single-instance database, 388–389
    enabling database password file authentication, 387
    granting additional roles and privileges to DBSNMP, 398–399

PERFSTAT user, 397–398
setting up segment findings, 399–401
starting database listener, 387
stopping and disabling Database Control, 405–407
database home, 613
database point-in-time recovery, 602
database software, installing, 107–108
database storage subsystem, 7
database templates, 109
Database tier (Tier 3), 7
DB Control. *See* Database Control
dba, creating OSDBA group, 62–63
DBMS Job Processing Time, 720
DBMS Job Status, 719
DBPITR, 602
DBSNMP, granting additional roles and privileges to, 398–399
dcmctl, 281
dcm-daemon, 282
default metrics, changing, 492–500
default policies, changing, 501–504
Dell Linux hardware monitoring, 410
demilitarized zones (DMZs), 679–680
deny from directive, 681
directories
    EMSTATE directory, 164
    Grid Control parent directory, 69
    monitoring, 410
    Oracle base directory, 65–66
    Oracle inventory directory, 66–69
disaster recovery, 624–625
    recommendations, 637–649
Disk Activity, 724
Distributed Configuration Management Control (DCM) utility, 281
DNS lookup, fully qualifying, 51–52
double colon (::), 169
downloading the software, 31–32
DR. *See* disaster recovery
DSA, 282
dynamic initialization parameters, 602
Dynamic Monitoring Service (DMS), 16

Index 741

E

EBS. *See* Oracle E-Business Suite
EM, 268
    startup/shutdown, 292
    startup/shutdown with standby OMR database, 293
    Windows EM Services, 290–291
EM CLI. *See* EM Command Line Interface
EM Command Line Interface, 13, 244
    installation, 263–264
    options for setting Preferred Credentials, 326
EM Java Console, 13, 14, 244
    Change Manager, 245–246, 249, 250–262
    creating a database user for Change Manager, 254–260
    creating a standalone repository for Change Manager, 260–262
    creating a tablespace for Change Manager, 251–254
    determining need for, 245–246
    installation steps, 246–249
    login, 272
    required shared library path environment variable settings, 249
    starting, 249–250
    vs. Web Console, 245
EM_MAX_RETRIES, 637
EM_RETRY_WINDOW, 637
EM2Go, 13
e-mail notification, configuring, 95–96
emctl, 238–240, 281, 284–285
    controlling OMA using, 286–287
emkey, securing, 150–151
emoms.properties, 270, 359
EMSTATE directory, 164
encryption. *See* data encryption
Enterprise Manager. *See* EM
Enterprise User Security, 683
environment variables, 85–86
equal sign (=), 139
Error Lookup Tool, 692–693

events, following from trigger to notification, 303–305
excessive reclaimable space, 399
Execute Host Command feature, 288–289
exports, setting up OMR Database exports, 594–597

F

failover, configuring OMS for, 138–139
Fast-Start Failover, configuring Data Guard Observer for, 643
fault-tolerant network devices, 626
FGAC. *See* Fine-Grained Access Control
files, monitoring, 410
Fine-Grained Access Control, 109
Flash Recovery Area, 109, 602
Flashback Database, 602
    complete recovery, 607
    incomplete recovery, 607
Flashback Drop, 606
flashback logs, 606
Flashback Query, 606
Flashback Table, 606
Flashback Technology, 606–608
Flashback Transaction Query, 606
Flashback Version Query, 606
Forms Application service model, 443
Framework Security, 656
    enabling between Agents and OMS, 656–663
    enabling for Agents, 659–662
    enabling for OMS hosts, 657–659
    locking OMS hosts to prevent Agent HTTP uploads, 662–663
FSFO. *See* Fast-Start Failover

G

GC. *See* Grid Control
GC Console. *See* Grid Control Console
GC_VIEWALL, 308, 309, 311
General preferences, 302, 306, 317, 318–319
Generic service model, 443

grid, the, 4–5
grid computing
    defined, 4–7
    features of, 5
    tiers, 6–7
    typical architecture, 6
Grid Control
    agents for, 26
    components, 10–13
    and AS Control, 22, 24–27
    data flow between components, 19–22
    and Database Control, 22, 23–24, 26–27
    defined, 7–9
    features of, 8
    monitoring itself, 22
    versioning, 9
Grid Control Console, 11, 13–14
    vs. EM Java Console, 245
    improving login performance, 727–728
    limiting access to certain clients, 680–682
    logging into, 114–115
    login as a user other than DBSNMP, 271
    login to GC, 269–271
    security, 675–676
    URL variations to login, 269
Grid Control environments
    deciding how many to build, 35–36
    vs. Grid Control sites, 34
Grid Control parent directory, 69
Grid Control sites
    vs. Grid Control environments, 34
    how many are required, 36–41
groups
    characteristics and advantages of, 430–431
    creating, 431–435
    group pages, 435–436
    Oracle inventory group (oinstall), 61–62
    OSDBA group (dba), 62–63
    OSOPER group, 63
    out-of-box reports, 437–438
    redundancy group, 431
    System Monitoring Dashboard, 436–437
guaranteed restore points, 610

H

HA. *See* high availability
hardware requirements, 30, 55–56
    Agents, 166–168
    commands to check, 59
    installation, 56–57
    operating, 57–59
Healthcheck metric, 498
high availability, 624–625
    recommendations, 625–637
host name constraints, 52–53
host name references, fully qualifying, 49–52
host name resolution, setup, 49
host-based access control, 675–676
    schemes, 680–682
hostname command, fully qualifying, 50–51
HTTP_Server, 282
hyperthreading, 735

I

Identity Management service model, 443
indexes, rebuilding, 695–698
initialization parameters, 109–110
installation
    additional OMS, 116–117
    Adobe SVG Viewer, 262–263
    Agent Cloning, 223–225
    ASM software, 105–106
    bugs, 77–83
    Clusterware software, 105
    configuring/tuning a database for the OMR, 110–111
    creating the Oracle database, 108–110
    database software, 107–108
    EM Command Line Interface, 263–264
    Grid Control security certificate, 115–116
    information need for, 77, 78–81
    installing the OMS on a RAC node, 113
    interactive installation, 158, 160, 228–236

log files, 90–91
logging into the Web Console, 114–115
nfsagentinstall, 202–213
OAS, 667–672
the OMR database, 105–111
Oracle database manuals, 104
Prereqchecker, 48, 73
running the installer, 111–113
running time, 90
setting/unsetting OS environment variables, 85–86
silent installation method, 87–88, 158, 160, 225–228
standalone Agents on dedicated OMR nodes, 114
starting and monitoring, 89–91
static ports feature, 88–89
types, 42
using an existing database, 103–114
using GC New Database option, 91–103
X Server setup, 83–85
See also preinstallation
instance recovery, 600
integrity. *See* data integrity
Interactive Agent, 158, 160
    installation, 228–236
    package requirements, 228–229
    parameters, 231
inventories, 376–381
inventory directory, specifying, 93–94
I/O bottlenecks, tuning, 723–726
*i*SQL*Plus, 108
    configuring access in GC, 142–148, 149
    login, 274–275
*i*SQL*Plus Server, 291
    controlling, 278–279
*i*sysman/emd/directory, 552

J

Java Console. *See* EM Java Console
Job Dispatcher Performance, 722
job purge policy, modifying, 150

K

known_hosts file, 183

L

language, selection, 93
latency
    specifications between GC tiers, 37, 38
    WANs, 39
listeners
    discovery and configuration, 407–409
    stopping database listeners using port 1521, 70
load balancing
    configuring OMS for, 138–139
    confirming that listeners load-balance across RAC nodes, 141–142
loader backlog, 715–719
locking OMS hosts, 662–663
log files
    generic log file monitoring, 409
    reducing GC logging, 130–134
logical backups, 594
logical processors, 735
login
    AS Console, 273
    EM Java Console, 272
    Grid Control Console, 114–115, 269–271
    improving Console login performance, 727–728
    *i*SQL*Plus, 274–275
    Metrics Browser, 276–278
LogLoader, 282, 283

M

mail servers, 331–332
maintenance, 686
    addressing OMR Database alert log errors, 691–693
    analyzing metric rollup tables, 693
    clearing Critical and Warning alerts, 690–691

maintenance (*cont.*)
    clearing OMS and OMR system errors, 687
    fixing target metric collection errors, 688–690
    monthly offline tasks, 693–698
    performing OMR partition maintenance, 693–695
    rebuilding tables and indexes, 695–698
    starting or removing targets in DOWN status, 690
    weekly online tasks, 687–693
    See also tuning
Management Agent. *See* Oracle Management Agent
Management Connectors, 302, 414, 419–420
management options, 109, 414–415
Management Packs, 415–418
    access, 301, 414
master Agent, 203
    installing on shared storage, 209–210
master Agent host, 203
Mean Time To Recovery (MTTR), 516, 600
media failure, 604
media managers, 569
Media Pack, ordering, 31–32
media recovery, 573, 600
menus
    navigation order, 305–307
    Preferences, 302–303
    Setup, 299–302
MetaLink. *See* OracleMetaLink
metric baselines, 504–506
    adaptive thresholds, 508–509
    baseline period, 506–507
    creating, 509–514
    disabling, 514–515
    normalized metric values, 508
    time groups, 507–508
    trimmed maximum, 508
metric snapshots, 529–530
    applying, 532–533
    creating, 530–532
    why snapshot computed thresholds require editing, 534

metrics
    adding corrective actions, 518–521
    changing defaults, 492–500
    creating UDM or service test, 704
    description, 704
    evaluating, 708–709
    evaluating job, notification, and alert metrics, 722
    informational, 699
    Management Services and Repository metrics, 699–702
    metric subtype, 704
    metric type, 704
    OMS and OMR host metrics, 699, 702
    Perl syntax for alert log metric thresholds, 500–501
    recommended thresholds, 704
    specific metrics to gather, 703–708
    UI navigation for historical data, 704
    UI page performance metrics, 699, 702–703
    See also User-Defined Metrics
Metrics Browser
    fixing target metric collection errors, 688–690
    login, 276–278
MGMT_ECM_DEPOT_TS, 18
MGMT_TABLESPACE, 18
Monitoring Agent, 381–382
monitoring templates, 301–302, 533–536
    applying, 541–544
    comparing settings between targets and the template, 544–547
    creating, 536–539
    maintaining custom metrics when applying, 540
    multiple threshold settings, 541, 542
    single threshold settings, 541
mount commands, 32
multi-inventory support, enabling, 376–381

N

navigation order, 305–307
network configuration, 30
Network Interface Combined Utilization, 723

Index

Network Interface Total, 723
NFS Agents, 160, 195–196, 198
    installing, 211–213
    installing master Agent on shared storage, 209–210
    typical installation using nfsagentinstall, 204
    variables and values, 205
    See also Agents; nfsagentinstall
nfsagentinstall, 158, 195–196
    configuring shared storage, 205–209
    installation, 202–213
    installation options to prevent Agent startup and auto-discovery, 210
    installing master Agent on shared storage, 209–210
    mounting the NFS share on target hosts, 207–209
    typical installation using, 204
    See also NFS Agents
nobody, creating an unprivileged user, 64
nonproduction environment, 626
normal restore points, 610
notification methods, 300, 306
    defining, 330–334
Notification Performance, 722
Notification Processing Time, 722
notification rules, 303, 306, 339–340
    aligning with target metrics/policies, 353–355
    Availability page, 343–345
    General page, 342–343
    Jobs page, 350–351
    Methods page, 351–352
    Metrics page, 345–348
    Policies page, 348–350
    recommendations, 355–356
    steps to create, 340–342
    subscribing to, 352–353
notification schedules, 303, 306, 334
    defining, 335–337
    suspending, 338
notifications, 490
    configuring for the OS UDM, 526
    enabling if Grid Control goes down, 381–383
    following events from trigger to, 303–305
    out-of-band (OOB) notification, 382
    suspending, 338

O

OAS, 107
    configuring for all OMR data transmissions, 668–669
    configuring for monitoring/local OMR Agents, 670
    configuring for OMR Database users, 670–671
    configuring for the OMR Database, 671–672
    configuring for the OMS, 669–670
    data encryption and integrity, 663–666
    installation, 667–672
    testing client connections with the OMR database, 672–673
    testing for monitoring/local OMR Agents, 674–675
    testing for OMR Database users, 673–674
    testing for the OMS, 673
OC4J
    home, 282
    logging, 133–134
    OC4J_EM, 283
    OC4J_EMPROV, 283
OCM. *See* Oracle Configuration Manager
OCS service model, 443
OHS
    enforcing HTTPS between the browser and, 677–678
    logging, 132–133
oinstall, creating Oracle inventory group, 61–62
OMA. *See* Oracle Management Agent
OMF. *See* Oracle-Managed Files
OMR. *See* Oracle Management Repository
OMR Database backup and recovery, 560–561
    backup reports, 608, 611–612
    choosing recovery settings, 599–603

OMR Database backup and recovery (*cont.*)
   complete recovery, 604–605
   configuring backup settings in GC, 567–573
   current backups, 608, 609–612
   Customized Backup, 574, 575, 583–588
   disaster recovery recommendations, 639–644
   implementing a GC backup method, 573–577
   incomplete recovery, 604, 605–606
   managing backups in GC, 608–612
   media failure, 604
   Oracle-Suggested Backup, 573, 575–583
   performing incomplete recovery of OMR Database, 605–606
   recovery, 597–598
   restore points, 608, 610–611
   restoring GC after OMR Database complete recovery, 604–605
   restoring GC after OMR Database incomplete recovery, 605
   restoring GC operations after OMR Database failure, 603–606
   RMAN Script job, 574, 575, 588–594
   setting up exports, 594–597
   software backup and restoration, 613–615
   using a recovery catalog in GC, 561–567
   using GC to recover target databases, 598
OMR Private Interconnect Transfer Rate (MB/Sec), 724
OMS. *See* Oracle Management Service
OMS Configuration Manager, 291
OMS hosts
   logging, 131–132
   setting up Oracle user environment on OMS hosts, 135–138
OMS loader backlog, 715–719
one-off patches, 125–129
   See also patches
ons.log, 130–131

Open Grid Services Architecture (OGSA), 5
oper, creating OSOPER group, 63
opmnctl, 281–284
ORA-600 errors, 691
   Error Lookup Tool, 692–693
ORA-7445 errors, 692
   Error Lookup Tool, 692–693
oracle, creating Oracle software owner, 63–64
Oracle Advanced Security. *See* OAS
Oracle Application Server Containers for J2EE (OC4J), 15–16
Oracle base directory, creating, 65–66
Oracle Cluster Registry (OCR), 614–615
Oracle Configuration Manager, 121
   installing, 152–153
   registration, 99–100
Oracle Configuration Manager client, 13, 244
   installing the OCM client, 264–266
Oracle Data Guard, 639–644
Oracle Database, 290
Oracle database home, 613
Oracle Database Instance, 291
Oracle E-Business Suite, monitoring as a service, 487
Oracle EM Advanced Configuration, 682–683
Oracle Enterprise Manager Console DB, 107
Oracle HTTP Server (OHS), 16
Oracle inventory directory, 66–69
Oracle inventory group (oinstall), creating, 61–62
Oracle Management Agent, 12, 14–15, 290
   controlling in the Console, 288–290
   controlling using emctl, 286–287
   data flow between components, 21
Oracle Management Repository, 12, 17–19
   choosing a new or existing database for, 43–47
   clustering, 634–635
   configuring and tuning the OMR database, 152
   configuring automatic OMR Database startup on boot, 636

Index **747**

configuring iSQL*Plus access in GC, 142–148, 149
confirming that listeners load-balance across RAC nodes, 141–142
controlling, 279
data flow between components, 21
database datafiles, 552
high availability recommendations, 633–636
installing the OMR database, 105–111
installing the Oracle Configuration Manager, 152–153
modifying the data retention policy, 148–150
modifying the job purge policy, 150
patches, 123–124
platforms supporting VLM and Reference Documentation, 735
securing the emkey, 150–151
setting up GC auditing, 151
setting up Oracle user environment on OMR nodes, 141
software home, 553
See also OMR Database backup and recovery
Oracle Management Service, 12, 15–17, 290
adding a SLB at the standby site, 646
adding OMR alias to OMS tnsnames.ora, 140
backing up critical OMS files, 140–141
configuring automatic OMS startup on boot, 627, 633
configuring for failover and load balancing, 138–139
controlling, 280–285
data flow between components, 21
disaster recovery recommendations, 644–648
enabling Framework Security between Agents and OMS, 656–663
enabling Framework Security for OMS hosts, 657–659
files in receive directory, 552
high availability recommendations, 627–633

installing additional OMSs at standby site, 644–646
installing an additional OMS, 116–117
installing multiple OMSs, 627, 628–631
installing on fault-tolerant storage, 627, 631
locking OMS hosts to prevent Agent HTTP uploads, 657, 662–663
modifying the default console timeout, 138
OracleMetaLink connectivity, 55
reducing GC logging, 130–134
replicating the shared receive directory at the standby site, 646–647
running an active OMS from a DR location, 40
server load balancers, 627, 632–633, 646
setting up Oracle user environment on OMS hosts, 135–138
Shared Filesystem Loader, 627, 631–632, 647
software backup and restoration, 615–618
triggering standby OMS startup from standby OMR startup, 647–648
tuning OMS thread pool size, 139–140
Oracle Partitioning, 107
Oracle software owner, creating, 63–64
Oracle Wait Interface, 509
OracleAS
Application Server bugs, 412
configuring OracleAS instances, 411–412
discovering OracleAS instances, 411
OracleAS Backup and Recovery Tool, 616
OracleAS Single Sign-On, 683
OracleAS Web Cache, 17
data flow between components, 20–21
security considerations, 679–680
Oracle-Managed Files, 109
OracleMetaLink, 357
configuring, 97
connectivity, 55
downloading one-off patches in, 126

Oracle-Suggested Backup, 573, 575–583
order deny directive, 681
ordering the Media Pack, 31–32
OS scripts, 332
OS User-Defined Metrics, 522
    creating, 522–527
    monitoring script, 523–524
    See also User-Defined Metrics
OSDBA group (dba), creating, 62–63
OSOPER group, creating, 63
out-of-band (OOB) notification, 382
out-of-box reports, 437–438
OWI, 509

P

partition maintenance, 693–695
Patch Cache, 357
Patch Validation feature, 358
patches, 120–121
    applying latest database patch set
        certified for OMR, 123–124
    applying required GC one-off patches,
        125–129
    applying the latest EM Critical Patch
        Update, 124–125
    applying the latest GC patch set,
        121–123
    GC Patch Installer, 121
    Patch Wizard, 127–129
    RefreshFromMetalink job, 362
    See also bugs
Patching Setup, 300, 356–357
    MetaLink & Patching Settings, 357–358
    Offline Patching Settings, 360–362
    Proxy & Connection Settings, 358–360
PERFSTAT user, 397–398
Perl, 16
    syntax for alert log metric thresholds,
        500–501
permissions, verifying Agent user permissions,
    188–189
PL/SQL procedures, 332
PL/SQL Server Pages, 16
PL/SQL stored procedures, 16

point of recoverability, 573
policies
    adding corrective actions, 518–521
    changing defaults, 501–504
policy rules, 501
    count by target type, 502
portlist.ini, 89, 269
ports
    involved in securing traffic, 655
    static ports feature, 88–89
    stopping database listeners using port
        1521, 70
Preferences menu, 302–303
Preferred Credentials, 303, 306, 318,
    320–321
    EM CLI options, 326
    entering, 322–327
    normal vs. privileged, 321
    programmatically setting, 325–327
    target vs. default, 321–322
    verifying default credentials, 328–330
preinstallation
    architectural design, 30, 33–47
    bandwidth and latency specifications
        between GC tiers, 37, 38
    broadband transmission rates, 39
    confirming platform-specific software
        requirements, 72–73
    connectivity checks, 53–55
    creating required directories, 65–69
    creating required OS groups and users,
        61–64
    downloading or ordering the software,
        31–32
    fully qualifying host name references,
        49–52
    hardware requirements, 30, 55–59
    host name constraints, 52–53
    host name resolution setup, 49
    how many GC environments to build,
        35–36
    how many regional sites are required,
        36–41
    installation types, 42
    LAN bandwidths for different LAN
        devices, 39

Index **749**

making required installation choices, 41–47
minimum network requirements between GC components, 38
mounting commands, 32
network configuration, 30, 47–55
software requirements, 30, 60–73
static IP addresses, 53
stopping database listeners using port 1521, 70
synchronizing OS timestamps/time zones, 70–72
verifying certification information, 60–61
Prereqchecker, 48, 73
production environment, 626
product-specific prerequisite checks, 94–95
program resource utilization monitoring, 410
properties files, 185, 186
provisional views, 19
proxy, configuring, 97

R

RAC, 634–635
RAC Management Repository, controlling, 279
RCA. *See* Root Cause Analysis (RCA)
rebuilding tables and indexes, 695–698
receive directory, OMS files in, 552
recovery. *See* backup and recovery (B&R)
recovery catalog, directing the OMR Database to use a recovery catalog in GC, 561–567
recovery configuration, 109
Recovery Time Objectives (RTOs), 550
recovery window policy, 573
redundancy group, 431
  See also groups
Reference Documentation, 735
RefreshFromMetalink job, 362
registration passwords, 301
release notes, 77–82
remote systems, staging the software on, 32
Repository. *See* Oracle Management Repository

resetting Agents, 554–557
restore points, 608, 610–611
retention policy, 573
RMAN, 561–562
RMAN Script job, 574, 575, 588–594
role reversal procedures, 643–644
roles, 299, 306
  creating, 307–313
rollup delays, minimizing, 719–722
Root Cause Analysis (RCA), 483–486
Root Key, 657
root passwords, 327
root.sh, bugs, 212
rotation frequency, 334
row chaining, 399

S

Search for Extra-Terrestrial Intelligence At Home (SETI@home), 5
Secure Sockets Layer. *See* SSL
security
  enforcing, 626
  Framework Security and OHS, 654
  goals, 653
  installing the Grid Control security certificate in your browser, 115–116
  methods and ports involved in securing traffic, 655
  Oracle HTTP Server (OHS), 16
  secure Console connections, 675–682
  secure Repository data transmissions, 663–675
  specifying options, 97–98
  See also Framework Security; OAS
semicolon (;), 169
server load balancer
  adding at a standby site, 646
  changing the response file if using an SLB, 170
  configuring, 165
  installing Agents through, 161
  modifying Agent Deploy properties file for SLB, 187
  using with OMS, 627, 632–633

Server Side Include, 16
service level agreements (SLAs), 439–441
    Recovery Time Objectives (RTOs), 550
services, 438–439
    adding service tests, 461–467
    Aggregate service model, 443
    beacons, 442, 467–469
    changing availability definitions, 473–475
    changing system configuration, 469–470
    configuring, 445–446
    creating, 445, 450–460
    creating a system, 445, 446–450
    editing service level rules, 481–483
    enabling Management Pack access for Web application, 461
    Forms Application service model, 443
    Generic service model, 443
    Identity Management service model, 443
    managing Web Server data collection, 470–473
    Monitoring Configuration tasks, 460–461
    monitoring Oracle E-Business Suite as a service, 487
    OCS service model, 443
    performance metrics, 440–441, 475–476
    performing monitoring configuration tasks, 445
    Root Cause Analysis (RCA), 483–486
    service test-based availability, 439–440
    system-based availability, 440
    test types, 443–444
    transactions, 442–443
    types of service models, 443–445
    usage metrics, 441, 476–478
    Watch List, 478–481
    Web Application service model, 443
Setup menu, 299–302
Shared Filesystem Loader, 627, 631–632, 647
Silent Agent, 158, 160
    installation, 225–228
single-instance Repository Database, controlling, 279

SLB. *See* server load balancer
SMTP Server connectivity, 53–54
snapshots. *See* metric snapshots
SNMP traps, 332
software
    confirming platform-specific software requirements, 72–73
    downloading or ordering, 31–32
    requirements, 30, 60–73, 168–173
software configuration management (SCM), 626
space character, 169
SQL User-Defined Metrics, 522
    creating, 527–529
    See also User-Defined Metrics
SSH Server (SSHD), setting up on Windows, 179
SSH user equivalence, configuring, 178–184
sshConnectivity script, 180–184
SSL, 655
standalone Agents, 33, 156
    changing the response file if using an SLB, 170
    choosing to secure the Agent or not, 166
    configuring a server load balancer first, 165
    confirming no existing agent is installed, 169–170
    cross-platform agent installation, 171–173
    hardware requirements, 166–168
    initializing the oracle user environment, 168–169
    installation methods, 158–162
    installing a local or cluster Agent, 163–164
    installing on dedicated OMR nodes, 114
    platform names in Agent download directory, 171
    preinstallation steps, 157
    preparing targets for discovery, 173
    required Agent installation information, 173
    software requirements, 168–173

using an existing user or creating a separate Agent user, 162–163
using an LDAP or local Agent user, 165
See also Agent Deploy; Agents
static initialization parameters, 602
static IP addresses, 53
static ports feature, 88–89
staticports.ini, 88–89
storage options, 109
See also Automatic Storage Management (ASM)
SVG Viewer. *See* Adobe SVG Viewer
system errors, clearing, 687
System Monitoring Dashboard, 436–437
System Monitoring Plug-ins system, 302, 414, 418–419

T

tables, rebuilding, 695–698
tablespace locations, 113
Target Subtabs, 303, 317
  customizing, 319–320
targets
  ASM discovery and configuration, 383–386
  discovering and configuring, 370–376
  properties, 413–414
  removing from Grid Control, 368–370
  removing from monitoring, 367–370
  starting or removing targets in DOWN status, 690
  types auto-discovered, requiring configuration, or additional setup, 371
  See also database discovery and configuration
targets.xml, 367
template accounts, 313
templates
  database, 109
  See also monitoring templates
tiers, 6–7
time groups, 507–508
time zones
  setting up time zone for SSH Server, 184–185
  synchronizing OS time zones, 70–72
timestamps, synchronizing OS timestamps, 70–72
TNS Listener, 290, 291
transactions, 442–443
  See also services
trimmed maximum, 508
tuning, 490–491, 709–710
  changing default metrics, 492–500
  changing default policies, 501–504
  corrective actions, 515–521
  evaluating job, notification, and alert metrics, 722
  improving Console login performance, 727–728
  increasing UI page performance, 728–734
  I/O bottlenecks, 723–726
  leveraging monitoring templates, 533–547
  metric baselines, 504–515
  metric snapshots, 529–533, 534
  minimizing rollup delays, 719–722
  platform-specific recommendations, 734–735
  reducing high CPU utilization, 710–714
  resolving loader backlog, 715–719
  User-Defined Metrics, 521–529
  See also maintenance

U

UDM. *See* User-Defined Metrics
UI page performance metrics, increasing UI page performance, 728–734
UNIX
  scripting technique to modify configuration files, 130
  shell variations, 84
  syntax differences between Windows shell scripting and, 137
  X Server setup, 83–85
unprivileged user, creating, 64

user permissions, verifying, 188–189
User-Defined Metrics, 521–522
    OS UDMs, 522–527
    SQL UDMs, 522, 527–529

V

variables, setting/unsetting OS environment variables, 85–86
versions, Grid Control, 9
views, 18–19
virtual directories, 16
VLM, 735
voting disk, 614
    backup and restoration, 615

W

WAN, broadband transmission rates, 39
Web Application service model, 443
Web Cache. *See* OracleAS Web Cache
Web Console. *See* Grid Control Console
WebCache, 283
WebCacheAdmin, 283
Windows EM Services, 290–291
Windows shell scripting, syntax differences between UNIX and, 137

X

X Server, setup, 83–85